CETACEAN SOCIETIES

CETACEAN SOCIETIES

Field Studies of
Dolphins and Whales

Edited by

Janet Mann,

Richard C. Connor,

Peter L. Tyack, and

Hal Whitehead

THE UNIVERSITY OF CHICAGO PRESS

Chicago and London

The University of Chicago Press, Chicago 60637
The University of Chicago Press, Ltd., London
© 2000 by The University of Chicago
All rights reserved. Published 2000
Printed in the United States of America

27 26 25 24 23 22 21 20 19 18   5 6 7 8 9

ISBN-13: 978-0-226-50341-7 (paper)
ISBN-10: 0-226-50341-0 (paper)

Library of Congress Cataloging-in-Publication Data

Cetacean societies : field studies of dolphins and whales / edited by Janet Mann . . . [et al.].
    p.   cm.
    Includes bibliographical references and index.
    ISBN 0-226-50340-2 (cloth : alk. paper); — ISBN 0-226-50341-0 (paper : alk. paper)
    1. Cetacea—Behavior.   2. Social behavior in animals.   I. Mann, Janet.
QL737.C4 C39 1999
599.5′156—dc21   99-045607

This book is printed on acid-free paper.

We dedicate *Cetacean Societies* to the late Kenneth S. Norris, 1924–1998.

KEN NORRIS

Teacher, pathfinding researcher, mentor, friend, and colleague to many who have sought to understand the lives of whales and dolphins.

CONTENTS

CONTRIBUTORS

Robin W. Baird
  Biology Department
  Dalhousie University
  1355 Oxford Street
  Halifax, Nova Scotia B3H 4J1
  Canada

Phillip J. Clapham
  Northeast Fisheries Science Center
  166 Water Street
  Woods Hole, MA 02543
  U.S.A.

Jenny Christal
  18 Nightingale Court
  North Road
  Hertford
  Herts SG14 1PD
  England

Richard C. Connor
  Department of Biology
  University of Massachusetts-Dartmouth
  285 Old Westport Road
  North Dartmouth, MA 02747
  U.S.A.

Janet Mann
  Department of Psychology & Department of Biology
  Georgetown University
  37th and O Street, NW
  Washington, DC 20057
  U.S.A.

Andrew Read
  Duke University Marine Laboratory
  135 Duke Marine Lab Road
  Beaufort, NC 28516
  U.S.A.

Randall R. Reeves
  27 Chandler Lane
  Hudson, QC J0P 1H0
  Canada

Amy Samuels
  Department of Conservation Biology
  Chicago Zoological Society
  3300 Golf Road
  Brookfield, IL 60513
  U.S.A.

  Biology Department
  Woods Hole Oceanographic Institution
  Woods Hole, MA 02543
  U.S.A.

Peter L. Tyack
  Biology Department
  Woods Hole Oceanographic Institution
  Woods Hole, MA 02543
  U.S.A.

Linda Weilgart
  Department of Biology
  Dalhousie University
  Halifax, Nova Scotia B3H 4J1
  Canada

Hal Whitehead
  Department of Biology
  Dalhousie University
  Halifax, Nova Scotia B3H 4J1
  Canada

Randall S. Wells
  Mote Marine Laboratory
  1600 Ken Thompson Parkway
  Sarasota, FL 34236
  U.S.A.

  Department of Conservation Biology
  Chicago Zoological Society
  3300 Golf Road
  Brookfield, IL 60513
  U.S.A.

Richard Wrangham
  Peabody Museum
  Department of Anthropology
  Harvard University
  11 Divinity Ave
  Cambridge MA 02138
  U.S.A.

FOREWORD

RICHARD WRANGHAM

THE SOCIAL lives and evolutionary ecology of cetaceans have long been mysterious, but for those lucky enough to glimpse their societies in detail, the potential is obvious. My introduction to them came in Shark Bay, Australia, in 1987. From small boats, Richard Connor and Rachel Smolker called out the names of dolphins surfacing anywhere from one to fifty meters away. Close encounters brought moments of revelation. A female, being herded by a trio of cooperating males, moved half a meter away from her mate-guarders, only to be warned by a popping call that she risked being attacked. Later, when the males were distracted by feeding, she bolted. She was forty meters away before the males began their chase, enough to allow her a full escape. It was like watching primates. The promise of a rich behavioral ecology of cetaceans was clear.

Such experiences of individual strategizing have lit the imagination of a new generation of cetologists determined to bring behavioral studies of dolphins and whales into the modern era. The difficulties of getting the data are huge. But creativity and determination have fostered a range of techniques for opening up the private lives of these elusive species. Cetacean species are now understood in far more detail than previously and are beginning to challenge generalizations from the terrestrial world. *Cetacean Societies* is the first major report of this new wave of studies.

Admittedly, even the most basic facts about taxonomy, distribution, foraging, and reproduction remain to be established in some of these huge, appealing creatures. How do they mate? Land-bound behaviorists will be startled to discover that not a single adult copulation has been definitely seen by any student of sperm whales (or, for that matter, humpback or killer whales). And of course, no one has ever been down for the hundreds of meters needed to watch sperm whales feed. But such daunting obstacles have not deterred the new generations of marine explorers. *Ceta-*

cean Societies shows how patience and ingenuity can allow observers to recognize individuals in the most unlikely of circumstances. And individual recognition is the breakthrough that allows the all-important description of social relationships.

As a result, it is already clear that cetaceans have as much diversity in life history and mating systems as is found in any taxon. To any land-bound ecologist, furthermore, the emerging account includes strikingly novel features. Living largely without physical refuges from predators, cetaceans might be expected to have high mortality rates and therefore to senesce early. To the contrary, there are species so long-lived that females challenge conventional wisdom either by breeding into their sixth decade (sperm whales) or by following the human pattern of a long postmenopausal life (short-finned pilot whales, and maybe killer whales). With no territoriality or female-defense polygyny yet reported, social species might have been expected to be under little pressure for philopatry. In fact, however, there are cetaceans in which both sexes are philopatric, a pattern unknown on land. Large testes in some species show that sperm competition is intense, which might suggest that females are concerned to elicit benefits from males such as investment in young or protection from infanticide. But no such benefits are known, so variation among species in the intensity of sperm competition is unexplained. The young are precocial in their locomotion and in their independence from their mothers. Yet some receive more parental investment than any terrestrial species.

The promise heralded by these sorts of puzzles, richly outlined in *Cetacean Societies,* is intensified by their significance for cognitive evolution. Large brains, learned vocal outputs, echolocation abilities, and complex social relationships are all found in cetaceans. But how they fit together is still not well understood. Dramatic variations among species, such as the twofold difference in brain size between freshwater and marine dolphins, should help to unravel the causal links.

Part review, part testament to extraordinary dedication, and part call to get involved, *Cetacean Societies* highlights the achievements of behavioral ecologists inspired by the challenges of cetaceans and committed to the exploration of a new world. Their greatest reward will be a book in twenty years' time that makes this one redundant. The high standard set by the present volume makes that goal a truly worthy challenge.

Richard Wrangham

ACKNOWLEDGMENTS

WITH A volume of this size and scope, there are many to thank for their inspiration, critical evaluation, and logistical or financial support. Richard Wrangham played a central role in stimulating this project and encouraging us to bring cetacean studies to the larger scientific community. The Center for Advanced Study in the Behavioral Sciences (CASBS) in Stanford, California, selected the development of this book as a special project. In 1994, we began writing this volume while three of us (Janet Mann, Richard Connor, and Peter Tyack) were fellows at the center. At the center, the support staff were unfailingly helpful in every way: from extensive library assistance to statistical counsel, they offered an ideal work environment that kept us motivated and on task. CASBS director Neil Smelser and assistant director Bob Scott were most supportive in all respects and were constant sources of stimulating conversation and humor. Lynn Gale and Lincoln Moses were wonderful statistical consultants. Patrick Goebel helped with computer problems, networking, and other logistical issues. The library staff helped us find obscure materials, and Shannon Greene sustained us with fabulous lunches.

Richard Connor, Janet Mann, and Peter Tyack were sponsored at The Center for Advanced Study in the Behavioral Sciences with support from the National Science Foundation (grant SBR-9022192). Janet Mann was also supported while at the center by the Helen V. Brach Foundation. Peter Tyack was also supported while at the center by a Mellon Independent Study Award from the Woods Hole Oceanographic Institution.

Primate Societies and its editors have served as role models and mentors for us. Barbara Smuts was Janet Mann's mentor in graduate school, and Richard Wrangham was Richard Connor's—both then at the University of Michigan. Barbara Smuts provided extensive guidance along the way. Thomas Struhsaker was a welcome and rare voice from the field when Peter Tyack was a graduate student at Rockefeller University. Peter Tyack would also like to acknowledge Dorothy Cheney and Robert Seyfarth as helping to form his views on animal communication and how to study it.

Many of our colleagues commented extensively on chapters, and we are grateful for their input. Special thanks to our contributors, particularly Robin Baird, Amy Samuels, Linda Weilgart, and Richard Wrangham, for commenting on, as well as contributing to, this volume. Thelma Rowell's keynote address at the 1994 meeting of the Animal Behavior Society inspired Amy Samuels to initiate chapter 1. Other colleagues deserving our thanks for detailed feedback on one or more chapters are Vincent Janik, Anne Pusey, Susan Dufault, Jonathan Gordon, Larry Dill, Michael Heithaus, Donald Dewsbury, Aleta Hohn, Michael Scott, Bill Watkins, and especially Michael Pereira and Fritz Trillmich.

Many institutions and granting agencies have supported our field research over the years. We are indebted to the American Cetacean Society, Center for Whale Research, Chicago Zoological Society, Dolphin Quest, Dolphins of Shark Bay Research Foundation, Eppley Foundation for Research, Getty Foundation, National Geographic Society, Georgetown University, Green Island Foundation, International Whaling Commission, Office of Naval Research (grant N00014-93-1-1181, chapter 1), National Institutes of Health, National Science Foundation, Natural Sciences and Engineering Research Council of Canada, the University of Michigan, the University of Western Australia, Waikoloa Marine Life Fund, Whale and Dolphin Conservation Society, and Woods Hole Oceanographic Institution.

Flip Nicklin (Minden Pictures), Robin Baird, Jeff Jacobsen, Sascha Hooker, James Porter, Phil Clapham, Bill Everett, Randall Reeves, Rusty White, Steve Leatherwood, and Andrew Wight also generously donated photographs. Lynne Barre, Christian Johansen, Jana Watson, and Krista Irwin provided assistance with organizing the references. Susan Dufault provided reference assistance for chapters 3 and 9. Flip Nicklin gave us information on remote vehicles for chapter 3. William Schevill and W. A. Watkins offered their extensive library for the research, and C. Hurter

(Woods Hole Oceanographic Institution) and M. Rabb (Brookfield Zoo) assisted in library research for chapter 1. Cynthia Flaherty and Trevor Spradlin also helped with hard-to-find documents for chapter 1. Deirdre E. Valentine and Lynne Barre helped with the appendices. Lynne Barre and Sharon Hughes helped prepare the indexes. Christina Henry, Charles Clifton, Anita Samen, and Theresa Biancheri at the University of Chicago Press and copy-editor Norma Roche were always cheerful, helpful, and supportive at all stages of book preparation. We also thank Dennis Anderson for his care with the design.

The Animal Behavior Society has agreed to establish an international "Cetacean Behavior and Conservation" fund to support graduate students undertaking fieldwork on these topics. The proceeds from this volume will go to that fund.

INTRODUCTION

The Social Lives of Whales and Dolphins

RICHARD C. CONNOR, JANET MANN,
PETER L. TYACK, AND HAL WHITEHEAD

WHALES AND dolphins are long-lived, reproduce slowly, and spend most of their lives hidden below the water's surface. It is therefore not surprising that cetaceans have rarely been favorite subjects of the "question-focused" approach to animal behavior that has dominated behavioral ecology for the past thirty years. Instead, many cetologists began their studies with a "taxon-focused" orientation that is the necessary prelude to question-focused science (Gans 1978). They were often motivated by the kinds of "big" questions that are typically beyond the scope of individual research projects (for example, why do dolphins have big brains? Why do some cetaceans learn their vocalizations? How does the social organization of a fully aquatic social mammal compare with terrestrial species?). The results of these initial efforts are exciting and not only bring the study of whales and dolphins into mainstream behavioral ecology, but also provide incentive for further basic research on the majority of cetaceans that remain unstudied.

Like cetology, behavioral primatology initially focused on "big questions," such as "How similar are humans and nonhuman primates?" "Can nonhuman primates serve as models for human evolution?" The evolution of primatology from taxon-oriented to question-oriented science is illustrated by comparing the two primate volumes supported by the Center for Advanced Study in the Behavioral Sciences, *Primate Behavior: Field Studies on Monkeys and Apes* (DeVore 1965) and *Primate Societies* (Smuts et al. 1987). *Cetacean Societies: Field Studies of Dolphins and Whales* is closer in spirit to the first primate volume in its focus on a few well-studied species. However, long-term studies on cetaceans generally followed the early primate work by some twenty to thirty years and have been conducted at a time of much greater comparative knowledge and theoretical sophistication. Thus, we are able to place the early findings on cetacean societies in a broader comparative framework and generate more hypotheses for future testing. We hope this volume will help accelerate the transition of cetology from a taxon-focused to a question-focused science and that, in another twenty years, a volume on cetaceans will be produced that can stand with the data-rich *Primate Societies*.

Intended Audience

Cetacean Societies is directed at two general audiences. For students of cetology, we present results from field studies in a methodological, comparative, and theoretical context to facilitate the framing and testing of hypotheses. The novel and convergent features of cetacean societies (Connor et al. 1998), along with innovative methods developed for studying cetaceans, may be attractive to noncetologists interested in broadening their comparative perspective or in gaining new insights into areas of theoretical interest.

Although there are a number of outstanding papers on research methods in animal behavior, the reviews of methods in *Cetacean Societies* fill a need for an up-to-date presentation that is sympathetic to the unusual difficulties facing students of cetology. How does a researcher collect systematic behavioral data on an animal that spends only a small amount of time visible at the surface? If the "ideal" methods are not practical, how can an observer still obtain data that reveals a species' social structure? While the presentations on methods in this volume do not provide an answer to every problem facing a student of cetacean behavior, we hope they will at least point toward some solutions.

While it is true that every species is interesting, it is also true that nothing is interesting except in comparison with everything else. Any cetacean behavior or life history pattern discovered ought to be compared with similar (if there are any) phenomena in other taxa. The chapters in *Cetacean Societies* strive to place the phenomena described in a broad comparative perspective. The discovery that both male and female killer whales remain with their mothers as adults would not be so remarkable if bisexual natal philopatry were commonplace among mammals. Likewise, the observation that many aspects of sperm whale social structure and life history closely match those found in elephants—but are rare elsewhere—is of great interest and brings immediately into focus the question of what selection pressures favored such convergence.

We have also endeavored to interpret cetacean behavior in a modern theoretical framework, focusing on individual selective benefits and on the ecological factors that shape social systems and relationships. Cetaceans are ripe for more detailed studies on kin selection, altruism, mating systems and strategies, life history evolution, and so forth. What are the costs and benefits of philopatry that permit male and female killer whales to remain in their natal groups? What kind of social system could produce the life history pattern of Baird's beaked whale, in which males appear to outlive females by thirty years? Why has vocal learning evolved so often in marine mammals and birds, yet so rarely in terrestrial mammals? Noncetologists may enjoy pondering such questions; we hope that cetologists will be inspired to go find the answers.

Organization and Synopsis

Cetacean Societies: Field Studies of Dolphins and Whales is divided into three parts. Following this introduction, part 1 continues with a chapter on the history of cetacean behavioral research and two chapters on methods. The history of cetacean behavioral research is eclectic and provocative, with its disparate origins in the whaling industry, commercial "sea world" exhibits, and students interested in the natural behavior, ecology, and cognitive capabilities of whales and dolphins. Chapter 2 reviews methods developed outside of cetology for quantifying social behavior and is thus directed primarily at cetologists. These methods have been incorporated slowly into studies of the social behavior of whales and dolphins and are hard to use on some species. Although social behavior is the foundation for social structure, it is possible to study social structure in other ways. One of the brightest aspects of modern cetology is the development of a wide range of innovative research techniques. Chapter 3 describes these innovative methods of studying social structure and should be of interest to cetologists and noncetologists alike.

In part 2 of *Cetacean Societies,* four of the best-studied species (see Mann 1999) are reviewed: the bottlenose dolphin (*Tursiops* sp.), the killer whale (*Orcinus orca*), the sperm whale (*Physeter macrocephalus*), and the humpback whale (*Megaptera novaengliae*). Important discoveries are evident in each chapter, demonstrating the rewards of long-term cetacean research while at the same time pointing out many promising avenues for future investigations. Part 3, the final section of the book, presents reviews of major topics: group living, male and female reproductive strategies, communication, and conservation. These chapters broaden the discussions of particular species begun in part 2, bringing together observations from a much wider range of species and placing these findings in a comparative and theoretical context.

Rather than presenting an encyclopedic list of all the topics covered in each chapter, the following synopsis selects one or a few highlights from each chapter that we feel convey the flavor and excitement of cetacean behavioral research.

Part 1: History and Methods

As Samuels and Tyack explain in chapter 1, on the history of cetacean behavioral research, the inherent difficulty of observing the animals can only partly explain why cetology lags behind the study of primates and other terrestrial taxa by some twenty or more years. For example, the study of primate societies was given tremendous momentum by the influence of a few individuals who supported and pioneered the early naturalistic behavioral studies, with Louis Leakey, Sherwood Washburn, and Jane Goodall being prominent examples. Roger and Katy Payne played a similar role for young researchers interested in large whale—and in some cases dolphin—behavior. However, instead of reading about naturalistic observations of wild dolphins (as in Goodall 1967), many of us in cetology can recall a youthful time of seeking information and even inspiration from the books of John Lilly—a memory we have become almost embarrassed to admit. Samuels and Tyack explore Lilly's largely negative impact on cetacean behavioral research, following his transition from a neurobiologist interested in dolphin brains to a pseudoscientific mystic with a large popular audience. Lilly was introduced to dolphin vocalizations by one of the pioneers of cetacean research,

2

Forrest G. Wood, who later wondered whether history would have been different had he not been so generous with his tape recordings.

Chapter 2, "Unraveling the Dynamics of Social Life," is an explication of behavioral research methods directed largely at students of cetology. Methods developed mostly by primatologists and used widely by ethologists for gathering quantifiable data on individual social interactions and relationships may be imported into cetacean studies to a greater degree than has previously been the case. Many studies have relied almost exclusively on "snapshot" samples of individual associations and behavior during brief encounters with whale or dolphin groups. In a typical study, after photographing individually distinctive marks and recording a variety of ecological and behavioral data, researchers move on to search for other groups to "survey." A richer perspective on the dynamics of social life can be obtained by following a recognizable individual and systematically recording the behavior and interactions of that "focal" individual or by sampling the activities of group members. If focal individuals travel in reasonably small groups, observers may also include or focus on any occurrence in the group of a few behaviors that are considered especially illuminating. Surveys can be combined with "follows," and we urge those beginning new projects to integrate follows into their studies as early as possible, if only to begin habituating the animals to the continuous presence of observers. However, such methods may require subjects that are more consistently and continuously observable and identifiable than are many cetaceans.

When confronted with such elusive subjects, rather than folding up the tent and going home, cetologists have often developed new gadgets or methods that allowed them to learn about cetacean social structure. Chapter 3 focuses on such methods developed by cetologists, some of which may be useful exports to other fields. For example, faced with the difficulty of localizing sound underwater, cetologists have developed methods using hydrophone arrays to determine which whale or dolphin makes which sound (Clark 1980). Whitehead (1997; Whitehead and Dufault 1999) developed analytical techniques that reveal temporal patterning of social relationships through repeated "snapshot" sampling of individual associations. We hope this chapter will serve as an encouragement to innovative students of cetology as well as a source of ideas for the noncetologist audience.

Part 2: Four Species

In general, easy access in coastal waters has been a key factor in determining which species of whales and dolphins have

been most closely examined. However, our inclusion of a chapter on the open-ocean, deep-diving sperm whale shows that this need not be the case. Recognition that each of these four species has yielded important and exciting discoveries should stimulate and justify more "taxon-driven" investigations on lesser-known cetaceans. We find remarkable examples of convergence between cetaceans and terrestrial mammals as well as traits that appear to be uniquely cetacean, discoveries that promise to broaden our understanding of the relationship between ecology, life history, and social evolution in mammals that live underwater as well as on land.

The bottlenose dolphin (chap. 4), found widely in temperate and tropical coastal waters, is the most commonly studied cetacean. Our chapter compares results from the two longest-running *Tursiops* studies, in Sarasota Bay, Florida, and in Shark Bay, Western Australia. Bottlenose dolphins live in fission-fusion societies highly reminiscent of those of chimpanzees and spider monkeys, in which individuals travel in small groups that often change in composition. Some females are largely solitary, while most associate preferentially with several other females in loose "bands." In Sarasota, such bands are often, but not always, composed of maternal relatives. The strongest bonds among adults are found among males, which form stable alliances of two or three individuals. Some alliances in Shark Bay are similarly stable, while others are remarkably labile. In Shark Bay, teams of two or more alliances cooperate against other alliances in contests over females. Such "alliances of alliances" are not an obvious feature of bottlenose dolphin social structure in Sarasota, where it appears that more males are found alone or in pairs but rarely in the trios common in Shark Bay. This contrast between the two sites is strengthened by the apparent lack of strong bonds among adults at another study site in Scotland. Understanding the evolutionary basis of such variation in social structure should become a focal point for future studies.

The most intriguing aspect of killer whale society is how different it is from that of any terrestrial mammal (chap. 5). Killer whales have been most extensively studied in the waters off southwestern Canada, where two sympatric populations are found: *residents,* which feed primarily on fish, and *transients,* which prey on other marine mammals. *Residents* live in stable matrilineal groups of up to four generations of whales from which neither males nor females disperse. Matrilineal groups travel with other related matrilineal groups in "pods," which are thought to form by a gradual process of pod fissioning. Each pod has a distinctive repertoire of calls, and the degree of overlap in call repertoire has been used to construct pod pedigrees. *Transient* killer whales exhibit a fundamentally similar matrilineal social structure, but the energetics of feeding on their primary prey, the harbor seal, appears to constrain group size such that second-born sons and daughters of reproductive age disperse from their mother's pod to form new groups.

A comparison of the sperm whale and the African elephant (chap. 6) offers the most striking example of convergence between cetaceans and terrestrial mammals, beginning with their relatively large body and brain size, marked degree of sexual size dimorphism, life span, and age of maturity. Both the sperm whale and elephant live in stable matrilineal groups of about ten individuals that are often found in temporary associations with other matrilineal groups. Female bonds in both species center on the cooperative care of young, which, being equipped for neither fleeing nor hiding, are extremely vulnerable to predation. Both male sperm whales and male elephants graduate from spending time in bachelor groups to roving between female groups in search of estrous females.

Most baleen whales engage in extensive seasonal migrations in which they travel from high-latitude summer feeding areas to warm-water winter breeding grounds, where they give birth. Mating often occurs on breeding grounds or during migration. With a few notable exceptions (see chap. 8), descriptions suggesting strong social bonds among adult baleen whales are lacking. Thus, research on baleen whales, and the humpback whale in particular, (chap. 7), has focused on mating systems, feeding ecology, and life history. As is often the case for species in which female mating strategies are subtle and male strategies are not, most research has focused on male behavior. The discovery of complex song in humpback whales stimulated a large body of research on male mating strategies, of which there appear to be several kinds. Males may attract females and/or repel rival males by solitary singing, they may fight for access to individual females in "competitive" groups, and they may escort single females, but the relationship between these particular strategies and their relative payoffs remains elusive. Song in humpback whales provided the first evidence of vocal learning in a mammal outside of our own species. Thus, rather than comparisons with social mammals such as primates, humpback whale song invites comparisons with songbirds. Interestingly, despite literally thousands of observation hours during the mating season, no researcher has ever witnessed copulation in humpback whales. This should not be taken simply as a reflection of the difficulty of whale research, but as a revealing indication of the cryptic and probably brief nature of humpback mating. Humpback whales have relatively short penises and small testes, suggesting that females do not copulate with many males. In contrast, matings are often observed in right whales (even simultaneous copulation with two males and one female), which have long penises and nearly one-ton testes, the largest on Earth.

Part 3: Comparative Studies, Theory, and Conservation

Perhaps no other group of mammals has evolved in a habitat so lacking in refuges from predators. The open ocean does not offer the equivalent of a tree or burrow in which cetaceans can hide or to which they can flee. Individuals of some species may seek refuge from particular predators by swimming close to shore or diving deep, but for most cetaceans the only refuge from predators is found in group living, the subject of chapter 8. The cost of group living may be ameliorated by cetaceans' ability to move efficiently through their dense but buoyant medium. Relatively low costs of locomotion may reduce the costs of foraging competition in cetaceans compared with terrestrial species and even contribute to our understanding of the remarkable pattern of natal philopatry in *resident* killer whales (chap. 8).

Cooperation is often invoked to explain apparently coordinated behavior in cetacean groups, particularly in foraging and feeding contexts. In many such cases, possible selfish explanations have been overlooked. On the other hand, observations of coordinated feeding behaviors in stable groups of humpback whales can be used to infer greater social complexity than was previously thought to be the case. In general, the relationship between food distribution and group formation in cetaceans is poorly understood and points to the need for more fine-scale studies on prey distribution.

Chapters 9 and 10 review reproduction and reproductive strategies in female and male cetaceans. Compared with most mammals, cetaceans live long, mature late, and reproduce slowly, producing one calf at a time, in which they invest heavily. Maximum age estimates for bowhead whales are well over one hundred years, which would make them the oldest animals in the world (chap. 10). Reproduction in most mysticetes is closely adapted to their seasonal migratory cycle: females migrate to low latitudes to give birth, lactate for approximately six months while feeding at high latitudes, become pregnant during the following migration to the tropics, and after a one-year gestation period, give birth again. By cetacean standards, some small odontocetes follow a "live fast, die young" strategy. For example, female harbor porpoises are mature at a few years of age and give birth every year or two before death occurs in their early teens. Other odontocetes (for example, the killer whale) don't mature until their mid-teens, may live

RICHARD C. CONNOR, JANET MANN, PETER L. TYACK, AND HAL WHITEHEAD for the better part of a century, and typically reproduce only once every four to five years. The short-finned pilot whale *(Globicephala macrorhynchus)* and killer whale are the only nonhuman species with clear evidence for menopause: female pilot whales cease to produce calves by age thirty-six, even though they may live into their sixties—one female was lactating at age fifty. Sperm whales may continue to suckle for up to thirteen years. Such extended periods of parental care provide ample opportunity for young odontocetes to learn how to hunt and catch prey, how to avoid becoming prey, and how to cope in a complex society.

In general, odontocetes exhibit stronger and more complex bonds among adults than do mysticetes. The type of bond and its functional nature differ widely among species. In sperm whales, the strongest bonds are among females and, as noted above, are thought be based on protection of young from predators. Male-male alliances are the strongest bonds among adult bottlenose dolphins and are based on conflicts with conspecifics (male or female). Strong associations among adult males have been detected in other species, suggesting that male alliance formation might be common in odontocetes. The killer whale appears to be one of the few mammals to exhibit strong adult son-mother bonds.

We are only beginning to explore how particular attributes of the marine environment influence cetacean mating strategies. The three-dimensional underwater habitat rewards maneuverability in combat and may reduce selection for large size in males of some species. Likewise, male control of access to females may be more difficult underwater than on land. On the other hand, greater female maneuverability may provide an additional factor favoring male cooperation compared with terrestrial mammals. Low travel costs underwater may also reduce the cost of alliance formation in males—or group formation in females. They also enable cetaceans to range more efficiently over large distances in search of their often dispersed and patchy resources; such a ranging pattern may not only render territoriality less profitable, but may also, as noted earlier, allow such unusual phenomena as adult male killer whales traveling with their mothers in search of mates.

Cetaceans have evolved remarkable abilities of echolocation and long-distance communication, taking advantage of the remarkable physics of sound propagation in the ocean. In chapter 11, Tyack reviews the fundamental role sound plays in mating systems and a variety of other aspects of cetacean life. There is little evidence for vocal learning among nonhuman terrestrial mammals. In contrast, many cases of vocal learning and vocal imitation have been reported for marine mammals (Janik and Slater 1997).

Humpback whales singing on the same breeding ground converge on the same song, which nevertheless continues to change each month during the breeding season. Bottlenose dolphins are superb vocal mimics, either of computer-generated sounds or of each other's "signature" whistles. The individually distinctive pattern of frequency modulation in signature whistles is thought to allow individual dolphins to recognize each other. Tyack (Tyack and Sayigh 1997; Tyack, chap. 11, this volume) points out that the use of "voice" cues for individual recognition may be problematic in an environment of large pressure changes. As animals dive below the surface, increasing pressures tend to distort internal air-filled cavities involved in sound production, and hence any sounds they produce. The frequency-modulated design of signature whistles should be impervious to such pressure effects.

Cetaceans face a bewildering variety of threats from humans, and many species are threatened or endangered. Chapter 12 demonstrates how the magnitude of anthropogenic threats to cetaceans can be reasonably gauged by how close they live to shore, how large they are (and thus how attractive to whalers), and whether they are subjected to by-catch in human fishing operations. At one extreme is the Chinese river dolphin, or baiji *(Lipotes vexillifer),* which lives in a restricted habitat (the Yangtze River) heavily populated by humans. The baiji has been driven to the verge of extinction from being incidentally caught on fishing hooks and harmed by boat traffic, dams, pollutants, and explosions from construction projects. At the other end of the spectrum, the hourglass dolphin *(Lagenorhynchus cruciger)* has the good fortune to be too small to have attracted the attention of whalers and lives far from shore in the open Southern Ocean, where it does not encounter hazards from human fisheries. But even the hourglass dolphin may be feeling the effects of expanding fisheries, ozone depletion, and other global ecological changes caused by human activities. The transmission properties of sound underwater, which led natural selection to gradually mold cetaceans into acoustic specialists par excellence, have become a curse to cetaceans assaulted by anthropogenic noise. Noise from underwater explosions may be instantly lethal to cetaceans, while shipping may increase ambient noise levels to the point where large baleen whales, already reduced in numbers from whaling, are unable to locate mates over the long distances they did in preindustrial times.

Our ability to accurately assess threats to cetacean populations is limited (even basic population estimates are very difficult), but cetologists must not restrict their input on conservation issues to those cases in which the data are convincing. From public education to challenging assumptions about the burden of proof (do opponents of a development have to demonstrate a high probability that a cetacean population will be harmed or does the developer have to demonstrate that it will not?), researchers must expand their roles in cetacean conservation. If present trends continue, it is clear that cetacean conservation in the twenty-first century will be largely an effort to protect the critical habitats in which whales and dolphins live.

We value the presence of whales and dolphins in our oceans, seas, and rivers for many reasons. Some cetacean species show remarkably complex social behavior. It is encouraging that the number of important discoveries made in a species appears to correlate roughly with the intensity with which it has been studied. And our studies of cetaceans are just beginning. So, if we lose the cetaceans, we lose not only the animals and their societies, but also any chance of discovering what "society" really means to a whale or dolphin.

RICHARD C. CONNOR, JANET MANN, PETER L. TYACK, AND HAL WHITEHEAD

Part 1

HISTORY AND METHODS

1

FLUKEPRINTS

A History of Studying Cetacean Societies

AMY SAMUELS AND PETER TYACK

CETOLOGY IS at a turning point with respect to studies of the sociality of whales, dolphins, and porpoises. Building on elaborate analyses of social structure, cetacean behavioral biologists are beginning to examine details of social behavior, and in doing so, they are shifting their methods from qualitative natural history narratives to focused, quantitative analyses and hypothesis testing. This chapter recounts the events and ways of thinking that delivered cetology to this pivotal juncture.

To tell the story properly requires a cetacean-centric perspective. Little can be learned of the roots of cetacean behavioral biology by reviewing a history of the broader field of animal behavior, whose focus has primarily been terrestrial animals. Until recent years, the two endeavors have had only sporadic direct contact and intellectual exchange. In older animal behavior textbooks, cetaceans appear as exemplars of anthropomorphism (e.g., the dolphin's smile: Tavolga 1969) or untested sociobiological theory (e.g., cooperation: Wilson 1975). More recent texts are virtually devoid of new information about cetacean sociality (e.g., Hinde 1982; Gould 1982; Dewsbury 1984; Manning and Dawkins 1992), although cetacean studies are now emerging in animal behavior anthologies (e.g., Connor et al. 1992a; Tyack and Sayigh 1997).

Traditionally, cetacean biologists placed considerable emphasis on deciphering the social systems of their subjects; that is, determining species-typical grouping patterns on broad temporal and spatial scales. These efforts faced formidable obstacles imposed by marine conditions, to which cetologists responded with resourcefulness and inventive techniques. As a result, long-term research on individually recognized whales and dolphins is now well established, social structure on an ecological scale is known for many species, and extensive demographic, reproductive, and kinship information exists for numerous cetacean communities.

In stark contrast to the elegant analyses of social systems stand the anecdotal descriptions of social behavior that abound in the cetacean literature. Although careful analysis of social structure lays the groundwork for sophisticated investigations of social behavior, it is only recently that cetacean behavioral biologists have begun to examine in any systematic way the intricacies of social behavior that occur on more immediate time scales. Thus, within these community-level frameworks, little is known about the details of social interactions between individuals that occur on a minute-by-minute basis, or about the nature of social relationships between individuals that may change on a weekly, monthly, or seasonal basis.

Research on pilot whales highlights the inconsistency in what is known about cetacean sociality. Considerable effort and ingenious methods have been applied to identifying the characteristic group structure of pilot whales. Through meticulous determination of the age, sex, morphometry, reproductive status, and genetic relations of each member of pilot whale groups captured in drive fisheries, it was established that pilot whales (both long- and short-finned) have a common mammalian social organization. Related females travel together in stable groups that include their immature offspring, whereas males leave the natal group as they reach maturity, thereafter moving between groups of females in search of mating opportunities. These analyses also revealed a phenomenon that is rare among mammals: matrilineal groups often include elderly postreproductive females. However, despite the fact that this exciting discovery was made more than ten years ago, the observational study that focuses on the behavior of postreproductive females and their role in pilot whale society has yet to be conducted.

Such incongruity in what is known about cetacean sociality is perplexing. Many whale and dolphin societies are arguably among the most complex in the mammalian world—why have their social interactions and relationships been given short shrift? Cetaceans have been subjects of extensive and intensive scientific inquiry—why is knowledge of their social behavior often no more than a by-product of other nonbehavioral investigations? Cetacean sociability and intelligence have engendered enormous popular interest and countless myths—why hasn't this attention translated into more systematic, quantitative inquiries into their social behavior?

The customary answer is that the social behavior of difficult-to-see, difficult-to-follow marine animals is simply difficult to study. There are innumerable logistical hurdles that must be surmounted to learn about animals whose lives take place primarily beneath the water and whose behavior is typically viewed from unstable platforms at the water's surface. Many cetaceans are fast-moving, wide-ranging, elusive, and difficult to observe closely at sea. Indeed, studying the social behavior of these animals in the wild is exceedingly hard work.

Nevertheless, numerous nonmarine species are similarly resistant to direct observation because they fly or burrow, are nocturnal or nomadic, live in forest canopies or on ice floes; yet many of these animals are subjects of sophisticated animal behavior research. A good example is the naked mole-rat, whose subterranean existence has not prevented considerable scrutiny into its social organization and behavior (e.g., Sherman et al. 1991). In addition, primatologists—who are customarily thought of as striding across the open savannah trailing highly visible, ground-dwelling monkeys—have long bemoaned "the extraordinary obstacles in the way of primate research. In the often remote areas where primates still survive in nature, there are truly formidable difficulties involving logistics, disease, language, culture, and even violence" (Hamburg 1987: viii). Hardships notwithstanding, field studies of primate social behavior have flourished and are an integral part of primatology, featuring not only the accessible, terrestrial baboons and macaques, but also the more cryptic forest monkeys (e.g., Cords 1987; Strier 1990).

In some cases, the coastal marine environment can actually be more accessible than many terrestrial habitats. Large concentrations of cetaceans live within close reach of urban centers and research institutions. The marine habitat, however, is more than merely a hindrance to direct viewing of underwater activity. The sea is additionally hostile to tangible signs of lifestyle such as tracks, scats, or nests—a factor with which few land-based biologists have had to contend. Thus, cetacean biologists, handicapped by impaired viewing conditions and imperceptible clues, must be singularly inventive in their efforts to detect and decipher the social lives of their subjects. This history, therefore, is in part a celebration of the successes in finding ways to decode cetacean social systems.

The sheer adversity of fieldwork at sea does not entirely explain the present status of research on cetacean social behavior. Other factors have also come into play. Human interests often dictate that certain taxonomic groups, like the cetaceans or the primates, be viewed through special lenses. Consequently, cetology and primatology have each come from different backgrounds, with terminologies, methodologies, and emphases so dissimilar they might well have come from separate cultures. For example, the focus of primate studies was essentially shaped by a desire to know more about the behavior of hominid ancestors (Washburn and DeVore 1961). In concert with input from evolutionary theory, natural history, comparative psychology, and ethology, a humanistic emphasis in primate studies and a resultant interest in individuality (e.g., Rowell 1994) set the standard for behavioral research on many terrestrial mammals (e.g., spotted hyenas: Frank 1986; African elephants: Moss and Poole 1983) as well as studies of marine mammals while on land (e.g., elephant seals: Cox and Le Boeuf 1977).

In contrast to an anthropocentric fascination with primates, attitudes toward cetaceans had a wholly different derivation. Scientists cared little about these animals as individuals because they scarcely considered that the behavior of such strange beings would reveal anything about ourselves. To the contrary, whales and dolphins are sufficiently alien to have been considered appropriate stand-ins for preparations to communicate with extraterrestrial life (e.g., Wooster et al. 1966). Indeed, until a decade ago, it was debated whether cetacean behavior had any resemblance at all to that of terrestrial mammals (e.g., Darling 1988)!

Instead, for centuries, human interest in large cetaceans has been a commercial one. Through the whaling eras of open, human-propelled boats and sailing ships to the modern factory vessels, any knowledge acquired about the social behavior of the great whales was employed to increase whaling harvests or to manage a lucrative but diminishing resource. As a result of this exploitative association, knowledge about whale social behavior has been built upon foundations of population biology, stock management, and analysis of whale carcasses and whaling statistics—avenues of inquiry whose methods, philosophies, and vocabularies imposed idiosyncratic ways of thinking about whale behavior. As whale numbers and habitats declined, the consumptive attitude shifted to a more conservation-oriented perspective, emphasizing population dynamics, social systems, and life history parameters of living whales at sea.

The relationships of humans with small whales, dolphins, and porpoises had a different origin. The existence of the smaller cetaceans was hardly noticed, except as fishermen's pests or mythological beasts, until trained dolphins became the star entertainers of early oceanarium collections. The visibility and accessibility of small cetaceans at aquaria provided opportunities for close-up viewing and hands-on experimentation, thus attracting many scientists to investigate the intricacies of cetacean social behavior, sensory systems, and communication. Early descriptive studies form the basis of much of what is known today about the social behavior of small cetaceans. This prolific period of research was unfortunately short-lived. Anti-captivity sentiments, changes in the general character of zoo-based research, and sensationalized reports of human-dolphin communication all worked together to discourage further captive research on social behavior.

Until recently, studies of the social behavior of whales and small cetaceans operated largely in isolation from the broader field of animal behavior. This seclusion left cetacean biologists unschooled in modern methods of behavioral research, and it was the unusual cetologist who received formal training in behavioral biology. In this respect, cetacean behavioral biology presents another example of the detrimental effects of "carving up science along phyletic lines . . . [as] exemplified in studies of non-human primates. Through limiting their vision by phyletic boundaries, primatologists have too often tackled issues with which ornithologists were already highly experienced" (Bateson and Hinde 1976: 529). Although primatologists with behavioral interests long ago joined forces with the broader field of animal behavior, full integration of cetologists is a more recent phenomenon.

The following narrative outlines the sources of information for what is presently known about cetacean social systems and social behavior. Beginning with the hunting tales of the early whaler-naturalists, this account describes the attempts to infer behavior and social structure from studies of whale carcasses; the intimate observations of the social lives of small cetaceans at early oceanaria; the pseudoscientific explorations of human-dolphin communication; the decline of zoo-based research on cetacean social behavior; and the evolution of present-day, longitudinal field studies with their emphasis on conservation. As chroniclers, we

have tried to introduce and explain the central events, viewpoints, and concerns in the words of the principal players themselves, a task made all the easier by a literature rife with commentary.

The Early Whaler-Naturalists

It is hardly necessary to say, that any person taking up the study of marine mammals, and especially the Cetaceans, enters a difficult field of research, since the opportunities for observing the habits of these animals under favorable conditions are but rare and brief. My own experience has proved that observation for months, and even years, may be required before a single new fact in regard to their habits can be obtained.
Charles Melville Scammon (1874: 11–12)

A long time ago excellent possibilities really existed for observing whales from sailing boats and row boats in immediate proximity. In addition, one should bear in mind that before the invention of the harpoon gun, whalers were forced to study the peculiarities of the whales' behavior in more detail and more scrupulously not only for the success of the whaling but also for their own safety.
Alexey V. Yablokov (1972: 261)

Early whalers of the eighteenth and nineteenth centuries, like predators of any species, had extensive knowledge about the habits of their quarry. Their targets were large, slow-moving cetaceans—such as right, gray, sperm, bowhead, and humpback whales—that could be approached by rowboats. Quiet, open-boat whaling brought whalers and their prey into such close quarters that whalemen were able to describe secretive behaviors of sperm whales, including nursing and copulation (e.g., Bennett 1840; Bullen 1902).

Some early whalemen were self-styled naturalists who wrote about whales with intellectual as well as professional interest. The whaling captains Scammon (1874) and Scoresby (1820) and ship surgeons Beale (1835) and Bennett (1840) published observations of the social behavior, school composition, and natural history of large whales. Beale presented scholarly papers on sperm whale behavior to the Eclectic Society of London. Scammon was a contributor to the *Proceedings of the Academy of Natural Sciences of Philadelphia* and *The American Naturalist*, and published a monumental volume, *The Marine Mammals of the Northwestern Coast of North America* (fig. 1.1). Of his book, Scammon (1874: 11) said: "The chief object of this work is to give as correct figures of the different species of marine mammals . . . as could be obtained from a careful study of them from life, and numerous measurements after death. . . . It is also my aim to give as full an account of the habits of these animals as practicable." Scammon's book was

FRONTISPIECE

WHALING SCENE IN THE CALIFORNIA LAGOONS.

Figure 1.1. Whaling scene in the California lagoons. Frontispiece to the book *The marine mammals of the Northwestern coast of North America together with an account of the American whale-fishery* by Charles M. Scammon, a captain of American whaling and sealing ships (Scammon 1874).

praised by his contemporaries as one that "only a naturalist who combined his scientific knowledge with the experience of a whaleman" could have written (Allen 1874: 632–33).

"The experience of a whaleman" predisposed whalers to be best informed about those behavioral patterns likely to affect the outcome of the hunt. For example, whalers' knowledge of maternal behavior often condemned mothers and calves as easy targets. Whaleman Nordhoff (1895: 174–75) proclaimed the humpback as "the most stupid of whales [because it] clings obstinately to the [calving] place it has once chosen . . . [a fact which is] taken advantage of by whalemen, and great numbers of the old fish are slain annually." Similarly armed with the knowledge that "the right whale mother is very careful to choose a retired and unfrequented roadstead for the scene of her maternal labors" (Nordhoff 1895: 175), whalemen readily preyed upon mothers and calves of that species. Mother sperm whales were known to remain close by "so long as the young showed signs of life. For this reason, whalers, when harpooning calves, tried merely to wound and not kill them, so that both mother and young could be secured" (Caldwell and Caldwell 1966: 759; fig. 1.2).

Not only did whalers exploit bonds between mothers and calves, but "the literature of the eighteenth and nineteenth centuries reveals that many whalemen . . . of that era were aware of the succorant behavior that cetaceans displayed toward their wounded schoolmates, and [used] the

knowledge . . . to increase the whale catch" (Caldwell and Caldwell 1966: 757). Whalers' tales about whales that "hove to" when a schoolmate was distressed or injured (e.g., Beale 1835) formed the basis for present-day hypotheses about "epimeletic" or altruistic behavior of cetaceans (e.g., Caldwell and Caldwell 1966; Connor and Norris 1982). Whalers' recognition of strong bonds between individuals of certain species sometimes enabled the capture of entire social groups. Female sperm whales in particular were known to be "remarkable for their strong feeling of sociality and attachment to one another, and this is carried to so great an extent, as that one female of a herd being attacked and wounded, her faithful companions will remain around her to the last moment or until they are wounded themselves" (Beale 1835: 36).

On occasion, however, whalers' plans went awry when there was "active intervention by sperm whales in the fate of a 'comrade in distress'; for instance, sperms have dived under the ship in order to reach a wounded animal and pull it away from a dangerous spot; in several cases they have broken harpoons, bitten through harpoon lines to free their 'comrade,' or even attacked boats and destroyed them" (Berzin 1972: 256). Beale (1835: 48–49) confirmed that "these enormous creatures are sometimes known to turn upon their persecutors with unbounded fury, destroying every thing that meets them in their course, sometimes by the powerful blows of their flukes, and sometimes at-

Figure 1.2. Whalers took advantage of the tendency of sperm whale mothers to stay close to their young calves as long as the calves were alive. The whalers attempted to wound but not to kill a calf so that the mother could also be struck with a harpoon and killed. Reproduced from a watercolor in the collections of the Old Dartmouth Historical Society– New Bedford Whaling Museum (published in Purrington 1955).

tacking with the jaw and head." California gray whales were considered so dangerous that whalemen regarded them as "a cross between a sea-serpent and an alligator" (Scammon 1874: 272), and the hunt was "appropriately named 'devil-fishing'" (Scammon 1874: 260). Open-boat whaling clearly provided excellent opportunities for close-up viewing of, and even direct participation in, the whales' antipredator responses.

Early whalers were aware of distinctively marked individuals and the locations where such whales could be found "even before Melville transformed the true story of an unusually light-colored Sperm Whale named Mocha Dick into his epic novel, Moby Dick" (Katona and Whitehead 1981: 439). Southwell (1898: 403) reported: "[Captain Gray] states that whalers come to know strongly-marked individuals, and recognize them from time to time . . . In 1867 [Captain Gray] chased a [balaenid] whale 'with a growth like a beehive on the left side of its tail'; in 1872 he killed this same whale, and almost on the same spot. Writing in 1886, [Captain Gray] said that in 1880 he chased a whale with a large white splash on its back, and that he had seen it every year since."

Despite the vast knowledge of whalers concerning the habits of their prey, Southwell (1898: 397–98) cautioned against an inevitable bias: "[I]ntelligent as some of our whalemen have been . . . it must be borne in mind that their main object is the capture of these valuable prizes, and not for the study of their habits, except in so far as such a knowledge would conduce to that result." Indeed, early whalers were less familiar with those aspects of social behavior that did not directly influence hunting success. Thus, for example, the belief that the basic social unit of sperm whales consisted of a "schoolmaster" and his "harem" of females (e.g., Bennett 1840) could perhaps be credited more to whalers' longings in a woman-less society than to their keen powers of observation. Recent evidence refutes the schoolmaster theory, suggesting instead that roving males in search of mating opportunities are short-term visitors to stable matrilineal groups of females (e.g., Best 1979; Whitehead and Arnbom 1987; Whitehead and Waters 1990; Whitehead and Weilgart, chap. 6, this volume).

Smaller cetaceans—the dolphins, porpoises, and small toothed whales—were also captured for food, oil, and skins. For example, "Schools of [long-finned pilot] whales, known in the Faeroes as grind, . . . are hunted at every opportunity by the Faeroese, among whom the grind has a long and venerable history as a source of food" (Williamson 1945: 118). "The first record of whales being put to good use in the Faeroes is dated 1584. . . . Doubtless the whaling

is of much greater antiquity . . . and we may safely take it that that year marks merely the beginning of the written records, which thereafter were kept fairly regularly by the Danish Treasury, since a certain income was derived in tithe" (Williamson 1945: 130–31). In Great Britain, "the [common] Porpoise . . . formed the royal dish even so recently as the time of Henry VIII" (Norman and Fraser 1937: 310). Later, in the nineteenth century, a "species of Delphinus [sic], usually called Bottle-nose, . . . [was] occasionally driven on shore by the inhabitants of Shetland, Orkney, [Faeroe], and Iceland" (Scoresby 1820: 11), and there was "a fishery for the capture of the Bottle-nosed Dolphin . . . carried on from Cape Hatteras, North Carolina . . . [where] between the 15th November 1884 and the middle of the following May, no less than twelve hundred and sixty-eight of them were caught" (Norman and Fraser 1937: 328).

Fisheries for the smaller cetaceans tended to be seasonal, land-based operations, and behavioral information was less extensive than what could be obtained in the course of lengthy voyages in search of large whales. Caldwell and Caldwell (1972a: 149) lamented the loss of "tremendous amounts of good data" because serious biological study was rarely a component of the small cetacean fisheries. Nevertheless, these fishermen had some ideas about the social behavior of their prey (e.g., True 1890). Faeroese whalemen recognized the tendency for pilot whales to "behave . . . very much as though they were a flock of sheep" (Williamson 1945: 121) and used this observation to herd whales onto the beach (fig. 1.3). However, the whalers also claimed that escaped whales came back to be captured because they "'return to the blood,' as though this exerted some hypnotic influence" (Williamson 1945: 123). Williamson rejected this idea, proposing instead that a whale separated from its flok [school] was more likely drawn back by strong social bonds. Cape Hatteras bottlenose dolphin fishermen made similarly erroneous observations of their prey: "[W]hen very young [the calf] . . . is raised to the surface by [the mother] each time she rises to breathe" (reported by Townsend 1914: 299).

Although whalers' accounts were often rendered with a predatory point of view and spiced with stories of "castaways, mutinies, desertions, floggings, women stowaways, drunkenness, . . . hostile natives, barratry, brutal skippers . . ." (Sherman 1965: 22), the wealth of natural history narratives told by whalemen has proved to be a valuable source of behavioral information (e.g., Caldwell et al. 1966; Best 1983; Mitchell 1983; Wray and Martin 1983). In many respects, the early whalers' observations and their interpretations of what they saw form the cornerstone to un-

Figure 1.3. Fishermen in small boats herd a group of pilot whales into shallow water in Trinity Bay, Newfoundland, where they can be killed with lances and hauled onto the beach for flensing. Note how the group of pilot whales, in the center of the photograph, is closely bunched. Photograph by D. Sergeant (Sergeant 1962: 4).

derstanding the natural history and social behavior of the large cetaceans.

Modern Whaling

"Units rather than Whales"[1]

Since whaling is a marine enterprise, most patterns of thought that have been devoted to the harvesting of whales have been derived from fisheries biology, where it is customary to think in terms of populations and aggregates rather than individual animals.

George A. Bartholomew (1974: 295)

The modern whaling era "goes back to the invention of the harpoon gun and explosive harpoon head by the Norwegian Svend Foyn in the 1860s [fig. 1.4]; but it was the development of the floating factory in 1903, and especially of the factory ship stern ramp in 1925 . . . which made expansion into all Antarctic seas possible" (McHugh 1974: 321; fig. 1.5). Technological advances enabled whalers to hunt such fleeter species as blue, fin, sei, Bryde's, and minke whales in addition to the sperm whales that had been accessible to the early whalers. However, "the rapid and more intensive catching methods using noisy, propeller-driven catchers gave less time and opportunity for observations of undisturbed behavior. . . . For a long time after the end of open-boat whaling, therefore, first-hand observations and new data on social behavior . . . were slow in coming" (Best 1979: 228). As an example, Best (1979: 251–54) noted that "although eyewitness accounts of battles between individual sperm whales exist in the literature of open-boat whaling (see Caldwell et al., 1966), only one modern account of such behavior exists (Zenkovich, 1962). This may be because the presence of screw-driven vessels can be detected by sperm whales at a distance of up to eight miles, when their behavior usually changes markedly (Gambell, 1968[a])."

Crews of spotter aircraft sometimes came upon rare spectacles, such as the calving behavior of sperm whales (Gambell et al. 1973) or the protective behavior of hump-

---

1. McVay 1974: 374.

Figure 1.4. The harpoon gun of a Norwegian whaling catcher vessel from the first quarter of the twentieth century. From Commander Christensen's Whaling Museum, Sandefjord, Norway (published in Ellis 1991).

Figure 1.5. Illustration of a British whaling factory ship from the middle of the twentieth century, operating in the Antarctic with spotter planes and catcher vessels. From the Kendall Whaling Museum, Sharon, Massachusetts (published in Ellis 1991).

backs toward their calves during a killer whale attack (Chittleborough 1953). Scientists on whale-catching and marking expeditions also recorded behavioral events (e.g., fin and humpback whales: Andrews 1909; fin whales: Gunther 1949; sperm whales: Berzin 1972; gray whales: Bogoslovskaya et al. 1982). Frederick True (1903), aboard a catcher, produced some of the first photographs of living whales at sea (fig. 1.6). As in the open-boat era, it was the whales' defense against their human predators that was most commonly seen from the whalers' vantage point. A well-known example is "a very strange habit" of sperm whales described by Nishiwaki (1962: 2): "A group of sperm whales, about 20–30 individuals swimming leisurely, was found. The whale catcher approached very slowly and then shot the biggest whale. The instant the whale was hit all individuals of the herd made a circle like a marguerite [daisy] flower centering around the biggest whale. These radially gathered whales put their heads together and made many splashes with their tail flukes."

Although mechanized techniques reduced opportunities for first-hand observation, modern whaling did provide an incentive to better understand the social behavior and social structure of whales: "effective management of heavily exploited wild species obviously requires that harvesting procedures be based on accurate knowledge of their natural history" (Bartholomew 1974: 294; see also, e.g., Schevill 1974; Winn and Olla 1979). With establishment of the International Whaling Commission (IWC) in 1946, it was mandated that regulatory decisions be based on scientific findings (McHugh 1974). Unfortunately, despite "the large body of scientific data about the biology of whales, almost the only aspect of this knowledge that has been used by the whaling industry is information on the abundance of whales and where they can be found" (Bartholomew 1974: 294). The mandate was further undermined when equal consideration was accorded to nonscientific factors (McHugh 1974), which meant that IWC decisions also weighed the financial concerns of the whaling industry and the preservationist attitudes of environmentalists (Peterson 1992). An egregious example of the controlling influence of economics in whaling management was the "blue whale unit," defined as "an arbitrary expression intended to equate different whales on the basis of the amount of oil produced from them. In its later form 1 blue whale was considered equivalent to 2 fin whales, or to 2.5 humpbacks, or to 6 sei whales. This, while convenient for the whalers, was an unfortunate idea for conservation" (Schevill 1974: 414).

Scientific input did prevail in IWC decisions during the 1970s. As a first step, Bartholomew (1974: 295) urged cetologists to "use as our point of departure not fishery biology, but the ecology and social behavior of the large mammals—which, to say the least, is what whales are." With establishment of the New Management Procedures,

the Scientific Committee [was made] far more important . . . [by raising] the level of scientific argumentation that went into decision making. Before 1974, the members of the Scientific Committee had given the IWC a unanimous "best estimate" resting as often on political as scientific grounds . . . Spurred by pressures from outside cetologists and from members of the IUCN [International Union for the Conservation of Nature] and the FAO [Food and Agricultural Organization] . . . the Scientific Committee used the adoption of the new management procedures to establish a more open process in which papers were published, commentary was sought, and the scientific basis of conclusions was made explicit. (Peterson 1992: 164–66)

Figure 1.6. Photograph of a finback whale taken by Frederick True in 1899 from the bow of a Canadian whaling steamship in Notre Dame Bay, Newfoundland—one of the first published photographs of a living whale. (From True 1903.)

As a result, "the Twenty-fourth Meeting of the Commission [in 1972] was notable in a number of respects [including] the end of the blue whale unit as a method of regulating catches" (IWC 1974: 6) and its replacement with biologically relevant quotas related to species and breeding populations.

Such changes paved the way for regulatory decisions based on a more refined understanding of behavior, including the idea that sperm whale "social structure is such that the simpler population models are not at all applicable" (Holt 1977: 133). Many agreed that "both the underlying biology and the manner of harvesting demand that any management model for sperm whales should distinguish between the sexes" (Beddington and Kirkwood 1980: 57). Thus, even though male mammals are seldom considered in demographic models, specific attention was given to adult male sperm whales, and "separate [catch] limits for male and female whales in this species were set" (IWC 1974: 6). Considerable effort ensued to determine which social and demographic factors were critical to developing accurate models (e.g., IWC 1980). For example, after reviewing the survivorship of long-lived mammalian species, Ralls et al. (1977: 241) rejected "the current assumption of equal mortality rates for males and females in [sperm whales] . . . on both theoretical and comparative grounds." In addition, the reproductive role of male sperm whales was a particular concern, stemming from the early whalers' belief that males were "harem masters." Cetologists such as Mitchell (1977: 224) worried that selective whaling for large males would have disastrous effects:

> Behavioural processes occurring during the rendezvous between bulls and schools of mature females are unknown, as is the possibility of replacement of breeding bulls during the mating season. It is not certain whether the "idle" bulls take turns at being harem master . . . or whether bulls, once they reach this "idle" status and assume lengthy, high-latitude migrations, are ever again candidates for "harem master" status. . . . If behavioural and distributional factors insured that after competition for harems, the successful harem master would be the only male servicing a harem . . . then the removal of the harem master could reduce pregnancy rates in that school drastically.

Although it is now known that the harem master concept is inaccurate (e.g., Best 1979; Whitehead and Arnbom 1987; Whitehead and Waters 1990), concerns remain that a reduction in the number of large males may affect reproductive rates (Whitehead and Weilgart, chap. 6, this volume).

Despite concessions to male sperm whales, the overall impact of behavioral considerations on whaling management was minor. In contrast, a strong management em-

phasis exerted a substantial influence on how cetologists thought about and conducted research on whale behavior. Perceptions of whale behavior were also clouded because whaling science embraced and perpetuated a popular social theory, "group selection," that disputed Darwinian natural selection at the level of the individual. An outspoken proponent of group selection, V. C. Wynne-Edwards, proposed "the specific hypothesis that animals voluntarily sacrifice personal survival and fertility to help control population growth . . . [and] that this is a very widespread phenomena among all kinds of animals" (Wilson 1975: 110). Wynne-Edwards (1962: 18–19) described circumstances in which

> the interests of the individual are actually submerged or subordinated to those of the community as a whole. [For example] the social hierarchy . . . is a common and important product of conventional competition, and its function is to differentiate automatically, whenever such a situation arises, between the haves and have-nots. . . . For those high enough in the scale the rewards—space, food, mates—are forthcoming; but when food, for instance, is already being exploited up to the optimum level, the surplus individuals must abide by the conventional code and not remain to contest the issue if necessary to the death. It is in the interests of survival of the stock and the species that this should be so, but it ruthlessly suppresses the temporary interests of the rejected individual, who may be condemned to starve while food still abounds.

For a time, this type of species-benefit reasoning influenced thinking about the social behavior of all animals, and many "early ethologists often assumed that natural selection would produce animals that sacrificed personal reproductive success for the general benefit of their species" (Alcock 1993: 10). Cetologists such as Caldwell and Caldwell (1972a: 57–58), like their counterparts studying terrestrial animals, used this rationale to explain behavioral phenomena: "The subadult male [dolphin] is biologically expendable to the herd, being lower in the social hierarchy than the herd bull and less likely to impregnate the females. Thus by acting as scouts [in times of danger], the subadult males help protect the herd without endangering its long-term social structure or reproductive potential, and thus help maintain the species."

A group-focused view was ultimately shown to be largely erroneous when applied to the evolution of social behavior. "Wynne-Edwards' advocacy of group selection . . . had the . . . effect of stimulating a careful examination of the issues involved, leading to the conclusion that group selection is unlikely to be of widespread importance" (Gadgil 1982: 490). As a result, Darwinian natural selection, with its focus on the individual, soon regained prominence in evolutionary biology and animal behavior (e.g., Lack 1966; Williams 1966; Wilson 1975). With this conceptual shift, traditional explanations were reframed at the level of the individual, resulting in more comprehensive interpretations of behavioral phenomena. Hrdy (1977) gave the example of infanticide by male langur monkeys, which was historically explained as a means of regulating population numbers in the face of overcrowding; infanticide was now better understood as a reproductive strategy that benefits individual males in a number of species, including lions, colobus monkeys, and ground squirrels. Many apparently altruistic acts, previously assumed to be sacrifices for the good of the group, were now better understood as aid to close relatives (Hamilton 1964).

However, group selection explanations had considerable and lasting effects on cetology because explicit links were made to whaling management. In developing his ideas, Wynne-Edwards "took as his starting point an analogy from the whaling and fishing industry. Noting that 'over-fishing reduces both the yield per unit effort and the total yield [Wynne-Edwards 1962: 7–8)]' he argued that animals are no different in principle from fishermen. They must manage their own number to prevent overkilling their own prey" (Le Boeuf and Würsig 1985: 134). These views were then produced as endorsement for whaling practices, in that the "notion of self-regulation of animal numbers by 'self-destruction' for the good of the group was in accord with management philosophy . . . and justified culling. Culling saved animals the trouble of having to do it themselves" (Le Boeuf and Würsig 1985: 134–35). Whaling managers were loath to part with these ideas because, after all, group selectionism supported their own view of how humans should manage an animal resource to maximize yield.

The favor these ideas found in whaling management inevitably spilled over into how cetacean biologists talked and thought about whale behavior. In particular, whaling terminology, coined to facilitate resource management, effectively discouraged thinking about whales in ways that were biologically significant. Not only was the blue whale unit "an unfortunate idea for conservation," but as McVay (1974: 374) pointed out, "by talking in 'units' rather than 'whales,' you make it arithmetic, not biology." Talking in terms of blue whale units, stocks, and barrels was not conducive to thinking in terms of species differences, breeding populations, social units, or individual whales. To promote a change in focus, Bartholomew (1974: 295) urged fellow cetologists: "Perhaps instead of thinking of whales in terms of aggregates, we should think about them as individuals operating in a social context that is maintained by complex individual social interactions." A decade later, Le Boeuf and Würsig (1985) emphasized again the importance of making this conceptual shift.

Today, most biologists concede that group selection models generally "apply to extremely specialized conditions and provide no explanation for the evolution of altruistic traits in vertebrates, except in groups of related individuals" (Clutton-Brock and Harvey 1978: 6; see also Maynard Smith 1976). Kin-based colonies of social insects that form "complex cooperative societies with such internal cohesion and division of labor that they resemble single organisms" (Trivers 1985: 171) still evoke a group-focused perspective (e.g., D. M. Gordon 1987). Hölldobler and Wilson (1994: 107) maintain that "One ant alone . . . is really no ant at all. . . . The amazing feats of the weaver ants and other highly evolved species comes not from complex actions of separate colony members but from the concerted actions of many nestmates working together. . . . The colony is the equivalent of the organism, the unit that must be examined in order to understand the biology of the colonial species." Thus, "to speak of a colony of driver ants or other social insects as more than just a tight aggregation of individuals is to speak of a superorganism, and therefore to invite a detailed comparison between the society and a conventional organism" (Hölldobler and Wilson 1994: 110).

Group-level ideas also play a role in contemporary interpretations of the behavior of some cetaceans. Jerison (1986: 163–64) noted that "information from echolocation can be sensed at the same time by several individuals," which led him to suggest that dolphins may experience "communal cognition," something akin to "an extended self . . . constructed (and experienced) by a group of several animals."

A long-term study of Hawaiian spinner dolphins led Ken Norris (1991b: 13) to conclude that, as with colonial ants, "a spinner dolphin alone is very much less than a whole animal." Norris (1991b: 13–14) elaborated:

> It was only after much looking that we began to understand another key feature of [the spinner dolphins'] lives: they are so thoroughly creatures of their schools that they have surrendered some aspects of normal mammalian individuality to the group. . . . [Spinner dolphins] live locked in the geometry of their schools, playing out a life-long cat-and-mouse game with their predators. . . . [The dolphins'] ultimate defense is to behave like schooling fish. In doing so, their individuality is suppressed in favor of the school.

At other times, with echolocation providing an early warning system to detect predators, these dolphins can "afford to express all the complexity and individuality of their mammalian heritage. . . . But should the predator swim close, they then must revert to the fish's strategy, the school, in which they become faceless ciphers, obeying without question a group strategy" (Norris 1991b: 180–81). Is this something more than Hamilton's (1971) "selfish herd"? This intriguing but controversial proposal, put forward by a scientist well known for provocative ideas that have inspired the careers of innumerable cetacean biologists, awaits its turn for further scientific scrutiny.

Shoot First, Ask Questions Later: Deductions about Behavior from Dead Animals

There is no firm evidence that the post-reproductive phase occupies a major portion of the total life span of the females of any wild mammal other than Globicephala macrorhynchus. . . . *One of the best ways to obtain this information is to use the carcass-salvage approach on a large sample of conspecifics obtained through a fishery.*

Helene Marsh and Toshio Kasuya (1984: 334)

A major contribution of modern whaling to the understanding of behavior comes from a paradoxical source: much of what is known about the social behavior and social structure of large cetaceans has been inferred from dead bodies. Making deductions about social structure and behavior based on studies of cadavers was not unique to cetacean research. Zoology had a long tradition of emphasis on "comparative anatomy and the study of corpses rather than the behavior of living organisms" (Dewsbury 1973: 8). Anthropological investigations of hominid evolution came from an anatomical perspective: measurement of monkey and ape bones preceded a shift to studying living primates (Washburn 1951). Carcass analyses were also a component of research on African elephants, carried out in the course of culls intended to preserve woodland habitats (e.g., Douglas-Hamilton and Douglas-Hamilton 1975). Elephant carcass studies, however, were the direct descendants of cetacean research conducted by cetologist Richard Laws, who modeled his methods for estimating the age, maturational status, and reproductive condition of elephants (e.g., Laws and Parker 1968) on his whale carcass studies (e.g., Laws 1956).

Making deductions about the behavior of difficult-to-see animals based on physical remains resembles the work of paleontologists, who make inferences about the behavior and social structure of never-seen, extinct animals based on fossils. By examining fossilized bones, tracks, and assemblage compositions, paleoethologists attempt to reconstruct the lives of such dinosaurs as carnivorous *Deinonychus*, which probably hunted in packs (Ostrom 1986), and duck-billed *Maiasaura*, which may have provided care for its young in colonial nests (e.g., Horner and Makela 1979; Horner 1982). However, whereas paleoethologists are usually limited to small pieces of a few specimens, cetacean carcass analyses have been based on an enormous sample of bodies and even intact social groups.

A staggering number of whale carcasses were available for study: during 1957–1961, for example, Mackintosh (1965) estimated the average annual world catch to be 64,308 whales, including 31,326 fin whales, 21,155 sperm whales, and 3,598 humpbacks (fig. 1.7). By processing entire bodies and analyzing whaling statistics, cetologists were able to conduct truly cross-sectional life history studies (for example, humpbacks: Chittleborough 1954, 1955a,b, 1958, 1959a, 1960, 1965; sperm whales: Best 1967, 1968, 1969a,b, 1970; Best et al. 1984).

Such a solid foundation of life history data provided a basis from which cetacean behavior, social structure, and mating systems could be deduced. For example, a leading theory about cetacean learning arose from carcass analyses. Brodie (1969: 312) observed that "the most striking difference between odontocete and mysticete reproductive cycles is the significantly longer nursing period of odontocetes . . . [which] is attributed to more sophisticated navigational training, kin-cooperation and complex social structure." Brodie was "one of the first to suggest that the prolonged period of lactation in odontocetes may be related to the importance of social learning" (Tyack 1986b: 145).

As another example, careful studies of cyamid infestations on whale carcasses enabled further decoding of the complex social structure of sperm whales in the Southern Hemisphere:

> The infestation of . . . sperm whales involved only two species, *Neocyamus physeteris* and *Cyamus catodontis*. . . . Females and small males both appear to be infested almost exclusively with *N. physeteris*, but . . . by a body length of 42 feet males are infested exclusively with *C. catodontis*. The point at which 50 per cent of males are infested with either cyamid species is at a length of 39 to 40 feet, and this stage must correspond to one at which male sperm whales become segregated from female schools. . . . [I]t would seem reasonable to assume that at this stage male sperm whales on average enter the Antarctic for the first time. Thus a significant change in the behaviour of the male occurs at the size corresponding to the attainment of puberty. (Best 1969a: 12)

Carcass analyses, combined with observations, were also used to infer mating strategies of male baleen whales. Brownell and Ralls (1986: 107) found that "the exceedingly large testes of the right whale, its longer penis and the apparently much less aggressive interactions between

Figure 1.7. A 30 m blue whale on the flensing plan of a whaling station in Grytviken, South Georgia, in October 1925. Processing of whales on a factory ship or at a shore station was done in conditions suitable for carcass analyses that could be used for life history research. (From Mackintosh 1965.)

males in mating groups suggest that this species has been selected primarily for competition through multiple matings and sperm competition," whereas contrasting characters of humpbacks indicated that "males of this species compete primarily by attempting to prevent mating by other males."

Carcass studies provided new insights into the behavior of small cetaceans as well. Drive fisheries capturing entire social groups have been an important source of information about the social organization of small whales and dolphins. For example, by assessing the age, sex, body length, and reproductive status of short-finned pilot whales captured in Japanese fisheries, Kasuya and Marsh (1984: 307–8) reconstructed the social structure of the species: "[T]he *Globicephala macrorhynchus* school is usually a breeding unit composed of adult males, adult females of various reproductive stages . . . . and immature and pubertal individuals of both sexes . . . Females probably attain sexual maturity in their mother's school. The association of females persists for a long time . . . presumably for life." Based on genetic data obtained in Faeroese fisheries for the closely related long-finned pilot whale, Amos et al. (1991b) confirmed the matrilineal nature of social groups, but disputed Kasuya and Marsh's notion that males are long-term members of female groups: "Upon

reaching maturity, males probably leave their natal pod and begin to visit other pods, mating with receptive females they encounter. . . . [Adult males] tend not to be associated with any one pod themselves for any great length of time" (Amos et al. 1991b: 267).

From carcass analyses, Marsh and Kasuya (1984, 1986) also concluded that postreproductive females appear to be an integral part of pilot whale society, a rare phenomenon among nonhuman mammals. This finding engendered considerable discussion about the roles of elderly females in animal societies (Whitehead and Mann, chap. 9, this volume). Austad (1994: 258) affirmed the importance of the cetacean example: "[T]he single convincing example of substantial female postreproductive life in nature is the short-finned pilot whale. . . . The potential parallel with the parental care strategy of humans seems clear."

Inferring social behavior from dead bodies is making the best of a bad situation. There are certainly drawbacks to making inferences about longitudinal behavioral processes from a cross-sectional carcass perspective, and these attempts have not always been successful. For example, the oft-cited notion that "baleen whales are generally believed to tend toward a monogamous mating system" (Lockyer 1984: 28) likely originated with Nemoto's (1964) analysis of catch statistics for baleen whales on the feeding grounds. Mackintosh (1965: 38) summarized the report: "From Nemoto's material it seems that schools of baleen whales contain an almost random mixture of whales of different sexes, ages, and sexual condition, though there might be a tendency for males and females to form pairs before departing for the breeding grounds." Recent studies of living animals indicate that monogamy is an unlikely mating system for most baleen whale species (reviewed in Tyack 1986b; see also Brownell and Ralls 1986; Whitehead and Mann, chap. 9; Connor et al., chap. 10, this volume).

Whereas paleoethologists will never have the luxury of confirming their theories about dinosaur sociality from first-hand observation, cetacean biologists have been developing ingenious ways to directly and indirectly monitor the behavior and social structure of their elusive subjects. Research on living cetaceans, discussed below, will be the ultimate test of ideas about cetacean sociality that were generated by carcass analyses.

Studies in Captive Settings

Marine Studios: "A Window in the Sea"[2]

Less than 50 years ago virtually nothing was known about the social . . . behaviors of dolphins. Their underwater activities were effectively hidden from view, and since scientists had little comprehension of the behavioral attributes of these small toothed whales, there was neither incentive nor guidance for undertaking field studies . . . This situation changed rapidly when the first oceanarium, Marine Studios . . . . opened in 1938. Here, for the first time, scientists, along with the public could observe bottlenose dolphins at close range and for extended periods from below as well as above the surface.

Forrest G. Wood (1986: 331)

Aquariums can take credit for first bringing dolphins and whales to the world's attention as remarkable mammals that have family life and social behaviour analogous to other mammals. Before this these animals were seen merely as sources of meat, oil and leather products.

Murray A. Newman (1994: 212)

Because small cetaceans were commercially less important than large whales, little was known about their behavior and natural history until studies were conducted in captive settings. Following on the heels of the Victorian home aquarium craze of the 1850s (Rudwick 1992), the earliest records of public cetacean exhibitions date back to the 1860s and 1870s, when beluga whales, bottlenose dol-

phins, and a harbor porpoise were displayed in aquaria in Great Britain and the United States, and a beluga displayed in New York City by celebrated circus man P. T. Barnum was probably the first cetacean trained to perform for the public (Caldwell and Caldwell 1972a; Wood 1973; Defran and Pryor 1980). As Victorian home aquaria "permanently changed our view of sea creatures" from the top-down vantage point of a shorebound observer to an eye-to-eye perspective (Gould 1997: 14), so did public aquaria change perceptions of small cetaceans.

In the early 1900s, aquarium director Charles Townsend (1914: 289) boasted that "New York Aquarium has a school of porpoises and lays claim to the world's best single exhibit of captive wild animals." Fascinated by "the naturally sociable and gregarious habit of porpoises," Townsend (1914: 291–92) provided the first scientific report on the behavior of captive dolphins, including a preliminary description of a "wuzzle":[3] "Frequently three or four [dolphins] will bunch together in the center of the pool, rolling and rubbing against each other in a ball-like mass suggestive of the tussling of puppies." By the early twentieth century, small cetaceans were viewed in aquaria throughout Western Europe and the United States (Defran and Pryor 1980). In these early collections, however, it was the rare animal that lived long, and none produced surviving offspring; thus, the scope of behavioral research was limited.

The establishment of Marine Studios in 1938 marked the beginning of a new era in the public display of small cetaceans. Longtime director of Vancouver Aquarium Murray A. Newman (1994: 81–82), recalled:

> The first successful cetacean exhibition was created not by aquarists but by the film industry. In the 1930s a group of movie producers constructed a large tank just outside of St. Augustine, Florida, filled it with marine life and named it Marine Studios. Their objective was to use it as a safe, convenient set for undersea adventure movies. However, the place excited so much local interest that they soon realized more money could be made by charging admission to their exhibit. In a quick change of strategy, they renamed the huge tank Marineland of Florida and presented the world with its first oceanarium. Marineland's undisputed stars in those days were a colony of bottlenose dolphins . . . maintained and observed by curator Arthur McBride and his successor, F. G. Wood.

"[W]hen the concept of oceanariums was new, no one realized what remarkable creatures [dolphins] were. To Arthur McBride, the first curator at Marine Studios, they were just another possibility for a passive exhibit" (Norris 1974: 56). However, McBride (1940: 16; fig. 1.8) quickly

---

2. Hill 1956.

3. Coined by W. E. Schevill (quoted in Johnson and Norris 1994: 250).

Figure 1.8. Marine Studios, circa 1945. Arthur McBride, the first curator of Marine Studios, is on the right in the white coat. (From Caldwell and Caldwell 1972a.)

realized he had something more than a static display, and he was soon "introducing the readers of *Natural History* to one of their most 'human' deep-sea relatives . . . an appealing and playful water mammal who remembers his friends and shows a strong propensity to jealousy and grief."

The unique underwater viewing opportunities and the stories of sociable dolphins attracted many behavioral scientists to Marine Studios, including an up-and-coming comparative psychologist, D. O. Hebb, from the nearby Yerkes Laboratories of Primate Biology, who collaborated with McBride in a pioneering study of bottlenose dolphin social behavior (McBride and Hebb 1948). Other studies of the social behavior of small cetaceans soon followed, at Marine Studios (e.g., McBride and Kritzler 1951; Wood 1953; Tavolga and Essapian 1957; Essapian 1962, 1963; Tavolga 1966) and at captive facilities elsewhere (California: Norris and Prescott 1961; Caldwell and Caldwell 1967; Hawaii: Bateson 1974; the former Soviet Union: Bel'kovich et al. 1970; Denmark: Andersen and Dziedzic 1964; South Africa: Tayler and Saayman 1972).

Interest in cetacean social behavior was further enhanced by improved survival and breeding success at early oceanaria (e.g., Wood 1977; Prescott 1977). The first of a succession of live captive births was recorded at Marine Studios in

1947, and this calf, a female bottlenose dolphin named Spray, lived to produce five calves of her own (Wood 1973, 1977). Captive colonies containing a diversity of life stages provided opportunities for studying many aspects of cetacean social life. Caldwell and Caldwell (1972a: 31) noted: "Only in oceanariums with communities of captive dolphins can . . . scientists . . . regularly and conveniently observe a semi-natural colony of these aquatic mammals as they play, fight, form deep bonds of affection, reproduce, rear their young, and perhaps even die of old age."

These early natural history accounts furnished the first ever—and in some cases, the only—glimpses into certain aspects of dolphin social life. However, not all observations from early oceanaria have endured subsequent scrutiny. For example, observations of wild dolphins led Wells et al. (1980: 303) to revise McBride and Kritzler's (1951) idea from their captive studies "that the basic social unit for Atlantic bottlenosed dolphins was a family unit consisting of a single adult male and three to five adult females with either first- or second-year offspring." Instead, Wells et al. (1980) found that the composition of wild bottlenose dolphin groups was fluid, with notable long-term associations among adult males or between mothers and their young, but not between adult males and females.

A few other examples illustrate the extent to which these initial descriptions have contributed to current understanding of the social relations of small cetaceans. David and Melba Caldwell (fig. 1.9) were among the first to recognize that "individual recognition and attachment . . . play a major role in cetacean behavior" (Caldwell et al. 1963: 9), and they emphasized the importance of "strong bonds of affection between individuals in captivity. . . . Not only do two animals prefer to associate more with each other than with others in the same captive colony, but these relationships are often retained for long periods of time even when the animals are separated" (Caldwell and Caldwell 1972a: 54). Pre-dating current awareness of the long-term bonds that exist between certain male bottlenose dolphins in the wild (Wells et al. 1980, 1987; Connor et al. 1992a), McBride (1940: 25–26) described a close relationship of two adult males at Marine Studios:

> Because the two males were captured together, apparently their social relationship had been determined previously. . . . There was practically no fighting between the two, and aside from occasional jaw-snapping on the part of the larger, the two were very peaceable. . . . When the [smaller] animal was released into the tank [after a several-week separation], the greatest amount of excitement on the part of the larger male was exhibited. No doubt could exist that the two recognized each other. . . . For several days, the two males were inseparable.

Captive research also refined perceptions of cetacean helping behavior, initially shaped by whalers' impressions.

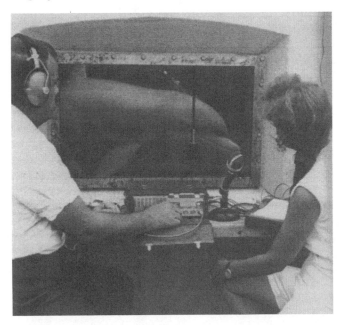

Figure 1.9. David and Melba Caldwell recording the sounds of a captive bottlenose dolphin at Marineland of Florida. This photograph illustrates the eye-to-eye perspective available at captive dolphin facilities. (From Caldwell and Caldwell 1972a.)

New information about the succorant behavior of small cetaceans was obtained not only during captures for aquaria (e.g., Brown and Norris 1956; Siebenaler and Caldwell 1956; Norris and Prescott 1961), but also from detailed observations afforded by captive settings (e.g., McBride 1940; Brown and Norris 1956; Tavolga and Essapian 1957; Norris and Prescott 1961; Lilly 1963b; Caldwell et al. 1963; Caldwell and Caldwell 1966). Helping behaviors such as supporting another and standing by could be closely examined and sometimes better understood within the context of known relationships of participants. Caldwell and Caldwell (1964) even carried out experimental studies that led them to conclude that supporting behavior was a social response, not merely an action elicited by floating objects (as suggested in Slijper 1962), based on dolphins' differential reactions to stimuli that were "inanimate" (a log) versus "animate" (a "life-like" vinyl calf and a thawed carcass!). A subsequent review of helping behavior both in captivity and the wild (Caldwell and Caldwell 1966) resulted in some general conclusions about the cooperative behavior of odontocetes: "[I]t appears that usually only young of either sex or adult females are aided. Adult males do not commonly receive help. . . . It is also much more likely that support will be given to an animal familiar to the group than a stranger—even if the criteria for proper age and sex are met" (Caldwell and Caldwell 1972a: 59). This assessment is now better understood in light of recent indications that there is a matrilineal basis to social groups of many odontocete species in the wild (e.g., bottlenose dolphins: Scott et al. 1990a; killer whales: Bigg et al. 1990b).

Contemporary summaries of bottlenose dolphin agonistic behavior (e.g., Shane et al. 1986) are still largely derived from reports that are many decades old (e.g., McBride 1940; Tavolga 1966). Despite the qualitative nature of those early studies, some of their findings have been confirmed by recent quantitative analyses (Samuels and Gifford 1997). For instance, there is general agreement that adult males are dominant to adult females, and that the agonistic dominance of adult males is distinct from the role of certain adult females as the social foci of their groups (e.g., Tavolga 1966; Tayler and Saayman 1972; Samuels and Gifford 1997). In early studies, however, dominance relations among adult females were typically dismissed as inconsistent or nonexistent (McBride and Hebb 1948; Tavolga 1966), whereas recent quantitative analysis suggests that, although agonism among females is infrequent, females may have stable dominance relationships (Samuels and Gifford 1997).

The accomplishments of the early oceanaria have had long-lasting effects. An important early discovery was that small cetaceans could be readily trained to perform complex behaviors (fig. 1.10). As Caldwell and Caldwell (1972a: 14) remembered:

> Dolphins had been jumping for food and doing simple . . . tricks at Marineland [of Florida] for some time, but one night Cecil M. Walker, Jr. (then a night pumpman and now assistant general manager), happened to notice that one of the bottle-nosed dolphins seemed to be tossing a pelican feather toward him. Walker retrieved it and with patient coaxing developed this behavior pattern until the dolphin was tossing not only the feather but also such substantial objects as pebbles, rubber balls, and small inflated rubber inner tubes. . . . Step by step this simple game developed into the highly trained dolphin shows that can be found in widely scattered corners of the world today.

Soon, Marineland was proudly exhibiting Flippy, the world's first "educated porpoise," tutored by Barnum and Bailey circus man Adolf Frohn (Hill 1956: 181).

It was not only the public that benefited from the cetacean responsiveness to training. "[P]ublic oceanariums have focused . . . scientific attention upon the remarkable attributes of the smaller odontocete cetaceans, such as bottlenose porpoises and pilot whales. As a corollary to this new interest, the biologist now finds that he can deal directly with a porpoise as an experimental subject" (Norris 1966: v). This realization led to productive areas of research on cetacean sensory systems, communication, and cognition, initiated in the United States during the 1950s and 1960s (e.g., Lilly and Miller 1961b; Lilly 1965; Bastian 1967)

Figure 1.10. Trained bottlenose dolphins performing synchronized leaps through hoops at Marineland of Florida. (From Caldwell and Caldwell 1972a.)

Figure 1.11. William E. Schevill and Barbara Lawrence training a bottlenose dolphin for echolocation studies in the early 1950s. (Photograph by James Moulton; courtesy of William A. Watkins.)

and in the former Soviet Union by the 1970s (reviewed in Linegaugh 1976).

Norris (1991c: 295) recalled that "the first conditioned response experiment on a dolphin" was conducted by Lawrence and Schevill (1954) (fig. 1.11), who demonstrated that a bottlenose dolphin could hear sounds above the hearing threshold of humans. In a subsequent experiment, Schevill and Lawrence (1956: 13) showed that a dolphin could find food in water so murky that sight was precluded, "thus supporting the widespread supposition (for example, Kellogg, Kohler and Morris 1953) that . . . cetaceans hunted [by means of echolocation]." Norris et al. (1961) went on to train a bottlenose dolphin to find underwater targets while blindfolded, providing more evidence for an ability to navigate using echolocation.

In the "creative porpoise" experiment, Karen Pryor (Pryor et al. 1969) trained two rough-toothed dolphins to display novel behavior spontaneously (fig. 1.12). Pryor (1975: 236) described the initial trials with the dolphin Malia: "She thought of things to do spontaneously that we could never have imagined, and that we would have found very difficult to arrive at by shaping. . . . Malia seemed to have learned the criterion: 'Only things which have not been previously reinforced are reinforceable.' She was deliberately coming up with something new."

Highly trained dolphins enticed the U.S. Navy to become a principal player in captive cetacean research. "Notty, the Navy's first porpoise" was acquired in the 1960s as part of an effort to improve the hydrodynamics of torpedoes (Wood 1973: 185). Soon thereafter, the Navy embarked on the man-under-the-sea program after losing to the Air Force in a bid to supply life-support systems for the man-in-space program (Wood 1973). Some Navy scientists "thought the exploration of the ocean was just as important as venturing into space, and they had begun plans to study dolphins. They wanted to know how dol-

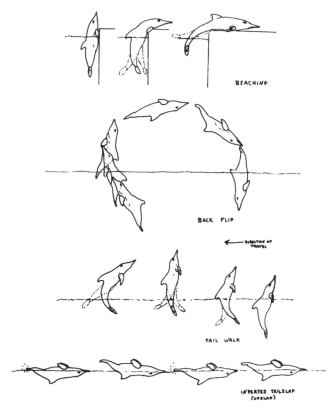

Figure 1.12. Drawings of some of the novel behaviors produced by a rough-toothed dolphin, *Steno bredanensis,* during Karen Pryor's "creative porpoise" experiment. These drawings were sent to trainers to judge how common the behaviors observed during this experiment were prior to training a *Steno* to produce novel behaviors. (From Pryor 1975.)

"The Mind of the Dolphin"[4]

I invite you to entertain some new beliefs about dolphins . . . [that] these Cetacea with huge brains are more intelligent than any man or woman.

John C. Lilly (1978: 1 [emphasis Lilly's])

Individual dolphins and whales are to be given the legal rights of human individuals. . . . Research into communication with cetaceans is no longer simply a scientific pursuit. . . . We must learn their needs, their ethics, their philosophy, to find out who we are on this planet, in this galaxy. The extraterrestrials are here— in the sea.

John C. Lilly (1976: 68)

Since the early captive studies, the notion of a "mind in the waters" (McIntyre 1974) has pervaded research on cetacean social behavior. The individual who popularized this concept was John Lilly, a medical doctor with expertise in neurophysiology. Like many scientists of the day, Lilly gained entry to cetology via Marine Studios in Florida. Former curator Forrest G. Wood (1973: 3) remembered that "Dr. Lilly had first visited the laboratory in 1955 as a member of what we called the 'Johns Hopkins Expedition' . . . [a group of] distinguished neurophysiologists. . . . [T]heir purpoise [sic] in coming to Marineland was to map the cortex of the bottlenose dolphin. . . . But they did not foresee the difficulty they would encounter in anesthetizing a porpoise." After much trial and error, Lilly (1958, 1961b) worked out methods to study brain functioning using less problematic local anesthetic and electrical stimulation (fig. 1.13).

His brain research led Lilly to champion the theory that "the absolute size of the mammalian brain determines its computing capability . . . ; the larger the computer, the greater its power" (Lilly 1967b: 33; see also Lilly 1963a). His ideas rekindled a turn-of-the-century debate about the link between brain size and intelligence (e.g., Gould 1981; Jerison 1986; Klinowska 1988). His brain research, however, also caused Lilly to "become a special target of antivivisectionists" (Wood 1973: 13).

While at Marine Studios, Lilly (1961b) listened to Wood's (1953) recordings of dolphin sounds and became fascinated by the then unfamiliar calls. Lilly provided some of the first structural descriptions of dolphin vocalizations (e.g., Lilly and Miller 1961a,b; Lilly 1963b), many of which are still considered valid today (e.g., Caldwell et al. 1990). He also called attention to the dolphins' facility in imitating sounds (Lilly 1961b, 1965), leading to studies of vocal mimicry, an important area of contemporary ceta-

phins could swim so fast and silently beneath the ocean's surface, how their sonar worked, how deep they could dive . . . [reasoning that] the answers to those questions might be useful to humans trying to live and work under the sea" (Ridgway 1987: 10).

Although largely disinterested in cetacean social behavior per se, the Office of Naval Research (ONR) sponsored the earliest symposia on cetacean research in 1963. Published proceedings from those meetings (Tavolga 1964; Norris 1966) have been key references for those interested in cetacean social behavior, cognition, and communication. ONR also played a major role in promoting early captive research on sensory systems, sponsoring, for example, many of the studies on echolocation, communication, and cognition listed above (e.g., Lawrence and Schevill 1954; Norris et al. 1961; Bastian 1967; Pryor et al. 1969). ONR was also a sponsor of the early work of John Lilly (e.g., Lilly and Miller 1961a,b). However, it was Lilly's later work and ideas, described below, that had widespread influence on studies of cetacean social behavior.

_____

4. Lilly 1967b.

Figure 1.13. Dr. John Lilly causing a captive bottlenose dolphin to vocalize by stimulating its brain electrically. (From Einhorn 1967.)

cean research (e.g., Richards 1986; Tyack 1986b; Janik and Slater 1997). However, Lilly's "attempt to attach a particular whistle to a definite situation" (Caldwell and Caldwell 1965: 434) was never validated. "Lilly . . . may have been the first to hypothesize about the functions of specific dolphin whistles when he wrote of a distress call and an attention call. . . . [With respect to the distress call] Lilly accurately described a widely generalized one-looped whistle of the species, but he assigned to it a specificity and uniqueness of context which was not tested. The evidence for both a context-specific distress call and a complex repertoire of context-specific whistles is weak" (Caldwell et al. 1990: 206).

Brain and acoustics research brought Lilly respect in the 1960s: he was an invited participant at the First International Symposium on Cetacean Research (Lilly 1966), and his reports were published in such prestigious journals as *Science* (e.g., Lilly and Miller 1961a,b; Lilly 1963b, 1965). In the heyday of space exploration, Lilly was an acknowledged expert on communication with extraterrestrial life (Wooster et al. 1966), and his ideas about interspecies communication received funding from federal agencies such as the National Science Foundation, the National Institute of Mental Health, and the Air Force Office of Scientific Research (Lilly 1967b). Ridgway (1987: 10) recalled the persuasiveness of Lilly's argument "that scientists should

learn how to communicate with dolphins to prepare for communication with intelligent life in outer space. Knowing of the dolphin's large, highly convoluted brain . . . many scientists were taken with Lilly's ideas . . . in the beginning."

These, however, were not the accomplishments that so profoundly influenced studies of cetacean social behavior. "These useful, early contributions were followed by a series of books in which Lilly . . . extended his real findings into claims that dolphins possessed a language and that some, such as the sperm whale, possessed an intelligence whose complexity far exceeded our own. [These claims] extended the hope of interspecies communication between humans and dolphins" (Norris 1991c: 298). Cetologists became increasingly critical as Lilly failed to produce tangible evidence to support his ideas (e.g., Caldwell and Caldwell 1965; Wood 1973). Extensive research by Caldwell et al. (1970: 12–13) refuted Lilly's claims, demonstrating that "the message content of dolphin whistles is simple and redundant rather than complex and specific. We found no evidence indicating a 'song patterning' or 'language.' The level of information content in the whistle may . . . even exceed that of other advanced social animals but is much inferior in specificity to even a rudimentary language." Prescott (1981: 130–31) added that "Dr. Lilly's initial results were no more than mimicry. . . . Nearly simultaneously, utilizing the same training techniques . . . an obscure dolphin trainer [at Ocean World, Florida] stumbled upon the ability of dolphins to mimic human sounds. . . . Unlike Lilly, this trainer realized that he had shaped a dolphin's behavior . . . and incorporated the result into a basic animal performance, leaving only the audience to misinterpret the results." Cetacean biologists became all the more dismayed when Lilly's focus shifted to altered states of consciousness, including experimentation with the effects of a psychedelic drug on dolphin behavior (Lilly 1967a).

Disapproval also came from biologists studying other taxa. Sociobiologist E. O. Wilson (1975: 474) denounced Lilly's books, *Man and Dolphin* (Lilly 1961b) and *The Mind of the Dolphin: A Nonhuman Intelligence* (Lilly 1967b), as "possibly the most widely read books on sociobiology and therefore . . . extraordinarily misleading to both the general public and a wide audience of scientists." In his critique of *Man and Dolphin*, Wilson (1975: 474) wrote:

Although Lilly never states flatly that dolphins and other delphinids are the alien intelligence he seeks, he constantly implies it. . . . Anecdotes are used to launch sweeping speculations. . . . Objective studies of behavior under natural conditions are missing, while "experiments" purporting to demonstrate higher intelligence consist mostly of anecdotes lacking quantitative measures and controls. Lilly's writing differs from that of Herman Melville and Jules Verne not just in its more modest literary merit but more basically in its humorless and quite unjustified claim to be a valid scientific report.

Wilson (1975: 474) said that he "dealt frankly with these two books [because a] noncommittal attitude only serves to perpetuate the myth that Lilly helped to create."

If Lilly helped to create a myth, it was eagerly embraced and promoted by many others. Scientific disclaimers did little to diminish the public fascination with possibilities for communication with "an alternate sentient being—benign, philosophical, and gifted with the patience and wisdom of the sea" (Parfit 1980: 73). Lilly charmed the public with his popular books and his self-portrayal as the one "willing to stick his neck out" in defiance of narrow-minded scientists (Lilly 1961b: 135). Captivation with the promise of cetacean language and intelligence was also fanned by numerous nonprofessional publications, including pseudoscientific books (e.g., Stenuit 1968; Fichtelius and Sjolander 1972) and feature articles in such magazines as *Life* and *The Saturday Evening Post* (Lilly 1961a; Schulke 1961; Appel 1964). A blockbuster novel, *The Day of the Dolphin* (Merle 1969), which, according to Lilly (1978), was loosely based on his life and ideas, was widely misconstrued as factual (Wood 1973). "[I]n recent years the authors of one popular book after another have started from the basic premise . . . that the cetaceans represent a high order of . . . intelligence. Human nature and the press being what they are, some of these accounts have received wide publicity . . . to the extent that complex dolphin sociology and high cetacean intelligence have joined motherhood and apple pie in the public mythology" (Gaskin 1982: 115).

The status of cetaceans as "floating hobbits" (Pryor and Norris 1991c: 2) was clinched with the 1963 hit movie *Flipper,* which resulted in a long-running television series. "Even the Soviet government embraced an idealized image of the porpoise . . . [announcing in 1966] a ban on the catching and killing of porpoises . . . [because] extensive research both in Russia and abroad had shown that the porpoises' brain power makes them 'marine brothers of man'" (Wood 1973: 7). Public infatuation went so far as to promote dolphins as "the status pet . . . you can enjoy in your own swimming pool" (Ciampi 1964: 22). Deploring the huge impact of Lilly's ideas, Wood regretted having been the one to introduce Lilly to dolphin vocalizations: "I'm not sure if I hadn't played him the tapes the world would be a different place now" (quoted in Parfit 1980: 74).

The impact extended to obstruction of scientific prog-ress. For example, "the important discovery that each dolphin has a unique acoustic signature, first reported [in Caldwell and Caldwell 1965], languished while many embraced the more fashionable view that dolphins had a complex language and that it was only a matter of time before researchers could decode it" (Leatherwood 1991: 98). Reeves (1983: 709) worried that negative reactions to Lilly's ideas might be counterproductive: "I wonder if [Gaskin (1982)] hasn't over-reacted to some of the anthropomorphisms and sentimentalisms, in the process becoming not only provocative but defeatist. After all, . . . serious questions about cetacean intelligence and social structure do not deserve to be dropped entirely just because a few investigators have approached them irresponsibly." Lilly's ideas about "the mind of the dolphin" flew in the face of a long-standing opposition among animal behaviorists to anecdotal studies of the animal mind and consciousness (Galef 1996). Biologists studying other taxa who looked askance at Lilly's work sometimes regarded with suspicion the cetologists whom they considered Lilly's colleagues, asking, "Can you be a serious scientist if you work with dolphins?" (Norris 1991c: 298).

Few contemporary discussions of cetacean intelligence credit (or even mention) Lilly's ideas, and few cetologists care to be linked with the name of John Lilly. Nevertheless, Lilly's initial work set the stage for productive, legitimate scientific inquiries into cetacean communication and cognition (e.g., Richards et al. 1984; Richards 1986; Tyack 1986a; Caldwell et al. 1990). Unfortunately, Lilly's influence has also lived on in an unwavering public mythology that continues to bias perceptions of scientific studies of cetacean behavior.

The Decline in Captive Behavioral Research

[I]nterpretation of behavior observed in captivity must be approached with great caution. Moreover, captive bottlenose dolphins display a marked propensity to learn complex behavior sequences by imitation . . . and thus studies of their behavioral repertoire are fraught with further possible pitfalls of misinterpretation.
Graham S. Saayman and Colin K. Tayler (1979: 166)

After several decades of productive research in zoo and aquarium settings, there has been a virtual hiatus in captive studies of cetacean social behavior since the 1970s (but see, e.g., Overstrom 1983; Wells 1984; Tyack 1986a; Östman 1991; Samuels and Gifford 1997). This decline cannot be attributed to any single, definitive cause; rather, the evidence points to a suite of factors. Le Boeuf and Würsig (1985: 143) suggested an economic explanation, that "conducting marine mammal research in the lab is much more costly today [than a few decades ago] in large part because of the legal husbandry requirements. . . . Consequently, outside of commercially self-sustaining oceanaria, not much behavioral work is being done on captive animals." This interpretation fails to explain why primate behavioral research has not been similarly affected, why research on cetacean social behavior has declined in oceanaria as well, or why captive studies of cetaceans now emphasize sensory systems over social behavior.

A transformation of the general character of zoo and aquarium research may have been a contributing factor. At the time when behavioral studies of captive cetaceans were flourishing, behavioral research in general was thriving in zoo environments (e.g., Morris 1966; Kummer and Kurt 1965; Rabb et al. 1967). Indeed, "many influential biologists owe much of their interest in biology to early experiences of animals in zoos, and this is particularly true of ethologists" (Robinson 1991: 120). Since the 1980s, however, there has been a decline in appreciation of the value of zoo research that focuses on social behavior (Kleiman 1994). Nowadays, zoo research is typically driven by the needs of collection management and wildlife conservation, with an emphasis on "high-tech" applied research, particularly in genetics (Kleiman 1992; Thompson 1993; Wemmer and Thompson 1995). Kleiman (1992: 310) worried that "zoos will be making a major mistake if they totally abandon more classical descriptive behavioral research and basic behavioral research." This loss has already been felt in studies of cetacean social behavior.

For cetaceans, it seems likely that a decline in zoo-based behavioral research is also part of the Lilly legacy. In the late 1960s Lilly became a staunch opponent of maintaining cetaceans in captivity and closed his dolphin lab, saying, "I began to see the ethical implications of my beliefs about dolphins. If what I believed about dolphins was true, I had no right to hold them in a concentration camp for my scientific convenience" (Keen 1971: 77). Lilly was inspirational to the animal activist movements that proliferated in the 1970s (e.g., O'Barry and Coulbourn 1988), and the burgeoning anti-captivity stance was no doubt strengthened by the wide publicity accorded to the beliefs of someone of Lilly's stature (e.g., Keen 1971; Hussain 1973).

On the one hand, public attraction to cetaceans was engendered in part by oceanarium displays. For example, the killer whale's fearsome image was rendered more benign as the public gained close proximity to whales in captive exhibits. Whereas previously "Navy training films portrayed killer whales as dangerous vermin that might attack lifeboats and swimmers [and] some military fliers reportedly used them for bombing practice" (Pryor and Norris 1991a: 383), Newman (1994: 160) later reported "how quickly attitudes towards killer whales have changed over the decades since we captured Moby Doll in 1964. After years of people seeing . . . [killer whales], there is a distinct social revulsion against shooting them." This positive change in public opinion had the additional effect of fostering anti-captivity sentiments: "Many ordinary citizens became uneasy about [live captures of killer whales] and began to press for regulation of the numbers captured" (Newman 1994: 155; see also Bigg and Wolman 1975).

The public also became critical of cetacean research conducted in captive settings. In the United States, distrust of the scientific profession came from a coupling in the public eye of captive dolphin research with military efforts and the much-hated Vietnam War. The public was told that "[dolphin] research has passed from those with broad interests and a love of the animals [like Lilly] to scientists with narrow interests and US military money. . . . The dolphin has become just another experimental animal, but one which conveniently can be trained to perform military tasks and dolphinaria tricks" (Hussain 1973: 182). Forgotten were Lilly's own failures to resuscitate his dolphin subjects (Lilly 1961b), the military dollars that sustained his early research (Lilly 1967b), and his willingness to dose dolphins with psychedelic drugs (Lilly 1967a). Even when Lilly later softened his position to resume research with captive dolphins (Lilly 1978), biologists working with captive cetaceans continued to be viewed with suspicion by animal rights groups.

These sentiments came on top of an already uneasy partnership between scientists and members of the marine mammal public display community. Pryor (1975: 2) admitted that, even in early oceanarium days, "Public exhibits and private research didn't mix well. Experiments sometimes detracted from exhibits, and the scientists on the staffs of these oceanariums told horror stories of precious research animals being pressed into public shows just when the data collecting was getting good." As the first curator of Marineland of the Pacific, Norris (1974: 99) recalled having conflicting views about his own research: "[T]he housekeeping for a porpoise is expensive; their tanks must be kept clean and supplied with running sea water, and they eat a dozen or more pounds of fish every day, so it is wasteful to have nonproductive [i.e., nonperforming] porpoises at an oceanarium."

Even when it was possible to gain access to dolphins for research purposes, some scientists worried that aspects of captive conditions—such as atypical group composition, human-animal bonds, training, sensory deprivation, or restricted space—might distort the animals' natural social be-havior (e.g., Gaskin 1982). Those beliefs were part of a broader debate between animal psychologists and etholo-gists: "Ethologists have traditionally supposed the effects of captivity to be distorting [whereas] psychologists have supposed them to be innocuous or helpful" (Boice 1981: 407). Wood (1986: 332) responded to this controversy by noting that "these are, of course, well-founded concerns" and providing guidelines for assessing the suitability of a given captive situation: "[T]he competent student of dol-phin behavior will take into consideration the quality of the captive environment. Does it provide some simulation of natural conditions . . . ? Is the dolphin colony relatively stable and do births occur regularly? Does the behavior of the animals appear unstressed and natural, as opposed to stereotyped and with indications of boredom?"

Wood's common-sense checklist suggested that behav-ioral observations made in captivity could be cautiously in-terpreted by taking the specific captive circumstances into account (see also, e.g., Saayman and Tayler 1979; Johnson and Norris 1986; Östman 1991). Many cetacean biologists nevertheless remain unconvinced of the validity of captive studies, some apparently misinterpreting cautious expla-nations as denials of worth. Newman (1994: 199) recalled that "a schism had developed among marine mammal sci-entists, and it was very visible at [the Sixth Biennial Confer-ence on the Biology of Marine Mammals in 1985]. The split was over the issue of aquariums and captive cetaceans. . . . Ken Norris . . . represented the faction in the society that appreciated the value of captive cetaceans to the acqui-sition of scientific knowledge . . . [and particularly] in be-haviour studies."

One outcome of this long-standing controversy has been that areas other than social behavior have taken precedence in captive cetacean research. Specifically, "since the discov-ery of the echolocating capabilities of dolphins, most of the [captive] behavioral work has concentrated on the acoustic modality" (Tavolga 1983: 19). In the Navy's psychophysi-cal approach to studying echolocation (e.g., Au 1993), the presumed artificial effects of captivity are no cause for con-cern. To the contrary, a controlled setting is ideal for exper-imental tasks designed to reveal the extent of odontocete echolocational abilities rather than to determine how ani-mals actually make use of these capabilities. Such studies have demonstrated, for example, that "trained bottlenose dolphins can detect the presence of a 2.54 cm solid steel sphere at . . . nearly a football field away (Murchison 1980) . . . [and] can discriminate targets that are identical in terms of shape and differing only in composition (e.g., Kam-minga and van der Ree 1976)" (Tyack 1998).

Nevertheless, a stalwart few maintained the belief that a dialogue between field and captive efforts holds "the key to understanding wild dolphins" (Pryor and Norris 1991b: 291; see also, e.g., Saayman and Tayler 1979; Wells et al. 1980; Pryor and Kang Shallenberger 1991; Norris et al. 1994; Samuels and Gifford 1997). Saayman et al. (1973: 229–30) explained that "preliminary studies demonstrate the importance—indeed, the necessity—of simultaneously conducting complementary studies of animals under con-trolled captive conditions, where details of behaviour can be determined at close range, as well as under free-ranging conditions, where behaviour observed in captivity can be seen functioning under the appropriate socio-ecological cir-cumstances." Norris (1985: 7) voiced a similar viewpoint in a plenary paper at the Sixth Biennial: "A full understand-ing of the behavior of marine mammals requires studies both at sea and in captivity. Each provides a different view of behavior, and by working in both ways one may check and correct interpretation made in each."

These scientists appreciated the unique benefits of work-ing in a controlled environment, including the ability to view cetaceans underwater and at close range, to observe entire sequences of behavior, and to monitor the social rela-tionships of known individuals over the long term. Such detailed, close-up observations have permitted investiga-tions difficult to achieve in the wild, such as studies linking social behavior and hormones (e.g., Wells 1984) or evaluat-ing the functional significance of specific social interactions (e.g., male-female sexual behavior: Puente and Dewsbury 1976; male-male sexual behavior: Östman 1991; agonis-tic behavior: Samuels and Gifford 1997). In addition, glimpses of social behavior at sea have sometimes been bet-ter understood in light of intensive viewing of the same interactions in a captive setting (e.g., Saayman and Tayler 1979; Pryor and Kang Shallenberger 1991). For instance, in their behavioral study of dolphins trapped in tuna nets, Pryor and Kang (1980: 74–75) noted that "experience with spotters and spinners in captivity was fundamental to ob-servation in the nets. Virtually all of the behavioral events . . . were well-known to us from captive animals. . . . Our familiarity with individual behavioral patterns, as evi-denced by the preparation of a very adequate 'dictionary' before going to sea, allowed us to identify actions which might be indecipherable to a novice observer."

Whatever the reason, it remains true that, since the 1970s, aquarium-based studies of cetacean social behavior have been virtually nonexistent. As unfortunate as this lapse is for a better understanding of the social behavior of small cetaceans, captive research played a significant role in pro-moting another major phase of cetacean research: field studies. In the 1970s, some cetologists took their cue from the primatologists, many of whom were shifting toward field research. At the time, Evans and Bastian (1969: 470–71) noted that

> the current state of [cetacean behavioral research] is very reminiscent of the recent history of primate behavior studies. Popular interest has long supported public display of captive primates in much the same way that the cetacean . . . displays now enjoy the public's fancy. But although much was written about primate social behavior based on close observation of these captive groups, a large part of the ideas that resulted from these efforts has been forced to be drastically revised. The recent flourishing of ecologically sophisticated studies of free-ranging populations that has been the happy lot of behavioral primatology has provided a much deeper and fuller understanding of the social life of these animals. . . . Our fervent hope is that the same history will unfold in the study of the social behavior of marine mammals.

Norris (Pryor and Norris 1991a: 385) recalled that it was not only the limitations but also the exciting discoveries of captive research that inspired cetologists to take the next step into fieldwork: "From the first few captives in oceanariums, we began to understand that these cetaceans were complex mammals many of whose behavior patterns bore a startling resemblance to those of terrestrial mammals. So a few people . . . began to grapple with learning about dolphins at sea."

Field Studies

"No Longer Must We Kill Whales to Study Them"[5]

These new [passive observational] approaches are bound to bring new understanding. Far from nurturing the growth of knowledge in whale biology, I feel that the availability of large numbers of corpses, and thus the possibility of more years of the old study methods, has actually held back the growth of this branch of science.
Roger Payne (1983: 3)

By the 1960s and 1970s, many cetologists had concluded that only limited deductions could be made about cetacean social systems and behavior without observations of live animals in their natural environment. Those who embarked on field studies of small cetaceans were inspired by findings from oceanarium studies or spurred by their concerns about captivity. Those who went to sea to study large whales often did so in opposition to the traditional reliance on carcass analyses or in abhorrence of killing whales. McVay (1974: 381) felt strongly that "cetology has for a long time been a 'dead' science. . . . [T]he bulk of the

scientific reports are based on data taken from dead whales and those data consequently are industry dependent. This means that, wittingly or unwittingly, the whale scientist may often be in a parasitic relationship to the whaling industry. . . . [W]hat has been missing from the equation has been any systematic study of the whole organism and its relation to group and environment." In the same vein, Payne (1983: 2) rejected the idea that "serious science cannot be done without dead whales," a stance he believed to be fostered by the whaling industry to garner continued support for whaling. As an alternative, Payne (1983: 1) presented an edited volume, *Communication and Behavior of Whales*, the studies in which were "all based on passive observation techniques. There is no result in this book that was derived from killing, capturing, confining, or even touching a whale . . . [which demonstrates] that basic science can be done at a useful level of rigor . . . without resorting to intrusive techniques or commercial whaling operations."

The disdain for studying the corpses of whales was long preceded by similar sentiments in other animal studies. At the beginning of the twentieth century, ornithologist Edmund Selous "declared war on all previous ornithological writing" (Stresemann 1975: 342), stating that the "zoologist of the future should be a different kind of man altogether: the present one is not worthy of the name. He should go out with glasses and notebook, prepared to see and to think. He should stalk the gorilla, follow up the track of the elephant, steal up on the bear . . . but it should be to biographise these animals, not to shoot them."

> Some men have strange ambitions. I have one:
> To make a naturalist without a gun.
> (Selous 1905: 323)

By the 1920s, many ornithologists reacted against carcass studies: "We are concerned here not with the study of skins . . . but rather with subtleties of behavior, with growth and development, with molting, with instinctive actions and mental abilities—in short, with matters that up to now have been scarcely considered" (O. Heinroth, quoted in Stresemann 1975: 348). Allee (1933: 320) made a similar remark in his review of Bingham's (1932) monograph, *Gorillas in a Native Habitat*: "[A]t last such field studies have been put on a sound basis which should result in the hunting of information rather than of specimens."

Admonishments like these led to "a new generation of Dutch and German ornithologists [and zoologists] that soon became the leading investigators of behavior" (Stresemann 1975: 348). These classical ethologists of the 1930s and 1940s were "careful observers who were more con-

---

5. Darling 1988: 872.

cerned with the observation and description of behavior under natural conditions than with the formulation of complex theories. To use Tinbergen's label (1958), ethologists were 'curious naturalists'" (Dewsbury 1984: 10). By the 1950s, modern ethology emerged as the classical form blended with such disciplines as ecology, comparative psychology, and physical anthropology (Hinde 1966; Dewsbury 1984).

Perhaps because of the lucrative influence of whaling, or perhaps because "at first, we cetologists literally did not know whether behavioral studies . . . in the wild were possible" (Pryor and Norris 1991a: 385), it was not until somewhat later that cetacean biologists were able to replace carcass studies with research on live animals in the natural environment. Caldwell (1955) and Schevill and Backus (1960) set the stage for studies of free-ranging cetaceans by demonstrating that a bottlenose dolphin and a humpback whale, respectively, could each be identified over a period of days while alive and at sea (fig. 1.14). These studies established that it was possible to obtain information about the behavior and ranging patterns of whales and dolphins

in the wild. Many cetologists followed this lead, and cetology increasingly came to emphasize nonlethal methods for studying free-ranging animals. Norris (1991a: 9) recalled early field studies of small cetaceans: "[B]y the late 1960s, a few Western naturalists had hitched up their field pants and begun to seek out the best means and the best places to observe wild dolphins. They chose sea cliffs, they developed little radios that could be affixed to dolphin fins, and they began to watch dolphins underwater."

Calling the movement toward field research "a fresh breeze," McVay (1974: 381) singled out other exemplary efforts: "While attention to the natural history of cetaceans is not new, the beginnings of a stronger orientation toward living cetaceans are found in such work as the phonograph [record] . . . produced by Schevill and Watkins in 1962. Scientists are now determined to know the whale in its natural habitat of the sea."

The incentive for many early field-workers was more than scientific or moral. Whitehead (1989b: vii) described his 1981–1984 voyage to the sperm whales of the Indian Ocean as "an immensely powerful experience. . . . It intro-

Figure 1.14. William E. Schevill observing a right whale at sea. (Photograph by William A. Watkins, Woods Hole Oceanographic Institution.)

duced me to one of the largest and most unusual animals on earth in more intimate terms than I had ever known any other. . . . We encountered other unusual and intriguing animals, sailed to strange and exciting countries and fought our way through an almost endless sequence of obstacles."

"[I]ndividual Identification Has Indeed Become a Staple of Field Research"[6]

Megaptera novaeangliae is a species in which minor individual variations are often sufficiently conspicuous and distinctive to enable even a shipboard observer to recognize individual whales out of small groups. . . . [O]ur subject was readily distinguishable by its larger size, by the shape of the dorsal fin or hump (especially variable in this species), and by the distinctive color pattern of the underside of the flukes (markedly unlike any of the others with it).

William E. Schevill and Richard H. Backus (1960: 279–80)

Early field-workers not only launched studies of free-ranging cetaceans, but perhaps more importantly, demonstrated that it was possible to repeatedly find and recognize naturally marked individual cetaceans on separate occasions. Ornithologists had been the first to appreciate the significance of recognizing individual animals for the study of behavior and social structure. In the late nineteenth century, bird banding—enabling recognition of individual birds without their recapture—was pioneered as an aid to migration research (Delany 1978). By the 1930s, "important insights into the structure and dynamics of societies were provided by the colored band. . . . For the first time it was possible to follow the fate of individual birds from birth to death, determine exactly their fertility rate, examine their relation to other members of the same population, and obtain much other information about which previously there has been only the vaguest notions" (Stresemann 1975: 359). For some time thereafter, for most species, "artificial marking and tagging was considered almost a prerequisite for behavioral work" (Würsig and Jefferson 1990: 43).

Although artificial tags continue to be used for such difficult-to-distinguish species as birds (but see, e.g., Scott 1978), by the 1960s and 1970s, biologists came to realize that individuals of many marine and terrestrial mammalian species could be identified by natural marks. For example, it became well known in primate field studies that "monkeys and apes tend to show so much variation in their facial and other features that numerous individuals can be recognized" (Schaller 1965: 628). Thus, Schaller (1963) kept track of individual mountain gorillas by making a collection of

"nose-print" diagrams, and Goodall (1971) was able to identify individual chimpanzees by unique facial characteristics. Animals other than primates were also found to be individually distinct: the Douglas-Hamiltons (1975) recognized African elephants by looking at tusk shapes and ear outlines, and Pennycuick and Rudnai (1970) discovered that lions could be precisely discriminated by patterns of vibrissa spots. The unique striping patterns of plains zebras were detected by the Klingels (1965), who went on to lead the development of photo-identification techniques (Moss 1975).

Since the Discovery Investigations of the 1920s, individual whales had been monitored by means of artificial tags to obtain estimates of population parameters and detect migration patterns. Unfortunately, the Discovery "marks" were shot into the whale and could be recovered only after a tagged whale was killed and its body rendered for oil (Brown 1978). Many cetologists felt it was time to replace these lethal tags and "consider other means of carrying out research on large whales . . . without killing large numbers of animals" (Brown 1978: 73). The findings of Caldwell, Schevill, and Backus offered promise that for some cetacean species, individual recognition might be accomplished by noninvasive means.

The idea caught on quickly. "[T]he extensive use of natural marks [to identify individuals] began for four odontocete species in five widely separated projects all within a two- or three-year period . . . [at the same time as individual] recognition of humpback whales . . . and Southern Hemisphere right whales" (Würsig and Jefferson 1990: 43). Payne (1995: 63) remembered preliminary stages of the southern right whale project at Península Valdés, Argentina: "I guessed that by photographing the heads of all the whales from the air we could create a 'head catalog' of known callosity patterns and thereby keep track of individual whales over long periods. In 1971 we demonstrated that this was indeed feasible but more time had to pass before we were finally sure that the patterns were constant enough to be used in identifying right whales throughout their lives" (see also Whitehead and Payne 1981; R. S. Payne et al. 1983). Bigg (1994: 14; see also Bigg 1982) described a similar realization in the 1970s about photographing killer whales of the Pacific Northwest: "The pictures revealed several individuals with distinctive nicks and gouges on their dorsal fins. This provided us with natural identification tags. . . . We had now discovered a method to study killer whales and could begin documenting the life histories of many individuals" (fig. 1.15).

Thereafter, many cetacean field-workers learned to discriminate the often subtle natural markings that distinguished individual whales and dolphins (reviewed in

6. Würsig and Jefferson 1990: 43.

Amy Samuels and Peter L. Tyack

Figure 1.15. Mike Bigg and Graeme Ellis photo-identifying individual killer whales from a small boat. (Photograph by Flip Nicklin/Minden Pictures; from Hoyt 1984.)

Hammmond et al. 1990). These markings included color patterns on ventral surfaces of humpback whale flukes (Katona et al. 1979; Katona and Whitehead 1981); marks on the trailing edge of sperm whale flukes (Whitehead and Gordon 1986); dorsal fin markings and back pigmentation of minke whales (Dorsey 1983; Dorsey et al. 1990); dorsal fin nicks and shapes of bottlenose dolphins (e.g., Irvine and Wells 1972; Würsig and Würsig 1977; Shane and Schmidly 1978; fig. 1.16), humpback dolphins (Saayman and Tayler 1979), and spinner dolphins (Norris and Dohl 1980a). For some species, individual recognition was enhanced and its validity confirmed by nonlethal artificial marking techniques such as visual tags and freeze-branding (e.g., bottlenose dolphins: Irvine et al. 1982).

Thus, in the 1970s, the ability to recognize individual animals repeatedly over periods of years ushered in a new era of long-term field research for marine and terrestrial species alike. Wells (1991c: 201) recalled that "when our research program [on bottlenose dolphins in Sarasota waters] began in 1970, it was not planned with the intention that it become a long-term study." However, Sarasota researchers learned that "conclusions based on short-term data tend to be simplistic and transitory. Collecting data for only 2 or 3 years is unlikely to give a complete picture of a complex society of long-lived animals" (Scott et al. 1990a: 242).

Longitudinal studies of many terrestrial mammals and marine species that periodically come to land were initiated in the 1960s and 1970s, with some continuing for decades

(e.g., chimpanzees: Goodall 1965, 1986; Pusey et al. 1997; baboons: Altmann and Altmann 1970; Altmann 1991; lions: Schaller 1972; Packer et al. 1988; elephants: Moss 1977, 1988; elephant seals: Le Boeuf and Peterson 1969; Le Boeuf and Reiter 1988; many other species: Clutton-Brock 1988). Long-term field studies of cetaceans also began at the same time, with many ongoing since their inception (e.g., bottlenose dolphins: Irvine and Wells 1972; Scott et al. 1990a; Wells 1991c; killer whales: Bigg 1982; Balcomb et al. 1982; Ford et al. 1994; spinner dolphins: Norris and Dohl 1980a; Norris et al. 1994). Payne (1995: 102) exulted in the returns from twenty-five years of studying right whales: "[W]e now know over twelve hundred individual right whales. We are expecting our third generation of calves—descendants of mothers we first met back in 1970, many of whom are still alive and still in their calf-bearing years."

During the same period, the field of animal behavior came into its own: the Animal Behavior Society was organized in 1964, a number of professional journals were established (e.g., *Animal Behaviour, Aggressive Behavior, Hormones and Behavior, Behavioral Ecology and Sociobiology*), and many textbooks were published (e.g., Dewsbury and Rethlingshafer 1973; Alcock 1975; Wilson 1975; Colgan 1978). In 1973, the contributions of ethology were honored when Konrad Lorenz, Niko Tinbergen, and Karl von Frisch were awarded the Nobel Prize, an "event [which] provided inspiration for all animal behaviorists" (Dewsbury 1984: 11).

Figure 1.16. Repeated photographs of three individual bottlenose dolphins, showing the stability of natural markings over a year in the wild. (From Würsig and Würsig 1977.)

Kodachrome, Hydrophone, and a "Semisubmersible Seasick Machine"[7]

It is obvious that no matter where and how it is studied, the whale requires the application of a wide range of innovative methodologies and techniques.

Howard E. Winn and Bori L. Olla (1979: xii)

Cetologists had to devise ingenious methods to monitor their elusive subjects (reviewed in Read 1998; Whitehead et al., chap. 3, this volume). Many investigative techniques that would become standard in cetacean fieldwork were added to the toolkit during these early efforts. With the

7. Norris and Wells 1994: 58; coined by W. E. Schevill (W. A. Watkins, personal communication).

near-simultaneous proliferation of several field projects in the 1970s, it is difficult to pinpoint who first developed or applied which technique. However, it is widely acknowledged that Roger and Katy Payne (fig. 1.17) and their colleagues were especially influential in introducing, adapting, and validating a number of methodologies for studying cetacean behavior at sea; therefore, a review of procedures implemented during the initial fieldwork at Peninsula Valdés provides a sampler of techniques routinely employed in cetacean field biology today.

In studies of bottlenose dolphins at Peninsula Valdés, the Würsigs were among the first to apply to cetaceans photographic techniques for recording individuals by their natural markings (see fig. 1.16; Würsig and Würsig 1977), a method now used in nearly all cetacean field research (e.g., Hammond et al. 1990 and references above). Photographic methods were devised to interpret social structure by measuring group stability and preferential associations of individuals (Würsig and Würsig 1977; Würsig 1978). These procedures have been replicated or adapted in studies of the social systems of, for example, killer whales (e.g., S. L. Heimlich-Boran 1986), bottlenose dolphins (e.g., Wells et al. 1987), and sperm whales (e.g., Whitehead and Arnbom 1987; Whitehead et al. 1991).

Peninsula Valdés researchers also experimented with ways to observe cetaceans from afar to eliminate reliance

Figure 1.17. Roger and Katy Payne studying southern right whales at Peninsula Valdés, Argentina. (Photograph by J. and D. Bartlett, Bruce Coleman Photo Library.)

on seagoing vessels, which can be disruptive to animals or restrictive to research budgets. For instance, Roger Payne adapted the use of a surveyor's theodolite to monitor movements of nearshore cetaceans from clifftop vantage points, a method put to use in studies of dolphin behavioral ecology (Würsig and Würsig 1979, 1980) and southern right whale communication (Clark and Clark 1980). The theodolite has become the tool of choice when precise records of coastal movements of cetaceans are required (e.g., Tyack 1981; Würsig et al. 1991). In addition, the observation of Roger Payne et al. (1983) that southern right whales rarely reacted to circling aircraft led to the use of small planes to obtain an overhead, big-picture view of whale behavior (see also Watkins and Schevill 1979). This technique is still commonly used in remote areas to examine, for example, the behavior of Arctic bowhead whales (Würsig et al. 1984, 1985, 1993) and the school structure of pelagic dolphins (Scott and Perryman 1991).

For longer-range monitoring of movement patterns, radiotelemetry devices were adapted for cetacean research. William Evans "was the cetologist most responsible for developing the dolphin radio tag that now allows us to follow dolphins at sea" (Norris 1991a: 9). Evans did not work at Peninsula Valdés, but his early tags were used there to monitor dusky dolphins (Leatherwood and Evans 1979; Würsig and Würsig 1980). Development of radio tags for large whales began in 1961 (Schevill and Watkins 1966), with many refinements since that time (e.g., Watkins and Schevill 1977a; Leatherwood and Evans 1979). Technological advances have made it possible to monitor a cetacean's "environment (e.g., water temperature, salinity . . .), behavior (e.g., diving depth, swimming speed, sound production), or physiological state (e.g., heart rate, body temperature) as a function of time and location" (Leatherwood and Evans 1979: 2). Tagging and biotelemetry have been applied to learning about the lives of even the most elusive cetacean species (reviewed in Norris et al. 1974; Scott et al. 1990b; Würsig et al. 1991; see also, e.g., radiotelemetry: Ray et al. 1978; Watkins et al. 1981; Read and Gaskin 1985; time-depth recorders: Kooyman et al. 1983; acoustic telemetry: Watkins et al. 1993; satellite telemetry: Watkins et al. 1996; Mate et al. 1997).

Focal-animal behavioral sampling techniques (Altmann 1974) were first introduced to cetacean field studies in observations of southern right whale mothers and calves from the cliffs of Peninsula Valdés (Taber and Thomas 1982; Thomas and Taber 1984). There, too, sound playback techniques borrowed from investigations of bird song, grasshopper calls (reviewed in Falls 1992), and seal sounds (e.g., Watkins and Schevill 1968) were applied to studies

of cetacean communication (Clark and Clark 1980). The prediction that this method would prove "useful in determining the biological function of the sounds in a whale's acoustic repertoire" (Clark and Clark 1980: 664) has been confirmed many times over, as in evaluating the functions of humpback song (Tyack 1983; Mobley et al. 1988) and bottlenose dolphin signature whistles (Sayigh et al. 1999).

Innovative ideas came to fruition at other sites as well. In their studies of Hawaiian spinner dolphins, Norris and Wells (1994: 54) had long felt that "[a] major challenge of the study of dolphin natural history is to place an effective observer under the water in the ocean where dolphins live out their life patterns." Underwater observations have been crucial in, for example, deciphering the sexes and roles of singing humpbacks and their associates (Glockner 1983). Norris (1991b: 215), however, had dreams of more extended observations below the surface, of being "like Captain Nemo sitting before his underwater picture window . . . [looking out on the dolphins'] lives from the comfort of [an] air capsule." Norris and his colleagues built several incarnations of underwater viewing chambers. The prototype, with the stomach-churning nickname of "semisubmersible seasick machine" (Norris and Wells 1994: 58), enabled them to be the "first scientists to study wild [dolphin] societies underwater where their lives are truly spent" (Norris 1991b: 13; fig. 1.18). "[The] underwater observational dimension . . . allowed us to observe [spinner dolphins] in the context of a wild school, complete with predators, food sources, and the physical world of the sea" (Norris 1994: 2).

Others had a different approach to "seeing" underwater: "In order to reach below the surface and try to assess the behaviors of submerged whales, we utilized underwater sound" (Watkins 1981: 84) (fig. 1.19). Techniques such as a "non-rigid three-dimensional hydrophone array" (Watkins and Schevill 1972) made it possible to track the movements (e.g., Watkins and Schevill 1977b) and record the vocalizations (e.g., Watkins and Schevill 1977c) of individual cetaceans underwater. "The sounds from finback whales . . . provided the stimulus for much of the early progress in design of equipment and techniques for acoustic observations at sea" (Watkins 1981: 84) because these whales turned out to be the source of the mysterious 20-cycle pulses (Schevill et al. 1964) that had long puzzled underwater listeners, including geophysicists and the military. Watkins et al. (1987: 1901) later suggested that "direct association of the [signals] with the reproductive season for this species points to the 20-Hz signals as possible reproductive displays by finback whales."

Nearly all these methodologies exemplify the promi-

A

B

Figure 1.18. Ken Norris (A) riding on the Semisubmersible Seasick Machine (B), which he developed for underwater viewing of wild dolphins. (A, from Norris 1974; B, photograph courtesy of W. E. Evans.)

nence of "employing a team approach" (Scott et al. 1990a: 243) in cetacean fieldwork. Those studying bottlenose dolphins in Sarasota waters "learned the value of simultaneously pursuing multiple lines of investigation. . . . The study has become more and more of a corporate affair, uniting biologists interested in behavior, life history, genetics, acoustics, reproduction, and population biology" (Scott et al. 1990a: 243). Examples of special collaborative efforts in cetacean field research include partnerships forged by scientists with members of the public or with whale-watch operators to locate, census, and photograph killer whales of the Pacific Northwest (e.g., Bigg et al. 1990b; Ford et al. 1994) and humpback whales of the Gulf of Maine (e.g., Clapham and Mayo 1990; Katona and Beard 1990; Lien and Katona 1990; Clapham 1994).

Save the Whale[8]

To avoid [extinction of] right whales, we need to know much more about them. In spite of 30 years of nominal protection, they have not undergone the rapid recovery in numbers that gray whales have. . . . We have no idea why this may be so, so little is known about the basic biology of this species. Our aim . . . was to study the basic biology of the right whale and to develop estimates of its population . . . [and] to apply what we had learned to preserving the species.

Roger Payne (1980: 551)

Whereas previous research had often been geared toward better management or exploitation of whales, the new era of field research was founded upon the preservation of these animals. For Payne (1980, above) and many other cetologists, conservation was a particular incentive for field studies of cetaceans. Bigg and colleagues echoed these sentiments in describing the initiation of their field research on killer whales:

> The study began in 1970 [when] biologists in British Columbia and Washington State were faced with an urgent request. . . . The questions posed concerned whether the removals were endangering the local killer whale population and what restrictions should be introduced if more whales were to be taken. This required knowing how many killer whales were in the region; whether the whales taken in Washington State were from the same stock as those taken in British Columbia; what the productivity of the population was; and whether the removal of one particular age or sex was detrimental to productivity. Little was known about these topics. (Bigg 1994: 13)

Thus, the new generation of field cetologists wore several hats: not only were they scientists collecting facts about

---

8. The motto of the 1970s anti-whaling movement (see, for example, Day 1987).

AMY SAMUELS AND PETER L. TYACK

Figure 1.19. William A. Watkins recording the sounds of a right whale in Vineyard Sound, July 1964. (Photograph by James Moulton; courtesy of William A. Watkins.)

their chosen study subjects, but many also became ambassadors for their whale of choice. To better serve as advocates for the cetacean they championed, many early field-workers focused their research on the species rather than on a research question. Methods conformed to a conservation ethic: these cetologists were adamant that, despite difficult study conditions at sea, their research would be conducted without "even touching a whale" (Payne 1983: 1).

In the United States, conversion to a conservation focus involved more than merely a change in ideals; it became the law: "The Marine Mammal Protection Act of 1972 makes the United States government responsible for long-term management of marine mammal populations. This means conserving and protecting these populations and doing research on them to see that it is done wisely. . . . The positive effects, from a scientific point of view, are that money is allocated for applicable research on phenomena important in managing populations" (Le Boeuf and Würsig 1985: 139).

Where research funding came from "greatly influenced whom we cetologists were talking to and what kinds of question we were asking" (Pryor and Norris 1991a: 385). Funds earmarked for estimating cetacean population parameters led to research priorities with a population focus. From a conservation perspective, such research was badly needed. And, as Le Boeuf and Würsig (1985: 139) pointed out, "although the research mandated appears to be closely tied to management's charge of keeping populations near optimal levels, the information gained is likely to be of general interest." Thus, for example, significant contributions to knowledge about bowhead whale behavior were by-products of studies funded by the U.S. Minerals Management Service "to obtain information directly useful for management decisions regarding potential disturbance of whales through oil and gas industry activities" (Würsig and Clark 1993: 157; see also Würsig et al. 1984, 1985, 1993).

The challenge for field researchers was to show that the new passive techniques could provide accurate estimates of population size and reproductive rates, statistics previously derived from carcass analyses. Whitehead and Gordon (1986: 163) demonstrated that "benign [noninvasive] research can duplicate the kinds of data provided by commercial whaling, as well as investigat[e] some areas of . . . whale biology for which catch data could not provide information" (see also, e.g., Whitehead and Payne 1981; Payne 1983). Once "it became clear that data on resightings of individuals could provide information on the abundance, survivorship, reproductive rates, and population differentiation of whales" (Hammond et al. 1990: v), the newfound ability to distinguish individual cetaceans was widely applied to calculations of population parameters.

These applications led to considerable interest in developing capture-recapture methods to approximate population and life history parameters based on resightings of naturally marked cetaceans (e.g., Hammond 1986). An entire special issue of the IWC *Reports* was devoted to "shooting whales (photographically) from small boats" (Mizroch and Bigg 1990: 39), with articles detailing photographic methods, field protocols, modeling and statistical techniques, and the advantages and disadvantages of using natural versus artificial marks to identify animals for population-focused analyses (Hammond et al. 1990). The high-level investment was due to the urgency of the conservation-based effort and the perceived novelty of the enterprise: "Although the recognition of individual animals from natural markings is a common practice in behavioural studies, these data are rarely used for the estimation of population size. Apart from a study of alligators . . . . I know of no others of this kind except for those on whales" (Hammond 1986: 254).[9]

Conservation concerns forced a partnership between cetologists interested in behavior and those with an interest in population dynamics. In 1982, they came together at an IWC-sponsored workshop dedicated to "the behaviour of whales in relation to management" (IWC 1986). "Prior to that meeting cetologists studying behaviour and cetologists studying population dynamics had kept themselves to themselves, apart from the occasional complaint from the behavioural people that modellers took no account of their work and the riposte from the modellers that until the behavioural observations were quantified it was impossible to incorporate them into models" (IWC 1986: iii). The workshop "served two important purposes: (1) it showed management scientists how current behavioural knowledge could assist them in their work; (2) it suggested new areas of behavioural research which would materially assist future management of whale stocks" (IWC 1986: iii). Thus, for example, behavioral research on right whale calves (Taber and Thomas 1982; Thomas and Taber 1984) was used to illustrate the value of longitudinal studies for accurate recruitment estimates: "Without callosity identification we might have mistaken the yearlings returning to Golfo San Jose with their mothers as infants born out of the normal calving season . . . [because] these yearlings acted superficially like infants, and did not appear to our eyes to be much larger than infants we had watched depart six months earlier. Without certain identification, we would have made incorrect inferences about the range of the calving season and about mother-infant behaviour" (Thomas 1986: 118). From the other camp, Hammond (1986) provided behavioral biologists with a clear explanation of the proper techniques for collecting data that could be used in population analyses and modeling.

Even though the threats have changed over time (Whitehead et al., chap. 12, this volume), conservation persists as the principal research thrust of cetology. Some risks to large cetacean populations have diminished since the IWC adopted a worldwide moratorium on commercial whaling, which took effect in 1986 (IWC 1983). Similarly, the effects of live capture on small cetacean populations in U.S. waters have declined since federal capture quotas were instituted in the 1970s (e.g., in 1977 for bottlenose dolphins: NMFS 1990), and especially since zoos and aquaria discontinued live capture of the most commonly exhibited species, the bottlenose dolphin, in 1990 (NMFS 1993). "New threats have emerged, however, that are more subtle in their expression but perhaps no less significant. These include: incidental take during fishing operations; entanglement in lost and discarded fishing gear; disturbance by boats engaged in whale-watching and other activities; and habitat degradation and destruction due to fishery development, dumping, dredging, offshore oil and gas development and other human activities" (Hofman and Bonner 1985: 116).

"Unraveling the Structure of a Cetacean School"[10]

Individual recognition of animals is essential in a detailed study of social behavior.
George B. Schaller (1965: 628)

Conservation efforts took advantage of the take-home message from earlier efforts to adapt whaling models to the sperm whale's unconventional social structure: population parameters could not be accurately modeled unless the complexities of cetacean social organizations were better defined. Thus, armed with photo-identification techniques and other innovative methods, field-workers set off to determine the characteristic grouping patterns of their subjects. These efforts resulted in a wealth of data that revealed the framework of many cetacean societies (chapters 4–7 of this volume).

Despite the logistical difficulties encountered by cetacean field-workers, investigations of cetacean social structure have been shown to compare favorably both in quantity and sophistication with those of terrestrial mammals such as ungulates and primates (H. Whitehead and S. Dufault, unpublished data). These analyses have revealed unique aspects of killer whale society (Baird, chap. 5, this

---

9. This claim to distinction refers to the special problems associated with estimating population parameters from mark-recapture models based on naturally marked animals.

10. Norris and Dohl 1980b: 212.

volume) as well as a number of parallels between societies of cetaceans and those of terrestrial mammals (Connor et al, chap. 4; Whitehead and Weilgart, chap. 6, this volume). The alliances of male bottlenose dolphins, for example, have been likened to those of chimpanzees (Connor et al. 1992a) and lions (Wells et al. 1987), the bottlenose dolphin fission-fusion society to that of chimpanzees or spider monkeys (Tayler and Saayman 1972; Würsig 1978; Smolker et al. 1992), and the matrilineal bands of female bottlenose dolphins to prides of lions (Wells et al. 1987). Sperm whale society also appears to bear many strong resemblances to that of African elephants, including a matrilineal basis for stable groups of females, the temporary coalescing of female groups, and solitary, roving males in search of mating opportunities (Weilgart et al. 1996).

The result of these efforts is that social structure on a broad ecological scale is now known for many cetacean species. These analyses include such detailed information about individual community members as their demographic, reproductive, and kinship histories. Within this community-level framework, however, the particulars of the daily social lives of individual whales, dolphins, and porpoises remain largely unknown. What is the role of postreproductive females in pilot whale groups? What is the nature of the relationship of a female sperm whale with members of her own kin group versus with members of other matrilineal groups? What is the social significance of a bottlenose dolphin producing its own signature whistle versus that of a companion? Do wild bottlenose dolphins have agonistic dominance relationships? What factors determine the mating success of a male sperm whale or a bonded pair of male bottlenose dolphins? Questions like these, and many more, are only beginning to be addressed.

In studies of social interactions and the dynamics of social relationships between individual animals, cetacean behavioral biology lags far behind its nonmarine counterparts. Field studies of cetacean social behavior are only now emerging from the "curious naturalist" phase, with its emphasis on "description of behavior under natural conditions . . . [rather than] the formulation of complex theories" (Dewsbury 1984: 10). Only recently have cetacean behavioral biologists begun to apply hypothesis testing and systematic methodologies to examine the intricacies of cetacean social behavior—that is, the social interactions between individuals that occur on a second-by-second or minute-by-minute basis and the social relationships between individuals that may change on a weekly, monthly, or seasonal basis.

Again, it was ornithologists who first realized the significance of individual identification not only for analyses of social structure, but also for studies of social behavior: "At first the metal ring seemed merely to be a new aid to migration research. But then the marking of individuals proved to have a much more comprehensive significance, because it helped the study of behavior on the breeding grounds" (Stresemann 1975: 338). By the 1970s, the individual had become the unit of theoretical interest in animal behavior studies because of the "growing acceptance of the evidence that the potency of natural selection is overwhelmingly concentrated at levels no higher than that of the individual" (Alexander 1974: 325). As a result, "in the past twenty years . . . there has been a theoretical revolution in evolutionary biology, leading to reexamination of early work . . . and a dramatic increase in studies designed to clarify how factors such as kinship, reciprocity, sexual selection and life history affect the evolution of behavior" (Cheney et al. 1987: 2). Following the principles of maximizing individual reproductive success (Williams 1966) and kin selection (Hamilton 1964), "animals came to be viewed as individuals who were armed with many behavioral options in their struggle for maximizing either their own reproduction, or that of their relatives. . . . By analyzing the behavior of individuals, the foundations for a comprehensive theory of social behavior were laid" (Rubenstein and Wrangham 1986: 4). From this theoretical perspective, studies of social behavior came to emphasize long-term monitoring of known individuals to assess factors that influence lifetime reproductive success.

Further sophistication in observational studies of social behavior came with the development of systematic sampling techniques to minimize biases due to the conspicuousness of certain behaviors or age-sex classes (e.g., Altmann 1974; Mann, chap. 2, this volume). "Asked to prepare a short piece on different ways of analyzing [behavioral] data, [Jeanne Altmann] . . . decided the problem was not really how people analyzed data. . . . The problem was how people collected data. . . . The embarrassing truth was that many of the regularly cited field studies especially before the mid-1970s both gathered and analyzed data in a way that did not justify the conclusions reached" (Haraway 1989: 307). Altmann (1974: 229) suspected "that the investigator often chooses a sampling procedure without being aware that he is making a choice," and proposed that judicious use of "sampling decisions . . . in observational studies of social groups can increase the validity of comparisons both within and between studies, whether observational or experimental, field or laboratory" (Altmann 1974: 231). To make sampling choices explicit, Altmann (1974) wrote what has come to be regarded as a preeminent handbook of observational methods for studying social behavior.

One such method focuses on the behavior of a single individual at a time (focal-animal sampling: Altmann 1974). This technique routinely takes the form of following a known individual for a specified period of time while recording specific behavioral information in a systematic fashion. Repeated over weeks, months, or years, these "follows" furnish a comprehensive, longitudinal perspective on each individual's social life. Whereas repeated, brief sightings of individuals provide information about characteristic grouping patterns, repeated, protracted follows of individual animals are integral to understanding their social relationships within those groups (Mann, chap. 2, this volume).

Attention to the social relationships of primates brought further conceptual advances in animal behavior. In particular, "the new thing that students of primate behavior did was to recognise the individuality of their animals" (Rowell 1994). By the 1980s, "groups of individually recognized [primates had] been studied continuously for 10 years or more. . . . [These] long-term primate studies were among the first to show the critical importance in mammals of kinship, social relationships, and individual variations in behavior" (Smuts et al. 1987: ix). In particular,

> [W]idespread existence of long-term social relationships among primates has frequently forced primatologists to approach the evolution of behavior with a [different] perspective. . . . [T]raditional ethological research has concentrated on interactions between individuals and has examined the function of single acts by measuring their immediate and long-term consequences. In primates, however, an interaction like grooming clearly has consequences beyond both its immediate function of ectoparasite removal and the longer-term function of making subsequent grooming bouts more likely, since grooming can also contribute to the maintenance of a relationship that may have important reproductive consequences. . . . [Thus] primate studies have begun to document the importance of analyzing behavior at the level of social relationships. (Cheney et al. 1987: 4).

These conceptual and methodological advances in the field of animal behavior are largely lacking from accounts of the social behavior of wild cetaceans (but see, e.g., Taber and Thomas 1982; Slooten 1994; Mann and Smuts 1998). Instead of judicious sampling decisions, hypothesis testing, and a focus on the interactions and relationships of individuals, descriptions of cetacean social behavior are all too often based on what Norris (1991a: 9) described as Saayman's 1960s style of recording "whatever one can see" (i.e., the ad libitum technique common to early field studies of primate behavior). In a discipline that has gone to extensive lengths to ensure the precision of population-level appraisals (e.g., Hammond 1986; Hammond et al. 1990) and analyses of social structure (see references above), it is ironic

that no comparable degree of rigor has been applied to field studies of social behavior.

In particular, although cetacean biologists acknowledge that the ability to identify individuals "lay[s] the foundation for modern behavioral studies of . . . whales" (R. S. Payne et al. 1983: 373), individual cetaceans are rarely the subjects in present-day investigations of their social behavior. This is true despite the fact that recognition of individuals is now common practice both in studies of animal behavior and in analyses of cetacean social structure. Having borrowed a tool from social behavioral research—individual recognition via natural markings—cetologists have been slow to return this device to its original purpose: studies of social behavior.

No single factor suffices to explain the delayed development in social behavioral research on cetaceans. When primatologists in the 1970s and 1980s introduced the "relatively new approach" of focused studies and hypothesis testing to field research on primates, they acknowledged that "general descriptive natural history . . . is a necessary prerequisite for problem-oriented studies" (Sussman 1979: v). Some cetologists similarly maintained that focused studies of social behavior must necessarily be preceded by population- or community-level analyses, and that it was only a matter of time until behavioral studies of cetaceans would catch up to those of terrestrial mammals. Fieldworkers studying Sarasota bottlenose dolphins realized that "our [early] focal animal observational efforts pointed to the need for more detailed background information on the community members. All too often, our known animals interacted with identifiable animals of unknown age or sex" (Wells 1991c: 205). "[W]e recognized that the validity of conclusions about the behavior of these animals depended on the quality of the background information that we collected for each member of the cast of characters. . . . Thus a major thrust of the [early] research program has been to refine our abilities to identify individuals positively and to obtain the most accurate information possible on sex, age, reproductive condition, and genetic relationships" (Wells 1991c: 201). Focused behavioral studies of Sarasota dolphins now coexist alongside community-level investigations (Wells 1991c), and include focal follows of individual calves to identify the social factors that influence signature whistle development (Sayigh et al. 1995a).

In addition, research on social behavior is typically considered to be tangential to conservation efforts. This general devaluation of social behavioral studies is not restricted to cetology: "The spurious view that much basic behavioural research is of no practical use—and therefore a waste of time—is more prevalent than in the case of, say, physiological or biochemical research. This is at least partly because people have insight into their own actions and may also be familiar with the behaviour of some animals" (Martin and Bateson 1994: 2). Nor are cetologists the only ones wearing "intellectual blinders" to the value of behavioral biology in conservation efforts: "Behavioral biologists rarely recognized behavioral contributions to conservation other than captive breeding and reintroduction programs of endangered species. The conservation biologists typically claimed that behavioral research was mostly 'descriptive natural history'" (Clemmons and Buchholz 1997: xi).

In cetology, this attitude flies in the face of the lessons for management derived from deciphering the social structure and mating system of the sperm whale and from the successful IWC workshop that brought together behavioral biologists and population modelers (see references above). In part, the separation between behavioral and conservation efforts stems from the conflicting demands of population-level analyses and social behavioral research: "[B]ehavioural investigations emphasizing focal animals . . . require sampling strategies that are not ideal for providing data useful in estimating population size" (Hammond et al. 1990: 7) because population statistics require maximizing sample size. However, "behavioral diversity . . . is an important but often overlooked component of biological diversity. Failure to identify and preserve endangered behavior will undermine species' potential for not only survival but evolutionary change, and hence, long-term persistence. Efforts aimed solely at conserving genetic diversity do not automatically subsume the survival of behavioral diversity" (Clemmons and Buchholz 1997: xii).

In cetology, the devaluation of behavioral research may have also emanated from a perception that social behavioral research on cetaceans is "mostly descriptive natural history," which, excepting the analyses of cetacean social structure, is largely true. The potential for anecdotalism and subjectivity in cetacean social behavioral research has been an ongoing concern: "Behaviorists especially . . . have had to struggle to escape viewpoints based on their own lives rather than the lives and environment of the animals under observation" (Norris 1991b: 218). Gaskin's commentary no doubt reflects the frustration of many cetologists:

No research on Cetacea has attracted more public attention in recent years than work on their behaviour, communication and intelligence. . . . Yet surely no aspects are more difficult for the scientist to study effectively. The biologist, educated to respect the "hard data" of the numbers and weights of population samples or the calculated values from studies of cellular enzymatic reactions and blood chemistry, usually views behav-

ioural work as occupying the "soft" fringe of biology (meaning that area which abuts on psychology, and is therefore barely respectable) (Gaskin 1982: 112).

For many cetologists, the customary explanation for the arrested development of cetacean social behavioral research is that it is simply too difficult to conduct focused, problem-oriented studies of the social behavior of marine animals that are fast-moving, wide-ranging, elusive, difficult to observe closely at sea, and whose lives take place primarily beneath the water's surface. How, then, does one account for those cetologists who overcame such obstacles to conduct sophisticated studies of cetacean social behavior? Following the pioneering efforts of the 1980s (e.g., spotted and spinner dolphins: Pryor and Kang 1980; right whales: Taber and Thomas 1982; Thomas and Taber 1984), systematic methods and hypothesis testing are becoming more widespread in field research on social behavior (e.g., bottlenose dolphins: Connor et al. 1992a; Smolker et al. 1993; Sayigh et al. 1995a,b; Mann and Smuts 1998, 1999; see also Mann, chap. 2, Whitehead et al., chap. 3, this volume). The discrepancy between the convictions of many cetologists and the accomplishments of a handful of cetacean behavioral biologists bears a closer look.

"History as a Vehicle to Effect Change"[11]

As this narrative arrives at present-day studies, it becomes more difficult to remain impartial. A stated goal in compiling this review was to promote the transition in studies of cetacean social behavior from qualitative, descriptive natural history to focused, quantitative analyses of the social interactions and social relationships of individuals. Like animal behavior historian Donald Dewsbury before us, "[our] goals are best described as a cross between recording history and advocacy"; we, too, "hope to escape some criticism by making [our] objectives and biases explicit" (Dewsbury 1984: ix–x). It was only after compiling this history that we, the chroniclers, realized how close cetacean behavioral biology is to making the transition.

We, therefore, cautiously venture the opinion that many of the stated obstacles to focused social behavioral studies have not been fully put to the test. Some objections to new ideas may turn out to be old habits that die hard. Focal animal sampling, for example, has been considered an appropriate technique only for a handful of species with special characteristics and under special circumstances, but has this assumption been tested? Real-time identification of individuals, a prerequisite for this method, is believed impos-

---

11. Dewsbury 1984: ix.

sible for most cetacean species. Elephant researcher Ian Douglas-Hamilton began with a similar viewpoint in the 1970s: because "many elephants looked similar to others and were distinguishable only by minute differences, . . . I therefore believed that photography was probably the only method of recording details with sufficient accuracy." However, a colleague insisted that he sketch distinctive features instead, because "the problem of making notes while taking photographs would inevitably lead to greater muddles" than the occasional misidentification (Douglas-Hamilton and Douglas-Hamilton 1975: 43). Douglas-Hamilton soon learned to recognize many individual elephants by eye while continuing to maintain a catalog of photographs for confirmation. This method has worked well in behavioral studies of bottlenose dolphins (e.g., Smolker et al. 1993). The successes of the few attempts to follow individual baleen whales while confirming their identity photographically (e.g., minke whales: Dorsey 1983; blue whales: A. Samuels, P. L. Tyack, and C. Carson, unpublished data) suggest that the technique may be more broadly applicable than previously assumed.

Where species' attributes fall short, innovative solutions can fill the gap. A descendant of the semisubmersible seasick machine enabled Östman (1994; Norris and Wells 1994) to be the first to conduct underwater focal follows of free-ranging spinner dolphins. In addition, the continued reliance of bird behavioral researchers on colored legbands to identify individuals suggests another solution for cetacean behavioral biology: new developments in noninvasive tagging methods for cetaceans (e.g., Stone et al. 1994) are likely to facilitate focused behavioral studies of species that cannot be readily identified by eye.

The perception that protracted follows are impossible to achieve at sea (e.g., Ohsumi 1971) or disruptive to the animals' behavior (e.g., Würsig and Würsig 1980) also needs to be revisited. The problem of habituation to observers (and their vessels), a key component of behavioral research, has often been approached by cetologists as an all-or-nothing proposition rather than a process that takes time. After all, it took eight months before the chimpanzees stopped running away from Jane Goodall, and another ten before they permitted sufficiently close approaches that she could observe social behavior (Goodall 1986)! For deep-diving sperm whales, lengthy follows were deemed indispensable for deciphering their social structure and mating system because "most observations were of sub-groupings of a larger school, whose members were never all seen together at the surface at the same time" (J. C. D. Gordon 1987a: 214).

Some methods of studying cetacean social behavior may have trade-offs, and to evaluate these, the field-worker's equivalent of Wood's (1986) common-sense checklist for captive behavioral studies is needed. The potential for hidden drawbacks is evidenced in Burley's (1981) demonstration that colored legbands influence zebra finch reproductive success. However, had bands never been used to identify individual birds, it is unlikely that ornithology would have paved the way for scientists studying other species.

This history reveals that the groundwork has, in fact, been laid for sophisticated studies of cetacean social behavior to follow in the footsteps of the elegant analyses of cetacean social structure. Long-term research on individually recognized cetaceans is well established; extensive demographic, reproductive, and kinship information has been obtained for many individual whales and dolphins; captive conditions are much improved; behavioral sampling techniques are available from behavioral biology; and natural history observations have generated numerous research questions. As a result of hard work, perseverance, and the development of ingenious technologies, cetacean behavioral biologists now have the option of looking at individual cetaceans within the context of demographic factors, familial relationships, and social associates. It is time to overcome the historical impediments: to explore innovative solutions to the challenges posed by problem-oriented behavioral research, to appreciate the contributions of complementary captive and field studies, and to acknowledge the significance of social relationships between individuals. With the hard work already accomplished, the stage is set to usher in focused, quantitative studies of cetacean social behavior.

2

UNRAVELING THE DYNAMICS OF SOCIAL LIFE

Long-Term Studies and Observational Methods

JANET MANN

A dark gray fin breaks the water surface and a slight spray catches the light. The boat driver steers west toward the animal steadily, at moderate speed, hoping to get close without disturbing it. The researcher raises the camera, poised to shoot, trying to predict exactly where the next breath will occur, but the animal has turned, and dives into the glare. It's too late. Four minutes later, the animal resurfaces, this time 200 m to the south. The boat is never close enough before the animal dives. Down again.

CETOLOGISTS CAN spend hours trying to get close enough to animals to photograph them and identify them, let alone collect demographic, reproductive, association, and behavioral data. So many species, especially deep-diving pelagic animals, are quite difficult to study. But these difficulties are tempered by the fact that all cetaceans surface regularly to breathe and can spend considerable periods of time at the surface. Some species may not be that much more difficult to study than forest creatures, which also go in and out of view, or elusive animals occupying less dense habitats. However discouraging initial attempts to study dolphins and whales may be, the chapters in this volume show that it is both feasible and rewarding to study the social relationships and social organization of wild cetaceans.

This chapter, directed largely to students of cetology, is about the benefits of long-term study, the significance of social bonds in cetaceans, and the appropriate observational techniques for unraveling cetacean social relationships. Long-term behavioral studies are key to understanding the nature of cetacean sociality and the selective forces that shaped patterns of cetacean communication, cognition, life

history, behavior, and ecology. Studies of behavioral variation and its contribution to an individual's reproductive success are at the heart of understanding the constraints, trade-offs, and opportunities that cetaceans face. Even in species in which there are few stable relationships, such as humpback whales, long-term studies and follows of individuals and groups help us identify the functions of brief social encounters (e.g., Tyack 1981, 1983; Tyack and Whitehead 1983) and examine important reproductive and life history parameters (e.g., Wiley and Clapham 1993; Baraff and Weinrich 1993).

Why Long-Term Studies of Behavior?

Long-term behavioral studies of terrestrial mammals have greatly influenced the development of new concepts and hypotheses concerning the evolution of social behavior. This has been especially true of primatology, which has shifted from short-term to long-term studies of relationships (Cheney et al. 1987). These studies have depended on detailed behavioral observations and on knowing the relationships between individuals. Complex patterns of social behavior such as cooperative lethal raiding in chimpanzees (e.g., Goodall 1986) weren't revealed until after nearly fifteen years of observations at Gombe, but stimulated provocative theories on the evolution of human warfare (see Manson and Wrangham 1991; Boehm 1992). Long-term observations of the stability of cercopithecine female dominance hierarchies, combined with female philopatry and patterns of intergroup conflict, contrasted with the relative rarity of male philopatry, and inspired the groundwork for modeling social organization in complex mammalian groups (e.g., Wrangham 1980; van Schaik 1983; Cheney 1987). Dominance, reconciliation, alliance formation, and social intelligence are all concepts derived from patterns observed or predicted in social relationships through repeated encounters (e.g., see Harcourt and de Waal 1992; Byrne and Whiten 1988; de Waal 1987). None of these concepts can be understood on the basis of one encounter between individuals. The meaning of acts is determined through the context and history of interactions between the participants.

In the last few years, several long-term studies of cetaceans have similarly begun to link complex patterns of behavior and social organization into models of social structures and mating systems (see chaps. 4–7, this volume). Most long-term cetacean studies are based on annual surveys or periodic censuses of populations, rather than following individual animals for short or long periods of time. The advantages and disadvantages of the types of studies vary, but complementary approaches are highlighted here.

Cross-sectional and Longitudinal Studies

Cross-sectional research involves the sampling of individuals (or cases) *once,* at a specific point, or for a specific time period (e.g., sample all individuals born between 1970 and 1979 once). Longitudinal studies require that the same individuals be sampled for at least two distinct time periods or points, and that the analysis allow for comparisons of the same individual between those time periods. A one-time census of animals, whether they are individually identified or not, alive or dead, is a cross-sectional data set. Censuses, whaling/fisheries data, and stranding data provide a cross section of individuals or groups of animals, which can be used to assess population and demographic parameters. Censuses or whaling/fisheries by-catch data may be accumulated and compared across years, but because different individuals are typically sampled each time, this is a repeated cross-sectional design, not a longitudinal design. Data from cross-sectional studies are subject to certain drawbacks because this method offers few tools for

explaining variation between individuals. However, some life history data can be acquired only from cross-sectional studies, thus providing much-needed information.

Some of the largest databases for any mammal are those on dead cetaceans. These data are often used to estimate population size or changes in population growth in response to anthropogenic activities. Whaling and fisheries data are analyzed from a management perspective, targeted toward estimating sustainable harvests of the animals. The methods used involve relating age to reproductive, demographic, and physiological features. Baleen whales can be aged by ear plugs (reviewed in Lockyer 1984), and odontocetes can be aged by growth layers in the teeth (e.g., Perrin and Myrick 1980). Ovarian scars, pregnancy rates, lactation, and testes maturation are then correlated with age and body size to determine age of sexual maturity, interbirth intervals, gestation length, growth rates, and other parameters.

Whaling and fisheries data on dead animals can be particularly useful in identifying onset of sexual maturity (either first ovulation or first conception), reproductive senescence, and breeding season (seasonal variation in sperm production and testes size). The study of live animals may reveal little about such patterns. However, these data are less valuable for other reproductive estimates. For example, ovarian scars (from the corpora albicantia) have been used to estimate ovulation rates in many species, including spinner and spotted dolphins, common dolphins, humpbacks, sei and fin whales, pilot whales, and sperm whales (e.g., see Perrin and Donovan 1984). Corpora counts alone tell us little, however, about birth rates, interbirth intervals, age at weaning, survivorship, or population growth. For example, Gambell (1973) calculated a mean ovulation rate for sperm whales at one corpus per 2.33 years, yet sperm whale infants are probably not weaned until four or five years of age (Whitehead and Weilgart, chap. 6, this volume). Corpora counts obfuscate several important issues, such as spontaneous abortion rate and interbirth intervals for mothers with surviving versus nonsurviving infants. Perrin and Reilly (1984) calculated calving intervals for nine delphinid species as the reciprocal of the annual pregnancy rate (percentage of mature females that are pregnant divided by gestation period). Bottlenose dolphin (*Tursiops* sp.) calving intervals were 1.3–1.5 years, which underestimates values from longitudinal studies by one-third (Connor et al., chap. 4, this volume). These discrepancies are rooted in the difference between the cross-sectional approach and the longitudinal approach. Cross-sectional models of interbirth intervals do not discriminate between mothers with surviving and nonsurviving infants. The pregnancy rate may be high because females have lost their offspring, not because they have weaned them. Some of the problems associated with estimating demographic, reproductive, and population parameters from census and fisheries data are described in several sources (Perrin and Donovan 1984; Goodman 1984; DeMaster 1984; Hester 1984; Olesiuk et al. 1990).

Cross-sectional studies of dead animals can also reveal intriguing life history patterns. Kasuya and Marsh (1984; Marsh and Kasuya 1984, 1986) studied the reproductive biology of over eight hundred short-finned pilot whales (*Globicephala macrorhynchus*) and documented menopause in that species (reviewed by Whitehead and Mann in chapter 9 of this volume). In addition to providing some of the best data available on menopause in nonhumans, such cross-sectional studies can be used to examine sex-biased investment in whales. Some male pilot whales still had milk in their stomachs into their teens, but no females had milk in their stomachs after the age of seven, suggesting that mothers may invest more in sons than in daughters, a pattern predicted by Trivers and Willard (1973): if male variance in reproductive success is much greater than female variance in reproductive success, and if maternal investment can significantly influence the reproductive outcomes of offspring, then mothers who are in good condition (can invest more) should invest in sons, the sex with the greatest variance in reproductive success.

Female pilot whales begin ovulating between seven and twelve years of age, but rates of first-born survivorship are not known. Although typical weaning age was estimated to be three to six years, males (presumably sons) were sometimes nursing into their teens, when females (presumably daughters) were already producing their own offspring. Older mothers (after thirty-five years) were thought to be nursing their infants to older ages than younger mothers. If sons are being nursed for longer than daughters, and older mothers are nursing their offspring longer than younger mothers, then older mothers may be more likely to produce sons than daughters, and younger mothers may be more likely to produce daughters than sons. The hypothesis that sex ratio bias changes with female reproductive age in pilot whales remains to be tested.

Sex, age, relatedness, lactation status, and stomach contents can be retrieved from carcasses; long-term observations of individual animals could not provide this kind of data. However, long-term observations can provide insights into the adaptive function of male-biased investment, menopause, and other intriguing life history patterns. A longitudinal approach is necessary to understand adaptive function and the dynamics of complex social life.

Survey and Follow Data

The longitudinal approach involves sampling individuals repeatedly over periods of time. Longitudinal data, such as infant survivorship, are important for looking at which factors contribute to an individual's reproductive success. A longitudinal study typically includes surveys and sometimes group or individual follows. Surveys are useful for calculating coefficients of association, determining mean and range of group sizes and ranging patterns, and acquiring a wealth of demographic and reproductive data. Every long-term study needs survey records to keep track of many or most members of the population or group. Surveys are less useful for behavioral data because the sampling period is so brief. Follows are a better approach for investigating behavior and social bonds.

Annual resightings of individuals using photo-identification have been used to estimate demographic parameters in a variety of species, including bowhead whales (*Balaena mysticetus:* Miller et al. 1992; Rugh et al. 1992), gray whales (*Eschrichtius robustus:* Jones 1990), right whales (*Balaena glacialis:* Hamilton and Mayo 1990; *B. australis:* Bannister 1990; R. S. Payne et al. 1990), humpback whales (*Megaptera novaeangliae:* Clapham and Mayo 1990; Glockner-Ferrari and Ferrari 1990; Barlow and Clapham 1997), bottlenose dolphins (*Tursiops* sp.: Wells 1991c; Defran et al. 1990; Würsig and Jefferson 1990; Connor et al., chap. 4, this volume), and killer whales (*Orcinus orca:* Olesiuk et al. 1990; Bigg et al. 1990b; Brault and Caswell 1993). Interbirth intervals, patterns of association, ranging, and mortality have been estimated from these long-term studies. Wells and his colleagues (Wells et al. 1980, 1987; Wells and Scott 1990; Duffield and Wells 1991; Connor et al., chap. 4, this volume) have provided the most detailed picture of a cetacean community from a longitudinal study using survey methods. Killer whale research by Bigg and colleagues (see Baird, chap. 5, this volume) has provided similarly detailed life histories.

In contrast to survey records, data from individual follows, or focal animal sampling (Altmann 1974), tend to be detailed, with repeated sampling on a restricted number of individuals, usually within one age or sex class. Focal animal data are optimal for studying behavior, interaction, and social relationships, but are less useful for gaining broad measures for the rest of the population or group. One illustration of the differences between survey and focal behavioral data comes from our Shark Bay studies of bottlenose dolphin mothers and infants. In survey data (Smolker et al. 1992), mother-infant association coefficients are 100% until weaning at age four. Focal data, using the same

10 m chain rule (any group member who is within 10 m of any other group member is in the group), show that some infants are rarely apart from their mothers (3% of the time), but others spend up to 30% of their time away from their mothers (Mann 1997; Smolker et al. 1993). How do we reconcile these differences? During surveys, animals may join or leave a group unnoticed. For example, if a young calf is not in the presence of her mother, but arrives fifteen minutes after the survey begins, the observer may believe the calf must have been there all along. Because mothers and infants typically separate for about thirteen minutes at a time (J. Mann, unpublished data; Connor et al., chap. 4, this volume), the infant is bound to appear if the observer waits for the typical duration of survey sightings. In later years, we reduced this bias by including animals who were present only during the first five minutes of the survey. Other methodological differences may have contributed to these discrepant results. Single animals and all foraging groups were excluded from the analyses. In focal studies, infants were most likely to separate from their mothers during maternal foraging (Mann and Smuts 1993, 1998); however, the exclusion of foraging groups cannot completely account for the different results from surveys and focal follows. Mothers and infants separate and rejoin over periods of minutes or hours, changing with social and ecological context. The survey "snapshot" shows the stable association between mother and calf, but not the dynamic, fluid changes in their pattern of association over the day, weeks, or years.

Similarly, pairs and triplets of *Tursiops* males have very high coefficients of association in Shark Bay (Smolker et al. 1992), but without focal follows on males, Connor and colleagues could not have discovered why: males form stable alliances that cooperate to aggressively herd cycling females and keep other males away from them (Connor et al. 1992a,b). The details, the behavioral sequences, who initiates and who terminates interactions, who stays with whom and how close, are all intricate parts of the cetacean social world. Survey and focal data are both important sources for any field study, but to understand social bonds, one must go beyond the survey snapshot and aim for the full-length film.

Developing Research Questions: The Value of Comparisons Within and Across Taxa

The comparative method is particularly powerful for highlighting potential mechanisms and adaptive functions of behavior. For example, Whitehead and Weilgart's comparison between sperm whales and African elephants in chapter 6 of this volume shows remarkably similar life history strategies and patterns of kinship and mating. These comparisons suggest further questions for later study. The roving behavior of large males in both species poses problems for estrous females, who need to attract males over long distances, perhaps to induce male-male competition or provide more mating options. Female elephants solve this problem by emitting a very low frequency estrous call. Males approach calling females from over tens of kilometers (Poole et al. 1988; Poole 1989). Do estrous sperm whales give calls as well? We don't know, but if mature males are in short supply, the ocean is an excellent medium for broadcasting reproductive state acoustically.

Cetaceans are suitable for comparisons across species because they show such diversity in behavior, ecology, and life history. For example, theories about what selective pressures favored larger brains have been difficult to test because of methodological problems in comparing animals of different body sizes and/or across different taxa. However, the natural variations in brain size in odontocetes of similar body size, such as baiji (*Lipotes vexillifer*, which is small-brained, and bottlenose dolphin (*Tursiops* sp.), which is large-brained, provide convenient examples for comparing hypotheses regarding social versus ecological selective pressures in brain evolution (see Connor et al. 1992b). Within-species variation can also be substantial in cetaceans. The twofold variation in body size within *Tursiops* sp. may help identify factors linked to male-male alliance formation (Connor et al., chap. 4, this volume). The marked differences between *transient* and *resident* killer whales in group size and hunting behavior can be used to test optimization and ecological models (Baird, chap. 5, this volume). Variation within populations can pose conceptual and statistical problems, however. Once can't assume that the mean or median is representative of a population, particularly where age or sex class variations may be important.

Comparisons with terrestrial mammals and/or pinnipeds may also be appropriate. With the remarkable adaptations cetaceans have made to marine life, convergence of their strategies with those of terrestrial mammals can be striking. Cetacean foraging ecology may be similar to that of felids (e.g., compare killer whales with lions and cheetahs, or bottlenose dolphins with primates); life history strategies may be similar to those of bovids (e.g., compare humpback whales with wildebeest, buffalo), or primates (compare bottlenose dolphins with chimpanzees: Whitehead and Mann, chap. 9; Connor et al., chap. 10, this volume). Baleen whale milk composition is comparable to that of seals and bears, which also fast during most of lactation and wean quickly (Oftedal 1993, 1997). Weaning age in most odontocetes is comparable to that of some primates (Whitehead and Mann, chap. 9, this volume). The newborn cetacean's motoric precociousness and following response is ungulate-like, but the intensity and duration of some odontocete mother-infant bonds are primate-like (Whitehead and Mann, chap. 9, this volume). Evidence for elaborate vocal behavior and learning invites comparisons with bats, humans, and birds (Janik and Slater 1997; Tyack, chap. 11, this volume). Cetacean diving behavior, breathing patterns, and predation risk may be shared with some pinnipeds. Coalitionary behavior in *Tursiops* has some striking similarities to coalitions in some primate species (Connor et al., chap. 10, this volume). These comparisons help identify patterns and direct future research.

With research questions in hand, some sampling decisions become self-evident, such as which animals to sample, what behaviors to record, and what sampling regime to use. We know so little about the behavior of most dolphin and whale species that initial research questions tend to be very broad, such as, are there sex differences in adult behavior and patterns of association? For many research questions, observational methods are not the most appropriate tool. For example, if one is interested in kinship bonds among male kin, it would take many years of observation to document genealogical relationships. Genetic sampling is a much quicker and more reliable means of determining relatedness, especially for paternity studies. If one is interested in diving depths during different activities, then tagging the animal with a time-depth recorder would be more effective than observing how long the animal stays submerged between breaths. (Such technological innovations are discussed in detail by Whitehead et al. in chapter 3 of this volume.) Analyses of stomach contents from dead animals may reveal much more about diet and foraging behavior than behavioral observations. However, many research questions can be answered with appropriate observational sampling methods. The next section reviews the practical issues involved in observing cetaceans, including identification of animals, habituation to boats, defining behavior (developing an ethogram), defining association, deciding whom to follow (groups vs. individuals), sample length, observational sampling methods (e.g., continuous, point, scan sampling), and how to cope with difficult sampling conditions.

Observational Methods

Identification

Observational techniques often depend on identifying individuals, or at minimum, being able to distinguish be-

tween one animal observed and the next. Identification is typically not possible without markings that are visible above water, such as distinctive dorsal fins, tail flukes, or other reliable dorsal markings (figure 2.1). For some large whales that have distinctive tail fluke markings, observers must wait until a "fluke-up" dive to identify the animal they've been watching. Not all dives are "fluke-up," or at the proper angle to photograph, and it can potentially take hours of the observers' time to identify a whale. (Photo-identification techniques and tagging are covered in more detail by Whitehead et al. in chapter 3 of this volume.)

Fortunately, the majority of cetacean species have dorsal markings. At least thirty species of cetaceans have been studied using individual identification (table 2.1), although very few studies have focused on behavior. Over time, many of these animals can be sexed, either by direct views of the genital area, persistent presence of a young calf next to a (female) cetacean, adult body size, or genetic sampling. Body size may be useful only for discriminating mature males in highly dimorphic species (e.g., sperm whales), since smaller animals can be younger males or females. In some species, such as Pacific white-sided dolphins and killer whales, dorsal fins are sexually dimorphic.

Relatively unbiased estimates of behavior rates can still be acquired when individuals are not readily identifiable. If the animals are reasonably solitary (e.g., some baleen whales), then the observer may follow an individual animal for a day and record its behavior, or just observe one animal for as long as possible. Richardson and colleagues successfully followed individual bowhead whales (*Balaena mysticetus*) for two to eight hours (Richardson et al. 1995a). Others have successfully followed blue whales (*Balaenoptera musculus*) (A. Samuels, personal communication), suggesting that several balaenopterid species might be systematically sampled. If the animals are migrating, or if there are many animals, it may be unlikely that the same animal will be sampled on two different days, allowing researchers to consider each follow independently. Mallonée (1991) employed this technique for observing gray whales (*Eschrichtius robustus*) during their migration off the California coastline. The likelihood of resighting and sampling the same individual should be tested with identifiable individuals before assuming that each sample is independent. If the animals are in groups, the observer may use scan sampling to instantaneously sample each group member's behavior in some predetermined sequence (discussed in greater detail below). Alternatively, observers may stay with particular types of groups, such as the humpback whale competitive groups described by Tyack and Whitehead (1983). Here, observers may be able to distinguish between individual

A

B

Figure 2.1. Identification photographs for bottlenose dolphins (A–C) and blue whales (D–E), showing how the dorsal fins and tail flukes show reliable markers. (A–C, photographs by Janet Mann; D, E, photographs by Robin Baird.)

C

D

E

Table 2.1. Markings and other morphological characteristics used for individual identification in cetaceans

Common name	Species	Markings[a]	Vessel[b]	Animals[c]
Mysticetes				
Right whale, N. Atlantic	*Balaena glacialis*	Callosities, scars, lip crenulations	B, A	257
Right whale, S. Hemisphere	*Balaena australis*	Callosities, scars	B, A, S	4–850
Bowhead whale	*Balaena mysticetus*	Dorsal pigment	A	1,400
Blue whale	*Balaenoptera musculus*	Flukes, pigment, scars, mottling	B	32–220
Fin whale	*Balaenoptera physalus*	Dorsal fin, pigment, scars	B	149–200
Sei whale	*Balaenoptera borealis*	Dorsal fin, scars	B	60
Bryde's whale	*Balaenoptera edeni*	Dorsal fin, pigment, scars	B	50–160
Minke whale	*Balaenoptera acutorostrata*	Dorsal fin, pigment, scars	B	6–30
Gray whale	*Eschrichtius robustus*	Pigment, scars	B	10–701
Humpback, N. Pacific	*Megaptera novaeangliae*	Flukes (dorsal fin, pigment)	B	50–6,000
Humpback, N. Atlantic		Flukes (dorsal fin, scars)	B	550–3,700
Humpback, S. Hemisphere		Flukes (dorsal fin, pigment)	B	10–400
Odontocetes				
Sperm whale	*Physeter macrocephalus*	Flukes, dorsal fin	B	40–580
Killer whale	*Orcinus orca*	Dorsal fin, saddle patch, scars	B, S	10–350
Beluga whale	*Delphinapterus leucas*	Scars	B	?
Bottlenose dolphin	*Tursiops* spp.	Dorsal fin, scars	B	400–700
Atlantic spotted dolphin	*Stenella frontalis*	Dorsal fin, fluke, spotting pattern	B	?
Hump-backed dolphin	*Sousa chinensis*	Dorsal and dorsal fin scars	B	50
Hector's dolphin	*Cephalorhynchus hectori*	Dorsal fin, scars (pigment)	B	300
Risso's dolphin	*Grampus griseus*	Dorsal fin	B	60–250
Spinner dolphin	*Stenella longirostris*	Dorsal fin	B	220
Short-finned pilot whale	*Globicephala macrorhynchus*	Dorsal fin, saddle mark	B	100
Baiji	*Lipotes vexillifer*	Dorsal fin, face pigment	B	?
Amazon river dolphin (boto)	*Inia geoffrensis*	Dorsal and dorsal fin, pigment	B	44
Tucuxi	*Sotalia fluviatilis*	Dorsal and dorsal fin, pigment	B	40
Dall's porpoise	*Phocoenoides dalli*	Dorsal fin pigment	B	63
Pacific white-sided dolphin	*Lagenorhynchus obliquidens*[d]	Dorsal fin, pigment	B	?
Atlantic white-sided dolphin	*Lagenorhynchus acutus*	Dorsal fin, pigment	B	?
White-beaked dolphin	*Lagenorhynchus albirostris*	Scars, pigment	B	10
Dusky dolphin	*Lagenorhynchus obscurus*[d]	Dorsal fin, pigment	B	?

Source: Adapted and summarized from Hammond et al. 1990 and Würsig and Jefferson 1990; additional information from Corkeron 1990; Gonzalez 1994; and the author

Note: This is not a complete list of all species with identifiable individuals.

[a]Parentheses indicate that markings of this type are less reliable.

[b]B, boat; A, aircraft; S, shore

[c]Ranges for number of animals are presented if more than one field study was conducted.

[d]LeDuc et al. (1999) recently reclassified these *Lage norhynchus* as *Sagmatias*.

animals long enough to identify their "roles" in the group and collect behavioral sequences (see the discussion of sequence sampling below).

Habituation and Effects of Observation Vessels on Behavior

If observers are using boats to study behavior, then habituation to the presence of boats is necessary. Killer whales in British Colombia and bottlenose dolphins in Shark Bay, Western Australia, allow boats to remain at close range, but at other sites it may be difficult to get close to the animals and stay there. Even if dolphins or whales allow a boat to remain close, that does not mean they are not affected by the boat's presence. With coastal species, shore-based observers can test the effects of the boat by contrasting cetacean behavior when boats are present and absent. Janik and Thompson (1996) did such a comparison and found that bottlenose dolphins dove for longer periods when a dolphin-watching tour boat was near than when non-tour boats passed through and when no boats were present. Even without land-based observations, researchers can use sonobuoys to compare vocal communication with and without boats present, or can compare group size and behavior patterns observed at a distance (e.g., 100 m) with those observed at close range (using gross-level categories, such as speed, dive times, direction changes, degree of clustering—choosing only those that can be reliably measured at varying distances). Indirect effects of the boat should also

Figure 2.2. Researchers following dolphins in a small boat. (Photograph courtesy of Andrew Wright.)

be considered; for example, boats may reduce or increase foraging success by affecting fish behavior or deter predation by affecting the behavior of large sharks.

As any field researcher can attest, experience matters in learning how to avoid disturbing the animals and how to stay with them. Rapid habituation depends on sensitive boat driving. Experienced observers avoid sudden turns, accelerations, and decelerations, approaching the animals head-on, or zooming up from behind; such maneuvers are prohibited by most whale-watching guidelines designed to reduce harassment. Maintaining a steady speed so as to keep pace with the animals helps to maximize their visibility and minimize disrupting their behavior (fig. 2.2). Depending on behavioral context or how the animals are identified, observers may try to stay parallel to (when identification is based on dorsal fins) or behind (when identification is based on tail flukes) the animals during much of the observation period. Researchers may take their cues from the animals and adjust their driving according to the animals' activities. For example, when dolphins are resting, observers may be able to cut the motors. Similarly, during foraging, when animals make rapid accelerations and turns, it is easy for boats to get in the way, and maintaining some distance may reduce interference. Sailing vessels are often used to study sperm whales (Whitehead and Weilgart, chap. 6, this volume), and although sailing is impractical for studying some cetaceans, using four-stroke, electric, or diesel motors or rubber-mounted engines can reduce noise levels for the animals. (The effects of sound pollution on cetaceans are discussed in more detail by Whitehead et al. in chapter 12 of this volume.) From the observer's perspective, good data collection depends on the observers' ability to situate the boat in the right place *before* the animals resurface.

Developing an Ethogram

The sounds and gestures of cetaceans are difficult to interpret compared with those of the terrestrial mammals with which we are more familiar. Thus, an ethogram—a list of defined behavior categories—may be challenging to develop. The meaning or function of a behavior pattern can be derived from examining the contexts and consequences of its occurrence and is not normally part of an ethogram. For example, the ethogram might contain the behavior "petting": one animal actively moves its pectoral fin up and down the body part(s) of another animal (fig. 2.3). Although the observer may believe this is an affiliative behavior, s/he wouldn't write down "being friendly" or "affiliation" when the behavior occurred. If petting is affiliative, one might expect it to occur between animals who approach each other frequently, show preferential association, and do so before and after petting occurs. Other behaviors may occur in a variety of contexts and may not serve any one function. For example, "belly-up swim" may occur during play, during hunting, during agonistic encounters, or during mating. The more specific the description and definition of the behavior, the better, and the more one watches animals, the easier it becomes to make refined distinctions between different behaviors. One can always lump different types of social behaviors into one category for analysis later, but if all social behaviors are just called "social" during data collection, the data cannot be subdivided later into sexual play, agonistic interaction, affiliative

Figure 2.3. Shark Bay bottlenose dolphin mother and calf petting. Petting (active rubbing of the pectoral fin or fluke on the body of another) is a common expression of affiliation among bottlenose dolphins and especially between mothers and calves. (Photograph by Janet Mann.)

interaction, and so on. Some behaviors are difficult to identify; for example, traveling and foraging are sometimes obvious, sometimes ambiguous. Dolphins, for example, may appear to be traveling, but periodically echolocate or inspect a weed patch for fish. Sound recording may be particularly helpful in resolving ambiguous situations. Otherwise, the observer may opt to have a combined travel/foraging category for occasions when finer discriminations are not possible.

The ethogram helps ensure that the observer defines and records the behaviors the same way whenever they occur. However, it does not substitute for reliability measures within and between observers. Few ethological studies present intra- and inter-observer reliability coefficients. Reliability measures identify the sources of error in behavioral sampling, including intra-observer variation (from day to day, sample to sample, or subject to subject) and inter-observer variation (stable differences between observers). Reliability scores include percentage of agreement, Phi correlation coefficients (Bakeman and Gottman 1986), and generalizability scores (Shavelson and Webb 1991), the latter accounting for the most sources of error. Of seventy-four cetacean field studies of behavior reviewed by Mann (1999), none reported observer reliability. This is understandable, given that for most field studies, only one observer is typically available to record at any one time. All the more reason, then, for the ethogram and sampling protocol to be unambiguous and well-defined. Other observers should be trainable and the study replicable. Reliability measures are difficult to obtain, since they require two observers who simultaneously but independently code behaviors—both individuals should be on the same vessel and have the same vantage point. This is difficult when observers can see or hear each other. Field researchers may compensate by co-observing periodically and reaching consensus on behavior coding. Videotapes are valuable to train and test field observers when on-the-spot reliability testing is impractical.

Events and States

Altmann (1974) distinguished between events (short-duration behaviors) and states (long-duration behaviors) because this distinction has numerous implications for how one samples behavior. In developing an ethogram, the researcher quickly realizes that some behaviors are brief and some have appreciable durations. Behaviors brief enough to be classified as events may include single behavioral displays (breaches, head slaps, spyhops, lobtails: see plate 3), vocalizations, or breaths. Other behaviors, such as approaches and leaves, can be turned into brief events by defining them in terms of a criterion distance. The initiations and terminations of behaviors (transitions between states) may also be considered events. When the animals join, that is an event; how much time they spend together is a state.

For events, frequency, converted into a rate per time unit, is important. High-frequency behaviors, such as breathing, may be difficult to record if one is also collecting other types of data. In such circumstances, it is often convenient to consider event clusters as bouts and score the onset and offset, or initiation and termination, of the bout instead. The researcher might record the duration of a surfacing bout, rather than count every breath within the bout.

States are ongoing behaviors, such as the "big four": socializing, resting, traveling, or foraging. These are expressed as proportions or percentages of time spent engaged in the behavior (time budgets). Time budget data can help identify important developmental, behavioral, or reproductive phenomena. For example, the proportion of time infants spend foraging may be a good measure of their relative independence from the mother. Or seasonal changes in the proportion of time adults spend socializing may indicate mating, even if copulations are not observed.

It is often difficult to determine the exact onset or offset of a behavior. Biases against the offset can be an even greater problem. Since the intensity of behavioral states may wane over time, the offset is less noticeable. Such bias will be minimal if states are mutually exclusive because when resting starts, socializing is, by definition, over. Fortunately, observers have a number of sampling options for estimating time proportions, which are described in greater detail below.

Association

One of the fundamental questions facing a student of social behavior is "who is with whom?" or conversely, "who is *not* found with whom?" For animals to interact, they must make contact: vocal, visual, chemical, or physical. The most distal level of contact (acoustic range) may help to broadly define group membership because it allows individuals to coordinate their movements to stay roughly "together," while more proximal contact, such as touching, can define a closer level of association. Levels of association can also vary in frequency, duration, and intensity over different time scales (e.g., fleeting, seasonal, lifetime) and depend on life history factors (e.g., emigration from the natal group at maturity). Sperm whale units (Whitehead and Weilgart, chap. 6, this volume) and killer whale pods (Baird, chap. 5, this volume) may coordinate their movements over kilometers. Yet, even within those groups that are stable for life, some animals will consistently stay closer together than others or form preferential associations. Furthermore, the flexibility of cetacean associations can vary enormously, from the relatively stable pods of killer whales (Baird, chap. 5, this volume) to the more fleeting associations of humpback whales (Clapham, chap. 7, this volume).

Association indices (e.g., Cairns and Schwager 1987; Ginsberg and Young 1992) are critical measures of dyadic or between-group relationships. Repeated measures of who tends to be in the same group (coefficients of association or proportion of time spent together) provide basic data on social life. Association patterns have been documented

in a number of cetacean species, and the closeness and temporal stability of such associations can be modeled to provide a cohesive picture of social structure (Whitehead 1997). This section focuses on the conceptual and practical problems of defining who is "together."

Defining Groups

Most cetacean studies define groups based on "coordinated activity" or distance measures (Mann 1999). Distance measures assume nothing about the individuals' activities. "Coordinated activity" definitions of groups consider animals as members of the same group if they are engaged in the same activity or are traveling in the same direction as others (e.g., Shane 1990a; Ford 1989). This definition makes implicit assumptions about distance, since observers must be close enough to monitor the activities of all potential group members. Observers cannot tell if an animal a kilometer away is coordinating its activities with those of others equally spaced. How long must they be "traveling in the same direction" to be coordinated? How close must they be, and does sea state influence one's ability to identify such associates? Similarly, does that mean that animals closer to one another, but engaged in different activities, are not members of the same group? Different weight may be given to different members of varying age or sex classes. For example, if an infant is playing while its mother and other adults are foraging, isn't the infant still part of the group? Does the observer need to document the activity and direction of travel for all possible animals within a large range to determine group membership? Although coordinated movements and activities may be important to the animals for maintaining group membership, they are impractical to quantify reliably. Spatial or proximity-based definitions are recommended for association measures because they are meaningful, more easily quantified, and make fewer assumptions about what a group member must do to qualify for membership. Some researchers have opted to combine proximity and "coordinated activity" definitions (e.g., Weinrich and Kuhlberg 1991; Weinrich 1991; Whitehead et al. 1992; Mattila et al. 1994).

If behavioral or group definitions are too broad, important details are sacrificed. For example, if all individuals sighted in the bay that day are classified as associates, closer association patterns may be missed. Furthermore, the animals sighted and called "associates" may, in fact, be avoiding one another, thus rendering association categories biologically insignificant to the animals themselves. So what distances are meaningful? The time it would take to harm or help another is important. For a dolphin or whale mother and calf, a separation of 10 m versus 100 m could change the course of a predation attempt. Animals may relate to one another differently if they are close enough for immediate physical contact. The definition of a group might also reflect the functional aspects of association. For example, although Smolker and colleagues (1992) used a 10 m chain rule to identify dolphin group membership, they chose to eliminate foraging groups from their association analysis because their proximity may be caused by a common interest in a food source, not in each other. In addition, complete group size may not be reliably recorded during some types of foraging. This strategy is not equivalent to defining groups based on whether or not they are engaged in the same activity, but it does analyze association based on whether or not the majority of group members (determined by proximity) are engaged in a particular activity. Some studies consider animals to be together if they are within one or two body lengths (e.g., humpback whales: Weinrich 1991) or five body lengths of each other (bowhead whales: Dorsey et al. 1989). Species body sizes should be taken into consideration in defining association because it reflects the animals' vulnerability (i.e., to predators) and possibly other factors (e.g., swimming speed, detectability). Twenty meters of separation may mean little to a fin whale mother and infant, but a lot to a harbor porpoise mother and infant. Researchers can experiment with several definitions and compare the results. The definition of association should capture the fluid or stable nature of these relationships, and how context affects their patterns.

Once a protocol is developed to define "who is with whom," researchers might turn their attentions toward considering why some animals are *not* together. For example, on the humpback breeding ground, groups with two mothers and calves are seldom seen, and mothers with calves avoid one another (Tyack 1982; Herman and Antinoja 1977). On a larger geographic scale, young male sperm whales remain at high latitudes, feeding, rather than entering the breeding grounds during the breeding season; in contrast, large males move toward warmer waters to mate (Whitehead and Weilgart, chap. 6, this volume). Immature males may associate with mature males during nonbreeding times of the year. Even if young males virtually never associate with mature females, is it fair to say that they are avoiding them? Immature males may not be avoiding females; instead, they are probably avoiding competition with large mature males.

Avoidance relationships are more difficult to study, but acoustic data can provide clues. To test the hypothesis about sperm whale avoidance behavior, adult male sounds could be played to immature males to see whether they provoke different reactions than playbacks of female codas.

Playback experiments are particularly valuable in examining avoidance relationships. In another example, Tyack (1983) found that most humpback whales on the breeding grounds avoided playback of humpback song, which is usually produced by lone males, in contrast to playback of "social sounds" of males competing for access to a female, which often attracts other males. These patterns match naturalistic observations of how whales respond to other whales producing these sounds (Clapham, chap. 7, this volume). Kroodsma (1989) provides an excellent review of playback techniques, and Sayigh et al. (1999) illustrate the value of such experiments for cetaceans.

Proximity measures are critical tools for investigating the nature of relationships (Carpenter 1964). Close bonds can be revealed by measures of who tends to be next to, or close to, specific others within groups. The "nearest neighbor" method, which involves identifying the closest animal(s) to a focal animal at regular intervals, has been widely applied in ethology since the 1960s (e.g., Kummer 1968). We may not know why the animals are together, yet, but we can develop some hypotheses (e.g., close associates are kin or alliance partners). Scan sampling of associates within a well-defined radius can yield clusters or subgrouping patterns within groups. Scott and Perryman (1991) used the nearest neighbor approach with aerial photogrammetry to examine age-sex class association and approximate age of weaning for large schools of over two thousand spotted dolphins. Similarly, patterns of preferential association in killer whale pods (S. L. Heimlich-Boran 1986) were determined by examining neighbors in photographs. The stability (duration over days, months, or years) of associations is another measure of "preference."

From Preferential Association to Social Relationships

Why is animal A with animal B? To determine *why* animals associate, one must observe what they actually *do* together. By observing individual behavior and interactions, we gain the individual's perspective on social relationships. Proximity, affiliative behaviors, agonistic interactions, approaches and leaves (who initiates), vocal communication, and the context-dependent nature of these interactions inform us of the nature of social relationships. For many displays, it is possible to identify both who produced the display and who is the "intended" recipient. Such measures are widely used to study dyadic interactions in ethology. Long-term, repeated observations of dyadic interactions are critical to building from the level of interactions to the level of relationships. For example, the Hinde index (Hinde and Atkinson 1970) has been used to determine who is responsible for maintaining proximity in a particular dyad by comparing the asymmetries in who approaches and who leaves the other most often. Two field studies have employed the Hinde index for studies of cetacean mothers and infants (southern right whales: Thomas and Taber 1984; bottlenose dolphins: Mann and Smuts, 1999) and have found results similar to those found for other mammals, and consistent with predictions generated by Trivers's (1972) theory of parental investment.

To move from preferential association (who is closer to whom within groups) to social relationships, behavioral definitions and functional categories must be quantitatively assessed. For example, agonistic interactions, as originally defined (Scott and Fredericson 1951), are any acts that encompass either aggressive or submissive behaviors. Repeated observations of dyadic encounters are necessary to identify dominance relationships and their stability (Samuels and Tyack, chap. 1, this volume). Aggressive, submissive, and affiliative behaviors may be difficult to define (see Samuels and Tyack, chap. 1, this volume). Observers may initially come up with categories that describe the acts, but ascribe these to larger conceptual categories later. For example, a "lateral head jerk" may be described, but in the analyses, the researcher decides it should be subsumed with other aggressive behaviors because of its context and the responses of others—that is, based on the observation that it is accompanied by obvious aggressive behaviors, such as biting or hitting. Mounting, on the other hand, may occur in aggressive, playful, mating, or affiliative contexts, and may not be placed in a superordinate functional category. Without such categories and definitions of behavior patterns, we cannot examine the functional significance of social behavior and define relationships.

Even if it is difficult to observe affiliative, aggressive, and mating behaviors among cetaceans, proximity is sometimes visible at the surface, and a record of who is responsible for maintaining proximity, and for how long, provides a wealth of information. Is there an asymmetry? Does the pattern change when others are present? Are there seasonal changes? I now turn to specific problems that confront the field observer in sampling behavior.

Follow Protocol

Once research questions are defined and an ethogram developed, the ethologist develops a protocol and method for measuring behavior. One choice concerns which subject(s) one watches and for how long; the other concerns the details of how behavior is recorded (Martin and Bateson 1986). These decisions may be distinguished as "follow protocol" and the "sampling method." "Follow protocol" refers to how long an observation extends and to whether researchers follow a group or an individual animal. "Sampling method" refers to the procedures used to "sample" the behavior of individuals or groups (Mann 1999). Most ethological research is based on the study of individuals—the focal animal protocol. However, with large or fast-moving groups of cetaceans, it may not be feasible to stay with one particular individual and record its behavior. For historical, theoretical (Samuels and Tyack, chap. 1, this volume), and practical reasons, most studies of cetacean behavior employ a "group follow protocol" (Mann 1999).

Group follows. Group sizes for cetaceans can range into the hundreds (e.g., Rose and Payne 1991: Shelden et al. 1995) or thousands (Scott and Perryman 1991). Large groups may be easier to find and keep track of, but are more challenging settings for minimizing observational bias. If cetologists choose to follow groups, sampling protocols must specify a strategy for sampling all individuals, or a restricted subset of individuals, and/or obvious behaviors (those less vulnerable to bias—i.e., that are unlikely to be recorded at different rates because of group size, spread, activity, or speed). Observers cannot record the activities of all group members at once, but they can develop systematic sampling procedures, including the use of video or photography, for scanning the behaviors and spatial relationships of some or all group members (e.g., Scott and Perryman 1991). Obvious behaviors, such as breaches and surface displays (hitting the tail, pectoral fins, or other body parts on the water surface), may be visible enough to score reliably for all group members (e.g., Waters and Whitehead 1990). Depending on the size and other characteristics of the group, sampling decisions necessarily entail restricting the amount of information that can be consistently and reliably recorded. For example, Baird and Dill (1996) followed small and stable groups of *transient* killer whales and could consistently document the rate of kills and prey sharing. If groups fission or individuals leave, observers may be tempted to stay with the larger or more active group, or may not notice the departures of less active group members. Because this is likely to bias data collection, a decision rule for following animals under changing conditions must be developed a priori.

Individual follows. Individual follows (focal animal sampling) enable the observer to focus on an individual animal's "perspective." What is a day in the life of that animal like? Who does it approach, avoid, stay close to, interact with, mate with, and fight with? Observing the continuous stream of individual behavior in different contexts can be instrumental in elucidating the dynamics of social relationships.

One of the first focal animal studies with wild cetaceans was a study of southern right whale *(Balaena australis)* mothers and infants (Taber and Thomas 1982; Thomas and Taber 1984). This study quantified patterns of behavior, development, and spatial proximity between mother and calf from birth to weaning. Its detailed observations revealed distinct stages of early development, changes in the mother-calf relationship, and weaning conflict (Thomas and Taber 1984). More recently, focal animal observations of bowhead whales *(Balaena mysticetus)* revealed activity budgets, detailed foraging information, and intriguing patterns of association and social behavior (Richardson et al. 1995a). Focal animal studies of bottlenose dolphins in Shark Bay, Australia, have shown how mothers and calves use whistles during separations (Smolker et al. 1993) and the development of infant behavior patterns and relationships during the first months of life (Mann and Smuts 1998).

Sampling Methods

The following section scrutinizes quantitative sampling methods. Quantitative systematic observations of repeated interactions between individual animals in varying contexts help reveal the functions of relationships.

Sample Length

Ideally, the sample length or period should be as long as possible so that the timing of intervals between rare behaviors is preserved in the record (see Rogosa and Ghandour 1991). However, it is sometimes impossible to stay with one animal for more than five or ten minutes, either because its movements are too fast or unpredictable or because groups are so large that animals are easily lost in the crowd (e.g., spinner dolphins: Östman 1994). One solution to this problem is to use very short focal sample periods, such as five minutes. If loss of focal animals is dependent on their behavior (as it often is), then the sampling length should be short enough that all behaviors have an equal probability of being sampled when they occur. For example, if the observer typically loses track of a foraging animal within five to seven minutes, but can remain with a socializing or resting animal for twenty minutes or more, then samples should be five minutes long. Otherwise, if the researcher just stayed with the focal animal for as long as possible, or even for fifteen minutes, s/he would have lots of resting and socializing data and very little foraging data. The cost is that long sequences of behavior (over five minutes) are lost. If behavioral states have long durations,

or the intervals between events are long, then short sample lengths will censor and bias the data. In the case of behavioral states, they will result in sample periods that end before the behavioral state has terminated. In the case of rare behavioral events, this would mean that short sample lengths would seldom include two of the events, and those that did would represent a biased sample. If loss of the focal animal occurs occasionally, and is not tightly linked to major behavioral states, then long follows would still be appropriate. When sample length is cut short because of the animal's behavior, observers are obliged to document the cause, if known.

To avoid having too much censored data (e.g., if bouts of behavior are very long), or if there are important behavioral sequences that require long periods of sampling, the ethologist may use a different protocol to sample those incidents. Courtship, for example, may go on for extended periods of time in some species or among some individuals. If observers are interested in how different types of courtship strategies result in potentially conceptive matings, then they may target the onset of such male-female interactions (e.g., among humpback or right whales) and stay with the pair until they animals disperse for some predetermined period of time. In primate studies of reconciliation, de Waal (1993), started sampling dyads as soon as an agonistic interaction was observed. He would then stay with that dyad for a prescribed period of time to see if they "reconciled." This strategy is a form of "sequence sampling," in which the first occurrence of a particular obvious behavior observed in a group is selectively sampled from beginning to end (Mann 1999).

Sample length, hours of observation, and number of subjects observed are critical but rarely reported elements in cetacean studies (Mann 1999). In my survey of studies, data were often pooled across subjects, or each event, rather than by each subject. This approach creates a serious problem because some individuals were overrepresented in pooled samples. In other words, samples were not independent. In general, measures of behavior should be pooled by individual, so that each individual has a rate or proportion of time spent in given activities for a time period or field season. In general, the sample size ought to refer to the number of subjects, not the number of samples or observations taken (e.g., see Martin and Bateson 1986; Martin and Kraemer 1987).

Choosing Whom to Sample

Decision rules concerning whom to sample are essential for limiting the biases of behavioral sampling. Such "rules" involve which group or which individuals to search for, and within a given group, which individuals to sample and in what order.

Because of the search costs of finding focal animals or groups, remaining with one animal or one group per sampling day is often advisable. If observers are following large groups, then only a subset of the animals can be sampled. The decision rule might involve sampling all individuals who surface within a randomly selected area, or sampling only animals who can be readily identified—because of distinct markings or size. For small groups, scanning all individuals from left to right or in a preselected order is feasible. Each decision can either increase or diminish the biases of data collection depending on the characteristics of the animals and how observers follow them.

If observers are following individuals, random sampling of focal subjects is usually not possible, either because of weather conditions or because time constraints may prohibit searching for animals near the boundaries of the study area. The frustration of spending time searching during valuable calm days may make it difficult to continue searching for the top-priority animal on your list while passing up animals 2 and 3. The main point is to have some clear decision rules, and stick to them, even if there is something interesting going on elsewhere. It is critical to avoid following individuals just because they are doing something more exciting than the animals that you are supposed to be following. Data will otherwise be biased toward certain events. In my bottlenose dolphin mother-infant study, all infants were ranked each day by sampling priority. The rank depended on how many hours of observation had already been collected on an individual and when it was last observed. The decision rules concerned weather, search time, and time of day. This flexible but standardized system was developed as a compromise between minimizing sampling bias and maximizing focal data collection.

Types of Sampling

Jeanne Altmann's (1974) classic paper on sampling methods identifies different sampling techniques and examines the biases inherent in each method. Because of the methodological challenges posed by cetacean studies, I reexamine sampling issues here with special consideration of the cetologist's sampling conditions. Table 2.2 indicates what follow protocol (group vs. individual) and sampling method (continuous, point, scan, predominant activity, etc.) is most appropriate depending on the species' characteristics. I also provide specific examples of cetacean and noncetacean behavioral studies to illustrate the kinds of data such

methods yield. I do not cover every sampling method. Interested students should consult other works (Martin and Bateson 1986; Dunbar 1976; Altmann 1974; Rogosa and Ghandour 1991). I do not claim that systematic observational sampling is simple or completely free of bias. I suggest techniques that are likely to reduce sampling bias, and recognize that all of the challenges faced by researchers cannot be addressed with these generalizations.

Continuous sampling. Continuous sampling creates a systematic record of frequencies and durations of behaviors for one animal or a pair. Surfacing bout durations, breath frequencies, dive types, surface display rates, and synchronous surfacings between the focal animal and others may be recorded on a continuous basis for many dolphin and whale species. Activities at the surface (during surfacing bouts) may also be recorded continuously. If the activities are brief, or difficult to time, they can be scored as events (frequencies) rather than states (onset-offset). Continuous data can be the most difficult to collect and requires that all target behaviors be recorded under all conditions. For example, if the researcher examines breath rates as a function of activity, then breath counts are likely to be accurate when the animal is resting or traveling, but less accurate during socializing or foraging. The ethogram and protocol must specify what behaviors can and cannot be recorded in all situations.

Continuous data collection does not require that animals be in continuous view. Some behaviors can always be seen, such as fluke-up dives or surface displays (e.g., lobtails), some are occasionally seen (e.g., social interactions and directed displays), and some are virtually never seen (e.g., deep-water foraging). The distinction between events and states is especially critical for continuous sampling. Event rates depend on the proportion of time that the animal is in view. For cetaceans, unbiased event rates can be calculated only for behaviors that, by definition, always occur at or near the surface (e.g., breaches) unless underwater observation techniques are applied (see Whitehead et al., chap. 3, this volume). If, by measuring the correspondence between surface and subsurface behavior, the researcher can determine the likelihood of event occurrence in relation to the animal's location in the water column, then surface observations can be used to calculate rates (event frequency divided by the proportion of time the animal is in view or prorated proportional to the relationship between surface and subsurface behavior). In sum, researchers can calculate rates for events that always occur at or near the surface, or prorate event rate estimates based on correspondence

Table 2.2. Recommended uses for different sampling methods

Method	Rapidly identifiable?		Group size		Rate of change in group membership		Dive time	
	Yes	No*	<10	>10*	<12 hr*	>12 hr	<10 min	>10 min*
Individual follow approach								
Continuous	Yes	No	Yes	?	Yes	Yes	Yes	?
Point[a]	Yes	No	Yes	?	Yes	Yes	Yes	?
PAS	Yes	No	Yes	?	Yes	Yes	Yes	?
Group follow approach								
Scan	Yes	Yes	Yes	Yes	?	Yes	Yes	?
Sequence[b,c]	Yes	?	Yes	?	Yes	Yes	Yes	Yes
Incident[b]	Yes	Yes	Yes	?	?	Yes	Yes	Yes

Source: From Mann 1999.

Note: Asterisk indicates more challenging sampling conditions. Yes indicates that this sampling method can probably be used if there are no other problems. No indicates that this method is not likely to be applicable unless animals are solitary.

? indicates that it may be possible to use this sampling method, depending on other conditions.

[a]Recommended for spatial proximity measures and behaviors that are rapidly identifiable.

[b]Recommended for infrequent and obvious behaviors, such as breaches or lobtails.

[c]Sequence parameters must be carefully defined.

between surface and subsurface observations. If the relationship between events observed at surface and subsurface is not known, then researchers may still examine other aspects of each event, such as the event rate at the surface and behaviors or individuals associated with the event.

States can continue over periods during which the animals move in and out of view (e.g., foraging, traveling, socializing, resting). If the "out-of-sight" periods tend to be short relative to the bout lengths of the behavior, then one convention is to designate the activity for the out-of-sight period as either the last activity seen before the subject disappeared or the first activity seen when the animal reappeared (Lehner 1996). Alternatively, the observer may consider a state continuous only if the last activity seen before and the first activity seen after a brief out-of-sight period are the same. This strategy is not effective if behaviors, such as foraging or hunting, consistently occur out of the observer's view. In such circumstances, cetacean observers may detect foraging using other cues, including bubble patterns at the surface, echolocation clicks, or intermittent sightings of fish catches or chases. Observers may use fleeting observations of events to define states, or indirect cues (rapid surfacing, changing direction) to define "unobservable" states. One may also tag an individual to track behaviors at depth for comparison with cues visible at the surface (see Whitehead et al., chap. 3, this volume).

Continuous sampling is one of the richest sources of information on social behavior and relationships because it focuses on details, sequences, and actual rates and dura-

tions of behaviors of individual animals. It is easiest to conduct when animals are readily identifiable, group sizes are small (making it easy to find the focal animal quickly), dive times are short, or animals are in continuous view (see table 2.2). Even when continuous sampling is not possible for all members of a population, it can be useful for studying subsets of behaviors (e.g., singing bouts of humpback whales).

Point sampling. Point sampling entails scoring activity or other data as a "snapshot" at a given moment (e.g., every thirty seconds). In direct behavioral observations, distance to others, activity, or other information may be scored on a point sampling basis. For example, the observer may score who is the focal animal's nearest neighbor at the first surfacing after each five-minute interval. Activity can be more difficult to score. In reality, it takes several seconds or more to "decide" what the animal is doing. It is important that the observer decide as quickly as possible, so that s/he does not inadvertently wait until the animal does something that is more interesting or easier to score. For cetaceans, the "point" often happens when the animals are submerged. If the observer samples behavior at the first surfacing after the interval point, then the data are biased toward behaviors that occur at the surface. For example, sperm whales forage at depth and rest during surfacing bouts (Whitehead and Weilgart, chap. 6, this volume). Other behaviors may not change at the surface (socializing, resting, traveling). One option is to define the breaks between foraging dives as part of the continuous state of "foraging" and simply define it as such. In this example, if the animal is submerged at the point, the activity may be designated as foraging, and if the point occurs when the animal is at the surface, resting is scored. As with continuous sampling, recording "unobserved" behavior is problematic because the researcher must rely on indirect cues. Observers may observe their focal animal as continuously as possible, constantly reassessing activity state and using that information to decide activity state at the "point."

Scan sampling. Scan sampling is useful for cetacean studies because it can be applied under a diverse set of sampling conditions (see table 2.2). Scan sampling entails taking a "point" or "instantaneous" sample of an individual's behavior or location before moving to the next animal and doing the same. Observers may use a random scan sampling schedule, scan right to left, or move ahead of a group and scan animals as they pass an arbitrary location. Whichever scan sampling protocol is used, individuals must be sufficiently discriminated to avoid sampling the same individual more than once per scan sample. The usefulness of this protocol depends on how large groups are, how they tend to move and surface, and how distinctively marked individuals are.

Scan sampling can be used in a variety of ways, but is best applied to associations or states. Here are some examples. First, the observer may scan the group every ten minutes to assess who is still present, or who is within 10 m of the focal animal. Second, the observer may scan the group to get a snapshot of group activity by identifying (as quickly as possible) each animal's activity. This is difficult because of the time it takes to identify the animal's activity. One strategy is to observe each individual for a predetermined brief interval (e.g., three minutes) to determine activity, with the midpoint of the observation taken as "activity." If animals tend to surface synchronously (most individuals near the surface at the same time) regardless of activity, then more rapid scans might be possible. Third, the observer might use scan sampling to assess nearest neighbors for each animal to examine preferential association. If focal data and group scans are combined, it is important that the scans be completed very quickly in order to prevent the observer from being distracted from the focal animal. With scan sampling, cetacean researchers can broaden their data set to look at coordination of group activities and refined measures of association for a number of animals of different age and sex classes. With large groups, it is often necessary to scan a subset. No matter what strategy is used, the likelihood of biasing scans toward

individuals because of their behavior needs to be assessed. This may be achieved by comparing different scan sampling protocols (e.g., comparing scans conducted at different intervals, comparing scans of identifiable and unidentifiable individuals, and comparing scans with focal individual follows). Scan samples are preferable to focal group activity sampling for surveys (see below), so that the activity of each or most group member(s) is explicitly noted.

Predominant activity sampling. Predominant activity sampling (PAS), developed by Hutt and Hutt (1970), refers to scoring a behavior as the predominant activity over some interval (e.g., thirty seconds) only if that behavior occupied 50% or more of that interval (fifteen seconds). This technique accurately represents the proportion of time in which state behaviors occur, but not event frequencies (Tyler 1979). Very short behaviors or displays will not be represented in PAS data unless the observer uses very short intervals. PAS should not be confused with predominant group activity sampling (see below), which is an estimate of what activity the majority (50% or more) of the animals in a group are engaged in. PAS refers to the proportion of a time interval that an individual engaged in a behavior.

PAS is useful for sampling animals that go in and out of view for brief periods. In our sample of Shark Bay *Tursiops* infants, some types of data were collected using focal scans and focal PAS simultaneously. Thus we could compare infant activity budgets based on sampling type. For example, data on "infant position" (when the infant is under the mother and in contact with her) were collected using scan samples at regular two-and-a-half-minute intervals (first surface after the interval), but the infant's predominant activity could also be calculated for the preceding interval based on the observed onsets and terminations of infant position. When we compared proportion of time in infant position for nineteen infants in our longitudinal sample over a total of forty infant years (counting each year as independent) and over 750 hours of focal data, the data were highly correlated ($r = 0.974$). The mean error rate for PAS was 3.2%. It is important that the interval length be short enough to capture the shortest states of interest. No cetacean studies I have surveyed used PAS (Mann 1999).

Sequence sampling. In sequence sampling, the observer systematically records behavioral sequences or interactions of a certain type, maintaining the sequence of the behaviors in the record (Altmann 1974). Parameters defining how an interaction begins and ends must be specified in sequence sampling. For example, sequences of killer whale beach hunting of pinnipeds might begin when one or more killer whales swims within 50 m of the shoreline and ends when the prey are consumed or the whales are all over 200 m from shore. Sequence sampling is distinct from incident sampling because during sequence sampling, multiple events may occur in a group, but the observer focuses on the start of the first event s/he sees and records that particular event or interaction to completion (even if other group members engage in similar behaviors at the same time). With incident sampling, all behavioral events of a specific type must be scored for the entire group. Sequence sampling can be applied during individual or group follows (see table 2.2) and is useful for following chains of interactions. The main limitation depends on the observer's ability to identify and follow through on each sequence.

One-zero sampling. One-zero sampling entails scoring whether or not a behavior occurs during an interval (e.g., thirty seconds), rather than scoring how frequently or for how long the behavior occurred. For example, an observer may score whether or not an animal vocalized during thirty seconds, or whether or not it groomed, rather than the rate of vocalizing (an event) or the duration of grooming (a state). This method does not accurately estimate true rates or proportions of behavior, and the results from one-zero data can provide different results from those of continuous data (Altmann 1974; Mann et al. 1991).

Researchers sometimes use one-zero events to define states. That is, one-zero data may inform the development of an ethogram, but not be part of the sampling method. For example, if two animals are seen in contact during an interval, this may "define" socializing, regardless of the duration or frequency of contact. Or, one might define foraging based on whether or not the animal fluke-ups at the dive. Observers may use one-zero events to define group membership—for example, whether or not animals are observed together during a one-hour period. It is important, then, to distinguish between the use of one-zero (categorical) information to help define behaviors that are presumed to be continuous, and treating one-zero scores as state behaviors themselves.

Incident sampling. Incident sampling entails scoring all behavioral events of a specific type in a group. The observability of the events is key to the success of this method. The behaviors themselves must be obvious and attractive enough to alert the observer (e.g., breaching). In addition, the observer must be able to record all the events regardless of how many animals are present. Thus, for incident sampling to be successful, the behavior must be sufficiently infrequent to allow complete recording for the group. As Altmann (1974) points out, incident sampling is a form of continuous sampling of a group for a restricted set of behaviors. However, with continuous sampling, the individual animals are identified. During incident sampling, distinctions between subjects may or may not be made.

Baird used incident sampling to look at marine mammal kills by focal groups of killer whales (Baird, chap. 6, this volume; Baird and Dill 1996). Waters and Whitehead (1990) used this method to record breaches and lobtails (percussive displays) for sperm whale groups. Although individual identification is not a requirement of incident sampling, observers should still discriminate between the actors of behaviors to avoid the same biases of pooling across subjects (i.e., a minority of subjects would disproportionately contribute to the data set).

Focal group sampling. Focal group sampling is recording group behavior by assessing what activity all or most members of the group are engaged in. Assessments may be continuous (e.g., the group foraged from 10:30 until 11:14) or scored at regular intervals (e.g., predominant group activity assessed over five-minute intervals). After ad libitum sampling, focal group sampling is the most common method of behavioral sampling among cetacean biologists (Mann 1999). It is used with both group follows (staying with a group for prolonged periods of time) and surveys (brief assessments of group composition, location, activity).

In practice, observers informally scan the group to determine what activity most of the group members are engaged in. The informal aspects of such assessments of group activity make group sampling problematic. Predominant group activity (defined as the activity that 50% or more of the group members are simultaneously engaged in) *can* be assessed, but by *explicit* scans of individual group members (scan sampling method). The distinction between scan sampling and focal group sampling to determine group activity is that scan sampling specifically and explicitly samples the behavior of individuals in rapid sequence, yielding a count of *how many* or what proportion of individuals are engaged in a given activity, while focal group sampling involves a global assessment, one that is not based on precise sampling of each group member or some subset of the group.

Altmann (1974) has been cited in support of focal group sampling (e.g., Shane 1990a). However, Altmann described focal (sub-) group sampling as appropriate only under a very restricted set of conditions, "in which all individuals in the sample group are continuously visible throughout the sample period. . . . if one is working with observational conditions that are less than perfect, focal animal sampling should be done on just one focal individual at a time, or at most, a pair (e.g., mother and young infant)" (243–44). These conditions are virtually never met for wild cetaceans.

Altmann (1974) prescribed these conditions for focal group sampling because of the many potential biases inherent in watching groups (as opposed to individuals). Fragaszy et al. (1992) empirically demonstrated the considerable inaccuracy of group sampling in a study of forest primates (see also Mann 1999 for an analysis of group sampling error rates). This alone should give pause to anyone considering using this method.

Biases from group sampling are likely because of a number of factors. First, and most important, the observer's attention is naturally drawn to certain events, such as socializing. No observer can continuously monitor and record the behavior of all group members simultaneously. Sampling one individual is difficult enough. Second, the visibility of group members can change depending on their activity. If foraging animals stay submerged for long periods, but socializing animals remain at the surface, then the predominant group activity may appear to be socializing, because the foraging animals are less visible. Third, group activity may be easy to determine if all members of the group are doing the same thing, but difficult to determine if some animals are doing different things. If most of the animals are resting at the surface, group activity may be easy to identify accurately. If they are engaged in several activities, observers must use a different method of assessment. Fourth, observers are often making implicit assumptions about the relative importance of behaviors of different age and sex classes. If two mothers are hunting, but their infants are traveling, is the group hunting or traveling? In sum, if groups are small and cohesive (e.g., five animals or fewer), then the observer may be able to monitor their behavior and estimate what the predominant activity is for over half the group members, but then sampling is dependent on group size, cohesiveness, and the activities of the animals, and these factors introduce many biases into data collection. Brief surveys of group activity are subject to the same biases.

Researchers do not have to forsake following groups, but they can record their behavior in a different way. Researchers may track a group (stay with them) and record all occurrences of certain very obvious behaviors, thus using inci-

dent sampling. Alternatively, they may stay with a group and repeatedly scan the behavior of all individuals or a subset (scan sampling). Sequence sampling is also appropriate for group follows (see table 2.2).

Ad libitum sampling. Ad libitum samples are "typical field notes" in which the observer writes down what s/he finds of interest, or simply as much as possible (Altmann 1974). Ad libitum sampling is recording what is most readily observed when watching a group of animals. The majority of cetacean behavioral studies are still of this type (Mann 1999). Ad lib data are useful for certain kinds of comparisons, but they are not suitable for looking at rates of behavior, or for comparing behavior patterns of different age or sex classes (Altmann 1974). As Altmann points out, some animals may be more noticeable because of their behavior (more active), size, or level of habituation. Ad lib data are valuable for looking at the direction of interactions, or at what happens within a dyad. For example, ad lib data are commonly used to assess dominance relationships (e.g., see Samuels and Tyack, chap. 1, this volume), in which one animal "wins" and the other "loses" agonistic interactions. They can also be used to look at who pets or grooms whom within a dyad, but not the rate of petting or grooming for those individuals.

Combining Methods

It is practical and advisable to use more than one sampling method. With marine mammals, observers cannot use continuous sampling for all behaviors for the focal animal or group members. Compromises must be made, and it is important that researchers both recognize and minimize the biases inherent in different methods. Sampling schedule, length, method, interval, choice of subject(s), and field conditions contribute differentially to data biases. For example, if the observer follows one focal animal, s/he may use PAS to score activity states, point sampling to assess nearest neighbors, and scan sampling for group composition and activity, water depth, location, and so forth. For all of these methods, the observer must decide on the appropriate sampling interval (e.g., sample group composition once per hour vs. every five minutes) based upon some prior knowledge of the stream of behavior.

Rare Events

Our emphasis on systematic sampling does not completely preclude the option of dropping a focal sample in order to follow something unusual. If something extraordinary comes up, it would be foolish to miss it. Rare events such as predation, a birth, or a lethal fight can provide critical information and insights into your study species that a few hours of focal data could not. For example, Pack et al. (1998) recently reported the first documented death of a male humpback whale during competitive group interactions. Early in any field study, the observer may not know what is exceptional behavior and what is not, making such decisions difficult. Researchers must be explicit when they go "off sample" and must be careful to exclude such observations from the systematic sample data set.

Sampling Problems with Cetaceans

Table 2.2 presents a guide for choosing sampling methods under different observation conditions. If animals are not rapidly identifiable, it is difficult to do focal animal sampling unless animals are solitary. When group sizes are small, many sampling methods are possible, but if groups are large (ten animals or more), then scan sampling of the group or a subset may be the most appropriate method. Long dive times present some of the greatest challenges for both practical (easy to lose subjects between surfacings) and methodological reasons.

Individuals in large groups, such as *Lagenorhynchus, Delphinus,* and *Stenella* spp., seem impossible to follow because they move so quickly and are difficult to identify. However, Östman (1994) has successfully completed short focal follows on *Stenella* in Hawaii using an underwater observation booth built into the vessel, suggesting that it is possible to stay with individuals long enough to get a five-minute sample. Scan sampling is another possible approach with large groups of unidentified animals. In such circumstances, researchers would need to scan a subset of the group, possibly by randomly selecting subgroups or particular age-sex classes. For example, Pacific white-sided dolphin males have distinctive dorsal fins when they are fully mature. Even in large groups (over seventy animals) it is possible to scan behavior and nearest neighbor (identifying approxi-

mate age-sex class or body size) for mature males (J. Mann, personal observation).

The quality of observations varies greatly according to typical diving depth, water clarity, and sea conditions. Researchers may alter their protocol slightly (e.g., switch from continuous sampling to point sampling) under poor observational conditions. They may still be able to get activity budgets in poor conditions, but not behavioral events. Thus, the researcher might report the proportion of time spent resting, traveling, foraging, or socializing under poor conditions, and could use data collected under good conditions to get more refined estimates of time spent in different types of socializing, partners, and so forth. This strategy, however, is based on the assumption that the observational conditions affect the observer, not the animals. Comparisons of activity budgets in different observational conditions could provide an estimate of this bias.

Conclusions

Studies of cetacean behavior have increased in number and scope in recent years. These studies hold the promise of elucidating complex patterns of social behavior, relationships, and organization in dolphin and whale species. Long-term studies of known individuals provide the most detailed and cohesive picture of cetacean social lives. Theoretical and empirical advances in evolutionary biology offer the perspectives necessary to test hypotheses and interpret data. The integration of cetacean research into the field of behavioral ecology depends on how research questions are asked and how animals are sampled. Although dolphins and whales are not easy to observe, there are reliable techniques for doing so. Alternative methods of sampling the behavior of cetaceans under a variety of conditions are available. Evolutionary theory, long-term study, and improved quantitative measures are key to unraveling the dynamics of social life among cetaceans.

3

STUDYING CETACEAN SOCIAL STRUCTURE IN SPACE AND TIME

Innovative Techniques

HAL WHITEHEAD, JENNY CHRISTAL, AND PETER L. TYACK

Distant splashes on a calm sea first indicate the presence of the dolphins. From a crow's nest 10 m above the surface, the observer trains her binoculars on the splashes. Using her years of experience of the subtle cues that are all the ocean usually provides, she deduces that the splashes are indeed made by cetaceans, and not sailfish, manta rays, or tuna, which also leap from these waters. She estimates their range at 7 km and directs the vessel toward the dolphins. At 3 km, she can see surfacing animals in five clusters spread out ahead. Each cluster is a few hundred meters across, and, although only about ten animals are seen at the surface at any time in a cluster, the observer has the impression that each may contain a few hundred animals. She looks behind and sees two more clusters a few kilometers back, near the position of the boat when she first saw the distant splashes. Her documentation of the spatial patterning of the animals is complicated by an apparent split in one cluster, a spreading out of another, and the disappearance of a third. Another crew member, using more powerful binoculars from the steadier platform of the deck, tentatively identifies them as short-beak common dolphins (Delphinus delphis).

Five animals leap as they rush toward the research vessel. Two minutes later, four dolphins are weaving and turning beneath the bow. The crow's-nest observer has a beautiful view of the dolphins through the calm surface: the snouts, the eyes, the sweeping color patterns, the scrapes and scars. It is easy to identify the species— bottlenose dolphins (Tursiops spp.), not short-beaks—and she would soon be able to recognize individuals. But twenty seconds later, they are gone. The crew member on deck still believes the majority of the animals present are short-beaks, but there are clearly some exceptions. However, twenty-five minutes after the first sighting, the animals have sped up, leaping synchronously as they outpace the vessel toward the horizon.

THIS DESCRIPTION recounts a fairly typical sighting of oceanic dolphins in good weather, calm seas, and with experienced observers. But even in these almost ideal conditions, the observers were unable to reliably record the number of animals present, their behavior, or even their species.

This sighting illustrates that in most circumstances we cannot use visual observations to deduce the "content, quality and patterning of relationships," as Hinde (1976) defines "social structure" (a term we use synonymously with "social organization"), among cetaceans. There are many other mammals whose social behavior is difficult to monitor, including bats and arboreal primates. However, studying the social behavior of cetaceans poses particular challenges. These animals are often hard to see, many live in quite inaccessible habitat (the deep ocean), they usually cannot be handled and marked, and most do not survive captivity for any length of time. They also live their lives, and hold their social relationships, over spatial scales of up to thousands of kilometers and temporal scales of decades. Compared with most large terrestrial mammals, cetaceans have many more potential conspecific associates, often numbering in the thousands.

These constraints mean that passive observational methods that can be so effective in uncovering social structure (see Mann, chap. 2, this volume) do not always work well for cetacean species. Even where they do, they are not appropriate at all relevant scales. (Here and in the rest of this chapter, we use the unqualified term "scale" to refer to both spatial and temporal scales.) Thus students of cetacean societies have needed to develop innovative techniques. These have come from many directions: from technological and analytical advances, and as by-products of methods developed for the study of physiology, population ecology, genetics, and other areas of biology (see Read 1998). Cetaceans are particularly suited to some of the new techniques. Generally large and neutrally buoyant, they can carry bigger, more sophisticated tags with less effect than many terrestrial or aerial species. The importance and prevalence of cetacean vocalizations (see Tyack, chap. 11, this volume), coupled with excellent sound transmission underwater, makes acoustic techniques especially useful.

In this chapter we first discuss the information we need to build models of cetacean social structure, then review those innovative techniques that have the most potential to contribute toward our understanding of the societies of whales and dolphins. We think that some of these new techniques developed by cetologists offer creative solutions for studying the social structures of other animals as well as cetaceans.

Scales of Social Experience for Cetaceans

In Hinde's (1976) framework for the examination of social structure, the fundamental elements are behavioral interactions between pairs (or possibly larger numbers) of animals. The content, quality, and patterning of these interactions, then, defines the relationship between the animals. Finally, the nature, quality, and patterning of pairwise relationships form much of the social structure of the population (the "surface structure" in Hinde's terminology).

Behavioral interactions between animals often take place over time scales of milliseconds to seconds. For instance, two sperm whales *(Physeter macrocephalus)* may precisely interleave their communicative click patterns, or codas, giving rise to "echocodas" that consist of two overlapping codas, identical in temporal pattern plus or minus a few milli-

seconds, offset by about 50–100 milliseconds (Weilgart 1990). Interactions such as coda exchanges can mediate associations (Watkins and Schevill 1977c) and group formation (Weilgart and Whitehead 1993). Temporal scales of group stability range from minutes in the case of some baleen whales and dolphins (see, for example, Connor et al., chap. 4; Clapham, chap. 7, this volume) to lifetimes as is common among large odontocetes (Baird, chap. 5; Whitehead and Weilgart, chap. 6, this volume).

Spatial scales of cetacean social experience also range widely, with important interactions occurring when animals touch (see, for example, Connor et al., chap. 4, this volume), as well as over at least 10 km when humpback whales *(Megaptera novaeangliae)* respond to the sounds of conspecifics (Tyack 1983), and possibly over thousands of kilometers, as suggested by Payne and Webb (1971) for the low-frequency sounds of finback whales *(Balaenoptera physalus)*. In the case of migratory animals, scales of thousands of kilometers and the annual cycle are important facets of social structure. For instance, humpback whales and male sperm whales have different associates and types of relationships at high and low latitudes (Whitehead and Weilgart, chap. 6; Clapham, chap. 7, this volume).

Each communication channel used by cetaceans has a characteristic range of scales (fig. 3.1). In addition, animals may form social bonds based upon remembered experiences going back a lifetime. Animals that live in groups that cohere for more than one generation may also develop behavioral traits influenced by the culture of the group that may form over generations (J. R. Heimlich-Boran and S. L. Heimlich-Boran, in press; Whitehead and Mann, chap. 9, this volume). These patterns may also drive social

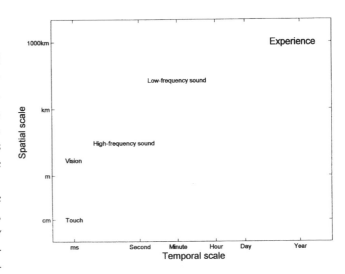

Figure 3.1. Scales of communication and experience of cetaceans.

behavior over time scales much longer than those traditionally studied by observational methods. We need a suite of methods to examine interactions between animals over small time periods, to integrate them into relationships over much larger scales, and finally to build a model of social structure (Whitehead 1997).

Interactions, Associations, and Groups

To follow Hinde's framework literally, we must observe and measure interactions. This can be done quite well with captive cetaceans (Samuels and Tyack, chap. 1, this volume), but it poses a severe problem for those studying whales and dolphins (as well as for students of other hard-to-observe animals) in the wild. The best known circumstances for observing the social behavior of cetaceans in the wild probably exist in Shark Bay, Australia (Connor et al., chap. 4, this volume). Here a small population of individually identifiable bottlenose dolphins can be observed almost continuously when they are at the surface, and sometimes when they are beneath it (e.g., figs. 9.16 and 9.18). However, even in Shark Bay, not all interactions between a focal animal and conspecifics can be documented. It can be difficult to keep up with fast-moving males, and observations are limited by daylight and good weather (R. C. Connor, personal communication). With the many other cetaceans that live in murky waters (such as river dolphins), large groups (such as pelagic dolphins), or far from shore (such as most beaked whales), only rarely is it possible to observe with the naked eye, or through binoculars, anything that might reasonably be termed an "interaction" between two individually identifiable animals.

If we know the circumstances in which interactions between animals usually take place (spatial ranges, behavior types, etc.), then we can use records of the presence of pairs of animals in such circumstances—often termed "associations"—as substitutes for records of actual interactions in analyses of relationships and social structure (Whitehead 1997). For instance, if detailed observation on a subset of animals shows that when two or three males travel in a coordinated fashion less than 10 m apart, they are always part of a long-term cooperative alliance, then observations of such behavior among less well studied individuals may be used to infer alliances. With detailed information relating rates and types of interaction to types of association—for instance, "at what spatial ranges do animals interact?"—this is a legitimate and useful procedure that shortcuts the need for extensive records of interactions (Whitehead and Dufault 1999).

Unfortunately, information on the circumstances under which interactions take place is also unavailable for many animals. So, consciously or unconsciously, many ethologists studying social organization make the "gambit of the group" (Whitehead and Dufault 1999). The "gambit of the group" assumes that when animals are clustered (spatially or temporally), they are interacting with one another (see Connor, chap. 8, this volume). Membership in the same cluster, sometimes called the "group," is then used to define association (fig. 3.2). This allows measures of association to be calculated and social structure to be analyzed. The "gambit of the group" has been a primary method for studying social structure in cetaceans, in part because so little information on social interactions is available for most cetacean species.

The "gambit of the group" may or may not be justifiable. Whitehead and Dufault (1999) suggest that it is probably reasonable if most interactions take place within groups. This assumption can often be defended if groups are separated by distances greater than the maximum range of communication by the animals. The definitions of grouping used by cetologists, however, often do not meet this criterion. Common definitions for grouping involve separations from tens to hundreds of meters, yet most cetaceans can communicate acoustically over ranges of kilometers. Another concern is that animals might be scored as belonging to the same group because they happen to be in the same place, even if they are not interacting ("non-mutualistic" groups: Connor, chap. 8, this volume). This could happen if the clustering of individuals is not social, but related to external factors, such as the presence of prey. With a few exceptions (some feeding baleen whales, phocoenids, platanistids perhaps), clustered cetaceans do seem to interact strongly with other members of their group, often coordinating behavior and movements with each other rather than some external stimulus. In these cases the gambit of the group would seem to be at least roughly justified. However, the more it can be bolstered, the more legitimate the analysis of social organization.

One way to ameliorate these problems is to define two or more types of groups and thus have more than one measure of association (Connor, chap. 8, this volume). Alternatively, a measure of association derived from common group membership can be used in conjunction with another measure of association defined using other means (e.g., nearest-neighbor). Whitehead (1997) suggests methods for analyzing such multivariate association data.

Given this theoretical framework for our attempts to describe cetacean social structure, we need technical input

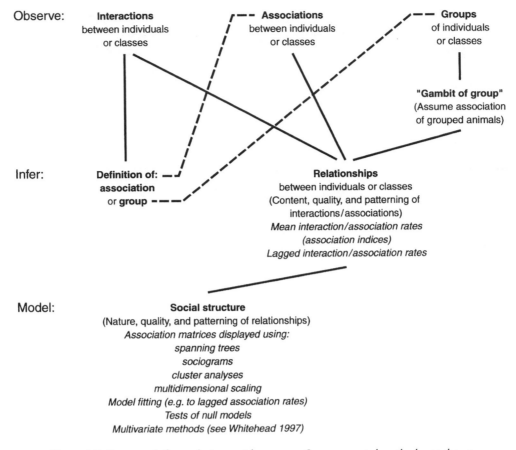

Figure 3.2. Framework for analyzing social structure. Some suggested methods are shown in italics. "Classes" are classes of individuals categorized by age, sex, and so forth. (Adapted from Hinde 1976; Whitehead 1997.)

in a number of areas to find and/or justify a procedure for deriving a model of social structure (Whitehead and Dufault 1999):

- General data on types of interactions between individuals
- How do interaction rates and types vary with circumstance, so that associations may be defined?
- How are animals clustered in space and time, so that groups may be defined?
- If groups are defined, then what proportion of interactions occur within groups?
- Do interactions occur randomly within groups?

Once a framework (which, for instance, might include a definition of group and a justification that most interactions take place within groups) has been developed, the next step is to gather the data that will be used to derive a model of social structure. This task normally includes one or all of the following:

- Identify individuals, and/or classify them by age, sex, or in other ways

- Observe and note interactions, associations, and/or groupings among identified individuals and/or classes of individuals (e.g., defined by sex or age)

After they have been collected, the data on interactions, associations, and groupings need to be summarized into relationships between individuals or classes of individuals. Measures of relationship can then be related to other factors (e.g., age difference, genetic relatedness) and a model of social structure developed (Whitehead 1997; see fig. 3.2).

Some important areas of the study of social structure are largely outside Hinde's (1976) framework. These include information on reproductive success, which may be used to assess the functions of different types of behavior, and details of events that are rare, but are closely tied to fitness. Mating is the most obvious of these, but parturition, intrasexual combat, weaning, and predatory incidents may also involve rare but significant interactions.

Thus, to make a reasonably rigorous analysis of the social organization of cetaceans, we need four general types of methods: methods that give information on interactions (and usually concern temporal scales of minutes or less); methods for collecting large amounts of data on the associations and groupings of individual animals (usually over temporal scales ranging from minutes to one day); analytical techniques to process this information (over temporal scales ranging from minutes to decades) so that we can measure relationships, relate them to other factors (such as age, sex, relatedness, and reproductive success), and thus build a model of social structure (Whitehead 1997; see fig. 3.2); and methods to infer the details of rare but important events (such as mating). In all these areas, cetologists are developing new, and often innovative, techniques.

Photographic Identification of Individuals

The identification of individual whales and dolphins using photographs of natural markings has been key to all long-term and reasonably detailed studies of cetacean social structure (figs. 3.3 and 3.4). The technique was principally developed in the 1970s and came to fruition in the 1980s (see Samuels and Tyack, chap. 1, this volume, for a history of this development) with the publication by the International Whaling Commission of the large and important volume *Individual Recognition of Cetaceans: Use of Photo-identification and Other Techniques to Estimate Population Parameters* (Hammond et al. 1990). Although the International Whaling Commission was principally interested in using the technique for studying population biology, the volume contains important papers on social structure, most significantly the work of Bigg et al. (1990b) on the social structure of killer whales *(Orcinus orca)*. In the 1990s, established photo-identification data sets have grown in longevity and size, studies of new species have been initiated, and techniques for managing and analyzing the data have improved. Summaries of techniques for photo-identifying cetaceans are given by Hammond et al. (1990) and in box 3.1.

To our knowledge, photo-identification has proved successful on every species for which it has been seriously attempted (see table 2.1, this volume), with the exception of the Ganges river dolphin, or susu *(Platanistida gangetica)*, which has no dorsal fin, surfaces in an undramatic manner, and is uniformly gray (A. Smith, Whale and Dolphin Conservation Society, personal communication). However, the ease of use and efficiency of photo-identification varies considerably between species. In some species, such as killer whales and humpback whales (Baird, chap. 5; Clapham, chap. 7, this volume), virtually all animals within a population can be identified reliably. In others, such as short-finned pilot whales *(Globicephala macrorhynchus),* current techniques allow only a proportion of the population (those most clearly and permanently marked) to be reliably identified (Shane and McSweeney 1990). If the population is over about 10,000, then managing a photo-identification database becomes extremely unwieldy.

Photo-identification of cetaceans using natural markings can provide information on movement patterns and on population size and dynamics. Long-term photo-identification studies can help to improve our knowledge of basic life history parameters such as age at sexual maturity, calving intervals, and reproductive and total life span (Hammond et al. 1990). If the identifying photographs are collected together with sufficient data for associations and/or groups to be defined (which is often the case), then they also have the potential to provide a model of social structure. Some photo-identification databases that have to date been used only to examine questions of population biology contain considerable potential for uncovering information on social structure.

Analytical methods for converting long-term photographic identification databases into models of social structure have not been standardized, although Whitehead (1997) has produced an analytical framework (summarized in fig. 3.2) to guide such analyses. The general procedure is to define and calculate association indices between all pairs of identified animals (see Cairns and Schwager 1987; Ginsberg and Young 1992), which together make up an association matrix. The association matrix can then be displayed using methods such as cluster analysis or a sociogram (e.g., Wells et al. 1987; Bigg et al. 1990b). An important and often overlooked element in such analyses is the temporal patterning of relationships, which can often be described effectively using lagged interaction/association rates (Whitehead 1995a). This technique has been applied to a few cetacean populations (Whitehead et al. 1991; Slooten et al. 1993; see fig. 6.10, this volume), and a few other species (e.g., Underwood 1981). Another analytical technique that can be very revealing in some circumstances, and should be used more often, is to test for preferred companionships using permutations of association or interaction measures (e.g., Slooten et al. 1993; Bejder et al. 1998; Whitehead 1999).

Photographs can provide other important information in addition to identity. When two or more animals are photographed in the same frame, this evidence can be used directly as a measure of association (e.g., S. L. Heimlich-Boran 1986). Animals can also be measured using photographs taken from aircraft (Cubbage and Rugh 1982) or boats (Gordon 1990; Dawson et al. 1995).

Figure 3.3. Photographing northern bottlenose whales for individual identity from shipboard in the Gully, a submarine canyon off Nova Scotia, August 1990. (Photograph courtesy of H. Whitehead laboratory.)

Tagging

Since the first "Discovery" tags were shot into large baleen whales in the 1920s, tagging has provided important information about some aspects of cetacean biology. Discovery tags are simply metal cylinders marked with a serial number that are recovered when a whale is killed. A recovered Dis-covery tag gives information on the movement of a whale from the date and place where it was tagged to the date and place where it was taken by whalers. The rate of return of Discovery tags can also be used to estimate population size (Brown 1978). Most modern whale tags are less intrusive and much more informative than those simple metal cylinders, although generally they have shorter lifetimes.

Figure 3.4. Individual identification photographs for northern bottlenose (upper pair), finback (second pair), humpback (third pair), and sperm whales (bottom pair). (Photographs courtesy of H. Whitehead laboratory.)

Box 3.1

Methods of Photo-identifying Cetaceans

Field Methods

A variety of research platforms have been used for photo-identification research. Shore-based studies are inexpensive and avoid disturbance to the subject animals, but are suitable only for cetacean populations that are regularly found within 500 m of the shore (Würsig and Jefferson 1990). Aerial photography is suitable for larger whales, whose identifying marks are visible on the head or body (Payne 1986; Best and Underhill 1990; Rugh 1990), but less so for smaller species or for those whose identification is based primarily on dorsal fin or fluke markings. Air time can be prohibitively expensive, and low-flying aircraft may affect the behavior of marine mammals (e.g., Richardson et al. 1985b). The use of airships generally causes less disturbance (Hain 1991), but they are much more expensive to operate. In some cases, airplanes are used as spotters, to find animals and direct boats to their locations.

Boats are the most widely used platforms for photo-identification (table 2.1), enabling researchers to find and travel with cetaceans and to maneuver for the best photographic angle (see fig. 3.3). Boats and engine noise are potentially disruptive, and care must be taken in approaching animals to be photographed. The specific tolerances of different species, and the angles required to photograph identifying marks, will dictate the most effective way to approach the animals.

Different types of markings are used for identifying different species (see fig. 3.4; table 2.1). Callosity patterns on the heads of right whales are individually distinctive (R. S. Payne et al. 1983), as are markings on the ventral surface and trailing edge of the flukes of humpback and sperm whales respectively (Katona et al. 1979; Arnbom 1987). But for most species, individual identification is made from photographs of the shape, markings, and coloration of the dorsal fin and surrounding area (Würsig and Jefferson 1990; see table 2.1).

The sampling regime used in collecting photo-identifications varies depending on the behavior of the subjects and the study priorities. For many rarely encountered species, or species that are not the specific target of a particular study, photo-identification may be opportunistic, with individuals photographed as they are encountered. Where the primary goal of the research is population censusing, efforts are usually made to photograph as many individuals as possible within a study area. However, in studies in which behavior or group membership are of particular interest, it may be considered preferable to stay with specific individuals and perhaps not collect photographs of other individuals that are present in the vicinity. Some researchers photograph only those individuals that appear to have identifiable markings (e.g., Würsig and Jefferson 1990), whereas others photograph all animals, regardless of apparent identifiability (Whitehead et al. 1997c).

Thirty-five millimeter single-lens reflex cameras are the primary tools of photo-identification researchers. Motor drives are widely used, particularly by those studying animals that surface frequently, repeatedly presenting their identifying marks (e.g., dorsal fins) to be photographed. Databacks, which imprint date and time information onto the film, are useful, although as a rule, information such as date, time, location, behavior, and group size is recorded for each frame taken, either on datasheets or on audiotape, for later transcription. Optimal lens choice depends on the characteristics of the species being studied and the platform. Researchers studying dolphins often use variable focal length (zoom) lenses (e.g., 80–200 mm), which enable them to photograph animals that are close to or farther away from the boat (e.g., Würsig and Jefferson 1990). Fixed-length telephoto lenses (e.g., 300 mm) are often used for photographing larger animals, which may be more distant, and for taking fluke photographs (e.g., Arnbom 1987). Lenses are selected to be as fast (low "f-stop") as possible, given constraints of weight and cost. Some researchers recommend the mounting of cameras and long-lenses onto gun-stocks or shoulder braces to improve stabilization during boat work (Würsig and Jefferson 1990).

Most researchers use black-and-white negative film, although color slide film is also used. Fast films are preferred by people studying larger whales, with ISO 400 film commonly chosen, sometimes pushed to ISO 1,600 to permit fast shutter speeds and maximal depth of field

Hal Whitehead, Jenny Christal, and Peter L. Tyack

(Hammond et al. 1990). The small, sometimes subtle, identifying markings of dolphins and porpoises require-film with small grain size and maximal resolution. Faster films may be too grainy, and ISO 100 film has proved to be adequate for small cetacean photo-identification under most light conditions (Würsig and Jefferson 1990).

Laboratory Methods

Initial analysis of identification photographs may be by observing negatives on a light table using optical magnifying loupes or dissection microscopes (Dufault and Whitehead 1993), by examining contact sheets (Carlson et al. 1990), or by projection of slides (Würsig and Jefferson 1990). The best photograph of each identifiable individual from each encounter is then printed and compared with a catalog of existing individual photo-identification records. This comparison and matching procedure may be performed manually, but where large numbers of records must be compared, a computer-assisted retrieval and matching system (Mizroch et al. 1990; Whitehead 1990b) can greatly improve efficiency. With all current computer-matching systems, a human operator is still required to make decisions about potential matches suggested by computer programs. Experience and time available are important factors in the accuracy of matching procedures. More experienced individuals make fewer mismatches, and the number of errors made is inversely proportional to the amount of time spent matching (Carlson et al. 1990). Cataloging of individual records requires some form of categorization of mark types; for example, the amount and patterning of white on the ventral surface of humpback whale flukes (Katona et al. 1979).

Variations in photograph quality are unavoidable as researchers attempt to focus on moving animals from a moving platform under differing light conditions, and these variations can cause problems. Arnbom (1987) found that the certainty of identifying a sperm whale from a fluke photograph depended on various physical features of the image: the resolution available (a factor that relates to the sharpness of focus and the size of the fluke within the frame), the vertical orientation of the fluke with respect to the camera, and the tilt of the fluke away from the vertical. All of these factors are independent of the number or quality of an individual's identifying marks. The inclusion of poor-quality photographs of well-marked individuals and exclusion of good-quality photographs of individuals with subtle identifying characteristics introduces biases into population analyses using photo-identification data (Hammond 1986), and also may be detrimental to studies of social relationships. In an attempt to circumvent these problems, researchers may develop quality categories for their identification photographs and restrict specific analyses to photographs of particular qualities (Hammond 1986).

Variation in identifiability can limit the information that is available for some individuals, and therefore has consequences for comprehensive studies of social structure. Some animals may behave in such a way as to reduce the chances of a good-quality identification photograph being obtained. For example, particular individuals may be more wary of boats than others, and may therefore stay at distances that prohibit identification. Particular age or sex classes may be less amenable to photo-identification. Humpback whales of different age groups exhibit different fluking behavior, which affects the probability of obtaining a usable photograph (Rice et al. 1987). Identifiability is dependent on characteristic individual markings, and the predominance of these markings varies between species and populations. Over 90% of sperm whales possess identifying characteristics (Arnbom 1987), but other species contain lower proportions of identifiable individuals (e.g., short-finned pilot whales: 45% (Hawaii), 35% (California), Shane and McSweeney 1990).

The permanence of identifying marks affects researchers' ability to recognize resighted individuals and follow their patterns of association over the long term. Although some marks are present at birth and are retained throughout life, marks may be added or changed during an animal's lifetime, and may disguise or obliterate others that were previously important identifying characteristics. The severity and depth of wounds determines the length of time for which scars persist and may be used for identification in bottlenose dolphins (Lockyer and Morris 1990).

They can collect a range of data on dives, movements, physiology, oceanography, and behavior and relay it back to scientists in a variety of ways (box 3.2). Because of their size and neutral buoyancy, cetaceans can support larger, and more informative, tags than many terrestrial or aerial animals.

Some of the greatest advances in tag technology have been initiated on pinnipeds, with the principal objective of studying underwater behavior and diving physiology (e.g., Le Boeuf et al. 1988; Kooyman 1989; Fletcher et al. 1996; Read 1998). Cetologists have been able to capitalize on these developments, as their study animals share the same habitat as pinnipeds and they are interested in similar questions. However, attachment is usually more difficult for whales and dolphins (fig. 3.5) than for seals. Seals have a pelage suitable for gluing attachments, and they can be temporarily captured and restrained while they are hauled out on land or ice.

Tagging studies have much to contribute to the study of cetacean social structure. They permit the collection of continuous data from identified individual animals throughout a dive cycle. The kinds of data that tags can sample in this fashion include vocalizations, heart rate, breathing, swim stroke, orientation, depth, and swim speed, to name a few. Tags provide a powerful complement to observational methods, which often are limited to biased sampling of the most visible behavior at the surface. Detailed information on interactions among animals underwater may be obtained from tags that record visual and acoustic behavior (Fletcher et al. 1996; Marshall 1998). This information will help provide a secure basis for studies of social structure. For instance, video and acoustic tags can record vocal interchanges and can indicate how animals react to signals from nearby conspecifics (both behaviorally and physiologically), and perhaps how their reactions change with the range of the signaler or with other characteristics of the received signal, such as loudness. This information would greatly help attempts to define associations and groups in a way that is meaningful to the animals.

Over longer time scales, when two or more members of a population are tagged simultaneously and the tags record position, temporal patterns of association may be deduced by comparing tracks (e.g., Butler and Jennings 1980). A tag might potentially record information from animals surrounding the carrier of the tag, which could allow these associates to be identified, giving a detailed temporal record of associations. An active acoustic telemetry system could be designed to detect which tagged animals are in the vicinity of another tagged animal and to measure the distance between the two tags. The system would preferably operate at a frequency and source level that the tagged animal could not hear. The tag itself could be used to verify lack of response to the pulses. At a regular interval, say, 5 minutes, each tag could emit a query pulse. If another tag was within range, it could respond with a tag-specific coded sequence after a precise delay. The initial tag could then respond with its own code. By comparing the round-trip travel times, each tag could estimate the distance to each other coded tag.

Passive tags might also collect less systematic information on the neighbors of a tagged animal. For example, if attributes of vocal repertoire were individually distinctive, then an acoustic tag could provide information on the identity of associates. Similarly, a video-recording tag could show distinctive morphological features of nearby animals, allowing their individual identification.

So far tags have not been much used to study social organization. Exceptions include Ohsumi's (1971) discovery of long-term bonds between female sperm whales using Discovery tag data and Butler and Jennings's (1980) inferences about the cohesion of dolphin schools from the tracks of animals simultaneously carrying VHF tags. An important limitation on such work has been that researchers (except those using Discovery tags) have generally tagged only a few animals, rarely simultaneously. However, this problem may be overcome by reusing a number of short-term suction cup tags (Baird 1998), as Giard and Michaud (1997) have done in their study of finback whales in the St. Lawrence estuary. Tagging also has ethical, veterinary, and logistical difficulties (see box 3.2). However, as tags become better collectors, archivers, and transmitters of data, the potential for using them to learn about interactions and relationships among animals will increase dramatically (Read 1998).

Acoustic Techniques

Sound is, in most ways, the most efficient information channel underwater (Tyack, chap. 11, this volume). Therefore it is not surprising that acoustic methods have become vital tools in the study of marine animals.

To study cetaceans using sound we need an acoustic source. Three possibilities are background sounds, artificial sources, and the animals themselves. Of these three, background sounds have not been used, and such studies, although perhaps theoretically feasible (Buckingham et al. 1996), would seem to have limited benefits in the immediate future. Artificial sources are used in three general ways. Acoustic tags, in which the source is placed on the animal, are discussed briefly in box 3.2. To investigate the signifi-

Box 3.2

Types of Tags and the Information They Can Gather

There are a range of ways to attach tags to animals, which range from highly invasive and long-lasting to noninvasive and temporary.

- Shooting the cylindrically shaped body of the tag into the blubber or muscle, as with Discovery tags and the tags used by Watkins and his coworkers (e.g., Watkins et al. 1993)
- Bolting, or otherwise attaching, the tag to the dorsal fin of animals that can be temporarily restrained (e.g., Mate et al. 1995; Westgate et al. 1995)
- "Barnacle"-type tags in which the body of the tag is outside the skin, but barbs hold it in place, often for periods of weeks (e.g., Mate et al. 1983; Goodyear 1993)
- Noninvasive suction cup tags, which usually stay on the animal for periods of hours (Stone et al. 1994; Baird 1998)

Except when an animal is temporarily restrained, the tag must be attached remotely. For penetrating tags, this requires considerable inertia, and these tags are shot into whales using guns (Brown 1978; Watkins and Tyack 1991). "Barnacle"-type and suction cup tags require less force and may be attached using crossbows or poles (e.g., Mate et al. 1983; Goodyear 1993; see fig. 3.5).

There are a range of ways of retrieving the data from tags. The simplest is to archive the data in the tag and retrieve the data with the tag. This is how Discovery tags, as well as some temporary tags, work (e.g., Westgate et al. 1995). Discovery tags are detected using metal detectors on flensing platforms, or later when the whale is being processed (Brown 1978). Suction cup tags and other short-term tags are positively buoyant and have radio transmitters that transmit continuously when above the surface, allowing them to be located and retrieved when the tag falls off (e.g., Westgate et al. 1995). Generally, tags designed to stay on cetaceans for more than a few days are not retrieved. Their data can be transmitted to the researchers using high-frequency acoustic transponders (e.g., Goodyear 1993; Watkins et al. 1993), VHF radios (e.g., Goodyear 1993), or satellite radio links (e.g., Mate 1989).

Among the kinds of information that tags can provide are:

- Identity. (Even Discovery tags managed this!)
- Position. This can come from the noted locations in which the tag was attached and recovered, sightings of the tagged animal, satellite-derived positions, usually when satellites pick up tag signals (e.g., the Argos system: Mate 1989), or geolocation (measuring the times of sunrise and sunset: Hill 1994).
- Depth (usually from a pressure sensor). It is also possible to record whether the tag has emerged from the surface using a conductivity switch (although this does not work very well for some small cetaceans that break the surface only briefly).
- Speed and movement. Speeds can be measured using flow meters or inertial systems.
- Oceanographic environment. Sensors can measure temperature, salinity, pressure, light level, and other measures of the ambient environment.
- Visual environment. Small video (or still) cameras can record the visual environment (Marshall 1998).
- Acoustic environment. Hydrophones can record sounds from the tagged animal, nearby conspecifics, or other natural or anthropogenic sources (e.g., Fletcher et al. 1996). They can also record heart rate, breathing, and swim stroke (Fletcher et al. 1996).
- Physiology, including heart rate (e.g., Kooyman 1989) and internal body temperature (Watkins et al. 1996).

Some forms of behavior (for instance, blows, breaches, vocalizations) may be recorded using hydrophones or other sensors.

Problems with Tagging

Tagging is not without its difficulties. Some feel uneasy about the act of attaching devices to the body of a sentient animal, especially when the attachment is invasive. There has been concern that animals could be harmed during the procedure of attaching a tag. Discovery tags, which can penetrate well into the whale, have injured or killed some animals, especially smaller ones (Brown 1975). In contrast, suction cup tags are unlikely to cause veterinary problems except in the most unusual circumstances; for instance, if during attachment a tag were accidentally shot into the eye of the whale.

Another concern regarding the validity of tagging data is that tagging may affect the behavior of the tagged animal, or its companions. This concern has grounds both ethical (especially in the case of long-term, invasive attachments that might change the energetics of swimming or feeding, mating, or predation rates) and scientific (if the tag affects the behavior of the animal, then the data it collects are not valid representations of the natural behavior of the species). Cetaceans frequently react when tagged by remote devices such as guns and crossbows. However, in most cases, these reactions seem to last little more than a few minutes, after which behavior appears to return to "normal" (e.g., Watkins and Tyack 1991; Goodyear 1993). In contrast, Schneider et al. (1998) found that bottlenose dolphins in New Zealand almost invariably reacted to suction cup tagging with persistent dramatic changes in behavior.

Tagging of cetaceans also has logistical limitations. Tags are expensive, costing about U.S.$500–1,500 for a simple VHF or acoustic tag (e.g., Goodyear 1993) and U.S.$3,000–5,000 for a long-term satellite tag (Mate 1989). Placing the tag on the animal is not always easy, particularly if it must be on a particular area of the body (as with VHF or satellite tags, which must regularly break the surface for substantial periods). Remotely deployed tags, and especially suction cup tags (Baird 1998), frequently fail to attach, and invasive satellite tags are rarely recovered. Power limitations and attachment failure limit the usefulness of most satellite and VHF tags to periods of a few weeks or months (Watkins et al. 1996), and, although data storage capability is increasing dramatically, video and audio storage is currently limited to hours.

All techniques of retrieving data have problems. Discovery tags may be shed, archival tags may not be recovered, satellite tags may fail to transmit, acoustic transponder tags have only short ranges (a few kilometers at best, and transmission may sometimes be erratic: Watkins et al. 1993), and VHF tags are often difficult and expensive to track for more than a few hours (Mate 1989; Watkins and Tyack 1991).

Despite these limitations, tags have provided substantial insights into some aspects of the lives of whales and dolphins (e.g., Watkins et al. 1993; Westgate et al. 1995; Winn et al. 1995). They have the potential to revolutionize our view of these animals, particularly those aspects of their lives that occur below the water surface.

---

cance of certain sounds to animals, researchers can play recordings or artificially constructed sounds to animals whose behavior and movements are being recorded. Playbacks to cetaceans have used sounds of predators (e.g., Cummings and Thompson 1971; Fish and Vania 1971), anthropogenic noise (see Richardson et al. 1995b), and sounds of conspecifics. Conspecific playbacks can provide significant information on how vocalizations mediate interactions (e.g., Tyack 1983) and as such have an important and underutilized role in the study of cetacean societies. Sonars, in which a sound is produced (usually from a ship, and preferably in a frequency range that is not detectable by the animal) and its echoes recorded and processed, have been widely used in studies of cetacean diving behavior (e.g., Lockyer 1977; Dolphin 1987a; Papastavrou et al. 1989; Ridoux et al. 1997). However, such studies have provided little (e.g., Papastavrou et al. 1989) or no information on social behavior.

In contrast, studies of the sounds produced by the animals themselves are keystones in research on the social organization of cetaceans. In these studies, the sounds of whales and dolphins are recorded on single hydrophones or arrays of hydrophones and then processed, either in real time or later, to give a variety of information about the signaler, its location, and its behavior (box 3.3). Acoustic data can be especially valuable when they are linked to visual observations, preferably using simultaneous visual and acoustic recordings (e.g., Tyack 1981; Frankel et al. 1995). Another useful technique is to link regular scan samples and acoustic recordings (Whitehead and Weilgart 1991). Sound travels so well underwater that vocal behavior is the one behavior that can be routinely sampled continuously throughout the dive cycle. An ideal, not yet achieved, would be recording a number of interacting animals on a hydrophone array in such a way that the source position and identity of each vocalizer can be determined. Some studies have come close to this, however, and have provided us with very useful information on how animals interact

Figure 3.5. Attempting to attach a suction cup tag to a killer whale, San Juan Islands, Washington State. (Photograph courtesy of S. Hooker.)

over small scales (e.g., Watkins and Schevill 1977b). Such research can inform us about interactions, associations, and groupings over a wide variety of spatial scales.

Of all the sensory modalities available to cetaceans for communication, sound is the only one capable of operating at ranges of greater than a few tens of meters. This means that when cetaceans are separated by this range or more, acoustic techniques ought to be able to provide biologists with all of the information available to a cetacean about conspecifics and other aspects of its environment. Making sense out of this information remains an enormous challenge.

Studies of vocalizations, and other cetacean sounds, can provide many types of information:

- the presence of animals
- numbers of animals present, from rates of hearing sounds (e.g., Whitehead and Weilgart 1990) or formal acoustic censuses (Ko et al. 1986; Leaper et al. 1992)
- the bearing of an animal or a group, information that is useful for acoustic tracking (Whitehead and Gordon 1986) or censuses (Leaper et al. 1992)
- location, and thus spatial arrangement, of animals, from arrays of three or more hydrophones over scales of meters to thousands of kilometers (e.g., Watkins and Schevill 1977b; Ko et al. 1986; Clark 1995; Frankel et al. 1995)

- individual identity, when animals have distinctive vocalizations (e.g., Caldwell et al. 1990)
- group identity (e.g., Ford and Fisher 1982)
- sex or age class when these have distinctive vocalizations, as is the case for mature male sperm and humpback whales (Whitehead and Weilgart, chap. 6; Clapham, chap. 7, this volume)
- size of animals, at least for sperm whales, in which the interpulse interval of the echolocation click is correlated with body size (Gordon 1990; Goold 1996)
- class of interaction, such as threats (Connor and Smolker 1996), or behavioral mode, such as foraging or socializing (e.g., Whitehead and Weilgart 1991)
- vocalization sequences by one or more animals (e.g., Watkins and Schevill 1977c)

The most informative studies give several of these types of information continuously. Studies of cetacean vocalizations have many roles in research on social structure, but three are particularly important: they allow us to examine interactions underwater and often in detail, indicating the circumstances in which animals interact and thus providing bedrock for definitions of association; they can show the spatial patterning of animals over a range of scales, and thus help us to define groups; and they allow us to trace patterns of affiliation and disaffiliation.

Box 3.3

Techniques of Acoustic Recording and Analysis

Deployment of Hydrophones

Terrestrial ethologists can rely upon their own ears to detect the calls of animals, and can use their own abilities of acoustic localization to find a vocalizing animal. Humans cannot hear very well underwater, so marine ethologists usually must use underwater microphones or hydrophones to listen to marine animals. Hydrophones are currently readily available, relatively inexpensive, small, and easy to use. The miniaturization revolution in electronics has also created a variety of acoustic recording systems that are portable enough for any field situation, including attaching the recorder to an animal (Fletcher et al. 1996). Recording of low-frequency sound and calibrated recording are facilitated by digital acoustic recording systems.

Cetacean ethologists have been in the forefront of developing systems to locate vocalizing animals. The simplest solution for animals that make high-frequency sounds is to use a directional hydrophone, which functions like the parabolic reflector microphones often used for terrestrial animals. For example, Whitehead and Gordon (1986) used a reflector hydrophone to determine the direction of sperm whale clicks. This allowed them to follow groups of sperm whales for periods of several days.

Other methods of localization rely upon an array of hydrophones. One solution is to place hydrophones less than half a wavelength apart and measure differences in phase for the same signal arriving at adjacent hydrophones. For example, Clark (1980) used a small rigid array to estimate the bearings of the calls of southern right whales in real time. If the hydrophones are spaced more widely, then one can measure differences in the time of arrival of a sound at each hydrophone in order to estimate the source location (e.g., Watkins and Schevill 1972).

The earliest technique commonly used for recording sounds of cetaceans involved deploying a single hydrophone from a boat. Typically the vessel maneuvers near whales, stops, and shuts down its engines, and underwater sounds are recorded while the vessel is drifting. When a vessel is under way, water flowing past hydrophones generates flow noise, which is particularly intense at low frequencies. High-frequency vocalizations can be recorded while under way by using filters that block this low-frequency noise (e.g., Sayigh et al. 1993a). Even better is to tow an array of hydrophones in a straight line. If the signals from these hydrophones are combined with no delays, the array will be most sensitive to sounds coming perpendicular to the array and will be least sensitive to sounds, such as the engine noise, that are coming from a bearing parallel to the array (Urick 1983). This kind of deployment is well suited to acoustic census techniques (Thomas et al. 1986; Leaper et al. 1992), but is less useful for studying behavior. As long as the spacing of the hydrophones is appropriate, signals from this array can also be input to a beamformer, which can construct the signal that is arriving at any bearing for the array (Johnson and Dudgeon 1993). As beamformers become smaller, cheaper, and more practical for field biologists, they offer great promise for following the vocalizations of individual cetaceans (Miller and Tyack 1998).

The two primary problems of recording from vessels are the noise from the ship's propulsion system and flow noise as the vessel moves through the water. One solution to both problems is to record sounds from a small quiet buoy that drifts with the water to minimize flow noise. This is such an effective solution that there is a commercially available device, called a sonobuoy, that can be deployed from a ship or aircraft, drops a hydrophone to a preset depth, and uses a radio transmitter to broadcast sounds, including the vocalizations of cetaceans, received by the hydrophone. Sonobuoys can also be deployed in a large enough array to allow localization of vocalizing whales by measuring time delays between the arrival of a sound at different hydrophones (e.g., Frankel et al. 1995).

In polar regions where there is a solid ice cover over the sea, it is often possible to deploy hydrophones from the ice. Ice-covered seas have little wind or wave noise; if the ice itself is not moving and making noise, there may be relatively low noise levels near the surface in this environment. As with sonobuoys, it is often cheaper and easier to broadcast the signals via radio rather than running cables to all of the hydrophones (Clark and Ellison 1988).

HAL WHITEHEAD, JENNY CHRISTAL, AND PETER L. TYACK

The variation of sound speed with depth in many polar regions causes sound energy to refract upward; this often creates an advantage for hydrophones deployed at shallow depths. However, sound energy tends to concentrate at depths of a kilometer or more in temperate and tropical seas. Noise from wind and waves also often originates at the sea surface. These considerations mean that it is often advantageous to mount hydrophones on the seafloor in such environments.

The U.S. Navy has devoted considerable resources to using bottom-mounted hydrophones to locate and track ships. Few other institutions have the financial resources required to pay for cabling arrays of hydrophones tens of kilometers back to the land, and for integrating data from many of these arrays. Tracks of continuously vocalizing finback whales were obtained in the early 1960s using data from bottom-mounted hydrophones made available by the U.S. Navy (Patterson and Hamilton 1964). These arrays have proved capable of detecting whales at ranges of hundreds of kilometers, as was predicted by the acoustic models described by Payne and Webb (1971) and Spiesberger and Fristrup (1990), and have been used to track a blue whale for 1,700 km over forty-three days (Costa 1993). If there is no need to locate animals in real time, then a cost-effective alternative to cabling hydrophones to shore is to deploy independent recorders on the seafloor and recover the data later. Hydrophones placed on the seafloor to record seismic activity have also been used in this way to track blue and finback whales (McDonald et al. 1995). These units can be reused and redeployed wherever is optimal for a particular study.

Analysis of Acoustic Data

Detection of Vocalizations

A variety of promising signal processing techniques have been developed for detecting cetacean vocalizations. Some of these detectors automate the process humans use when examining spectrograms. For example, Mellinger and Clark (1993) developed an automatic detector for the endnote from songs of bowhead whales, which compares the spectrogram of a transient signal to a stored spectrographic model. Neural net processing can improve the performance of this kind of detector (Potter et al. 1994). Even a simple detector that tests for a threshold signal-to-noise ratio in a specific frequency band for a specific duration can reduce bias and

facilitate replication of the initial stages of analyzing animal sounds.

Classification of Vocalizations

When ethologists set out to categorize signals from animals, their goal is to establish categories that are perceptually relevant to the species under study. The most common analysis used in such categorization is the sound spectrogram (e.g., fig. 11.4), which plots the energy of sounds as a function of frequency and time. Spectrograms have the advantage that they can present acoustic information in visual figures that may roughly match the way sounds are processed by the peripheral auditory system in mammals. Spectrographic analysis can help aid the ear in such recognition, but often the biologist makes spectrograms of only a few examples of each sound type, and categorizes sounds by ear. The vocal repertoire is usually then presented as counts of the different signal types.

There are several problems with humans subjectively detecting and categorizing the signals of another species. Just as we may be more likely to attend to more dramatic visual displays (Altmann 1974), we may have perceptual biases in what auditory signals we detect and record, which may bias our counts of signals. If we view our task as assigning discrete categories, we may miss graded or continuous elements of a communication system. Even if the graded signals are perceived categorically (Harnad 1987) by other cetaceans, there is no guarantee that humans will draw the perceptual boundaries in the same place as our study animals. For example, Japanese macaques easily categorize their own calls, but a closely related macaque species does not categorize these calls in the same way as the species that makes them (Zoloth et al. 1979). If such close relatives differ in how they categorize species-specific vocalizations, how much less likely are humans to match the appropriate patterns of distantly related cetaceans? Any categorization of vocalizations must ultimately be validated by testing with the species producing the calls—a difficult procedure with most cetaceans.

There are also several alternatives to the subjective categorization of sounds. One is to measure specific acoustic features from a set of sounds, often taking the measurements from spectrograms by hand, and then analyze them using multivariate statistics (e.g., Clark 1982). Computerized digital signal processing facilitates making more robust measurements, measuring more features, and selecting features for their relevance to the animals. Another approach for classifying sounds is to create an explicit model of which acoustic features are most salient to an animal. For example, Buck and Tyack (1993) developed a method to compare similarity in frequency contour of whistles while allowing timing to vary. This kind of similarity index makes explicit assumptions that the pattern of change of absolute frequency of the fundamental is a more salient feature for whistles than small changes in timing. These assumptions can be tested by comparing variation in these features among natural signals, but the ultimate test must use playbacks to test how whales and dolphins perceive the similarity of different signals.

Identifying Individuals

The ability to determine the identity of a vocalizing animal is of great importance, allowing more detailed studies of interactions, permitting the use of acoustic data for studies of association, and spinning off into population analyses. This is a serious challenge for studies of cetaceans.

Sometimes acoustic location methods (described above) can be used together with visual observations to determine which individual produced a sound. However it is rare that all visually observable animals are individually identifiable, and this method does not work when animals swim close together.

There are many situations where it is possible to approach a cetacean for a short time, but where it is difficult to reidentify and follow the animal or to monitor its behavior in detail. In these cases, it may be advantageous to attach a tag that can record or telemeter acoustic data recorded at the animal along with other data (see box 3.2). This kind of tag can be used to identify vocalizations from the tagged animal. Such tags have been used with captive dolphins (e.g., Tyack

1991b) and wild elephant seals (e.g., Fletcher et al. 1996).

Better, but even harder, is to determine individual identity from the sounds themselves. Unfortunately, we cannot yet reliably identify all (or even most) members of a wild population of cetaceans using acoustic means. Bottlenose dolphins are reported to produce individually distinctive signature whistles (Caldwell et al. 1990), but other individuals may produce precise imitations of the signature whistle of a dolphin (Tyack 1986b). Recognition of individual vocalizations among cetaceans may depend more on stable acoustic features similar to the voice cues we use to identify other humans rather than on the category of the vocalization. Voice cues in humans depend upon the shape of gas-filled cavities in the vocal tract. These may change shape in diving animals; as an animal dives, the increased pressure reduces the volume of such a cavity, and this may change the acoustic properties of vocalizations. Further work on the mechanisms by which cetaceans produce vocalizations may help identify stable acoustic features of individuals. For example, Norris and Harvey (1972) have hypothesized that the interpulse interval in sperm whale clicks may result from reflections within the spermaceti organ. Goold (1996) has shown that these interpulse intervals are stable within six-minute sequences of clicks, presumably from one whale. These kinds of measurements show promise not just for estimating the size of the whale (Gordon 1991), but also perhaps for tracking the clicks of different individuals, although more information would be needed to distinguish an individual reliably from within a population of more than a few animals. Testing for individually distinctive features of vocalizations will usually require use of either the acoustic location or tagging methods just discussed.

Genetic Techniques

Social behavior, like all other aspects of the biology of animals, results from evolutionary processes, which cause changes in gene frequencies. Therefore, genetic analyses theoretically have the potential to examine the roots of social evolution: how social structures affect gene flow, and how genotypes influence social structure. While the answers to such fundamental questions may perhaps be better sought in better-understood systems, genetic analysis is an important tool in the investigation of social structure in cetaceans, and in many other species. Genetic techniques have the potential to provide detailed information on identity, sex, kinship, philopatry, dispersal, mating systems, and reproductive success, much of which is difficult or impossible to obtain by other means. Methods of collecting genetic

HAL WHITEHEAD, JENNY CHRISTAL, AND PETER L. TYACK

samples and details of the different types of genetic markers are summarized in box 3.4, together with some of the problems that may limit the effectiveness of these techniques.

Genetic analyses can inform us about sampled individuals in several ways. The most fundamental of these, although technically not the easiest, is individual identification (Amos and Hoelzel 1990). In some cases, biopsies or sloughed skin samples (fig. 3.6) may be collected in circumstances in which they cannot be assigned to specific individuals, but may be linked to a specific social group or cluster of animals. Techniques such as minisatellite DNA fingerprinting or multilocus microsatellite analysis allow identification of individuals and the exclusion of duplicate samples, unknowingly collected from the same animal at different times, from data sets (Amos and Hoelzel 1990; Richard et al. 1996b). These genetic identifications can be used exactly as photo-identifications are, for the analysis of associations, local and migratory movements, and population size (Palsbøll et al. 1997).

One of the most basic pieces of information that is required for studies of social structure is the sexes of the animals involved (Whitehead and Dufault 1999). Yet this can be hard to determine by observation for many species of free-ranging cetaceans. Even in species with marked sexual dimorphism, it may be impossible to distinguish between juveniles of one sex and adults of the other. Simple genetic techniques using the sex-linked ZFY and SRY markers can overcome these problems (Baker et al. 1991; M. W. Brown et al. 1991a; Berubé and Pallsbøll 1996).

Knowledge of the parental origins of individuals can provide important information on philopatry and dispersal (with respect to social groups or geographic locations) and the patterns of relatedness within social groups. For example, low mitochondrial (mt) DNA diversity within groups may be indicative of female philopatry and suggests a matrilineal group structure. In contrast, the presence of more than one mtDNA haplotype within a social group, as has been found for bottlenose dolphins (Duffield and Wells 1991) and sperm whales (Richard et al. 1996a), indicates that group members sometimes come from different natal groups. Differences between the mtDNA haplotype distributions of male and female belugas (Delphinapterus leucas) in different summering areas indicate sex-biased locational dispersal (O'Corry-Crowe et al. 1997).

From the perspective of the evolution of social and mating systems, a particularly important relationship is paternity, yet this cannot be inferred reliably by observation even in apparently monogamous species (Burke 1989). Both minisatellite and microsatellite techniques may be applied to questions of paternity. Paternity analysis is simplified when the mother is sampled, since bands or alleles that are present in the offspring but absent in the mother must derive from the father. This set of bands or alleles is thus a diagnostic paternal subset with which possible fathers can be compared. Any male that possesses this paternal complement must be considered a putative father, and any male that could not have contributed this set of alleles or bands can be excluded. Paternity testing where all possible fathers have been sampled may permit positive identification of the father (e.g., captive killer whales: Hoelzel et al. 1991). Paternity exclusion data may indicate the presence or absence of breeding within social groups, and patterns of paternity among fetal cohorts (Amos et al. 1991a, 1993) may reflect the mating strategies of males. Female mating strategies have been investigated by assessing the paternity of successive calves born to female humpback whales (Clapham and Palsbøll 1997).

A knowledge of the patterns of genetic relationships among social group members gives an insight into the origins, stability, and structures of cetacean groups (Dizon et al. 1991; Duffield and Wells 1991; Amos et al. 1993; Richard et al. 1996a), as well as mating systems and the reproductive success of individuals. Genetic relatedness between individuals can be estimated using either minisatellite or microsatellite techniques. The proportion of shared fingerprint bands in minisatellite analyses gives an indication of the relatedness of individuals. Parent-offspring pairs (and full siblings) will share 50% of bands, whereas more distant relatives will share smaller percentages. Since microsatellite alleles are inherited in a standard Mendelian fashion, any two individuals that share an allele at every locus can be considered a putative parent-offspring pair. Full siblings may also follow this pattern of allele sharing, but due to the single births (Whitehead and Mann, chap. 9, this volume) and generally polygamous mating systems of cetaceans (Connor et al., chap. 10, this volume), full siblings are not expected to be common. With sufficient numbers of highly variable microsatellite markers, more distant relationships between individuals, such as grandparent-grandchild or aunt-niece, can also be discriminated with some accuracy. A comparison of allele frequency distributions between groups may provide evidence for kinship within groups, given the assumptions that kin groups are expected to demonstrate a nonrandom distribution of allele frequencies and that homogeneity will be greater within rather than between groups (Hoelzel and Dover 1989; Amos et al. 1991a). The exact methodology employed in the analysis of group genetic structure will reflect the nature of available data. Where a large number of individuals from different groups within a population have been sampled,

Box 3.4

Genetic Methods

Collecting Samples

Tissue samples for genetic analysis may be collected in a variety of ways. DNA is an incredibly resilient molecule, and recent experiments have proved that restrictable DNA can be obtained from archived whaling samples, such as baleen plates, as much as 30 years old (Kimura et al. 1997) and from dolphin blood stains on fisheries observer reports (Eggert et al. 1998).

Stranded animals provide an easily accessible source of tissue. Although immediate sample collection is preferable, skin samples collected from animals that have been dead for several weeks may be of adequate quality for DNA analysis (Amos and Dover 1990). While samples obtained from lone stranded animals may provide valuable information on population structure or geographic distribution, it is mass strandings, which may represent whole social groups of cetaceans, that are of greatest use in the systematic study of social structure. Not only can samples be collected for genetic analysis, but the size, sex, and relative positions of individuals, as well as vocalizations from living animals, may be recorded.

Collections of tissue samples that represent whole, or large parts of, social groups may also be obtained from species that are exploited, whether the animals are incidental by-catch (e.g., Dizon et al. 1991) or targets of dedicated fisheries (e.g., Amos et al. 1991b). Although some researchers may have ethical concerns over using samples from such sources, some of the most urgent questions for which genetics may provide answers relate to exploited populations.

Although it is frequently difficult to obtain tissue samples from living cetaceans, there are distinct advantages to such nonlethal sampling. The extensive capture-and-release study of bottlenose dolphins in Sarasota, Florida, allows researchers to obtain blood samples for genetic analysis from living and well-documented individuals (Duffield and Wells 1991).

Biopsy sampling allows small samples of skin and underlying tissue to be collected from animals in the wild. Specific equipment designs vary considerably, but most use a punch sampling dart with a bore of about 1 cm, fired from a crossbow or air gun at short range (<20 m)

(International Whaling Commission 1991; Barrett-Lennard et al. 1996b). Since DNA analyses require tiny quantities of DNA, these small plugs of skin may provide sufficient DNA for a number of different techniques. Biopsy sampling has proved more successful on the larger, slower-moving baleen whales than on smaller, faster cetacean species, although as expertise rises, this should be less of a problem. Individuals' reactions to biopsy sampling vary widely (M. W. Brown et al. 1991b; International Whaling Commission 1991; Weinrich et al. 1991), and for species in which photo-identification is dependent on a normal "fluking" dive (humpback and sperm whales), the reaction to biopsy sampling may prevent a skin sample being linked to a photographic identification. Less invasive than biopsy darting is the use of skin swabs to obtain small samples for genetic analysis (Whitehead et al. 1990; Harlin et al. 1989). These techniques seem particularly useful and appropriate for bow-riding dolphins (Harlin et al. 1999).

Some species of cetaceans naturally slough visible pieces of skin, which can readily be collected using a dip net and provide a noninvasive means of DNA sampling (Amos et al. 1992; see fig. 3.6). Samples are routinely stored in a 20% DMSO saturated salt solution (Amos and Hoelzel 1991), which preserves DNA at ambient field temperatures until freezing is possible. Although the degraded nature of the samples may reduce DNA quality, restrictable DNA is readily obtainable from sloughed skin samples (Amos et al. 1992). Because skin may be most easily collected following social or aerial behavior (breaches, lobtails) (e.g., Clapham et al. 1993c) or may be collected from a group of animals, it may not always be possible to link samples to photographic identifications of specific individuals (Richard et al. 1996a).

Feces has recently been recognized as a useful source of DNA. Since fecal matter contains cells from an animal's digestive tract, DNA can be isolated and genetic analyses performed (Kohn and Wayne 1997). Although collection of fecal matter may be more problematic for aquatic species than for terrestrial animals, genetic analysis of dugong DNA isolated from feces has been successful (Tikel et al. 1996).

Types of Markers

Genetic analyses use "genetic markers," segments of DNA that may differ between individuals. Two of the most commonly heard genetic terms are "DNA fingerprinting" and "microsatellites." Both of these terms relate to Variable Number of Tandem Repeat (VNTR) markers, single-copy DNA sequences that are scattered throughout the nuclear genome. VNTR markers consist of multiple copies of a short DNA sequence motif arranged in a tandem array. They are highly variable, and have the potential to positively identify close relatives (rather than simply excluding individuals from being close relatives). Most of the genetic variation revealed by VNTR markers is due to differences between individuals in the number of copies of the repeat motif.

DNA fingerprints are generated using minisatellite markers, VNTRs that have repeated sequence motifs of 10–50 nucleotide base pairs (bp). A size-fractionated and Southern-blotted sample of restriction-digested genomic DNA is screened using a probe that detects several similar minisatellite loci simultaneously. This generates a complex multilocus pattern of bands (a DNA fingerprint), which is unique for each individual (Jeffreys et al. 1985a,b). Multilocus minisatellite fingerprinting is a powerful tool for the examination of first-order relationships (parent-offspring, full siblings), but the analysis of more distant relationships is problematic for a number of reasons (Lynch 1988; Tautz 1990). Techniques that may overcome this limitation of DNA fingerprinting (Wong et al. 1986, 1987) have not been widely applied to nonhuman studies because of difficulties in isolating probes.

Microsatellites are VNTR markers with a di-, tri-, or tetranucleotide repeat motif (2–4 bp). The microsatellite DNA sequence can be amplified using the polymerase chain reaction (PCR: Mullis et al. 1986; Saiki et al. 1988; White et al. 1989) and primers that target the nonrepetitive unique sequence that flanks the repetitive array. PCR products are then size-fractionated on a polyacrylamide gel.

The mitochondrial DNA genome exists as multiple copies in eukaryote cells. This small, circular DNA molecule has a mutation rate estimated to be 5–10 times that of single-copy nuclear DNA (W. M. Brown et al. 1979), and therefore has the potential to be a highly variable DNA marker within populations. In general, inheritance of mtDNA is thought to be strictly matrilineal, without recombination (Lansman et al. 1983; Gyllensten et al. 1985; Avise and Vrijenhoek 1987; but see Kondo et al. 1990; Gyllensten et al. 1991; Zouros et al. 1992; Wallis 1999). Many mtDNA studies focus on the most variable part of the mitochondrial genome, the control region (also known as the displacement loop or d-loop) (e.g., Wada et al. 1991; Baker et al. 1994). The control region is amplified using PCR and the nucleotide sequences examined for base changes, which indicate different mitochondrial haplotypes and thus different matrilineal origins (e.g., Mullis et al. 1986; Saiki et al. 1988; White et al. 1989; Richard et al. 1996a; Lyrholm and Gyllensten 1998).

Molecular sexing techniques use genetic markers located on the sex chromosomes. Recent work has focused on two genes found on the Y chromosome, *ZFY* (the zinc finger protein gene: Page et al. 1987) and *SRY* (the sex-determining region Y gene: Sinclair et al. 1990), which are highly conserved in many different mammals. A clone of the human *ZFY* gene has been used as a probe to identify male-specific restriction fragments on Southern blots of genomic DNA for a number of mammal species, including several cetaceans (Baker et al. 1991; M. W. Brown et al. 1991a; Palsbøll et al. 1992).

Other genetic techniques that have proved useful in studies of cetacean social structure are protein electrophoresis and chromosome heteromorphism analysis (e.g., Duffield and Wells 1991).

Problems

The genetic analysis of relationships between individuals is dependent on the existence of variability in DNA markers. Cetaceans as a group were found by protein electrophoresis to be relatively poor in genetic variation (Hoelzel and Dover 1989), and studies using other genetic markers confirm this finding. DNA fingerprints for pilot whales were found to be considerably less variable than those of humans or birds (Jeffreys et al. 1985a; Burke and Bruford 1987; Amos and Dover 1990), and Hoelzel and Dover (1991a) found the rate of cetacean mitochondrial control region evolution to be an order of magnitude lower than that suggested by human studies. Lower levels of DNA variability result in similarities between individuals that may not be closely related, and widen the confidence limits around any analysis of patterns of relatedness.

Some early genetic studies of cetaceans were limited in their scope because of small sample size. Although this may still be a problem for species that are particularly inaccessible or difficult to approach, developments in biopsy technology and expertise (e.g., Barrett-Lennard et al. 1996b) and the use of sloughed skin are increasing the number of samples available. The expense of genetic techniques is certainly an important consideration, but prices are decreasing considerably as methods become more widely used.

Figure 3.6. Collecting sloughed skin from sperm whales at sea (insert: collected skin). (Photograph courtesy of H. Whitehead laboratory.)

relatedness within a group can be estimated by comparing the similarity of allele frequencies between grouped individuals with that in a random sampling of the gene pool (Queller and Goodnight 1989). Where population allele frequency data is not available, group genetic structure may be investigated by examining the proportion of putative parent-offspring pairs that are present within groups and the extent to which particular alleles remain with, and are propagated within, a group (Amos 1993). Thus genetic data can provide information on the patterning of genetic relationships between groups, within groups, and between individuals.

One of the most important attributes of genetic markers is that they carry information about past events, over time scales ranging from the life span of an animal to the evolution of a species. For instance, Amos et al. (1991b) estimated the year in which a pilot whale group split by analyzing changes in the correlation between allele frequencies within age classes among the two resulting groups. For rigorous use of the comparative method for distinguishing be-

HAL WHITEHEAD, JENNY CHRISTAL, AND PETER L. TYACK

tween hypotheses of social evolution, we need a valid phylogeny (Harvey and Pagel 1991). Genetic techniques have contributed vital information to the question of evolutionary relationships between cetacean species, sometimes suggesting that traditional phylogenetic groupings are inaccurate (e.g., Milinkovitch et al. 1993; Le Duc et al. 1999).

Genetic techniques are most informative about social structure when the results of different molecular methods are combined (Duffield and Wells 1991; Richard et al. 1996a) or linked to other sources of information. For instance, the combination of genetic data with observational or photo-identification studies enables researchers to link the sex and/or relatedness of individuals to their behavior and patterns of association (e.g., Clapham et al. 1992).

The Power of Inference

When scientists cannot directly study the subjects in which they are interested, it is common to make inferences from observations of linked phenomena. Two factors have encouraged this practice in research on cetacean social structures: the difficulties of observing behavior directly, and the existence of large quantities of data on dead animals. Thus information on anatomy, morphology, life history, and parasitology has given us good working hypotheses about the social structures of cetaceans, particularly those of the species that are hardest to study.

In some instances, environmentally induced attributes can be used to deduce elements of social structure. For instance, the scarring patterns on beaked whales of the genus *Mesoplodon* strongly suggest that there is conflict between adult males using their erupted teeth (Mead 1989; MacLeod 1998), and similarities in the scarring patterns on female sperm whales in the same group indicate that there is communal defense against predators (Dufault and Whitehead 1995b; fig. 3.7). Parasites are also useful clues. Best (1979) used information on the species of cyamids found on sperm whales to deduce that relationships between large males and groups of females were transitory.

If we make the assumption that most traits are largely adaptive, then inferences about behavior can also be made from studies of anatomy or life history. For instance, following Harcourt et al.'s (1981) demonstration that the relative testis size of male primates is closely related to their methods of competing for mates, Brownell and Ralls (1986) examined relative testis size in baleen whales. They predicted that females of species with relatively large testes (such as the right whale, *Balaena* spp.) were likely to mate with many partners and so induce sperm competition, whereas females of species whose males have small testes

Figure 3.7. Similar marks and scars on members of a sperm whale group, photographed off Peru in April 1993. (Photographs courtesy of H. Whitehead laboratory.)

(such as the blue whale, *Balaenoptera musculus*) were likely to mate with few males during any mating season. One of the most interesting conjectures in the study of cetacean societies has been Kasuya and Brownell's (1989) postulated social system for Baird's beaked whales *(Berardius bairdii):* based on life history, anatomical, and scarring information, they suggest that males are primary caregivers for the off-spring (see Connor et al., chap. 10, this volume).

These examples demonstrate the potential of inferential techniques for developing hypotheses about cetacean social organization. Although inferior to direct observation of behavior, and unlikely to definitively establish significant new knowledge about social structure, they can be reliable. For instance, Best's (1979) conclusion about the short duration of associations between male sperm whales and groups of females from analysis of ectoparasites was later confirmed by observation (Whitehead 1993). Inferential methods are particularly important for animals that are very hard to observe, such as the beaked whales of the genus *Mesoplodon*, and when events, such as male combat and communal defense against predators, are rare but important. They are particularly useful for targeting promising species and issues for later focused direct research.

Where Next?

In the coming years we expect that there will be dramatic advances in the techniques we use to study cetacean social organization (see Read 1998). Many of these advances will be improvements in our current methods.

The powerful observational techniques described by Mann in chapter 2 of this volume are often limited in their use on cetaceans by our inability to see the animals and their interactions clearly. Small vessels and shore sites often provide poor viewing platforms for behavior, and large vessels are expensive, unmaneuverable, and intrusive. Aircraft are also expensive and often disruptive to the animals. Small, remotely controlled or autonomous surface vessels, submarines, and aircraft, especially blimps (Hain 1991; Read 1998), which are currently under development for many purposes, should greatly improve our ability to view cetacean behavior in many circumstances. If such vehicles do not need to carry humans, they can be made much smaller, and so more maneuverable, less intrusive, and cheaper. Video or audio records of the behavior of the animals can be transmitted by radio or acoustic modem from remotely operated vehicles to an operator who controls the vehicle (again by radio or acoustic modem). Autonomous vehicles usually archive information for later retrieval, but process some of it internally in real time to make autonomous decisions about movements and sampling. Autonomous vehicles can be programmed to follow acoustic or radio tags. These vehicles are likely to greatly extend the potential of techniques such as focal animal sampling for cetaceans.

In just a few years, the majority of photographs used for individual identification of animals will be taken using digital cameras, which will make cataloging and automated matching simpler and cheaper. Current data sets will grow both in longevity and detail. Analysis of these large and multivariate data sets should benefit from improved software and hardware, allowing more flexible analyses. For instance, faster computers and better software should make it much more feasible to use permutation tests to examine the significance of apparent structure in social data, or bootstrap methods to obtain robust confidence intervals around parameter estimates.

The most important acoustic developments are also likely to involve information processing. Two goals stand out: fast (and preferably real-time) source location from hydrophone arrays when several animals are vocalizing together, and reliable (again, preferably real-time) individual identification from sound recordings. If both of these could be achieved, we could gain a much better understanding of cetacean communication as well as interactions between animals.

Tag designs and capabilities are improving rapidly as digital electronics become cheaper, smaller, and more powerful. From the perspective of understanding cetacean social structure, we look toward the evolution of two rather different types of tag. Tags that intensively sample the visual or acoustic environment of an animal as well as its movements, behavior, and physiology (e.g., heart rates and possibly hormone levels) should tell us a great deal about social interactions. Because of data storage limitations, these would be short-term tags. In contrast, long-term tags that stay on animals for months or years and send back (probably via satellite) information on position, movements, and behavior could, if deployed on a number of members of a population, tell us much about scales of relationships.

Genetic techniques are advancing especially fast, as they have such wide applicability. We look forward to developments that will cut the costs of large-scale analyses involving many individuals with multiple markers and allow us to overcome some of the limitations imposed by low levels of genetic variability among some cetacean populations. There will undoubtedly be new and better markers (probably some that are male-specific and so can complement the use of maternally inherited mitochondrial DNA) as well as new analytical techniques for estimating relatedness among individuals.

We anticipate that techniques for integrating different types of data will improve. For an example at small scales, locations of sound sources derived from hydrophone arrays could be overlaid on video images (see Dudzinski et al.

Table 3.1. Techniques and their uses (or potential uses) in studying cetacean social structure

	Types of information			
	Interactions	Associations and groups	Processing interaction/group data	Rare but important events
Scales				
Spatial	cm–1,000 km?	m–km	10–10,000 km	
Temporal	ms–min	min–day	min–decades	
Methods				
Photo-ID		***		*
Processing ID data			***	
Observational	**	**		**
Tagging	**	**		*
Acoustics	***	**		*
Genetics		*	***	***
Inferences			*	**

Note: *** = very important; ** = fairly important; * = occasionally useful

1995 for an attempt at this). Over large scales, geographic information systems allow oceanographic data to be linked easily with tracks of individuals obtained from tags or acoustic arrays.

In the next few years we expect that a range of new techniques will be available to provide new insights into cetacean social structure. Some, such as using satellite imagery to examine group formations, can be anticipated, but there will be others that would seem as strange to us now as genetically identifying individual cetaceans from sloughed skin might have in 1980.

Conclusions

In table 3.1, we summarize how the types of techniques that we have discussed, as well as those described in chapter 2, relate to the four categories of methods needed to study social structure: methods of obtaining information on interactions, methods of collecting data on associations and groupings, analytical techniques for building a model of social structure, and methods that allow us to infer the details of rare but important events. It can be seen that none of the techniques that we have considered covers all areas. Each category of techniques requires data over a certain range of temporal and spatial scales, and the effectiveness of each available category (whether photo-identification, tagging, or genetics) is also scale-dependent. We need to use a coordinated suite of methods to uncover the societies of cetaceans, and the greater the degree of coordination (for instance, between photographic identification, genetic and acoustic methods), the more profitable the study.

Progress in studies of cetacean social organization depends on continual innovation. Recent technological innovations, such as suction cup tags and microsatellite analyses, are currently producing a wealth of high-resolution data on the lives of cetaceans. As these data are analyzed, new windows open into the animals' social lives. The pace of innovation, and the rate of discovery, are accelerating. For instance, we now know something about the social structure of large pelagic dolphin schools like those described at the start of this chapter from a combination of genetic (Dizon et al. 1991), photogrammetric (Scott 1991), VHF tagging (Butler and Jennings 1980), and other techniques.

Some innovations pioneered by cetologists should be valuable for studying the social organization of other groups of animals. For instance, the genetic methods pioneered by Amos (1993) could help unravel the social structures of other inaccessible species; techniques of looking at temporal patterning in social relationships, which have been most fully developed in studies of whales and dolphins, could be applied usefully to many species with fission-fusion social structures (Whitehead and Dufault 1999); and the use of acoustic arrays to observe interacting animals (e.g., Frankel et al. 1995) has clear potential for studying interactions in other social and vocal species.

Part 2

FOUR SPECIES

4

THE BOTTLENOSE DOLPHIN

Social Relationships in a Fission-Fusion Society

RICHARD C. CONNOR, RANDALL S. WELLS,
JANET MANN, AND ANDREW J. READ

FOR MOST people the word "dolphin" conjures up an image of the bottlenose dolphin, *Tursiops truncatus*. This comes as no surprise, given that bottlenose dolphins are the most common cetaceans in aquaria (Defran and Pryor 1980) and given their history of association with humans in coastal waters (Lockyer 1990). Found in temperate and tropical waters worldwide, bottlenose dolphins are common in pelagic as well as coastal waters, where they are often found in bays and tidal creeks, and are even known to travel up rivers (Caldwell and Caldwell 1972a). Its accessibility in nearshore waters has also made the bottlenose dolphin the best-studied cetacean. Since the early 1970s, studies of individually recognized bottlenose dolphins have been conducted in a number of countries (e.g., Argentina, Australia, Bahamas, Costa Rica, Croatia, Ecuador, Mexico, Portugal, Scotland, South Africa, and the United States), representing a variety of inshore habitats. The authors of this chapter have conducted research as part of the two longest-running field studies on bottlenose dolphins, Sarasota Bay, Florida, and Shark Bay, Western Australia (fig. 4.1). We will focus on observations from these two sites, incorporating data from other locations where possible.

One characteristic that all populations of *Tursiops* appear to have in common is a fission-fusion grouping pattern, in which individuals associate in small groups that change in composition, often on a daily or hourly basis (e.g., Würsig and Würsig 1977; Wells et al. 1987; Smolker et al. 1992). Patterns of sex-specific bonds within the bottlenose dolphins' fission-fusion society suggest intriguing parallels with the social organization of common chimpanzees *(Pan troglodytes)* and spider monkeys *(Ateles paniscus)* (Connor et al. 1992a; Smolker et al. 1992, chapter 10).

In a fission-fusion society such as the bottlenose dolphin's, even studies of group composition can reveal important insights about social relationships. At a given time, an individual may have the opportunity to associate in a number of small groups or to travel alone. Unlike animals that live in groups of constant composition, social relationships in a fission-fusion society may depend strongly on the social context: who is there and who is not. As more details become available from bottlenose dolphin studies, a fascinating picture is emerging of variation in social strategies within and between populations. Strong male-male bonds are found in Shark Bay and Sarasota, but possibly not in the Moray Firth, Scotland (Wells et al. 1987; Smolker et al. 1992; Wilson et al. 1993). Some males form strong bonds with other males in Sarasota, but some males travel alone (Wells et al. 1980, 1987). In Shark Bay and in Sarasota, some females are relatively social while others are more solitary. Ultimately, it may be possible to link such differences between populations, and in some cases within populations, to differences in predation risk or the availability or use of resources.

Evidence for intra- and interpopulation variation in foraging techniques abounds (reviewed in Shane 1990a). For example, a small proportion of Shark Bay dolphins hunt with marine sponges on their rostra, representing the first report of tool use for any wild cetacean (Smolker et al. 1997; tool use has been reported previously for dolphins in oceanaria: Caldwell and Caldwell 1972a). Such foraging specializations within habitats may be strongly influenced by learning (below). Together, the intra- and interpopulation variation in foraging and in social bonds found in bottlenose dolphins provides an important opportunity for

Figure 4.1. A young male bottlenose dolphin in Shark Bay, Western Australia. (Photograph by Richard Connor.)

testing hypotheses about the ecological determinants of dolphin social relationships.

General Appearance and Systematics

The bottlenose dolphin is a medium-sized delphinid, varying in color from slate grey to charcoal, with noticeably lighter ventral pigmentation (Leatherwood et al. 1983; Wells and Scott 1999). In many areas of the Indian Ocean and the tropical western Pacific, including Shark Bay, both males and females develop ventral speckling as they mature (fig. 4.2; Ross and Cockcroft 1990). Speckles appear to increase in coverage and density with age after sexual maturity, but individuals also vary in the extent of speckling (Ross and Cockcroft 1990; Smolker et al. 1992).

The genus *Tursiops* exhibits striking regional variation in body size, with the largest mature individuals measured in some populations (e.g., 220–230 cm in the Spencer Gulf, South Australia, and Shark Bay: Ross and Cockcroft 1990) dwarfed by the largest individuals measured elsewhere (e.g., 350–410 cm in the northeastern Atlantic: Fraser 1974; Lockyer 1985). In Sarasota Bay, female bottle-

nose dolphins reach their asymptotic length at about 250 cm (Read et al. 1993). In contrast, three parous females in Shark Bay were measured at 186, 210, and 213 cm, the size of typical two- or three-year-old female calves in Sarasota. Larger body size is generally associated with colder water temperatures (e.g., Ross and Cockcroft 1990), but whether this relationship reflects a direct adaptation to thermal requirements or is more closely related to dietary differences, or other factors, is unclear.

The systematics of the genus *Tursiops* is presently unclear. Two questions predominate. The first is whether there are several species or subspecies of *Tursiops* or a single species, *T. truncatus,* with marked geographic variation, including a large range in body size and proportions, tooth count, and coloration (Mitchell 1975b). Phylogenetic analysis of samples identified as *T. truncatus* and those identified as the nominal species *T. aduncus* (sensu Ross 1977) indicate that the two are separate species (Curry 1997; Curry and Smith 1997; LeDuc et al. 1999).

The picture is further complicated by the presence in many regions of two forms of bottlenose dolphins, labeled "inshore" and "offshore," recognized on the basis of gross

Figure 4.2. Numerous speckles cover the ventrum of this deceased Shark Bay female, Holey-fin. For over twenty years Holey-fin delighted tourists with her daily visits to the beach at Monkey Mia, where she gently accepted dead fish and stroking. Her death, at approximately age thirty-five, resulted from a stingray spine that migrated slowly through her body until it penetrated her heart. (Photograph by Richard Connor.)

morphology, hematology, cranial morphology, and parasite faunas (eastern North Pacific: Walker 1981; southeastern United States: Hersh and Duffield 1990; Peru: van Waerebeek et al. 1990; northeastern United States: Kenney 1990). The two forms may differ somewhat in their latitudinal distribution, as in the northwestern Atlantic, where the nearshore form extends only as far north as New Jersey while the offshore form continues north to Nova Scotia (Gowans and Whitehead 1995).

Hoelzel et al. (1998b) have found nuclear and mitochondrial genetic distinctions between nearshore and offshore populations in the western North Atlantic. Similarly, mtDNA sequence data analyzed by Curry (1997) indicate that inshore bottlenose dolphins from the western North Atlantic and Gulf of Mexico are reproductively isolated from offshore populations sampled from all ocean basins.

A second question regarding the systematics of *Tursiops* is whether the genus, as currently recognized, is monophyletic. A molecular investigation of the systematics of the Delphinidae provided evidence that the *T. aduncus* type is not a sister taxon to *T. truncatus,* but is more closely related to several species of *Stenella* as well as *Delphinis delphis* (LeDuc and Curry 1996; LeDuc et al. 1999). Anticipating further morphological and genetic analyses, LeDuc et al. (1999) retain the generic name *Tursiops* but recognize *T. aduncus* as a species. It is likely that these aspects of *Tursiops* taxonomy will take several years to resolve, and we offer this information so that our interpretations regarding the social ecology of bottlenose dolphins can be assessed. Following LeDuc et al. (1999), the comparisons we make here between Shark Bay bottlenose dolphins and other populations are interspecific and may eventually be determined to be intergeneric.

Field Studies

Longitudinal studies of individual wild bottlenose dolphins began in the 1970s (Irvine and Wells 1972; Würsig and Würsig 1977, 1979; Wells et al. 1980; Irvine et al. 1981). Prior to this, field reports were limited to general descriptions of the activity and size of groups (e.g., Norris and Prescott 1961), sightings of a single identifiable dolphin (on three occasions over a four-and-a-half-month period: Caldwell 1955), or specific accounts of unusual observations, such as individuals charging partway out of the water to feed on fish pursued onto mud banks (Hoese 1971) or supporting the remains of dead calves (Hubbs 1953; Moore 1955).

With the development of photo-identification techniques (Würsig and Würsig 1977; Wells et al. 1980; Samuels and Tyack, chap. 1; Whitehead et al., chap. 3, this volume), studies of bottlenose dolphins ushered in a new era of cetacean research. Recognition of individual bottlenose dolphins is based almost entirely on scars and the shape of the dorsal fin, which is conveniently projected above the water surface each time a dolphin breathes (fig. 4.3).

Figure 4.3. Nicks on the trailing edge of the dorsal fin, as well as overall fin shape, reveal that the dolphin on the left is the female "Uhf." The dolphin on the right, "Natural Tag," has more subtle markings but has nonetheless been well known to researchers for fourteen years. (Photographs by Janet Mann.)

Determining the sex of individuals is difficult because of the lack of obvious dimorphism in adults and because the ventral genitalia and mammary slits are usually hidden from researchers. Thus in many studies sex determination has been limited to individuals judged to be females because of the consistent presence of a small calf. Sex determination in Sarasota is carried out by physical examination of individuals that are briefly captured, given veterinary examinations, and released. More than 90% of the resident Sarasota dolphins are of known sex and age. In Shark Bay, clear water, male erections, and the dolphins' penchant for bow riding upside down has allowed researchers to determine the sex of a large number of individuals, paving the way for behavioral studies (Smolker et al. 1992). However, the value of such opportunistic observations has now been largely marginalized by more efficient genetic sex determination (M. Krützen, unpublished data).

Behavioral studies of known individuals have been carried out on nearshore populations of bottlenose dolphins in eastern and western Australia (Corkeron 1990; Smolker et al. 1992), Argentina (Würsig and Würsig 1977, 1979), Croatia (Bearzi et al. 1997), Portugal (dos Santos and La-

cerda 1987; dos Santos et al. 1995), Ecuador (Felix 1994), Scotland (Wilson et al. 1992, 1993, 1997), and western North America along the coast of California (Hansen 1990; Defran and Weller 1999; Defran et al. 1999), in the Gulf of California (Ballance 1990), along the coasts of Florida (Shane 1990a,b; Wells et al. 1987; Wells 1991a) and Texas (Shane 1980, 1990a; Bräger et al. 1994; Fertl 1994), and in the Bahamas (Rossbach and Herzing 1997). Most of these studies have focused on activity patterns, habitat use, and association patterns among individuals.

The longest study of bottlenose dolphins was initiated in 1970 in Sarasota Bay, Florida (Irvine and Wells 1972; Scott et al. 1990a; Wells 1991c). Built around the long-term study of identifiable individual dolphins across four generations (Wells et al. 1987), the Sarasota project has included studies on social associations, social development, mating system, and reproductive patterns (Wells et al. 1980, 1987; Wells 1991b,c, 1993b; Urian et al. 1996; Moors 1997); ranging patterns and habitat use (Irvine and Wells 1972; Irvine et al. 1981; Waples 1995); life history and population dynamics (Hohn et al. 1989; Wells and Scott 1990; Read et al. 1993; Tolley et al. 1995); genetic and behavioral studies of population structure, relatedness, and paternity (Duffield and Wells 1986, 1991; Duffield et al. 1991, 1994; Wells 1986); comparisons with dolphins in adjacent waters (Wells et al. 1995a,b, 1996); energetics, health patterns, and the effects of environmental contaminants (Costa et al. 1993; Wells 1993c; Lahvis et al. 1995; Wells et al. 1995a,b; Vedder 1996); and vocal communication (Sayigh et al. 1990, 1995a,b).

Long-term observations in Shark Bay, Western Australia, were initiated in 1984, following a pilot study in 1982 (Connor and Smolker 1985). Observations included several individuals that were provisioned in a shallow-water area (Connor and Smolker 1985; Mann and Smuts 1999), but most observations in Shark Bay focus on non-provisioned dolphins (Smolker et al. 1992). In Shark Bay, studies have focused on social relationships, including male-male relationships and mating strategies (Connor et al. 1992a,b, 1996, 1999; Connor and Smolker 1995); female-female relationships and behavior (Richards 1996); mother-infant relationships and behavioral development and life history (Mann and Smuts 1998, 1999; Mann et al. 1999); and vocal communication (Smolker et al. 1993; Connor and Smolker 1996; Smolker and Pepper 1999).

The use of different methods complicates our attempt to compare bottlenose dolphins across study sites. This difficulty is clearly illustrated by comparing group size in different areas, a problematic issue anyway given the fission-fusion nature of bottlenose dolphin grouping patterns (Smolker et al. 1992). The mean group size reported for bottlenose dolphins ranges from 5 to 140 individuals, but much of this variation can be attributed to differing definitions of "group" (table 4.1). For example, the Shark Bay study used one of the most conservative definitions: individuals were considered to be in a group if they were within 10 m of any individual in the group (a 10 m "chain rule"). The Sarasota Bay study considered a group to include all of the individuals that were moving in the same general direction, interacting with one another, or engaged in similar activities. Typically these animals were within an area of about 100 m diameter, and this cluster of individuals often accounted for all dolphins in sight at any given time. In contrast, other studies have considered any dolphins moving along the shore at the same time, or even in the same bay, to be in the same group (table 4.1). To some degree, differences in group size definitions reflect the regularity with which dolphins are observed at a study site. In Shark Bay, dolphins are always observed during surveys, usually in multiple, spatially discrete subgroups. In Golfo San José, Argentina, dolphins were sighted on only 44% of the observation days, and all dolphins passing the shore-based observation area were included in the same group (Würsig and Würsig 1979).

Life History and Reproduction

Bottlenose dolphins are long-lived: females in Sarasota may live for more than fifty years and males for more than 40 (Hohn et al. 1989; Wells and Scott 1999; table 4.2). Female bottlenose dolphins typically begin breeding at five to thirteen years of age (Sergeant et al. 1973; Perrin and Reilly 1984; Wells et al. 1987; Mead and Potter 1990; Mann et al. 2000). Estimates of the age of maturity for males range from eight to thirteen years (Sergeant et al. 1973; Perrin and Reilly 1984; Harrison and Ridgway 1971; Wells et al. 1987). In Sarasota, females reach an asymptotic length of 2.5 m at about age ten, while males continue to grow, especially in weight and girth, for another few years, reaching an asymptotic length of slightly over 2.6 m when they are 33–39% heavier than females (Read et al. 1993; Tolley et al. 1995). Differences in girth in the thorax and abdomen account for most of the mass difference between the sexes (Tolley et al. 1995). In smaller *Tursiops* off South Africa, males are only 11% heavier than females (Cockcroft and Ross 1990a). Sexual size dimorphism may vary with body size in bottlenose dolphins (Read et al. 1993) in a fashion predictable by ontogenetic scaling (e.g., Shea 1984). In Shark Bay, where adults are even smaller than in South Africa, sexual size dimorphism has not been measured, but is clearly not pronounced. The ontogenetic scaling hypothesis could be tested with further measurements from Shark Bay and from populations with dolphins larger than in Sarasota (e.g., North Atlantic).

After a twelve-month gestation period (Schroeder 1990; Schroeder and Keller 1990), bottlenose dolphins give birth to a single offspring that remains with its mother for several years (Wells et al. 1987, 1993b; Smolker et al. 1992; Mann et al. 1999). Typically young attempt to nurse up to age three to five in Shark Bay (Mann and Smuts 1998; Mann et al. 1999). In Sarasota, it is not uncommon to find lactating females with calves of three to four and a half years of age, and some lactating females have been accompanied by calves of seven to nine years (Wells 1993b; Vedder 1996). The duration of dependency can be considerably longer; one undersized and anemic male remained with his mother in Sarasota for eleven years, at which time he was the size of a typical four-year-old (Wells 1993b). In Shark Bay, a female juvenile, aged seven years or more, who was clearly not small for her age, was still attempting to nurse and regularly traveling with her mother in the position typical of dependent infants (Richards 1996). The minimum in-

Table 4.1. Group sizes and definitions of groups used in seventeen studies

Country	Location	Average group size	Group definition	Source
Argentina	Chubut Province, Golfo San José	14.9	Individuals passing the shore at the same time	Würsig 1978
Australia	Western Australia, Shark Bay	4.8	10 m chain rule; singletons excluded	Smolker et al. 1992
Australia	Queensland, Moreton Bay	10.4	Individuals within 100 m of the boat	Corkeron 1990
Croatia	Cres and Losinj Islands	7.4	From Shane 1990b: dolphins in apparent association, moving in same direction, usually engaged in same behavior	Bearzi et al. 1997
Mexico	Kino Bay, Gulf of California	15.0	All dolphins in Kino Bay (5.5 × 13 km)	Ballance 1990
Pacific Ocean (offshore population)	Off North and South America	57 (median 10)	Not stated	Scott and Chivers 1990
Portugal	Sado estuary	13.7	All animals that could be seen even when spread over a wide area	dos Santos and Lacerda 1987
Scotland	Moray Firth	6.3	Individuals within 100 m of each other	Wilson et al. 1993
South Africa	Eastern Cape	140.3	Generally dispersed over several square kilometers	Saayman and Taylor 1973
United States	California	18	Any aggregation of one or more dolphins	Hansen 1990
United States	California	19.8	Dolphins observed in close proximity and usually moving in the same direction and engaged in similar behavior	Defran and Weller 1999
United States	Florida, Sanibel Island	2.4–7.4 (reported by activity)	Dolphins in apparent association, moving in same direction, usually engaged in same behavior	Shane 1990a,b
United States	Florida, Sarasota Bay	4.8	Individuals within 100 m of the boat	Irvine et al. 1981
United States	Florida, Sarasota Bay	7.0	Individuals within a 100 m radius	Wells et al. 1987
United States	Texas, Aransas Pass	3.8–6.9 (reported by activity)	Dolphins in apparent association, moving in same direction, usually engaged in same behavior	Shane 1990b
United States	Texas, Galveston Bay	4.4	From Shane 1990b: dolphins in apparent association, moving in same direction, usually engaged in same behavior	Bräger et al. 1994
United States	Texas, Galveston Bay	2.9/3.2 groups w/ and w/o mother-calf pairs, respectively (calves excluded)	One or more individuals engaged in same activity within 5 body lengths of each other or for individuals feeding near shrimp boats, all dolphins within 10 m of boat	Fertl 1994

Table 4.2. Life history parameters of bottlenose dolphins

	Males	Females	Source
Length at birth	110–134 cm	95–132 cm	Hohn 1980a
Length at physical maturity	265 cm	249 cm	Read et al. 1993
Mass at physical maturity	259 kg	194 kg	Read et al. 1993
Minimum age at weaning	1–2 years	1–2 years	Wells and Scott 1999
Calving interval		2–6 years	Wells et al. 1987
Age at sexual maturation	8–12 years	5–10 years	Wells et al. 1987
Longevity	40+ years	50+ years	Wells and Scott 1999

Note: Values are given for the morphotype inhabiting coastal waters of the Gulf of Mexico and western North Atlantic. Parameter values vary widely among different morphotypes.

terbirth interval for females with surviving calves is three years in Shark Bay (Mann et al. 2000) and two years in Sarasota (Wells 1993b), but three- to six-year intervals are more common in Sarasota (Wells 1993b) and four- to six-year intervals in Shark Bay (Connor et al. 1996; Mann et al. 1999). Differences in body size or nutrition levels may account for variation within and among habitats (Connor et al. 1996), as has been suggested for primates (Lee 1987). In captivity, with liberal feeding and no need to learn a variety of methods for procuring prey, bottlenose dolphins nurse their young for only eighteen to twenty months, and exhibit correspondingly short interbirth intervals (two to three years) (e.g., McBride and Kritzler 1951; Tavolga and Essapian 1957; Tavolga 1966).

Health differences may also explain variation in the age at which females first give birth. Females as young as seven to ten years old have been reported to give birth in captivity (Tavolga and Essapian 1957; Schroeder 1990), and three females in Sarasota have given birth at ages six to seven (Wells et al. 1987; R. S. Wells, unpublished data). In Shark Bay, the youngest female to give birth was twelve years old (Mann et al. 2000). Mortality among calves is high in Sarasota and Shark Bay. Overall, 19% of Sarasota infants die in their first year (Wells and Scott 1990) and about 46% die before separation from their mothers. Approximately 29% of infants in Shark Bay die before age one and 44% die before weaning (Mann et al. 2000). Mortality may be especially high for infants of primiparous females. Few first calves of young females have survived in Sarasota, but first calves of older females usually die as well (Wells 1993b). The calves of at least two primiparous females in Shark Bay survived to weaning but four have not (Mann et al. 2000). Given the high mortality of first infants, one can ask why some females attempt to reproduce earlier than most. It is possible that a female that can afford to reproduce early benefits by producing a surviving calf earlier than most, even if her first calf is more likely to die than the first calf of older primiparous females.

Like many delphinids for which information is available, bottlenose dolphins exhibit diffuse seasonal reproduction (Perrin and Reilly 1984; Urian et al. 1996), with a single or bimodal birth season centered in spring/early summer and fall (Harrison and Ridgway 1971; Ross 1977; Wells et al. 1987; Mead and Potter 1990; Connor et al. 1996; Mann et al. 2000). Studies of plasma progesterone levels in captive dolphins have established that *Tursiops* is a spontaneous ovulator and seasonally polyestrous, producing two to seven ovulations per year (Kirby and Ridgway 1984; Yoshioka et al. 1986; Schroeder 1990; Schroeder and Keller 1990). Cycle length was estimated by Yoshioka et al. (1986) to be about thirty days. Testis size in bottlenose dolphins is large compared with most mammals, but small for a delphinid (Kenagy and Trombulak 1986; Connor et al., chap. 10, this volume). Captive males have been found to exhibit seasonal peaks in plasma testosterone levels and sperm concentration (Schroeder and Keller 1989). Males in Sarasota also exhibit seasonal changes in testosterone concentrations in blood and in testis size and diameter as determined by ultrasound (Wells et al. 1987; R. S. Wells, unpublished data).

Ecology

Feeding and Foraging

Bottlenose dolphins feed on a wide variety of fish, as well as some cephalopods (squid and octopus), and occasionally shrimp (Gunter 1951) and small rays and sharks (Mead and Potter 1990). As the following examples illustrate, they pursue schooling and solitary prey throughout the water column as well as into the air above, into the sand below, and even onto the shore. Bottlenose dolphins are often reported to circle around fish schools, with one or a few individuals at a time darting into the school to feed (Morozov 1970; Leatherwood 1975; Hamilton and Nishimoto 1977; Bel'kovich et al. 1991). In the Bahamas bottlenose dolphins dive into the sand up to their eyes after some still unknown prey, leaving areas of the seafloor resembling a miniature moonscape of small sand volcanoes (Rossbach and Herzing 1997). It is unclear whether buried prey are detected with echolocation or visually by some surface disturbance in the sand, although intense echolocation is typically heard during these feeding episodes. Individuals are often reported to swim upside down while chasing fish near the surface (e.g, Leatherwood 1975; Bel'kovich et al. 1991). In Shark Bay, dolphins often swim belly up in pursuit of fish that jump or skip along the surface for considerable distances before the fish is grabbed in the air. In tidal creeks of Georgia and South Carolina, dolphins pursue prey onto mud banks and slide back into the water (Hoese 1971; Rigley 1983; Petricig 1993; see plate 5). Bottlenose dolphins may also stun or kill fish by "whacking" them up to 9 m into the air with their flukes (Wells et al. 1987; Shane 1990b). In one common foraging pattern, dolphins arch their peduncles or lift their flukes out of the water as they dive, presumably to increase the angle of descent as they travel toward the bottom (Shane 1990b). In a few 8–12 m channels in Shark Bay, dolphins engaging in this behavior carry conical sponges over their rostra. Observations of sponge use are incomplete, but Smolker et al. (1997) suggest that the sponges may function as tools to protect the dolphins' rostra as they probe around the bottom for prey (fig. 4.4).

There are three points worth emphasizing here. First, dolphins employ a diverse repertoire of behaviors during feeding that reflect variation in prey strategies to avoid detection or capture. Dolphins are adept at capturing fish in schools as well as finding solitary, possibly cryptic prey resting on the bottom or even hiding in the sand or seagrass.

Second, dolphins may use both social and solitary feeding strategies. Group assaults on schooling fish may allow them to counter the antipredator schooling mechanisms of prey species (Major 1978; Norris and Dohl 1980b; Pitcher and Parrish 1993). Particular feeding techniques may be performed in groups or alone. Bottlenose dolphins that pursue fish onto mud banks in South Carolina usually do so in groups, with individuals beaching simultaneously side by side (Petricig 1995), but in Shark Bay dolphins that pursue fish onto the beach usually do so alone (Berggren 1995).

Finally, the variety of feeding methods underscores the potential for studies of individual foraging specializations in bottlenose dolphins, especially the role of learning of foraging and feeding techniques (early observations of captive dolphins suggested the importance of learning to Norris and Prescott [1961]). Evidence for learning among wild bottlenose dolphins is strongest in the accounts of cooperative feeding associations between humans and dolphins that are found from the first century to modern times (Busnel 1973). In the town of Laguna, in southern Brazil, fishermen line up in murky thigh-deep water, holding weighted throw nets, in pursuit of fish they cannot see. One or two dolphins, facing seaward several meters offshore of the fishermen, occasionally submerge, moving seaward, only to reappear moving toward the men a few seconds later (Pryor et al. 1990). The dolphins then come to an abrupt halt 5–7 m from the fishermen, diving with a surging roll that cues the fishermen to toss their nets. Both humans and dolphins likely do better fishing together than either would working alone (Busnel 1973; Pryor et al. 1990). Cooperative fishing with humans in Laguna is practiced by a subset of dolphins who may learn the behavior as calves (Pryor et al. 1990).

The most complete record of the feeding habits of bottlenose dolphins comes from the analysis of stomach contents of individuals caught incidentally in nets. Cockcroft and Ross (1990b) examined the stomach contents of 165 nearshore bottlenose dolphins that had become trapped in nets set to protect swimmers from sharks along the Natal coast of South Africa. Overall, fish accounted for 75% and squid 25%, by mass, of the diet. Of over seventy prey species recorded, six accounted for 60% of the diet, including two cephalopods and four fish species. A range of habitats is found even among the four predominant fish species; one was a benthic species that inhabits inshore reefs, another was a benthic species that prefers sandy-bottom areas, and the remaining two were pelagic shoaling fish. Similar

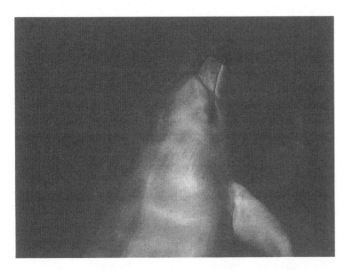

Figure 4.4. Tool use in dolphins? This dolphin is "wearing" a sponge on its rostrum. The sponges used vary considerably in size, but don't usually have an intact holdfast, as this rather small sponge does. (Photograph by Michael Heithaus.)

Richard C. Connor, Randall S. Wells, Janet Mann, and Andrew J. Read

large-scale studies of stranded or captured dolphins yield broadly similar results: a large percentage of the diet comprises several species, followed by a large number of less frequent and "occasional" items (Mead and Potter 1990; Barros and Odell 1990; Van Waerebeek et al. 1990). Geographic shifts in diet have been linked to changes in relative prey abundance (Barros and Odell 1990).

Predators

At the northern extreme of the range of Atlantic bottlenose dolphins, a population in the Moray Firth, Scotland, is thought to live relatively free from shark predation (Wilson et al. 1997). If so, then the Moray Firth is probably the exception to the rule. Dolphins that survive shark attacks bear evidence of the encounter in the form of distinctive crescent-shaped scars (fig. 4.5). In Morton Bay, Queensland, 36% of 334 individually identified bottlenose dolphins had shark scars (Corkeron et al. 1987). Of the 145 bottlenose dolphins caught incidentally in nets set to protect bathers from sharks off the Natal coast of South Africa, 15 (10%) had clear shark bite scars; another 13 had scars that may have been produced by shark attack (for a possible total of 19%). Based on a series of admittedly crude calculations, Cockcroft et al. (1989) suggest that sharks might kill over 2% of the Natal dolphin population annually.

In Sarasota, 22% of 64 non-calves examined during brief captures bore shark bite scars; none of the 22 calves in the same sample were scarred (Wells et al. 1987). The majority of shark victims in Natal were young dolphins, suggesting that the relatively low incidence of scars on young dolphins might be due to a greater proportion of fatal attacks (Cockcroft et al. 1989). However, in Shark Bay, 34% (12 of 35) calves in a focal study on infants had shark bite scars (Mann and Barnett 1999). Mothers of young calves may also be frequent victims (fig. 4.6). Corkeron et al. (1987) noted a preponderance of fresh shark wounds (5 of 21) on females with calves less than eighteen months old.

Nine species of sharks caught in the Natal nets contained the remains of dolphins in their stomachs; the four most common predators on dolphins were the bull shark *(Carcharhinus leucas)*, dusky shark *(Carcharhinus obscurus)*, tiger shark *(Galeocerdo cuvieri)*, and great white shark *(Carcharodon carcharias)*. Some or all of these four species have been implicated in dolphin predation elsewhere (McBride 1948; Wood et al. 1970; Wells et al. 1980; Corkeron et al. 1987). A recent decline in the population of white sharks has been implicated in the paucity of shark bite scars on Croatian bottlenose dolphins (Bearzi et al. 1997).

Direct observations of shark predation on wild dolphins are lacking. A five-month-old infant clearly died from wounds received from a shark in Sarasota (Wells 1993b). In Shark Bay, a seven-month-old infant died from shark bite wounds, and tourists observed a 2 m tiger shark killing a 4-month-old infant of one of the provisioned females

Figure 4.5. The large shark bite scar behind the male "Wow"'s dorsal fin was present when he was first photographed in 1982. This picture was taken fifteen years later in 1997. (Photograph by Richard Connor.)

A

B

Figure 4.6. Mother with severe shark bite wound (probably from tiger shark, *Galeocerdo cuvier*) within a week of the attack *(top)* and again several weeks later *(bottom)*. Although calves in Shark Bay are most vulnerable to predation, mothers may also be at increased risk. (Photograph by Janet Mann.)

(Mann and Barnett 1999; fig. 4.7). Sharks have often been observed close to bottlenose dolphins, whose reactions varied, not surprisingly, with species and circumstance. In captivity, bottlenose dolphins exhibited a stronger negative reaction to tiger sharks than to other species, avoiding an adult tiger shark and attacking and killing two smaller ones (McBride and Hebb 1948). A bottlenose dolphin trained to repel sandbar *(Carcharhinus milberti)*, lemon *(Negaprion brevirostris)*, and nurse *(Ginglyostoma cirratum)* sharks became agitated and refused to respond to commands when

Figure 4.7. Holey-fin nudges in toward her dead 4-month-old calf, Hobbit, who moments earlier had his tail bitten off and was disemboweled by a 2 m shark. Holey-fin chased the shark away when it returned. (Photograph by Janet Mann.)

bull sharks were introduced into its pool (Irvine et al. 1973). Bottlenose dolphins have been observed avoiding hammerhead sharks *(Sphyrna zygaena)* and a great white shark *(Carcharodon carcharias)* in South African coastal waters. When a 2.5–3.0 m white shark cruised into a resting group of six females with three older dependent calves (ages three and a half to five years) in Shark Bay, the dolphins immediately dove, came up leaping away, and did not cease leaping until they had covered 3 km ten minutes later (Connor and Heithaus 1996). The young calf killed in Shark Bay was tens of meters from his mother when killed, but other dolphins rushed to the scene, chasing the shark away. The mother, who remained with the partially eaten body of her calf, chased away a 2 m tiger shark when it approached about forty-five minutes after the initial attack (Mann and Barnett 1999). In other cases, dolphins and sharks in captivity and in the wild have been observed swimming in close proximity with little or no apparent reaction by either party (e.g., Wood et al. 1970; Leatherwood 1977). Mutual tolerance was the rule between bottlenose dolphins and sandbar sharks that lived in the same tank until one of the female dolphins began giving birth. The other dolphins formed a tight school below and beside the female, and herded away the sharks, which were apparently attracted by the blood in the water (McBride 1948).

Killer whales *(Orcinus orca)*, which enjoy an even wider distribution than bottlenose dolphins, prey on a variety of marine mammals and have been implicated in attacks on the other two species featured in this volume (Whitehead and Weilgart, chap. 6; Clapham, chap. 7, this volume). Although there have been no documented attacks on bottlenose dolphins, killer whales may occasionally threaten coastal and offshore bottlenose dolphins in all locations. Würsig and Würsig (1979) observed scarring on one bottlenose dolphin in Golfo San José, Argentina, that appeared to have been made by killer whale teeth, and on two occasions saw dolphins leaping away from killer whales at a high rate of speed. A 288 cm male bottlenose dolphin from offshore waters in Peru exhibited tooth scars that were likely made by a false or pygmy killer whale (A. J. Read, unpublished data).

Interspecific Associations

Bottlenose dolphins are commonly found in mixed-species groups in pelagic waters. Scott and Chivers (1990) found a significant increase in mixed-species groups with increasing distance from shore in the eastern tropical Pacific. Over 40% of mixed-species groups were made up of bottlenose dolphins and short-finned pilot whales *(Globicephala macrorhynchus)*. The bottlenose dolphin-pilot whale association is common but poorly understood (Faeroe Islands: Kraus and Gihr 1971; the North Pacific: Norris and Prescott 1961). Kenney (1990) reported that bottlenose dolphins off the northeastern United States were with other

THE BOTTLENOSE DOLPHIN

101

species in about 10% of sightings, and about half of these involved pilot whales. Bottlenose dolphins were present in at least seven out of twenty-seven pods of short-finned pilot whales caught in a drive fishery near Japan (Kasuya and Marsh 1984). Pilot whales feed extensively on squid, and Kraus and Gihr (1971) found scars from squid suckers on one of two bottlenose dolphins captured in a drive fishery with a school of 101 pilot whales in the Faeroe Islands.

Herzing and Johnson (1997) observed a variety of interactions between Atlantic spotted dolphins *(Stenella frontalis)* and bottlenose dolphins in the Bahamas. Interactions between individuals of the two species included affiliative contact behavior as well as sexual and aggressive behaviors. Aggressive behavior by the much smaller spotted dolphins was performed by males in groups. Sexual behaviors were initiated by adult female bottlenose dolphins toward young male spotted dolphins as well as by male bottlenose dolphins toward male and female spotted dolphins.

Contrary to their popular image, bottlenose dolphins are also one of the very few mammals known to direct lethal, nonpredatory aggression at other mammalian species. In the Moray Firth, Scotland, bottlenose dolphins have been observed to attack harbor porpoises *(Phocoena phocoena)* on several occasions (Ross and Wilson 1996). In detailed postmortem examinations between 1991 and 1993, forty-two stranded porpoises from the Moray Firth exhibited evidence of blunt trauma consistent with attacks from bottlenose dolphins. The reasons for such attacks are unclear—possibilities include play (as "practice" infanticide, Patterson et al. 1998) and competition for prey. Subsequent observations in Cardigan Bay, Wales (Jepson et al. 1997), and Virginia Beach, USA (A. J. Read, unpublished data), indicate that such lethal aggression may be a common behavioral feature of bottlenose dolphins wherever their ranges overlap with those of harbor porpoises. These interactions may serve to limit the geographic ranges of harbor porpoise populations.

Habitat and Group Size

Dolphin species that inhabit more open, pelagic habitats generally form larger groups (Norris and Dohl 1980b), and similar results are often reported for bottlenose dolphins (reviewed by Shane et al. 1986). For example, in Sarasota, Wells et al. (1980) found larger groups in the deeper waters of the passes and open Gulf than in the shallow areas inland of the barrier islands. Similar patterns have been reported from Tampa Bay, immediately to the north of Sarasota, and Charlotte Harbor to the south (Wells et al. 1995a,b, 1996). Factors contributing to this trend are thought to be the risk of predation in open habitats and a transition from

solitary, individual prey items on reefs or shallow seagrass beds to schooling fish in the pelagic environment (Norris and Dohl 1980b; Wells et al. 1980). However, while bottlenose dolphins in the eastern tropical Pacific are occasionally found in aggregations numbering in the thousands, the median group size (ten) is very similar to that found in the adjacent coastal habitat (twelve) (Scott and Chivers 1990). Even the very large groups are usually aggregations of smaller groups spread out over several miles. Scott and Chivers (1990) point out that the prey of pelagic bottlenose dolphins probably do not occur exclusively in large schools, and that the large body size of pelagic bottlenose dolphins may substantially reduce their predation risk compared with that of smaller delphinids, such as spinner and common dolphins (Scott and Chivers 1990).

Day Range

There are few data on the day range of individual bottlenose dolphins, but those that are available suggest that day range is highly variable. Radiotracking data from Sarasota indicate that dolphins typically travel within a 40 km long home range at 2–5 km per hour. North-south movements of 30 km in a day have been observed, but are not typical (Irvine et al. 1981). In nearby Tampa Bay, a dolphin whose movements were monitored by a satellite for twenty-five days traveled a minimum of 581 km, or 23 km per day, but was recorded moving 50.2 km in one day (Mate et al. 1995). These figures are minimum figures, as an average of only four satellite locations were obtained each day (Mate et al. 1995). Lynn (1995) tracked the movements of five radio-tagged male bottlenose dolphins (ages eight to nineteen years) and five females (ages eight to thirty-one years) along the central Texas coast for periods ranging from twelve to sixty days. Individuals ranged widely on some days (one female traveled 55 km in a twelve-hour period) while remaining in limited areas of 1–2 km^2 on other days (Lynn 1995).

Home Range and Migration

Variation between populations. The Sarasota Bay bottlenose dolphins are long-term, year-round residents of a home range of about 125 km^2 (Wells et al. 1980; Irvine et al. 1981; Wells 1986, 1993a; Scott et al. 1990a). Through 1995, 75% of the dolphins first identified in the same waters during 1970–1971 were still under observation, as were numerous calves and grandoffspring. The Sarasota dolphins' home range includes shallow estuarine bay waters and channels inshore of a north-south series of barrier islands, narrow, deep passes leading to the Gulf of Mexico, and Gulf coastal waters within several kilometers of shore.

From April to September the dolphins concentrate in the inshore waters, including shallow seagrass meadows, in their daily movements. During the remainder of the year the dolphins move into the deeper inshore waters and spend more time in the Gulf coastal waters. Such seasonal habitat shifts are common in nearshore populations (Shane et al. 1986), and are apparently related to water temperature changes, either through the thermal requirements of the dolphins or changes in the distribution of their prey or predators (Irvine et al. 1981; Wells et al. 1980; Shane et al. 1986).

At the other end of the spectrum, some populations with larger seasonal range shifts are described as migratory. An example is found off the mid-Atlantic United States, where the northern extent of the range of the coastal population is Long Island in the summer and Cape Hatteras in the winter (about 400 km) (Mead 1975; Mead and Potter 1990; Blaylock et al. 1995). The offshore population off the northeastern U.S. coast also exhibits north-south seasonal migrations (Kenney 1990).

No seasonal range extensions have been reported for inshore bottlenose dolphins along the coast of California, yet individuals in this population are highly mobile, moving as much as 286 km in fourteen days (Defran et al. 1999). Ranging patterns of individual dolphins along the east coast of the United States have not been fully documented, but a few resightings of individuals identified along the coasts of North Carolina, Virginia, and New Jersey have been reported (Hohn 1997).

Climate changes can contribute to large-scale nonseasonal range extensions. A 600 km northward range extension of some members of the coastal California population from San Diego to Monterey Bay was linked to warming of the water during the 1982–1983 El Niño event (Wells et al. 1990). Bottlenose dolphins have continued to inhabit Monterey Bay for at least ten years since their arrival (Feinholz 1995).

Individual variation within populations. Individuals in Sarasota differ in their ranging patterns, and much of this variation can be explained by sex differences. Females preferentially use smaller core areas along the 40 km long study area, which they share with other females. Males more frequently range to either end of the study area; on occasion they may even disappear from the area for days to months at a time, and have been observed interacting with dolphins in adjacent areas (Wells et al. 1987, 1996). Much less frequently, Sarasota females may disappear from the home range for weeks to months (R. S. Wells, unpublished data). Home range variation is also found among individuals of

the same sex. One unusually small dolphin in Sarasota was found exclusively (99 sightings) in a deep-water hole (200 m × 75 m × 7 m deep) for three and a half years (Wells et al. 1990).

Dispersal

There is mounting evidence for natal philopatry in bottlenose dolphins. Overall annual rates of immigration and emigration are very low in Sarasota, no more than 2–3% (Wells and Scott 1990). Of twenty-two female calves (born to sixteen Sarasota mothers) and sixteen male calves (born to eleven Sarasota mothers) known to have survived through separation from their mothers, all have remained within the Sarasota dolphin community, for up to twenty-three years (Wells 1991c; R. S. Wells, unpublished data). Eight of the female offspring have given birth to their own calves within the community home range; one of these births represents the fourth generation of one maternal lineage observed in Sarasota (R. S. Wells, unpublished data).

Two of the provisioned females that visited the Monkey Mia shallows in Shark Bay produced female calves that became provisioned, remained in the area, and produced calves of their own. These two cases are suspect because provisioning could alter dispersal patterns; however, another calf of one of these females remained in the area and reproduced although she did not become provisioned. Of thirty-two calves (eleven males, seventeen females, four sex unknown) born to eighteen non-provisioned Shark Bay females who survived to weaning, all were sighted frequently in the study area post-weaning for up to ten years. One female has given birth and remained within her natal range. Over twenty juveniles first sighted in the mid- to late 1980s have remained in the same home range for at least ten years, into adulthood.

Thus, both females and males may be philopatric. Some adult males may range more widely than females in Sarasota, disappearing from the area for months at a time, and do not maintain a high level of association with their mothers. However, evidence suggests that most males continue to include their mother's range as part of their home range, as described above. Although interactions may be infrequent, males may maintain a social relationship with their mothers into adulthood. One female's fourteen-year-old son was seldom with his mother, but was with his mother during the first sighting of her with a new male calf (Wells 1991a).

Do Bottlenose Dolphins Live in Closed Groups?

There is no evidence that the four hundred or more identified dolphins in the Shark Bay population live in a "closed" or "semi-closed" community in which community members are generally hostile to nonmembers (Smolker et al. 1992). Within the 250–300 km² study area there appear to be no sharp discontinuities, but rather a continuous mosaic of overlapping home ranges for both males and females.

The long-term residents of Sarasota Bay are considered to be a community rather than a population, in the strictest sense of the term (Wells 1986; Wells et al. 1987). Bands of females in Sarasota form the core of a hundred-member community whose members interact with each other much more than with members of similar units in neighboring areas (Wells et al. 1987). About 83% of the sightings involving Sarasota community members include only members of the Sarasota community. Occasional agonistic interactions near the Sarasota community boundary pitting pairs of Sarasota community males against pairs or trios of large, heavily scarred individuals (likely males) from adjacent areas suggest that males may defend the community territory (Wells 1991c), although other explanations are also possible. Though the social association and daily movement patterns suggest a somewhat closed unit, genetic characterization based on mitochondrial DNA and results of paternity analyses indicate that significant genetic mixing occurs (Wells 1986; Duffield and Wells 1986, 1991). The Sarasota dolphin community is part of a larger mosaic of slightly overlapping home ranges of different dolphin communities along the central west coast of Florida (Wells 1986, 1991a; Wells et al. 1987, 1995a,b, 1996).

Social Behavior and Communication

Social relationships, the focus of this book, are revealed through behaviors expressed in social interactions, repeated over days, months, and years (Mann, chap. 2; Whitehead et al., chap. 3, this volume). Much of what we know about social behavior in bottlenose dolphins comes from ad lib observations in captive settings during the 1940s through the 1960s (Samuels and Tyack, chap. 1, this volume) and some recent focal studies of captive dolphins (e.g., Eastcott and Dickinson 1987; Cockcroft and Ross 1990c; Peddemors 1990; Östman 1991; Reid et al. 1995). These descriptions, together with more recent observations of free-ranging individuals, allow us to summarize how bottlenose dolphins conduct their social relationships. Their acoustic communication system is less well understood (Tyack, chap. 11, this volume).

Affiliation

Bottlenose dolphins express affiliation by proximity, physical contact, and synchronous movement. Strongly bonded males in Shark Bay often swim side by side less than a body length apart and surface in precise synchrony (fig. 4.8; R. C. Connor et al., unpublished data). Young calves and their mothers also maintain close proximity and swim synchronously (Mann and Smuts 1999). Similar patterns are observed in Sarasota Bay (Wells et al. 1987). Two dolphins may swim in close proximity with one dolphin resting its pectoral fin against the side of another (Tavolga and Essapian 1957), a behavior referred to as contact swimming (Samuels and Tyack, chap. 1, this volume). While contact swimming is one expression of affiliation, there are more active forms of affiliative touching. Gentle stroking with the pectoral fin or rubbing against another individual is a common expression of affiliation in bottlenose dolphins (Tavolga and Essapian 1957; Samuels and Tyack, chap. 1, this volume). In Shark Bay, rubbing and petting are commonly observed between closely bonded individuals, such as mother and calf (Mann and Smuts 1998, 1999). Rubbing occurs when one individual actively rubs a body part on another individual (fig. 4.9). Among adults in Shark Bay, affiliative behavior involving contact with the flukes or pectoral fin of another individual can be distinguished from more intense forms of body-to-body contact, which produce splashing at the surface, often include socio-sexual contact (below), and may express more than affiliation. We do not know how acoustic signals may mediate affiliative behavior.

Socio-Sexual Behavior

It is difficult to distinguish "sexual" behavior from "socio-sexual" behavior unless "sexual" behavior is, by definition, considered potentially conceptive based on the age-sex classes and reproductive states of the actors (e.g., see Wrangham 1993; Wells 1984). In Shark Bay, males mount females or other males by sliding the ventrum along the side of the other dolphin or over the back. During a mount, the penis of the male is hooked under the ventrum of the other animal (fig. 4.10). Intromission is difficult to see, having been observed only once between adults in Shark Bay, but is frequently seen between immature animals in both Shark Bay and Sarasota Bay. This side-to-dorsal mounting approach differs from the ventrum-to-ventrum mating described previously, in which the male is stationed below the female (Tavolga and Essapian 1957) or alongside her at the surface (dos Santos and Lacerda 1987).

Socio-sexual behavior, which is nonconceptive, may mediate social relationships or serve a communicative function in bottlenose dolphins, as proposed for other animals (e.g., bonobos, *Pan paniscus*, Wrangham 1993; spinner dolphins, *Stenella longirostris*, Wells 1984). Like that of bonobos and

Figure 4.8. Males often swim and dive in perfect synchrony with their alliance partners. (Photograph by Richard Connor.)

spinner dolphins, socio-sexual behavior among bottlenose dolphins may involve almost any age-sex class combination of individuals, and does not appear in exclusively affiliative or agonistic contexts. Mounting of one male by another may be used to express dominance (Östman 1991), but when newborn infant males mount their mothers, other interpretations are warranted (Mann and Smuts 1999). In Shark Bay, we (Connor and Mann) have observed male-male mounting with erections in clearly agonistic contexts that included head-to-head behavior and biting, as well as in more relaxed interactions between strongly bonded males. We have observed male calves mounting older males, older males mounting calves, and reciprocal mounting between a large, heavily speckled adult male and a smaller, non-speckled, subadult male. Pregnant and immature females have been observed mounting other females, although female-initiated mounts are much less common than those initiated by males. On one occasion we observed a female mounting an adult male. Immatures spend long periods of time engaged in socio-sexual exchanges. Another behavior commonly associated with socio-sexual behavior is rostro-genital contact ("goosing"), in which one individual moves its rostrum into the genital area of another, gently or roughly. "Genital inspections" are frequently observed in Shark Bay, in which one animal brings the rostrum close to, but does not contact, the genital area of

another. Echolocation clicks are sometimes heard during these "inspections," possibly providing information about reproductive state.

Aggressive Behavior

Posture, movement, and sound are also used to express aggression on a continuum that ranges from vocal and physical threats to violence. A bottlenose dolphin may threaten another with a distinct posture, arching the head and flukes down (e.g., Östman 1991). One dolphin may charge at another, accelerating rapidly and veering off suddenly before contact (Connor et al. 1996). The "jaw clap," a rapid opening and closing of the jaws, and the "head jerk," which involves a rapid lateral or vertical jerk of the head, are well-known physical threats that include an acoustic component (Overstrom 1983). Connor and Smolker (1996) identified an acoustic threat, "pops," that does not appear linked to a prominent visual display. Physical violence includes biting, the less severe tooth-raking, and strikes with the rostrum, pectoral fins, or peduncle and flukes (McBride and Hebb 1948; Norris 1967a; Östman 1991). An aggressive interaction may incorporate a number of these behaviors. Two dolphins may face each other head-to-head in an aggressive interaction, sometimes jerking their heads up and down while producing low-frequency vocalizations. Head-to-head interactions sometimes escalate into a fight includ-

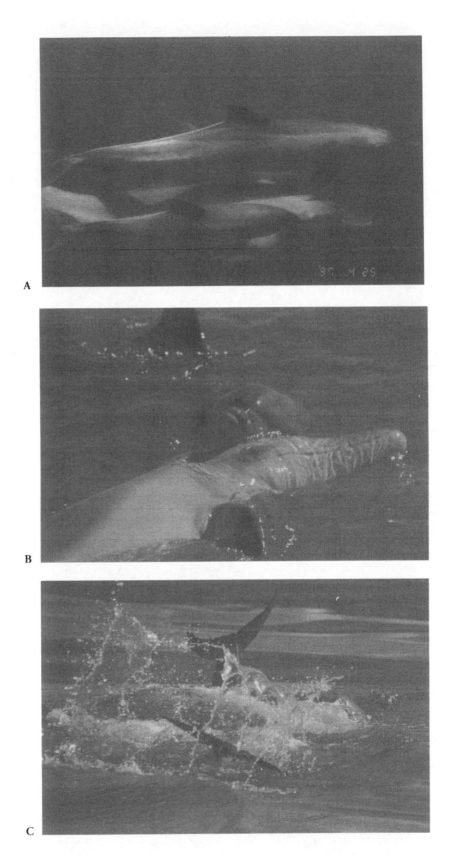

Figure 4.9. Affiliative behavior varies in intensity from (A) swimming side by side with gentle intermittent touching to (B) body-to-body rubbing and (C) more intensive "playful" contact that may include splashing. (A, photograph by Janet Mann; B, C, photographs by Richard Connor.)

Figure 4.10. Sex often occurs in nonreproductive social contexts in bottlenose dolphins. Here one male in Sarasota attempts intromission with another male, who also has an erection.

ing bites and strikes. A dolphin may ram another with its rostrum. In captivity, bottlenose dolphins have killed small sharks and conspecifics by striking and ramming them (McBride 1940; McBride and Hebb 1948; Caldwell et al. 1968). Young conspecifics and harbor porpoises *(Phocoena phocoena)* have been battered to death by free-ranging bottlenose dolphins in this manner (Ross and Wilson 1996; Patterson et al. 1998; Dunn et al. 1998).

Dominance

Agonistic interactions and dominance relationships are not equivalent (see Samuels and Gifford 1997). Furthermore, it is important to distinguish, as primatologists have done (e.g., de Waal 1987), between formalized dominance ("rank"), in which repeated encounters between individuals reaffirm asymmetries in the relationship, and "power," which may depend on alliances between individuals, the influence of individuals under particular circumstances (reproductive condition, access to resources, presence of a newborn infant, etc.), or other individual characteristics. Dominance relations have not been studied in wild bottlenose dolphins, but were examined recently in captivity by several researchers (Samuels and Gifford 1997; Östman 1991). Dominance was determined most reliably by the "loser" in dyadic agonistic interactions, who, in response to aggressive or neutral behavior by the other individual, displayed submissive behavior ("flinch" or "flee": Samuels and Gifford 1997), but no aggressive behavior. The two males in the study were always dominant over females

(eight altogether, but two to four females were in the group at any one time). Male dominance persisted in spite of the illness of one male and the larger size of some females. Female dominance relations were stable over years and rarely contested. During the same period, the males punctuated periods of calm with episodes of intense rivalry, resulting in several role reversals. Nothing is known about dominance relationships between immature animals, or whether the dominance concept can be appropriately applied to young animals.

In Shark Bay, dolphins that have captured a fish often toss it several times up to 3 m away in the presence of other (often larger) dolphins, who do not attempt to take the fish (R. C. Connor and J. Mann, unpublished data). Thus, at least in the context of prey capture, "ownership" may be respected in wild bottlenose dolphins.

Displays

Many of the behavioral displays of bottlenose dolphins may not fall along the affiliative-agonistic axis. Individuals of either sex may perform a range of displays, many of which are performed at the surface where they are easily observed. A common display is to slap part of the body, such as the rostrum, flukes, or belly, side or back, on the water surface. Such displays appear to reflect excitement, but are not clearly either affiliative or agonistic. An affiliative or agonistic function is even less obvious for the elaborate synchronous or solitary displays males perform around females in Shark Bay (Connor et al. 1992a). A common display that

Figure 4.11. Two Shark Bay male bottlenose dolphins perform synchronous belly slaps as they travel in opposite directions on either side of a female. This particular display, like several others, has never been observed again. (Photograph by Richard Connor.)

may be performed by single males or simultaneously by two or three males traveling side by side is the "rooster strut," in which the head is arched above the surface and bobbed up and down. "Butterfly" displays are a common class of synchronous displays performed by two males around a female. The males execute a series of figure eights, forming the "wings" on either side of the female, who constitutes the body of the "butterfly." Butterfly displays are not stereotyped, but vary considerably based on the size and shape of the figure eights and the order in which they are performed. A number of displays that appear quite complex have been observed only once. An example is the display pictured in figure 4.11, in which two males performed synchronous belly slaps while moving in opposite directions on either side of the female. The variation in displays renders unlikely the possibility that males have a set repertoire of displays. Although we cannot explain why males perform such a variety of displays, we suggest that they accomplish this end by having one individual closely following the lead of another. This hypothesis finds support in captive observations on individuals trained to perform synchronous behaviors (Braslau-Schneck 1994). Synchronous displays may function to advertise health and vigor to females, but may also have agonistic or affiliative functions directed to the female, to other males, or to each other (Connor et al. 1992a). Females have not been observed engaging in synchronous displays, but synchronous surfacings or breaths between females are common.

Association Patterns and Social Relationships

In many habitats, determining the sex of individuals is difficult; thus sex-specific association patterns have been studied only in Sarasota and Shark Bay. Comparisons between these two sites, combined with general results from several other locations, show the potential for comparative socioecological studies of bottlenose dolphin populations.

Association patterns typically have been studied using the survey method, in which researchers travel through the study area and record the composition of groups during brief encounters, as well as their predominant activity and relevant environmental data (e.g., Bräger et al. 1994; Hansen 1990; Wells et al. 1987; Würsig and Würsig 1977; Shane 1980; Smolker et al. 1992). Repeated over a number of days and years, surveys can provide an accurate short-term or long-term picture of individual association patterns, presented as coefficients that estimate the percentage of time that any two individuals are found together (see Cairns and Schwager 1987; Ginsberg and Young 1992; Mann, chap. 2, this volume). However, surveys do not illuminate the social dynamics or ecological contexts of changes in group composition, which occur, on average, three or four times per hour in Shark Bay (box 4.1; R. C. Connor and J. Mann, unpublished data on focal follows of male and female dolphins).

Both the Sarasota and Shark Bay studies employed the

Box 4.1

Observing Bottlenose Dolphins in the Wild

Bottlenose dolphins live in fission-fusion societies in which individuals associate in small parties that frequently change in composition and behavior. The following account of "a day in the life"—or about six and a half hours, to be exact—of "Poindexter," a Shark Bay female, captures the dynamic nature of dolphin associations and behavior. Poindexter was followed by R. Connor on 19 June 1986, which is early winter in Shark Bay, when the dolphins feed extensively on schooling fish.

Poindexter was encountered at 11:00, foraging in 5 m of water near the female Square, her calf Squarlet, and Double-hook (sex unknown). As they moved north, still foraging, they encountered the females Joy and Joysfriend. Poindexter left the others by 11:20, moving in and out of feeding groups of cormorants, porpoising and leaping occasionally as she pursued schooling fish. She joined Square and Squarelet at about 11:35, and they traveled slowly together for fifteen minutes, then Poindexter left them, but rejoined them briefly around noon. Traveling alone by 12:10, Poindexter began foraging (milling) in 4.5 m of water again at 12:25, in the vicinity of the female Holey-fin, her female calf Holly, Square and Squarelet, the young male Pointer, and a couple of others we did not identify. By 12:40 Poindexter had settled into a slow traveling and resting group with Square, Squarelet, and the female ED. At 13:41 Poindexter left the others, porpoising rapidly toward feeding groups of dolphins. She entered the feeding area at 13:50, leaping and porpoising in an area with at least fifteen or twenty dolphins, including males and females, all in pursuit of schooling fish. The leaping and porpoising began to wind down to milling by 14:30, and by 14:45 dolphins began to coalesce into large socializing groups. Poindexter emerged at 15:05 in a smaller group including the females Uhf, Joysfriend, Crooked-fin, and her female calf, Puck. They met up a few minutes later with another group including the male Chop and the females Surprise, Blip, and her female calf, Flip. Nearly as soon as the two groups met they began a bout of intense socializing, but Poindexter left with Crooked-fin. They traveled for the next twenty minutes before rejoining the group, which had by now ceased socializing, of Joysfriend, Puck, Pointer, and Uhf (in the interim, Blip, Flip, and Chop had left). Poindexter and

Crooked-fin were on their own again within five minutes, and joined Joysfriend, who was now with the female Merapi, at 15:50. Poindexter and Crooked-fin parted ways at this point, and Poindexter joined the males Slash and Crinkles. Surprise joined them briefly at 16:01 as they rested. At 16:15 they joined the group of Joy, Joysfriend, Tongue-fin, her female calf, Lick, and the male Shave. This group slowly broke up, leaving Poindexter alone by 16:30. She traveled for the next fifteen minutes before joining Crooked-fin, Joysfriend, and Merapi at 14:46. Merapi and Joysfriend moved away within a few minutes, leaving Poindexter alone with Crooked-fin again. At 17:06 they met up very briefly with five members of the "Crunch-bunch," a group of males. For the last thirty minutes, until we left them at 17:40, Poindexter and Crooked-fin traveled to the edge of a deeper channel, where they began diving, and possibly foraging.

Poindexter's day is not intended to reflect a typical "day in the life"; indeed, there probably is no such thing. Groups may be stable for longer periods, and rest for hours; individuals may forage alone for hours or may spend longer periods in socializing groups. But the six and some hours of Poindexter's activities and group changes clearly illustrate just how dynamic bottlenose dolphin society can be.

Our ability to observe Poindexter and other dolphins depends on weather, water clarity, and how habituated the dolphins are. Even clouds may render clear, calm water opaque. With habituated dolphins, relatively clear water, and occasionally calm weather, observation conditions are rarely better than in Shark Bay. Rubbing and other social interactions can be observed to at least 3 m below the surface on clear, calm days. On windy days, observations are largely limited to surface behaviors. In other locales, murky water may prevent subsurface observations even on calm days. Behaviors that always occur at the surface, such as synchronous surfacing, can be recorded on a consistent basis. Behaviors that occur largely out of view underwater are not possible to record reliably. Even if the duration of a behavior is substantial, such as a several-minute petting bout, an observer might see it only in brief intermittent glimpses. Some important behaviors, such as intromission, are very difficult to see at all.

half-weight coefficient of association (COA) (Cairns and Schwager 1987; for rationale see Smolker et al. 1992), in which the association between two individuals A and B = $2N_{ab}/N_a + N_b$, in which N_{ab} is the number of times A and B were found in the same party and N_a and N_b are the total number of party sightings for A and B. We multiply this coefficient by 100, as in Smolker et al. (1992), yielding COAs ranging from 0 for two individuals that are never found in the same group to 100 for two individuals that are always sighted together.

Before we compare results from the two sites, we offer the caveat that definitions of group size differed between Sarasota and Shark Bay. In the Sarasota study, all individuals clustered in an area (up to but typically much less than 100 m), engaged in similar activities, or moving in the same direction were included in the group, but in Shark Bay, association data were limited to resting, traveling, and socializing groups, and inclusion required that individuals be within 10 m of any other individual in the group. These differences may tend to offset each other, as in Shark Bay individuals often spread out more than 10 m apart during foraging, a behavioral category that was excluded because the ephemeral nature of feeding associations rendered them difficult to document (Smolker et al. 1992).

Female-Female Relationships

The association patterns of females in Shark Bay and in Sarasota are similar in three respects (Wells et al. 1987; Wells 1991c; Smolker et al. 1992). First, females have a large network of associates and are linked to most other females in an area either through mutual associates or occasional occurrence in the same subgroup. Second, within this extensive social web, most females associate most strongly with a subset of other females in "bands"; and third, a minority of females at both sites do not belong to any particular band.

In Sarasota, females in bands of thirteen, seven, and two shared COAs greater than 30, and core ranges centered on shallow areas such as a small bay, a sound, and a river mouth (Wells et al. 1987). Five females in Sarasota did not associate preferentially with particular bands. The habitat in Shark Bay is not nearly so subdivided as in Sarasota, but bandlike association patterns are still apparent for some females, though not for others (Smolker et al. 1992). Over twenty-five years of observation in Sarasota suggest that bands may maintain their basic structure for many years, but can change over time as the female composition of the community changes (R. S. Wells, unpublished data). At least eight females have given birth in their natal bands.

Within bands, females with calves of similar age tend to associate with each other, as do females without calves (Wells et al. 1987).

Limited genetic evidence supports the long-term observations in Sarasota in suggesting that at least some females in bands are closely related (Duffield and Wells 1991). Five females of the thirteen-member "Palma Sola" band share a unusual chromosome marker that has not been found in sixteen other adult or subadult females that reside in Sarasota Bay or in forty-nine other dolphins that live in adjacent or outside areas. On the other hand, mtDNA analysis indicates that multiple generations of more than one maternal lineage may be present in a band (Duffield and Wells 1991).

One of the most striking features of female-female relationships in Shark Bay and Sarasota is the variation among females in "sociability." Simply, some females tend to be rather solitary, while others are usually found in groups (Smolker et al. 1992; Wells et al. 1987; Mann et al. 1999). As a class, females in Shark Bay were found alone more often than males (Smolker et al. 1992).

Why females differ in sociability is unclear, but may be related to feeding strategies. "Sponge-carriers" in Shark Bay spend most of their time foraging and are rarely found in social groups (Smolker et al. 1997). Sponge carrying may be a low-yield, solitary strategy that requires nearly constant foraging, leaving little time for associating in parties.

A solitary lifestyle may carry a high price. In Sarasota, five females do not belong to particular female bands. One of the most solitary of these, "Hannah," was the only female that did not share a COA of at least 20 with any other female. Hannah spent most of her time near structures such as bridges in regions used less frequently by other dolphins, foraging alone or in the brief company of small numbers of other individuals. In 1989 Hannah's five-month-old calf was attacked by a shark; the calf died within two weeks (Wells 1991a). In general, calf survivorship is related to group size and stability in Sarasota; females raising their offspring within bands have a significantly higher probability of successfully rearing their calves than do non-band members (Wells 1991b, 1993b).

In addition to protection from sharks, females may enjoy other benefits from bonds with other females. Females may benefit by cooperating against harassing males (Connor et al. 1992a, Richards 1996). Somewhat surprising is the lack of observations of females forming cooperative alliances against other females. We will address the socioecology of female-female bonds more extensively below.

RICHARD C. CONNOR, RANDALL S. WELLS, JANET MANN, AND ANDREW J. READ

Male-Male Relationships

Male bottlenose dolphins in Shark Bay exhibit two levels of alliance formation (Connor et al. 1992a; fig. 4.12). First, males in pairs and trios, "first-order alliances," cooperate to form coercively maintained consortships with individual females (Connor et al. 1992a,b, 1996). Second, teams of two or more alliances, "second-order alliances," cooperate in attempts to take female consorts from other alliances or to defend against such attacks (Connor et al. 1992a,b).

Bonds between males in some Shark Bay first-order alliances are extremely strong, with association coefficients ranging from 70 to 100, and stable, having been observed for periods ranging up to thirteen years (Connor et al. 1992b, 1999). Each pair or trio maintained moderately strong associations (COAs of 20–60) with one or two other pairs or trios (Smolker et al. 1992; Connor et al. 1992b). In striking contrast, first-order alliance bonds between males in one unusually large second-order alliance of fourteen individuals were highly unstable, with individuals switching alliance partners often, but always within the group of fourteen (Connor et al. 1999; fig. 4.13). During 1995–1997, males in this "super-alliance" herded females with five to eleven other members of the group and participated in four to eleven different alliance combinations. While it was rare to observe more than two stable alliances together in a social group, ten or more males were present in nearly a quarter of the social groups containing members of the super-alliance (Connor et al. 1999).

Males in Sarasota also form strongly bonded pairs (Wells 1991c; Wells et al. 1987); however, cooperative associations between alliances similar to those in Shark Bay have not been identified to date. Male pairs crystallize at sexual maturity in Sarasota and have been observed to last as long as twenty years, until the loss of one of the pair members (Wells 1991b,c; R. S. Wells, unpublished data). In Shark Bay, two new stable trios were formed when a single male joined an existing pair. Single males have not been observed to join existing pairs in Sarasota, but in four cases, a male whose partner died formed a tight bond with another male.

Alliance formation may not be as prevalent in Sarasota as in Shark Bay (Smolker et al. 1992). First, there appear to be more solitary adult males in Sarasota than in Shark Bay, and no long-term adult male trios have been recorded (Wells 1991c). In 1987, eighteen males in Connor's study in Shark Bay associated in four trios and three pairs. Fifteen of these individuals were well speckled and likely mature (one trio included two non-speckled males and one pair included one non-speckled male). Although alliance relationships within the super-alliance were labile, the vast majority of consortships involved trios rather than pairs (95% of 100 consortships: Connor et al. 1999). Solitary males are rare in Shark Bay compared with Sarasota, and appeared limited to old males who had lost their partners (Smolker et al. 1992). After 1987, two of the eighteen males became solitary for one year each (1989 and 1991), after their alliance partners disappeared and before they disappeared. Thus, there is a suggestion that adult males in

Figure 4.12. A "second-order" alliance in Shark Bay composed of one trio and one pair of males. (Photograph by Richard Connor.)

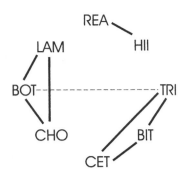

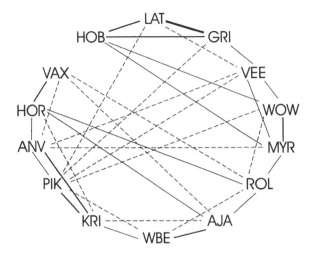

Figure 4.13. Contrasting patterns of alliance formation are represented by sociograms depicting the strength of alliance associations among *(top)* males in three stable alliances who formed second-order alliances together and *(bottom)* males in the fourteen-member super-alliance. Data were restricted to observations of males in an alliance with a female consort. Dotted lines represent the weakest associations (COA of 10–29) and thick bars the strongest (>70). (Reprinted by permission from *Nature* 397:571–72, 1999, Macmillan Magazines Ltd.)

Shark Bay tend to form more triplets and travel alone less often than adult males in Sarasota. Second, joint attacks by two or more alliances against one or two other alliances over access to a female have not been observed in Sarasota. Single alliances have been observed attacking single alliances on rare occasions.

The strongest contrast to Shark Bay may be found in an isolated population of about a hundred identified individuals in the Moray Firth, Scotland. The sex of most individuals has not been determined in the Moray Firth, but a four-year study has failed to reveal any high-level associations between dolphins of adult size that are in the range of the association coefficients found between males in Shark Bay and Sarasota (Wilson et al. 1992). Association coefficients were calculated for eighty-four clearly recog-

nizable individuals. Only eight COAs were above 50 (using the simple ratio index: Ginsberg and Young 1992), and these were between calves and adults and thus likely mother-calf pairs. No COAs greater than 46 were observed for any pair of adults. The average group size in the Moray Firth was 6.3 individuals (Wilson et al. 1993), which excludes the possibility that the differences result from differences in activity patterns (e.g., individuals spending more time alone foraging).

The evolutionary basis for cooperation between males in Shark Bay and Sarasota may include nepotism and/or some form of mutualism (Connor 1995b). In Sarasota, observations and genetic testing indicate that the members of male pairs tend not to be closely related, having different mothers and fathers (R. S. Wells, unpublished data). Given low female reproductive rates and the near ubiquity of male alliance formation in Shark Bay, it is unlikely that most alliance partners could be maternal siblings or even maternal first cousins. It is even less likely that males that share second-order alliance relationships are mostly close relatives. The super-alliance represents the largest second-order alliance known, but shifts in second-order alliances over time mean that even males in stable first-order alliances may have a large number (up to ten) of second-order alliance relationships.

Observations suggest that males sometimes participate in attacks on rival alliances to help another alliance obtain a female rather than to obtain the female for themselves (as in savanna baboons: Bercovich 1988; Noë 1992). Altogether, in only one of 350 consortships did an alliance consort with two females simultaneously. However, in three "thefts" observed in Shark Bay, one of the two attacking alliances already had a female, who remained with them before, during, and after the attack. In two of these cases the alliance that already had the female appeared to be recruited by the other alliance to participate in the theft (Connor et al. 1992a).

Why should one alliance help another acquire a female? Connor et al. (1992a) suggested that either reciprocal altruism or pseudo-reciprocity could be operating. In reciprocal altruism individuals exchange altruistic acts, but in pseudo-reciprocity altruism is performed to increase the chance of receiving benefits that are a by-product of selfish behavior of the recipient (see Connor 1986, chap. 8, this volume). In our present example, pseudo-reciprocity could operate if the assisting alliance were dominant to the alliance they assisted, so that, even if they did not herd the female themselves, they would still be able to mate with her when the alliances were together. Reciprocity would require that the assisting alliance be repaid by the other alliance. We cannot distinguish between these hypotheses presently.

It is important to stress that relationships between alliance partners may be simultaneously cooperative and competitive. Indications of conflict are found in both levels of alliance formation in Shark Bay. There is typically an "odd-male-out" in stable trios, consistent across years, whose COA with the other two males is lower than they share with each other (Smolker et al. 1992). The odd-male-out also participates less in the synchronous surfacing that commonly occurs between alliance partners (R. S. Connor, unpublished data). In one group of three males, relationships were unstable, and only two males consorted with a female together at a given time. The identity of the odd-male-out not only shifted in this case, but the frequency of shifts showed a strong seasonal component (Connor and Smolker 1995). The frequent partner switching in first-order alliances in the super-alliance may be a mechanism by which cooperative relationships are maintained within the larger group (Connor et al. 1999).

Conflict between alliances is suggested by the shifting allegiances of the stable pair RH, which has been in existence for thirteen years. In 1985, R and H were members of two different but very closely associating pairs and trios TBC and CBL were strongly bonded (table 4.3). In 1986, one of the partners of R and H was attacked by the other three and the trio CBL. After both of their partners disappeared in 1986, R and H formed a very tight pair and began associating with CBL and TBC, but not at the same time. This "triangle" continued until C and L vanished in May 1989, and the remaining male from that trio, B, joined RH as a distinct odd-male-out. RHB continued to associate with TBC until TBC disappeared after 1989. RHB then began traveling with the pair LP. This association lasted until 1994, when B and P formed a pair, apparently at the expense of L. Through 1999, L is an infrequent associate of his former partner P, and has not acquired a new adult alliance partner, but instead travels mostly alone or in female or juvenile groups, and is sometimes paired with an immature male.

We predict that as genetic information on relatedness

and paternity emerges, the picture will be one of variation, and that this variation will prove useful not only for testing the relative importance of kinship in male-male cooperation, but may provide an interesting application of reproductive skew and concession models (Clutton-Brock 1998; Connor et al., chap. 10, this volume). Males in some alliances are more strongly bonded than males in others. Thus, the COAs between alliance members might be correlated somehow to their genetic relatedness. However, cooperation with an unrelated male may be a better option than traveling alone (e.g., Packer and Pusey 1982; Packer et al. 1991).

Male-Female Relationships

Associations between males and females were tied strongly to female reproductive state in Sarasota and Shark Bay. Associations between particular females and males were much higher during years when a female was cycling than during years when she was pregnant (Wells et al. 1987; Smolker et al. 1992). Adult males and females associated much more often during the mating season in Sarasota, where Wells et al. (1987) recognized two patterns of male-female association. Solitary males ranged in a smaller area heavily used by particular female bands and were observed more often in association with receptive females. Males in pairs ranged far beyond the areas favored by resident female groups and were rarely observed with receptive females. In contrast, associations between individual cycling females and male pairs and trios in Shark Bay were extensive, and were particularly noticeable for males in the most tightly bonded pairs and trios (Smolker et al. 1992).

Not surprisingly, not only the frequency but also the nature of male-female interactions is strongly affected by the female's reproductive state. In Shark Bay, males in pairs and trios form consortships with single females that may last anywhere from a few minutes to over a month (Connor et al. 1992a,b, 1996). Many consortships were maintained by coercion, or "herding," and often began with a chase, aggression, and displays by the males. The males may either capture a female when she is alone or isolate her from a group of females. Females may try to escape from the males by rapidly bolting away from them. A consortship ends when the female leaves the males, either via a successful escape or simply by swimming away with no apparent reaction from the males.

Males occasionally produce a distinctive, low-frequency "pop" vocalization that induces their female consort to remain close (Connor and Smolker 1996), as do escalated threats such as head jerks, charges, and overt aggression

Table 4.3. Coefficients of association among three alliances in Shark Bay during a five-year period (1985–1989)

	1985	1986	1987	1988	1989
CBL & TBC	65–68	55–71	9–16	7–14	—
CBL & RH	10	15–19	34–38	45–48	—
TBC & RH	14–15	16–27	31–42	31–39	51–56

such as hitting (Connor et al. 1992b, 1996). During consortships the males typically travel abreast behind the female or flank her on either side and slightly behind (fig. 4.14). Aggression is not observed during some consortships, suggesting that coercion may not be important in some cases. However, once a coercive consortship is established, aggression may occur infrequently; one consortship was observed for eleven hours over two days before aggression was seen (Connor et al. 1996). The males and the female need not synchronize their activities during consortships; it is not uncommon to find the female foraging and the males just following behind her. Females were consorted much more often during years when they were not pregnant than when they were pregnant (Connor et al. 1992a, 1996). Consorting is the only documented mating strategy in Shark Bay, but other less obvious strategies may exist (Connor et al. 1996).

The picture in Sarasota is somewhat different. Both solitary males and pairs of males are often observed in the company of receptive females, and both have been identified as sires of Sarasota calves (Wells et al. 1987; Duffield et al. 1994; R. S. Wells, unpublished data). Pairs of males are commonly observed flanking or following females during the breeding season in Sarasota. However, agonistic interactions are observed infrequently during these associations, so it is not yet clear whether such consortships are coerced, as in Shark Bay (Connor et al. 1992b, 1996), or reflect a form of mate guarding (Moors 1997). Male-female asso-

ciations are not limited to within-community interactions. In many cases, pairs of males are observed with females who are not residents of the Sarasota community. About 40% of the calves born to Sarasota females during the 1980s were sired by non-Sarasota males (Duffield et al. 1994).

Although males do not usually associate strongly with particular females except when they are receptive, mixed-sex groups are not uncommon in Shark Bay. About 50% of the groups recorded in Shark Bay contained both males and females (excluding calves) (Smolker et al. 1992; Mann et al. 2000). Only 31% of the groups recorded in Sarasota were of mixed sex (Wells et al. 1987). In focal follows of mothers with newborn calves, adult and subadult males were present only an average of 6% of the time. However, very old and subadult males accounted for 85% of the petting bouts that the mother engaged in (outside of those with her infant). Males and females may forage together, engage in occasional affiliative contact, and often travel and rest together. Such mixed-sex parties provide an opportunity for the development of male-female relationships beyond those exhibited in consortships (Mann and Smuts 1999; see also Connor et al. 1996).

If males and females continue to use part of their natal range as adults, then males may continue to maintain social bonds with their female kin. Given the low rate of association between the sexes, however, it is doubtful that bonds with adult kin of the opposite sex play an important role

Figure 4.14. Two male bottlenose dolphins in Shark Bay flank a female *(center)* consort. (Photograph by Richard Connor.)

RICHARD C. CONNOR, RANDALL S. WELLS, JANET MANN, AND ANDREW J. READ

in the mating strategies of male or female bottlenose dolphins, as they may in killer whales (Connor et al., chap. 10, this volume). Wells's (1991a) observation of a fourteen-year-old male briefly associating with his mother and newborn brother suggest, however, that appropriately timed associations, such as when the mother has a vulnerable new offspring, may be significant.

Mother-Calf Relationships and Development

Not surprisingly, calves associate strongly with their mothers for their first few years; COAs remained near 100 for most mother-calf pairs during the calf's first three years in both Sarasota and Shark Bay (Wells et al. 1987; Smolker et al. 1992). The majority of calves are weaned at age three or four (Mann et al. 2000), which is considered to be the typical age of separation in Sarasota as well (Wells et al. 1987; Wells 1993b). A sharp decline in the mother-calf COA is associated with the mother becoming pregnant in some, but not all, cases (Wells et al. 1987; Mann et al. 1999). A mother may actively reduce the close association with her calf before the next calf is born. Maternal aggression may occasionally play a role, as suggested by the appearance of extensive tooth-rakes on one three-and-a-half-year-old female calf immediately prior to separation, and on a four-and-a-half-year-old male who continued to swim alongside his mother for seventeen months after she gave birth again (Wells et al. 1987).

Paradoxically, bottlenose dolphin calves exhibit precocious locomotion, but prolonged dependence on their mothers. Shark Bay females are virtually intolerant of separations from their calves in the presence of others during the first week of life, and will chase and retrieve the calf frequently (Mann and Smuts 1998, 1999). The calf has a strong following response from birth, but tends to follow any rapidly moving object in front of it. Other females and calves appear to capitalize on this tendency, and occasionally attempt to bolt with the infant (Mann and Smuts 1998). Mothers sometimes chase and threaten the would-be kidnapper with charges, jaw claps, and head jerks. Maternal chases wane by the end of the first week, and the animal who attempted to bolt with the infant during the first week may escort the calf away from the mother during the second week without incident. Mann and Smuts (1998) suggest that the chases in the first week and the tolerance of newborn escorts during the second week are indicative of an imprinting period. Allomaternal care is discussed in greater detail by Whitehead and Mann in chapter 9 of this volume.

The newborn period is characterized by echelon swimming (infant close, nearly parallel, swimming just above the mother's midline), respiratory synchrony, nearly constant contact, and frequent rubbing with the mother (Mann and Smuts 1999; fig. 4.15). All of these behaviors decline over the first two months, as infants become more independent and begin associating with other individuals. Infants increase the time spent swimming in infant position (fig. 4.16), in which they have access to the mammary slits, over the course of two months (from 0% during the first week to an average of 18% by the second month: Mann and Smuts 1999), and maintain this position approximately 30–45% of the time until weaning (Mann 1997; see fig. 9.4). Infants continue to spend more time farther away from their mothers as they age, but some infants rarely venture 20 m or more from their mothers for several years (Mann 1997).

Newborn development is rapid. In the first two weeks, most of the infant's social interactions occur with the mother, and involve primarily rubbing (infant-initiated) on the mother's head region. Mothers pet their infants as early as the first week, but infants do not pet their mothers for the first time until they are at least into the second or third month of life. By three weeks of age, infants produce a diverse repertoire of displays (chin slaps, spyhops, jaw claps, belly slaps, etc.); sometimes the infant displays with no one nearby (closer than 5 m), but most displays occur with social partners close by, typically other infants or nulliparous females. Playful chases, including those with role exchanges (infant chases partner, then partner turns and chases infant, then the sequence is repeated) occur within the first month of life, even involving the exchange of objects, such as sea grass (Mann and Smuts 1999). While social behavior develops rapidly, most of it involves locomotor play, and there is little body contact, outside of rubbing with the mother. Socio-sexual play and petting are uncommon until the latter part of the first year (J. Mann, unpublished data). Foraging skills develop slowly. It appears to take several months of practice before the infant actually catches a fish and years before it can be nutritionally independent (J. Mann, unpublished data). Infant time spent foraging increases steadily with age (averaging approximately 30% by the infant's third birthday), while social behavior increases rapidly during the first six months of life, and levels off to about 12% of the time (see fig. 9.4).

Although Shark Bay newborns are, on average, in larger groups than older calves (6.3 vs. 4.5–5.5, Mann et al. 2000), newborns and their mothers spend 44% of their time alone with each other (SD = 7.24, range = 39–53%, $N = 4$ mother-infant pairs) (Mann and Smuts 1999). Sayigh (1992) found that one of four Sarasota newborn-mother pairs spent 50% of their time alone. The other

Figure 4.15. Three weeks old, the infant male "Cookie" stills swims alongside his mother's head in echelon position. (Photograph by Richard Connor.)

Figure 4.16. Calf surfaces, breaking infant position only to breathe. (Photograph by Janet Mann.)

three pairs spent less than 20% of their time alone. Average group size for Sarasota mother-infant pairs was between three and seven individuals (Sayigh 1992). Mothers and infants associate with thirty or more different animals during the first few months, primarily adult and subadult females and offspring (Mann and Smuts 1999). "Female" groups, consisting of mother, infant, and one or more adult females (Mann and Smuts 1999), accounted for approximately 45% of the time during the first few months of life, and about 19% of the time during the infant's third year. Adult males associate with mothers and offspring approximately 6% of the time from birth to weaning (Mann and Smuts 1993), and interactions between infants and adult males are rare. Group size tends to decline as calves age (over the course of three or four years), both in Shark Bay (Mann et al. 2000) and in Sarasota (Wells et al. 1987), but group sizes were much larger in Sarasota than in Shark Bay.

During the first few years of life, mothers and calves spend the majority of their time together (within 10 m of each other), mostly with the calf in infant position, but calves also spend 3–30% of their time away from their mothers. Infants are often alone during these separations, at distances of up to several hundred meters. Given the possible risk to isolated calves from sharks, one must ask why such separations occur at all. Mann and Smuts suggest that early separations between Shark Bay mothers and infants are necessary because of the dolphins' foraging behavior. Separations, even during the newborn period, occur primarily when the mother is foraging (Mann and Smuts 1998). Mothers do not cache or carry their offspring during foraging, and accelerated chases during hunting enforce repeated separations with the infant.

During the majority of separations, infants are, like their mothers, foraging alone. However, during roughly 20% of their separations, infants are with other immatures (Mann 1997; Mann and Smuts 1993; J. Mann, unpublished data). Although infants form long-term bonds with other infants, we still know little about the nature of such relationships. One pair of male infants, Coo and Smo, developed a preferential association independent of their mothers during the first year of life, which continued well after they were both orphaned at 43–44 months of age. They often traveled several hundred meters away from their mothers to join, pet, play, hunt, or just swim together. After their mothers disappeared, they spent most of their time together, occasionally engaging in prolonged mutual petting (up to twenty-two minutes, the longest bout recorded in our Shark Bay sample) (J. Mann, unpublished data).

Clearly, learning to negotiate separations and reunions with the mother and specific others is one of the critical tasks of infancy, and sound must play a critical role. Mother-calf communication and signature whistle development is one of the best-studied aspects of vocal communication in *Tursiops* (e.g., Caldwell et al. 1990; Sayigh et al. 1990, 1995a; Smolker et al. 1993). Yet, we still know little about how signature whistles develop and how they are used in dolphin communication. We do know that *Tursiops* produce frequency-modulated whistles, ranging from 4 to at least 20 kHz, that become individually distinctive and stereotyped by the end of the first year (reviewed in Caldwell et al. 1990). Using the focal follow method in Shark Bay, Smolker, Mann, and Smuts (1993) examined the use of whistles during naturally occurring contexts. They found that infants often whistled during separations from the mother, and that the distance (not the duration) of the separation predicted best the probability of calf whistling. (Whistles were attributed to the calf using behavioral cues, such as an increase in whistle amplitude as the calf approached the hydrophone.) However, whistles were not used to maintain continuous acoustic contact (i.e., contact calls). Infants were silent for the first half of the separation, but were likely to whistle during the latter half of the separation, before returning to the mother, suggesting that they used whistles to indicate an interest in reuniting with the mother. Mothers rarely whistled; the infant seemed to already "know" where the mother was, and would whistle and then swim toward her. However, due to equipment limitations, distant whistles may not have been detected.

Under stressful (involuntary) separations, both mothers and infants whistle more often. For example, during herding by males, mothers often become separated from their infants, and both whistle almost continuously (Smolker et al. 1993). Mothers and calves in Sarasota also whistle almost continuously when they are briefly removed from visual contact during temporary capture-release efforts by researchers (Sayigh et al. 1990). In captivity, when mothers and calves are involuntarily separated, both whistle at very high rates (Caldwell et al. 1990). Whistles are likely to convey the identity, location, and motivational state of the whistler.

Vocal learning appears to play an important role in signature whistle development (reviewed in Janik and Slater 1997; Sayigh et al. 1995a; Tyack 1986b). Some male calves in Sarasota produce signature whistles similar to those of their mothers, whereas female calves tend to produce whistles distinct from those of their mothers (Sayigh et al. 1990, 1995b), and may use acoustic models to develop their signatures in the early months (Sayigh 1992). Because females continue to associate with related females after weaning (Wells et al. 1987), selection may favor the development of highly individualized signatures among females to avoid confusing identity (Sayigh et al. 1995b). Because males do not continue to associate consistently with their mothers after weaning, there may not be selection against males adopting some features of their mother's signature (Sayigh et al. 1995a).

Developing Relationships in a Fission-Fusion Society

The several years bottlenose dolphins spend with their mothers may be a critical time in which they learn foraging techniques and social skills they will need as independent juveniles and adults. During this time striking variation between individual calves in both acoustics and behavior emerges. These individual differences, such as tendency to separate from the mother (Mann 1997), remain stable over time. This variation predicts infant survivorship: calves who separate more often are more likely to survive to weaning than calves who stay close to the mother in "infant position" (J. Mann, unpublished data). However, it may not be that calf separations contribute to infant survivorship per se; rather, calves who are in good health may be able to better afford forays away from the mother than calves in poor health. Habitat differences also help explain these patterns. Shallow-water habitats (less than 4 m) correlate with calf survivorship, although adventurous calves are no more likely to live in shallow water than deep (J. Mann, unpublished data). Calves who separate more often may gain more hunting and social experience than those who stay close to their mothers (Mann 1997). This is consistent with findings that female reproductive success in Shark Bay is predicted by water depth, not group size (Mann et al. 2000). The underlying causes of such individual differences are unknown, but it is likely that maternal experience, sociability of the mother, and ecological (i.e., predator and fish distribution) and other factors interact with calf characteristics to shape the course of development.

In Sarasota, juvenile males and females associate extensively with male and female peers after they leave their mothers (Wells et al. 1987). In Shark Bay and Sarasota, male-male bonding seems to be under way even before male calves leave their mothers (J. Mann, unpublished data; R. S. Wells, unpublished data). Females in Sarasota often, and perhaps always, return to join their natal bands. In both Shark Bay and Sarasota, females develop and maintain social relationships with a large number of females outside their bands as well.

Immature dolphins spend considerable time in large groups engaging in social play. We suspect that immature animals are cultivating important social relationships in these groups and practicing social skills for the present and future. Thus, these social interactions may have short-term and long-term benefits. In Shark Bay, we often observe apparent "play" herding by immature individuals. For example, during one hour of intense socializing, a group of twelve immature dolphins frequently split up into two or three smaller groups. In each of the smaller groups, two or three individuals performed nearly continuous chasing, sexual behavior, displays, and in one case pops toward another individual. Three aspects of the behavior in this group differed from adult consort herding associations. First, the composition of these smaller groups was highly labile: when the groups came together, the males would form new "makeshift" alliances in pursuit of different females. During the one-hour observation period, one nulliparous female was observed engaging in herding-like associations with three different "alliances." Second, one individual of unknown sex was observed in both herded and herder roles during the socializing. Third, when the socializing ended, the herding-like behaviors ceased and the group dispersed. Herding by adults continues through all activities, including foraging, traveling, and resting.

Even though both males and females spend up to ten years as juveniles/adolescents (post-weaning, prereproductive), over twice the duration of infancy, this period remains virtually unexplored. As in primates, this period represents roughly 25% of their post-weaning lives. Even after several decades of detailed primate behavioral studies, juveniles have been largely neglected (Pereira and Fairbanks 1993). Why animals should delay reproduction is not well understood. But the trade-offs between growth and reproduction may provide a selective advantage to those who delay reproduction in favor of reaching a certain body size (e.g., Pereira 1993; Pagel and Harvey 1993; Whitehead 1994). Once delayed maturation is favored, extended learning of social and foraging skills during this phenotypic limbo may further reinforce this life history strategy (Pagel and Harvey 1993).

Mating Strategies

In most mammals, most or all parental investment is performed by the female (Clutton-Brock 1989). Reproduction in females is thus limited by resources that can be translated into offspring, whereas reproduction in non-investing males is limited by access to females (Trivers 1972). Where males vary in some trait that may benefit the female's offspring, females are expected to discriminate among possible mates so long as such choice is not too costly (e.g., Trivers 1972; Andersson 1994). Female bottlenose dolphins invest heavily in each offspring and should attend strongly to such differences among males. Male bottlenose dolphins, like other mammals that do not participate in parental care, are predicted to channel their reproductive effort toward obtaining matings.

Our understanding of male and female mating strategies in bottlenose dolphins has advanced considerably in the last decade but is still very far from complete. We will begin our consideration of bottlenose dolphin mating strategies with the well-documented use of coercion by males in Shark Bay to establish and maintain consortships with individual females. We include here some observations from Sarasota on consortships between male pairs and individual females, although it is not yet clear whether such consortships are maintained with coercion in Sarasota.

Male coercion of females was elevated by Smuts and Smuts (1993) to the status of a third kind of sexual selection, to stand beside mate competition and mate choice. Male coercion represents mating strategies in conflict. We will address the nature of this conflict and how it affects the relationship between female choice and male mating strategies. We will conclude this section by examining evidence for noncoercive male mating strategies in bottlenose dolphins.

What kind of male mating strategy is herding? Connor et al. (1996) suggest three possibilities. First, males might be able to force copulation with females during herding, either directly or by threatening aggression. Second, herding might be a "war of attrition" (Parker 1970; Connor et al. 1992a,b, 1996) in which females mate with their male consorts because they are prevented from mating with preferred males. A war of attrition could also be waged by males simply by guarding females rather than also herding them, as may occur in Sarasota (Moors 1997). Third, females prefer to mate with multiple males. In the "war of attrition" and multiple male mating models, herding can be viewed as much as a strategy to prevent females from mating with rival males as to prevent rival males from having access to the female.

Herding may be a strategy to monopolize females, but it is not an entirely successful one. Females in Sarasota and Shark Bay were observed in consortships with up to twelve or more males during the season when they conceived (Connor et al. 1996; Moors 1997). One way females gain access to other males is to escape from the males they are with (Connor et al. 1992a). However, females may be unable to escape, or males may effectively guard them. Moors (1997) found that receptive females in Sarasota were typically accompanied by a pair of males for an extended period at the probable time of conception. However, by increasing the number of times they cycle, females may effectively "escape" from males that monopolize them during nonconceptive cycles (Connor et al. 1996).

Two features of consort formation are puzzling. First, Shark Bay females that conceive during the breeding season (September–December) are often herded extensively during the previous winter (May–August), when only two of forty (5%) recorded conceptions occurred (based on a twelve-month gestation period) (Connor et al. 1996). Thus, even though females sometimes conceive in the winter, the extensive herding during this period suggests that nonconceptive cycles are more prevalent at this time, possibly as part of a female strategy to mate with multiple males and confuse paternity (Connor et al. 1996). Second, although the majority of consortships last less than a week, which accords with the five- to seven-day period of rising estrogen levels reported from captivity (Schroeder 1990), others in both Sarasota and Shark Bay last considerably longer—up to five weeks (Connor et al. 1996; Moors 1997). Borrowing an explanation offered for a similar phenomenon in chimpanzees (Goodall 1986), Connor et al. (1996) suggest that males may initiate consortships prior to a female's period of maximal attractiveness to avoid being preempted by other alliances.

How can females choose mates in this system? The elaborate synchronous displays that males perform around female consorts have obvious parallels with coordinated displays in birds (e.g., Trainer and McDonald 1993). In these species females may use male displays as a choice criterion (e.g., Foster 1981). Female choice in this context may be used to discriminate between males within or between alliances, which is compatible with both the war of attrition and the multiple male mating models of the herding strategy (Connor et al. 1996). However, herding by males may impose significant constraints on female choice by limiting a female's ability to seek out favored males. Multiple cycling may allow females to retain choice of mating partners: if they are monopolized by undesirable males during one cycle, they may be able to mate with preferred males during the next cycle (Connor et al. 1996). However, multiple cycling may also be a strategy to reduce the risk of infanticide, as has been suggested for multiple-male mating in some primates with a similar mating system (Hrdy 1979; Wrangham 1980b; Struhsaker and Leland 1985; Hiraiwa-Hasegawa 1987). The infanticide hypothesis received support with the recent discovery of apparent infant killing by conspecifics in the Moray Firth and in coastal Virginia (Patterson et al. 1998; Dunn et al. 1998).

Are there male mating strategies besides herding in bottlenose dolphins? Mating strategies based on affiliative interactions between females and males in or outside of consortships may be important in bottlenose dolphins, but remain largely undescribed. This may be due to the greater visibility of agonistic interactions compared with affiliative behavior (Connor et al. 1996). If other strategies exist, and individuals vary in the extent to which they use particular strategies, we might expect herding to be more prevalent among a subset of males. In fact, data suggest that herding may be an age-related strategy in Shark Bay. During 1987–1989, Connor et al. (1992a,b) recorded fifty-eight herding events involving males. The seven focal alliances in the study were responsible for forty-nine of those events. The herding totals for each alliance are listed in table 4.4, together with the total number of party sightings, focal follows, and follow hours for members of each alliance during 1987–1989.

Alliances A, B, and C participated in more herding events than alliances E, F, and G. Herding events were recorded during all phases of offshore data collection, including party sightings and focal individual follows. While exact controls for observation time are lacking (e.g., alliances may travel with each other for varying periods during follows), it is clear that differences in opportunity to observe herding cannot account for the large differences in herding among the alliances.

Alliances D and G were the only two alliances with members that had no ventral speckling. Alliance G, for which we documented no consortships, had two members that did not have any ventral speckling through 1989. Until August 1988, alliance D consisted of one male with ventral speckling and one that did not develop speckles until

1989 (suggesting he was a subadult). In August 1988 a third male with ventral speckles began associating with alliance D. We documented the first cases of herding by alliance D in 1988, coinciding with a substantial increase in cross-alliance herding by members of alliance D with the three provisioned males (Connor et al. 1992b).

All members of alliances A, B, C, E, and F were speckled. Alliances E and F often associated with each other as second-order alliance partners, as did alliances A, B, and C (Connor et al. 1992a,b; Smolker et al. 1992). All members of alliances E and F were thought to be old males because (1) they were all very heavily speckled, (2) another heavily speckled male who associated with them before becoming emaciated and disappearing in 1987 had extensive tooth wear, and (3) they exhibited general behavioral lethargy (R. C. Connor, unpublished data). Thus, some alliances herd females more than others, and our limited data suggest that herding may be age-related and primarily the domain of mature, but not old, males.

Alternative mating strategies such as "friendships" between male and female savanna baboons (*Papio anubis:* Smuts 1985) might have parallels in dolphins. For example, males in alliance E were often seen petting with one old female even during late pregnancy (Connor et al. 1996). Old or juvenile, but not "prime" males, have been observed petting with mothers of new calves (Mann and Smuts 1998). Observations in which one alliance appeared to prevent females from being herded or harassed by other males also hint at differentiated male-female relationships (Connor et al. 1996). Wells et al. (1987) found that resident single males in Sarasota were found more often in the company of receptive females than were male pairs, but subsequent observations and the results of preliminary genetic testing indicate that both single males and male pairs are successfully breeding within the community (Duffield et al. 1994). Further focal behavioral observations combined with genetic paternity testing should help us uncover the complete range of mating strategies in Shark Bay and Sarasota.

Table 4.4. Herding events recorded for each of seven alliances of non-provisioned males at Shark Bay, 1987–1989

Alliance	A	B	C	D	E	F	G
Herding events	16	16	10	6	1	0	0
Sightings[a]	78	81	57	64	57	25	107
Follows	24	29	21	20	20	7	22
Follow hours	87	113	79	81	62	21	71

Note: Also given are the total number of sightings (occurrences in resting, traveling, or socializing groups: Smolker et al. 1992) of the most commonly sighted member of the alliance and the number and total duration (hours) of focal follows on alliance members. Two of the three members of alliance C disappeared during May 1989, after which the third member joined alliance A. Sighting and focal totals for this individual were thus included with alliance C prior to 1 May 1989 and with alliance A afterward. Only one member of the pair of males in alliance F was a focal male, thus the low number of follows for this alliance. These data indicate that observation time does not account for differences in herding among alliances.

The Ecology of Social Relationships in Bottlenose Dolphins

The pattern of male and female social relationships should be determined by ecological and social influences (Wrangham 1987). As mentioned above, female reproduction is limited more by access to resources while males are limited more by access to mates (Trivers 1972; Bradbury and Vehrencamp 1977; Wrangham 1980a). Female grouping patterns and relationships are, therefore, usually considered to be more directly affected by ecological parameters, such as predation and resource distribution, including conflicts with other females over access to resources (Wrangham 1980a). The distribution of females then determines the options available to males, whose strategies (e.g., infanticide) can, in turn, influence female grouping patterns (e.g., Wrangham 1987; van Schaik 1989; van Hoof and van Schaik 1994). Further, ecological parameters (e.g., predators) can affect the relative value of alternative mating strategies available to males (Wrangham 1987). Such complex interactions between ecological and social influences on female and male strategies makes difficult our task of understanding the evolution of any species' social structure.

The ubiquitous fission-fusion grouping pattern of bottlenose dolphins is a good place to begin our inquiry into ecological influences on their social organization. The patterns of grouping and sex-specific bonds in bottlenose dolphins are rare in mammals, but strikingly similar to those in common chimpanzees and spider monkeys (Wrangham and Smuts 1980; Symington 1990; Smolker et al. 1992), as was noted in some of the earliest bottlenose dolphin field studies (Irvine and Wells 1972; Saayman and Taylor 1973; Würsig 1978). Würsig (1978) suggested that a similar patchy and irregular distribution of food might explain the similarities in dolphin and chimpanzee grouping patterns. Now, over twenty years later, can we make this comparison more precise?

A patchy and irregular distribution of food is common to many species with widely varying social organizations. For example, the distribution of food for the bonobo, *Pan paniscus*, is also similar to that of common chimpanzees in being patchy and irregular, yet the social organization of bonobos is quite different from that of chimpanzees. Strong female-female and extended mother-son bonds, and occasionally strong male-male bonds, characterize bonobo social relationships (Kano 1992; Furuichi 1989; Ihobe 1992), in contrast to strong male-male and weak female-female and mother-son bonds in common chimpanzees (Goodall 1986). Wrangham (1986) suggested that the difference in social organization between bonobos and chimpanzees reflects an adaptation in the chimpanzee to prolonged periods when individuals are forced to travel in small parties or alone, due either to seasonal shortages of fruit trees or less reliance on evenly distributed terrestrial herbaceous vegetation (Wrangham 1986; Chapman et al. 1994).

We can make a crude analogy between chimpanzee fruit trees and the schools of fish dolphins sometimes feed on. Fish schools are rich patches that can feed a number of individuals. If large fish schools were consistently available,

we might expect dolphins to travel in consistently large groups (Wells et al. 1980), which could lead to stronger female-female bonds. In addition to feeding on schooling fish, however, bottlenose dolphins often feed on solitary prey. Dolphins in both Sarasota and Shark Bay, for example, spread out to feed on prey that live near or on the bottom. In Shark Bay, mothers with newborns, who might benefit most from associating in larger groups, spend approximately 44% of their time alone, and all maternal hunting observed was solitary (Mann and Smuts 1998). We suggest that the fission-fusion grouping pattern in bottlenose dolphins reflects a requirement to spread out to reduce feeding competition.

For animals that must disperse to feed, the cost of forming groups and maintaining social bonds is also affected by the cost of locomotion (Connor, chap. 8, this volume). A low cost of locomotion reduces the cost of spreading out, and individuals can more easily afford to regroup between feeding bouts. In comparison to terrestrial mammals such as chimpanzees, bottlenose dolphins enjoy a low cost of locomotion (Connor, chap. 8, this volume). Thus, with a distribution of food identical to that found in common chimpanzees, we might still expect a different pattern of social relationships in bottlenose dolphins.

The benefits of social relationships must be considered as well as the costs (Connor, chap. 8, this volume). Different species and different populations of the same species may differ in the degree to which individuals are threatened by predators or conspecifics. The threat from conspecifics, for example, will vary with the rate at which individuals encounter each other in competitive circumstances, which may correlate with, among other factors, population density (Connor et al., chap. 10, this volume). As population density increases, individuals are more likely to encounter conspecifics when foraging alone, which will increase the benefit of traveling with an ally.

In the following sections we ask why bottlenose dolphins form social bonds, what benefits they accrue, and why individuals and populations vary in their patterns of bond formation.

Female-Female Relationships

The ecological basis for female-female bonds in bottlenose dolphins is not well understood. Up to four generations of kin associate in female "bands" in Sarasota, and similar moderately strong female-female bonds have been reported in Shark Bay. Additionally, one of the striking features of female-female associations in Shark Bay and Sarasota is the large number of occasional associates (Wells et al. 1987; Smolker et al. 1992). These lower-level associations are not merely females congregating at resources, as associations in Shark Bay are limited to traveling, resting, and socializing groups. Thus models for female-female bonds must take into account not only persistent associations with a relatively small number of associates that may be mostly kin, but the relatively large number of associates that are unlikely to be kin. Here we consider as causes for female-female bonds vigilance, cooperative defense against predators, male harassment, and competition with other females over resources.

The high incidence of shark bite scarring at several sites suggests that vigilance and predator defense might be the primary selective agents behind female-female bonds. Calves are likely the most vulnerable individuals, but dolphins do not "outgrow" their predators as some large ungulates such as elephants might. The fission-fusion nature of dolphin grouping likely relates to the distribution of food: individuals most often disperse in small groups or singly to forage efficiently. As in other mammals, efficient foraging and predator defense may often come into conflict in bottlenose dolphins (e.g, see Mann and Smuts 1999). Female foraging requirements shift during the reproductive cycle. In the eastern tropical Pacific, pregnant spotted dolphins (*S. attenuata*, but possibly closely related to *Tursiops*: Perrin et al. 1987) fed predominantly on squid while lactating females fed more on flying fish (Bernard and Hohn 1989). Lactating females may switch to feeding near the surface to avoid leaving their vulnerable calves during deep, prolonged dives for squid (Bernard and Hohn 1989). Different foraging requirements may exacerbate conflicts between foraging and predator avoidance for females in different stages of the reproductive cycle. Females in phase reproductively might not only have similar predator defense requirements, but similar foraging requirements as well.

Cycling females might also benefit from associating with other cycling females. Connor et al. (1992b) reported two possible cases of females cooperating against males. In one case, six females chased three males who were chasing a (probable) female. The males gave up the chase when the larger group of females closed to within 10 m behind them. None of the six females had a calf, and four had been in consortships with males in the previous four months, indicating that they were cycling. In several other cases, females have moved closer together and more synchronously around the time that male harassment ceased. All of these cases involved younger males, rather than the three alliances of mature males (A, B, C) that were responsible for most herding by non-provisioned males. These preliminary observations suggest that female cooperation is most suc-

cessful against alliances containing younger males. In mammals as diverse as orangutans and sea lions, males that are unable to compete with prime males may employ tactics that include aggression toward females (Mitani 1985; Campagna et al. 1988). Young males might impose costs on female bottlenose dolphins, for example, by attempting to herd them during periods when they would not be attractive to prime males.

Together, these observations suggest that females may benefit by associating with other females in similar reproductive states who share similar requirements for food and defense against males and predators. Opportunities to affiliate with kin in similar reproductive states may be limited given the high and unpredictable mortality of calves. We suggest that the advantages to females of associating with other females in similar reproductive states may be the primary reason females associate (and possibly maintain bonds) with a larger number of unrelated females (see de Waal and Luttrell 1986).

We have not observed females cooperating against other females to defend resources. This is somewhat surprising, given the extensive observations of resource (or territory) defense by female-bonded mammals (Wrangham 1980a; Isbell 1991). The highly mobile nature of prey in an aquatic medium may render them less defensible than terrestrial resources. Another possibility is that dolphins don't defend home ranges or individual food patches, but instead defend whatever part of their range they are in at a given time. Wilson et al. (1997) observed seasonal movements by dolphins in the Moray Firth that were stratified. Individuals that commonly inhabited the area during the winter months were found deeper in the Moray Firth in summer, when their winter locations were occupied by dolphins moving in from areas outside of the inner Moray Firth. Wilson et al. suggested that "area defense" by dolphins could account for the lack of mixing between dolphins that use the inner Moray Firth seasonally and those that are present year-round.

Another important issue is variation among females in their tendency to be found in groups. If differences are not age-related, then solitary females might be those who do not have close relatives to associate with or those whose foraging specialization does not favor grouping (e.g., sponge carrying). Since overall group size does not predict female reproductive success in Shark Bay (Mann et al. 2000), a more detailed look at the context of grouping is needed to understand female bonds.

Variation in female-female bonds across sites is not as evident as variation in male-male bonds. The highest association coefficient between two adults in the Moray Firth

RICHARD C. CONNOR, RANDALL S. WELLS, JANET MANN, AND ANDREW J. READ

is in the same range as the strong female-female COAs in Sarasota and Shark Bay. If predators are important for female grouping and predators are relatively rare, then we would expect to see lower female-female COAs in the Moray Firth. A site with low predator pressure and lack of herding by males could be used to test our hypothesis that bonds with large numbers of nonrelatives are maintained to ensure that females have a selection of social partners appropriate to their reproductive state.

It would also be of interest to investigate the relationship between resource distribution, group size, and the pattern of female-female bonds. Coastal California bottlenose dolphins may provide an opportunity to explore this issue. Although small group sizes occur commonly (22% of groups contained 2–5 individuals) among California bottlenose dolphins, the average group size (19.8) is larger than reported from most nearshore populations (Defran and Weller 1999). If this difference reflects a greater preponderance of large prey schools, and individual females in California associate on a more consistent basis, the pattern of female-female bonds, and thus possibly that of males as well, may differ from that in populations with smaller group sizes.

Male-Male Relationships

Males form alliances of two or three individuals in Shark Bay. Trios are rare in Sarasota, where males more often form pairs or travel alone. Association data from the Moray Firth show that males do not form long-term stable alliances, but do not rule out the possibility that males form temporary coalitions. How can we understand this variation? To answer this question, we need to understand why males form alliances—and why they don't.

Coalition and alliance formation in male birds and mammals is universally associated with competition over access to females, either directly or indirectly (e.g., resource defense or rank acquisition) (Connor et al., chap. 10, this volume). Cooperative herding of females suggests that male alliance formation in bottlenose dolphins might be driven at least partially by conflicts with females. Further, predators and resources, the primary determinants of group formation (Alexander 1974), may also factor into the relative success of solitary versus group-based alternative mating strategies.

To understanding variation in bottlenose dolphin alliance formation, research exploring ecological, social, and genetic factors is needed. Here we attempt to focus these future studies by framing hypotheses that can potentially explain variation in alliance formation within and across habitats. We consider separately the roles of male-male and male-female interactions, as well as the effects of predators and resource distribution, on male alliance formation.

Individual variation. The first problem we address is why we find some males in alliances and some males alone within a study site. Wells et al. (1987) suggested that solitary males with home ranges that coincide extensively with particular female bands might be engaging in a different mating strategy than males in alliances. In theory, there are three mechanisms that might maintain alternative mating strategies (see Gross 1996): (1) frequency-dependent selection acting on genetic polymorphisms for alternative phenotypes that yield equal fitness payoffs (e.g., "fight" or "sneak"); (2) genetic monomorphism in which alternative tactics are expressed in each individual according to probabilities determined by frequency-dependent selection; and (3) genetic monomorphism in which alternative tactics yield different payoffs, but individuals select the tactic that, based on their condition or "status," maximizes their fitness. Of these three possibilities, the first and second are considered rare, so we will focus here on the third category, "conditional strategies."

Within the framework of a conditional strategy with the alternative tactics "travel alone" and "form an alliance," being solitary may have a lower or a higher fitness payoff than participating in an alliance. First, solitary males may be "making the best of a bad job"—for example, the old males in Shark Bay that became solitary after their alliance partners died. Other males in Shark Bay have successfully established themselves in new alliances after losing their partners. In Sarasota, four males that lost alliance partners joined other solitary males to form new pairs. These observations suggest that males in alliances prefer to be in alliances rather than solitary (probably the rule in Shark Bay), but do not rule out the possibility that traveling alone is the preferred option for some prime males in Sarasota.

Following Noë's (1994) model of alliance formation in male baboons, pair formation in Sarasota might be based on dominance rank and limited to mid- or low-ranking males that can monopolize females only in alliances. Solitary males might be high-ranking individuals that can monopolize females by themselves or low-ranking males that cannot monopolize females either alone or in an alliance. This model can also be extended to incorporate variation in the size of alliances within habitats. It predicts that lower-ranking males in Shark Bay would be more likely to form triplets and higher-ranking males pairs, and provides a potential explanation for the larger groups formed by young males that are old enough to reproduce, but not old and large enough to compete with prime males. The super-alliance is clearly competitive with other prime males in Shark Bay (Connor et al. 1999), and has won the several encounters observed, but it is possible that the males in this group would not be competitive operating in pairs of alliances.

Interpopulation variation. Differences between populations in male alliance formation cannot be attributed to age-specific reproductive tactics or rank-related strategies. It is also unlikely that availability of kin could differ that markedly between habitats. Here we consider three factors that might help us to explain interpopulation differences in alliance formation: rate of male-male interactions, predation risk, and resource distribution. A fourth factor, the quality of male-female interactions, is considered in the following section.

Alliance formation may correlate with the rate of interaction between rival males over access to receptive females (Connor et al., chap. 10, this volume). We consider four possible parameters that might vary sufficiently between populations to produce differences in the rate of male-male interaction: population density, operational sex ratio, day range, and "sphere of communication."

Differences in population density should correlate with encounter rates and alliance formation in dolphins. Shark Bay appears to have a higher density of dolphins than the Moray Firth. Wilson et. al. (1992) identified about 100 individuals during 92 surveys along 40 km of coast. During 172 survey days in Shark Bay (1994–1996), 378 dolphins were identified along approximately 20 km of coast and ranging several kilometers offshore, but the actual area surveyed may be at least as large as in the Moray Firth, where researchers followed a predetermined survey route. A more careful comparison controlling for survey method and effort is needed. This comparison also ignores irregularities in the distribution of dolphins in a habitat where local densities can vary significantly (Wilson et al. 1997).

At a given population density, a greater day range would increase the interaction rate of males. The dolphins in the Moray Firth may have larger daily ranges than in Shark Bay, perhaps due to their larger size (1.5–2 times longer), or a different resource distribution. B. Wilson (personal communication) reports that individuals in the Moray Firth routinely travel distances of 20 km along the coast during several hours of observation. Over a similar time period it would be unusual for dolphins in Shark Bay to move more than 10 km.

Smolker et al. (1992) suggested that because a male's sphere of communication is restricted in subdivided habitats, a solo strategy might be favored relative to more open habitats. Sarasota is much more subdivided than Shark Bay or even neighboring Tampa Bay, offering a possible test of this hypothesis. However, the most direct route to testing this hypothesis is to follow males and measure the rate at which they encounter and avoid other males.

Another factor that may influence the rate of male-male interaction over females is the operational sex ratio. The level of male-male competition among bottlenose dolphins will be affected by the proportion of adult females that become receptive in any given year. Females in Shark Bay typically begin cycling during the third year of nursing a calf (Connor et al. 1996). Interbirth intervals of less than four years are unusual, while in Sarasota two-year intervals have been recorded and three-year intervals are not uncommon. A longer interbirth interval would increase the operational sex ratio and produce more intense competition for receptive females, and perhaps favor larger alliances in Shark Bay. However, the proportion of females that are cycling will be strongly influenced by infant mortality before weaning, which is higher in Shark Bay (Mann et al. 2000).

Wilson et al. (1993) suggested that food distribution or the low number of predators might account for the lack of stable first-order alliances in the Moray Firth. Male alliances in Shark Bay clearly serve to increase males' access to females; how can predators or resources influence alliance formation? It is possible that traveling alone is a better male mating strategy than alliance formation, but is disfavored in high-risk habitats because solitary males are more vulnerable to predation. This hypothesis is weakened somewhat by observations of solitary males in groups of mostly females. However, given that males must still travel between groups to locate receptive females, two or three males may be able to detect or deter potential predators well enough that the lower predation risk males experience in alliances outweighs the loss of mating opportunities. To test for the influence of predators, populations could be compared for the proportion of individuals with shark bite scars. Wells (1991b) found that males in pairs acquired more shark bite scars than did single males, but pair members tended to live longer than single males, suggesting that pair members were able to recover from injuries that singles were not, presumably due to the protection afforded by a partner.

Similarly, differences in prey distribution may affect the cost of grouping in different habitats and favor or disfavor alliance formation. Males in some Shark Bay alliances routinely spread out to forage near the bottom, possibly on solitary prey that are evenly distributed but widespread. It is possible that resources are even more widely distributed in the Moray Firth, to the extent that alliance formation

RICHARD C. CONNOR, RANDALL S. WELLS, JANET MANN, AND ANDREW J. READ

124

is disfavored altogether. This possibility seems unlikely, given the similarity in average group size in the Moray Firth (6.3), Shark Bay (4.8), and Sarasota (4.8 and 7.0).

Male-Female Relationships

Why do male bottlenose dolphins cooperate to herd females? One possibility is suggested by observations of two males chasing a female in Shark Bay. Rather than pursuing the female directly from behind, the males angle out and pursue her from either side, effectively closing the distance if the female subsequently turns in either direction. A male pursuing a female alone might have a much more difficult time catching her, suggesting that a female's ability to maneuver underwater may have favored cooperative herding partnerships among males in the shallow waters of Shark Bay. The greater depth of the Moray Firth provides what is essentially another possible escape route for females, likely making the task of controlling female movements even more difficult for males.

If female maneuverability is critical for the formation of cooperative male relationships, we would not expect to find populations in which single males herd females. By extension, it seems reasonable to suggest that male alliances should be more important in populations with greater sexual size dimorphism (unless the disparity in maneuverability is such that even cooperation is no longer effective), but this does not appear to be the case.

On the other hand, observations of single males herding females in populations with greater sexual size dimorphism would not eliminate the male-female conflict hypothesis for alliance formation in Shark Bay. Physical size, rather than maneuverability, might be the critical factor in herding. Shark Bay males, at best only slightly larger than females, may require assistance in coercing a female. Relatively larger males in other populations, such as Sarasota, where males have allometrically larger propulsion features and weapons (Tolley et al. 1995), may be able to coerce females without help.

Observations of interactions that suggest affiliative relationships between male and female bottlenose dolphins were discussed above, and remain an open area for future investigation.

Future Directions

The bottlenose dolphin, because of its wide distribution in the warm coastal waters favored by our own species, will likely continue to be one of the most frequently studied cetaceans. The differences in social and foraging behavior reported here suggest that new study sites, especially in new habitats, will be highly productive. With appropriate meth-ods, and if maintained for a sufficient length of time, new studies will add to the diversity of behavior and social relationships reported in this chapter and provide tests of the tentative hypotheses offered here. We suspect that bottlenose dolphin researchers, in the not so distant future, will be contributing to an edited volume titled *Bottlenose Dolphin Cultures,* after the recently published volume on chimpanzees (Wrangham et al. 1994). In addition to highlighting differences in social relationships and foraging behavior, we expect that such a volume will describe cultural differences in vocal behavior (see Tyack, chap. 11, this volume).

Conservation

Their proximity to human activity exposes bottlenose dolphins to numerous threats, such as pollution, boat traffic, interactions with commercial and recreational fisheries (through entanglement and competition for fish), habitat alteration, and other forms of human interference with their lives (Mann et al. 1995; Wells 1993a; Wells and Scott 1997). Although not endangered at the genus level, specific populations (e.g., the Moray Firth, mid-Atlantic seaboard of the United States: Wang et al. 1994) might be considered to be threatened.

In addition to being of academic interest, knowledge of behavior and social relationships is important to the conservation of bottlenose dolphins. The discovery of individual foraging specializations in bottlenose dolphins suggests that they might be more vulnerable to habitat alteration than would be indicated by a population-level analysis of diet. Our understanding of bottlenose dolphins at many coastal sites supports the consideration of social relationships along with ranging patterns to provide for the identification of population units that can form the basis for management. In Florida, patterns of social association facilitate the partitioning of continuously distributed resident bottlenose dolphins into biologically meaningful, geographically based management units (e.g., "communities": Wells 1986). The frequency of associations between individuals inhabiting overlapping or adjacent ranges can help to define the extent of population units; this approach is strengthened when ranging patterns and social associations are correlated with data on genetic patterns of distribution (Wells 1986; Duffield and Wells 1991; Duffield et al. 1991). Because many threats to coastal bottlenose dolphins are geographically localized (e.g., fisheries, point-source pollution, coastal development), definition of management units makes it possible to relate specific threats to particular population units, allowing evaluation of potential impacts and providing a focus for mitigation efforts.

On the other hand, researchers must be careful not to extrapolate between study areas. Distinct communities have not been detected in an area of Shark Bay that is twice as large as the home range for the Sarasota community. The continuous mosaic of overlapping home ranges that is found in Shark Bay suggests that localized threats in some populations of bottlenose dolphins may have ripple effects that are not impeded by any kind of community boundary. Foraging specializations, widely documented in bottlenose dolphins, may limit an individual's ability to compensate for damage to even small areas. Further, for wide-ranging dolphins such as those in California, the effects of localized threats may be felt hundreds of miles away.

With long-term studies of individually recognized dolphins, it becomes possible to monitor trends in population dynamics and health (Wells and Scott 1990; Wells et al. 1995a,b). In addition to the obvious direct conservation applications for the dolphins themselves, monitoring the health of bottlenose dolphin populations may provide an important indicator of the overall health of their habitat and the ecosystem it supports. Studies of large "charismatic" terrestrial mammals such as chimpanzees have been important in recognizing and challenging threats to their habitats (Tyack et al., Epilogue, this volume). With their widespread distribution and high public profile, bottlenose dolphins are well suited to a role as environmental sentinels for coastal habitats.

Richard C. Connor, Randall S. Wells, Janet Mann, and Andrew J. Read

5

THE KILLER WHALE

Foraging Specializations and Group Hunting

ROBIN W. BAIRD

AMONG THE cetaceans, killer whales (*Orcinus orca:* fig. 5.1) exhibit several unusual features related to social organization, ecology, and behavior. Perhaps the most striking of these features are their dispersal patterns. For two so-called *resident* populations in the eastern North Pacific (numbering about two hundred and eighty-nine individuals, respectively, as of 1998), neither sex has been recorded dispersing (neither locational nor social dispersal—cf. Isbell and van Vuren 1996) from their natal groups over a twenty-one-year period, nor has immigration into a group been recorded (Bigg et al. 1990b). Natal philopatry by both sexes has not been positively documented for any other population of cetacean or, for that matter, for any other species of mammal. Individuals from *resident* populations feed on fish, and individuals from another, sympatric population, *transients,* specialize on marine mammal prey. These two forms were termed *resident* and *transient* based on research in the 1970s (Bigg et al. 1976; Bigg 1982). These names have been shown subsequently not to be particularly descriptive of the movement patterns and site fidelity of the two forms (Guinet 1990; Baird et al. 1992), but they have been retained as the common names. One apparent consequence of the differences in diet is the differences in dispersal patterns. *Resident* killer whales travel in long-term stable groups made up of several maternal lineages (Bigg et al. 1990b). However, among *transients,* all female offspring and all but one male offspring seem to disperse from their maternal groups (social dispersal), but dispersing offspring continue to use their natal range (locational philopatry) (Baird 1994). Besides the difference in diet, *resident* and *transient* killer whales also differ in behavior, acoustics, morphology, pigmentation patterns, and genetics (table 5.1; fig. 5.2).

Foraging specializations appear to occur in killer whale populations elsewhere, though research efforts have been generally insufficient to determine whether, as in the North Pacific populations, sympatric forms specialize on different prey types. Individuals of some Southern Ocean populations feed almost exclusively on marine mammals (Hoelzel 1991a; Guinet 1991b; Baird et al. 1992). Predation on marine mammals makes the study of foraging behavior easier than perhaps for any other species of cetacean because the prey are large, breathe at the surface, and are often captured close to, or even on, shore. Several interesting findings have come from these studies, including apparent teaching of hunting skills to offspring (Lopez and Lopez 1985; Guinet 1991a; Hoelzel 1991a) and a strong relationship between group size and foraging success in one population (Baird and Dill 1996). Other studies have demonstrated features for killer whales that appear to be unusual among mammals in general, including the presence of some females who live twenty or more years beyond the birth of their last known offspring (Olesiuk et al. 1990) and the occurrence of group-specific vocal dialects within killer whale populations (Ford and Fisher 1983; Strager 1995). In this chapter I review the general biology of killer whales, focusing on several longitudinal studies on free-ranging animals. Information on feeding habits, ranging patterns, and social organization and behavior is emphasized.

Taxonomy

The killer whale is a member of the suborder Odontoceti, family Delphinidae, subfamily Orcininae. The member-

Figure 5.1. An adult female *transient* killer whale porpoising off Victoria, British Columbia. (Photograph by Robin W. Baird.)

ship of the subfamily Orcininae has not been agreed upon, however (Heyning and Dahlheim 1988); some include only the genus *Orcinus* and the false killer whale *(Pseudorca crassidens)*, others include only *Orcinus* and the Irrawaddy dolphin *(Orcaella brevirostris)*, while still others include the pilot whales *(Globicephala* spp.) and pygmy killer whale *(Feresa attenuata)* in addition to *Orcinus, Orcaella,* and *Pseudorca.* At present, only one species in the genus *Orcinus* is generally recognized, *O. orca.* Several authors have suggested recently that there is more than one species in the genus. Based on animals killed in Soviet whaling operations in the Antarctic, Mikhalev et al. (1981) and Berzin and Vladimirov (1983) independently described new species in the genus (*O. nanus* and *O. glacialis* respectively), both of which seem to refer to the same population, with a smaller average body size than *O. orca* (Heyning and Dahlheim 1988). Berzin and Vladimirov (1983) also noted differences in morphology, group size, and diet, with the species

they described, *O. glacialis,* feeding primarily on fish and being found in large groups (150 to 200 individuals), while *O. orca* fed mainly on marine mammals and was found in smaller groups (10 to 15 individuals). While the proposed new species differed somewhat in habitat from *O. orca* (*O. glacialis* was found in among the ice floes while *O. orca* was found in open water), their ranges, based as well on the data from Mikhalev et al. (1981), did overlap. Neither of these new designations have been generally accepted (Perrin 1982; Heyning and Dahlheim 1988). Based primarily on behavioral and ecological data, Baird (1994) has argued that the two sympatric forms in the nearshore waters of the eastern North Pacific (one feeding on fish and the other feeding on marine mammals) are reproductively isolated and, thus, should be considered separate species. However, the use of a morphological species concept by cetacean taxonomists, rather than a reproductive one, makes such a suggestion unlikely to be accepted (Baird 1994).

Table 5.1. Differing characteristics of *resident* and *transient* killer whales in nearshore waters of the eastern North Pacific

Morphology/genetics

    Shape of dorsal fin (Bigg et al. 1987; Bain 1989)

    Saddle patch pigmentation (Baird and Stacey 1988)

    Possibly eye patch pigmentation (D. Ellifrit, personal communication, cited in Baird 1994)

    Mitochondrial and nuclear DNA (Stevens et al. 1989; Hoelzel and Dover 1991a; Hoelzel et al. 1998a)

Behavior/ecology

    Diet (Bigg et al. 1987, 1990a; Morton 1990; Baird and Dill 1996)

    Travel patterns/habitat use (Heimlich-Boran 1988; Morton 1990; Baird and Dill 1995)

    Respiration patterns (Morton 1990)

    Vocalizations (Ford and Hubbard-Morton 1990; Morton 1990)

    Echolocation (Barrett-Lennard et al. 1996a)

    Amplitude of exhalations (Baird et al. 1992; Baird 1994)

    Possibly diving patterns (Baird 1994)

    Group size (Bigg et al. 1987; Morton 1990; Baird and Dill 1996)

    Pattern and extent of natal philopatry (Bigg et al. 1987; Baird and Dill 1996)

    Seasonal occurrence (Guinet 1990; Morton 1990; Baird and Dill 1995)

    Geographic range (Bigg et al. 1987)

Figure 5.2. Adult male *resident (top)* and *transient (bottom)* killer whales from British Columbia, showing differences in dorsal fin shape (typically *resident* dorsal fins are more rounded). Saddle patch pigmentation patterns also differ, with some *residents* (including the example shown here) having black pigmentation intruding into the grayish-white saddle patch, while *transients* do not (see Baird and Stacey 1988). (Photographs by Robin W. Baird.)

Ecology and Social Organization

Field Studies

Prior to the 1970s, little was known of the biology of killer whales. Occasional observations of stranded animals or animals taken in whaling operations, as well as anecdotal observations of the behavior of animals in the wild and a few captive individuals, were recorded (see, for example, Carl 1946; Backus 1961; Caldwell and Brown 1964; Newman and McGeer 1966; Rice 1968). Detailed field studies on killer whales were first initiated in the early 1970s and have been undertaken in several nearshore locations around the world (Bigg et al. 1990b; Guinet 1990; Lopez and Lopez 1985; Lyrholm 1988). These studies have relied on photo-identification of individuals, based on distinctive acquired and congenital characteristics of the dorsal fin and the saddle patch, a lightly pigmented area just below and behind the dorsal fin. All killer whales in an area can be identified with high-quality photographs. Focal sampling is relatively easy with killer whales because their morphological characteristics (large size, distinctive dorsal fin and saddle patch, sexual dimorphism) allow for rapid individual identifica-

tion. Photo-identification of killer whales has been utilized in many localities (e.g., Sigurjónsson et al. 1988; Black et al. 1997), but in four areas in particular, long-term studies of identified individuals have been undertaken (e.g., Bigg et al. 1990b; Hoelzel 1991a; Guinet 1991b; Bisther and Vongraven 1995; Similä et al. 1996). I discuss each of the four areas in detail below.

British Columbia, Washington, and Alaska. Spong et al. (1970) initiated the first field study of killer whales in British Columbia, but this early work did not incorporate photo-identification of individuals. Its primary goal was to examine the behaviors of wild individuals, following on Spong's earlier research on captive animals (e.g., White et al.

1971). Population studies utilizing photo-identification, initiated in response to a live-capture fishery (in which animals were taken for captivity), began both in British Columbia and in the state of Washington in 1973 and have continued to date (Balcomb et al. 1982; Bigg 1982; Bigg et al. 1990b; Olesiuk et al. 1990). The proximity of these areas to large human population centers and the ease of working in the calm inshore waters attracted numerous investigators to work there. Most of the research focused on *residents*, which were predictably found in particular locations at certain times of the year. In early years, research efforts were focused in two specific areas: Johnstone Strait, off northeastern Vancouver Island, and Haro Strait, an area that straddles the border between the United States and Canada off the southern tip of Vancouver Island. These areas were the focus of virtually all research on killer whales in British Columbia from 1973 through the late 1980s. Only in the early 1990s did research begin to encompass a broader geographic range within British Columbia, with further work in the Strait of Juan de Fuca, off the west coast of Vancouver Island, and in areas in northern British Columbia and the Queen Charlotte Islands. Studies in British Columbia and Washington have provided the most detailed information on killer whales anywhere in the world. The early photo-identification studies provided a basis for numerous behavioral and ecological studies. Some of these later studies were initiated in the late 1970s, but most were started in the 1980s, covering a diverse range of topics, including foraging and feeding (J. R. Heimlich-Boran 1986; Felleman et al. 1991; Nichol and Shackleton 1996), vocal traditions and vocal behavior (Hoelzel and Osborne 1986; Ford 1989, 1990), habitat use (Heimlich-Boran 1988; Hoelzel 1993), life history and population dynamics (Olesiuk et al. 1990; Brault and Caswell 1993), and social behavior and social structure (Haenel 1986; S. L. Heimlich-Boran 1986; Bigg et al. 1990b; Rose 1992).

Residents in British Columbia appear to be divided into two geographic populations, one termed the "northern" *residents*, usually found off northern Vancouver Island and in southeastern Alaska (Dahlheim et al. 1997), and the other termed the "southern" *residents*, usually found off southern Vancouver Island and in Washington. Considerably less research has been undertaken on the *transient* population. Commercial production of a catalog of known individuals, both *residents* and *transients*, in 1987 (Bigg et al. 1987), greatly facilitated matching of individuals between areas and comparisons between studies. An updated version containing only individuals from the *resident* population was produced in 1994 (Ford et al. 1994), and a catalog of all individuals found in southeastern Alaskan waters, including both *resident* and *transient* individuals found farther south in British Columbia, was produced in 1997 (Dahlheim et al. 1997). Comparisons between areas for *transients* documented since the 1987 catalog are undertaken primarily through the exchange of photographs by researchers working at various institutions (e.g., Baird 1994; Dahlheim et al. 1997).

Largely in response to the potential for live-capture fisheries in Alaska, research on killer whale population size and dynamics was begun in Prince William Sound in 1983 and in southeastern Alaska in 1984 (Leatherwood et al. 1984, 1990) and has continued to date (Matkin and Saulitis 1994; Matkin et al. 1994; Dahlheim et al. 1997). Production of catalogs of known individuals from these areas (Heise et al. 1991; Dahlheim et al. 1997) has facilitated studies within each area as well as comparisons between them.

Argentina (Patagonia). Lopez and Lopez (1985) initiated behavioral research on killer whales in the nearshore areas of Punta Norte, Peninsula Valdés, southern Argentina, in 1975. Southern elephant seals *(Mirounga leonina)* and southern sea lions *(Otaria byronia)* utilize beaches in that area and give birth to their pups in September and October and from January to February respectively. These are also the time periods when killer whales are most commonly seen swimming close to the coast, preying mainly on pups and juveniles, but sometimes on adults. In this area, intentional stranding on shore to capture hauled-out prey is the primary foraging tactic (Lopez and Lopez 1985; Hoelzel 1991a). Research efforts at this site have been largely land-based examinations of the behavior of individuals foraging in nearshore areas.

Norway. Photo-identification studies in Norway were initiated in 1983, largely in response to management needs for population estimates in the area, where extensive hunting of killer whales had been conducted for over forty years (Lyrholm 1988). Boat-based photo-identification studies have been conducted both around the Lofoten and Vesteralen islands, in northern Norway, and around the More area in southwestern Norway (Lyrholm 1988). Killer whales are most abundant in those areas between October and January, when they follow the movements of herring *(Clupea harengus)* into nearshore areas (Bisther and Vongraven 1995; Similä et al. 1996). Studies on acoustic and feeding behavior have also been possible in the protected waters of enclosed fjords (Similä and Ugarte 1993; Strager 1995; Similä et al. 1996).

Crozet Archipelago. Killer whales are the most frequently seen cetacean in the nearshore waters of Possession Island, Crozet Archipelago, located in the southwestern Indian

Ocean. They can be found in that region year-round and can be seen daily from October to December at Possession Island, where they feed on a wide variety of prey, including elephant seals, penguins, fish, and other whales (Guinet 1991b). Guinet studied this population from 1987 through 1990, primarily to examine behavior and ecology and to make comparisons with other killer whale populations (Guinet 1991b, 1992). Research in the area has been largely land-based. Marine mammals are the most frequent prey observed taken, and intentional stranding to capture prey hauled out on shore is regularly recorded.

Comparisons between sites and data collection. As noted, methods at these different field sites have varied. Such differences are due both to site-specific conditions and to differing numbers of researchers working at each site. Research in the Crozet Archipelago and in Patagonia has been primarily shore-based and undertaken by a very small number of investigators. Studies in all areas have been largely restricted to small geographic areas for only a few months of the year, when weather conditions are best (British Columbia and Washington) or when killer whales move into nearshore areas (Norway). Seasonal differences in the occurrence of pods that use extremely nearshore areas (within a few hundred meters of shore), combined with pod-specific and seasonal differences in behavior (Baird and Dill 1995), all bias data collected at different sites as well as comparisons between sites.

Encounters with killer whales at shore-based sites are typically limited in duration because of the small geographic area covered. Killer whales are detected visually (Punta Norte) and/or acoustically (Crozet Archipelago) with hydrophones as they move toward these land-based sites. Primarily boat-based research sites, such as British Columbia, Washington, Alaska, and Norway (although there are several land-based studies in British Columbia and Washington running concurrently with boat-based studies), have typically covered wider geographic areas. Detection of killer whales utilizing these areas has relied on a combination of land-based observers, radio reports from commercial fishermen or whale-watching charter operators, hydrophones deployed from shore, and boat-based surveys looking for killer whales. Virtually no research has been undertaken at night in any of these areas, due to the difficulty in tracking and observing killer whales at night. For larger groups (more than ten individuals), boat-based tracking can be undertaken for extended periods of up to ten or twelve hours. For smaller groups (fewer than four individuals), extended tracking can be more difficult. This is especially true for *transient* killer whales, which travel

in smaller groups, dive for longer periods, and follow less predictable routes (Morton 1990; Baird and Dill 1995). For both photo-identification and behavioral studies, killer whales are typically tracked from distances ranging from 10 to several hundred meters. For behavioral studies, most researchers have relied on following particular groups or, less frequently, focal animal sampling.

Distribution and Seasonal Occurrence

Killer whales are cosmopolitan, having been observed in all oceans of the world. However, concentrations generally occur in colder regions and in areas of high productivity (Bigg et al. 1987; Heyning and Dahlheim 1988; Guinet and Jouventin 1990). No clear evidence of seasonal north-south migrations is available. Based on sightings from whaling vessels in the Southern Hemisphere, Mikhalev et al. (1981) described seasonal migrations from high-latitude areas (most south of 50°S) in the summer months to lower-latitude areas (most north of 50°S) in winter. However, no information was presented on potential seasonal biases in effort, and the conclusions were based on densities of whales recorded in particular areas, not on movements of individual animals, so it is difficult to judge the validity of such conclusions (Perrin 1982). In polar areas, the occurrence of killer whales may be limited by the presence of pack ice in the winter months, thus some north-south movements would have to occur (Reeves and Mitchell 1988a). However, a recent sighting of killer whales deep in Antarctic sea ice in winter indicates that not all individuals move away from the poles (Gill and Thiele 1997). The extreme seasonal biases in effort could be partly responsible for the perception that all killer whales move toward lower latitudes in the winter months (Gill and Thiele 1997). Killer whales are present year-round in many areas. Evidence suggests that individuals occupy very large ranges (see, e.g., matches between California, Oregon, British Columbia, and Alaska presented in Black et al. 1997), and the proportion of time spent in different parts of their ranges may vary seasonally.

In the British Columbia and Washington study areas, both *resident* and *transient* killer whales are present year-round. Several authors have suggested that *residents* are rare in the core study areas during the winter months due to the decreased presence of salmon (*Oncorhynchus* spp.), one of their primary prey species (J. R. Heimlich-Boran 1986; Bigg et al. 1987; Nichol and Shackleton 1996). Several seasonal biases in effort are present, however. In general, less effort has been extended in the winter months. Inclement weather conditions and low daylight hours during the winter months also likely decrease the probability of visually detecting killer whales when they are present. Seasonal comparisons of vocalization rates among *resident* killer whales in Johnstone Strait suggest decreased presence during the winter months (Nichol and Shackleton 1996). However, possible biases include a decreased rate of vocalizations by *residents* during the winter months (D. E. Bain, personal communication) and use of travel routes farther from shore during these times (R. W. Baird, personal observation), thus affecting visual or acoustic detection from shore. Methods for measuring salmon abundance have also been indirect, relying on sports fishing catches, commercial catches, and the number of salmon arriving at spawning rivers (J. R. Heimlich-Boran 1986; Nichol and Shackleton 1996). Until these issues are resolved, correlations between the seasonal presence and abundance of killer whales and prey availability remain unclear. During periods when effort is relatively high, however, there is a general correlation between the presence and/or number of killer whales using an area and the relative abundance of salmon (J. R. Heimlich-Boran 1986; Nichol and Shackleton 1996). Such a correlation implies that availability of prey may limit the number of killer whales that use an area. Different northern *resident* pods may correlate with different runs and/or different species of salmon (Nichol and Shackleton 1996), suggesting that pods have different primary foraging areas within their overlapping home ranges and that temporal segregation may occur for some pods that share primary foraging areas.

Seasonal influxes into nearshore areas where pinnipeds are abundant have been noted for killer whales around Marion Island, the Crozet Archipelago, and Punta Norte, Argentina (Condy et al. 1978; Guinet 1992; Hoelzel 1991a). For *transient* killer whales in southern British Columbia, a strong seasonal peak occurs, coinciding with the period when harbor seal *(Phoca vitulina)* pups are being weaned (Baird and Dill 1995). This peak was not due to a general increase in the visitation by *transients* of this area; rather, some pods of *transients* appeared to use this area preferentially during pup weaning while others were seen there regularly year-round. Those pods that used the area year-round also tended to travel farther from shore (typically more than 1 km). The general result was that pods present during fall through spring typically spent more time away from the shoreline; therefore, land-based observers or spotters were less likely to notice them. Due to this seasonal difference in use of nearshore areas, many shore-based studies may be biased when examining seasonal presence (Baird and Dill 1995). Even boat-based studies typically have focused on nearshore areas (within 20 km of

shore); because of this bias, it is unknown where individuals go when not in nearshore areas.

Social Organization

Killer whale groups vary in size from single animals to as many as several hundred individuals (Perrin 1982). However, larger groups appear to be temporary associations of smaller, more stable groups. In all areas where longitudinal studies have been carried out, evidence suggests that there are long-term associations between individuals and limited dispersal from maternal groups (Lopez and Lopez 1985; Bigg et al. 1990b; Guinet 1991b; Similä and Ugarte 1993; Baird 1994; Baird and Dill 1996). Such evidence is most conclusive for the British Columbia and Washington study areas, and this area also has the best data for variability in group size, structure, and stability, showing differences in these characteristics between the sympatric *residents* and *transients*.

Groups of killer whales have been defined and categorized based on spatial associations, synchronization of respirations, acoustic dialects, and coordination of activity. Categorizations of groups and associations have varied between studies. Associations within groups have generally been based on distance between individuals during observations (Rose 1992) or on presence together in the same photographic frames during photo-identification studies (S. L. Heimlich-Boran 1986; Bigg et al. 1990b). Groups have been defined by general presence in an area (S. L. Heimlich-Boran 1986; Bigg et al. 1990b) or as all individuals swimming within 100 m of each other (Lopez and Lopez 1985). Baird (1994) defined group membership for *transients* as when all whales, within visual range of observers, acted in a coordinated manner during an observation period. This definition could include individuals up to a kilometer or more apart, when coordination of activities was made apparent by individuals converging on a single prey item discovered by one member of the group. Such a definition of group membership typically required an extended period of observation; determination of group membership was often not possible for short-duration encounters (for example, those of less than fifteen minutes).

Resident pods, subpods, and matrilineal groups. Over a fifteen-year period, Bigg et al. (1990b) studied the social organization of two populations of *resident* whales (totaling about 260 individuals at the end of the period). Multiple encounters each year with all three pods in one population and most of the pods in the other population allowed for detailed examination of *resident* social organization. Information was collected year-round, though there was a seasonal bias, with most data collected between June and September. *Resident* social organization was classified as a series of progressively larger groups, with each category showing no changes in membership, either seasonal or long-term (Bigg et al. 1990b). The smallest group is termed a matrilineal group or intrapod group. The researchers found that an individual only very rarely separated from its intrapod group for more than a few hours. These groups are of mixed age and sex and range in size from two to nine individuals (mean = 4). They appear to contain a single matriline of from one to four (mean = 3) generations, with both male and female offspring found in association with the oldest female in the group. Subpods comprise one to eleven (mean = 2) matrilineal groups and are defined as matrilineal groups that spend more than 95% of their time traveling together. These matrilineal groups are thought to be more closely related to each other than to other matrilines within the pod (Bigg et al. 1990b). Pods comprise from one to three (mean = 2) subpods and are defined as groups of subpods that travel together more than 50% of the time; pod memberships were determined with repeated observations over a period of years. Average pod size in the two populations combined was approximately twelve individuals (range of three to fifty-nine individuals).

Resident *dispersal.* Evidence for lack of dispersal from natal groups by *residents* is conclusive, based on the long-term stable associations observed, the lack of resightings of any individuals outside of their natal range or away from their natal pod, even with considerable research effort over a twenty-one-year period, and adult survival rates. One hundred and thirty-four individuals were born into the two populations during the duration of the study. Fifty-nine individuals disappeared, and none of those individuals were ever sighted again in the company of other whales (all were thought to have died). Additionally, none of the individuals that have disappeared have been sighted in groups of *resident* killer whales observed in the adjacent waters of southeastern Alaska. Individuals that disappeared ranged in age from newborn calves to old adults, with both sexes represented. Disappearance rates (= presumed mortality rates) did vary with age and sex (Olesiuk et al. 1990), but the individuals that disappeared at the highest rates (calves younger than five years, males older than twenty-five years, females older than fifty years) were those least likely to have dispersed, according to current thought regarding mammalian dispersal patterns. Additionally, if more than a couple of the older individuals that disappeared had actually dispersed and survived, the already high survival rates of adults (see below) would be unrealistically inflated. Concurrent

studies by other investigators on the same populations, an additional six years of data collected subsequent to the study reported by Bigg and his colleagues (Ford et al. 1994), and research on *resident* populations in Alaska (Matkin et al. 1994) also support the lack of dispersal noted.

Cohesion and splitting of resident *pods.* As noted, *resident* pods are defined as groups of subpods that spend more than 50% of their time together over a period of years. Such a definition does not do justice to the wide range of circumstances in which pods can be encountered. Members of one or more subpods can be encountered in extremely close proximity to one another (as many as fifty-nine individuals clustered within less than 100 m of each other) or spread out over several kilometers. On some occasions, subpods may join and leave each other over the space of hours or days, while during other periods, the entire pod can be seen together repeatedly for days on end, with all individuals present. Some evidence also exists of seasonal trends in pod spacing: during the winter months individuals within a pod often appear to be spread out over much larger areas (D. E. Bain, personal communication; R. W. Baird, personal observation). Such variability means that definitions of "pods" in other study areas may require numerous repeated observations of groups, both within and between years. *Resident* pods are thought to form due to the splitting of a single pod into two or more over a period of many years, perhaps decades (Ford 1990; Ford et al. 1994). Based on the "50% rule," two pods of *residents* (one northern and one southern) appear to have split since the British Columbia study began in 1973. No quantitative analysis is available in either case, but in both cases, three subpods were seen together for the majority of sightings in the 1970s, but gradually spent less and less time together during the 1980s, and in the 1990s are usually not seen together (Ford et al. 1994; R. W. Baird, personal observation; D. K. Ellifrit, personal communication). Ford et al. (1994) speculated that the death of the oldest female in the group, from whom many of the individuals are usually descended, can destabilize a pod or subpod and begin the process of pod splitting.

Acoustic clans as a measure of resident *grouping.* Ford's research on *resident* killer whale acoustics demonstrated the existence of stable pod-specific dialects and showed that some pods shared a number of calls (Ford 1990; Ford and Fisher 1983). He suggested that shared calls between pods reflect common ancestry. Taking into account both association patterns and pod-specific vocal repertoires, within British Columbia and Washington, *resident* social organization can be further categorized for groups greater than the level of pods. Acoustic clans can be defined as pods that share one or more calls (Ford 1990). Four acoustic clans have been identified for British Columbia and Washington *residents*. Three of these clans share a common range and regularly associate with each other off northern Vancouver Island, northern British Columbia, and southeastern Alaska (the "northern" *residents*). A fourth clan, whose members have not been observed associating with individuals in the three northern clans (the "southern" *residents*), is usually found off southern Vancouver Island and in Washington. The southern *resident* community contained eighty-nine individuals as of 1998. The northern *resident* community has been reported to contain about two hundred individuals (Ford et al. 1994), but some pods within this community have been seen interacting with *resident* killer whales in southeastern Alaska (Dahlheim et al. 1997), which in turn have been observed interacting with *residents* in Prince William Sound, Alaska (Matkin et al. 1997); thus the population size of northern *residents* is likely much greater.

Resident *interpod and intrapod associations.* Pods within a community are frequently observed associating with one another, particularly during the summer months. The presence of high prey concentrations during these months could simply result in an aggregation of whales in an area of high food availability, or, conversely, it could allow whales to congregate for social purposes without the cost of increased competition for prey. One unusual behavior observed among southern *resident* killer whales when meeting during these periods has been termed a "greeting ceremony" (Osborne 1986). A greeting ceremony occasionally occurs when two or more pods meet after having been separated for more than a day. When this happens, individuals within each pod have been observed in a line abreast formation at the surface, facing the other pod. They approach each other, and at a distance of 10–20 m, they remain motionless for approximately ten to thirty seconds. Both pods have then been observed to submerge and swim toward the other, so that when they resurface, they form tight mixed groups, a common feature of intermingling behavior (Osborne 1986). Social and sexual behavior is frequently observed during greeting ceremonies and in associations of up to a hundred individuals. The increased social and sexual behavior in large associations suggests that pods come together for social interactions when competition for prey is reduced. There is no clear evidence of territoriality between pods within either the northern or southern *resident* communities. If killer whales are not territorial, it would

no doubt relate to their large home ranges (cf. Mitani and Rodman 1979).

Since no dispersal of either sex occurs from *resident* pods, breeding is likely to occur between pods rather than within a pod (although cf. Hoelzel et al. 1998a). The presence of reproductively active females within pods containing no adult males (e.g., Bigg et al. 1987) also suggests that breeding must occur between pods. The increased frequency of sexual behavior in multipod associations also supports this supposition. Copulation between an adult male and an adult female has rarely been positively documented in the wild, however, and genetic data are not yet available to positively confirm that mating occurs between *resident* pods.

It has been erroneously reported in the literature that the northern and southern *resident* communities have nonoverlapping ranges (for example, Bigg et al. 1990b; Felleman et al. 1991). In fact, their ranges overlap by over 120 km on both the east and west coasts of Vancouver Island. Behavioral interactions have not been observed between individuals from the different *resident* communities, although relatively little research has been undertaken in the areas where the populations overlap. Core areas have been identified for each community, and these are separated by about 390 km (2.5 days of travel at 3.5 knots: Bigg 1982). Differences in pigmentation patterns suggest that the communities may be reproductively isolated (Baird and Stacey 1988). Genetic analyses have been undertaken using samples collected from stranded, captive, and free-ranging animals (Stevens et al. 1989; Hoelzel and Dover 1991a; Hoelzel et al. 1998a). Mitochondrial DNA comparisons within the northern and southern *resident* killer whale populations have demonstrated no variability (Hoelzel et al. 1998a). A comparison between the two populations found a small (one base pair) but fixed haplotype difference (Hoelzel et al. 1998a), suggesting that these populations may have arisen due to separate founding events.

Transient *social organization.* Less investigation into the social organization of *transient* killer whales has been undertaken. New adult and/or subadult *transient* individuals are regularly documented in the British Columbia/Washington study area (Baird and Dill 1995). In 1986 only 79 individuals had been identified in the population, while by 1993 a total of 170 individuals had been documented (Bigg et al. 1987; Ford et al. 1994). Some individuals are seen numerous times each year, but the long resighting intervals for others (up to ten years) have made positive recording of dispersal (and, for that matter, of deaths) much more difficult than for *residents*. Using Bigg et al.'s (1990b) 50% association rule to define pods (that is, individuals that travel together more than 50% of the time over a period of years), *transients* clearly associate in distinct pods (Baird and Whitehead 1999). Each pod appears to be equivalent to a single *resident* matrilineal group, with from one to two generations present (Baird 1994; Baird and Dill 1996). Average pod size in Baird and Dill's (1996) study was two individuals, with a range from one to four individuals.

Transient *dispersal.* Unlike *resident* pods, definite social dispersal (Isbell and van Vuren 1996) from *transient* pods has been recorded, although only on two occasions (one of each sex: Bigg et al. 1987; Baird and Dill 1996). The two dispersing individuals were resighted within their natal range but did not associate with their natal group (locational philopatry). Extensive indirect evidence of dispersal also exists: maximum pod size appears to be four individuals, while pods containing only a single individual made up 31% of the pods recorded by Baird and Dill (1996), and *transient* pods usually contain only one adult male and/or one reproductive female. By contrast, the smallest *resident* pod recorded is three individuals, and *resident* pods often contain more than one adult male and/or more than one reproductive female. For *transients,* based on the 50% association rule, it appears that either a male or a female could be considered a pod of size one. However, this is not to say that these individuals are always found alone. All lone individual *transients* documented have been adult or subadult males (Baird 1994). Female *transients* with no offspring that appear to have dispersed from their natal group seem to travel with a variety of other *transient* pods for temporary periods. Male *transients* that appear to have dispersed spend some time alone and some time traveling temporarily with other *transient* pods (in Baird's 1994 study, of five different "lone male" pods, five of the twelve sightings of these individuals involved other pods present, while the remaining seven sightings involved only the single individual).

Transient pods are fairly stable, with some close associations (that is, they are virtually always seen together) between individual *transients* documented over fifteen years or more. Based on both direct and indirect evidence, it appears that female *transients* disperse from their matrilineal group either when they reach sexual maturity or when they give birth to their first calf, effectively forming their own pod (Baird 1994). Females without dependent offspring, as noted, and females with a young calf do not appear to travel alone. Instead, they temporarily associate with a variety of other *transient* pods. Dispersal of male *transients* occurs, but not all males seem to disperse. In one

of the two cases of dispersal recorded, a subadult male left his natal group, which also contained an adult male thought to be his maternal sibling. Based on this observation and the occurrence of, at most, one adult male in a *transient* pod, Baird (1994) suggested that all males other than the first-born disperse before the onset of sexual maturity.

Transient *interpod associations.* All *transient* pods documented in British Columbia that are larger than one individual in size contain an adult female, unlike those off Punta Norte, where a stable group made up of a pair of males has been observed in a study that identified thirty individuals (Lopez and Lopez 1985; Hoelzel 1991a). *Transient* pods often associate with one another; group size of *transients* noted by Baird and Dill (1996) ranged up to fifteen individuals, with a mean group size of about four individuals. No evidence of *transient* communities, as noted for *residents,* has been found, although not all *transient* pods are equally likely to occur in any particular area (Baird and Dill 1995; cf. Nichol and Shackleton 1996). Associations between *transient* pods do not appear to be completely random, however; they depend in part on pod size and the age and sex of all pod members (see below) and in part on the predominant foraging tactics exhibited by the pod. Baird and Dill (1995) demonstrated pod-specific foraging specializations in *transients,* which exhibited two general types of foraging, nearshore and offshore. Pods tended to associate more frequently with others that shared similar foraging tactics than with those that differed in terms of foraging specialization. Such associations may be due in part simply to foraging in similar habitats, but Baird et al. (1992) and Baird and Dill (1995) suggested a possible functional explanation for this pattern: individuals should associate more with others that share their foraging abilities when it is advantageous to forage cooperatively in very small groups.

Interactions between residents *and* transients. Bigg (1982) defined the *transient* population as a third community, in addition to the northern and southern *residents.* Early ideas regarding *transients* were that they were individuals rejected from *resident* pods (M. A. Bigg, personal communication; cited in Baird 1994) and that they had been relegated to a less desirable lifestyle with low productivity (Bigg 1979). *Residents* and *transients* have subsequently been described as "races" by several investigators (Bigg et al. 1987; Ford et al. 1994), though "race" is usually defined in a geographic sense, implying geographically isolated populations, which are typically given subspecific designation

(Mayr and Ashlock 1991). While the two forms are sympatric, behavioral evidence suggests that they remain socially isolated. Observations of *residents* and *transients* near each other have been reported on only twenty-one occasions (Jacobsen 1990; Morton 1990; Baird and Dill 1995; Barrett-Lennard et al. 1996a). Whether the two groups are on an intersecting or nonintersecting course seems to be an important determinant of reactions, although this has not always been reported. On eight occasions no change in direction of travel was recorded for either form as they passed by each other on a nonintersecting course within a couple of kilometers (Jacobsen 1990; Baird and Dill 1995). On eight occasions when on intersecting courses, *transients* have been seen changing their direction of travel, effectively avoiding the *residents* (Morton 1990; Baird and Dill 1995), while *residents* have been recorded changing their direction of travel in the proximity of *transients* three times (Morton 1990; Barrett-Lennard et al. 1996a). Relative group sizes have not been presented for all of these observations, but *residents* are typically observed in much larger groups than *transients* (Bigg et al. 1987; Morton 1990; Baird 1994). Since *residents* vocalize more frequently than *transients* (Ford and Hubbard-Morton 1990), it is likely that *transients* detect the presence of *residents* much sooner and much more frequently than the other way around (Baird and Dill 1995). Thus the cases of *residents* showing no reaction when near *transients* may be due simply to their being unaware that *transients* were nearby. *Transients* have been recorded avoiding *residents* more frequently than vice versa. One observation of aggression between the two forms (Baird and Dill 1995) involved a large group of *residents* (approximately thirteen individuals) chasing and apparently attacking a small group of *transients* (three individuals). This observation suggests a functional reason for *transient* avoidance of *residents*—the typically larger groups of *residents* may be a threat to *transients* (Baird and Dill 1995). Why *residents* would attack *transients* remains unclear; with their tendency to feed on other marine mammals, one possibility is that under some circumstances (e.g., at times of the year or in areas where other marine mammal prey are unavailable and where *transient* group size is larger than *resident* group size), *transient* groups might prey upon lone, injured, or young *residents*.

We still know little about killer whale social organization at other sites. Lone animals have also been observed off Punta Norte, and groups of as many as twelve animals have been documented (Lopez and Lopez 1985). Stable groups off Norway range from six to thirty individuals, with a median group size of fifteen individuals (Bisther and Vongraven 1995; Similä et al. 1996), while group size off

the Crozet Archipelago ranges from two to seven individuals (Guinet 1991b). In general, group size for populations that feed on fish tends to be larger than for populations that feed on marine mammals (cf. Jefferson et al. 1991). Evidence from other sites suggests long-term associations between individuals (Guinet 1991b; Hoelzel 1991a; Bisther and Vongraven 1995; Similä et al. 1996).

Habitat Use and Ranges of Pods

Several authors have discussed habitat use by *residents* and/or *transients* (Heimlich-Boran 1988; Felleman et al. 1991; Morton 1990; Hoelzel 1993; Baird and Dill 1995). Both *residents* and *transients* frequent a wide range of water depths. Both use deep areas (>300 m), but *residents* tend to spend more time in deeper water than *transients*. *Residents* occasionally move into water less than 5 m deep, but some *transient* pods spend considerable time in even shallower depths, often foraging in intertidal areas at high tides (fig. 5.3). To quantify relative distribution patterns and habitat use in the Haro Strait region, Heimlich-Boran (1988) divided the area into 4.6 × 4.6 km quadrats. While over twenty times the observational hours were collected for *residents* than for *transients*, he noted that for 34% of the quadrats that *transients* were recorded in, *residents* had not been documented. Baird and Dill (1995) found considerable variability in habitat use between *transient* pods, with some spending significantly more time foraging in very nearshore areas than others. Heimlich-Boran (1988) noted an increase in *resident* foraging behavior in areas of high-relief subsurface topography, along the major routes for salmon migration. Using a different measure of foraging behavior, Hoelzel (1993) found no such correlation between feeding and bottom topography for *resident* killer whales in the same area. The time frames for these two studies did not overlap, however, so it is unclear whether the different conclusions reflect the different methods or a change in whale behavior between the two time periods.

Travel routes of *transients* and *residents* differ, with *residents* more typically traveling in straight lines while *transients* often follow the contours of the shoreline and frequently enter small bays (Morton 1990; Felleman et al. 1991). More extensive data on *transients* indicate that this may be true for *transient* pods that specialize in foraging in nearshore areas (cf. Baird and Dill 1995), but not for pods that usually forage in open water.

Individual killer whales have been documented to move over very large areas. Perhaps the widest movement documented is that of a number of individuals seen both in central California and in southeastern Alaska, a linear distance of 2,660 km (Goley and Straley 1994; Black et al.

Figure 5.3. A *transient* killer whale foraging in shallow water at Race Rocks, a sea lion and harbor seal haul-out site off southern Vancouver Island, British Columbia. (Photograph by Robin W. Baird.)

1997). Actual home range sizes are unknown, primarily because virtually no photo-identification work has been done in offshore areas. For killer whales in British Columbia and Washington, utilizing the northernmost and southernmost sightings of particular pods, combined with the limited knowledge of east-west (onshore-offshore) movements, the largest documented range for a *transient* pod is 140,000 km^2, while the largest documented range for a *resident* pod is approximately 90,000 km^2. Both *resident* and *transient* individuals have been documented to move up to 160 km in one twenty-four-hour period, but groups of both types also spend extended periods in very small areas. *Resident* pods around southern Vancouver Island may be repeatedly seen off Victoria or San Juan Island for days or even weeks in a row during the summer months. At this time of the year salmon are extremely abundant, and individuals may not have to move far to find concentrations of food. *Transients* may spend several hours milling in one small area (usually a harbor seal haul-out site) and

may repeatedly visit a larger area several times in the space of a week. Saulitis (1993) studied a group of *transient* individuals in Prince William Sound, Alaska, and compared their behavior with published studies of *transients* in British Columbia. She found that the Prince William Sound individuals spent more time traveling than *transients* in British Columbia, and suggested that this may be due to a lower abundance of marine mammal prey in her area, which could force the animals to make wider-ranging movements (there were, however, differences in behavioral definitions used in the different areas, potentially confounding such comparisons).

While killer whales are regularly observed in offshore areas off British Columbia, evidence suggests that offshore populations are distinct from both the inshore *residents* and *transients*. Recent research has documented the occurrence of several large groups that have been termed "offshore" killer whales (Ford et al. 1994; Walters et al. 1992). These offshore whales appear to differ from both *residents* and *transients* in several ways. Group sizes of offshore whales (ranging from two to seventy-five, but usually thirty to sixty individuals: Ford et al. 1992, 1994; Walters et al. 1992) appear most similar to those of *residents.* Range sizes of offshore whales seem to be larger than those of *residents* (see e.g., Black et al. 1997), though precise limits are unknown at this time, and offshore whales seem to differ acoustically from both *transients* and *residents* (Ford et al. 1994). Offshore whales appear to share morphological characteristics with both *residents* and *transients,* with their dorsal fins being more similar to those of *residents* and their saddle patches more similar to those of *transients* (cf. Bigg et al. 1987; Baird and Stacey 1988). While few measurements are yet available, offshore whales appear to be smaller in body size than either *residents* or *transients* (Walters et al. 1992; Ford et al. 1994). Mitochondrial DNA comparisons of these offshore animals with *residents* and *transients* suggest that they are closely related to southern *residents,* sharing the same mtDNA haplotype (Hoelzel et al. 1998a).

Diet and Foraging Behavior

Information on feeding in killer whales has come from a variety of sources. Stomach contents from animals taken in whaling operations or from stranded individuals have provided extensive information on species consumed, particularly for fish and squid. In many cases, stomach content data are the only information documenting the occurrence of different species in the diet. In recent years extensive studies have also been undertaken on killer whale feeding behavior, and anecdotal information continues to accumulate on the range of killer whale prey (e.g., Dahlheim and Towell 1994; Florez-Gonzales et al. 1994; Constantine et al. 1998). Killer whales are top predators, with an extreme range in food items reported taken, including squid, octopus, bony and cartilaginous fish, sea turtles, seabirds, sea and river otters, dugongs, pinnipeds, and cetaceans, as well as occasional reports of terrestrial mammals such as deer, moose, and pigs (Heyning and Dahlheim 1988; Guinet 1992; Jefferson et al. 1991). Extensive stomach content data (Nishiwaki and Handa 1958; Rice 1968), numerous scattered published reports (reviewed by Hoyt 1990), and observations of behavioral interactions with other marine mammals, both predatory and nonpredatory (reviewed by Jefferson et al. 1991), provide a wealth of data on prey types. Individual populations of killer whales appear to specialize in particular types of prey (Felleman et al. 1991; Jefferson et al. 1991; Baird et al. 1992). Although most prey are taken in the water, killer whales regularly beach themselves to take prey in the Crozet Archipelago and at Punta Norte, Argentina.

Resident killer whales in British Columbia, Washington, and Alaska appear to feed primarily on fish (Bigg et al. 1990a; Matkin and Saulitis 1994). In an observation study, 95% of the fish kills observed were salmonids, and of these, 50% were one species, chinook *(Oncorhynchus tshawytscha),* the largest species occurring in that area (Bigg et al. 1990a). It seems likely, however, that salmon, particularly chinook, are disproportionately recorded by such observational methods. One reason is that salmon are typically found in the upper portions of the water column (see discussion and references below) and thus are more likely to be seen then deeper-water fish during prey chases or if prey handling is of short duration. Chinook, being the largest species of salmon, may take longer to consume and may be more likely to be shared between individuals as well. Observations of predation on bottom fish have been rare in the behavioral studies undertaken, probably because killer whales are unlikely to bring prey caught at the bottom to the surface, but stomach contents from the occasional stranded animal suggest that predation on bottom fish occurs regularly (Baird 1994; Ford et al. 1994). During their migration toward breeding rivers, salmon tend to be found primarily in schools, while juvenile salmon and overwintering adults tend to be more solitary. *Residents* have occasionally been observed feeding on herring (Ford 1989), a schooling species, but the species of bottom fish recorded are typically solitary.

Diving behavior. Little has been published on the diving behavior or diving capabilities of killer whales. An early report of an animal entangled in a submarine cable brought up from 1,030 m in depth (Heezen and Johnson 1969) suggested that killer whales can dive deeply; however, it is unclear whether this animal was entangled in the cable while diving to that depth or if it had become entangled after death or while the cable was being brought to the surface. The U.S. Navy used two killer whales in their deep object recovery program and trained one of the animals to dive to a depth of 260 m (Bowers and Henderson 1972). The use of killer whales in that project was discontinued suddenly (one animal became sick and the other escaped), so the depths recorded were not considered to be the maximum depth attainable. Baird (1994; Baird et al. 1998) used recoverable time-depth recorders (TDRs) on free-ranging killer whales and collected data from twenty-one individuals (twenty *residents* and one *transient*), for periods ranging from fifteen minutes to over thirty hours. Whales were usually tracked visually while tagged; thus information on behavior was collected simultaneously with diving data. Diving patterns of *residents* were extremely variable while foraging. Some long-duration dives were very shallow (e.g., less than 10 m), some were to midwater, and some were to the bottom, in depths up to 260 m (fig. 5.4). While regularly diving deeply, the *residents* spent the vast majority of their time (more than 70%) in the top 20 m of the water column. The limited information available on the depth distribution of salmon in that area suggests that most species spend the vast majority of their time in the upper levels of the water column (that is, less than 30 m: Quinn and terHart 1987; Quinn et al. 1989; Ruggerone et al. 1990; Olson and Quinn 1993; Baird 1994), so the time spent by *residents* in the upper levels of the water column may reflect their regular predation on salmon. Foraging *residents* did dive regularly to the bottom, however (Baird 1994; Baird et al. 1998), and such behavior may reflect predation on bottom fish, as has been noted from stomach contents.

Diving behavior has been recorded for only a single *transient* (Baird 1994). Some obvious differences from *resident* diving behavior were apparent, but clearly more data are necessary to confirm whether such differences consistently occur between the two forms. It is worth discussing the diving behavior of the *transient,* though, as it may be relevant to prey detection. The diving pattern of the *transient* was extremely regular, with all long-duration dives (between one and seven minutes) being to depths between 20 and 60 m. The individual spent more than two-thirds

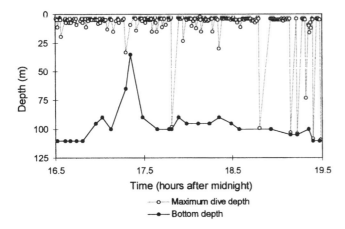

Figure 5.4. An example of the diving pattern of a *resident* killer whale. This animal (a thirteen-year-old male) was followed while tagged for the three hours shown here and was considered to be foraging. The bottom depth was recorded periodically during this period (shown by solid line). The distance traveled over this period was approximately 20 km, thus the steepness of bottom contours is exaggerated. Only a single point is shown for each dive, representing the maximum depth. On some occasions the animal dove to the bottom in up to 100 m of water, similar to all foraging *residents* in the study by Baird (1994; Baird et al. 1998). (From Baird 1994.)

of its time between 20 m and 60 m in depth. The behavior of the animal was classified as foraging during this time, so it appears unlikely that the observed pattern differed from that of the *residents* due to a difference in behavioral state. Baird (1994) suggested that swimming at these depths may allow *transients* to detect prey visually, using the silhouettes of prey against downwelling surface light.

Prey detection. A variety of differences in the echolocation patterns of *residents* and *transients* have been reported (Barrett-Lennard et al. 1996a). Barrett-Lennard et al. (1996a) noted that echolocation click trains produced by *transients* are of shorter duration than those produced by *residents,* and that *transient* click trains have unevenly spaced clicks while *resident* click trains are made up of evenly spaced clicks. *Transients* were also reported to use isolated clicks more frequently and to produce click trains less frequently than *residents.* Barrett-Lennard et al. (1996a) suggested that all of these differences arise from the differences in prey taken, since marine mammals can hear echolocation clicks and potentially evade capture while fish generally cannot. Several authors have suggested that killer whales do use passive listening to detect marine mammal prey (Barrett-Lennard et al. 1996a; Guinet 1992). Observations of marine mammals becoming silent and motionless in response to the presence or sounds of killer whales also support the use of passive listening in prey detection by killer whales (Jefferson et al. 1991). Even so, the relative roles of vision and passive listening in prey detection by *transients* remain unclear. Vision may be important in prey detection for several other species of marine carnivores; namely, sperm whales (*Physeter macrocephalus*) and white sharks (*Carcharhinus carcharias*) (Fristrup and Harbison 1993; Klimley 1994). Energy intake rates of *transients* hunting in areas with relatively high noise levels (namely, noises produced by a research vessel as well as other vessels in the area powered by outboard motors), which might mask the sounds of potential prey, were more than sufficient to meet the energetic needs of the whales (Baird and Dill 1996). Based on these observations and the diving pattern of the single *transient* studied, Baird (1994) suggested that vision may be regularly used to detect prey. *Residents* appear to locate prey underwater using a combination of echolocation and passive listening, and both vision and echolocation are probably important during prey capture (Barrett-Lennard et al. 1996a). *Residents* appear to spend much less time foraging at night than during the day; such a difference in activity state could reflect the importance of vision in prey detection and capture (Baird et al. 1998). Killer whales occasionally use aerial vision to detect prey on beaches or on floating ice (Smith et al. 1981; Guinet 1992), where echolocation is ineffective.

Cooperative hunting and prey handling. Killer whales often forage in groups. Foraging in groups could occur due to an immediate benefit from group hunting or due to longer-term benefits of group living. Social functions of grouping in killer whales are discussed below. Immediate benefits of group hunting could involve an increase in the rate at which prey are encountered, an increase in prey capture success, a decrease in prey handling time (or reduced risks of prey capture), or an increase in the ability for groups to defend prey during intergroup conflicts.

In general, larger groups should have higher prey encounter rates (cf. Pitcher et al. 1982; Connor, chap. 8, this volume). If the prey (or the prey patch) is larger than can be consumed by a single individual, such an increase in encounter rates should favor larger group sizes. In theory, for prey that could be consumed by a single individual, larger groups would have to increase encounter rates beyond simply the additive effect of several individuals hunting together. One possible way in which encounter rates of marine mammal prey may increase in a multiplicative fashion is that potential prey may detect one whale in a foraging group and move away from its path of travel, effectively "blindly" entering the path of another forager while preoccupied with avoiding the first. Increases in prey encounter rates for larger groups have not been demonstrated with killer whales, but numerous authors have suggested that both *residents* and *transients* may benefit from cooperative food searching (Hoelzel and Osborne 1986; Ford 1989; Felleman et al. 1991; Hoelzel 1993; Baird and Dill 1996). Baird and Dill (1996) noted that for larger groups hunting pinnipeds, encounter rates could theoretically begin to decrease, as the likelihood that potential prey might detect the foraging group and escape onto shore or into underwater hiding sites should increase with predator group size.

In terms of an increase in prey capture rates with group size, more information is available, both for populations feeding on marine mammals and for those feeding on fish. Some mammalian prey, such as harbor seals, appear to "hide" at the bottom, perhaps in underwater caves or crevices (Baird and Dill 1995). In these cases, several individual *transients* appear to coordinate their surfacing patterns so that one whale is always at the bottom while the other replenishes its air, effectively waiting for the seal to run out of air (Baird and Dill 1995). Lone whales would have to

return to the surface to breathe, possibly allowing a seal to escape. For killer whales feeding on schooling herring off Norway, Similä and Ugarte (1993) describe whales circling under and around schools, apparently keeping the herring in a tight school and near the surface, where individual whales would strike the school with their tail flukes and eat individual fish. While no information was presented on whether larger groups of whales were more successful at corralling schools or at maintaining larger herring schools with fewer fish escaping, it appears unlikely that single or very small groups of whales could prevent the dispersal of herring schools.

Decreasing handling time or decreasing the risk of prey handling is likely to be important only for potentially dangerous prey, such as adult male sea lions or large whales. Killer whales appear to spend considerable time in the process of killing such prey, possibly to minimize the potential for injury (Baird and Dill 1995), and at least one case of a large sea lion escaping after having been caught has been reported (Bigg et al. 1987). For prey the size of harbor seals, no effect of group size on handling time was found; individual whales or whales in very small groups appear to be able to capture, kill, and eat harbor seals very quickly (e.g., in less than two minutes), although handling time is often extended for other, unknown reasons (Baird and Dill 1995).

Transients hunting primarily harbor seals had significantly higher food intake rates in groups of three than in groups of other sizes (either larger or smaller: Baird and Dill 1996) (fig. 5.5). The increase in foraging success with group size likely resulted from the synergistic effects of several individuals hunting together, by increasing both prey encounter and capture rates. For prey of a relatively constant size, as group size increases, competition over the carcass also increases; thus for larger prey, the optimal group size would likely be larger than three individuals (Baird and Dill 1996). Similarly, for dangerous (e.g., adult male Steller sea lions, *Eumetopias jubatus*) or more difficult to capture prey (e.g., Dall's porpoises, *Phocoenoides dalli*), an increase in group size should decrease costs of prey capture or increase prey capture rates (Baird and Dill 1996).

Large groups should be better able to defend prey patches or prey carcasses from smaller groups. Such intergroup conflicts over prey have not been observed for *resident* or *transient* killer whales in British Columbia, Washington, or Alaska but they have been reported elsewhere. Bisther and Vongraven (1995) observed occasional "feeding patch takeovers" off Norway, apparent competitive interactions in which one group of killer whales rapidly approached another that was feeding on a herring school,

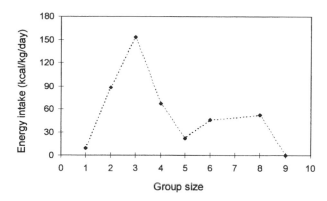

Figure 5.5. Energy intake (measured in kcal/kg/day) versus group size for *transient* killer whales, showing the clear peak in energy intake rates for individuals in groups of three. (From Baird and Dill 1996.)

with the original group leaving. No information was presented, however, on the relative size or usual association patterns of the respective groups. Hoelzel (1991a) noted similar behavior in killer whales feeding on sea lions off Argentina: a larger group of whales (seven individuals) appeared to actively displace a smaller group (two individuals) from the most productive foraging area. C. Guinet (personal communication) has also observed similar behavior on one occasion in the Crozet Archipelago, where a group of seven individuals appeared to take an elephant seal pup that had been killed by a group of five individuals.

Sharing of prey is occasionally observed among *residents* feeding on fish. Prey sharing, and even provisioning, is observed much more frequently among killer whales feeding on marine mammals (Hoelzel 1991a; Guinet 1992; Baird and Dill 1995). Baird and Dill (1995) suggested that prey sharing occurred in virtually all of the marine mammal attacks they observed. Hoelzel (1991a) observed prey sharing in all three groups he studied, and noted that one whale in each group did the majority of the hunting and provisioned the other whales in the group. Guinet (1992) observed prey sharing in twenty-seven of twenty-nine cases. One of the two cases in which sharing did not occur involved four consecutive elephant seal captures, in which the last seal in the series was consumed by only one of the whales present. The remaining case involved a single whale apparently searching to determine the location of the remainder of her group after capturing an elephant seal pup, then moving in the opposite direction to consume the prey alone (Guinet 1992).

Among *transients* in British Columbia, apparent cooperative foraging occasionally occurs between pods (that is, all members of two or occasionally three pods forage together and share prey captured). Multipod foraging associations

are likely to be most beneficial when *transients* are hunting prey that are difficult to capture or dangerous, yet they do occur when *transients* are hunting harbor seals, which are neither difficult to capture nor potentially dangerous (Baird and Dill 1996). Multipod foraging associations have not been observed among killer whales hunting pinnipeds off the Crozet Archipelago (Guinet 1992) or off Punta Norte (Hoelzel 1991a). When intentional stranding is the primary hunting tactic (as in these latter two localities), it is unlikely that additional individuals could increase the capture rate or decrease the risk. Off the Crozet Archipelago, where killer whale group size is typically small, larger groups of killer whales have been observed attacking large baleen whales (Guinet 1991b). Combined with such occurrences elsewhere (Silber et al. 1990; Jefferson et al. 1991), Guinet (1991b) suggested that temporary feeding associations do occur. For killer whales cooperatively feeding on herring off Norway, Similä and Ugarte (1993) noted that only one group would feed on a school of herring at a time.

Killer whale calves and young juveniles often are not involved in capturing prey that may be somewhat dangerous, such as during intentional strandings or attacks against large baleen whales (Guinet 1991a; Hoelzel 1991a). In general, though, killer whales of all ages and both sexes participate in marine mammal attacks and subsequent feeding (Jefferson et al. 1991). However, some sex-specific differences in hunting tactics have been noted for both mammal-eating and fish-eating killer whales. In British Columbia, male *resident* killer whales often forage in deeper water than do females, or in peripheral positions in a group. Since *residents* do not appear to cooperatively herd or chase prey, it has been suggested that such differences may be due to different diving capabilities of the sexes (which should correlate with size differences: Bain 1989). However, time-depth recorders deployed on both subadult and adult female *resident* killer whales in southern British Columbia did demonstrate their abilities to regularly use bottom areas in relatively deep (>150 m) waters (Baird 1994; Baird et al. 1998). J. K. Jacobsen (personal communication) observed *resident* killer whales occasionally attempting to capture fish (large salmon) that were trying to hide in rock crevices along the shoreline; only females and subadults were observed in this behavior. Due to the high degree of sexual dimorphism, adult males may be limited in their range of behaviors associated with prey capture in situations requiring extreme maneuverability or travel in shallow waters. Only females engage in intentional strandings to capture pinnipeds in the Crozet Archipelago (Guinet 1991a), while both males and females participate in intentional strandings to capture pinnipeds at Peninsula Valdés, Argentina

(Lopez and Lopez 1985). This difference may be due to the lower-grade slopes of the beaches in the Crozet Archipelago compared with Peninsula Valdéz: larger males may be unable to beach themselves successfully, or to do so without risk, at Crozet (Guinet 1991a, 1992). During attacks on a Bryde's whale *(Balaenoptera edeni)* and on Pacific white-sided dolphins *(Lagenorhynchus obliquidens),* males have been noted to play a minimal role, if any, in the attack (Silber et al. 1990; Dahlheim and Towell 1994).

Among *transients* hunting harbor seals, males and females seem to play similar roles in finding and capturing seals, but occasionally differ in how they handle prey (Baird and Dill 1995). During several seal attacks, as a group of females and subadults would pass by a seal at the surface, striking it with their tails or pectoral flippers, an adult male in the group would appear to prolong its dive time, possibly staying beneath the seal to keep it from escaping (Baird and Dill 1995). Among killer whales searching for marine mammal prey along beaches in the Crozet Archipelago, in two different pods, individuals maintained specific foraging positions relative to other individuals, both within and between years and between different bays (Guinet 1992). Such differences, like the differences noted above with *transients* handling harbor seals, might be considered "division of labor" (cf. Stander 1992 for examples with lions), with specific individuals repeatedly taking the same role in hunting or prey handling. No apparent division of labor was observed for killer whales cooperatively feeding on schools of herring off Norway (Similä and Ugarte 1993).

Foraging tactics vary between populations, depending on habitat, type of prey taken, and factors such as prey abundance. Intentional stranding behavior to capture hauled-out pinnipeds is frequently observed in Argentina and the Crozet Archipelago. Such behavior is extremely rare for mammal-eating *transients* in British Columbia and Washington (Baird and Dill 1995). Two likely reasons exist for its rarity in those areas. One is that there appears to be a cost associated with the behavior. For example, in one case a juvenile killer whale intentionally stranded and was unable to return to the water on its own (it was pushed into the water by researchers) (Guinet 1991a). Six other stranded killer whales had been recorded on the Island over a twenty-seven-year period (five juveniles and one adult), implying that these individuals may have died as a result of intentional stranding (Guinet 1991a: he did not report, however, the sexes of the stranded individuals, whether any evidence was available that they had stranded intentionally, or whether all died). While mortality associated with intentional stranding may be extremely infrequent, the benefits would presumably have to be high to warrant this behavior.

The benefit in areas of high prey abundance, such as with harbor seals around southern Vancouver Island, is low, in that prey can be captured relatively easily without engaging in this behavior.

Killer whales off Norway use bubbles, lobtailing, and body pigmentation to herd and encircle herring, force them to the water's surface, and prevent them from escaping (Similä and Ugarte 1993). Whales were repeatedly observed swimming with their ventral surface toward the herring schools (which presents a white pattern bordered with black anteriorly, and a more complex black-and-white patterning posteriorly), often in response to movements of herring away from the school. Such behavior appeared to result in herring moving back into the school. The exact mechanism of such herding is unclear, as the white anterior area might function to tighten the herring school while the black-and-white posterior area of the ventral surface would be more likely to disrupt the herring schooling (Wilson et al. 1987).

Predation and Parasitism

No predators on killer whales have been recorded. Young or sick whales are likely potentially at risk from attacks by large sharks in some areas, but no observations of individuals with scars from failed shark attacks have been reported, as is seen with other dolphin species (e.g., Corkeron et al. 1987; Cockcroft et al. 1989; Connor et al., chap. 4, this volume). Scars from intraspecific interactions are frequently observed (Scheffer 1969; Visser 1998; fig. 5.6), but it is unknown whether such interactions ever result in mortality.

A variety of endoparasites have been recorded from killer whales, including trematodes, cestodes, and nematodes (reviewed in Heyning and Dahlheim 1988). Transmission of such parasites is primarily through ingestion of infected food items. External parasites are rarely seen, but some killer whales have been seen with barnacles on the rostrum, the trailing edge of the tail flukes, or the trailing edge of the dorsal fin, and with a species of cyamid ectoparasite. The probable cause of transmission of ectoparasites is body contact between individuals, both during social contacts and from mother to offspring.

Interspecific Associations

Killer whale interspecific associations are primarily thought to involve predation upon other species (fig. 5.7), although a variety of nonpredatory associations have also been observed. Unfortunately, many reports in the literature document interspecific associations but without the necessary behavioral detail to specify the form or type of association;

Figure 5.6. Fresh tooth rakes, most likely caused by another killer whale, on the side of a ten-year-old female "southern" *resident* off Victoria, British Columbia. (Photograph by Robin W. Baird.)

namely, predatory versus nonpredatory (e.g., Mikhalev et al. 1981). At least twenty-six species of cetaceans and seven species of pinnipeds have been observed associated with killer whales in nonpredatory contexts (Jefferson et al. 1991). Often such interspecific associations have involved the co-occurrence of killer whales and other species of marine mammals in the same area at the same time, with little or no behavioral interaction. Other times, however, both large and small cetaceans and pinnipeds have often apparently deliberately approached and interacted with killer whales. In some cases, such as those in British Columbia and Washington, these instances have involved *resident* killer whales, which are typically little threat to other marine mammals, but some have involved *transients*. Groups of sea lions occasionally jump into the water when *transients* are near and often follow the whales (R. W. Baird, personal observation). In large groups, the risk to an individual sea lion should be reduced due to dilution, and killer whales may be reluctant to attack a large group due to in-

creased risk of injury, thus the threat may be not be great in such situations. Harbor seals seem to be the preferred prey of *transients* in this area (Baird and Dill 1996), thus sea lions are at low risk of attack. Such behavior may function as a "pursuit invitation," alerting the whales that they have been detected and that the element of surprise is lost (Jefferson et al. 1991; Smythe 1970; Connor, chap. 8, this volume). It is also possible that sea lions benefit from such behavior by learning more about their predators; such information may be valuable at a later date if an individual is attacked, since such attacks are not always successful (Bigg et al. 1987).

Life History

Birth

The most detailed information on life history is for British Columbia *residents* (table 5.2); many of the characteristics

Figure 5.7. A *transient* killer whale throwing an adult male Dall's porpoise into the air, Chatham Strait, Alaska. (Photograph by Robin W. Baird.)

Weaning and Interbirth Intervals

Precise age at weaning is not known, but killer whale infants begin taking solid food at a very young age. Heyning (1988) noted solid food and numerous parasitic nematodes (whose first hosts were fish) in the stomach of a 2.6 m long animal. No milk was visible in the stomach of that animal, but the contents were not tested for the presence of milk lactose. In other species (e.g., sperm whales), the presence of solid food in the stomach is known to be a poor method for estimating the age at weaning. Using the ages at which killer whales begin spending more time away from their mothers, as well as when they are observed taking fish, Haenel (1986) estimated weaning to occur at between one and a half and two years of age. Weaning in at least one other species of odontocete, the bottlenose dolphin (*Tursiops* spp.), does not occur until about three or four years of age, even though calves begin capturing their own fish by six months of age (Connor et al., chap. 4, this volume), so it is likely that nursing in killer whales may continue beyond the ages suggested by Heyning (1988) or Haenel (1986). Guinet and Bouvier (1995) noted that killer whales appear to first be able to successfully capture elephant seal pups by intentional stranding at about six years of age, and thus suggested that they are still somewhat dependent on adults at that age.

Olesiuk et al. (1990) noted that calving interval, defined as the interval between births of surviving calves, ranges from two to twelve years in British Columbia (mean = 5).

likely vary between populations. Some information—for example, gestation period—has been best established with captive animals. Gestation periods in captive animals, measured using hormone levels, ranged from 468 to 539 days (average of 517 days; SD = 20 days) (Duffield et al. 1995). Length at birth in British Columbia and Washington ranges from at least 218 to 257 cm (Olesiuk et al. 1990). One animal was born in captivity at a length of 206 cm (Duffield and Miller 1988), while the largest fetus recorded worldwide appears to have been 270 cm in length (Nishiwaki and Handa 1958). A single calf is usually born, though Olesiuk et al. (1990) reported two cases of twins, one of which was subsequently determined to be a case of mismatching (J. K. B. Ford, personal communication). In the remaining case, both animals (a male and a female born in 1980) survived to at least thirteen years of age (Ford et al. 1994). If this situation is not a case of adoption (no adult females in that pod went missing that year: Olesiuk et al. 1990), then killer whales appear to be the only species of cetacean in which viable multiplets have been recorded. Calving occurs year-round in British Columbia, but there appears to be a peak in births between fall and spring (Olesiuk et al. 1990). Definitions of the age at which individuals are no longer "calves" vary between studies.

Table 5.2. Life history characteristics of killer whales

Maximum body size: female	7.7 m
Maximum body size: male	9.0 m
Gestation length[a]	Mean = 517 days; range = 468–539 days
Weaning age	??
Interbirth interval[b]	Mean = 5 years; range = 2–14 years
Calf mortality (to 6 months)[b]	37–50%
Length of estrous cycle[a]	Mean = 42 days; range = 23–49 days
Calving season[b]	Year-round with winter peak
Age of female sexual maturity (first birth)[b]	Mean = 15 years; range = 11–20 years
Age of male sexual maturity (asymptotic growth)[b]	20 years
Maximum life span: female[b]	80–90 years
Maximum life span: male[b]	50–60 years

[a]Data from Robeck et al. 1993.
[b]Values derived from studies of *resident* killer whales from British Columbia and Washington (Ford et al. 1994; Olesiuk et al. 1990).

ROBIN W. BAIRD

Subsequent to their study, one fourteen-year calving interval was noted in the same population (Ford et al. 1994). The occurrence of two-year calving intervals implies that females are able to become pregnant while still nursing a calf. In Prince William Sound, calving interval has been observed to range between four and ten years (Matkin and Saulitis 1994). Calving interval increases slightly with age, but there is extremely high variability (Olesiuk et al. 1990).

Mortality and Life Expectancy

Mortality rates for British Columbia and Washington *residents* vary with age and, for older individuals, with sex (Olesiuk et al. 1990). Neonatal mortality, defined as that which occurs between birth and six months of age, is very high. Olesiuk et al. (1990) estimated neonatal mortality of *residents* in two ways, using survival rates of calves first encountered during winter and using the discovery of stranded animals. These estimates were 37% and 50%, respectively. Bain (1990) independently estimated neonatal mortality in the population of *resident* killer whales off northern Vancouver Island at 42%, based on the distribution of calving intervals. The causes of this high neonatal mortality remain unclear, but similar neonatal mortality levels have been noted for bottlenose dolphins (Connor et al., chap. 4, this volume). Predation does not appear to be a significant cause of mortality in the *resident* populations; no individuals appear to have scars associated with failed shark attacks, and stranded animals that have been found show no evidence of predatory attacks. Unfortunately, detailed postmortem examinations of stranded neonates from these populations have been undertaken on only one individual, and the cause of death was not determined (R. W. Baird, unpublished data).

After six months of age, mortality rates decline steadily for both sexes. Mortality rates are lowest around twelve to thirteen years of age for males, and around twenty years of age for females, after which mortality rates begin to increase steadily with age (Olesiuk et al. 1990). Maximum longevity has been estimated at about fifty to sixty years for males and eighty to ninety years for females. At birth, average life expectancy is about twenty-nine years and seventeen years for females and males respectively. From six months of age (excluding the high mortality rate during the first six months), average life expectancy increases to about fifty years for females and about twenty-nine years for males. Life expectancy at sexual maturity (assumed to occur at about fifteen years of age for both males and females: see discussion below) is about sixty-three years for females and thirty-six years for males (Olesiuk et al. 1990). The causes of the shorter life expectancy of males are unknown, but are presumably somehow related to sexual selection.

Growth and Age at Sexual Maturity

Based on captive data, growth rates are similar for males and females. Growth rates tend to be linear for the first nine to twelve years for females and for the first twelve to sixteen years for males, after which both sexes show a decrease in growth rate (fig. 5.8; Bigg 1982; Duffield and Miller 1988). However, growth rates vary between individuals as well as between and potentially within populations (Duffield and Miller 1988). The mean annual growth rate for six eastern North Pacific killer whales was 38 cm per year (range 26–53 cm/yr). Growth rates of North Atlantic killer whales fell into two distinct categories, those that grew at about 21 cm per year (range 17–25 cm/yr) and those that grew at about 39 cm per year (range 31–48 cm/yr) (Duffield and Miller 1988). Based on whaling data, Christensen (1984) suggested that growth rates of wild killer whales off Norway are not linear with age, and that males may show a secondary growth spurt associated with adolescence. However, lengths reported were estimates rather than precise measurements; thus, the validity of these conclusions is unclear. The maximum lengths recorded for males and females are 9.0 m and 7.7 m respectively (Heyning and Brownell 1990), although average maximum sizes attained by both sexes appear to be much smaller (Duffield and Miller 1988).

Age at sexual maturity for females has been reported in a variety of ways, including age at first ovulation, age at first pregnancy, and age at first parturition. Olesiuk et al. (1990) defined age at sexual maturity for females as the age at which they first give birth to a viable calf, and noted

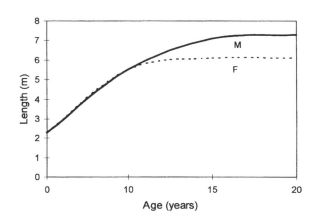

Figure 5.8. Theoretical average growth curves for eastern North Pacific killer whales based on measurements from captive animals. (After Duffield and Miller 1988.)

that it varies between twelve and sixteen years (mean = 14.9). Since their study, one female in the same population has given birth at eleven years of age (Ford et al. 1994). Ford et al. (1994) show one female giving birth at ten years of age, however, the age of that female was estimated plus or minus one year (Olesiuk et al. 1990). Similarly, the upper limit for the age at which females give birth to their first viable offspring has risen, with two females first giving birth at twenty years of age (Ford et al. 1994). Age at first parturition (for viable calves) in captive animals has averaged 12.7 years (range 8–15: Duffield et al. 1995). It is unclear, however, whether such pregnancies of younger animals would be unsuccessful in wild individuals, or just do not occur, being suppressed either behaviorally or physiologically. Olesiuk et al. (1990) also noted that fecundity rate (defined as the proportion of mature females that give birth to viable calves each year) declines linearly with age. However, this conclusion is sensitive to age estimation techniques for older females. Some females in their study that were adult-sized at the beginning of the study period (in the early 1970s) and were not recorded to have given birth during the study were assumed to be older individuals. Females appear to become reproductively senescent at a mean age of about forty years (though one female reproduced at approximately fifty years of age) (Olesiuk et al. 1990). This supposition is based on several findings. About seventeen females that were adult-sized at the beginning of the study, and which were still alive as of 1993, have never been observed with calves (Ford et al. 1994). An additional seven adult females died ten or more years after first being encountered without ever having been recorded with a calf (Ford et al. 1994). However, a very small proportion of individuals may be infertile, due perhaps to disease or congenital problems. Evidence for this conclusion comes from one southern *resident* individual of known age (+/− two or three years), who matured in the mid- to late seventies but has not been known to have given birth to a calf as of mid-1999, despite being observed regularly each summer every year since she matured. Regardless, the relatively small proportion of such infertile females among those born into the population in the last thirty years makes it likely that some females do live twenty or more years after reproductive senescence (that is, are postreproductive).

Appendage size differs between the sexes, with adult males having larger pectoral fins and tail flukes than adult females. Adult males have a tall, triangular dorsal fin that may reach up to 1.8 m in height, while in juvenile males and adult females the fin reaches 0.9 m or less and is generally more falcate (fig. 5.9). Onset of sexual maturity for

males has been defined by an increase in the growth rate of the dorsal fin (Olesiuk et al. 1990). Using this criterion, onset of sexual maturity for males ranges from ten to seventeen and a half years (mean = 15) (Olesiuk et al. 1990). The use of such a criterion, however, needs to be tested using hormonal levels, perhaps with captive animals. Dorsal fin growth for males continues for at least six years after the onset of maturity, and physical maturity may be reached at the end of that period (Olesiuk et al. 1990).

Social Behavior and Communication

Preferential associations both within and between groups have been reported at different field sites; these include female-calf, male-calf, all-male, and multigroup associations. Seasonal changes in behavior have been noted for *transients* in British Columbia and for killer whales in the Crozet Archipelago. In both cases, an increase in social behavior has been demonstrated during the period when prey abundance is highest (Baird and Dill 1995; Guinet 1991b).

Interactions with Young

Observations of killer whale births in the wild have been documented on two occasions, and both involved the mother giving birth among a larger group of related individuals (e.g., Stacey and Baird 1997). In both cases, considerable percussive activity by other individuals present and rubbing and lifting the infant into the air was observed (Stacey and Baird 1997). The relationship between a mother and her calf is not abruptly interrupted as the latter matures (Haenel 1986). Calves do tend to spend less time with their mothers as they grow, because they start swimming with other members of the pod. Among *residents,* even though a juvenile decreases its dependence on its mother, they will never be completely separated for long periods. Associations between infants and individuals other than the mother fall into several categories. Multi-calf play groups occasionally form. Older individuals, including subadults and both adult females and males, associate with infants. Such behavior has often been referred to as alloparental or allomaternal care (Haenel 1986; Bain 1989; Rose 1992; Bisther and Vongraven 1995). It is unclear, however, whether and how such behavior benefits the calf or its mother. Data on young calves (those under one year of age) associating with individuals other than their mothers are rare, and the necessary detailed behavioral data (e.g., activity, travel speed, respiration rate) on mothers with and without calves, and on calves with their mothers, alone, and with other individuals, are lacking.

Plate 1. Pacific white-sided dolphin leaping in Juan de Fuca Strait, British Columbia. Robin Baird

Plate 2. Hourglass dolphin, Southern Ocean. S. Hooker

Plate 3. Shark Bay bottlenose dolphin calf spy-hops during a play bout. Janet Mann

Plate 4. Bottlenose dolphins around the world have found many unusual ways to catch their prey, including chasing fish onto mud banks in South Carolina. Flip Nicklin, Minden Photos

Plate 5. Mysterious mass stranding of long-finned pilot whales on the coast of Massachusetts. Phil Clapham/Center for Coastal Studies

Plate 6. Socializing Shark Bay dolphins. Janet Mann

Plate 7. Shark wound healing on a five-year-old bottlenose dolphin male, "Smokey," Shark Bay, Australia. Janet Mann

Plate 8. Male narwhals "fence" with 3 m long tusks. Flip Nicklin, Minden Photos

Plate 9. Studying a group of socializing sperm whales off the Galápagos Islands.
Flip Nicklin, Minden Photos

Plate 10. Humpback whale bubblenet feeding in Alaska. Andrew Wight

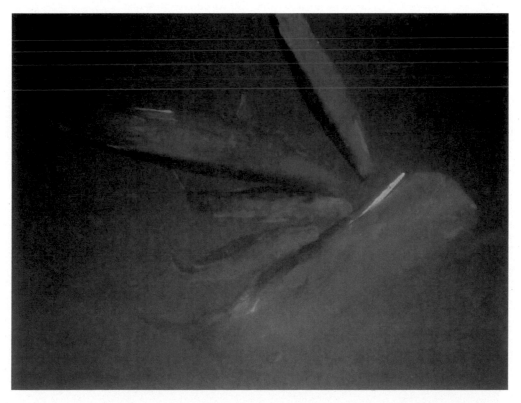

Plate 11. Three times heavier than the female he courts, this large male sperm whale displays an erection during an interaction with a group of females and young. Flip Nicklin, Minden Photos

Plate 12.   Schools of rough-toothed dolphins occasionally come close enough to shore to draw the attention of Japan's dolphin hunters. R. White

Plate 13.   By-catch of Dall's porpoise on the deck of a salmon vessel. Bill Everett

Plate 14. Aboriginal hunting: preparing to flense a pregnant female narwhal, Baffin Island, Canada. Randall Reeves

Plate 15. By-catch of dolphins, such as this spinner dolphin in Sri Lanka, has led to the development of a marketing system and directed hunts. S. Leatherwood

Plate 16. The stranding that wasn't. Minimal tidal movement allowed these false killer whales to remain in a protective formation around a large dying male for three days. James W. Porter

Figure 5.9. Adult female *(left)* and adult male *(right) transient* killer whales, showing the smaller and more falcate dorsal fin of the female. Adult males also have larger pectoral flippers and tail flukes than females. (Photograph by Robin W. Baird.)

Adults may teach young to hunt, based on observations of coordinated intentional strandings between an adult and a juvenile, in which the adult would capture a prey item and pass it to the juvenile (Lopez and Lopez 1985). In their review of teaching in nonhuman animals, Caro and Hauser (1992) noted that the evidence for teaching by killer whales was weak. They defined teaching as a modification of behavior only in the presence of a naive observer, which involves a demonstrated cost for the teacher (or lack of immediate benefit) and a demonstrated benefit for the observer that would not have occurred in the absence of teaching. Information collected subsequent to Lopez and Lopez's study provides further evidence for teaching by killer whales. Several types of costs associated with purported teaching events have been identified. One individual adult female had decreased capture success when hunting in close proximity to two juveniles (Hoelzel 1991a). As mentioned previously, Guinet (1991a) noted a risk of mortality associated with intentional strandings, apparently greatest for juveniles, though some risk may occur for adults as well. Evidence on benefits to young would need to compare success rates of calves that beach with potential adult "teachers"

versus those that beach alone, as well as those that beach frequently versus rarely. That some individuals are better hunters than others, and that even adults practice hunting techniques, has been demonstrated (Hoelzel 1991a; Guinet 1991a). Guinet (1991a) also provided additional information relevant to the issue of whether teaching occurs. Intentional strandings in the Crozet Archipelago appear to be of two types, those directly intended to capture prey, and coordinated group strandings (when no prey are on the beach), which appear to be a form of social play (Guinet 1991a). Only a few females (all adults) engaged in strandings to capture prey, while all juveniles and the other adult females engaged in coordinated group strandings (Guinet 1991a). Guinet (1991a) also noted association patterns between stranding individuals. Unlike the situation in Patagonia, group strandings did not involve the capture of prey or appear to represent failed attacks, thus adults received no energetic benefit of the behavior. The most unusual pattern was that one juvenile associated more frequently with an adult female that regularly hunted by intentional stranding than it did with its mother (who had not been recorded capturing prey by intentional stranding). It appears that the juvenile preferentially associated with the adult from whom it could benefit the most, in terms of learning successful hunting techniques, rather than with its mother, with whom it associated in all other contexts (traveling, resting, feeding). One of the primary problems with the killer whale data in terms of Caro and Hauser's (1992) definition of teaching is that the adult whales also exhibit the behavior (intentional stranding not associated with prey capture) in the absence of potential pupils (although they could have been doing so to practice the hunting technique themselves). Guinet (1991a) noted some modification of adult behaviors when calves or juveniles participated, however, in that adults returned to the water at the side of the calves, helping them roll back into the water. One instance of a female pushing her infant onto shore to strand it, then stranding beside it to assist it back into the water, also suggests that teaching is occurring.

Male Behavior and Relationships

For northern *resident* killer whales, a variety of individual and age-specific relationships between males and other age-sex classes have been described (Rose 1992). Subadult and juvenile males occasionally aggregate in "play" groups, in which frequent body contact, splashing, and sexual behavior is observed. The latter consists of penile erections and/or beak-genital orientation by one or both males. These activities may be performed by more than two males and with males of different age classes. To test whether these male-only social groups represented agonistic/dominance interactions or play, Rose (1992) examined age distribution, relatedness of participants, and reciprocity of physical contact, among other parameters. These groups most frequently involved animals of different age classes, and specific behaviors exhibited were usually reciprocated. Trios and quartets frequently occurred (one-third of all male-only groups observed), and individuals did not appear to actively try to avoid larger (thus potentially dominant) whales. As such, Rose (1992) suggested that these associations were more likely to represent play groups than agonistic interactions. Adolescents participated in such interactions four times as often as adults, and Rose (1992) suggested that these associations may help adolescents gain courtship skills. In comparing her observations of male social behavior with the results of other killer whale studies, Rose (1992) stated that males, and particularly young males (from twelve to twenty-five years), socialize more, and more vigorously, than females. However, methodological differences between studies suggest that this question warrants further attention, particularly quantification of female social behavior using focal animal sampling. Sex differences in patterns of play have been observed in some primates (with juvenile males of some species playing more than juvenile females) and may be related to development of fighting skills (Fagen 1993).

Vocalizations

Killer whale vocalizations have been grouped into three distinct categories: whistles, discrete calls, and clicks (Ford 1989). Research efforts have focused on the latter two types of vocalizations. Studies of the communicative functions of such vocalizations have been hampered by the difficulty of localizing underwater sounds and of recording high-frequency sounds, both of which require specialized equipment (see Miller and Tyack 1998). No published studies using sounds localized from specific individuals are available, but some higher-frequency work has been undertaken on echolocation clicks. Ford (1989) suggested that discrete calls produced by *resident* killer whales function as social signals between pod members, because production of calls seems to elicit calls from other individuals, and call and whistle rates are highest when whales are socializing. As for the communicative function of echolocation clicks, Barrett-Lennard et al. (1996a) noted that while production of click trains from one individual did not elicit click responses from other individuals, click trains were frequently produced during social interactions of *residents*. However, no detailed analysis of killer whale echolocation clicks and their potential communicative function has been undertaken (cf. Dawson 1991 for a study of Hector's dolphin, *Cephalorhynchus hectori,* clicks and communication). Based on a negative correlation between group size and echolocation use for *residents,* Barrett-Lennard et al. (1996a) suggested that information collected through echolocation is shared between individuals. However, their study did not appear to control for group size effects on behavior, which have been documented for both *residents* and *transients*. In a study of southern *residents,* group size and the occurrence of fast, nondirectional surfacings (which were interpreted as indicative of feeding) were negatively correlated (Hoelzel 1993). For *transients,* social behavior increased with group size, while foraging behavior decreased with group size (Baird and Dill 1995). Since feeding seems to occur less frequently in larger groups of both *residents* and *transients,* a decrease in echolocation would be expected for that reason alone. Thus it is unclear whether such sharing of information actually occurs.

Killer whales exhibit a variety of percussive behaviors, such as breaching (fig. 5.10), tail slaps, pectoral fin slaps, and dorsal fin slaps. Norris and Dohl (1980b) have suggested that such behavior may function as a means of com-

Figure 5.10. An adult male *transient* breaching, a behavior that usually occurs only after a prey capture or when multipod groups are engaged in social behavior. Such behaviors are probably infrequent at other times for *transients* due to their potential negative effects on foraging success, since marine mammal prey such as harbor seals could easily detect them. (Photograph by Robin W. Baird.)

munication between individuals, though this has not been rigorously tested with killer whales. The role of body posturing or touching between individuals has not been investigated as to potential communicative functions.

Socioecology

Resident killer whales are members of the only mammalian population in which no dispersal of either sex has been recorded. Male-biased dispersal and female philopatry is the most common pattern observed in mammalian populations (Greenwood 1980; Clutton-Brock 1989). When male killer whales stay within their natal group, it is likely because the costs of staying are low and/or because there are some benefits to staying. Direct costs of staying in the maternal group could include decreased opportunities for mating or energetic costs associated with increased competition for food. However, because of their wide-ranging movements and regular interactions between pods, oppor-

tunities for mating may not be lower for killer whales that remain within their maternal group (Baird 1995; Connor, chap. 8, this volume). Bain (1989) suggests that potential costs of competition for food for *residents* are reduced, as adult males should be able to feed at greater depths and on larger prey items, essentially dispersing ecologically, not geographically (although the time-depth recorder work of Baird et al. [1998] suggests that females and subadults can dive as deeply as adult males, at least in the inshore waters around southern Vancouver Island). Potential benefits of staying, in terms of inclusive fitness, could include assisting with the care of related calves within the group, assisting with group defense (as in *transients* defending against potential attacks by *residents,* or vice versa, or through competition between *resident* communities), or helping the pod to locate and capture prey. For *residents* in British Columbia and Washington eating fish, cooperating to capture prey may not be important, but among fish-eating whales elsewhere, such as off Norway, all individuals appear to cooperate in herding fish. Increased access to mates is an additional potential benefit of philopatry for males, as brothers might form coalitions (Bain 1989), or mothers might help their adult sons gain access to mates (Connor et al., chap. 10, this volume). The costs for a female *resident* of remaining philopatric are also likely to be low for the same reasons: mating likely occurs between pods and not within, so selection for dispersal to avoid inbreeding is probably low. Females may benefit for other similar reasons as well: the increase in inclusive fitness associated with the care of related calves within the group, group defense, and locating prey schools.

There is an extremely wide range in pod sizes for *resident* killer whales (from three to fifty-nine individuals). Some of this variability is likely due to chance demographic circumstances. Small pods in which the surviving offspring of a lone adult female are all males are destined to die off, and in fact, several northern *resident* pods or subpods appear to be doing exactly that (Ford et al. 1994). This wide variability, as well as the observation that these smaller pods or subpods do not join with other groups, suggests that selection pressure on group size for *residents* is not strong. This supposition is supported by the analysis of Brault and Caswell (1993), who found no significant demographic effects of pod size. This variability contrasts strikingly with the small variability observed for *transient* pod sizes (Baird and Dill 1996), from one to four individuals. Selective pressure favoring small pods, due to the energetic costs of foraging in large groups when hunting marine mammals, appears much stronger in this circumstance. Alternatively, a small pod joining with a larger pod for a long period might not be an option if the larger group actively worked to prevent such joining (although cf. Giraldeau and Caraco 1993).

For *transients*, remaining in a group larger than three has a direct cost in terms of a reduced energy intake rate (Baird and Dill 1996). When adult and subadult individuals are foraging, they are found most frequently in groups of three, the group size at which individual energy intake rates are maximized (Baird and Dill 1996; see fig. 5.5). Dispersal of both sexes from groups larger than three or four individuals likely occurs in response to the energetic costs associated with remaining in a large maternal group. Groups larger than three or four individuals are regularly observed nonetheless (the largest documented in Baird and Dill's study was fifteen individuals), and the benefits of the occasional formation of such large groups may outweigh the short-term energetic costs. Social behavior is more common than foraging in larger groups of *transients*, and these

large groups may provide opportunities for mating, allomaternal care, and/or learning mating or courtship skills (Baird and Dill 1995). Larger groups of *transients* also contain a disproportionately large number of calves and juveniles, and these groups may function to protect these more vulnerable individuals from attacks by *residents* (Baird and Dill 1996).

While the data are incomplete, a picture of the social system and social organization of *transients* can be suggested (fig. 5.11). The choice of social partners for a female *transient* varies according to a variety of factors, including her age, the number and age of her siblings, and the number, age, and sex of her offspring. Juvenile females appear to remain with their mothers until they reach reproductive age. At that time, two possibilities seem apparent. One is that her maternal pod will accept the presence of an adult male for a period of time, despite the energetic costs of an additional individual in the group, to allow mating to occur. How such males are chosen is unclear. The other possibility is that she will begin to travel for periods of time with one or more *transient* pods that contain an adult male. Despite the energetic costs to the individuals in the pod containing the adult male, such pods are probably willing to accept a reproductive female for the sake of his opportunity to mate. This latter scenario is more likely to occur when the female's maternal group has already reached a size of four individuals; thus, in terms of inclusive fitness, an energetic benefit exists to leaving (cf. Giraldeau and Caraco 1993). If a female who has dispersed loses her calf or is unable to conceive, she may return to her natal pod, assuming that returning does not increase its size above the optimum. Once a female gives birth, or if she has two juvenile or infant offspring, she appears to spend several years temporarily associating with one or more pods of *transients*. I suggest that these temporary associations function to protect the offspring of the female from attacks by *residents*. As noted, although one attack by *residents* on a *transient* pod has been recorded, it is unclear why *residents* would attack *transients* (though one possibility is that *transient* pods occasionally pose a threat to *residents*, particularly to small groups or sick individuals). Attacks on adults by other *transients* are unlikely. Unless a clear numerical advantage is available for an attacking group, the risk of injury associated with such an attack would probably be too high. As well, *transients* benefit from the occasional temporary foraging associations needed to subdue difficult-to-capture prey, and aggression between *transient* groups could disrupt their future ability to cooperate.

The presence of a female and her offspring brings an

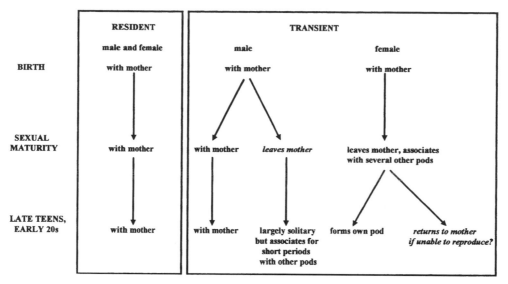

Figure 5.11. A diagrammatic representation of *resident/transient* differences in the patterns of association between mother and offspring, from birth through the early twenties. Those links labeled in italics are the most speculative, since the absolute occurrence of events to support such outcomes is rare, given the small population size, long periods between resightings of individuals, and long life span of these animals.

energetic cost to the members of the group she joins. Again, such pods may be willing to accept the female and her calf if the female gives adult males within the pod opportunities for mating. If there is a nonzero probability of a male fathering the female's next calf, or of being the father of her existing calf, other individuals in the group may also gain inclusive fitness benefits. Thus the increased probability of reproductive success for males within the group outweighs the short-term energetic costs. As her offspring age, and at least one (either a male or female) becomes old enough to assist the mother with group defense and foraging, she will then spend the majority of her time traveling only with those individuals.

Males appear to have two options for dispersal, depending on their relative position within their maternal group. A first-born male may stay with his mother his entire life, while other males disperse from their maternal group (while remaining within their natal range) sometime before sexual maturity is reached. For males, remaining philopatric is likely to be the preferred strategy. The first-born male, because of his larger size, is probably able to retain his position within the maternal group, from which he benefits in two ways: by increased energy intake rates associated with hunting in a group of two or three individuals, and by associating with his mother as an alliance partner (Connor et al., chap. 10, this volume) or in defense against *residents*. With the long interbirth interval, it is un-

likely that a second-born male will reach sufficient size and strength to be able to challenge an older brother for many years. Dispersing males appear to remain solitary for a large proportion of their time and thus suffer a direct energetic cost of dispersing (Baird and Dill 1996). Such males do associate for temporary periods with other pods, both those that already contain an adult male and those that do not. Although the data are limited, associations with pods that already contain an adult male appear to be of shorter duration than those with pods in which adult males are absent (R. W. Baird, unpublished data). Males are likely to be accepted into groups for longer periods when the group contains one or more young whales but lacks an adult male, as their presence may be valuable to the group in terms of defense against attacks by *residents*. Lone *transient* males do not appear to join with other lone males. One interaction observed between two lone males passing within 500 m of each other involved considerable percussive behavior by one individual, but both whales continued their routes of travel past each other (R. W. Baird, personal observation). It is surprising that lone males do not associate in pairs, considering the extensive energetic benefits associated with foraging in groups of two or three versus foraging alone (Baird and Dill 1996). Pair formation could theoretically confer benefits in terms of increasing mating success, if a pair of males were more able to mate with a female, or to prevent the female from mating with other males. The lack of such pair formation, especially considering its foraging advantages, suggests that any increase in the ability to sequester females by pair formation would be outweighed by the reduced likelihood of paternity (see Connor et al., chap. 10, this volume).

The Future of the Taxon: Conservation Status and Critical Research Issues

In general, most killer whale populations have probably been affected by human activities to a relatively small degree when compared with other marine mammal species. Killer whales have been hunted for oil and meat (for human or animal consumption, fertilizer, or bait) in many areas, but particularly the western North Pacific (off Japan), the eastern North Atlantic, and the Antarctic (Berzin and Vladimirov 1983; Bloch and Lockyer 1988; Oien 1988; Reeves and Mitchell 1988b; Hoyt 1990). These fisheries have been discontinued since the early 1980s, and only very small numbers are taken occasionally today, either directly or incidentally (IWC 1993). Culling of animals, because of their perceived or known threat to fisheries, has also occurred (Dahlheim 1988; Olesiuk et al. 1990). Live-capture fisheries for public display in oceanaria have been focused in three areas, British Columbia and Washington, Iceland, and Japan (Hoyt 1992). While only small numbers of animals were taken, these takes had a substantial effect on local population sizes in British Columbia and Washington (Olesiuk et al. 1990). While the population of *resident* killer whales off northern Vancouver Island has recovered to levels higher than those prior to the live-capture operations (Olesiuk et al. 1990), southern *resident* killer whale population growth has been more sporadic, and the population has recently been declining (Baird 1999). Considering the extensive whaling takes off Japan from the early 1950s through the early 1980s (summarized in Hoyt 1990), even the small number of animals (five) live-captured off Japan in 1997 could jeopardize the recovery of the local population.

In recent years, nonconsumptive utilization of killer whales—that is, whale-watching—has become particularly prominent in Washington and British Columbia. These activities have raised a variety of concerns among researchers and members of the public as to the potential for disturbance in what are generally considered important feeding areas (Kruse 1991; Duffus and Dearden 1992, 1993; Phillips and Baird 1993; Duffus and Baird 1995). The available evidence for effects of boats on whale behavior or occurrence is generally unclear, and research on these interactions is continuing.

Killer whales have been shown to have among the highest levels of contaminants of any cetacean worldwide (Calambokidis et al. 1990). Surprisingly, levels of mercury appear to be higher in the tissues of *resident* whales than in *transients,* an unexpected trend considering the relative trophic levels of the two forms. Presumably such levels reflect consumption of heavily contaminated prey, but such prey consumption has not become apparent in the observational studies of foraging undertaken to date. The deployment of TDRs has suggested more extensive foraging on bottom or midwater fish than previously expected, thus identifying a possible source for higher levels of heavy metals (Baird 1994). TDR deployments on larger numbers of individuals in a wider variety of circumstances, behaviorally, temporally, and geographically, may also help better elucidate the regions of the water column where *residents* forage. Further work on year-round habitat use, perhaps through the deployment of satellite tags, is also necessary to identify general movement patterns during the winter months. These two areas of investigation are necessary if a potential source of contaminants is to be identified (and would also be of great value in understanding other aspects of killer whale biology). Little is known about the potential effects of such contaminants, and further research is also needed on that subject. Toxins such as spilled oils probably have an acute effect (Dahlheim and Matkin 1994).

The potential for direct competition with humans for marine resources, or for indirect interactions through the food web, has been virtually unexplored (Baird et al. 1992). The major prey populations of both *residents* and *transients* in the nearshore waters of the eastern North Pacific (salmon and seals) have been substantially reduced in size in the last hundred years, yet nothing is known about the potential effects of these reductions on top-level predators.

One of the most interesting aspects of killer whales is their extreme variability in foraging tactics, behavior, and dispersal patterns, even within one geographic area. Perhaps the most interesting questions that can be addressed for killer whales concern some of the causes and consequences of this variability. Better documentation of killer whale populations in areas outside British Columbia and Washington, to show whether such variability occurs elsewhere and what variations exist depending on the particular ecological circumstances in each area, is needed to address these questions. In terms of better understanding the relationship between *resident* and *transient* killer whales, several areas of research should be pursued. One is the determination of a behavioral isolating mechanism, and the clear differences in underwater sounds produced by the two forms is the obvious candidate. The role that pod- or community-specific dialects may play in isolating these groups socially could be investigated by monitoring the reactions of *transients* to playbacks of *resident* sounds, and vice versa, as well as playbacks of the calls of northern and southern *residents* to each other. The diversity of playback circumstances would provide a clear experimental forum for understanding how *transients* and *residents* might interact upon meeting, a situation that is rarely observed in the wild, yet may have important implications for population structure. The possible differences in body size of off-shore killer whales from either *residents* or *transients* could be investigated using photogrammetry (as could potential morphological differences between *residents* and *transients*). Another approach is examining the consequences of reproductive isolation between the two forms in terms of any skeletal or other morphological differences that may exist.

Further investigation of genetic differences at all levels of social organization is needed. This approach could include more detailed studies of population-level differences between offshore, *resident,* and *transient* populations, between northern and southern *residents,* and between those whales and populations of *resident*-type whales identified in the contiguous area farther north in Alaska and off California and Mexico, as well as within-group genetic variability. Genetic analysis of paternity could be used to interpret social behavior and to determine male and female mating strategies. For *transients,* analyses of relatedness for groups not regularly observed together could be used to confirm more cases of dispersal and provide more concrete evidence for the rules of dispersal that *transients* follow. Genetic data could also be used to examine relatedness between individual *transients* that share similar foraging specializations (nearshore versus offshore foragers). In theory, *transients* from different pods that share similar foraging tactics should be more strongly related maternally than paternally, assuming that such foraging tactics are learned and passed on maternally.

While research on killer whales worldwide has been concentrated in the research sites I focus on here, efforts are being made to initiate studies of killer whales elsewhere. Further support for studies examining populations at the fringes of the well-established research sites, as well as more isolated populations, will greatly increase our knowledge of home range size and population variability respectively. At all sites, research on nighttime behavior has been virtually nonexistent, as has been work on offshore movements or populations. Research in virtually all areas has also had a strong seasonal bias, with research being undertaken when prey abundance is highest. Detailed studies of behavior, social organization, and ecology are needed during other seasons, when prey is likely to be more limiting. These studies could be combined with acoustic measurements of blubber thickness of known, free-ranging animals (M. Moore, personal communication) to examine how reduced prey availability, and thus presumably reduced energy stores, affect social behavior, as well as to correlate reproductive success with energy stores.

Even though many detailed behavioral studies have been undertaken, their reliance on methods of observing groups, rather than individuals, has limited the conclusions of many studies and has made comparisons with studies of other animals difficult, if not impossible (Mann, chap. 2, this volume). Further focal animal studies will be valuable. These could include studies of the relationship between infants and their mothers and other individuals in terms of the potential costs and benefits of allomaternal behavior, and of the relationship between postreproductive females and other individuals in the population, in terms of selective pressures favoring survival beyond reproductive senescence for females. The differing methods of researchers working on different populations of killer whales have also made comparisons difficult, and multipopulation behavioral research studies would be of value. Combined with the continuation of the long-term studies discussed here, such work will lead to a much greater understanding of the complexities of behavior and social organization of killer whales.

6

THE SPERM WHALE

Social Females and Roving Males

HAL WHITEHEAD AND LINDA WEILGART

THE SPERM whale (*Physeter macrocephalus* Linnaeus) is the preeminent vertebrate predator of the mesopelagic ocean—the waters at depths between about 200 and 1,000 m. In terms of the biomass it removes from the oceans, the species rivals humans (Kanwisher and Ridgway 1983). This remarkable ecological success is likely related to its large brain (the world's largest), slow maturation, long and deep dives, and peculiar acoustic system, which seems to include the extraordinary spermaceti organ (fig. 6.1), as well as to a complex social organization (Weilgart et al. 1996), the principal subject of this chapter.

The sperm whale is the largest of the toothed whales (Odontoceti), with the greatest sexual dimorphism of any cetacean (fig. 6.2). Sizes vary between ocean areas and with degree of exploitation (Kasuya 1991), but most mature females are between 9.5 and 11 m, and most mature males between about 13 and 18 m. When seen from above the surface, they usually look much like dark gray logs (fig. 6.3). The blowhole is at the apex, and on the left side, of the smooth head, which takes up one-quarter to one-third of the body and largely contains the spermaceti organ, a long barrel-shaped structure filled with spermaceti wax (Rice 1989). Its principal function seems to be in the production of loud clicks (Norris and Harvey 1972; Cranford et al. 1996). The rear part of the animal is covered with large "crenulations," or wrinkles. From underwater the sperm whale appears more graceful. The underslung, rod-shaped lower jaw, frequently outlined in white (see fig. 6.1), is distinctive.

Rice (1989) gives a general summary of sperm whale biology, and there are comprehensive, although sometimes dated, reviews of sperm whale behavior (Caldwell et al. 1966), social organization (Best 1979), reproduction (Best

et al. 1984), acoustics (Watkins 1980), diet (Kawakami 1980), and stock structure (Dufault et al. 1999) available. In this chapter we will integrate information from the whaling industry, mainly describing diet, reproductive biology, and life histories (usually citing these reviews), with more recent research on the behavior and social organization of living animals. We will suggest how the remarkable extremes achieved by the sperm whale in so many areas may be related, and how they contribute to special conservation concerns for this species.

Phylogeny

The sperm whale is quite distinct taxonomically. It is the only living species in its genus, and in the family Physeteridae there are only two other, much smaller, extant species: the pygmy sperm whale, *Kogia breviceps,* and the dwarf sperm whale, *Kogia simus.* Physeteroid fossils have been found from the late Miocene (ca. 5–10 million years ago) (Kellogg 1928). The sperm whales seem to have diverged from members of the suborder Odontoceti, the toothed whales, early in cetacean evolution. Using molecular data, Milinkovitch et al. (1993) argue that sperm whales may be a sister group to the baleen whales, suborder Mysticeti, and thus that the Odontoceti are paraphyletic. However, the analytical methods used in arriving at this result have been challenged, and morphological data strongly support sperm whales being members of a monophyletic Odontoceti (Heyning 1997).

Hunting and Studying Sperm Whales

The sperm whale has had a close relationship with humans since the eighteenth century. It was the subject of two mas-

154

Figure 6.1. The head of a sperm whale. The volume above the narrow lower jaw and in front of the eye is largely filled by an oil-filled sac, the spermaceti organ. (Photograph courtesy of H. Whitehead laboratory.)

Figure 6.2. Mature male with females on the Galápagos breeding ground (1989), showing the considerable sexual dimorphism present in adult sperm whales. (Photograph courtesy of H. Whitehead laboratory.)

sive hunts. The first of these, peaking in about 1840, greatly reduced sperm whale populations throughout the world (Whitehead 1995b) while providing the oil that lubricated the Industrial Revolution and stimulating Herman Melville (1851) to write the great novel *Moby-Dick*. Mechanized whaling killed sperm whales during the twentieth century until 1988, but the species was a particularly important target in the 1960s, when they were killed at a rate of over 20,000 per year (Best 1983).

Some aspects of sperm whale biology have been studied using material collected during the most recent hunt (especially between 1950 and 1980). These studies have provided very useful information on the reproductive biology, life history parameters, and diet of sperm whales (e.g., Clarke 1980; Kawakami 1980; Best et al. 1984). Inferences about behavior and social organization were also made from this material, but these were necessarily incomplete (Best 1979).

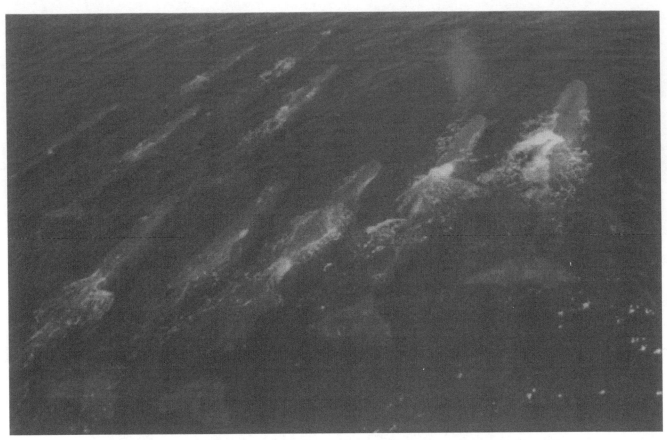

Figure 6.3. A group of female and immature sperm whales congregating during a social period at the surface off the Galápagos Islands. (Photograph courtesy of S. Hooker.)

Longitudinal studies of living animals began in 1982 off Sri Lanka (Whitehead and Gordon 1986; J. C. D. Gordon 1987b). There are current multi-year studies of females and their offspring off the Galápagos Islands (Whitehead 1990a), Dominica, the West Indies (Watkins et al. 1985; Gordon et al., in press), and the Azores (Gordon and Steiner 1992), and of males off northern Norway (Lindhard and Strager 1989) and New Zealand (Childerhouse et al. 1995). Shorter studies have been carried out, or are being started, in a number of other locations, including the Canary Islands (André 1997), the Seychelles (Kahn 1991), Nova Scotia (Whitehead et al. 1992), and mainland Ecuador (Whitehead and Kahn 1992). The studies of females mostly use 10–13 m ocean-going auxiliary-powered sailing vessels, but successful research has been conducted using larger vessels (e.g., Kahn 1991; Palacios and Mate 1996) and smaller, outboard-powered inflatables (e.g., Dawson et al. 1995).

The methods used to study living sperm whales are rather different from those that have become standard for other cetaceans, but the species is amenable to some kinds of nonlethal study. Sperm whales typically dive for periods of thirty to sixty minutes (e.g., Papastavrou et al. 1989; Watkins et al. 1993), but they can be located and tracked underwater by listening for their distinctive clicks (audible underwater to about 7 km: Leaper et al. 1992) with directional hydrophones (Whitehead and Gordon 1986). In daylight with good sighting conditions, they can be seen breathing at the surface at ranges of a few kilometers, but their blows are less prominent than those of similarly sized baleen whales, so sighting surveys are less efficient (e.g., Christensen et al. 1992). Individual males can be tracked, using a combination of acoustic and visual methods, for up to about twenty-four hours (e.g., Mullins et al. 1988), and groups of females for several days (e.g., box 6.1). However, in areas of reasonably high abundance, a tracking vessel may inadvertently transfer between individuals or groups one or more times per day.

During social periods, and the approximately eight-minute intervals when sperm whales are at the surface between deep dives, they can usually be approached to within about 30 m by a slow-moving boat. Individuals are identified principally from marks on the trailing edge of their flukes, which are photographed as the whale dives (figs. 3.4

Box 6.1

A Week in the Life of a Sperm Whale Group

JENNY CHRISTAL AND HAL WHITEHEAD

In 1995, we followed a group of sperm whales near the Galápagos Islands continuously for one week (except for one eight-hour break) (fig. 6.4). The group contained twenty-two individuals (all photo-identified) and had consistent membership throughout the week. Eighteen of these whales had been identified during previous years, and their patterns of association in those previous sightings (as well as a subsequent sighting in 1996) enabled us to divide the group into two distinct units (fig. 6.5). Table 6.1 gives sex, size, and age information for each individual, while table 6.2 summarizes the movements and behavior of the group.

Features of the behavior and characteristics of the group that are representative of sperm whales studied off the Galápagos include:

1. Size of group: Groups of female and immature sperm whales encountered in this area average about twenty animals.

2. Unit structure: Groups encountered in the field often represent the temporary (hours to days) association

of two (or sometimes more) separate units. In this case, the group consisted of one unit with five mem-

Table 6.1. Demographic information for members of the group

ID number	Sex[a]	Size (m)[b]	Age (in 1995)[c]	Female of reproductive age?[d]
235	—	—	>10*	?
255	F	—	>10*	Y
2242	F	—	>6*	?
2361	F	10.2	20	Y
2818	—	—	>4*	?
793	—	10.7	>30(F)/14(M)	?
795	F	—	>8*	Y
754	M	9.5	9	N
804	F	10.2	21	Y
806	—	11	>30(F)/15(M)	?
807	F	10.4	23	Y
809	F	10	18	Y
810	F	10	18	Y
811	F	9.7	15	Y
812	F	—	>8*	Y
814	F	—	>8*	Y
2935	—	—	>1*[e]	?
2942	F	8.2	7	N
3287	F	8.6	9	N
3290	—	10.1	18(F)/11(M)	?
3295	—	9.2	11(F)/8(M)	?
3303	—	9.4	12(F)/9(M)	?

[a]Sex was determined from sloughed skin samples, using a modification of the *SRY* molecular sexing technique developed by Richard et al. (1994).
[b]Size was determined using a photographic measurement technique (Gordon 1990). Where more than one measurement was available for a particular animal, the average value is given.
[c]Ages were determined in two different ways. For those individuals for which a length measurement was available, approximate age was determined using published growth curves (Best 1970). Where a length measurement but no sex was available for an individual, age estimates for both sexes are provided. Where no length measurement was available, the age given (and marked with an asterisk) represents the number of years since the first identification of that individual. These figures are likely to be underestimates of age, particularly since none of these individuals were noticeably small on first identification, and were therefore probably at least five years old at that time.
[d]Females were considered to reach reproductive maturity at nine years of age (Best 1968).
[e]Although individual 2935 was not photographically measured, no individuals other than 2942 and 3287 were noted to be small, and so 2935 is not believed to have been a juvenile.

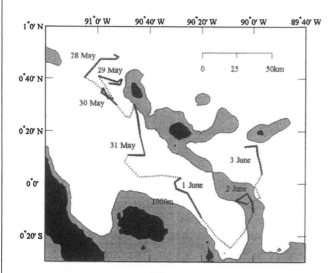

Figure 6.4. Map showing the track of a group of sperm whales near the Galápagos Islands, 28 May–3 June 1995 (daylight hours, solid line; overnight acoustic tracking, dotted line). Islands are black, and waters less than 1,000 m deep are shaded.

Unit	Whale	1985	1987	1988	1989	1991	1994	1995	1996
A	235	✓	✓		✓			✓	
	255	✓	✓	✓	✓	✓		✓	
	2242				✓	✓		✓	
	2361				✓	✓		✓	
	2818					✓		✓	
B	793		(✓)				✓	✓	✓
	795		(✓)	✓			✓	✓	✓
	754		✓				✓	✓	✓
	804		✓	✓			✓	✓	✓
	806		✓				✓	✓	✓
	807		✓				✓	✓	✓
	809		✓				✓	✓	✓
	810		✓	✓			✓	✓	✓
	811		✓	✓			✓	✓	✓
	812		✓				✓	✓	✓
	814		✓				✓	✓	✓
	2935						(✓)	✓	✓
	2942						✓	✓	✓
	3287						✓	✓	✓
	3290							✓	✓
	3295						✓	✓	✓
	3303						✓	✓	✓

Figure 6.5. Sighting histories based on photo-identification records for the twenty-two members of the group. A check mark indicates that an individual was identified during that year; units were delineated on the basis of members being identified together on the same day on several occasions (Christal et al. 1998). The two units were identified together only during 1995. Circled check marks indicate that although those individuals were identified during that year, they were identified with other units, not with unit B. These three whales are believed to have transferred into unit B at some point after their initial identification (Christal et al. 1998).

bers and another with seventeen members, which remained in association for at least seven days (fig. 6.5). Mean unit size off the Galápagos is twelve (Christal et al. 1998).

3. Lack of calves: The low reproductive rate of sperm whales in the Galápagos area is exemplified by this group, which contained a minimum of ten females of reproductive age, yet no animals of less than about six years of age (table 6.1).

4. Brief encounters with a mature male: The same mature male was seen in association with the group for brief periods on two consecutive days (table 6.1). On 31 May, the male was first seen at 16:30 at one end of a rank of approximately twenty group members. The male remained at the surface for twenty minutes after all members of the group dove, then he dove, and was not seen again that day. On the following day, 1 June, the same male was observed 400 m away from the group at 12:45, but was not seen to associate directly with any members of the group, and appeared to leave the area shortly after 13:00.

5. Daily distance moved correlated negatively with feeding success ($r = -0.71$): Feeding success, as indicated by defecation rates, was highest on 29 May, the day on which the group traveled the shortest straight-line distance, and lowest on 1 June, when the group traveled its greatest daily distance (fig. 6.4; table 6.2).

Rather unusual features of the behavior of this group included:

1. No encounters with other groups: Groups in the Galápagos have often been observed to associate with other groups to form temporary aggregations lasting

Table 6.2. Summary of daily movement and behavior

Date	Distance moved (km)[a]	Feeding success (%)[b]	Socializing periods	Encounters with male
28 May	31.9	21.2	13:30–18:30	—
29 May	21.1	30.2	None	—
30 May	37.9	7.4	12:50–17:00	—
31 May	44.2	13.9	15:00–17:00	Y
1 June	81.9	6.3	07:45–11:58	Y
2 June	47.9	17.1	10:48–14:20	—
3 June	28.0	15.5	None	—

[a]Distance moved was calculated using the straight-line distance between the GPS position of the group at midnight on successive days. The value for 28 May is for eighteen hours (06:00–24:00), as is the value for 3 June (24:00–18:00).
[b]Feeding success was estimated by the defecation rate: the percentage of all observed fluke-ups at which defecation was seen to occur (Smith and Whitehead 1993).

for periods of hours (Whitehead and Weilgart 1990). This group did not associate with any other groups during the seven days of observation. There was a very low density of groups of female sperm whales in Galápagos waters at the time, a symptom of a general decline in population size in the area (Whitehead et al. 1997a).

2. Incursion into waters less than 1,000 m deep: The group spent most of 2 June in waters shallower than 1,000 m (fig. 6.4), whereas sperm whales in this area generally stay in deeper waters. However, the depth in this area exceeded 850 m, and because of the underwater topography (fig. 6.4), the only way the group could have continued to travel in waters deeper than 1,000 m was to return the way they had come, through a region in which they had had poor feeding success.

3. Temporal patterns of socializing behavior: Sperm whales around the Galápagos Islands are usually observed to show social or resting behavior during the afternoon (Whitehead and Weilgart 1991). This group followed that general pattern on only three days, 28 May, 30 May, and 31 May (table 6.2). On two other days, no social/resting behavior was observed during daylight hours. Following one of those days (29 May), the group was lost during overnight acoustic tracking for eight hours. It is possible that the whales were silent at this time if they were socializing or resting following a day of unbroken foraging.

4. Presence of individuals that had transferred into a unit: One pair of individuals, and one single whale, that were identified as members of unit B over periods of years were found to have been associated with whales other than unit B at the time of first identification (fig. 6.5). This finding suggests that these individuals transferred into unit B at some time between their first and second identifications, and provides support for suggestions raised by genetic studies (Richard et al. 1996a; Christal 1998) that there is sometimes long-term association between individuals from unrelated matrilines.

and 6.6; Arnbom 1987; Dufault and Whitehead 1995a; Childerhouse and Dawson 1996). Living sperm whales are measured using stereophotography (Dawson et al. 1995) or photographs showing the whale and the horizon taken from a known height (Gordon 1990). Length can also be estimated from analysis of the interpulse interval of sperm whale clicks (Gordon 1991; Goold 1996). Diving behavior is studied using sonar or depth-recording tags (e.g., Papastavrou et al. 1989; Watkins et al. 1993). Squid beaks can be collected from feces (Smith 1992) to supplement the information on diet that formerly came from stomach contents (Kawakami 1980). Collections of sloughed skin or samples collected by biopsy dart have been used for genetic analyses of sex and relatedness (Whitehead et al. 1990; Richard et al. 1994, 1996a,b; Lyrholm and Gyllensten 1998; Lyrholm et al. 1999; Whitehead et al., chap. 3, this volume).

Behavior has been recorded in several ways. Focal animal follows, in which all blows, dives, and movements are recorded as they happen (sometimes with concurrent

Figure 6.6. A diving sperm whale showing its flukes. Markings along the trailing edge of the flukes are used to identify individuals. (Photograph courtesy of H. Whitehead laboratory).

acoustic recordings) are possible for single males over periods of hours (e.g., Mullins et al. 1988) and for females in groups over periods of minutes (e.g., Gordon and Steiner 1992). To date, the most successful method of systematically recording the behavior of grouped females and immatures over periods of many hours has been to use scan samples, in which the positions (relative to the boat), movements, and observable behavior of all visible whales are recorded every five minutes. These samples can be linked to regular acoustic recordings (Whitehead and Weilgart 1991). Observations of behavior at night have been few and haphazard, although some inferences can be made from regular acoustic recordings (e.g., Whitehead and Weilgart 1991).

Distribution and Ecology

Habitat, Diet, and Predators

Sperm whales may be seen almost anywhere in the deep ocean, although they are more frequently found in certain geographic areas spanning roughly 300–1,500 km, which the whalers called "grounds" (Townsend 1935; fig. 6.7). These grounds are often, but not invariably, coincident with regions of increased primary productivity caused by upwelling (Gulland 1974; Jaquet et al. 1996). At any time, sperm whales may form concentrations a few hundred kilometers across in waters characterized by increased secondary productivity (Jaquet and Whitehead 1996; table 6.3, fig. 6.7). These concentrations usually lie within a ground.

Females are found only rarely in waters less than 1,000

m deep. The males are also generally deep-water animals. However, in some parts of the world, such as the western North Atlantic, larger males quite consistently use relatively shallow waters 40–500 m deep (e.g., Mitchell 1975a; Whitehead et al. 1992; Scott and Sadove 1997). While females and immatures are usually restricted to latitudes less than 40° (except in the North Pacific, where they are found up to 50°N), males generally move to higher latitudes as they mature, with the largest males sometimes using waters close to the ice edge in both hemispheres (Rice 1989; fig. 6.8).

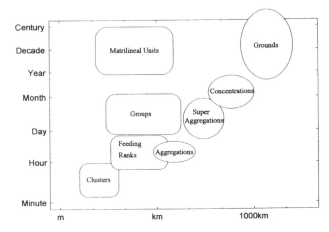

Figure 6.7. Approximate temporal and spatial scales of the spatial and social organization of sperm whales. The rounded boxes represent primarily social structures ("mutualist groups"); the ellipses indicate animals aggregated because of extrinsic factors, primarily the distribution of food ("non-mutualistic groups"). (Adapted from Jaquet 1996.)

Table 6.3. Levels of social and spatial organization of female and immature sperm whales off the Galápagos Islands and in the South Pacific

| Level | Approximate scale | | | Possible function |
	Number	Time	Space (distance across)	
Concentration	1,000	??	200 km	General area of food
Super-aggregation	?	days-months?	40 km	Prey patch
Aggregation	40	6 hours	15 km	Concentrated food
Group	20	10 days	1 km	Efficient foraging
Unit	12	>12 years	0.5 km	Calf protection
Cluster				
foraging	2	5 minutes	20 m	Social, predator protection
socializing	15	minutes-hours	200 m	

The world's population of sperm whales has a varied diet, including squid with mantle lengths ranging from a few centimeters to 10 m, midwater and benthic fish, and crustaceans (Kawakami 1980; Rice 1989). However, their staple diet in most oceans consists of mesopelagic squid with masses of about 1 kg. Together, the world's sperm whales consume a similar biomass to all human fisheries for all marine species, roughly 100 million tons per year (Kanwisher and Ridgway 1983).

Sperm whales possess a range of internal parasites, especially helminths, and external parasites, including copepods and barnacles, but none seem to be an important source of mortality (Rice 1989). They can be attacked by killer whales, *Orcinus orca* (Arnbom et al. 1987; Brennan and Rodriguez 1994; Pitman and Chivers 1999), and there are observations of apparent predation or harassment by large sharks (Best et al. 1984), pilot whales, *Globicephala macrorhynchus* (Weller et al. 1996), and false killer whales, *Pseudorca crassidens* (Palacios and Mate 1996). Their large size and methods of communal defense appear to keep healthy

adult sperm whales largely safe from mortality due to predators (Jefferson et al. 1991), although Pitman and Chivers (1999) recently observed killer whales killing at least one sperm whale, and severely injuring several others, in an attack on a group of females and immatures off the coast of California.

Ranging Behavior and Migrations

The home ranges of individual females seem to span distances of approximately 1,000 km (Best 1979; Dufault and Whitehead 1995c). However, occasionally females travel several thousand kilometers across large parts of an ocean basin (Kasuya and Miyashita 1988). These rare long-distance migrations may account for a lack of geographically based structure in the mitochondrial DNA of female sperm whales within ocean basins, or in nuclear DNA around the world (Dillon 1996; Lyrholm et al. 1996, 1999; Lyrholm and Gyllensten 1998). The Galápagos population on the equator shows no indication of seasonal migrations (H. Whitehead, unpublished data), but females at higher latitudes seem to move away from the poles during winter (Best 1979).

After dispersal from their natal family units, males move to generally higher latitudes as they age and grow (see fig. 6.8; Best 1979). Their annual home ranges seem to be larger than the females' (e.g., Martin 1982), and they appear to migrate seasonally toward the equator in winter (Best 1979). However, some mature or maturing males show strong site fidelity to high-latitude areas just a few kilometers across (Childerhouse et al. 1995). Males older than about twenty-five years migrate to tropical breeding grounds, but little is known about the frequency, the duration, or the seasonality of the visits of individual males to the tropics. Photographically identified mature males spent periods of at least two months on the Galápagos breeding ground (Whitehead 1993). We do not know the geo-

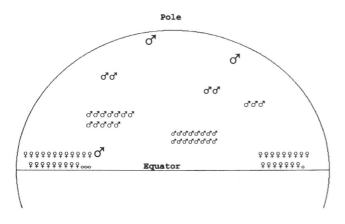

Figure 6.8. Social organization and latitudinal distribution of sperm whales in the Northern Hemisphere. A similar pattern is found in the Southern Hemisphere.

graphic relationships between the range of a male's natal family unit, his high-latitude feeding areas, and the locations of the breeding grounds that he visits.

The movement patterns of sperm whales depend very considerably on their foraging success. In the eastern tropical Pacific, foraging success in any area shows great variation over time scales of years and is negatively correlated with the presence of the "El Niño" oceanographic phenomenon (Smith and Whitehead 1993; Whitehead 1996b). When feeding success, as indicated by defecation rates, is high, sperm whales tend to stay within geographic areas a few tens of kilometers across, zig-zagging back and forth, whereas when there is little food available, they move in fairly straight lines, covering about 90 km per day (Whitehead 1996b; Jaquet and Whitehead 1999). Since at any time there are considerable differences in oceanic conditions, and so feeding success, between areas a few hundred kilometers apart (Whitehead 1996b), this strategy of moving away from areas with poor food supplies likely allows sperm whales to survive El Niño conditions, which can be so disastrous for less mobile animals (Arntz 1986). A home range spanning approximately 1,000 km should, most of the time, contain areas where food is reasonably abundant.

Foraging Behavior

Sperm whales have substantial energy requirements (Lockyer 1981a), but generally eat relatively small food items (ca. 0.01% of the mass of a sperm whale) with rather little nutritional value, which they encounter singly or in small schools (Clarke 1980). Off the Azores, about 77% of the food of sperm whales (by mass) consisted of small (<1,200 g), slow-moving, luminous and neutrally buoyant squid, and about 23% of larger, fast-moving, nonluminous cephalopods (Clarke et al. 1993). A general foraging pattern seems to consist of dives to about 300–800 m below the surface for periods of about forty minutes separated by about eight minutes of breathing at the surface (Papastavrou et al. 1989; Gordon and Steiner 1992; Watkins et al. 1993). However, foraging dives may last for over an hour, reach below 1,000 m, or be constrained by water depths of less than 200 m (Whitehead et al. 1992; Watkins et al. 1993; Scott and Sadove 1997). While at depth the whales swim at about 4 km per hour (e.g., J. C. D. Gordon 1987b; Mullins et al. 1988; Watkins et al. 1993).

The mechanisms by which sperm whales find and catch their food are unclear. Loud, regular trains of clicks (with about 0.5 seconds between clicks) are heard from both solitary males and aggregated females most of the time that they are foraging at depth (Mullins et al. 1988; Whitehead and Weilgart 1991; Goold and Jones 1995). Most scientists believe that these "usual clicks" are used for echolocation (e.g., J. C. D. Gordon 1987b; Goold and Jones 1995). Vision may also be important in finding food (Fristrup and Harbison 1993). Many of the squid that sperm whales eat are bioluminescent (Clarke et al. 1993). Gaskin (1967) has proposed that squid may be attracted by their white jaws as well as luminescent mucus from previous prey. The methods of prey capture probably vary with prey type. Clarke et al. (1993) suggest that smaller, slower, luminous squid may first be detected by echolocation, then seen and easily eaten, whereas larger, more agile, nonluminous prey probably need additional use of echolocation and may be actively chased. The sperm whale's teeth and lower jaw are not necessary for the ingestion of food, as well-nourished animals have been caught with broken or deformed lower jaws (Rice 1989). The whales probably suck their prey into their mouths (Rice 1989).

Mature and maturing males usually seem to forage independently (e.g., Gordon et al. 1992; Whitehead et al. 1992), although there may be coordination of movements over scales of a few kilometers (Christal and Whitehead 1997). Groups of females and immatures usually spread out while foraging (Watkins and Schevill 1977b; Whitehead 1989a). The dives of groups containing calves are staggered in such a way that there are shorter intervals with no adults at the surface than in groups without calves (Whitehead 1996a). There is considerable variation in the extent of dispersion while foraging, but the modal spread of a group of twenty or so females and immatures may be about 500 m in the horizontal plane. Off the Galápagos, groups of females and immatures frequently forage in a rank aligned perpendicular to the direction of movement (Whitehead 1989a). When breathing at the surface between dives, they are usually alone (no animals within 100 m) or in clusters of two animals (see fig. 6.7, table 6.3).

Most studies of diet have found no consistent diurnal pattern of feeding (Clarke 1980). The little information that does exist also suggests little difference in dive depths between day and night (Papastavrou et al. 1989; Watkins et al. 1993). However, off the Galápagos (Whitehead and Weilgart 1991) and in the western South Pacific (H. Whitehead, unpublished data) foraging behavior, as indicated by diving and regularly spaced click vocalizations (see below), was more common at night and in the morning than during the afternoon.

Life History

The sperm whale is the epitome of a "*K*-selected" animal, in the sense of a species that seems to have evolved with populations close to the carrying capacity of the environment (see Whitehead and Mann, chap. 9, this volume, for comments on the use of this term). Sperm whales have long life spans (sixty to seventy years), low mortality (estimated to be 5.5% per year for females and 6.6% for males: International Whaling Commission 1982), very low fecundity, large size, and prolonged parental care (Rice 1989; see table 6.4). Following a gestation period of about fourteen to sixteen months, sperm whales are born singly at about 4.0 m long, with a sex ratio of 1:1 (Best et al. 1984). Calves first eat solid food when less than one year old, but may continue to suckle for at least five more years. Lactose was found in the stomach of a thirteen-year-old male (Best et al. 1984), although the significance of this is debated (Oftedal 1997).

There is considerable variation in the age at which males disperse from their family units and begin their progressive relocation to cooler waters, but the mean seems to be about six years of age (Best 1979; Richard et al. 1996a). Males gradually become sexually mature in their teens, but do not seem to take an active role in breeding by entering the warm-water breeding grounds until their late twenties, when they are about 13 m long (Best 1979). However, they continue growing at least until their forties (fig. 6.9).

Females, which grow more slowly than males of the same age, become sexually mature between seven and thirteen years of age, at a length of roughly 8–9 m (fig. 6.9; Rice 1989). In each population studied there is a three-month period of peak conception, principally during spring, although successful breeding may occasionally take place at other times of year (Best et al. 1984). Within these locally preferred mating seasons, females within groups synchronize their estrous periods, possibly on account of their shared environment (Best and Butterworth 1980). The mean calving rate is thought to be about one every five years in a population near carrying capacity, and one every four years in a depleted population (Best et al. 1984), although there are some exceptions. During field studies of living animals and studies of animals caught by whalers, there have been some measurements of very low pregnancy or calving rates (about one surviving calf every ten years) (Clarke et al. 1980; Whitehead et al. 1997a). These may be due to a scarcity of breeding males caused by selective whaling resulting in reduced pregnancy rates. The International Whaling Commission's "Sperm Whale Model,"

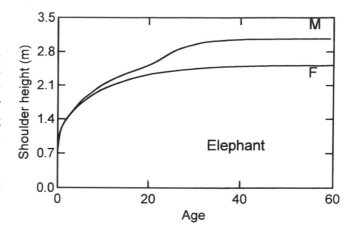

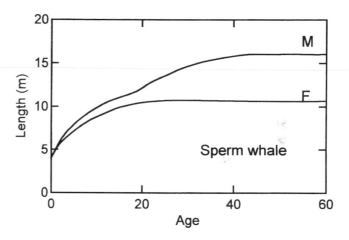

Figure 6.9. Growth curves for male and female sperm whales (data from Rice 1989) and elephants (data from Laws et al. 1975). Both curves are based on very large sample sizes.

which was used to study the dynamics of sperm whale populations, contains a component relating declines in pregnancy rates to the relative depletion of males (International Whaling Commission 1982). Female fecundity declines with age, with females over age forty being found pregnant about half as often as those aged between ten and thirty, but there is no indication of menopause, as indicated by a cessation of ovulation (Best et al. 1984).

Spatial Aggregation, Social Behavior, and Communication

Sperm whales are often seen in clusters of two to forty animals swimming in the same direction, a few meters apart, at the surface (see fig. 6.3). These clusters frequently change composition as animals converge, diverge, and dive. In the case of females and immatures, the clusters and their dynamic memberships are the most obvious visible manifestation of an apparently complex social structure.

Over spatial scales within home ranges and temporal scales within life spans of individuals, we recognize several hierarchical levels of social and spatial organization in female sperm whales. These range from the temporary clusters of animals seen at the surface together through the most durable associations, which we call "units" (typically about ten animals), "groups" of about twenty animals that are stable for periods of hours to days, temporary "aggregations" and "super-aggregations" of forty or more animals, to geographic "concentrations" of a thousand animals or more (Paterson 1986; Whitehead and Weilgart 1990; Whitehead et al. 1991; see fig. 6.7). The attributes of these levels vary considerably from the approximate medians presented in table 6.3, both geographically and temporally (Whitehead and Kahn 1992). Intermediate or additional, currently unrecognized, levels may exist, and some of the levels may not be represented in certain areas or at certain times.

Spatial Aggregation

On the broadest scale, during a survey of the South Pacific, female sperm whales were found in concentrations spanning approximately a thousand kilometers and separated by a few thousand kilometers. These concentrations correlated with areas of increased biomass at depths of a few hundred meters (Jaquet and Whitehead 1996), and acoustic censuses suggest that they contain on the order of a thousand whales (N. Jaquet and H. Whitehead, unpublished data).

Substantial numbers of sperm whales—aggregations—can sometimes be found within a radius of 15 km or so, particularly in productive waters, probably in response to temporary food concentrations. Off the Galápagos, an especially productive area, the mean aggregation size in 1985 and 1987 was about forty animals (Whitehead and Weilgart 1990). From South Pacific survey data, Jaquet (1996) found rather larger structures, about 10–70 km across, which she called "super-aggregations" and related to patches of prey.

Relationships between Females and Care of Calves

Within these aggregations, female sperm whales are most typically encountered in groups of twenty or so animals that move in a coordinated fashion. Some pairs of females can be found in the same group years after their first identification together, whereas other pairs seem to remain together only for periods of days or less (fig. 6.10; Ohsumi 1971; Whitehead et al. 1991). Analysis of several thousand identification photographs collected over a period of four

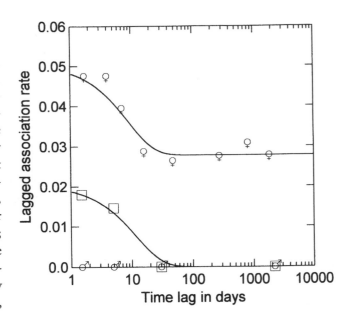

Figure 6.10. Standardized lagged association rates for associations among mature males (♂), between mature males and female/immatures (□), and among female/immatures (♀) off the Galápagos Islands, with fitted models of exponential decay in association rates. The standardized lagged association rate of lag d is the probability that if a particular animal X is identified within two hours of animal Y now, then X will also be identified within two hours of Y after a lag of d days, standardized by the rates of observing animals. (From Whitehead 1995a.)

years off the Galápagos indicated that the groups encountered often consisted of two or more stable units that had coalesced for periods of very approximately ten days (fig. 6.10; Whitehead et al. 1991).

The sizes of Galápagos units range between three and twenty-four animals, with a mean of twelve (Christal et al. 1998). Although there are occasional splits, mergers, and transfers between units, most members of a unit stay together for at least years (Christal et al. 1998). Genetic studies have mostly been carried out at the level of the group. The results generally suggest groups consisting of one or more matrilines (Dillon 1996; Richard et al. 1996a; Christal 1998; Lyrholm and Gyllensten 1998; Bond 1999; Lyrholm et al. 1999). However, there are suggestions of paternal relatedness between grouped matrilines, and of "missing" members of matrilines (Richard et al. 1996a; Christal 1998; Bond 1999). There has been only one study of genetic relationships within known stable units, and this study examined just two units, those described in box 6.1 (Christal 1998). One female in unit A possessed a distinct mitochondrial haplotype and so was unlikely to have been maternally related to the other members of her unit, and there was a lower level of relatedness among members of both units than would be expected in purely matrilineal units (Christal 1998). Nonetheless, these results, and those from the studies of groups, show that female sperm whales generally stay with at least some of their close relatives for long periods. Thus, it now seems reasonable to expand the terminology for a set of stable associates from "unit" to "family unit." We believe these units to be the fundamental elements of sperm whale society.

Whalers occasionally found groups apparently consisting entirely of immature males and small females together, suggesting some female dispersal from the family units (Best 1979). If female dispersal is present, then for the units to be made up of related animals, either a considerable proportion of the females must not disperse, or many females that do disperse must later rejoin their natal units. Perhaps the occasional sightings of groups of young males and females together were a result of the destruction of family units by whaling.

Calves associate with different adults and immatures within their units. As an older companion raises its flukes at the start of a long dive, a calf will swim to another nearby adult, or to the position where one is about to surface. Individual calves accompany a variety of adult or immature members of their unit, although they show a particularly strong association with one adult, presumably the mother (J. C. D. Gordon 1987a; Whitehead 1996a). J. C. D. Gordon (1987a) discusses observations of an identified calf suckling from different females, and two calves of similar size (so not siblings, as twins do not seem to survive: Rice 1989) simultaneously suckling from the same adult. Additionally, whalers often found more lactating females than calves in a group (Best et al. 1984). These results suggest communal suckling, although the data are not conclusive. Observation of the events subsequent to a sperm whale birth indicate that the communal life of a calf seems to begin soon afterward (Weilgart and Whitehead 1986).

When attacked by whalers, females cooperatively tried to assist their offspring and each other (Caldwell and Caldwell 1966). Members of the family unit were drawn to injured calves. The whalers made use of this by harpooning, but not killing, calves and then slaughtering the females who stood by them (Caldwell and Caldwell 1966). Females also stood by each other when injured, and there are reports of females attacking whaling boats that had fastened to another female, as well as attempting to break harpoon lines. Males sometimes appeared to aid females, but never each other (Caldwell and Caldwell 1966).

Similarly, when attacked by killer whales, female sperm whales take communal defensive measures. During an encounter between these species near the Galápagos, female sperms formed a very tight cluster, with individuals almost touching each other, around a centrally placed small calf, and maneuvered to keep their heads, and jaws, facing toward the attacking killer whales (Arnbom et al. 1987). An attendant large male was positioned behind the females. Using this formation, the females appeared to defend themselves, and the calf, successfully during a 2.7-hour attack. In other cases, sperm whales defend themselves communally by adopting the "marguerite" formation, with the animals appearing like the spokes of a wheel, their heads at the hub, and all animals swimming inward (Nishiwaki 1962). Sperm whales have several times been observed to adopt this formation in response to attack or harassment by pilot whales, *Globicephala macrorhyncha* (Weller et al. 1996), as well as false killer whales, *Pseudorca crassidens* (Palacios and Mate 1996). The sperm whales that were attacked and killed by killer whales off California had been adopting the marguerite formation (Pitman and Chivers 1999). The importance of communal defense in sperm whale groups is indicated by similarities in the type and degree of markings found on the flukes of sperm whales from the same group (Dufault and Whitehead 1998) (see fig. 3.7). Sperm whales that encounter and confront predators as a cohesive group might be expected to receive similar predator-produced scars.

Relationships between Males

After dispersal from their natal family unit, males are found in loose aggregations (which have been called "bachelor schools") usually consisting of animals of similar size (Best 1979). As males grow and move to higher latitudes, the sizes of these all-male aggregations become generally smaller (Best 1979). On the Scotian Shelf, 90% of the male sperm whales observed at the surface were alone (no other animals within 100 m); the rest were in pairings (clusters of two) of short duration, and no pairings between known individuals were observed over more than one day (Whitehead et al. 1992). However, male sperm whales often strand multiply (Rice 1989; Reeves and Whitehead 1997), which might suggest that they possess an underlying social organization that we have so far been unable to detect. In two all-male strandings in Scotland, the males were predominantly unrelated to one another (Bond 1999).

More than one male may attend a group of females at the same time, but we have found no consistent associations between breeding males over periods of days (see fig. 6.10; Whitehead 1993). Instead, there is evidence that breeding males avoid one another, with two or more males being found simultaneously with a female group less often than expected if their movements were independent (Whitehead 1993). The large males do fight with each other on occasion, and have the scars to prove it (Best 1979; Kato 1984), but such fights appear to be rare. Males within a particular breeding ground may assess one another using their loud and distinctive "slow clicks" (Weilgart and Whitehead 1988; Gordon 1991).

Mating

Mature male sperm whales stay resident within the concentration of females around the Galápagos for periods of at least several weeks (Whitehead 1993), and three males were observed off Dominica in successive years (Gordon et al., in press). However, while on the breeding grounds, they rove between groups of females, spending just a few hours with each at a time, but sometimes revisiting groups on consecutive days (see box 6.1; Whitehead 1993). While with the groups, the males behave in a manner quite similar to the females that they are accompanying, although their presence seems to increase the rates of social activities and vocalizations of the females (Whitehead 1993). Groups of females are sometimes seen accompanied by no mature males, sometimes by one, and occasionally by more than one (Whitehead 1993).

There are a few reports of apparent copulation in the literature (Di Natale and Mangano 1985; Ramirez 1988; Rice 1989), but none are detailed or particularly convincing. The low pregnancy rate of sperm whales implies that conceptive copulations are very rare events. Thus copulation itself is likely to be rare, unless it is very inefficient, as in lions *(Panthera leo)*, or used extensively for social rather than reproductive purposes, as in humans.

Social Behavior

Foraging at depth, interspersed with respiration periods at the surface, occupies roughly three-quarters of a sperm whale's time. The remaining one-quarter is generally spent in a social or resting mode of behavior at or near the surface (Whitehead and Weilgart 1991).

Female and immature sperm whales socialize near the surface in large clusters (see fig. 6.3). A group of twenty whales may form one to four clusters or aggregate with other groups to make a cluster of fifty or more whales (Paterson 1986). These clusters usually move slowly, with little directional consistency, but in other ways their behavior when socializing seems more varied than the quite stereotypical dive cycles of foraging sperm whales (Whitehead and Weilgart 1991). During social periods animals perform few fluke-ups (tail flukes raised into the air, usually before deep dives), but lobtails (flukes thrashed onto the water surface: fig. 6.11), sideflukes (half of flukes visible above water surface, as when the whale is turning), spyhops (head

Figure 6.11. A sperm whale "lobtailing"—thrashing its flukes on the water surface—off the Galápagos Islands. (Photograph courtesy of H. Whitehead laboratory.)

raised above the water), and breaches (leaps from the water: fig. 6.12) are often seen (Whitehead and Weilgart 1991). Sperm whales socializing near the surface often touch or stroke one another with their jaws or flippers. Sometimes they actively maintain physical contact with each other for periods of more than a minute by swimming gently into each other. On other occasions the tightly clustered whales lie very still and silently for periods of up to a few hours.

Off the Galápagos, periods of surface social behavior generally seem to last either about one hour or about five to six hours, and peak in the late afternoon (Whitehead and Weilgart 1991). The whales often end these social periods and start foraging dives at sunset. Likely functions of these periods at the surface include rest and the maintenance of social bonds after dispersion of the group during foraging (Whitehead and Weilgart 1991).

Mature males at high latitudes, which mostly appear to be solitary, also spend periods of an hour or more moving slowly at the surface (Gordon et al. 1992; Whitehead et al. 1992). However, breaching and lobtailing, activities that are thought to have a social function, are seldom observed (Waters and Whitehead 1990).

Most striking and unusual when compared with other social odontocetes is the strength of the correlation between broad categories of visually observable behavior and vocalization type in female sperm whales. In particular, "codas," which are short, patterned series of clicks (fig. 11.10; Watkins and Schevill 1977c), are associated with surface social

Figure 6.12. A sperm whale "breaching" off the Galápagos Islands. (Photograph courtesy of H. Whitehead laboratory.)

behavior (Whitehead and Weilgart 1991), whereas regularly spaced "usual clicks" (fig. 11.7) are not.

Communication

Like other cetaceans (Tyack, chap. 11, this volume), sperm whales probably rely on the acoustic channel for communication over most ranges. Visual, tactile, and possibly chemical communication are likely to be important only at close quarters.

The most obvious form of social communication in sperm whales is the exchange of codas (Watkins and Schevill 1977c). Codas can be easily classified, forming clear categories based on their temporal pattern and the number of clicks they contain (Weilgart and Whitehead 1993). Codas are often given as exchanges between individual whales (Watkins and Schevill 1977c), and there seems to be some degree of order in the sequential arrangement of these coda exchanges (Weilgart and Whitehead 1993). For instance, regularly spaced five-click codas often initiate exchanges, and seven-click codas often follow eight-click codas, but rarely does the reverse occur.

Although an individual whale may repeat one coda type several times within a short period of time, individuals also produce different coda types. Since many different whales can be heard using the same coda types, individual identification is not likely to be the primary function of variation in coda type. There is evidence of "dialects" in the differing proportional usage of coda types by different groups (using "group" in the sense defined earlier) as well as some geographic variation in coda repertoire (Weilgart and

Whitehead 1997). For instance, about twenty-five coda types are found in both the Galápagos and the Caribbean, but only some of these types are common to both areas (Moore et al. 1993; Weilgart and Whitehead 1993). In the most striking example of geographically distinctive coda repertoires, during eleven recordings made in the Tyrrhenian Sea, only one coda type ("click-click-click-pause-click") was ever heard (Borsani et al. 1997). When comparing coda repertoires from concentrations of sperm whales in the South Pacific, we found that, while there was some geographic variation, the greatest variation in coda repertoire was accounted for by the differences between groups (Weilgart and Whitehead 1997). Even neighboring groups differed in how frequently they used various coda types.

Codas are very rarely heard from mature or maturing males (e.g., Gordon et al. 1992; Goold 1999).

Codas probably function to mediate social interactions in groups that have coalesced after foraging periods. Similarly, in squirrel monkeys (*Saimiri* spp.), periods of resting, socializing, and caregiving are accompanied by vocal exchanges, which are thought to function in reaffirming social bonds following periods of dispersion while foraging (Symmes and Biben 1988).

"Slow clicks," called "clangs" by J. C. D. Gordon (1987b), which are distinctive loud clicks with repetition rates of only one click every six seconds or so, are heard only from mature males and mainly on the breeding grounds (Weilgart and Whitehead 1988). They may be used by males to assess or avoid each other (Whitehead 1993), or to announce their presence to groups of females. The slow clicks may contain structural elements that make them "honest" advertisements of the vocalizer's size, state of maturity, competitive ability, or physical fitness, as in the croaks of toads (Davies and Halliday 1978) or the roars of red deer (*Cervus elaphus:* Clutton-Brock and Albon 1979). However, no thorough analyses have been made of correlations between the structure of the slow click and any of these attributes. If some characteristic of the slow click is related to the size of the spermaceti organ, as is the case for usual clicks (Gordon 1991; Goold 1996), and it is used for assessment in interactions with females or other males, then there could be selective pressure for increases in the size of the spermaceti organ (J. C. D. Gordon 1987b; Clarke and Paliza 1988). The spermaceti organ is proportionally larger in males than in females, and in larger, more mature males than in smaller, immature males (20% of total length in 11 m female, 23% in 11 m male, and 26% in 16 m male: Nishiwaki et al. 1963).

"Creaks" (clicks emitted at high repetition rates, sounding like a rusty hinge) are heard in social contexts as well as from foraging whales. They are probably a form of echolocation, but may occasionally also be used as communication (J. C. D. Gordon 1987b; Weilgart 1990). Sperm whales also produce intermediates between, and composites of, creaks and codas (Weilgart 1990). There are a few non-click vocalizations, but these are uncommon and of low amplitude (J. C. D. Gordon 1987b; Goold 1999). Breaches and lobtails (see figs. 6.11 and 6.12), which frequently seem to display the extent of an animal's physical abilities, are likely also to have a communicative function (Waters and Whitehead 1990).

Socioecology

Scales of Experience

With their large ranges and long lifetimes, sperm whales show a diversity of social and spatial structures. The approximate spatial and temporal scales of those structures that we currently recognize are indicated in fig. 6.7. Structures that do not seem to be social ("non-mutualistic groups" in the terminology of Connor: see chap. 8, this volume) have positively correlated spatial and temporal scales (fig. 6.7), as is found in the space-time structures of other oceanic phenomena such as physical processes (Stommel 1963) and plankton communities (Haury et al. 1978). In contrast, truly social structures ("mutualistic groups") have durations ranging from minutes to at least decades, but usually seem to exist over ranges of less than a few kilometers (within acoustic communication range), and

show no such pronounced relationship between time and space.

The Family Unit

The central feature of sperm whale society is the family unit: females spending most, if not all, of their lives with their female kin. Best (1979) suggested two possible functions for these close, permanent bonds: care of calves, especially protection against predators, and cooperative feeding.

Over periods of hours to days, units in the same area may coalesce into groups, which forage in structured ranks. This observation suggests that there may be some advantages to coordinating foraging behavior with nearby animals. However, the small size, dispersed spatial organization, and lack of evasive ability of most sperm whale food all argue against much advantage for sophisticated cooperative foraging. Animals may gain useful information about gradients or concentrations of food from listening to the acoustic output of nearby animals, and they may forage side by side so that they do not swim through waters recently examined by another whale (Whitehead 1989a). However, there seems to be no need to form long-term, kin-based family units to reap these advantages while foraging.

In contrast, the care and survival of the calf is of vital importance to a female with such a low reproductive rate. Calves do not, and probably cannot, dive to the depths and for the durations at which their mothers feed (J. C. D. Gordon 1987a; Papastavrou et al. 1989). They are left at the surface, where they are vulnerable to predators such as killer whales and large sharks. In a family unit within which dives are staggered, apparently purposefully (Whitehead 1996a), at least one female or juvenile is likely to be present at or near the surface who can accompany a calf, be vigilant for predators, and defend against them. In the case of attacks by extremely dangerous cooperative predators—killer whales—cooperative defense may be vital for ensuring the survival of members of the group, especially calves.

Genetic data (Dillon 1996; Richard et al. 1996a; Christal 1998; Lyrholm and Gyllensten 1998; Bond 1999; Lyrholm et al. 1999) indicate that some permanent units contain two or more maternally unrelated matrilines, and recently Christal et al. (1998) discovered occasional splits in Galápagos units, or mergers between them, over time scales of five to ten years. These findings suggest that a unit's membership is determined by more than genealogy, and that there are circumstances that favor long-term associations between unrelated females, or the splitting of a unit. Perhaps, if babysitting is the primary function of permanent units, females who are members of numerically small matrilines improve their fitness by merging with others. Unusually large matrilines may suffer a feeding disadvantage and therefore be selected to split, as is suggested for *transient* killer whales (see Baird, chap. 5, this volume).

While vigilance for predators and the defense of calves would seem to be the most likely primary functions of communal kin-based living, this social system has other potential benefits for female sperm whales. Structured foraging formations and a combined memory of the distribution of temporally variable food sources over a large spatial range and long time span should both increase foraging efficiency and tend to smooth out temporal variations in food abundance (Whitehead 1996b).

Evidence is beginning to accumulate that sperm whale units have distinctive vertically transmitted cultures: functionally important information learned maternally or within matrilineal units. Units have characteristic coda repertoires, which are unlikely to be genetically determined (Weilgart and Whitehead 1997), and there are indications that units that share recent ancestry but whose ranges do not overlap also have similar coda repertoires (Whitehead et al. 1998).

The "Colossal Convergence": Sperm Whales and Elephants

Elephants (Elephantidae) and sperm whales have remarkably congruent life histories and growth patterns (see fig. 6.9), and the social organizations of both species are arranged into hierarchical levels, including cooperative matrilineal units and "kinship groups" (Best 1979; Weilgart et al. 1996). Elephants are also characterized by large size, large brains, sexual dimorphism, extensive home ranges, and ecological success (Laws 1970; table 6.4).

Table 6.4. Sexual dimorphism, life history parameters, and social organizations of African elephants and sperm whales

	Elephant	Sperm whale
Adult mass, male:female	2,700:4,700 kg[2]	49,000:15,000 kg[6]
Life span	~60 years[2]	≥60–70 years[6]
Age at maturity		
Females (sexual)	10–12 years[5]	7–13 years[6]
Males (sexual)	~17 years[5]	~20 years[1]
Males ("sociological")	29 years[5]	25 years[1]
Gestation length	22 months[2]	15 months[1]
Calving interval	4–5 years[5]	5 years[1]
Age at weaning (mean)	5 years[3]	2 years[1]
Age at weaning (max.)	8 years[3]	13 years[1]
Matrilineal family unit		
Females and offspring	~10[4]	~12[7]
Group	~2 units[4]	~2 units[7]

Sources: 1, Best et al. 1984; 2, Laws et al. 1975; 3, Lee and Moss 1986; 4, Moss and Poole 1983; 5, Poole 1994; 6, Rice 1989; 7, Table 6.3, this chapter.

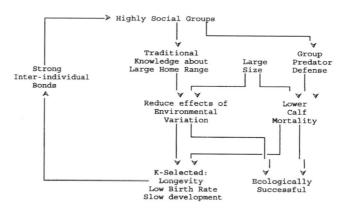

Figure 6.13. Some of the ways in which *K*-selection and sociality reinforce one another, resulting in slow-reproducing, altruistic, and ecologically successful animals, such as elephants and sperm whales.

In the evolution of sperm whales and elephants, a low reproductive rate, prolonged care of the young, large size, and cooperation among females have mutually reinforced one another (c.f., Horn and Bradbury 1984). The sperm whale's brain is a little larger than that of a blue whale (*Balaenoptera musculus*) which has over twice the body mass, and the largest terrestrial brain belongs to the elephant. Large brain size and advanced cognitive abilities are often linked to complex social organizations (Herman 1980; Byrne and Whiten 1988; Barton 1996). As reproductive success becomes increasingly tied to social ability, prolonged care of offspring will be favored, reinforcing the sociality/*K*-selection evolutionary cycle (fig 6.13). The result is a large, intelligent, wide-ranging, and very slowly reproducing animal. The sperm whales' (and elephants') large sizes and wide ranges, as well as (perhaps) a long communal memory, allow them to ride out considerable temporal fluctuations in their food supply and contribute to their trophic significance (Weilgart et al. 1996; see Millar and Hickling 1990).

Another remarkable parallel between elephants and sperm whales is in the breeding behavior of males. In both species, mature males are much larger than females (fig. 6.9; table 6.4) and rove singly between groups of females (Barnes 1982; Whitehead 1993). Competitive breeding is effectively restricted to older, larger males by dispersal to high latitudes in sperm whales and the phenomenon of "musth" in elephants. A male elephant in musth is dominant over non-musth males, but suffers considerable loss of condition (Poole 1987). Only the largest males enter musth during the peak breeding season of the females (Poole 1989).

The Behavior of Males

Why do male sperm whales disperse from their family units and then segregate to high latitudes? Other odontocete males are able to remain with their female kin and obtain mating opportunities when groups converge (Connor et al., chap. 10, this volume). Groups of female sperm whales are often within acoustic range of one another, and so a male would meet many unrelated females if he stayed with his family unit. He would also have the benefit of communal defense against predators, communal foraging, and the possibilities of assistance from his relatives in breeding competition and of increasing his inclusive fitness by helping them in turn (Connor et al., chap. 10, this volume). However, it appears that, in sperm whales, sociality is not an important determinant of feeding rates, and males are able to defend themselves against predators. Therefore a male sperm whale should leave his family unit if he can thereby enhance his potential reproductive success, and it seems that this is the case.

The extreme sexual dimorphism of sperm whales strongly implies that male mating success is size dependent because of intrasexual competition and/or female choice. Thus, lifetime reproductive success is likely to be closely tied to growth, emphasizing the importance of feeding for maturing males and the continuation of growth past sexual maturity (see fig. 6.9). The dominant vertebrate mesopelagic predator, the female sperm whale, is restricted to low latitudes, perhaps because of the thermal tolerances of her offspring. The males may experience better feeding opportunities and greater growth nearer the poles, where they are not competing with females for resources and where food items also tend to be larger and thus, perhaps, more suitable for a larger animal (Best 1979). When female sperm whales deserted equatorial Galápagos waters in the mid-1990s, they were partially replaced by aggregations of large, but apparently nonbreeding, males, behaving in a way similar to those in male aggregations at high latitudes (Christal and Whitehead 1997; Christal 1998). This episode may suggest that avoiding competition with females has been more important in the dispersal of males to high latitudes than attributes of the polar waters themselves.

With size-based competitive breeding, it will pay males to return to the warm-water breeding grounds and suffer decreased feeding opportunities, the energetic costs of migration and breeding, and, perhaps, greater mortality only when there are few larger males to compete with. A small male will generally have greater expected lifetime reproductive success by remaining at high latitudes, growing faster, and indulging in only the limited breeding opportunities

presented by latitudinal overlap with females and scarcity of larger males. In this kind of system delayed competitive breeding is favored (Pereira 1993; Whitehead 1994). Male sperm whales do not generally reenter the warm-water breeding grounds until their late twenties, a similar age to that at which male elephants enter the state of musth during the prime breeding season (Poole 1989). In both cases this is very approximately double the age of male sexual maturity (see table 6.4). Similarly, males of highly sexually dimorphic macropod species segregate from females while feeding and growing, and eventually return to rove among groups of females (Jarman 1989).

When a large male does return to the breeding grounds, he could either accompany an unrelated family unit, attempting to breed with its females, or rove between units searching for receptive females (Clutton-Brock 1989b). The relative benefits of these strategies depend on the distribution and accessibility of the groups of females, and in particular, on the relationship between the receptive period of individual females and the travel time for males between groups (Whitehead 1990c; see also Sandell and Liberg 1992). Moving between groups takes male sperm whales a few hours, almost certainly less than the receptive period of a female, so roving is theoretically favored (Whitehead 1990c). However, concentrations of females a few hundred kilometers across are generally ten or more travel days apart for a male, so the same theoretical model predicts that males should stay resident within concentrations, as is found off the Galápagos.

The role of females in pairing is almost completely unknown, but likely to be important. There is correlational evidence that artificially low numbers of mature males result in reduced female pregnancy rates (Clarke et al. 1980; Kahn et al. 1993). The relative abundance of adult males off the Galápagos has been reduced, presumably by selective whaling, to about 20% of that found in the same area during the early nineteenth century or expected from population models (Hope and Whitehead 1991), and the calving rate is considerably depressed (Whitehead et al. 1997a). However, females encounter mature males on about 75% of the days during the prime breeding season (Whitehead 1993), which might appear sufficient to ensure pregnancy. This paradox could be resolved if the presence of several males were necessary to induce mating behavior in either the males or females, or if females chose to mate with only some individual males, so that higher densities of males than those currently found off the Galápagos were required to ensure conception. If females within a family unit have similar preferences, then female choice could also explain genetic indications of shared paternity within groups

(Richard et al. 1996a), although a dominance hierarchy among males within a breeding ground, as suggested by Watkins et al. (1993), might also have this result.

Prospects for Future Studies

By combining results from extensive research on the hundreds of thousands of carcasses produced by the whaling industry with more recent longitudinal behavioral studies of living animals and groups of animals, we have developed a basic feel for how the ecology of the sperm whale may be related to its behavior. However, many aspects of the lives of these intriguing animals remain unknown.

Some of the most compelling gaps in our knowledge, such as a precise understanding of how sperm whales communicate using codas, are probably beyond current technical and conceptual expertise. However, the function of the males' slow click vocalization can probably be elucidated using playbacks and/or correlational analyses, and the obvious deficiency in our knowledge of the migrations and social behavior of mature and maturing males may soon be removed with the development of long-term satellite tags, analyses of DNA, and other techniques (Whitehead et al., chap. 3, this volume). Technological advances are also likely to increase our knowledge of diving behavior, and it should be possible to improve our methods of studying the diet of living animals, for instance, by improving fecal sampling methods or by analyzing fatty acids or stable isotope ratios in blubber samples. As long-term studies continue and develop, and especially as genetics becomes an integral part of such work, we will learn more about the structure and stability of female groups, increasing our insight into the functional significance of the social structure of sperm whales.

Conservation Status

Superficially, the sperm whale seems to be surviving rather better than most other large cetaceans. Their offshore habitat is infrequently used by humans and is farther from most sources of pollution than the continental shelves where many large cetaceans live, and their food has little overlap with the species currently being caught by human fishers. The global population estimates of a million or more sperm whales that are sometimes cited (e.g., Evans 1987; Rice 1989) are based on largely discredited analyses (Cooke 1986), but at least a hundred thousand sperm whales probably survived the most recent phase of whaling (Reeves and Whitehead 1997).

The little that we do know of the trends in sperm whale populations, and increasing reports of conflicts with humans, suggest that we should be concerned for the status of sperm whale populations. The population parameters most recently used by the Scientific Committee of the International Whaling Commission (1982) allow for a maximal rate of increase of a sperm whale population with a stable age structure of just 0.8% per year. These animals are not adapted to dealing with sudden population declines, and recover from depletion very slowly.

By the 1860s the open-boat whaling industry had probably reduced most sperm whale populations to less than 25% of their pre-exploitation size (e.g., Tillman and Breiwick 1983; Whitehead 1995b). Because of their slow rate of increase, and occasional whaling in the intervening period, they were still considerably depleted when modern sperm whaling reached its stride in 1946 (Whitehead 1995b). Modern whaling reduced populations substantially, but to an extent that is not well known because of technical difficulties with the data available (Cooke 1986). A particularly troubling problem has been the definition of the boundaries between populations for animals in which the two sexes show very different patterns of migration and ranging (International Whaling Commission 1982; Dufault et al. 1999).

The sperm whales' unusual segregation of the sexes, together with a concentration by modern whalers on the larger, more valuable and higher-latitude males, has necessitated the use of population models that consider the effects of changing numbers of breeding males on the pregnancy rates of the females (International Whaling Commission 1982). The modern whalers' preferential selection of the larger males radically changed the adult sex ratio on the breeding grounds, which in turn, in some areas, seems to have decreased the pregnancy rate of the females (Clarke et al. 1980) or the rate of observing calves (Whitehead et al. 1997a). These reduced pregnancy rates are likely persisting well beyond the end of whaling on account of the long delay before most males begin to enter the breeding grounds, and may partially account for a decline in the population of female sperm whales off the Galápagos (Whitehead et al. 1997a). Conversely, the rise in sightings of males in Antarctic waters during the 1980s reported by Butterworth et al. (in press) is also likely related to selection of large males by modern whalers. The Antarctic, from where mature and maturing males were largely removed in the 1960s and 1970s, is now being repopulated by maturing males from less decimated breeding populations at lower latitudes.

Whaling for females could also have population effects beyond the removal of the animals actually killed. In the similarly structured elephant societies, the removal of females by poachers increased the mortality of surviving animals and their calves as the family units lost members as well as the valuable survival skills and knowledge those members possessed (Poole and Thomsen 1989). Whitehead et al. (1997a) believe the principal cause of the rapid decline in the population of female sperm whales off the Galápagos during the 1990s to be movement into suitable habitat in the Humboldt Current off the west coast of South America, from which sperm whales had been removed by whalers operating from Peru between 1957 and 1981. Thus, for this species, the effects of whaling, which can often be unexpected and unusual, linger years after its cessation.

Other anthropogenic factors may be affecting sperm whales and their populations. Forty-one sperm whales were found dead in northern Europe during 1988–1989, and another twenty-one around the North Sea in 1994–1995 (Christensen 1990; Law et al. 1996). The reasons for these strandings are unclear, although some of the animals showed moderate levels of organochlorine pollutants (Law et al. 1996). Sperm whales generally have pollutant levels intermediate between the relatively uncontaminated baleen whales and the more highly polluted delphinids (Aguilar 1983).

In the Mediterranean Sea tens of sperm whales have been found dead in recent years following entrapments in gill nets set illegally for swordfish *(Xiphius gladius)* (di Natale and di Sciara 1994), and fisheries in other parts of the world take sperm whales incidentally (Barlow et al. 1994; Haase and Félix 1994). Throughout the world, sperm whales are in danger of death or injury from collisions with commercial shipping (e.g., André 1997; Reeves and Whitehead 1997) and from swallowing plastics (e.g., Viale et al. 1992).

Sperm whales seem to be especially easily disturbed by noise (e.g., Watkins and Schevill 1975; Bowles et al. 1994; Mate et al. 1994a). As sound seems essential to both their foraging and social behavior, increasing levels of underwater noise from shipping, military and industrial sonars, and oceanographic experiments are of particular concern for this species (Gordon and Moscrop 1996; Whitehead et al., chap. 12, this volume).

The actual or potential effects of these threats to sperm whale populations are largely unknown, and we need research into them. However, in the case of whaling, in which the level of the threat (numbers of animals targeted) and its effect on individuals (death) were both well known, results at the population level have been either unclear or unexpected. This unpredictability is partly the result of difficulties inherent in studying any wide-ranging pelagic animal, but special attributes of sperm whales, and perhaps especially their social structure, make predictions of population dynamics particularly uncertain. No research can assure us what are "safe levels" of the new and, compared with whaling, much less quantifiable threats. Therefore, the principal priority for conserving this species should be action to reduce by-catch, chemical pollutants, noise, and other anthropogenic effects on the ocean and to prevent the resumption of large-scale commercial whaling.

7

THE HUMPBACK WHALE

Seasonal Feeding and Breeding in a Baleen Whale

PHILLIP J. CLAPHAM

THE HUMPBACK whale, *Megaptera novaeangliae*, has been the most intensively studied of the mysticete (baleen) whales. Although it is one of the world's most wide-ranging mammals, its preference for nearshore habitats, its tendency to concentrate on traditional breeding grounds, and the ease with which individuals can be identified from natural markings have made it a relatively accessible subject for a number of long-term field studies over the last two decades. Furthermore, aspects of the humpback's biology and behavior (such as its highly complex song) have stimulated many researchers to adopt it as a principal study animal. The aim of this chapter is to identify ecological influences on the social behavior of this species, notably with regard to the distinctly seasonal nature of feeding and breeding.

Like most mysticetes, humpback whales maintain a strong annual cycle, which allows them to exploit the productivity of high-latitude habitats in summer and to mate and calve in tropical or subtropical waters in winter. These distinct patterns of temporal and geographic distribution, and the extensive migrations that link them, fundamentally influence many life history traits. For example, it has been argued that the extremely rapid fetal growth rate that is characteristic of mysticetes evolved as a response to metabolic pressures arising from the lack of a year-round food supply (Frazer and Huggett 1973). Similarly, the necessity for prolonged fasting while residing in unproductive tropical waters requires large energy stores, most notably for lactating females. Furthermore, the virtual absence of appropriate prey during the winter months has a profound effect on the distribution, and thus the social ecology, of this and perhaps other mysticete species.

The social organization of baleen whales in both high- and low-latitude habitats generally lacks the cohesive stability characteristic of many odontocetes, in which individuals or groups may remain associated for periods of years (e.g., Wells et al. 1977; Bigg et al. 1990b; Baird, chap. 5; Whitehead and Weilgart, chap. 6; Connor et al., chap. 10, this volume). While large aggregations are often observed, notably on a feeding ground, it is not clear whether these possess any underlying structure, or whether they simply reflect a concentration of animals brought together by a common interest in locally abundant prey. While a few long-term associations between individuals have been documented, small unstable groups appear to be the rule in all studied mysticetes.

The intensive exploitation of the great whales drastically reduced the abundance of most mysticetes. The advent of mechanized high-seas whaling operations in the 1920s led to the effective commercial extinction of some populations, notably in the once rich whaling grounds of the Southern Ocean. This broad elimination of a major group of predators must have significantly perturbed the marine ecosystems concerned. Furthermore, intensive human exploitation of fish stocks has had major effects on prey abundance in many areas (e.g., Brown and Halliday 1983; P. M. Payne et al. 1990). While the effects on the social ecology of humpback and other whales of these major changes in their population density and environment are unknown, it is important to keep in mind that the conditions and population densities under which the whales' behavior evolved may have been significantly different from those that we observe today.

Taxonomy and Morphology

The humpback whale (order Cetacea, family Balaenopteridae) is the sole species in the genus *Megaptera*. A recent species account is given by Clapham and Mead (1999), and selected life history parameters are summarized in table 7.1. Like all balaenopterids ("rorquals"), humpbacks possess a series of ventral pleats that expand to permit the whale to engulf a vast quantity of water and prey during feeding. The humpback is most obviously distinguished from other whales by its remarkably long flippers (proportionately the longest of any cetacean), which are approximately one-third of its body length. It is likely that the flippers provide the humpback with considerable maneuverability (Whitehead 1981), a factor that may be significant in both feeding and intrasexual competition.

The humpback is a moderately large baleen whale, with a maximum recorded length of 17.4 m (Chittleborough 1965). As with most baleen whales, females are larger than males, being typically 1 to 1.5 m longer (Chittleborough 1965). It is likely that the sizes of male and female humpbacks are the product of very different selective forces. The bioenergetic demands of reproduction may favor large size (and perhaps greater fat storage) in females (Ralls 1976). Among males (which in this species often compete aggres-

sively for females), absolute size may be less important in intrasexual competition than maneuverability (see Connor et al., chap. 10, this volume).

It is well known that Antarctic balaenopterids are on average somewhat larger than their Northern Hemisphere counterparts. This fact has been cited by Mayr (1965) as an example of Bergmann's rule; specifically, that larger size in the Southern Ocean has been selected for because of the lower water temperatures there. By contrast, Brodie (1975, 1977) adopts a more complex energetics approach: he argues that body size among balaenopterids represents a dynamic balance between short-term residency in cold, productive areas and lipid storage to sustain the animal while it is in the less productive but energetically less demanding waters of low latitudes. In other words, size relates primarily to fasting in warm water, not to residency in cold. That this translates into larger body size among Southern Hemisphere animals he attributes to a shorter feeding season and lower diversity of prey than in the north. As attractive as this hypothesis is, it cannot be tested without better energetics data than we currently possess.

Table 7.1. Life history parameters for the humpback whale, *Megaptera novaeangliae*

	Mean	Range/Max	Other	Reference/Notes
Adult body size (kg × 1,000)				Various sources, reviewed by Lockyer 1976. Sample = one 12.9 m male and five females (range: 12.5–15.0 m). Individual weight at a given length varies greatly by season and female reproductive condition.
Male			27.7	
Female		24.8–40.8		
Adult body length (m)				Matthews 1937; Chittleborough 1965. Other reported values up to 16.2 m (female) and 17.4 m (male), but reliability of measurement is unknown.
Male	13.0	14.8		
Female	13.9	15.5		
Gestation length (months)	11–12			Chittleborough 1958.
Weaning age (months)	10–12			Chittleborough 1958; Clapham and Mayo 1987, 1990; Baraff and Weinrich 1993. Independent feeding can occur at six months. A few calves (ca. 5%) remain associated with their mothers for a second year.
Interbirth interval (years)	2.4	1–>5	Mode = 2	Clapham and Mayo 1990; Barlow and Clapham 1997.
Calf mortality (first year)			0.125	Barlow and Clapham 1997. Figure estimated from modeling.
Length of estrus (days)	??			Estrus duration is unknown for any mysticete.
Breeding season				Kellogg 1929; Chittleborough 1958, 1965; Baker and Herman 1984a.
N Hemisphere		Dec–Apr		
S Hemisphere		Jun–Oct		
Age at first birth (years)	6	5–8		Chittleborough 1965; Clapham 1992.
Age at sexual maturity (years)				Chittleborough 1965; Clapham 1992.
Male	5	4–8		
Female	5	4–8		
Maximum life span (years)		>48		Chittleborough 1965. Likely significantly greater than this value given removal of older animals by whaling and known longevity of other balaenopterids.

Field Studies

The earliest formal studies of humpback whales were conducted in association with commercial whaling. The large-scale exploitation of baleen whale populations in the twentieth century provided an opportunity for the gathering of sometimes vast quantities of data on morphology, physiology, reproduction, diet, distribution, and movements (see Slijper 1962 for a comprehensive review). It is noteworthy that, despite the large gaps in our knowledge of these animals, some of the sample sizes concerned are among the largest available for any mammal.

Basic knowledge of humpback whales gathered by the whaling industry included the seasonal migration and the absence of feeding behavior during winter (Chittleborough 1965; Dawbin 1966), the seasonality of breeding behavior (Omura 1953; Symons and Weston 1958; Chittleborough 1965), and basic reproductive parameters. Tagging using so-called "Discovery" marks provided information on individual movements over periods of from days to years (Brown 1977; Chittleborough 1959b, 1965). A technique for aging dead balaenopterids by counting annual growth layer groups (GLGs, with one GLG equalling two layers) in waxy plugs in the ears (Laws and Purves 1956) permitted assessments of age at sexual maturity and other factors. The rate at which these growth layers are deposited in humpbacks remains in dispute, although data from Clapham (1992) support Chittleborough's (1958, 1959a, 1965) con-

tention that two GLGs are deposited each year (not one, as occurs in fin whales: Aguilar and Lockyer 1987).

The development of a technique for identifying individual humpback whales stimulated the initiation of several long-term studies of living whales in the mid-1970s. The considerable variation in the ventral fluke pattern (figs. 7.1 and 7.2) and its stability over time permitted unequivocal identification of individuals over many years (Katona and Whitehead 1981). Similar variations in dorsal fin shape and scarring provided additional means of recognition in both short- and long-term field studies.

In the North Atlantic, a central catalog for fluke photos of humpback whales (Katona et al. 1980) served to compare photographs of whales observed in locations as far apart as the Arctic and the West Indies, allowing assessment of the long-range movements of individuals (Katona and Beard 1990). In the early 1990s, this collaboration culminated in a seven-nation project known as Years of the North Atlantic Humpback (YONAH). Using standardized sampling protocols, YONAH photographed and biopsied (for genetic analysis) several thousand humpback whales in an intensive two-year effort to study the species across almost its entire North Atlantic range (Katona 1991; Palsbøll et al. 1997). Among the regional projects brought together under YONAH were two long-term studies in the Gulf of Maine, where almost two decades of daily field access during summer afforded by commercial whale-watching resulted in remarkably long and detailed sighting histories

Figure 7.1. Ventral fluke pattern of a mature female humpback whale (named "Beltane") from the Gulf of Maine. This individual, born in 1980, was the first female of known age recorded with a calf, at age six. (Photograph courtesy of Phil Clapham/Center for Coastal Studies.)

Figure 7.2. Researchers in an inflatable boat photograph the ventral fluke pattern of a humpback whale in the West Indies. (Photograph courtesy of Phil Clapham/Center for Coastal Studies.)

of individual whales (e.g., Weinrich and Kuhlberg 1991; Clapham et al. 1993a). The humpback whales of the Gulf of Maine represent what is probably the best-known population of baleen whales in the world.

In the last twenty years, regional projects were also initiated in the North Pacific, notably on the Alaskan feeding grounds, and in the Hawaiian breeding range (for an overview see Perry et al. 1990). Later, work was extended to the Pacific waters of Mexico (Urban and Aguayo 1987), the California coast (Calambokidis et al. 1990), and Japan (Darling and Mori 1993). As in the West Indies, much of the work in Hawaii was focused on population assessment; however, Hawaii became the focus for considerably more investigations of humpback whale behavior, including some focal animal studies (e.g., Tyack 1981). The recent inclusion of molecular techniques in population and behavioral studies of large whales (see Hoelzel 1991b) has added an exciting new dimension to current work on humpback whales in all oceans.

Behavioral studies of cetaceans have frequently been hindered by the inability of investigators to obtain data on a basic variable: the sex of a study animal. Studies of humpback whales were facilitated in this regard when Glockner (1983) noted that the sex of a humpback can be determined if the genital area is observed (females have a hemispherical lobe in this region; males do not). More recently, the development of molecular sex determination techniques has made the sexing of large numbers of whales a practical undertaking (Palsbøll et al. 1992).

Ecology

Distribution and Migratory Behavior

Humpback whales are found in all oceans of the world (fig. 7.3). In summer, they are found from temperate to coastal zones, generally close to coastlines or in continental shelf waters (CeTAP 1982), where they feed and where little or no reproduction occurs (Chittleborough 1965; Dawbin 1966). In winter, they are typically observed clustered around insular coasts or on offshore reef systems in the tropics (Dawbin 1966; Balcomb and Nichols 1982; Whitehead and Moore 1982). It is possible that a requirement of mothers with newborn calves for sheltered water may partly explain the aggregations of whales that occur in such areas; however, there is no obvious reason why other classes of whales should be similarly confined in their winter distribution. In fact, numerous animals spend at least some of this period in deep water offshore (Winn et al. 1975), as confirmed by recent acoustic data (C. W. Clark, personal communication). Furthermore, dependence of calves upon sheltered water is not known in any other rorqual, although it may be important to right whales (*Balaena* spp.; Kraus et al. 1986; Payne 1986).

Mating and calving take place in winter, and effectively no feeding occurs at this time (see Baraff et al. 1991 for an exception). Because of the seasonal opposition of the hemispheres, whales in boreal waters are feeding while their austral counterparts are breeding, and vice versa. There is evidence for overwintering in high latitudes by some in-

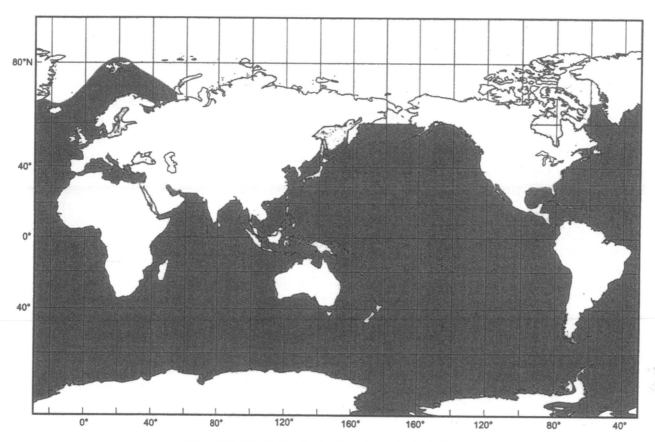

Figure 7.3. Worldwide distribution of humpback whales.

dividuals (Straley 1990; Clapham et al. 1993b; Swingle et al. 1993). Furthermore, M. R. Brown et al. (1995) have challenged the generally accepted view that the great majority of whales undertake the migration to the tropics, and have hypothesized that many nonbreeding females remain on feeding grounds.

Examination of whaling catches showed that both northward and southward migrations are characterized by a staggering of sexual and maturational class (Nishiwaki 1959; Chittleborough 1965; Dawbin 1966). Lactating females are among the first to leave the feeding grounds in autumn, followed by immature animals, mature males, "resting" females (those neither pregnant nor lactating), and last, by pregnant females. In late winter, this order is broadly reversed, with newly pregnant females among the first to begin the return migration to high latitudes.

Population Structure

The population structure of humpback whales varies between different oceans. Humpbacks in the Northern Hemisphere generally feed in relatively discrete subpopulations in summer, but mix on a common breeding ground in winter. In the North Atlantic, for example, humpback whales feed in the Gulf of Maine, off eastern Canada (Newfoundland, Labrador, the Gulf of St. Lawrence, and Baffin Island), and off western Greenland, Iceland, and Norway. These subpopulations are relatively discrete (Katona and Beard 1990), and fidelity to each is maternally directed (Clapham and Mayo 1987). Long-term studies have shown high annual return rates (up to 90% of identified individuals) to feeding grounds such as the Gulf of Maine (Clapham et al. 1993a). Interestingly, recent molecular work using mitochondrial DNA (which is maternally inherited) suggests that such matrilineal fidelity persists on an evolutionary time scale (i.e., for the thousands of years necessary for it to be reflected in the population's genetic structure) despite the lack of obvious barriers to movement in this aquatic environment (Palsbøll et al. 1995; Larsen et al. 1996). However, while there is little exchange among them in summer, whales from all these subpopulations migrate each winter to the West Indies. There, behavioral and genetic (both mitochondrial and nuclear DNA) data have shown that they mix and interbreed (Clapham et al. 1993b; Palsbøll et al. 1995, 1997).

In the North Pacific, the humpbacks that feed in Alaskan waters migrate primarily to the Hawaiian Islands (Darling and McSweeney 1985; Baker et al. 1986), while those from the apparently separate California stock (Calambokidis et al. 1996) breed primarily in coastal Mexican waters (Perry et al. 1990). The status and movements of humpback whales in the western North Pacific are unclear, but the recently documented movement of a photographically identified individual between Hawaii and Japan indicates that at least a few individuals are capable of considerable longitudinal movement (Darling and Cerchio 1993).

In the Southern Hemisphere, humpbacks feed in circumpolar waters surrounding the Antarctic continent. The traditional division of this population into six discrete stocks (numbered I–VI) was convenient for whaling management and was based largely upon Discovery tag returns (Mackintosh 1942). However, too few data exist to preclude the possibility that there is significant mixing during summer among these subpopulations. Chittleborough (1965), summarizing many years of whaling data collected in areas IV and V, noted that the two populations were generally separate, but that overlap occurred in some years due to temporary feeding range expansions. There is good evidence that the majority of whales from areas IV and V segregate to separate breeding grounds (Chittleborough 1965; Dawbin 1966; Gill and Burton 1995). Similarly, it is thought that humpback whales from other parts of the Antarctic migrate to a number of different low-latitude breeding grounds in the austral winter. These include Madagascar and eastern Africa (Findlay et al. 1994), tropical and equatorial waters of western South America (Stoneet al. 1990; Flórez-González 1991), and around islands and reef systems in the South Pacific (Abernethy et al. 1992; Hauser et al. 1999).

Habitat Use and Home Range

Our knowledge of habitat use and home range in humpback whales is biased by the fact that most observer effort has been concentrated in areas close to the coast. That said, known ranging patterns among humpback whales exhibit considerable variation in scale. Although movements by identified individuals between widely separated areas in both summer and winter have been recorded (Mattila and Clapham 1989; Katona and Beard 1990), repeated resightings and prolonged stays by individuals, notably in specific local feeding areas, are common (Clapham et al. 1993a). While feeding humpback whales may remain in specific areas (such as in bays or on banks) for periods of days or even weeks, they do not exhibit the rigid spatial fidelity characteristic of territorial animals. The absence of territoriality in a feeding area has been tied to the patchy, mobile nature of the whales' prey, which renders economic defense of specific areas unprofitable (Clapham 1993b); as such, humpbacks are typical of taxa confronted by heterogeneously distributed and spatially unstable resources (Gosling and Petrie 1981). In general, it is clear that humpback whale distribution in high latitudes is closely tied to the distribution of resources, and whales will undertake substantial movements and even abandon habitats in response to changes in prey abundance (P. M. Payne et al. 1986, 1990).

While there appears to be more of a trend toward continual turnover of whales in breeding areas than on feeding grounds (Baker and Herman 1981; Mattila and Clapham 1989), resightings of individuals over days or weeks suggest that not all whales "pass through" an area during winter. Furthermore, at least some whales return to specific habitats in different years. For example, 15.8% of individuals identified in Samana Bay (Dominican Republic) were observed on more than one day during a winter (maximum = five days), over periods ranging from one to thirty-three days, and 4.5% were seen in more than one year (Mattila et al. 1994). Glockner-Ferrari and Ferrari (1990) report multi-year sightings (up to ten different years) for male and female humpbacks identified during their long-term study in Hawaiian waters. A reasonable understanding of the movements and residency of humpback whales in winter requires greater field effort than has been invested to date.

There is evidence that females with calves preferentially use shallower or nearshore waters in both high- and low-latitude ranges. In the tropics, this may be to take advantage of calmer water, to minimize the possibility of predation by sharks (Whitehead and Moore 1982; Glockner and Venus 1983; Mattila and Clapham 1989), or to avoid harassment by males (Smultea 1994). On a feeding ground in Massachusetts Bay, Clapham and Mayo (1987) found that mature females were more likely to be found in this inshore region in years when they had a calf than in years when they did not, but the reason for this apparent preference is unclear.

Diet and Foraging Behavior

Prey species. Humpback whales feed on a variety of small schooling fish, including capelin, *Mallotus villosus* (Whitehead 1981, 1983), herring, *Clupea* spp. (Watkins and Schevill 1979; Baker et al. 1985), mackerel, *Scomber scombrus* (Geraci et al. 1989), and sand lance, *Ammodytes* spp. (Overholtz and Nicolas 1979; P. M. Payne et al. 1986, 1990). In some areas, the primary or exclusive prey are euphausiids of several genera, notably *Euphausia, Thy-*

sanoëssa, or *Meganyctiphanes* (Matthews 1937; Nemoto 1957; Slijper 1962). Central to a consideration of humpback foraging ecology is that all of these prey species are found in schools or (in the case of euphausiids) dense patches of varying vertical and horizontal extent.

Foraging behavior and energetics. There have been few quantitative studies on the diving behavior of foraging humpback whales. Dolphin (1987c) found that whales in Alaska invariably exploited the shallowest patches of prey, and that very few whales dive below 120 m despite the frequent availability of prey below this depth; this finding is consistent with predictions of cost-benefit models based on physiology and energetics (Dolphin 1987a,c).

Like all rorquals, humpback whales are considered "gulp feeders" in that they feed in discrete events, engulfing a single mouthful of prey at a time, rather than continuously filtering food in the manner characteristic of balaenid whales (Ingebrigtsen 1929; Watkins and Schevill 1979). Humpback whales are frequently observed feeding in the same area as confamilials such as fin or minke whales (Whitehead and Carlson 1988), but there is currently little evidence for interspecific competition among mysticetes (Clapham and Brownell 1996).

An unusual humpback whale foraging behavior, termed "bubble feeding," involves the production of bubbles in the form of clouds, nets, or curtains to corral or trap prey (Ingebrigtsen 1929; Jurasz and Jurasz 1979; Hain et al. 1982). Furthermore, there appear to be major differences in the specific techniques used between oceans: North Pacific humpbacks have never been observed using bubble clouds, which is the most common bubble structure used in the North Atlantic (personal observation; W. F. Dolphin, personal communication). Bubbles appear to be used primarily on certain species of schooling fish, but the exact way in which they serve to facilitate feeding is not clear.

Humpback whales often forage alone, but are also frequently observed to feed in coordinated groups (fig. 7.4). The term "coordinated feeding" is generally taken to mean synchronous behavior by two or more whales in which group members surface and dive together, and exploit the same prey patch, either for a single feeding event or repeatedly. Whitehead (1983) was the first to relate group size among foraging humpbacks to the horizontal extent of the prey patch being exploited. The relationship between prey type and foraging behavior has been noted by several observers. A pattern of fluid (occasionally coordinated) associations between whales feeding on swarming krill and stable, often coordinated associations between individuals preying on schooling fish was noted off Alaska by Baker and Herman (1984b) and by Perry et al. (1990). In particular, these authors reported two groups in which associations between core members persisted over all observations in several years; where it was possible to determine the associated prey type, it was always herring. Baker et al. (1992) described a similarly cohesive group of seven to nine whales feeding on herring in Icy Strait, Alaska, in 1986. This trend was also noted in Alaskan waters by D'Vincent et al. (1985), who reported observing distinct changes from uncoordinated to coordinated foraging among a particular group of

Figure 7.4. A group of humpbacks lunge-feeding on a large school of sand lance (*Ammodytes* sp.) in the Gulf of Maine. (Photograph courtesy of Phil Clapham/Center for Coastal Studies.)

whales as they encountered different patches of euphausiids; the whales concerned subsequently exhibited consistently coordinated behavior as they encountered herring schools.

In the Gulf of Maine, the predominant prey for over a decade was the American sand lance (Overholtz and Nicholas 1979; P. M. Payne et al. 1990). Stable associations among humpback whales in this area were rare, and never involved large groups (Weinrich 1991; Clapham 1993b). Foraging on sand lance was both coordinated and uncoordinated, with group size positively correlated with the size of the prey school (Clapham 1993b). Overall, it seems that coordinated feeding behavior is most likely to be effective when whales encounter fast-moving schooling fish such as herring, but is less useful when dealing with slower prey such as euphausiids. The rarity of stable groups anywhere (including Alaska) is discussed further below.

Learning. Truly learned behavior in whales is difficult to document. Clapham and Mayo (1987) observed unweaned calves blowing small bubble clouds next to their feeding mothers, possibly in an attempt to imitate this prey capture technique, but it is not clear whether this represents learning. The apparent spread by cultural transmission of a novel feeding technique ("lobtail feeding") in the southern Gulf of Maine (Weinrich, Schilling and Belt 1992) represents better evidence for a learned behavior, particularly since it was acquired by animals born after the technique first appeared in the population. This phenomenon may have been linked to an ecosystem shift that replaced herring with sand lance, thus requiring the development of a different feeding style specific to exploitation of the latter species.

Humpback calves have a longer period of dependence (up to a year, occasionally two: Clapham and Mayo 1990) than other balaenopterids; blue and fin whales, for example, appear to separate from their mothers at about six months (Lockyer 1984). It is possible that this relates in part to the necessity of learning more complex foraging skills such as bubble feeding, which is not practiced by other balaenopterid whales.

Predation, Parasitism, and Causes of Mortality

Predation. Given the potential effect of predators in shaping the social ecology of a species, it is unfortunate that the degree to which humpback whales are subject to predation remains unknown. The most frequently cited potential predators are killer whales *(Orcinus orca)*. While observations of fatal attacks on mysticetes are rare, records of such events involve several species, including Bryde's whales (Silber et al. 1990), gray whales (Baldridge 1972), blue

whales (Tarpy 1979), and minke whales (Hancock 1965). Dolphin (1987b) also suggests false killer whales *(Pseudorca crassidens)* as a potential predator of humpbacks, and there have been various references (notably in the whaling literature) to the idea that sharks may attack newborn calves (Budker 1953).

Humpback whale flukes (and sometimes other body parts) show abundant evidence of attack in the form of teeth marks or apparent bites. These marks are generally assumed to come from attacks by killer whales. Katona et al. (1980) noted that 33% of humpback whale flukes photographed in the western North Atlantic showed such marks. However, very few attacks by killer whales on humpback whales have been observed, and only one may have resulted in the death of an animal (Flórez-González et al. 1994). Dolphin (1987b) reviewed his data on humpback whales and killer whales in Alaskan waters and found no evidence for attacks; he also stated that he had "observed no situation that could unequivocally be labelled avoidance of killer whales by humpbacks." However, Dolphin did not state whether the killer whales in question were *transients* (and therefore fed on mammals) or *residents* (who would have fed on fish). Whitehead and Glass (1985) described two attacks off Newfoundland, neither of which resulted in serious injury or death to the humpbacks concerned.

It is frequently argued that just because predation is not witnessed does not mean that it occurs too seldom to represent a significant source of mortality. However, we are confronted with the fact that two decades of almost daily observations in the Gulf of Maine (a feeding ground) and sixteen winters of work in various parts of the West Indies breeding range have produced no record of killer whale attack on humpbacks; furthermore, there have been no observations of killer whales during the research in the West Indies, and only a handful of sightings in the Gulf of Maine. A similar dearth of killer whale sightings and absence of attacks applies to virtually all other sites where humpback whales have been intensively studied for many years.

Where, then, do the teeth marks on humpback whale flukes come from? Unpublished data from the Gulf of Maine suggest an explanation. Of the many humpbacks exhibiting teeth marks, all but one possessed those marks on the first occasion that the whale was recorded; in other words, only one whale was observed one summer without marks, but returned the following year bearing them (personal observation). I therefore suggest that attacks occur almost exclusively upon newborn calves at some point on the migration from the West Indies (where they are born) to the high-latitude feeding grounds where they spend their first summer. Chittleborough (1953) and Dolphin (1987b) make similar suggestions for other areas. The frequency with which such attacks occur on the migration is unknown; given that we can observe only the survivors, it must be acknowledged that they could represent a significant source of neonatal mortality.

Parasites. Humpback whales can carry a heavy parasite load. Whale lice (*Cyamus* spp.) infest humpback whales, although in considerably smaller numbers than is typical for right whales, whose dead skin they have been shown to eat (Rowntree 1996). Cyamids are found in sheltered locations on the body where water flow and drag are minimal; it seems unlikely that they are a significant threat to the health or well-being of the animal. Internal parasites, notably the giant nematode *Crassicauda*, probably represent a more significant problem for humpbacks and other balaenopterids (Lambertsen 1986), but related mortality is unknown.

Barnacles such as *Coronula* spp. are not true parasites, but rather use the whale as a substrate to which they attach (Baylis 1920). It has been suggested that barnacles may benefit at least male humpback whales through their use as weapons in intrasexual competition (Pierotti et al. 1985); a similar function has been proposed for the callosities of right whales (Payne and Dorsey 1983). However, there is no direct evidence for the barnacle hypothesis in humpbacks.

Other causes of mortality. Known causes of mortality include various anthropogenic sources, including entanglement in fishing gear and ship strikes (Whitehead 1987; Lien 1994). The apparent poisoning of many humpbacks by dinoflagellate toxins (a "red tide" event) has been documented only once (Geraci et al. 1989), but the long-term frequency and significance to population dynamics of such perturbations is unknown. It should be noted that this mass mortality involved numerous animals that died independently over several weeks; mass strandings of the type frequently recorded for some odontocetes (in which part or all of an entire group dies together) are unknown in baleen whales.

Life History

Birth, Growth, and Maturity

Gestation and birth. Humpback whale calves are born after a gestation period lasting between eleven and twelve months (Chittleborough 1958). No birth event has ever been observed in this species, but the abundance of females with very young calves in tropical waters during the winter makes it clear that the majority of calves are born in low latitudes (Chittleborough 1965).

There are no reliable records of a humpback giving birth to twins; all of the mothers in the various long-term studies cited here have been uniparous (Baker et al. 1987; Clapham and Mayo 1990; Glockner-Ferrari and Ferrari 1990), and whaling catch data show that very few humpback whale pregnancies (0.28–0.39%) involve multiple fetuses (Slijper 1962; Chittleborough 1965).

The peak birth month in the Southern Hemisphere, as determined from fetal length data, is early August (Matthews 1937; Chittleborough 1958, 1965). Studies of live whales suggest that births in the Northern Hemisphere peak in February (Herman and Antinoja 1977; Whitehead 1981). The mean length of calves at birth is between 13 and 15 feet (4–4.6 m: Clapham et al. 1999). Calves are precocious; they may begin the migration to the mother's high-latitude feeding grounds when only a few weeks old, where they continue to nurse and eventually feed independently.

Weaning and separation. Based upon the presence of milk in the mammary glands of mature females caught without calves, Chittleborough (1958) estimated that humpback calves were weaned at between ten and eleven months, and at a length of approximately 29–30 feet. Van Lennep and Van Utrecht (1953) believed that weaning was gradual, with a transition period during which the calf fed on both milk and solid food. Clapham and Mayo (1987) concluded from observations in the Gulf of Maine that most calves begin to feed independently at six months; however, apparent nursing behavior is observed for up to several months after this.

Although separation of calves from their mothers in the autumn of their birth year has been documented (Baraff and Weinrich 1993), the great majority remain with their mothers until some point during the calves' second winter. Sightings of mother-calf pairs still together, or apart, on the breeding grounds during the second winter indicate that separation does not consistently occur at a particular time or stage of migration (Baker and Herman 1984b; Clapham and Mayo 1987; Glockner-Ferrari and Ferrari 1990). Clapham (1993a) reported that 6 (three males, one female, and two of unknown sex) of 107 calves observed in the Gulf of Maine remained with their mothers for two years, but none remained into a third year.

Maturity, growth, and longevity. Different stages of maturity are recognized in humpback whales. Sexual maturity is defined by the presence of sperm in the testes of males or the occurrence of ovulation in females (Chittleborough 1954, 1955a,b). Physical maturity occurs later, and is generally defined by complete fusion of the vertebral epiphyses (Chittleborough 1955b). Social maturity is often defined in two ways: as the age at which a whale's association patterns become indistinguishable from those of other adults, or as the age when the whale begins to breed.

Female humpback whales grow faster and attain greater lengths than males at all stages of life (Nishiwaki 1959; Chittleborough 1965); this pattern is observed in most mysticetes. Growth curves for Antarctic humpback whales are presented in figure 7.5. Observations both from whaling (Chittleborough 1965) and from a long-term study in the Gulf of Maine (Clapham 1992) indicate that the mean age at attainment of sexual maturity in females is five years. That most males are also sexually mature by five is strongly suggested by Chittleborough's (1955a) records of testicular maturity at this age and by observations of breeding-related behavior among known-age males in the West Indies (Clapham 1992). As described further below, social matu-

rity among both males and females seems to occur at about this same age of five years (Clapham 1994). However, in species with polygamous mating systems, reproduction is frequently dominated by older males (Greenwood 1980; Johnson 1986). Clapham (1992) suggested that many male humpbacks may not be able to reproduce until they are large enough or experienced enough to successfully compete with other males. Physical maturity in male and female humpback whales is not reached until eight to twelve years after sexual maturity (Chittleborough 1965; Lockyer 1984).

The life expectancy of humpback whales is difficult to estimate, in part because whaling removed most "old" animals from the population. Chittleborough (1965) reports that the oldest whale he examined off western Australia was 48 years old. Reliable data on mortality are lacking; Chittleborough (1965) notes that an estimate for the natural mortality rate of 0.09 should be regarded as a maximum value in view of the fact that the data set upon which it was based took no account of probable unreported catches (a suspicion that proved correct: Zemsky et al. 1995; Yablokov et al. 1998). Recently, Barlow and Clapham (1997) used a birth interval model to estimate demographic parameters for the Gulf of Maine population; the results gave a non-calf mortality rate of 0.04 (SE = 0.008).

Reproduction

Working from whaling stations in eastern and western Australia, Chittleborough (1954, 1965) examined the ovaries of female humpbacks and determined that ovulation occurs from June to November. The peak was in late July, with 47.1% of sampled mature females ovulating at this time; however, "considerable numbers" of animals were in estrus in August and September. Chittleborough also determined from histological examination that most females ovulate only once during a winter, but that some (estimated at 16–28%) ovulate twice, and a few (0–8%) three times. Chittleborough also found that aseasonal (i.e., austral summer) ovulation and conception occurred, but only rarely. The frequency of ovulation among any nonmigratory females (see M. R. Brown et al. 1995) is unknown.

Reproductive rates have been variously reported (and defined). Reported pregnancy rates from whaling data of 0.37 (western Australia: Chittleborough 1965) and 0.39 (Aleutian Islands: Nishiwaki 1959) are generally comparable to the measure "calves per mature female per year" reported from some recent long-term studies, including 0.37 for Alaska (Baker et al. 1987) and 0.41 for the Gulf of Maine (Clapham and Mayo 1990).

Interbirth intervals among mature female humpbacks

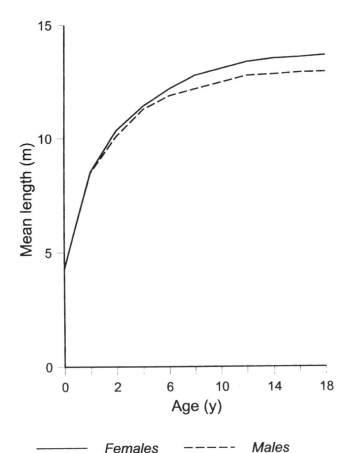

Figure 7.5. Estimated growth curves for male and female humpback whales from the Antarctic. (Adapted from Chittleborough 1965, Figure 19.)

PHILLIP J. CLAPHAM

vary from a single year to several years; the modal interval is two years (Chittleborough 1958). Chittleborough (1958, 1965) estimated that postpartum estrus occurred in a minority of cases; although he did not provide an estimate, 8.5% of females (reported by him from Norwegian whaling data) were simultaneously pregnant and lactating. In the Gulf of Maine, Clapham (1993a) reported that 63.2% and 25.3% of interbirth intervals ($n = 95$, all of them involving complete resighting histories) were of two and three years, respectively. Chittleborough (1958, 1965) suggested that three-year interbirth intervals resulted when a female lost a calf and subsequently conceived again the following winter. However, a three-year interval would also result if a female failed to conceive in two consecutive years, a possibility not addressed by Chittleborough. Data from the Gulf of Maine and southeastern Alaska suggest that females may reproduce at shorter intervals as they get older, or that their calves may be more likely to survive to be observed in feeding areas (J. M. Straley, P. J. Clapham, and W. Amos, unpublished data). However, age-specific fecundity and calf survival cannot be estimated from current data.

Neonatal Mortality

No good data exist with which to reliably estimate neonatal mortality in humpback whales. Clapham (1993a) compared the frequency of one-year birth intervals (6.3%) in the Gulf of Maine population with a much higher rate (14%) reported from the Hawaiian breeding range by Glockner-Ferrari and Ferrari (1990). He suggested that the difference in part reflected neonatal mortality prior to the mother's return to high-latitude feeding grounds, since any calf dying in the West Indies or on the migration north would not be recorded in the Gulf of Maine later in the year. Given the high cost of lactation in baleen whales (Lockyer 1981b, 1987), females with shorter interbirth intervals (notably one year) may have higher calf mortality.

However, the difference between the Gulf of Maine and the Hawaii data could also be attributed to sampling bias in the latter location: Glockner-Ferrari and Ferrari's sighting effort was strongly skewed toward mother-calf pairs, thus inevitably biasing their data set toward individuals with shorter interbirth intervals. This is further suggested by the discrepancy between their estimate of the proportion of females calving in consecutive years (14%) and the much lower figure (8.5%) given by Chittleborough (1958) for females who were simultaneously pregnant and lactating. However, Chittleborough's figure is almost certainly an underestimate, since it was derived from a Norwegian whaling catch sample in which lactating females were probably underrepresented. In practice, the difference in reported rates is probably due to a combination of both factors (mortality and bias).

Sex Ratio

Data from both whaling and long-term studies have shown that the sex ratio of humpback whale populations does not deviate significantly from parity (Chittleborough 1958, 1965; Matthews 1937; Clapham et al. 1995). However, that the operational sex ratio during the winter breeding season may be strongly biased toward males is suggested by several findings. First, Chittleborough (1965) noted that males made up a declining proportion of the catch off both coasts of Australia; a similar phenomenon was found by Budker (1953) off the coast of Gabon. Chittleborough linked this finding to heavy exploitation in low latitudes, with the (unstated) implication being that males were more abundant on the breeding grounds. Second, recent molecular sexing of free-ranging humpbacks on migration (M. R. Brown et al. 1995) and on a West Indies breeding ground (Palsbøll et al. 1997) showed a similarly strong bias toward males.

The question of whether the sex ratio of humpback whale calves varies by length of interbirth interval was investigated in the Gulf of Maine population by Wiley and Clapham (1993). They found that the sex ratio of offspring born after three-year intervals was significantly male-biased (seventeen male calves were born and only four females); the ratio associated with two-year intervals did not deviate from parity (twenty-two males, twenty females). The skew in these data is greater than in any other mammal that has been studied in this regard (see review by Clutton-Brock and Iason 1986). The authors interpreted the results as supporting Trivers and Willard (1973), who proposed that the ability of female mammals to vary the sex ratio of their offspring should be selected for in those species in which there is a greater variance in reproductive success among members of one sex (generally males), and in which the body condition of the mother during the period of parental investment affects the future condition (and competitive abilities) of the offspring.

This conclusion hinges entirely upon acceptance of the central assumption of the Wiley and Clapham (1993) study, that a longer calving interval is indicative of superior body condition. The authors argued that a female giving birth after three years would have less recently incurred the substantial cost of lactation, and cited supporting data from terrestrial mammals (Clutton-Brock et al. 1982; Rutberg 1986). However, the reverse argument can be made, that females calving at shorter intervals must be in better body condition to be able to do so (also, longer intervals may reflect the undocumented loss of a calf). Given that female humpbacks are larger and grow faster than males at all stages of life, it is possible that a female calf requires greater maternal investment; if so, mothers with shorter interbirth intervals might well be in better rather than worse condition. In light of these uncertainties, the functional significance of the Wiley and Clapham (1993) results, and the validity of the Trivers-Willard interpretation, remains unclear.

Social Organization

What Is a Group?

Balaenopterid vocalizations can potentially travel hundreds or thousands of kilometers (Payne and Webb 1971). Similarly, humpback whale song can be heard by large numbers of animals that the singer is not "with." Because of this, and the fact that visual contact underwater is limited, the term "group" is typically defined using proximity measures. Humpback groups have been defined as affiliations in which two or more animals swim side by side within 1–2 body lengths and generally coordinate at least their surfacing and diving, as well as their speed and direction of movement (see discussion in Clapham 1993b). This definition of "group" has also been employed (or is implicit) in both high- and low-latitude studies by Baker and Herman (1984a,b), Mattila and Clapham (1989), Perry et al. (1990), Weinrich (1991), and Weinrich and Kuhlberg (1991), among others. In some other studies, it is not clear whether "group" refers to animals associated in the sense above or merely observed in the same general area. Overall, it is important to remember that a proximity-based definition potentially underestimates the degree of association between animals that may be spatially distant but in acoustic contact.

Group Size, Structure, and Stability

Feeding grounds. With some interesting exceptions, the social organization of humpback whales in summer is strongly characterized by small, unstable groupings and by individual associations that last for periods of a few hours or less. Whitehead (1983) noted more variability in size among feeding than among nonfeeding groups off Newfoundland, with group size logarithmically correlated with the horizontal extent of prey patches. He also found that, while some pairs would maintain their associations within larger groups, there was no evidence for preferred companionships over more than a day.

Quantitative analyses of humpback whale group composition in high latitudes have been conducted only for the Gulf of Maine population. Clapham (1993a) compared sighting data for groups with and without a calf. More than half (58.4%) of the 22,672 recorded sightings without a calf were of singletons, and 30.1% were of pairs. Large groups were uncommon, and were invariably observed feeding on extensive schools of sand lance. Among groups containing a calf, the great majority (76.2%) included no whales except the mother and calf. For all whales, group size fluctuated, notably during feeding, most associations were brief, and individuals associated with many different conspecifics. Only six stable associations (defined as two whales recorded together in at least five consecutive observations over a period of two weeks or more) were recorded, and the sex and maturational class composition of these groups was varied. Furthermore, of 2,690 pairs containing animals known to have close maternal relatives in the population, only 12 consisted of related animals. Analysis by sex and maturational class of singletons and pairs (by far the most common group size) revealed that, among singletons, immature whales of both sexes were significantly overrepresented and mature females significantly underrepresented. In addition, male-female adult pairs were overrepresented, and pairings between adult males were underrepresented except during feeding. The prevalence of small, unstable groups was attributed to a lack of predation pressure and to the spatial characteristics of piscine prey (patches of constantly varying extent) favoring a foraging strategy involving frequent changes in group size. Clapham (1993a) further suggested that in such a system, kinship plays a limited or inconsequential role, since stable groups would be at a disadvantage unless they could consistently outcompete other whales for larger prey patches. Similarly, the absence of any obvious dominance interactions may be related to inability to defend the resource as well as to the difficulty of establishing dominance rankings among the large number of conspecifics that an individual will encounter during a summer.

Weinrich and Kuhlberg (1991) also analyzed data on short-term association patterns (i.e., the great majority of associations observed), and found that most groups were small and unstable. They concluded that the observed underrepresentation of male-male pairs reflected intrasexual avoidance as an extension of breeding competition. They also suggested that adult females were more likely to be social because of a need to cooperatively forage in order to meet the greater energetic costs of reproduction. Elsewhere, Weinrich (1991) studied stable associations, defined in two ways: "continuous" associations of whales that were observed together at least four times over a minimum of seven days; and "recurring" associations, in which two whales were recorded together for protracted periods, with occasions between on which they were separated. Weinrich (1991) noted that females made up the majority of stable associations; he hypothesized that such affiliations permit adult females to maximize net energy gain through cooperative feeding (notably when pregnant), and further, that they were composed of closely related animals or of whales with compatible feeding styles. However, if stable pair formation brought significant energetic benefits, it would presumably be adopted by other animals irrespective of need or relatedness.

The three Gulf of Maine studies (Weinrich 1991; Weinrich and Kuhlberg 1991; Clapham 1993a) are in agreement with regard to the small and unstable nature of the great majority of humpback whale groups. They are also in broad agreement that associations between males are less common than would be expected by chance, although their interpretation of this finding differs, as will be discussed further later. Weinrich's hypothesis that stable pairs are composed of related animals awaits molecular testing, but preliminary mtDNA analysis does not support the idea (P. J. Clapham and P. J. Palsbøll, unpublished data). Recent data on known relationships among members of the large stable groups reported from Alaska have indicated that these are not composed of related whales (C. M. Gabriele, personal communication).

Breeding grounds. As on the feeding grounds, the social behavior of humpback whales during their winter residency in low latitudes is characterized by small groups and brief associations (Mobley and Herman 1985; Mattila et al. 1994). However, interactions during this time are also marked by frequent agonistic encounters between males, and by other behaviors that are rarely observed during the summer feeding season. With very rare exceptions (Baraff et al. 1991), feeding is absent from the behavioral repertoire at this time, a fact that has been confirmed by whaling catch records showing empty stomachs and lower oil yields during winter (Chittleborough 1965; Dawbin 1966).

The transience of most affiliations during the breeding season was documented by Tyack and Whitehead (1983), Baker and Herman (1984a), and Mobley and Herman (1985). Off Hawaii, both females and males were resighted with different associates; in many cases, the associates were known to be of the opposite sex. Singletons were extremely common, as were pairs and trios (Mobley and Herman 1985). Many single whales were engaged in singing behavior (see below), an activity that (with a single, possibly aberrant exception) has been ascribed exclusively to males (Winn and Winn 1978; Tyack 1981). Mothers and calves were frequently accompanied by a third whale, termed an "escort"; all sexed escorts have proved to be male, a fact that has led virtually all observers to agree that escorts associate with lactating females because of the possibility of mating if the female comes into postpartum estrus (Tyack 1981; Glockner and Venus 1983; Baker and Herman 1984a; Glocker-Ferrari and Ferrari 1990).

Larger groups characterized by substantial surface activity, and sometimes high levels of aggression between members, are termed "competitive groups" (Clapham et al. 1992). Competitive groups (also referred to as "active," "surface-active," or "rowdy" groups by other observers) were first described in detail by Tyack and Whitehead (1983) and Baker and Herman (1984a), who suggested that they consisted of multiple mature males competing for sexual access to a single mature female (fig. 7.6). This was largely confirmed by Clapham et al. (1992), who used a molecular technique to determine the sex of all members of twenty-one competitive groups observed in the West Indies; the results generally supported the interpretation of intrasexual competition.

Tyack and Whitehead (1983) identified specific roles within competitive groups. A central whale that was generally less physically active was termed the "nuclear animal" (hypothesized and later confirmed to be female in most cases). A whale closest to the nuclear animal, who often aggressively fended off approaches to her, was called the "principal escort." Other whales in the group (excluding a calf when present) were termed "secondary escorts." Clapham et al. (1992) added the term "challenger" to distinguish any whale that actively challenged the principal escort for his position, and considered a secondary escort to be any whale in the group that did not adopt one of the other three (key) roles. Note that the term "escort" is used universally to refer to a whale accompanying a mother-calf pair on the breeding grounds. Escorts frequently become principal escorts when one or more other males join the group and challenge them for access to the nuclear animal (the mother).

Migration. With the exception of whaling studies, which found a temporal segregation of classes on migration (see above; Nishiwaki 1959; Chittleborough 1965; Dawbin 1966), the only study of group size and composition among migrating humpback whales has been that by Brown and Corkeron (1995), which encompassed the entire migration period (both south and north) off the Queensland coast of Australia. The social behavior of the whales was clearly related to breeding rather than feeding; this finding indicates that humpbacks continue

Figure 7.6. Researchers prepare to use a crossbow to biopsy members of a competitive group in the West Indies. The whale closest to the boat is the nuclear animal (which molecular sexing subsequently determined to be female). The individual with its head out of the water is a challenger "beaching" itself on the back of the principal escort (not visible), who is blocking his access to the nuclear animal. (Photograph courtesy of Phil Clapham/Center for Coastal Studies.)

to pursue mating opportunities while migrating, although over what portion of the migratory route this activity continues is not clear. Using skin biopsies and a molecular technique, Brown and Corkeron (1995) determined the sex of 134 individuals in sixty-three complete groups. They found that, while male-female was the most common combination among the twenty-seven sampled pairs, nine pairs consisted of two males, and two of two females. In addition, eight of seventeen groups of three or more whales consisted entirely of males (see discussion below).

Parental Care and Juvenile Development

Parental care of humpback whale calves is provided exclusively by the mother. It was once thought by whalers that a third whale frequently observed to accompany mother-calf pairs on the breeding grounds was the father, but it is now widely believed that the invariably male escort is unrelated. A mother provides her calf with food in the form of a fat-rich milk (Pedersen 1952). Beyond nursing, the nature of maternal care is unclear. Mothers certainly provide protection for their calves: they will often place themselves between the calf and an approaching boat, and steer the calf away (personal observation). Whalers would often exploit the maternal bond by injuring a calf in order to gain easier access to the mother (Allen 1916).

There is little doubt that calves learn from their mothers

the migratory routes that in future years will take them to the feeding grounds to which they will return each spring (Clapham and Mayo 1987; Palsbøll et al. 1995). The presumably large difference in potential swimming speed between mother and calf, in either local or migratory movements, may be compensated for by the calf swimming close to the mother and thus receiving some hydrodynamic benefit (Brodie 1977).

Once a calf has separated from its mother, there appears to be little contact between them (Baker et al. 1987; Clapham 1993a,b). The rate of reassociation between mothers and offspring in the Gulf of Maine is not significantly different from that between other whales (Clapham 1993a,b), although it is impossible to say whether such reassociations, when they occur, are qualitatively different from other associations. Juvenile humpback whales associate with many different conspecifics, but are observed alone significantly more often than adults. Clapham (1994) noted that, as juveniles age, the proportion of time spent with adults increases. By five years (the average age at attainment of sexual maturity), association patterns are essentially indistinguishable from those of adult conspecifics.

Unlike some other balaenopterids (e.g., minke whales: Jonsgård 1980; Wada 1989; fin whales: Rørvik et al. 1976), there is little evidence that humpback whale populations exhibit distinct segregation by sex or maturational class, except perhaps in migratory timing.

Social Behavior and Communication

Behavioral Displays

Humpback whales are well known for the frequency with which they engage in often spectacular aerial or other high-energy behaviors. These include breaching (the whale leaps headfirst from the water), lobtailing (the whale repeatedly slaps the surface of the water with its tail), flippering (the whale slaps the surface with one or both pectoral fins), spyhopping (part or all of the head is lifted out of the water), and tail breaching (the whale hurls its tail and caudal peduncle sideways out of the water). All of these displays are observed frequently in both high- and low-latitude habitats, and all are conducted by animals of both sexes and all age classes, from calves to adults. The function of such displays is generally unclear; Whitehead (1985c) proposed that all of them may act "like a physical exclamation mark, to accentuate other visual or acoustic communication."

The only directed study of breaching behavior was conducted by Whitehead (1985a). He noted that breaching was far more common in species that were more social and that had a tendency (like humpback whales) to congregate on breeding grounds. These observations suggest that breaching behavior is sometimes related to social activity or communication. This belief is reinforced by Whitehead's finding that breaches were more likely to occur during the joining or splitting of a group, and that breaching was seven times more common in breeding than in feeding areas. Many other suggestions for a function for breaching are critiqued by Whitehead (1985a); among the various ideas are the shedding of sloughed skin or parasites and a response to disturbance by vessel traffic. As Whitehead notes, there is probably no single function for this behavior, since it has been observed among all age-sex classes and in a variety of behavioral contexts. It is unlikely to be a sign of aggression, and when observed in competitive groups it is practiced only by peripheral animals, not by key participants.

Among other "active" behaviors, it appears likely that a tail breach (also known as a peduncle slap) is an aggressive or threat behavior in many cases, since it is frequently a response to harassment by boats (personal observation); it may therefore reflect annoyance. Neither lobtailing nor flippering has been closely studied. As with breaching, both behaviors are observed in a variety of social contexts. Whitehead (1983) found that, off Newfoundland, flippering was significantly more likely to occur when groups split than at other times, but he made no attempt to interpret the correlation. An idea voiced anecdotally by several observers that females use such displays to "call in" males, or to solicit competition in order to displace an unwanted companion, is perhaps supported by occasional observations of a female flippering prior to the formation of a competitive group (personal observation).

A further type of behavioral display involves the production of streams of bubbles (Tyack and Whitehead 1983; Baker and Herman 1984a). These are observed in competitive groups, and are generally made by the principal escort (Clapham et al. 1992). Bubble streams vary considerably in length, and when seen underwater, they resemble a curtain 2 or 3 m deep. Their function is unclear, but the most obvious possibilities involve threat or the creation of a screen to make it more difficult for challenging males to keep track of the female. It is also possible that bubble streams indicate physical condition to other males or to the female.

Inflation of the ventral pouch by animals (notably principal escorts) in competitive groups (Tyack and Whitehead 1983; Baker and Herman 1984a; Silber 1986) probably serves to create the impression of larger size and thus to intimidate challenging males; such a function has been ascribed to similar behaviors in other taxa and was noted by Darwin as early as 1871.

Song

In biology, a "song" is defined as any two or more notes that are repeated in a pattern. Humpback whale songs were first described by Payne and McVay (1971). As they describe, humpback songs are broken down into single units of sound that together make up a "phrase"; the repetition of a phrase is termed a "theme," and a "song" consists of a series of such themes sung in a specific order. Recorded songs vary in length from a few minutes to over half an hour; however, songs cycle without any obvious beginning or end, and whales may sing continuously for many hours. Singing behavior is heard throughout the winter in low latitudes, although it has also been recorded on migration (Stone et al. 1987; Clapham and Mattila 1990), on feeding grounds in spring and autumn (Mattila et al. 1987; McSweeney et al. 1989), and even in midsummer (D. K. Mattila, unpublished data).

Singers are always male (Glockner 1983; Baker and Herman 1984a) and are generally alone (Winn and Winn 1978; Tyack 1981), but occasionally escort conspecifics (e.g., Baker and Herman 1984a). The structure of a particular song changes progressively with time, yet all whales within a breeding population sing the same song and keep up with these changes (P. M. Payne et al. 1983; Payne and Payne 1985). Humpback whales from different ocean basins sing different songs, and song comparisons have been used to confirm hypothesized separation of populations (e.g., eastern and western Australia: Cato 1991; Dawbin and Eyre 1991) or to infer mixing among populations once thought to be discrete (e.g., Mexico and Hawaii: Payne and Guinee 1983).

It appears likely that one of the principal functions of song is as a powerful acoustic display with which to attract potential mates (Winn and Winn 1978; Tyack 1981). Other putative explanations include a spacing function among males (Frankel et al. 1995), a means of establishing dominance rankings (Darling 1983), and a threat display during intrasexual competition (Baker and Herman 1984a). That humpback song should serve more than a single function seems likely, notably in view of its multiple uses in other taxa.

Tyack (1981) studied interactions between singers and conspecifics in Hawaii, and found that singers frequently joined, or were joined by, other whales, at which point singing always ceased. Pairs resulting from a singer joining another animal remained together for periods ranging from a single surfacing to more than one and a half hours. Singers did not appear to be territorial; although most were relatively stationary (a finding subsequently contradicted by Frankel et al. 1995), they did not remain in a particular area from day to day. One of Tyack's most interesting findings was that singers sang for longer periods of time as the season progressed, and that the period of shortest song bouts coincided with the peak (hypothesized from whaling data) of estrus in females. Playback experiments conducted by Tyack (1983) found that most whales avoided song, while both singers and certain other whales responded to playback of "social sounds" (recorded from a competitive group) by charging the boat. In contrast, mothers with calves moved away from the sound source. Tyack concluded that both song and social sounds mediated responses of approach or avoidance between animals.

Mobley et al. (1988) also conducted playback experiments off Hawaii, using recordings of song and social sounds as well as vocalizations recorded from the Alaskan feeding grounds. They proposed that the various types of vocalization conveyed different information and levels of attractiveness to males prospecting for mates. Song indicated the presence of another male (low attraction), social sounds the presence of a female but accompanied by two or more competitive males (moderate attraction), and feeding sounds the presence of a probably unaccompanied female (high attraction). However, since many of the playbacks involved contextually improbable sounds (e.g., feeding on

a breeding ground), it is difficult to give a functional interpretation to the responses.

Darling's (1983) sexing as male of animals (number not given) who joined singers led him to propose that song served principally to mediate interactions among males rather than to attract females. However, both Tyack (1981) and Medrano et al. (1994) determined that some whales observed to join singers in nonagonistic interactions were female, lending support to the latter belief. Work by Frankel et al. (1995) suggested that singers were spaced farther apart than other whales, and that this spacing pattern was density-dependent. The authors suggested that, by maintaining as wide a spacing as possible, males would facilitate the location of individual singers by females. Chu and Harcourt (1986) and Chu (1988) proposed that song conveys information that permits conspecifics to determine the dive length, and therefore the potential condition (indicated by breath-holding ability), of the singer. Such a cue would be of obvious value both to females seeking potential mates and to potential competitors, but the considerable variation in the mean length of songs between years, and at different times within the same season, argues against this hypothesis.

Other Vocalizations

There has been little concerted effort to study non-song vocalizations in humpback whales. Perhaps as a reflection of this, all vocalizations that (unlike song) are variable through time and do not exhibit a consistent or continuous pattern have been lumped under the term "social sounds," irrespective of behavioral context. Silber (1986) found that, in the Hawaiian breeding range, social sounds were heard almost exclusively in groups containing three or more whales and were generally associated with intrasexual competition among males. He suggested that social sounds in this context reflected agitation and/or acoustic threat behavior by group participants. Beyond this general suggestion, however, there is no knowledge of the specific function, if any, of particular types of non-song vocalization during the winter.

From the increased frequency with which vocalizations occur in feeding groups, D'Vincent et al. (1985) hypothesized that such sounds served to coordinate lunging behavior. There have been no other directed (nonopportunistic) studies of humpback vocalizations in summer.

Interactions between Females

Data on interactions between female humpback whales are sparse, in large part because of the difficulty of reliably determining sex in the field. However, information on the association patterns of known females from long-term studies on feeding grounds provide some insights, as do two recent studies that have used molecular methods to determine sex in lower latitudes.

There is currently no direct evidence (in the form of observations of agonistic interactions) for intrasexual competition among females on either feeding or breeding grounds. However, it is possible that the rarity of female-female associations reflects avoidance, and thus some form of competition. During the winter, female-female pairs appear to be uncommon: there were none in a sample of thirty-four pairs biopsied in the West Indies in 1992 (D. K. Mattila, unpublished data), and Brown and Corkeron (1995) found only two out of twenty-seven sampled pairs during the migration past the Queensland coast. Neither Clapham et al. (1992) nor Brown and Corkeron (1995) found any competitive groups to contain more than one female. A similar result was reported from incompletely sampled groups in Mexican waters (Medrano et al. 1994).

In addition, lactating females appear to avoid each other in all areas. Groups containing more than a single calf are rare in summer (Clapham and Mayo 1987) and apparently absent in winter (Herman and Antinoja 1977; Baker and Herman 1984b; Mobley and Herman 1985). Females are generally unsocial when they have a calf; even in the Alaskan cases of stable core groups of humpbacks (Perry et al. 1990; Baker et al. 1992), mature female members of the groups did not participate in years when they had a calf (C. M. Gabriele, personal communication). Herman and Antinoja (1977) suggested that avoidance on the breeding

grounds allows time for the mother and calf to imprint on each other. However, the features of this avoidance are not consistent with an imprinting explanation because imprinting occurs primarily in gregarious species in which opportunities for confusion are high (Lorenz 1937; Bateson 1991). Imprinting periods are very narrow (a couple of days), but the humpbacks' avoidance of other mother-calf pairs lasts for weeks or months.

Interactions between Males

Interactions between male humpback whales show marked contrasts between summer and winter. In summer, males often associate and forage together, although as noted above, they tend to occur together less frequently than would be expected by chance (Weinrich and Kuhlberg 1991; Clapham 1993b). With the exception of some apparent late-season competitive group behavior (Weinrich 1995), agonistic encounters between males have not been observed on the feeding grounds. In winter, many interactions are agonistic and competitive. There is little doubt that most competitive groups represent intrasexual competition among males. Several observers (Tyack and Whitehead 1983; Baker and Herman 1984a) have noted the similarity of these groups to tending behavior in ungulates (Leuthold 1977), in which a male escorts an estrous or proestrous female and fends off other males. In such contexts, mating with the female is the goal, and it is generally assumed that this is also the case for humpback whales.

Competitive groups are often violent (figs. 7.6 and 7.7), with participants showing visible superficial wounds (Ty-

Figure 7.7. A principal escort slashes a challenger (foreground) with his tail in a competitive group in the West Indies. (Photograph courtesy of Phil Clapham/Center for Coastal Studies.)

THE HUMPBACK WHALE

ack and Whitehead 1983; Baker and Herman 1984a; Clapham et al. 1992). Serious injuries have not been recorded, although it would be impossible to detect significant internal injury unless it visibly compromised the behavior of the animal. Competitive groups vary considerably in their duration and dynamics. Many persist for several hours, often gaining or losing members. Some are notably and consistently violent, others relatively passive. Interactions between participants appear to be a matter of escalating actions, with visual (bubble streams, ventral pouch inflation) and probably acoustic threats followed by various degrees of physical contact. Principal escorts are successfully displaced by challengers in only a minority of cases: Tyack and Whitehead (1983) observed no displacements in Hawaii, and ten on Silver Bank, after groups had been followed for many hours. Clapham et al. (1992) recorded displacements in only three of forty-four groups in Samana Bay.

It is very likely that size and stamina are important determinants of the outcome of such contests, and perhaps the most violent groups reflect the presence of well-matched contestants. Maneuverability may also play an important role. It is clear from field observations that not all whales have an equal chance of attaining principal escort status. This finding is supported by a study of dorsal fin scarring in the West Indies, in which animals in key roles (i.e., principal escorts or challengers) within competitive groups were much more likely to be scarred than secondary members (Chu and Nieukirk 1988).

The discovery by Clapham et al. (1992) of pairs of males apparently working cooperatively in competitive groups led to the ineluctable conclusion that males may form coalitions to displace a principal escort. Pairs of males were observed entering or leaving competitive groups together; while in the group, these animals were often aggressive toward other males, but not toward each other. Similar observations were reported from eastern Australia by Brown and Corkeron (1995). Given that males are apparently always competing for single rather than multiple females, it is difficult to see the value of such affiliations unless either both males mate with the nuclear animal (and submit to sperm competition) or the coalitions consist of related animals. Although the latter seems unlikely in a species in which males from the same feeding ground are widely dispersed during the winter, microsatellite DNA analysis has recently found at least one example of related males within the same competitive group (Valsecchi 1997). Kin recognition is presumably difficult when most calves have been weaned before the birth of a sibling; the possibility that older animals might learn to recognize siblings via interactions with

their mother and her present calf seems unlikely given the low rates of reassociation observed after separation. On the other hand, a male with no chance of successfully displacing a principal escort on his own would have nothing to lose by forming a coalition and (if successful) submitting to the female's choice.

Male-Female Interactions

Male-female associations make up the greatest proportion of pairs observed on the feeding grounds (Clapham 1993b); these pairs, and female-female dyads, are more likely to remain together for periods of more than ten minutes than are male-male pairs (Weinrich and Kuhlberg 1991). Intersexual interactions during the winter fall into three broad categories: attraction/association, avoidance, and repulsion. Included in the first category are males joining females (e.g., singers or others moving to associate with females, whether alone, with a calf, or accompanied by one or more other males) and, more rarely, females joining males (e.g., females approaching singers). The large proportion of pairs that consist of a male and a female (Brown and Corkeron 1995) also fall into this category. Examples of avoidance include females actively swimming away from a singing male, or from playbacks of song (Tyack 1981).

There are only a few observations of apparent repulsion, all of them involving rejection of males by females. Clapham et al. (1992) reported two cases in which a female at the center of a competitive group displayed aggression toward one or more males. They suggested that these represented rejection of "unsuitable" males by females, and noted that this seemed particularly likely in one case, in which the male concerned was a known subadult.

Since reliable observations of humpback whale copulation do not exist, the frequency with which males and females mate is entirely unknown. Copulation at the surface has been reported in both southern and northern right whales (Payne and Dorsey 1983; Kraus 1986) and in gray whales (Swartz 1986); presumably the humpback whale copulates exclusively underwater. Humpbacks have a fibro-elastic penis, and it has been speculated that copulation is rapid (Slijper 1962), as it is in most ungulates (e.g., Clutton-Brock et al. 1982).

Socioecology

Why Do Humpback Whales Migrate?

The humpback's annual migration is one of the longest of any mammal (Stone et al. 1990; Palsbøll et al. 1997), and has a major effect on many aspects of the biology and behavior of this species. The migration takes whales from cold, productive waters in high latitudes to warm tropical environments that (with some localized exceptions) are largely devoid of appropriate prey. We do not understand all of the evolutionary compromises that this radical change has effected in the biology and behavior of the humpback, but there is no doubt that it has affected a wide range of characteristics and functions, from energetics to growth rate, from breeding strategy to social behavior.

All the more remarkable, then, that the function of this migration remains unclear. There has long been a popular intuitive belief that calving in warm tropical waters confers a survival advantage upon offspring. Yet many energeticists find this idea unconvincing, particularly since it fails to explain why males and nonpregnant females should undertake the trip. Attempts to model the energetic costs of migration versus "staying at home" (Brodie 1975; Kshatriya and Blake 1985) have produced conflicting results, a reflection of the fact that values for many of the variables involved are at best guesses.

That seasonal migration is somehow advantageous to humpback whales is certainly suggested by the considerable compromises that it appears to have exacted in their reproduction. Allometric comparisons of gestation time to body size highlight the extraordinarily rapid growth rate of the mysticete fetus compared with other mammals, which is twenty times that of some primates (Laws 1959). Frazer and Huggett (1973) hypothesize that this telescoping of fetal development into a twelve-month cycle has arisen to accommodate the necessity for females to incur the primary costs of lactation during a time when they can exploit the peak of summer productivity in high latitudes. This pattern contrasts with other large-bodied species such as sperm whales, which do not appear to be under such constraints. Their colonization of a niche in which they exploit deep-water prey gives them access to food year-round (Best 1979; Whitehead and Weilgart, chap. 6, this volume); consequently, their long gestation period (estimated at 15–16 months) is closer to what one would expect from allometric considerations alone.

Testing of the hypothesis that birth in cold water during winter would increase calf mortality requires better energetics data than currently exist, but this idea appears questionable in light of the adaptation of considerably smaller mammals (e.g., harbor porpoise, *Phocoena phocoena*) to life in such temperatures. An alternative version, that early development in warm water results in better condition then and later in life, and consequently in higher average reproductive success, is perhaps more tenable; however, while this would represent a powerful selective force for migration, it

fails to explain why whales of all classes should migrate if the primary benefit is to offspring. Indeed, in right whales, it is primarily parturient females that migrate to warm water in winter (Kraus et al. 1986).

In the absence of better data, the most plausible explanation for the humpback's migration is that migration to low latitudes is energetically more efficient for all members of the population despite the requirement for fasting. Brodie (1975) extends this argument to suggest that the optimal strategy for mysticetes should be to feed during pulses of high productivity in summer, but to reside in warm water when food availability in high latitudes is insufficient to offset energetic costs.

Energetic modeling will be crucial to the resolution of this important question if accurate data on key variables can be obtained. I suggest that a model that attempts to assess the costs and benefits of migration to individuals based upon a specific age or reproductive condition is preferable to a more general approach that does not distinguish such differences. Comparisons of ecological differences between humpback whales and other mysticetes, particularly those that do not migrate (i.e., Bryde's whales: Best 1977; and perhaps also Arabian Sea humpbacks: Whitehead 1985b; Reeves, Leatherwood and Papastavrou 1991; Mikhalev 1997) or in which only segments of the population do so (e.g., right whales: Kraus et al. 1986), represent another potentially valuable approach.

Selection Pressures Affecting Social Organization

Gosling and Petrie (1981) have noted that a social system must be viewed as the summation of the behaviors of the system's individual members, since it is upon individuals (rather than groups or societies) that selection operates. Group size is influenced by a variety of ecological factors, and the system may be further modified by the behavior of conspecifics. Major influences on group size and social behavior are predation pressure, the distribution of resources, and the need for parental care.

Predation. The extent to which predation influences humpback whale social structure is currently unknown. Unpublished data on scarring suggest that predation on young calves occurs during migration, where researchers have focused the least attention. Tactics to evade or defend against predators are unknown.

Distribution of resources. As with other mysticetes, humpback prey distribution and abundance varies greatly, both within a season and from year to year, requiring considerable movement within the feeding range. At the level of the patch, the widely varying size and speed of fish or euphausiid schools encountered by foraging humpback whales favors frequent changes in group size and foraging behavior.

On the breeding grounds, since defensible resources are absent and predation is probably minimal, group size and structure are determined by a limited number of individual mating strategies. Males can remain alone to display (sing) to attract mates (and probably also maintain distance from other males), or they can actively seek out and escort mature females, creating dyads (or trios when the female has a calf). Challenges by other males lead to intrasexual aggression in sometimes large, unstable groups. The absence of prey resources on the breeding grounds likely influences the widespread distribution and lack of association among females.

Parental care. Group care of offspring has been observed or inferred in many mammals (see Gittleman 1985). Allomothering behavior is frequently correlated with kin-based groups, and with prolonged infant development and concomitant protracted parental care. Humpback whales possess few, if any, of these characteristics. The offspring grow very rapidly (Chittleborough 1965), become independent early (Clapham and Mayo 1990; Baraff and Weinrich 1993), and attain sexual maturity relatively quickly (Chittleborough 1965; Clapham 1992). With few pressures to form groups, there are few opportunities or advantages for shared parental care. Again, avoidance of each other by lactating females could be interpreted as competition, but this is not clear.

The Mating System

It is now recognized that, in virtually all mammals, the spatial and temporal distribution of resources in large part determines a species' mating system by controlling the distribution of receptive females (Clutton-Brock 1989b). How do predictions derived from a knowledge of the humpback whale's winter environment fit with what we now know about the mating system of this species? First, because females are dispersed in feeding and breeding areas, males can defend neither resource-based territories nor groups of females. As we would predict, therefore, males are found competing for single rather than multiple females (Clapham et al. 1992). In addition, the rarity of female-female dyads, and the lack of observed aggressive interactions between females from any breeding ground, are expected consequences of their lack of need to cluster around resources, and of the apparent local sex ratio bias toward males. As noted by Berglund et al. (1993), intrasexual competition among females for mates will generally occur only when sex ratio is female-biased.

Second, as noted by Emlen and Oring (1977), communal display behavior by males (such as singing) is most likely to occur when males cannot control resources or mates. Third, because ecological conditions (notably resources, habitat type, and water temperature) are broadly similar in all humpback whale wintering areas, we would expect little intraspecific variation in the mating system between different populations (see Lott 1984), and none has been reported.

The humpback's mating system is characterized by singing behavior and intrasexual aggression among males, an apparent lack of male territoriality, male escorting (tending) of females, and lack of male parental care. Furthermore, although it is not clear whether females mate with more than one male per winter, we do know that females mate with different males in different years: molecular data from the Gulf of Maine show that the calves of individual mothers were fathered by several males (Clapham and Palsbøll 1997). Thus it is clear that humpbacks are broadly polygamous/promiscuous. Mobley and Herman (1985) have suggested that the humpback mating system may be characterized as a lek. This hypothesis is superficially attractive because of the presence of many singing (i.e., possibly displaying) male humpback whales in an area during the mating season; furthermore, it fulfills a prediction made by Emlen and Oring (1977) that leks should arise where resources are economically indefensible (in this case, absent). However, observed movements of individually identified males around and between breeding habitats (Baker and Herman 1981; Mattila and Clapham 1989; Mattila et al. 1989, 1994) indicate that rigid spatial structuring of the type observed in lek territories does not occur in this species.

Although similar to what Emlen and Oring (1977) have termed "male dominance polygyny," the mating system of the humpback whale does not easily fit existing classifications. Clapham (1996) has proposed the term "floating lek" to reflect the close similarities to many features of a lek and the apparent lack of rigid spacing among males. An important issue to be resolved is whether dominance rankings are established among males during the winter. With the possible exceptions of agonistic behavior within all-male groups (Clapham et al. 1992; Brown and Corkeron 1995) and of occasional apparently aggressive interactions between singers and conspecifics (Tyack 1981; Darling 1983), dominance sorting among male humpback whales has yet to be unequivocally documented. That some degree of such sorting would occur seems reasonable in light of the apparent residence characteristics of males documented in some locations (Clapham et al. 1992). However, the occurrence in the same wintering area, and often in the same competitive groups, of individuals of different high-latitude origins (Katona and Beard 1990; Clapham et al. 1993) means that most intrasexual competition for mates will occur among unfamiliar conspecifics who will have had no opportunity to establish dominance rankings prior to arrival. Furthermore, the lack of previously established dominance rankings among many males would explain the frequency with which aggressive interactions occur.

It is possible that the clustering of humpback whales in specific tropical habitats during the winter facilitates mate selection by both sexes. However, the large numbers of whales recently discovered singing in deep water (C. W. Clark, personal communication) cannot be ignored; it may be vain to assume that the majority of mating activity takes place among the well-known concentrations of whales on the banks, simply because they are much more accessible to study than those in unprotected water.

Mate choice. That females exercise mate choice is suggested by observations of apparent repulsion of certain males (notably juveniles) by females in competitive groups (Clapham 1992). Whether females incite competition among males (see Cox and Le Boeuf 1977) to aid in mate selection or to displace an unwanted escort is unknown; however, observations of accompanied females flippering or lobtailing prior to the formation of a competitive group suggest a parallel with the solicitation behavior of some other mammals (e.g., African elephants, *Loxodonta africana:* Moss 1983; Poole 1989). Glockner-Ferrari and Ferrari (1985) observed behavior that they interpreted as female attempts to influence the choice of an escort in a competitive group, but the behavior could also reflect a more general attempt to escape male harassment. It is not known whether female humpback whales assess the condition of males; possible criteria include body size, endurance, fighting ability, dive capability, or cues within songs.

Whether males exercise mate choice is an interesting issue, and there is now good evidence from some species for such selectivity on the basis of female reproductive potential (e.g., bison, *Bison bison bison:* Berger 1989). Fisher (1958) predicted that both sexes should discriminate among potential partners if the choice was likely to affect reproductive success. Anecdotal observations from competitive groups suggesting that some females are more "attractive" than others to males have been given by Tyack and Whitehead (1983) and Clapham (1996). In light of the costs to males of intrasexual fighting over several weeks or months (during which no feeding takes place), differential allocation of competitive effort based upon female condition is not unlikely.

Duration of estrus. A key question with regard to the humpback mating system concerns the duration of estrus. Estrus in mammals varies from less than an hour in red deer (Clutton-Brock et al. 1982) to eight or nine days in some canids (Marler and Hamilton 1966). Its duration in mysticetes is entirely unknown. This is unfortunate, since this factor has many ramifications for the mating strategies of both sexes. If estrus is short, for example, a male humpback who has inseminated a female would presumably have to practice mate-guarding behavior for only a short period before being free to seek other mates. Identified female humpbacks have been observed with the same male escort on consecutive days, and in different competitive groups over periods of several weeks (D. K. Mattila, unpublished data). These sparse data suggest a number of possibilities: (1) estrus lasts for two or more days; (2) the duration of estrus is short, but males associate with females during a proestrous period; or (3) some females are polyestrous (see Chittleborough 1958).

Male mating strategies. There appear to be three primary mating strategies among male humpback whales: (1) singing (displaying), presumably at least partly to attract females; (2) escorting/tending an estrous or proestrous female and awaiting the opportunity to mate with her; and (3) direct competition for a female with one or more other males. To these we may add a fourth: the possibility that males form coalitions to displace a principal escort (Clapham et al. 1992; Brown and Corkeron 1995).

It is not likely that all of these strategies are of equal value, nor that all are open to every male. Size and stamina are principal determinants of success in intrasexual competition in many other taxa (Le Boeuf 1974; Clutton-Brock et al. 1982), and humpback whales are unlikely to be different in this regard. Not all males will be large enough or strong enough to outcompete others in the physical combat of a competitive group (Chu and Nieukirk 1988). Males who cannot successfully compete in direct conflict may display (sing), or may adopt a roving male strategy by which they colonize low-density habitats containing fewer mates, but also fewer competitors. Davies and Halliday (1978) note that, in some taxa, this strategy is often as successful for a smaller or low-ranking male as searching for a mate in high-density areas.

Sperm competition. Brownell and Ralls (1986) reviewed the potential for sperm competition in baleen whales. They pointed out that there is considerable variation in relative testis weight between species: male right whales, which weigh up to 50 tons, possess testes whose average weight is 972 kg (Omura et al. 1969), the largest in the animal kingdom, while those of the blue whale, a species with a maximum weight in males exceeding 100 tons, have a mean weight of only 70 kg (Ohno and Fujino 1952). Male humpback whales (maximum weight approximately 40 tons) have relatively small testes for their body size (mean = 38 kg: Nishiwaki, 1959); consequently, we would predict that, while sperm competition may play a role in this species, it is not a primary male mating strategy (but should be in right whales). This consideration of relative testis size would also lead us to expect the aggressive interactions between male humpbacks that have been observed in competitive groups.

Summer. Although we know too little to reliably assess whether the behavior of humpback whales during their summer feeding season is to any extent determined by reproductive strategies, it is worth briefly considering this issue. Late autumn behavior resembling that described from competitive groups is occasionally observed (Weinrich 1995), and song is heard infrequently in spring and autumn (Mattila et al. 1987). These two phenomena suggest a residual effect close to the beginning or end of the seasonal migration. Indeed, while spermatogenesis is highly seasonal in humpbacks, Mitchell (1973) found high levels of sperm in two males killed off eastern Canada in mid-October and early November. The only hypothesis relating to singing in summer is that of Clapham (1996), who suggests that it may represent low-cost advertising by males with the possibility of a payoff among known individual females in the subsequent breeding season.

Other Mysticetes: A Comparison

A comparison of the socioecology of the humpback whale with that of other mysticetes exceeds the scope of this chapter, but it is useful to briefly consider some of the differences that exist within this taxon, and some examples of the kinds of questions that these differences raise. The baleen whales can be categorized according to various criteria, including distribution, migratory behavior, prey, feeding method, mating system, and life history parameters. Even a brief review of these characteristics highlights the considerable differences that exist between species. For example, mating systems vary considerably in both type and timing. Sperm competition appears very likely in right whales (and possibly gray whales and bowheads: Brownell and Ralls 1986), while the very small size of a blue whale's testes implies the existence of some other male mating strategy. Some species (humpbacks, gray whales, right whales) congregate on breeding grounds, and those that do so tend to be fatter and slower, with more aerial behavior (Whitehead 1981). Some species, like the humpback, are strictly seasonal breeders, while in the balaenids (right whales and bowheads) breeding is seasonal, but sexual activity is observed throughout the year (Kraus et al. 1986; Koski et al. 1993). In the coastal form of Bryde's whale, breeding takes place year-round (Best 1977); this observation argues strongly that, by remaining in low latitudes and exploiting a year-round food source, this species has freed itself from the constraints of migration. It is interesting that Arabian Sea humpbacks, which appear to remain in tropical waters throughout the year, are still seasonal breeders (Mikhalev 1997), although this strategy may be connected to low abundance of prey in winter.

Although all baleen whales appear to possess a social structure that is based upon small, unstable groups, the composition and dynamics of these units vary between species, and in many cases we know little or nothing about qualitative differences between them. Resource characteristics also differ widely among mysticetes. Prey varies widely, and includes copepods (right whales, bowheads, sei whales), small schooling fish (most rorquals), euphausiids (all rorquals, some balaenids), and benthic invertebrates (gray whales). These differences undoubtedly have major effects on social organization and behavior. Similarly, interspecific differences in life history parameters (see review in Lockyer 1984) prompt many questions. Why, for example, do humpbacks and right whales apparently invest more time in lactation than most other mysticetes? Why is age at sexual maturity (and perhaps life expectancy) much greater in bowheads than in other whales (Koski et al. 1993)?

One of the most obvious differences between some of the baleen whales is size. Body size is an important factor in consideration of life history (Western 1979; Schmidt-Nielsen 1984), yet it is frequently ignored in interspecific comparisons of cetaceans. Does the huge blue whale's virtually stenophagous diet of euphausiids reflect inherent biomechanical limitations that make foraging on fast piscine prey difficult? Do size and maneuverability affect the size of a prey patch that can be exploited, and therefore such factors as residency within an area, as has been suggested by Tershy (1992)? In addition, what do variations in sexual size dimorphism among species (Ralls 1976) tell us?

A comparative approach to understanding mysticete socioecology, in which interspecific variations in migration, behavior, and life history are linked to ecological differences between species, would be fruitful if sufficient data can be gathered.

Research Needs

Over the last two decades, a great deal has been learned about many aspects of the biology and behavior of humpback whales and other cetaceans from long-term studies of identified individuals (Hammond et al. 1990). While the continuation of such studies will undoubtedly yield further insights, it is important for field-workers to incorporate new technologies into their work, and in some cases to alter their focus.

Molecular genetics has demonstrated an ability to answer questions that in many cases simply cannot be addressed by other means. In particular, both mitochondrial DNA and microsatellite markers can provide an understanding of the extent to which social behavior is determined by kinship (Amos et al. 1993, 1995). Genetics can also provide information on reproductive success, although this is in practice more difficult for species such as humpback whales in which the number of candidates for paternity is potentially very large (and in which sample sizes are often small relative to total population size). However, assessing the reproductive success of particular classes of whales (e.g., principal escorts) may be feasible given a sufficiently robust sample size. The molecular approach is of particular value when applied to a well-known population, in which extensive sighting history data can be used to interpret genetic results.

Most behavioral work on humpback whales has been focused at the level of the group. We understand a good deal concerning the composition and dynamics of groups on both feeding and breeding grounds. While opportunistic observations and long-term investigations must continue, what is greatly lacking is information derived from focal studies of individuals. In such studies, animals of different sexes, ages, reproductive conditions, and behavioral roles are followed for extended periods of time to determine daily patterns of movement, behavior, and associations. There are many fundamental questions that can be addressed in this manner. In their frequency and duration of associations with animals of each sex, how do males differ from females? Juveniles from adults? Principal escorts from nuclear animals? What are the behavioral patterns of singers over the course of one or more days? Of mothers with calves? How much time do whales of different classes spend feeding and socializing? There is no shortage of questions,

and these cannot be answered without focal follows of many individuals.

Radiotelemetry (using either satellite or VHF transmitters) represents another important tool that should be incorporated into behavioral studies if logistical problems can be overcome. Long-term fieldwork and even focal follows will not tell us about the wider-ranging movements of individuals outside the relative safety of our small study areas. Do male humpbacks range farther in summer or winter than females? How do animals of different classes differ in their movements and occupancy patterns? Are males more likely than females to visit many potential breeding sites in the course of a winter, and do their migratory patterns differ significantly from those of female whales? Again, there are many questions, and they all require a means of tracking individuals over long distances and through often inhospitable conditions.

From a design and analytical viewpoint, we need to think more in terms of individual behavioral characteristics and strategies than has historically been the case. There is good evidence—and every predictive reason to believe—that behavioral decisions concerning whether to migrate, where to go, how much to eat, whom to associate with, and how to find a mate differ not just between the sexes, but with other factors such as age class and reproductive condition.

Conservation Issues

There is no doubt that commercial whaling had a devastating effect on populations of humpback whales worldwide: it is likely that as much as 95% of the pre-exploitation population was eliminated (Johnson and Wolman 1985). The recent release of formerly secret data on unreported Soviet catches shows that depletion of humpback and other large whales was considerably worse than once thought (Yablokov 1994; Zemsky et al. 1995; Yablokov et al. 1998).

The difficulty of obtaining reliable data on abundance and growth trends for any mysticete mandates a degree of caution when attempting to assess the status of previously exploited populations. Nonetheless, recent data from both sighting surveys and recapture of marked individuals strongly suggest that recovery is occurring in most humpback whale populations (see Best 1993 and Clapham et al. 1999 for a review). The growth rate of the Gulf of Maine population was recently estimated at 6.5% (SE = 0.012: Barlow and Clapham 1997).

Commercial whaling of humpbacks officially ceased in 1966, and although a few aboriginal hunts still occasionally take humpbacks, direct exploitation no longer represents a threat to this species. However, entanglement in fishing gear is today a significant source of mortality in some areas (Whitehead 1987; Perrin et al. 1994a). Because of the trophic level at which they feed, humpbacks should not be significantly affected by pollutants; existing data support this belief (O'Shea and Brownell 1994), although the possibility of effects resulting from transgenerational accumulation has yet to be addressed. The significance of other factors, including red tide events (Geraci et al. 1989), coastal development (Glockner-Ferrari and Ferrari 1990), and oil and gas exploration (Dawbin and Gill 1991) is unclear, but current data suggest no immediate cause for concern in these areas (Clapham et al. 1999).

These various potential threats collectively urge prudence in our view of the future of the species. However, while it is too early to suggest that the future of the humpback whale is assured, current data all point to the belief that most populations are recovering well from the long period of exploitation to which they were subject.

Part 3

Comparative Studies, Theory, and Conservation

8

GROUP LIVING IN WHALES AND DOLPHINS

RICHARD C. CONNOR

DURING THE summer of 1977, thirty false killer whales *(Pseudorca crassidens)*, floated in the shallows of the Dry Tortugas for three days (Porter 1979). A large male in the center of the tight group lay on his side, bleeding from his right ear. When a shark swam by, the whales flailed their tails. Individuals became agitated when people separated the whales to return them to deeper water, but became calm once back in physical contact with other whales. Despite the risk of stranding and growing blisters from exposure to the sun, the group stayed together and did not leave until the male died on the third day. (See plate 16.)

The behavior of these almost-stranded false killer whales reveals a strong dependence on group living. One population of killer whales *(Orcinus orca)* lives in the most stable groups known among mammals, so much so that individuals never disperse from their natal group (Baird, chap. 5, this volume).

In some other species of cetaceans, one or both sexes disperse from their natal group. After dispersal, the sexes may adopt different patterns of grouping. Female sperm whales *(Physeter macrocephalus)* live in stable matrilineal groups, but the oldest males are usually solitary (Whitehead and Weilgart, chap. 6, this volume).

In the fission-fusion societies of bottlenose dolphins *(Tursiops* spp.), individuals of the same sex vary in the degree to which they are found in groups (Connor et al., chap. 4, this volume). Individuals of species that do not appear to live in stable groups may join with others during particular activities, including migration (e.g., gray whales, *Eschrichtius robustus:* Swartz 1986), feeding (e.g., humpback whales, *Megaptera novaeangliae:* Clapham, chap. 7, this volume), and, obviously, breeding (e.g., humpback whales: Clapham, chap. 7, this volume).

In this chapter I review group living in whales and dolphins, focusing initially on the evolutionary and ecological factors that promote group formation. I begin with the question of what constitutes a "group." Individuals can aggregate without obtaining benefits from each other. Groups that are based on the exchange of benefits are referred to here as "mutualistic groups." This distinction is especially relevant for a volume such as this one in which the focus is on social relationships; only mutualistic group formation promotes the formation of social bonds.

Our current understanding of individual and sex-specific social relationships in cetaceans derives largely from detailed studies on a few species (see chaps. 4–7, this volume). Here I consider how their aquatic lifestyle affects the costs and benefits of grouping and social bond formation in whales and dolphins. This approach can help us understand the range of social bonds in cetaceans, from the relatively solitary minke whale to the extraordinary *resident* killer whales of British Columbia, in which adults of both sexes remain with their mothers.

The Costs and Benefits of Group Living

In his classic analysis of the evolution of social behavior, Alexander (1974) noted that "there is no automatic and universal benefit from group living but there are automatic and universal detriments." The universal detriments are increased parasite transmission and competition for resources. Other costs, such as an increased risk of detection by predators for individuals in large groups, may be important in particular cases. Group formation will be favored only when the benefits of being with others outweigh these costs.

Perhaps the major factor promoting group formation is predation (Alexander 1974; van Schaik 1983, 1989). Predators are a major source of mortality in most organisms. There are several ways, discussed in more detail below, in which individuals can reduce their risk of predation by joining groups (Alexander 1974; Inman and Krebs 1987). Group living can also reduce the risk to individuals from non-socially transmitted parasites (but not contagious parasites) in a manner similar to the way they reduce the risk of predation (Mooring and Hart 1992; Wcislo 1984; Cote and Poulin 1995).

Interference competition from conspecifics (Wrangham 1980a) or from members of other species (Buss 1981) can also promote group formation. Wrangham (1980a) suggested that cooperative defense of food patches can select for group living. Where food occurs in discrete, defensible patches that can support a limited number of individuals, groups of relatives may defend food patches against conspecifics (Wrangham 1980a, 1982). While females will join forces to defend food against other females (and sometimes males: Gompper 1996), males rarely do so. Females are the most important resource for males, who may cooperate to defend them against other males (Wrangham 1980a). Protection against aggression by males, in the form of harassment or infanticide, may also promote grouping among females, or of a female with a male "hired gun" (Wrangham 1979; Wrangham 1980a; Wrangham and Rubenstein 1986; van Schaik and Dunbar 1990; Campagna et al. 1992).

Finally, individuals may form groups to minimize their risks from challenges in the habitat. For example, individuals may huddle together to remain warm, or join others to reduce the cost of locomotion during swimming or flight (Lissaman and Schollenberger 1970; Hainsworth 1987; Partridge et al. 1983; Abrahams and Colgan 1985; but see Pitcher and Parrish 1993).

In the cases listed thus far, individuals join groups because of benefits they obtain from the presence or behavior of other individuals (mutualistic group formation). But other kinds of groups are possible. Groups may form independently of any benefits individuals obtain from other group members. Examples of such non-mutualistic group formation include aggregations where food is concentrated (Alexander 1974) or in a predator refuge (e.g., a hollow log) that can support more than one individual. Likewise, individuals may aggregate in areas that are energetically favorable (e.g., sunny spots on a forest floor) or provide shelter from extremes of climate (e.g., protection from wind and rain). For example, the social wasp *Polistes annularis* nests on cliff faces that serve as refuges from mammalian

predators. Within cliff faces, nests are aggregated in limited overhang areas that provide protection from the rain and sun (Strassman 1991).

For the rest of this chapter I will use the term "group" to denote mutualistic groups and refer to non-mutualistic groups as "aggregations." In a similar way, Norris and Schilt (1988) referred to all nonpolarized schools as "aggregations" and approximated the distinction made here between non-mutualistic and mutualistic groups with the terms "non-social" and "social" aggregations, "in which members are attracted to each other."

Recognizing Groups

Obviously, for groups to be mutualistic, individuals must be close enough to derive benefits from one another. For many kinds of benefits, such as predator protection, cooperative resource defense, or prey capture, this means that individuals will be closely spaced and groups will be easily identified as such by human observers. Other kinds of benefits do not require such close spacing and may be more difficult for observers to detect. For example, group formation in bowhead whales *(Balaena mysticetus)* may aid individuals navigating through the Bering Sea pack ice (Ellison et al. 1987; George et al. 1989; Clark 1989). By listening to the whales vocalize with an array of hydrophones off Point Barrow, Alaska, Clark (1991) estimated that individuals would pass by his listening station in groups of ten to fifteen, spread out over 4–8 square miles. Clark suggests that the bowheads can assess ice conditions from the echoes of other whales' vocalizations.

Non-mutualistic aggregations do not depend on individuals being close enough to benefit from each other and so occur on a wide range of scales. Many aggregations occur on a sufficiently large scale that they should not be easily confused with mutualistic groups. On the other hand, as the bowhead example suggests, some baleen whales could alter their movements in ways that increase the mutual exchange of benefits across large distances. Smaller aggregations are more likely to resemble mutualistic groups in scale and may even lead to the development of mutualistic groups if, for example, it becomes beneficial for two or more individuals to cooperatively defend resources (Wrangham 1980a, 1982). In cetaceans, such cooperative relationships are difficult to detect. For example, on breeding grounds, male humpback whales form competitive groups around single females and vie for the "principal escort" position next to her (Tyack and Whitehead 1983; Clapham et al. 1992; Clapham, chap. 7, this volume). The early descriptions of this behavior suggested that males competed

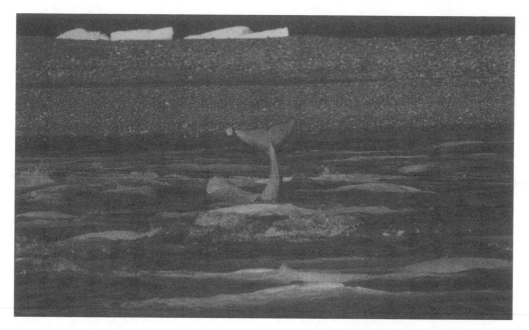

Figure 8.1. Beluga whales gathered in a shallow estuary. (Photograph by Flip Nicklin/Minden Pictures.)

individually for access to the female. However, more recent observations suggest that some competitive groups may contain males cooperating in pairs (Clapham et al. 1992; Brown and Corkeron 1995; Clapham, chap. 7; Connor et al., chap. 10, this volume).

A further point is that aggregations may consist of smaller mutualistic groups. Whitehead and Weilgart's study of sperm whales off the Galápagos Islands (chap. 6, this volume) illustrates nicely a hierarchical range of smaller mutualistic groups embedded in larger non-mutualistic aggregations. Whitehead and Weilgart report four levels of spatial organization in sperm whales: stable units of about ten related females with young, associations between two units lasting for periods of days, temporary aggregations of about forty individuals over a 15 km² area, and "geographic" concentrations of a thousand or so whales across a 1,000 km range. The smallest level of spatial organization is interpreted by Whitehead and Weilgart as a mutualistic group based on care and protection of dependent offspring. The second level, temporary associations between two units that move in a coordinated fashion, may be mutualistic and related somehow to improved foraging efficiency. The largest two levels are likely to be non-mutualistic aggregations, based on different spatial scales of food concentration.

Concentrations of food probably account for most non-mutualistic aggregations that might be confused with mutualistic group formation in cetaceans. Aggregations in predator refuges may also occur in cetaceans, but are un-

likely to be confused with mutualistic group formation. For example, gray whales gather to give birth in large lagoons that probably serve as refuges from predators such as killer whales (Swartz 1986). During the summer, beluga whales (*Delphinapterus leucas*) gather in shallow estuaries that are thought to serve as thermal refuges, minimizing the energetic costs for newborn calves (Sergeant and Brodie 1969; fig. 8.1).

Most or all cases of mutualistic group formation in cetaceans are likely to be associated with resource acquisition and threats from conspecifics or other species. The remainder of the discussion will be limited to those factors.

The Benefits of Group Living: Mutualism, Cooperation, and Altruism

Many cetacean behaviors, such as supporting a conspecific who is unable to swim, or joint herding of schooling prey, are claimed to be cooperative or altruistic (see Connor and Norris 1982; Bradbury 1986; Johnson and Norris 1986; Würsig 1986; Norris and Schilt 1988). Here I briefly review altruism and cooperation under the umbrella concept of mutualism before considering how they can contribute to our understanding of group living in cetaceans.

A mutualism is an association in which the inclusive fitness of each party is increased by the action of others (Janzen 1985; Pierce et al. 1987; Pierce 1989). Mutualistic interactions between conspecifics, including mutualistic group formation, are almost always based on by-product benefits, investment (altruism), or a combination of the two (Connor 1995b). Mutualism based on the exchange of by-product benefits was originally defined by West-Eberhard (1975) as "mutualism maintained by ordinary selfish behavior incidentally benefiting others." Connor (1995b) included as by-product mutualism any cooperative behavior that does not include altruism. In this view, there is a continuum of by-product mutualism ranging from acts performed irrespective of the behavior of others to highly coordinated actions. Thus, by-product mutualisms include interactions that are usually called cooperative in the cetacean literature. The concept of cooperation as used in the cetacean literature fits the general idea of an individual acting in concert with others to obtain a greater benefit than it would acting on its own (this has also been called "synergistic" or "simultaneous" cooperation: Maynard Smith 1983; Rothstein and Pierotti 1988). Here I will use the term *cooperation* in the way it is commonly used in the cetacean literature, to denote individuals doing better by acting in concert than alone, but not including exchanges of altruism (*sensu* Axelrod and Hamilton 1981).

Our use of the term *altruism*, for behaviors that benefit the recipient at a cost to the actor, also accords with its previous usage in the cetacean literature (e.g., Connor and Norris 1982). Group life offers many chances for individuals to behave altruistically. The return on an altruistic investment may take several forms. Kin-directed altruism serves to increase the inclusive fitness of donors, who therefore receive a genotypic, but not a phenotypic, return on their investment (Hamilton 1964). Altruistic acts may also be reciprocated (Trivers 1971), a behavior that likely originates among relatives (Alexander 1979; Axelrod and Hamilton 1981), but which may be extended to nonrelatives. Individuals may behave altruistically toward individuals from whom they receive by-product benefits, simply to increase the probability of receiving the by-product benefits. I have called this phenomenon "pseudo-reciprocity" (Connor 1986, 1995a). Finally, in stable social groups, individuals may dispense altruism or otherwise behave cooperatively in order to avoid punishment by more dominant group members (Clutton-Brock and Parker 1995; Frank 1996).

Partner Preferences, Mutualism, and Social Bonds

Under what conditions do group-living individuals develop social bonds? Differentiated social bonds are based on the differential allocation of benefits among individuals. Thus, for social bonds to develop, individuals must first discriminate among the potential recipients of benefits.

Altruists will discriminate among potential recipients that differ in the amount of return benefit they provide. Nepotism will be based on the recipient's reproductive value and degree of relatedness to the donor, and these factors will often vary enough to favor discrimination among potential recipients (Hamilton 1964) unless such discrimination is too costly to the potential altruist or recipient (Keller 1997). Partners in reciprocal altruism will be selected on the basis of the probability they will reciprocate, the amount they reciprocate, their projected value as a partner in future interactions, or even their gullibility (Trivers 1971). An individual engaging in pseudo-reciprocity will preferentially direct altruism toward those with the highest resulting probability of returning the most valuable by-product benefits to the altruist (Connor 1986).

By-product mutualisms can also favor partner preferences (Wrangham 1982), which are an expected outcome of individual variation in the ability to dispense or utilize by-product benefits (Connor 1996). Wrangham (1982) distinguished between two kinds of by-product mutualisms, "interference mutualism" (IM) and "non-interference mutualism" (NIM). In the former, two or more individuals benefit by acting together, but at the expense of others, as in cooperative defense of food (Wrangham 1982). An example of non-interference mutualism would be individuals joining groups to reduce the risk of predation, assuming that any individual can join a group. Because interference mutualism necessarily inflicts a cost on some individuals, it will typically favor the formation of stable kin groups (Wrangham 1982). On the other hand, factors such as competitive ability and reproductive value may strongly affect partner choice, especially in situations in which kin are not always available (Wrangham 1982).

In groups with more than two individuals, opportunities may arise to practice interference mutualism within the group. In groups in which individuals engage in interference mutualism on a recurring basis, differentiated social relationships should develop. In contrast, relationships should remain undifferentiated in groups in which opportunities for interference mutualism are absent or infrequent. For example, social relationships differ sharply between two closely related species of squirrel monkeys, *Saimiri oerstedii* and *S. sciureus* (Mitchell et al. 1991). Intergroup competition for resources appears weak in both species, and group formation in both is likely a response to predation risk (Mitchell et al. 1991). However, food patch size varies between the two species such that intragroup competition is rare in *S. oerstedii*, which has no dominance hierarchy or long-term affiliative bonds, but occurs often in *S. sciureus*, which, accordingly, has female dominance hierarchies and stable kin-based alliances.

The Influence of Dominance on Cooperation and Altruism

The circumstances in which individuals will engage in cooperative or altruistic behavior may be influenced strongly by dominance relationships (e.g., Noë et al. 1991; Clutton-Brock and Parker 1995; Frank 1996; de Waal 1996). Dominant individuals may play a primary role in "policing" groups, punishing individuals who do not behave cooperatively (Clutton-Brock and Parker 1995; Frank 1995, 1996). Even apparently altruistic behaviors may be coerced rather than offered in expectation of receiving return benefits (either altruistic or by-product). For example, rhesus monkeys *(Macaca mulatta)* may inform others of a food source they have located to avoid aggression that is directed at non-callers who are "discovered" (Hauser 1992). In some cases, lower-ranking rhesus monkeys who discover non-calling "cheaters" recruit others to participate in aggression against the non-caller (Hauser 1992); thus policing may be maintained by coalitions as well as by dominant individuals.

Dominance relationships can exert a strong influence on partner choice in by-product mutualisms. Once two individuals have collaborated to obtain a resource (e.g., food or a breeding site), a dominant individual may be able to exclude a subordinate from the resource. For example, unrelated male long-tailed manakins *(Chiroxiphia linearis)* cooperatively defend display territories, but only the alpha male mates with females attracted to the territory (McDonald and Potts 1994). The reward for the beta male lies in the future, in the chance that he will outlive the alpha and inherit alpha status. In other cases, the subordinate may have a reasonable chance of succeeding on its own and would be better off leaving if the dominant monopolized all of the resources. However, if the dominant individual expects to enjoy future by-product benefits from the subordinate, then the dominant should yield a portion of the resource to the subordinate as an incentive for it to remain (Vehrencamp 1983; Emlen 1984; Reeve and Ratnieks 1993; Jamieson 1997). Reproduction by subordinates, common in social insects, has also been reported in several cooperatively breeding mammals and birds (e.g., Keane et al. 1994; Craig and Jamieson 1990). In general, the amount of reproduction achieved by a subordinate is predicted to depend on such factors as the degree of control

the dominant has over reproduction in the group, the relative fighting abilities of the dominant and subordinate, their relatedness, how well the subordinate would fare by leaving, how well the dominant would fare if the subordinate left, and whether the dominant has to compete with others for the services of subordinates (Vehrencamp 1983; Reeve and Ratnieks 1993; Keller and Reeve 1994; Emlen 1997; Reeve 1998; Reeve et al. 1998).

Interference mutualism within groups may also lead to competition among individuals for alliance partners. An individual fortunate enough to be the preferred partner of two or more group members may play one off against the other and so enjoy a larger share of resources in mutualistic interactions than would be expected on the basis of relatedness or relative fighting ability (Wrangham 1982; Noë et al. 1991; Noë and Hammerstein 1994). Noë (1990) described such a case in which two lower-ranking male baboons *(Papio c. cynocephalus)* could defeat higher-ranking males only by forming a coalition with one middle-ranking baboon. The middle-ranking baboon evidently played the two lower-ranking males off against each other and enjoyed the majority of the benefits derived from their coalitions. Wrangham (1982) pointed out that in primate groups individuals may attempt to win alliance partners by grooming them, a case of pseudo-reciprocity in which altruism (grooming) is performed to increase the probability that the recipient will perform a selfish act (form an alliance) that benefits the altruist.

Interference mutualism between social groups can be an important factor elevating the value of subordinates to dominant individuals. Wrangham (1982) suggests that a victorious alliance may refrain from banishing or killing a rival defeated in intragroup conflicts if they benefit from the rival's efforts during intergroup encounters (see also Vehrencamp 1983). Similar influences may operate in social relationships among male bottlenose dolphins, which form alliances of alliances to compete for access to females (Connor et al. 1992a,b, 1999, chap. 10, this volume).

In spite of their near ubiquity among social animals, dominance relationships are surprisingly absent or muted in some cases. The absence of dominance relationships among lions *(Panthera leo)* is striking (Packer et al. 1988): female lions respect each other's "ownership" of feeding sites on carcasses, and males do not challenge the consortships of other males (Pusey and Packer 1998). To explain the lions' peculiar lack of dominance interactions, Pusey and Packer (1988) invoke the value of allies in intergroup conflicts and the dangers of escalated conflict for such evenly matched and well-armed contestants. From this "lethal weapons" perspective, killer whales are obviously interesting subjects for study. Escalated intragroup conflict might also be dangerous, or at least dangerously unpredictable, where contestants are members of different within-group alliances. Although dominance relationships have been documented in captive bottlenose dolphins (Samuels and Tyack, chap. 1, this volume), resource ownership may play an important role as well (Connor et al., chap. 4, this volume).

Group Formation in Whales and Dolphins

It is increasingly evident that many odontocetes have strong social bonds and long-term social relationships, which include such phenomena as competition for social partners (e.g., Connor et al., chap. 4, this volume). General aspects of sex-specific social bonds in cetaceans, including relevant models of social behavior and ecology, are discussed in chapters 9 and 10 of this volume. Other than those few species that have been studied in detail (chaps. 4–7), the sizes of stable social units are unknown for most cetaceans. Typical group sizes for a number of cetacean species, which may or may not correspond to important levels of social organization, are listed in table 9.1.

The remainder of this chapter will focus on the costs and benefits of group living in cetaceans. First I review how cetaceans living in groups might benefit from improved access to food or a reduced risk of predation, parasitism, and for females, male harassment. After reviewing the benefits of group living in whales and dolphins, I will consider how their relatively low cost of locomotion might reduce the cost of grouping and philopatry in cetaceans. Quantitative studies on the costs and benefits of group formation in cetaceans are largely lacking (but see Baird and Dill 1996; Baird, chap. 5, this volume), so it is hoped that this review of the largely anecdotal literature will encourage more systematic efforts. I include studies of group living in other organisms, both to illustrate particular phenomena and in the hope that a comparative approach will provide clues about the factors that promote group living in the Cetacea.

The Benefits of Group Living in Cetaceans: Predators, Parasites, and Prey

Cetacean Predators

Norris and Dohl (1980b) suggested that predation risk is the principal factor driving group formation in cetaceans. Two kinds of predators are most often implicated in pre-dation on cetaceans: sharks and killer whales. The killer whale, *Orcinus orca* (Baird, chap. 5, this volume), may be the most widely distributed cetacean and probably presents a threat to more cetacean species than any other predator (reviewed by Jefferson et al. 1991). Ranging from the tropics to polar waters, killer whales have been observed to attack the largest baleen whale (blue whale, *Balaenoptera musculus:* Tarpy 1979) and the largest toothed whale (sperm whale: Arnbom et al. 1987) as well as smaller dolphin species (e.g., Constantine et al. 1998). The only cetaceans that may be immune from killer whale predation are individuals that live in tropical rivers (Jefferson et al. 1991) or other protected coastal habitats that killer whales do not penetrate. Evidence for killer whale predation attempts comes not only from occasional observations of attacks, but also from scars on living cetaceans and from killer whale stomach contents. Scars on the flukes of 20–33% of young humpback whale calves suggest that predation might be focused on young whales, and that predation attempts may be much more common than indicated by direct observation (Clapham, chap. 7, this volume).

Predation by mammal-eating killer whales may also play an important role in determining group size in fish-eating killer whales. In the northeastern Pacific, two sympatric non-interbreeding populations of killer whales exhibit different dietary specializations: one on marine mammals and the other on fish (Baird, chap. 5, this volume). Baird suggests that the larger group size of the piscivorous killer whales might result in part from the threat of predation (on their calves) by the mammal-eating whales.

The killer whale is the only cetacean in which some populations specialize on marine mammal prey (Baird, chap. 5, this volume), but the false killer whale and, to a lesser extent, the short-finned pilot whale *(Globicephala macrorhynchus)* and pygmy killer whale *(Feresa attenuata)* have been observed in possible predatory attacks on spinner *(Stenella longirostris),* spotted *(Stenella attenuata),* and common dolphins *(Delphinus delphis)* during fishing operations by tuna boats (Perryman and Foster 1980).

Extensive evidence from wounds, scarring, and shark stomach contents indicates that sharks are important predators on many smaller tropical and temperate odontocetes (see Connor et al., chap. 4, this volume). Tiger sharks *(Galeocerdo cuvieri),* dusky sharks *(Carcharhinus obscurus),* great white sharks *(Carcharodon carcharias),* and bull sharks *(Carcharhinus leucas)* are the species most often implicated in attacks on nearshore odontocetes such as the bottlenose dolphin (McBride and Hebb 1948; Wood et al. 1970; Corkeron et al. 1987; Cockcroft et al. 1989; Long and

Jones 1996; Connor et al., chap. 4, this volume). White sharks have been implicated in predation on a wide range of odontocetes, from the small harbor porpoise to larger beaked whales (Long and Jones 1996). Sharks have been reported to accompany bottlenose dolphins (Leatherwood 1977) and sperm whales at sea (oceanic white-tipped sharks, *Carcharinus longimanus:* Best et al. 1979).

Evidence of shark predation on baleen whales is uncommon, although there is one second-hand report of a tiger shark attacking a young humpback whale calf, and a humpback calf stranded alive that had been attacked by sharks (Paterson et al. 1993; Clapham, chap. 7, this volume). White sharks have been observed approaching right whales in southern Australia, and right whale neonates often have what appear to be scars from such encounters (S. Burnell, personal communication) I have found no published reports of direct observations of sharks attacking cetaceans under natural conditions, but researchers have obtained eyewitness accounts shortly after fatal attacks in a few cases (e.g., Wood et al. 1970; Mann and Barnett 1999). Evasive reactions to the presence of a predator may indicate the degree to which dolphins perceive a threat (e.g., Connor and Heithaus 1996).

Until humans began to hunt cetaceans from boats, cetaceans were relatively free of predation from above the water surface. An exception to this is in the Arctic, where polar bears *(Ursus maritimus)* occasionally hunt beluga whales *(Delphinapterus leucas)* (Heyland and Hay 1976; Smith 1985; Lowry et al. 1987).

Before considering how group living can reduce the risk of predation on cetaceans, it is worth mentioning a few antipredator strategies that do not depend on group formation. Disturbed dwarf sperm whales *(Kogia simus)* may disappear, like squid behind ink, in a large cloud of reddish excreta (Scott and Cordaro 1987). Solitary antipredator strategies may even be preferable in group-living species when threatened by cooperatively hunting killer whales. The best way to avoid killer whales may be to hide from them or to travel to where they cannot follow (Jefferson et al. 1991). When killer whales are encountered near shore, dolphins and baleen whales may attempt to hide from them in the shallows of river mouths or the surf zone, or in kelp beds (Saayman and Tayler 1979; Jefferson et al. 1991). In offshore waters, deep-diving species such as sperm and beaked whales may escape killer whales by spending most of the dive cycle at depths greater than those to which killer whales dive. The yearly migrations to tropical breeding grounds by many baleen whales may be a strategy to reduce the risk of killer whale predation on vulnerable newborns

(Corkeron and Connor 1999). Although killer whales are found in tropical waters, they are much more common in high latitudes, where year-round populations of pinnipeds are likely the staple prey for mammal-eating populations.

Predation and Group Living

Group living may reduce an individual's risk of predation in several ways.

The dilution and encounter effects. Individuals in a group that has been detected by a predator enjoy a reduced attack rate compared with a solitary individual that has also been sighted, if the predator can consume only one or few individuals (Triesman 1975; Turner and Pitcher 1986). If a predator could consume the entire group, then the attack rate per individual would not decline with group size (Inman and Krebs 1987). Foster and Treherne (1981) observed this "dilution effect" during attacks on a marine insect, the ocean skater *(Halobates robustus),* by small juvenile fish *(Sardinops sagax).* The attack rate per individual decreased with group size such that a skater in a group of fifteen to seventeen individuals was sixteen times less likely to be attacked than a solitary individual.

The dilution effect depends on the "encounter effect"— the assumption that a predator's ability to detect prey does not increase in proportion to prey group size—a reasonable expectation in marine environments (Pitcher and Parrish 1993). If predators could consume only a single individual, but were ten times more likely to detect a group of ten individuals than one individual, then the lower per individual rate of attack by a single predator would be offset by the greater number of predators attacking (Triesman 1975; Turner and Pitcher 1986). Notice that for the dilution and encounter effects to be logically independent, we must assume that if the group is detected, all members of the group are detected (our hypothetical predator could not consume a whole group unless all members were detected).

Inman and Krebs (1987) argued that where grouping is favored for reasons other than predator deterrence, the encounter effect can be advantageous by itself. This may explain why Norris and Schilt (1988) suggest that, in addition to the dilution effect, schooling cetaceans might benefit from the encounter effect. To clarify, the encounter effect can never offer an advantage to group living on its own (i.e., without a dilution effect operating simultaneously). Without a dilution effect, the best the encounter effect can offer would be to render individuals in groups no more detectable to predators than solitary individuals, but they would be just as likely to be eaten once discovered.

The dilution effect may operate in the temporal formation of groups in response to predators or, in some cases, threats from conspecifics. Campagna et al. (1992) found that pregnant female southern sea lions *(Otaria byronia)* arriving early on breeding grounds suffer greater pup loss due to male harassment (causing infant death directly or indirectly via mother-infant separation resulting in starvation). The risk to females is diluted if they arrive when many other females are already present. A similar risk may explain why pregnant female humpback whales are the last to arrive on tropical breeding grounds (Clapham, chap. 7, this volume; Dawbin 1966). Male humpback whales compete vigorously for females, activity that may be hazardous to newborn infants. Female whales with newborns may be harassed by males if they have a postpartum estrus (some females calve in consecutive years) but, as with sea lions, young males may not discriminate. Female gray whales with young calves are sometimes pursued and harassed by groups of males (Swartz 1986). Unlike humpbacks, female gray whales are the first to arrive on the breeding grounds, but they usually remain in the interior of breeding lagoons, spatially separated from the courtship activity at lagoon entrances. In both species, females with newborns are the last to leave, probably to give their calves more time to grow and thus a better chance to survive the gauntlet of killer whales at higher latitudes.

An individual joining a group benefits from the dilution effect, but also contributes to the dilution effect enjoyed by others (a by-product benefit). Thus, we might expect partner preferences in group formation: kin may be favored in some circumstances (e.g., Young et al. 1994), but individuals may prefer to have more vulnerable nonkin join their group. Brooding convict cichlids *(Cichlasoma nigrofasciatum)* preferentially adopt fry smaller than their own, which are more vulnerable to predation (Wisenden and Keenleyside 1992). Some cichlids in Lake Malawi deposit their brood in the nests of bagrid catfish *(Bagrus meridionalis)*, reducing the predation risk for the young catfish (McKaye et al. 1992) and, presumably, also benefiting either the young cichlids or their mother. Adult catfish force the young cichlids to remain in more vulnerable positions on the periphery of the juvenile catfish group. A similar antipredator function may account for mixed schools of spinner *(Stenella longirostris)* and spotted dolphins *(S. attenuata)* in the eastern tropical Pacific (Würsig et al. 1994). When such mixed schools are captured in tuna nets, spinner dolphins remain in more peripheral positions and may actively avoid contact with spotted dolphins (Norris et al. 1978). It is possible that, under threatening circumstances, spinners are forced by the spotted dolphins to remain in more vulnerable peripheral positions.

The confusion and oddity effects. While the dilution effect reduces the attack rate per individual in groups, the "confusion effect" reduces the capture rate per attack (Miller 1922; Landeau and Terborgh 1986). The confusion effect occurs because the predator has greater difficulty tracking a single individual in a group (Milinski 1977). Synchronous movements of individuals in schools of fish or dolphins may enhance the confusion effect. This strategy exemplifies by-product mutualism, in which individuals increase the by-product benefits they receive from others by coordinating their actions. In addition to the dilution effect, ocean skaters in groups also enjoyed a relatively small confusion effect during predation by fish (Foster and Treherne 1981): the success rate of attacks on group-living skaters was three times lower than that of attacks on solitary individuals. Under laboratory conditions, largemouth bass *(Micropterus salmoides)* made successful attacks during 100% of five-minute trials with groups of one or two silvery minnows *(Hybognathus nuchalis)*, but only 17% of trials with groups of eight minnows (Landeau and Terborgh 1986). The bass also captured minnows in a higher percentage of trials when one or two fish colored differently were included in the minnow school (Landeau and Terborgh 1986). This "oddity effect" indicates that distinct individuals are more vulnerable or, from the opposite perspective, that conformity in color and size can enhance the confusion effect. A confusion effect, however, does not always accompany the dilution effect (e.g., Morgan and Godin 1985).

Where the confusion effect successfully reduces predator capture rates, predators may begin to avoid attacks on groups in favor of individuals (Pitcher and Parrish 1993). Thus, evidence of lower attack rates on groups may reflect a foraging decision by a predator faced with the confusion effect.

In response to disturbance, cetaceans have been observed to move closer together, making the group more compact (e.g., spinner dolphins caught in a tuna net: Norris et al. 1994; sperm whales: Best et al. 1979; humpback whales: Whitehead and Glass 1985). Compaction is common in schooling fish (Pitcher and Parrish 1993). Norris and Schilt (1988) suggested that such closer spacing in dolphin schools allows individuals to detect more subtle responses in their neighbors, and thus respond more quickly to threats (see also Pitcher and Parrish 1993). Dolphins may use vision to detect subtle movements of close neighbors (Norris and Schilt 1988), while fish can respond more quickly to the movements of close neighbors by detecting pressure waves, which fall off rapidly with increasing interfish distance (Gray and Denton 1991). Group compaction may also enhance the confusion effect by making it more difficult for a predator to single out an individual (Hobson 1978).

The selfish herd. Individuals in groups can also reduce their risk by selecting a location in the group that renders them less likely to be the target nearest to an approaching predator (Williams 1964; Hamilton 1971). Discussions of this "selfish herd" hypothesis often focus exclusively on the prediction that individuals should seek central locations in groups, and thus do not consider it as a possible explanation for cases of group compaction in cetaceans or fish, in which individuals do not change their *relative* positions (e.g., Norris and Schilt 1988; Pitcher and Parrish 1993). Certainly, some group-living aquatic organisms do compete for central locations. When threatened by sticklebacks, larger and faster water fleas *(Bosmina longispina)* escaped toward the center of swarms, leaving smaller individuals on the edge, where they suffered greater predation (Jakobsen and Johnsen 1988). However, the selfish herd concept is more general, and predicts that individuals should reduce their "domain of danger"; that is, that area that renders them closest to an approaching predator. Even if change in relative position is not possible, slight centripetal movement by a peripheral individual will reduce the size of the area in which it would be the nearest and most likely target for an approaching predator (Hamilton 1971). If others respond in a similar fashion, an overall compaction of the group will result.

A consideration of the confusion and oddity effects leads to a prediction of when individuals in Hamilton's (1971) "selfish herds" should attempt centripetal movement that changes their relative position and when they should limit themselves to reducing their inter-neighbor distances, resulting in group compaction. If coordinated movements in schools of fish or cetaceans generate a confusion effect, then attempts by individuals to move to the center of the group might not only negate the confusion effect, but render them more vulnerable as "odd" individuals (e.g., out of position relative to polarized individuals) in the group. We should therefore expect relatively more compaction in polarized schools, and centripetal movement with competition for central locations in nonpolarized groups.

Some predators employ methods that allow them to take advantage of the selfish herd behavior of prey (Pitcher and Parrish 1993). The great size of baleen whales enables them

to engulf entire schools of small prey. Group hunting strategies may also aid predators of schooling species. Jacks were more successful at capturing individual anchovies when they were in groups than when alone (Major 1978). Yellowtail *(Seriola lalandei)* were reported to herd schools of fish cooperatively (Schmitt and Strand 1982), with individuals making forays into a school of jack mackerel while the others maintained their peripheral positions, preventing the mackerel from escaping. The cooperative hunting techniques of killer and false killer whales may explain why dolphin schools are sometimes observed to scatter during attacks by these species (Perryman and Foster 1980; Acevedo-Gutiérrez et al. 1997).

Predator detection and vigilance. Group living not only decreases an individual's chances of being attacked and captured by a predator, but can also increase an individual's ability to detect a predator before an attack occurs (Kenward 1978; Pulliam and Caraco 1984; Uetz and Hieber 1994). Having alert group members around also enables individuals to reduce their own vigilance levels and devote more time to activities such as foraging (Pulliam and Caraco 1984; but see Lima 1995a,b). This predator detection benefit depends not only on a positive relationship between the number of alert individuals and the chance of the predator being detected, but also on group members learning that the predator has been detected. Group members that detect a predator may alert others passively by their behavior as they monitor or evade the predator (e.g., Uetz and Hieber 1994), or actively by giving an alarm call (Sherman 1977). Alarm calling is costly, at least in terms of energy expended, and should be considered altruistic in cases in which calling benefits the caller's kin or individuals that bestow by-product benefits on the caller (Trivers 1971; Connor 1986; Smith 1986). Alternatively, if a predator is zeroing in on a particular individual, that individual may generate a greater confusion effect with an alarm call that elicits a massive evasive response among group members, especially if they haven't seen the predator and don't know which way to flee (Charnov and Krebs 1975).

Sentinel behavior. "Sentinel" behavior, which is found in a number of avian and mammalian taxa, refers to "an alert non-foraging individual stationed in a prominent place while members of its group forage nearby" (Horrocks and Hunte 1986). While acting as a sentinel, a vervet monkey *(Cercopithecus aethiops sabaeus)* stations itself near the top of a tree and does not feed. Other troop members, foraging on the ground, need only monitor the behavior of the sentinel rather than scan a broad area. Despite appearances, sentinel behavior is not necessarily altruistic. The view afforded by their lofty position may yield a lower risk of predation for sentinels, whose rush to seek cover from an approaching predator may incidentally alert group members of the imminent danger (Bednekoff 1997). In Bednekoff's (1997) model, based on a trade-off between foraging and predation risk, individuals with sufficient energy reserves will act as sentinels only if others do not—a result that gives the appearance of coordination among individuals. In support of Bednekoff's model, Clutton-Brock et al. (1999) report that meerkat *(Suricata suricatta)* sentinels are not at greater risk of predation and increase their time on "lookout" if provisioned.

Norris and Johnson (1994) described behavior in three captive spinner dolphins that might reflect vigilance in the wild. Individuals would leave the group and engage in bouts of echolocation in which they appeared to thoroughly scan their captive environment. Norris and Johnson suggested that individuals took turns scanning, but a simpler alternative is that individuals scan at a rate determined by such factors as group size, predator threat, and costs of scanning (e.g., lost feeding time). An absence or failure of vigilance is indicated by observations of a juvenile white shark swimming slowly (about 2 km/hr) into a group of resting dolphins, whose subsequent flight reaction suggested they were unaware of the shark's approach (Connor and Heithaus 1996).

Predator inspection, pursuit, and mobbing. Once a predator has been detected, individuals may engage in evasive maneuvers, simply monitor the predator, or even approach it, a behavior called "predator inspection" (Pitcher et al. 1986). Approaching a predator is often risky (Curio and Regelmann 1986; Poiani and Yorke 1989; Pitcher 1992; FitzGibbon 1994), but may actually reduce risk in some cases (Godin and Davis 1995). Inspectors are thought to benefit from learning about the identity of the predator and the likelihood of attack (Pitcher 1992; FitzGibbon 1994). A predator being approached, having learned that it has lost the element of surprise, may be less likely to attack. Cheetahs are more likely to leave an area after they have been detected and followed by Thomson's gazelles *(Gazella thomsoni)* (FitzGibbon 1994). More vigorous "pursuit deterrent" behaviors, such as stotting by Thomson's gazelles, may further dissuade a predator by signaling that the stotting tommy is in good condition and won't be easy to catch (FitzGibbon and Fanshawe 1988; Caro 1994b).

A possible case of predator inspection in cetaceans is a behavior called "scouting," in which individual dolphins leave a group to inspect a human-made barrier (e.g., a net) (Evans and Dreher 1962; Caldwell and Caldwell 1964; but see Norris and Dohl 1980b). The term "scouting" carries the implication of an act performed to benefit others in the group and which is therefore altruistic. Although predator inspection carries some risk, it is thought to inform the inspector about the attack readiness of the predator and allow the inspector to shift to foraging in safer locations (Pitcher 1992). Pitcher (1992) considers whether inspectors behave altruistically or deceptively toward noninspecting group members. If individuals gain a net benefit from inspection, however, the behavior is simply selfish, and it is not necessary to invoke altruism or deception.

In some cases, the best defense is good offense. Some birds and primates persuade predators to leave an area by "mobbing" them, a strategy that includes chasing, harassing, and attacking behavior (Curio 1978; Cheney and Wrangham 1987). As with sentinel behavior, individuals engaged in mobbing may perform singular acts or take a lead role in rescue or defense. For example, the dominant male in a group of banded mongooses *(Mungos mungo)* led the charge that rescued a group member from the talons of an eagle (Rood 1983).

Saayman and Tayler (1979) described a case of mobbing by hump-backed dolphins *(Sousa* sp.), a species that reaches about 2.8 m, in South Africa. Two adults and a calf left a group of ten individuals, swam rapidly toward a 4–5 m shark, and herded it into a cove. The shark turned and lunged toward the dolphins, escaping the cove, but was pursued and cornered in another cove. Again the shark escaped and swam fast toward the open sea, with the dolphins in pursuit. Shortly, the dolphins broke off the chase and swam at a slower speed in the direction their group had been traveling.

Dolphins are reputed to cooperatively attack and even kill sharks, but there are only second-hand accounts of dolphins physically attacking sharks in the wild (e.g., Gunter 1942; Wood et al. 1970; Best et al. 1984). It is not clear from these accounts whether the attacks were in response to aggressive behavior by the shark. One female bottlenose dolphin whose calf was killed by a shark was observed chasing a shark away from her calf's body shortly afterward (Mann and Barnett 1999; see also Connor and Smolker 1990). In captivity, bottlenose dolphins will make unprovoked attacks on small sharks, ramming and even killing them (Wood et al. 1972).

If attacked, cetaceans may mount an aggressive defense,

but it is often unclear in such cases whether individuals coordinate their actions. Whitehead and Glass (1985) observed humpback whales violently thrashing their flukes, their major weapon, in response to an attack by killer whales. A group of three whales under attack maintained a closer interindividual spacing than usual. Arnbom et al. (1987) observed possible group defense by sperm whales during an attack by killer whales. The sperm whales formed a tight group and endeavored to face the killer whales. Facing outward may allow the sperm whales to use their teeth as a defensive weapon or simply make it easier to monitor the killer whales. Sperm whales surrounded by sharks or killer whales may form a tight bunch with calves in the middle (Best et al. 1979). The "marguerite" formation, in which sperm whales under attack form a circle with their heads pointed in and their thrashing tails out, often with calves in the center, is more clearly coordinated defensive behavior (Nishiwaki 1962; Palacios and Mate 1996; Weller et al. 1996). Several right whales were observed in this formation, with heads in and tails out, in the vicinity of one or more killer whales (K. Payne, personal communication).

In sum, protection from predators may be an important benefit of group membership for cetaceans, and, for at least some species (e.g., sperm whales: Whitehead and Weilgart, chap. 6, this volume), is a strong candidate for why groups form in the first place.

Cetacean Parasites

Grouping may also reduce the risk from non-socially transmitted parasites via the dilution effect (Mooring and Hart 1992). Duncan and Vigne (1979) found a higher per capita number of blood-sucking flies on horses in small groups of three individuals than in large groups of eight to thirty-two individuals. When horses from small groups joined larger groups they enjoyed a reduction in fly number (Duncan and Vigne 1979). Dolphins in tropical and subtropical waters are often attacked by what amounts to a foot-long biting fly: the "cookie cutter" shark *(Isistius brasiliensis)*, which removes circular or crescent-shaped chunks of flesh up to 7 cm across from its victims (Jones 1971). Cookie-cutters are thought to attack while dolphins are feeding (Jones 1971), suggesting a dilution benefit to dolphins hunting in groups. Group living may also reduce parasitism by the remora, which is found on tropical dolphin species. The spinning that gives the spinner dolphin *(Stenella longirostris)* its name may have been favored initially as a mechanism to eject hitchhiking remoras, although spinning now appears in other contexts as well (Hester et al. 1963; Norris et al. 1994).

Cetacean Prey

The defense of feeding areas or food patches may favor group formation (Wrangham 1980a). Individuals are expected to cooperate to defend resources when those resources are defensible and can support more than one individual. Even where groups form for other reasons (e.g., to reduce predation), there are a number of ways that individuals in groups may enjoy improved access to food. Before reviewing these food-related benefits of group living, it will be useful to have a brief overview of what cetaceans eat and how they feed.

Baleen whales are batch feeders, taking in large numbers of small crustaceans or fish in each feeding attempt, while toothed whales may consume solitary or schooling species, but pursue and capture their fish, squid, and occasionally crustacean prey on an individual basis. Some populations of killer whales specialize on mammalian prey, with the occasional avian offering, and marine mammals may appear in the diet of several other odontocetes (Jefferson et al. 1991; Perryman and Foster 1980). Odontocetes are found in rivers, shallow-water estuaries, and other near-shore environments; some are typically found offshore but over the continental shelf, and others (e.g., sperm whale) are found mostly in deep water. Many squid-eating species (Clarke 1986) are deep divers and are able to take advantage of prey that are abundant in the deep sea. The river dolphins, including the genera *Plantantista, Pontoporia, Lipotes,* and *Inia,* seize prey in the water with elongate pincer-like rostra. At the other extreme, blunt-rostrum odontocetes such as the monodontids (the beluga and the narwhal, *Monodon monoceros*), and even the sperm whale, (Whitehead and Weilgart, chap. 6, this volume) may rely primarily on suction for ingesting prey from the water or the bottom substrate (e.g., beluga) (Werth 1989, 1991; Heyning and Mead 1996).

The baleen whales can be subdivided into "skimmers," "swallowers," and "grubbers." The three species of balaenid, or "right" whales, are skimmers, swimming along the surface, mouths agape, trapping prey with long, fine baleen that provides the greatest filtering area per body length among the baleen whales (reviewed in Gaskin 1982). The balaenopterid family, or rorquals, are generally swallowers, but some species skim as well (e.g., the sei whale, *Balaenoptera borealis:* Horwood 1987). With the aid of ventral pleats, rorquals expand their mouths and gulp in huge amounts of water at a time, then expel it, trapping prey against their relatively short baleen.

Gray whales, the lone member of the family Eschrichtiidae, feed on the bottom by rolling on their sides and using suction to ingest their mostly amphipod prey. Plowing along the bottom abrades their snout and short, coarse baleen. Surprisingly, bowhead and humpback whales are also reported to engage in occasional bottom feeding (Würsig et al. 1989; Hain et al. 1995). Reports of substantial amounts of kelp and other vegetation in stomach contents, plus the presence of volatile fatty acids indicative of microbial fermentation, prompted Nerini (1984) to suggest that gray whales are partially herbivorous.

Food and Group Living

Cetacean prey, even mobile schooling species, should in many circumstances be defensible and worth defending, at least for short periods. There are, however, only two reports in the literature of an individual or group being displaced from food or a hunting area, and both accounts involve killer whales. In Argentina, Hoelzel (1991a) twice observed a pod of two adult females and five subadults displacing a pod of two adult males from a favored nearshore sea lion hunting area. In Norway, displacement occurred at herring schools when another group of killer whales rapidly approached the feeding group, which then left (Bisther and Vongraven 1995).

Territorial resource defense has not been reported in cetaceans and is obviously not an option for open-ocean populations. Even for populations that live in coastal habitats where boundaries can be defined, the mobility of their prey, their wide-ranging habits, and their three-dimensional environment all render territoriality a less inviting option than for most terrestrial species. The best hope for discovering territorial resource defense (permanent or seasonal; individual or community) in cetaceans may be found in those riverine or estuarine populations in which these factors might be minimized: populations that feed on abundant benthic or otherwise relatively nonmobile prey in shallow water with extensive land borders. Home ranges may also be defended, even in the absence of strict territoriality, when non-group members are encountered. This may explain why the two communities of *resident* killer whales in the coastal waters of British Columbia and Washington State have overlapping ranges, but have never been observed to mingle.

There are a number of ways individuals in the pursuit of food may benefit from the presence and behavior of others in their group. They may learn from others the location of food that occurs in unpredictable patches that are large enough to be shared (Ward and Zahavi 1973; Brown 1986). Pitcher et al. (1982) demonstrated experimentally that minnows *(Phoxinus phoxinus)* and goldfish *(Carassius*

auratus) detected patchy food faster in larger schools. A number of odontocetes spread out perpendicular to the direction of travel during what appears to be foraging (e.g., pilot whales: Norris and Prescott 1961; sperm whales: Whitehead 1989a; common, spotted, and spinner dolphins: Norris and Dohl 1980b; *resident* killer whales: Felleman et al. 1991). This behavior is often considered to reflect a cooperative effort to locate prey schools: "Delphinids in open water usually search for food in broad ranks which may be hundreds of meters wide, presumably to increase the chances of finding schools of fish or squid. Such ranks indicate that the animals are cooperating as a school in finding prey" (Würsig 1986). This may be true in cases in which individuals feed on large schools of fish (e.g., dusky dolphins in Argentina: Würsig and Würsig 1980), but individuals feeding on solitary items or small schools may spread out to reduce feeding competition. If groups form because of antipredator benefits, and in spite of increased costs of feeding competition, then foraging in ranks may be an effort to minimize this cost (e.g., sperm whales: Whitehead 1989a; Whitehead and Weilgart, chap. 6, this volume) and is not necessarily cooperative.

The cooperative feeding hypothesis predicts that cetaceans foraging in spread formation should coalesce into single large feeding groups once prey are located. The competition hypothesis predicts that individuals should not coalesce to feed. However, even species that forage in rank to avoid competition may occasionally coalesce to feed on large prey schools. Evidence is needed that establishes whether dolphins are feeding while they are spread out.

Individuals or small groups of dolphins are sometimes observed to leave larger nonforaging schools to hunt close to shore. As with the case of dolphins approaching human-made barriers, this behavior is often called "scouting" (Bel'kovich et al. 1991). Again, the term implies "sentinel"-like behavior, in which an individual takes on the burden of food location for the group. A simple alternative is that individuals leave the school to hunt alone.

In some species, individuals may actually feed better by informing others when they have located food. In the presence of insect swarms, colonially nesting cliff swallows *(Hirundo pyrrhonota)* give calls that attract other birds to feed. The additional feeding birds enable the caller to track the ephemeral swarms, and thus feed for a longer time (C. R. Brown et al. 1991). In this case, the value of byproduct benefits associated with feeding with others has selected for altruistic behavior by birds that discover insect swarms (C. R. Brown et al. 1991; Connor 1995a,b).

The question remains as to whether cetaceans that have located fish inform others. Bottlenose dolphins in the Moray Firth produce a low-frequency call, with most energy from 200-4,000 Hz, while feeding on salmonids. These feeding-associated calls produce rapid approaches by other dolphins in the area (Janik 1998). Recruitment may also be passive, as when a dolphin using echolocation to track prey inadvertently reveals the location of food to others nearby (Caldwell and Caldwell 1977; Dawson 1991).

Dusky dolphins *(Lagenorhynchus obscurus)* are a promising species in which to study recruitment to prey schools. Tayler and Saayman (1972) observed a single dusky dolphin discover and turn toward a school of fish, at which time approximately two hundred dolphins scattered over 8 km^2 converged on the area and began feeding. Würsig and Würsig (1979) made detailed observations of dusky dolphins feeding on schools of southern anchovy *(Engraulis anchoita)* in Argentina. Groups of six to fifteen dolphins apparently locate and herd schools of anchovy upward, trapping them against the surface (see also Norris and Dohl 1980b). While dolphins swim around and under the school, individuals rush through it, capturing up to five fish at a time (Würsig and Würsig 1979). Feeding bouts involving only one such group typically last only a few minutes, but feeding bouts may last for several hours when other groups join, leaping toward the feeding group from as far away as 8 km (Würsig and Würsig 1979). Dolphins may be attracted to feeding groups by the sight of feeding dolphins, which leap during feeding, feeding birds overhead, or food-associated sounds (Würsig and Würsig 1979). Würsig (1986) suggests that the relationship between dolphin feeding group size and feeding bout duration reflects a difference between the size of groups dolphins form to search for fish and the optimal group size for herding fish. Once fish are located by a searching group, the duration of feeding depends upon the number of dolphins that join. While it is likely that more dolphins are able to herd larger schools for longer periods, it is also possible that once large schools are discovered, they inevitably attract large numbers of dolphins irrespective of changes in herding ability. The observed relationship between the duration of feeding bouts and the number of feeding dolphins might be accounted for simply by positing a relationship between fish school size and the distance dolphins are willing to travel to feed. Small fish schools that can feed only a few individuals would likely be gone by the time individuals arrived from a distance. Further, the size of "searching groups" may be limited by prey school size, if the majority of prey schools encountered by the dolphins are small.

Aquatic predators such as cormorants and Atlantic bluefin tuna *(Thynnus thynnus)* form a feeding crescent as they pursue schools of prey fish (Bartholomew 1942; Partridge et al. 1983). Such spacing may simply reduce feeding competition, but individuals may also benefit when a fish attempting to escape their neighbor swims into their path. Whales and dolphins may also travel in close rank or crescent formation behind prey they are pursuing. Bottlenose dolphins often pursue fish in a crescent formation (Leatherwood 1975), and killer whales have been observed chasing sea lions in this manner (Norris and Prescott 1961). One group of killer whales swimming in tight formation generated a wave that tipped a crab-eater seal *(Lobodon carcinophagus)* from the safety of an ice floe (Smith et al. 1981). In the Beaufort Sea, bowhead whales often skim-feed in groups of two to ten or more individuals (Würsig et al. 1984). The whales were observed to skim-feed in a side-by-side formation as well as a staggered V formation. In the latter case, like the lead bird in a flock of geese, it seems unlikely that the lead whale would benefit from its neighbors. However, the lead position may be preferable when feeding on schooling prey (e.g., Major 1978).

Groups of dolphins often herd prey schools against barriers. The dusky dolphins observed by Würsig and Würsig (1980) trapped schools of anchovies against the surface. Atlantic spotted dolphins *(Stenella frontalis)* in the Gulf of Mexico were observed trapping schools of clupeid fish at the surface by swimming around and under them, tail-slapping, flashing the light-colored ventrum toward the school, and emitting bursts of bubbles around the fish (Fertl and Würsig 1995). Killer whales in Norway also emit bubble bursts while feeding cooperatively on schooling fish (Similä and Ugarte 1993; Similä 1997), but the most elaborate use of bubbles for herding prey is found in humpback whales (Jurasz and Jurasz 1979; Hain et al. 1982). Using sonar to "see" underwater, Sharpe and Dill (1995) found that the humpbacks dove below herring schools, then herded them up toward the surface while they wrapped the fish in a cylindrical bubble net.

The surface is not the only barrier against which dolphins can trap fish. At sea, bottlenose dolphin groups approaching a school of fish may split into two groups to attack the school from opposite sides (Tayler and Saayman 1972; Saayman et al. 1973; Würsig 1986). Bottlenose dolphins often herd prey schools against the shore (Morozov 1970; Leatherwood 1975; Bel'kovich et al. 1991; Petricig 1993). This behavior has repeatedly given rise to mutualistic interactions with humans (Busnel 1973; Hall 1985; Pryor et al. 1990). For example, on the coast of Mauritania, the local people fish in the shallows with hand-held nets for seasonally abundant mullet. Upon sighting a school of mullet, one of the fishermen enters the water and hits the surface repeatedly with a stick. This attracts bottlenose and hump-backed (*Sousa* sp.) dolphins, which feed in the shallows next to the fishermen. The net appears to serve as a barrier against which the dolphins herd fish, to the benefit of both dolphins and humans (Busnel 1973).

Cooperative hunting may allow individuals to capture larger or more elusive prey than they could hunting alone (see Caro 1994). Cooperatively hunting Harris's hawks (*Parabuteo unicinctus*) use several techniques to capture rabbits, including ambushes by several individuals converging from multiple directions and relays in which the lead position in a chase is taken up by a new individual following a failed attack by another (Bednarz 1988). The hawks were more successful in groups of five or six than in smaller groups, in terms of both capture rate and energy yield per hawk, which corresponds to the modal group size of five individuals (Bednarz 1988). Similarly, *transient* killer whales in British Columbia hunting harbor seals enjoyed a higher energy intake per individual in groups of three than in larger or smaller groups (Baird and Dill 1996; Baird, chap. 5, this volume). Again, groups of three were the most common during foraging. Larger group sizes in other mammal-eating *Orcinus* populations may result from their pursuing larger and more dangerous prey, such as whales (Baird 1994).

In lions, cooperative hunts may be highly coordinated. Some females stalk around to "wing" positions behind prey, from which they initiate chases, while others move short distances to "center" positions, where they wait to ambush prey that are flushed in their direction. Interestingly, individuals specialize in center or wing positions, and even whether they take the right or left wing (Stander 1992).

We might expect to find similar division of labor in cooperatively hunting delphinids. Würsig (1986) identified the observations of Morozov (1970) as a possible case. Morozov observed a group of fifteen bottlenose dolphins feeding on a densely packed school of fish close to shore. While two or three dolphins fed, the rest swam in small groups of two to four individuals in a semicircular pattern about 100 m seaward of the fish school, preventing the fish from escaping. Periodically, the feeding dolphins left the school, passing a new group that was moving in to feed. During a similar herding episode in South Africa, two large bottlenose dolphins maintained particular patrol routes farther offshore than most, leading Tayler and Saayman (1972) to suggest that the two individuals maintained a "supervisory function" over the herding operation. Neither report suggests a division of labor like that of lions, which exhibited individual roles during the pursuit of prey that

were consistent across hunting episodes (Stander 1992). The division of behavior in the dolphins was between feeding and patrolling (Morozov implies, but does not clearly state, that dolphins that had fed continued to patrol). Morozov suggested that feeding order might have been determined by dominance rank. In the South African observation, the two "supervisors" did occasionally capture fish, and may have simply preferred feeding farther offshore, as has been reported in larger killer whales (Baird, chap. 5, this volume). A division of labor might occur where individuals split up to attack schools from different positions. It would be necessary, however, to demonstrate that individuals took the same roles during repeated hunting episodes.

Possible division of labor during the hunting of harbor seals by *transient* killer whales in British Columbia was noted by Baird and Dill (1995). While females and subadults struck at the seal with their flukes or flippers near the surface, a male appeared to prolong its dive time, possibly preventing the seal from escaping (Baird, chap. 5, this volume).

Off southeastern Alaska, groups of up to eleven humpback whales feed in a coordinated fashion on schooling herring (Jurasz and Jurasz 1979; Hain et al. 1982; Perry et al. 1990). During "lunge-feeding," individuals gradually or rapidly break the surface with their mouths agape, rising up to a third of their body length out of the water as their throats distend with water and prey (see fig. 7.4). Groups of whales may perform repeated lunges side by side in unison (Jurasz and Jurasz 1979; Hain et al. 1982). Associations that are stable across years have been reported in a few such feeding groups (Perry et al. 1990). D'Vincent et al. (1985) reported that in 130 episodes of coordinated feeding by a group of eight humpbacks that they observed over a three-day period, all eight individuals maintained the same spatial position during lunges, both relative to each other and relative to the surface (some rose more vertically than others). Remarkably, although the composition of the feeding group has varied, the position of individuals relative to each other in the feeding group has remained constant across years (C. G. D'Vincent, personal communication). There are three possible interpretations of this behavior. First, it might reflect a division of labor. However, the different positions of the whales (see D'Vincent et al. 1985) suggest that lunging positions may not yield equal returns. Thus, an alternative explanation is that individual position is determined by position in a dominance hierarchy. Dominance hierarchies have not been reported for any baleen whale and are thought to be unimportant in humpback whales (Clapham, chap. 7, this volume). Both the dominance and division of labor explanations are strikingly at odds with the general view that long-term stable social bonds are weak or lacking in baleen whales.

The third hypothesis does not require social bonds or cooperation among the feeding whales. The group might contain a "producer," who herds the prey, and "scroungers," who take advantage of the producer's efforts. Producers and scroungers might coexist in the population in a mixed evolutionary stable strategy (Barnard and Sibly 1981). Group stability could result if it is more efficient for a scrounger to remain with a particular producer than to switch producers, and if producers are defended as a resource.

The Costs of Group Living: Travel, Feeding Competition, and Philopatry

Cetaceans almost certainly suffer the same costs of group living, namely, parasite transmission and resource competition, that afflict other taxa. Resource competition is strongly affected by travel costs, which are relatively low in cetaceans compared with most terrestrial mammals.

Costs of Locomotion and Group Living in Cetaceans

With their fusiform shape, cetaceans are well adapted for efficient locomotion in the aquatic environment (e.g., Hui 1987). However, Williams (1999) found that the total energy expended in moving a given distance is similar for animals that specialize in running, swimming, and flying. This total energetic cost of transport can be divided into basic maintenance costs and locomotion costs. Maintenance costs, which include energy expended for basal functions and endothermy (Williams 1999), are higher for marine than for terrestrial mammals, so locomotion costs are lower. Thus we can understand why marine mammals enjoy a relatively low cost of locomotion: the proportional increase in energy required for travel (relative to their maintenance level) is much less than for terrestrial mammals of comparable size. As travel costs normally decrease with increasing body size, we can expect larger cetaceans to enjoy even lower costs of locomotion. For a given body size, costs of locomotion are likely to vary with body shape and fluke morphology (Bose and Lien 1989; Bose et al. 1992; Curren et al. 1994).

I suggest that the relatively low costs of locomotion and concomitant extensive ranges of cetaceans have two important consequences for cetacean grouping patterns and social organization. First, other factors being equal, the low costs of locomotion lower the costs of grouping relative to terrestrial species with higher travel costs. Second, the low travel costs of cetaceans, combined with the fact that their off-spring are "followers" (so they are not tied to a breeding site), may significantly reduce the cost of natal group philopatry in *resident* killer whales and possibly some other odontocetes. Before exploring these issues further, I will review the data on travel speed and estimated daily travel distances for cetaceans.

Dolphins travel relatively fast for long periods of time. Unlike most terrestrial mammals, dolphins are usually on the move, stopping only briefly at the surface for rest periods. Estimates of the distance traveled in twenty-four hours, or "day range," can be made for a number of odontocetes. Three kinds of data were used for most of the estimates in table 8.1: extrapolations based on observations of diurnal travel speeds from fixed observation points, extrapolations from locations transmitted by radio and satellite tags, and individual or group follows. Radio tag and satellite data have the advantage of including nocturnal movements, but these methods will underestimate day range if they are based on infrequent locations of animals that frequently change direction. This bias is particularly strong for satellite data: the estimate for the radio-tagged Atlantic white-sided dolphin (*Lagenorhynchus acutus*) in table 8.1 (309 km during 64.3 hours, or about 115 km/day) is based on an average of 2.5 locations per day (Mate et al. 1994b). Extrapolations from speed estimates may be biased in either direction depending on how representative the observation site is in terms of travel speed and activity. For example, the twelve-hour estimate for hump-backed dolphin travel (24 km) is clearly a significant underestimate because it excludes the time (50%) in which the dolphins were socializing or foraging. Socializing and foraging dolphins may be moving at slower or faster speeds than traveling dolphins, but they are rarely stationary. The current speed relative to the dolphins' direction of travel can also affect estimates. Foraging bottlenose dolphins viewed by multidimensional sonar were moving at 6.5 km/hr relative to the seabed, but 7.9 km/hr relative to the water (Ridoux et al. 1997). If, for example, dolphins tend to forage into currents (as reported in Ridoux et al. 1997) but travel randomly with respect to currents, then day range estimates based on foraging observations will be low. Depth of the water may also influence travel speeds (e.g., Würsig and Würsig 1979). Estimates based on group or individual follows, given that they are of sufficient number and length, are less subject to such error.

In spite of these methodological difficulties, it is clear that dolphin day ranges often exceed 100 km/day, an order of magnitude above the day ranges of most primates and carnivores (Wrangham et al. 1993). The lowest estimate, of 15–20 km/day for radio-tagged harbor porpoises, may

Table 8.1. Travel speeds and estimated day ranges for several species of odontocetes

Case	Species	Size[a]	Method	Number of individuals	Observation duration	Average speed (km/hr)	Estimated daily travel (km) 24/12hr[b]	Reference
1	Harbor porpoise (Phocoena phocoena)	1.5	Radio tag	8	0.3–22 days		15–20	Read and Gaskin 1985
2	Harbor porpoise (Phocoena phocoena)	1.5	Radio tag	4	0.67–1.1 days		32–51	Westgate et al. 1995
3	Harbor porpoise (Phocoena phocoena)	1.5	Satellite	9	2–212 days		14–58.5	A. J. Read and A. J. Westgate, unpublished data
4	Dusky dolphin (Lagenorhynchus obscurus)	1.5–2.0	Theodolite, hourly estimates of group speed	Unknown, groups of unidentified individuals	14 hourly intervals (05:00–18:00)	7.7 (range about 5–12)	185/92	Würsig and Würsig 1980
5	Dusky dolphin (Lagenorhynchus obscurus)	1.5–2.0	Radio tag	2 individuals	3 days	2.3 3.2	55 77	Würsig 1982
6	Common dolphin (Delphinus delphis)	2.0–2.5	Radio tag				100	Evans 1974
7	Spotted dolphin (Stenella attenuata)	2.0–2.5	Radio tag	1	11 hr	9.1	219/109	Leatherwood and Ljungblad 1979
8	Spotted dolphin (Stenella attenuata)	2.0–2.5	Radio tag	2	3 + 4 days (16–24 hr/day)		119	Perrin et al. 1979 (data from two males tracked most consistently)
9	Hump-backed dolphin (Sousa sp.)	2.5	Reference aerial photographs of topographic features	Groups	Not reported	4.8 (50% diurnal activity)	>58/24	Saayman and Tayler 1979
10	Atlantic white-sided dolphin (Lagenorhynchus acutus)	2.5	Radio tag	1	64.3 hr		115	Mate et al. 1994b
11	Bottlenose dolphin (Tursiops truncatus)	2.5–3.5	Theodolite	Unknown, groups of unidentified individuals	10 hr (07:00–17:00)	6.1	146/73	Würsig and Würsig 1979[c]
12	Bottlenose dolphin (Tursiops truncatus)	2.5–3.5	Satellite	4	53–439 hr	1.4–2.7	Range = 33–64	Tanaka[d] 1987
13	Long-finned pilot whale (Globicephala melaena)	5–6.0	Satellite	1	14 days (18 hr/day)		74	Mate et al. 1987
14	Killer whale (Orcinus orca) (residents)	6–8.0	Theodolite	Unknown, individuals and groups	31 days	5.2	125/62	Kruse 1991
15	Killer whale (Orcinus orca) (transients)	6–8.0	Follows	Groups	NA		80–160	R. W. Baird, unpublished data
16	Sperm whale (Physeter macrocephalus)	10–12.0	Follows	Groups of females with calves	NA	2.7	89	Whitehead 1989a

[a]Sizes (in meters) are intended only as a general guide for adults of the species.
[b]Both 24- and 12-hour estimates are given for studies that included only diurnal speed estimates.
[c]Daily travel estimated from average speed. Average speed based on average of readings taken on unknown number of days during 21-month study.
[d]Based on four individuals Tanaka considered to have been tracked successfully.

have been low because Read and Gaskin (1985) were able to track individual movements only in nearshore areas (A. J. Read, personal communication). A later radio tag study on the same species found twenty-four-hour distances of up to 51 km. The lowest twenty-four-hour travel estimate for a delphinid, of 33 km/day for a bottlenose dolphin, was based on only four satellite locations, but still exceeds all carnivores but the spotted hyena (Crocuta crocuta) (40 km/day). Not surprisingly, the large day ranges of cetaceans are not always accompanied by equally impressive home ranges, and the geographic movements of some individuals may be quite limited (e.g., Connor et al., chap. 4, this volume). Generally, the straight-line distance between positions twenty-four hours apart will vary considerably be-

RICHARD C. CONNOR

tween species, populations, and individuals, and for individuals on different days.

Dolphins have large day ranges compared with terrestrial species not just because they travel fast, but because they travel during most or all of the day. Even in captivity, dolphins may travel continuously. Two common dolphins *(Delphinus delphis)*, an offshore species, were found to travel nearly continuously even in a large outdoor tank, averaging 5.8 km/hr and 136 km/day (Hui 1987). Behavioral observations of dolphins at night are limited. Using night-vision equipment, Petricig (1993) found that foraging activity of bottlenose dolphins in South Carolina was independent of the light cycle, but dependent on tidal stage.

Travel Costs, Feeding Competition, and Group Formation

One of the universal costs of grouping is resource competition (Alexander 1974). Variation in resource distribution can mean the difference between relatively low or high costs of grouping. Resource distribution is a relatively unstudied problem in cetacean biology, no doubt because it is difficult to measure the distribution of food and the rate of consumption by cetaceans. For a given resource distribution and consumption rate, the level of resource competition will be strongly affected by travel costs.

With increasing group size, the amount of food per individual at a given patch is reduced, so that the group must travel farther to find sufficient food (Clutton-Brock and Harvey 1977; Terborgh 1983; Isbell 1991; Wrangham et al. 1993; Janson and Goldsmith 1995). Other things being equal, a species with high locomotion costs will not be able to afford to travel in as large a group as a species that enjoys lower travel costs. Dolphins, with their relatively low travel costs, should therefore be able to travel in larger groups than terrestrial species of comparable body size.

The effects of travel costs on group formation are usually considered for cases in which group size is constant or foraging is continuous. However, many species of inshore delphinids live in fission-fusion societies characterized by variable group sizes and foraging behavior that occurs in distinct bouts. A model that incorporates bout feeding may give very different results than one based on continuous foraging. Imagine that the distribution of food and travel costs for a continuously foraging species disfavor group formation. Now consider that the same species forages in bouts. If the benefits of group formation are sufficient to outweigh the additional travel costs, then bout feeders will form groups periodically between feeding bouts where continuous feeders would not. Continuous foragers that decided to travel to join groups would suffer, in addition to

the extra travel costs, the loss of a proportion of their daily food intake.

Compared with a lumbering terrestrial mammal, a dolphin's low cost of locomotion further reduces the cost of grouping between feeding bouts. Consider two species that vary in locomotion costs but that otherwise share identical food distributions, durations of foraging bouts, and benefits of group formation. A higher cost of locomotion may tip the balance for one species toward a solitary existence while the other species, gliding efficiently through its realm, can afford to form groups between foraging bouts. Rodman (1984) makes such an argument to explain the difference in male mating strategies between chimpanzees and orangutans. Male orangutans are solitary, but chimpanzees engage in cooperative community defense. Chimpanzees enjoy much lower travel costs than the more arboreal orangutans, which allows them to spread out to forage apart, but to regroup when group defense is required.

Travel Costs, Ranging Patterns, and Dispersal

Natal philopatry by both sexes is rare among terrestrial mammals. In the *resident* population of killer whales in the northeastern Pacific, there is no dispersal from the natal pod by either sex (Baird, chap. 5, this volume).

A case for natal group philopatry by both sexes has also been made for the pilot whale. Two species of pilot whales are recognized, the long-finned pilot whale *(Globicephala melas)* and the short-finned pilot whale *(G. macrorhynchus)*. Using microsatellite DNA analysis on complete schools of 90 and 103 long-finned pilot whales captured in drive fisheries in the Faeroe Islands, Amos et al. (1993) found that mature males neither mate within nor disperse from their natal groups.

While it is tempting to speculate that pilot whales might have strong mother-son bonds like killer whales (Connor et al., chap. 10, this volume), it is important to point out that the large schools of pilot whales analyzed by Amos et al. (1993) may not be the social equivalent of killer whale pods. Sergeant (1962) reported that the average size of schools captured in drive fisheries of long-finned pilot whales is several times larger (mean = 85) than reported for pelagic schools of the same species (mean = 20). Kasuya and Marsh (1984) also found that larger aggregations of short-finned pilot whales tended to be captured in Japanese drive fisheries than was typical of offshore sightings. If these smaller groups represent a unit of social structure analogous to killer whale pods, the large schools analyzed by Amos et al. (1993) might represent a higher level of social structure, possibly analogous to a killer whale community, albeit a more cohesive one. Thus pilot whale philopatry within the large schools does not rule out dispersal between smaller, more stable, units. Further study of pilot whale social organization will allow us to distinguish between these possibilities.

A social organization based on stable pods with natal group philopatry by both sexes may occur in several members of the subfamily Globicephalinae, which includes pilot whales, the pygmy killer whale *(Feresa attenuata)* and the false killer whale (the position of killer whales in this group has recently been challenged: LeDuc et al. 1999). This would help explain otherwise puzzling behaviors such as Porter's (1979) observation of a pod of thirty false killer whales remaining for three days in tight formation around a large, dying male.

The enduring and consistent mother-offspring associations in *resident* killer whales focus our attention on their unusual pattern of natal philopatry, but the phenomenon may extend to bottlenose dolphins and other delphinids that live in fission-fusion societies as well. In Sarasota, Florida, female bottlenose dolphins associate in bands that include up to three generations of females. Males in Sarasota may range widely, but maintain their natal area in their adult home range (Scott et al. 1990a). *Transient* killer whales in the northeastern Pacific, whose range overlaps that of the *resident* population, travel in smaller groups than *residents,* and a few cases of dispersal from the natal pod have been documented (Baird, chap. 5, this volume). However, even though they no longer travel with their natal pod, dispersing individuals may continue to use their natal range.

Dispersal from the natal group by members of one sex is typically thought to be an adaptation to avoid inbreeding (Greenwood 1980; Pusey and Wolf 1996). If so, then either inbreeding is of little consequence for *resident* killer whales, or they are able to outbreed without emigrating from their natal pod. I suggest that the relatively low costs of locomotion and concomitant large day and home ranges of killer whales can contribute significantly to our understanding of the unusual pattern of natal philopatry in this species and possibly others. However, low costs of locomotion alone are insufficient to account for the unusual pattern of natal philopatry by both sexes. It is also critically important that they are not restricted in their movement by the need to remain at a breeding site and that resource distribution does not favor the defense of small territories (although community defense of large territories is possible). Flying, although not as efficient as swimming, is also less expensive than terrestrial locomotion (Schmidt-Nielsen 1990). Birds and bats are thus highly mobile, but anchored to breeding sites. In some colonial species, natal philopatry by both sexes may allow continued maternal associations by offspring of both sexes (e.g., Palmeirim and Rodriques 1995).

Amos et al. (1993) suggested that male pilot whales may encounter potential mates either by temporarily leaving the natal group to rove in search of mates or by encountering potential mates while traveling with their mothers. The cost of leaving for some period and returning is lowered if the cost of locomotion is low. Likewise, low costs of locomotion allow for large potential home ranges. At a given population density, having a larger home range means that males may encounter more potential mates, even while traveling with their mothers.

Resident and *transient* killer whales, bottlenose dolphins, spinner dolphins, and possibly many other species may exhibit geographic or "locational" philopatry (Isbell and Van Vuren 1996), in which members of both sexes maintain their natal range in their adult home range. For species that exhibit geographic philopatry, there may be a continuum in the degree to which members of both sexes maintain bonds with their natal group (social philopatry). *Resident* killer whales, in which both sexes remain with their mothers, represent one extreme in this continuum. Bottlenose dolphins, in which males only occasionally associate with their mothers, may represent another. The degree to which both sexes exhibit social (or, more precisely, maternal) philopatry should be limited by the cost of feeding competition. Baird (chap. 5, this volume) makes a convincing case that the smaller group size and the dispersal pattern of *transient* killer whales are products of feeding competition. For a discussion of the possible benefits of natal philopatry to male and female killer whales, see Baird, chap. 5; Connor et al., chap. 10, this volume.

Grouping Costs and Benefits: Seasonal and Maturational Changes

The factors favoring and disfavoring grouping may change during the course of development, seasonally, or moment by moment, and individuals often exhibit adaptive responses to those changes (reviewed by Lima and Dill 1990). Once male sperm whales have dispersed from their natal group, they associate in all-male groups of similar-sized individuals. As males continue to grow, they range into progressively higher latitudes and travel in smaller groups until they become solitary (Best 1979; Whitehead and Weilgart, chap. 6, this volume). Increasing male-male competition and possibly foraging competition may explain why mature males lead a mostly solitary life (but see Christal and Whitehead 1997), but it is less clear why smaller males aggregate (Whitehead and Weilgart, chap. 6, this volume). Possible factors include a relatively greater predation risk compared with larger males, reduced foraging competition (for smaller, more evenly distributed prey), the benefits of practice jousting with other males, or even cooperation against other males.

Migrating female baleen whales experience four rather distinct phases during a year, in which the costs and benefits of grouping vary considerably: feeding in high latitudes, breeding in low latitudes, and migrating twice between these destinations. The migration from the breeding to the feeding ground, when females are not feeding but have newborn calves in tow, is likely the time when predation pressure most strongly favors grouping (e.g., humpback whales: Clapham, chap. 7, this volume). Feeding competition disfavors grouping, but it is not clear when this factor might have its strongest influence. Competition may be less during the peak of the summer feeding pulse than before or after (and as the humpback whale observations suggest, group feeding may sometimes be beneficial). Resources may provide valuable if inconsistent opportunities to feed at other times of the year (e.g., near the breeding ground: Baraff et al. 1991), and competition for these limited resources might be strong enough to disfavor grouping.

Social Odontocetes and Asocial Mysticetes

Stable social groups and relationships appear to be widespread among odontocetes, but rare in mysticetes. Why is this so? The large size of mysticetes is clearly related to their feeding on abundant small fish and crustaceans. Their size also emancipates them from most predators that might threaten smaller odontocetes. There is, however, some overlap in size between odontocetes and mysticetes. Female sperm whales, which live in very stable groups, are similar in size (10 m) to minke whales and approach the size of other small mysticetes (the gray whale and Bryde's whale, *Balaenoptera edeni,* at about 14 m). Two ziphiids, the bottlenose whale (*Hyperoodon* spp.: 10 m) and Baird's beaked whale (*Berardius bairdii:* 13 m), are also in the size range of small mysticetes. The social system of the northern bottlenose whale, *Hyperoodon ampulatus,* resembles that of bottlenose dolphins, with strong male-male bonds and looser associations with some preferred companions among females (Gowans 1999). Baird's beaked whales are typically seen in groups, but whether these involve long-term stable bonds is unknown. Predation pressure should be highest on newborn whales. Here again we find an overlap in sizes: newborn sperm whales (4.0 m) are larger than minke calves (2.4–2.8 m) and slightly smaller than newborn humpback (4.5–5.0 m) and gray whales (4.9 m) (from Leatherwood et al. 1983). Thus, there appears to be no intrinsic difference in the vulnerability to predation of sperm whale calves and baleen whale calves.

Predation pressure is thought to be the primary factor underlying sperm whale social organization (Whitehead and Weilgart, chap. 6, this volume). Sperm whale calves are left near the surface when the mother dives to feed, as apparently they are not born with her diving ability. The threat of predation on sperm whale calves is thought to favor cooperative, stable social bonds among related females (Whitehead and Weilgart, chap. 6, this volume). There is evidence, however, to suggest that calves of smaller baleen whales born in tropical waters are also vulnerable to predation. In addition to the evidence of killer whale attacks on newborn humpback whales mentioned earlier, scars on newborn right whales in South Australia indicate that they are attacked by white sharks (Burnell, personal communication).

Having to leave their calves at the surface while feeding may not be the only factor that favors social living in large, deep-diving odontocetes. Differences in feeding competition between odontocetes and mysticetes may also be important (Gaskin 1982). In the Galápagos, female sperm whales feed at depths of 400–600 m on squid averaging 1 kg (Whitehead and Weilgart, chap. 6, this volume). If these squid are distributed in a temporally consistent fashion (uniform or very large patches capable of supporting twenty—spread-out—sperm whales during feeding bouts) that produces less feeding competition than the prey of baleen whales, sperm whales may be more able to afford to live in groups. It is also worth noting that even if sperm whale food were distributed in a continuous fashion that would otherwise favor continuous feeding (and perhaps solitary living: see above), sperm whales are "forced" into a bout feeding mode because the requirement to breathe necessitates a 1 km or more round trip to the surface as well as a period of surface "recovery."

A comparative study of feeding behavior and group size in Bryde's whales and fin whales *(Balaenoptera physalus)* suggests that feeding competition selects against the formation of stable groups in most baleen whales (Tershy 1992). The smaller Bryde's whale, which feeds primarily on fish, is more solitary than the euphausiid-eating fin whale. If predation risk were the most important factor, we would expect larger groups in the smaller, and presumably more vulnerable, Bryde's whale. Tershy (1992) suggests that baleen whales follow a pattern common to many terrestrial mammalian taxa in which smaller-bodied species feed on smaller patches of higher-quality food than larger species. Greater feeding competition in smaller patches prevents group formation in smaller-bodied baleen whales.

Although baleen whales appear to lack the stable social groups that are common among odontocetes, several observations suggest that long-term bonds might be more common in baleen whales than is currently thought to be the case. First, the majority of both sexes exhibit geographic philopatry (Clapham, chap. 7, this volume). Second, like odontocetes, large baleen whales enjoy relatively low costs of locomotion. Thus, even though stable groups may be disfavored, individuals can maintain fairly ready access to their maternal kin should conditions favor kin group formation. Third, baleen whales are thought capable of communicating over long distances; if vocalizations are individually distinctive, it should be possible for maternal kin to reunite after long-distance separations. In sum, social relationships may be expressed over a much larger spatial scale in large whales. We might, for example, expect to find kin bonds expressed during gray whale migrations from the breeding areas when females have new calves and are threatened by killer whales (e.g., Goley and Straley 1994). If individuals benefit from assisting each other to reduce the risk from predation (or from males, for that matter), then adult maternal kin may associate during migration. This assumes that feeding competition is unimportant during the return migration, an open question for this and other species (e.g., Swartz 1986). No data exist on associations during the whale migration or in the breeding lagoons, but some behavioral observations are suggestive. Four gray whales were observed supporting a fifth individual off southern California (Norris and Prescott 1961). On two occasions, adult gray whales were observed to assist a mother in freeing a stranded calf in a Mexican breeding lagoon (Swartz 1986). Assuming the adults to be unrelated, Swartz (1986) offered the behavior as a possible case of reciprocal altruism. It is quite possible, however, that the two adults were kin.

Brown and Corkeron (1995) reported that migrating humpback whale groups were larger during the return migration, when females had new calves. Groups of a single male and female, or two females, predominated during the southward migration. They suggested that females with calves might travel with an escort to increase the protection for their calves, and that males might accompany fertile females as a kind a mate guarding (Brown and Corkeron 1995). Again, it is possible that escorts might be older sons or daughters (see also Connor et al., chap. 10, this volume). The stable feeding associations that sometimes form among humpback whales in the northeastern Pacific also suggest longer-term relationships that are exhibited during times when food availability favors group feeding. Observations that individuals maintain particular positions during group feeding, possibly reflecting dominance relationships, are

especially provocative in this regard (Perry et al. 1990; D'Vincent et al. 1985).

Another possible factor differentiating odontocetes and mysticetes is the potential role that groups and social bonds may play in reducing the risk of infanticide. Recently, van Schaik and Kappeler (1997) argued that infanticide risk may explain the widespread occurrence in primates of year-round association between males and females. Female groups and year-round association between males and females may be common features of odontocete social organization (Whitehead and Mann, chap. 9; Connor et al., chap. 10, this volume). Further, two factors that are thought to favor infanticide (see Connor et al., chap. 10, this volume), a diffuse as opposed to a short, constrained breeding season and a lactation period that exceeds the duration of gestation, are characteristic of odontocetes, but not mysticetes.

Conclusions

Perhaps the most striking aspect of cetacean group living reviewed here is the phenomenon of natal philopatry by both sexes in killer whales and possibly other species. Natal philopatry by both sexes spans a continuum from the natal group philopatry of *resident* killer whales, in which both sexes remain with their mother, to "geographic" philopatry, in which one or both sexes leave their mother but continue to use their natal range. I have suggested that the option for natal philopatry by both sexes is made available by relatively low travel costs, a resource distribution that favors large ranges, and the fact that cetacean offspring are "followers" so that adults are not restricted to a breeding site.

Whether cetaceans disperse from their mother or their natal range is undoubtedly influenced strongly by the distribution of food, including the extent to which group formation is favored by cooperative feeding or disfavored by feeding competition. Feeding competition is a neglected element in studies of cetacean societies. Many accounts exist of apparent cooperative feeding; these events are easier to observe than competitive interactions, especially given the methods typical of cetacean studies.

Perhaps no other group of mammals has evolved in an environment so devoid of refuges from predators. Many cetaceans, especially smaller open-ocean species, have nothing to hide behind but each other. We are only beginning to understand the myriad ways in which this fact has shaped the parameters of social lives in cetaceans. Behavioral studies of known individuals, a focus of this volume, will go some distance toward enlarging our understanding of group living in cetaceans, as will studies on the distribution and habits of major predator and prey species.

9

FEMALE REPRODUCTIVE STRATEGIES OF CETACEANS

Life Histories and Calf Care

HAL WHITEHEAD AND JANET MANN

The newborn dolphin surfaces like a cork next to his mother. The deep creases, called fetal folds, and wobbly dorsal fin indicate that he has been born in the last day or so. His mother breathes and dives smoothly, but her new calf pops up suddenly, hesitates, breathes. He waits, bobbing slightly and flicking his tail on the surface, as if trying to figure out how it works. The bubble train from his blowhole and the overlapping whistles we hear through the hydrophones indicate that he and his mother are whistling. The calf, although newborn, is quite large—about 40% of his mother's length—and can follow the mother in his own awkward way, except when she dives deep to hunt. The mother surfaces again, and the calf moves alongside her. The mother slows, and the calf swims under her, nudging her mammaries to nurse. The calf is precocious in his motor skills and will be able to catch fish within six months. However, it will be at least four years before a younger sibling is born and he nurses no more.

THE PRODUCTION and care of young is a fundamental element of mammalian societies and the vigor of their populations. In most mammals, including cetaceans, only females actively parent (Clutton-Brock 1989b; Kleiman and Malcolm 1981). The female whale or dolphin, in her role as a mother, is the linchpin of cetacean population and behavioral biology. Thus an understanding of the reproductive behavior (in the broadest sense) of female cetaceans is crucial if we are to develop useful models of social evolution, socioecology, social behavior, or population dynamics.

In this chapter we use the adaptive paradigm to examine those areas of cetacean natural history that are largely specific to females: the timing of reproductive events, attributes of the newborn, calf development, maternal and infant behavior, lactation, and some aspects of female sociality and mating systems. In general, reproductive parameters (such as age at maturity and litter size) seem to be well adapted to an animal's environment, in the sense that changes in their values lower fitness. This is indicated by successful prediction of the values of some of these parameters by optimality models (e.g., Stearns and Koella 1986) as well as by experimental manipulations (e.g., Reznick and Bryga 1987). In species, such as cetaceans, with low adult mortality and fecundity but high infant mortality, we might expect other female traits that influence neonatal survival, such as nursing behavior and mother-infant communication, also to be closely related to maternal fitness, and so to be well adapted.

We therefore think it reasonable to examine the reproductive strategies of female cetaceans from an adaptationist perspective, while bearing in mind that there are reasons why traits may not be adaptive (Gould and Lewontin 1979). However, we neither test the hypothesis that the traits we consider are adapted to the animals' environment (e.g., by using the comparative method) nor use optimality models or techniques such as Charnov and Berrigan's (1990) dimensionless numbers to examine life history evolution in a rigorous quantitative manner. The principal reason for holding back from such endeavors, which can be so revealing with other taxonomic groups (e.g., Stearns 1992), is the poor quality of data on cetacean life histories (see below).

Instead, in each of the areas covered, we will give an overview of the range of behavior of female cetaceans and discuss its apparent adaptive significance. Our general procedure is to look for relationships between measures of the traits and the animals' environments, using theory developed for terrestrial mammals where appropriate, and suggest adaptive explanations for the patterns that emerge. We will principally examine differences between species and between the two suborders (mysticetes and odontocetes) of Cetacea, but some comparisons at other taxonomic levels (within species, between genera, and with noncetacean mammals) are also used. We also present correlations among measures of life history parameters, try to classify female cetaceans on the basis of their reproductive behavior, and compare their reproductive strategies to those of other large mammals.

Measuring Female Reproductive Parameters

To provide a general picture of female reproductive parameters, we compiled life history, environmental, and grouping data for the best-studied cetaceans (table 9.1). The general methods used to make these estimates, and their limitations, are described briefly in box 9.1.

Much of the information in table 9.1 should be viewed very cautiously, for the following reasons:

1. Some estimates, such as those for the river dolphins, are based on only a few specimens, which may have been acquired in a nonrandom way (e.g., strandings).
2. The methods used to derive some of the measures, especially aging techniques for mysticetes, have not been fully verified and could be biased by as much as a factor of two (see box 9.1).
3. Different methods, with different biases, have been used to estimate the same measure for different groups of species (e.g., interbirth intervals from longitudinal studies of identified living individuals and from age and corpora analyses of dead specimens).
4. Some measures, such as group size and habitat type, are not well defined and do not come from rigorous sampling procedures.

Almost all the measures show variation among populations, which in some cases is considerable. For instance, the mean adult size of female bottlenose dolphins (*Tursiops* spp.) ranges from 2 m in Shark Bay, Australia, to 4 m in the Moray Firth, Scotland (Connor et al., chap. 4, this volume), and some populations of spinner dolphins (*Stenella longirostris*) live in coral reef lagoons, whereas others inhabit the deep ocean (Norris 1991b).

Many areas of cetacean systematics remain unresolved. For the purposes of the analyses and discussion in this chapter, in which we often make comparisons between and within the two suborders of Cetacea, the most serious uncertainty concerns the taxonomic position of the sperm whales (family Physeteroidea). Some molecular analyses suggest that sperm whales are more closely related to the baleen whales than to the other odontocetes (Milinkovitch et al. 1993). However, since the sperm whales have teeth and much more closely resemble other larger odontocetes than the mysticetes morphologically (Heyning 1997), ecologically (they eat large animals singly), and in most of their life history parameters, in this chapter we continue to consider them odontocetes.

Because of these problems, table 9.1 should be viewed with care. In this chapter, we present general results derived from these data only when the effects are strong and clear, and try to use as examples results from large, reliable studies. Later in the chapter we take a broader view, comparing the ranges of female life history variables of cetacean superfamilies with those of other large mammal families and superfamilies (see table 9.3). Many of the same cautions about data quality apply to those discussions.

Reproductive Schedules

Sexual Maturity

Female cetaceans start to reproduce at three to fifteen years of age, when at about 85–95% of their mean adult body length. There is no obvious trend for odontocetes to start to reproduce at smaller or larger relative sizes than mysticetes (medians of 90.5% for both odontocetes and mysticetes), or for relative size at first reproduction to be related to the size of the species (table 9.2). In long-finned pilot whales (*Globicephala melas*), attainment of a certain length, rather than a particular age, seems to trigger first reproduction (Martin and Rothery 1993).

Presumably, reproduction when smaller than about 90% of mean adult body size is sufficiently inefficient that the potential benefits of offspring produced, whose survival may be very low, are outweighed by the demands on the mother's growth and the effects on her future offspring

(Reznick et al. 1990; Reiter and Le Boeuf 1991). Evidence for this conclusion includes the low success of first offspring. In mammals, first-born offspring are often smaller and more likely to die than later-born offspring (Clutton-Brock 1984). The proximate causes may be physiological or relate to maternal inexperience in parenting. In captive bottlenose dolphins, first-born offspring have higher infant mortality than later-born offspring (e.g., Tavolga and Essapian 1957; Cornell et al. 1987; Owen 1990; Connor et al., chap. 4, this volume). Unfortunately, in almost all other cetacean species, data are insufficient to compare offspring survivorship based on parity.

Age at first reproduction varies widely in odontocetes, both between and within species. Harbor porpoises (*Phocoena phocoena*) in the Bay of Fundy become mature at three or four years of age (Gaskin et al. 1984), whereas female killer whales (*Orcinus orca*) off British Columbia give birth to their first viable calf at a mean age of fifteen years (Olesiuk et al. 1990), although this varied from twelve to eighteen years. In Kasuya and Marsh's (1984) sample of short-finned pilot whales (*Globicephala macrorhynchus*), the youngest mature female (indicated by presence of corpora scars on the ovaries) was eight years of age, and the oldest immature was twelve years of age. In Shark Bay, Australia, bottlenose dolphin age at first birth ranges from twelve to fifteen years (Mann et al. 2000), but in Sarasota, Florida, females as young as six or seven years old have given birth (Wells et al. 1987; R. S. Wells, unpublished data).

Most mysticete females in unexploited populations are believed to mature at about ten years of age. However, this is often reduced for exploited populations, in which females presumably have increased access to food (Lockyer 1984). In addition, some of the best mysticete data on age at first birth indicate that humpback whale (*Megaptera novaeangliae*) females have their first calf at approximately five years of age (see Clapham, chap. 7, this volume). Disagreements concerning aging techniques for mysticetes remain to be resolved.

It is likely that most variation in age at first reproduction for female cetaceans is the result of differences in growth rates: some species, populations, or individuals grow faster, reach about 90% of their asymptotic size more quickly, and mature earlier. However, growing and surviving may not be the only functionally important activities during the juvenile-adolescent period. "Practice" in mating and (for species with intensive and extensive parental investment) parenting behavior during the juvenile and adolescent periods may contribute significantly to female reproductive success (e.g., primates: Fairbanks 1990; Epple 1978). Although such activities may be very important (see below), there is no indication in the cetacean data that they actually lengthen the period of sexual immaturity: the relative size at sexual maturity is slightly smaller for the slower-growing, and more social (see below), species (correlation between relative size at sexual maturity and age at sexual maturity, $r = -0.12$).

Seasonal Reproduction

Seasonal peaks in ovulation and birth have been reported among many cetacean species. These peaks are pronounced in the baleen whales (e.g., Mackintosh and Wheeler 1929), belugas (*Delphinapterus leucas*: Brodie 1989), harbor porpoises (Read and Hohn 1995) and botos—Amazon river dolphins (*Inia geoffrensis*: Best and da Silva 1989). In these populations the majority of births take place within a period just a few weeks long. Seasonal reproduction is least obvious in the offshore tropical odontocetes.

There are several reasons why particular seasons may be most suitable for birth or conception:

1. If food availability is seasonal, then females may have reproductive schedules arranged so that peak periods of food availability coincide with the times of greatest energy demand, especially mid-lactation (Oftedal 1984), before offspring begin to feed themselves consistently. However, depending on the energetics and physiology of gestation and lactation, lags between peak food availability and peak energy demands may be possible, or optimal, for the larger cetaceans.

2. Fasting in warm water during early and peak lactation may be more energetically efficient than feeding in colder water (as in hibernating mammals and phocids that fast during the first stages of lactation: Oftedal and Iverson 1995). Fasting in warm water may aid conversion of maternal body fat to high-fat milk and hence contribute to rapid calf growth.

3. Availability of easily caught prey at the time of weaning may be important for offspring (Kasuya 1995).

4. Newborn animals may have increased survival, and/or lower energy expenditure in warmer, calmer, or less predator-infested waters. In such cases females may give birth at those times of year that are warmer and calmer (usually summer), or when migrations to warmer and calmer waters result in less net energy loss for the mother (usually winter) (Brodie 1975; Lockyer 1987).

5. In some species, animals may congregate during certain periods of the year because of seasonal prey concentrations (northern right whales, *Balaena glacialis*: Gaskin 1987) or limited calving habitat (e.g., humpback whales: Whitehead and Moore 1982). Females that show estrus during these periods will increase the probability of finding a mate and have the greatest range of potential mates.

Table 9.1. Life history and other attributes of female Cetacea

Species	Common name[a]	Abbreviation[b]	Length[c]	Mass[d]	Birth length[e]	Birth mass[f]	First-year growth[g]	Maturity[h]
Mysticetes								
Balaena mysticetus	Bowhead W	Bmy	15	90,000	4.2		4	15
Balaenoptera acutorostrata	Minke W	Ba	8.8	6,600	2.8	200	5.5	8
Balaenoptera borealis	Sei W	Bbo	15	19,000	4.5	900		7
Balaenoptera edeni	Bryde's W	Be	13	12,000	3.4	900		11
Balaenoptera musculus	Blue W	Bmu	26.2	105,000	6.5	7,250	20	10
Balaenoptera physalus	Fin W	Bp	22	50,000	6.4	1,900	12	11
Caperea marginata	Pygmy right W	Cm	6.2	3,200	1.8			
Eschrichtius robustus	Gray W	Er	14.1	31,466	4.6	500		9
Balaena spp.	Right W	Eu	17.3	54,500	5.5			10
Megaptera novaeangliae	Humpback W	Mn	13.7	35,000	4.5	2,000	8	5.5
Odontocetes								
Berardius bairdii	Baird's beaked W	Bba	11	9,000	4.5			10
Cephalorhynchus eutropia	Chilean D	Ce	1.5	44				
Cephalorhynchus commersonii	Commerson's D	Cc	1.7	79	0.6	5	0.42	6
Cephalorhynchus heavisidii	Heavisides's D	Chv	1.7	72	0.8	9.5		
Cephalorhynchus hectori	Hector's D	Chc	1.3	41	0.6			8
Delphinapterus leucas	Beluga	Dl	3.6	400	1.6	79	0.56	7
Delphinus delphis	Common D	Dd	2.1	80	0.8			6.5
Globicephala melas	Long-finned pilot W	Gme	4.3	890	1.8	75	0.71	7
Globicephala macrorhynchus	Short-finned pilot W	Gma	3.6		1.9	60		10
Grampus griseus	Risso's D	Gg	3.3	320	1.3			12
Hyperoodon ampullatus	N. bottlenose W	Ha	6.9	3,000	3.5			
Inia geoffrensis	Boto	Ig	1.9	81	0.8	7.5	0.21	
Kogia breviceps	Pygmy sperm W	Kb	3.4	408	1.2	55		
Kogia sima	Dwarf sperm W	Ks	2.7	272	0.9	46		
Lagenorhynchus obliquidens	Pacific white-sided D	Lo	1.9	90	0.9	15		10
Lagenorhynchus acutus	Atlantic white-sided D	Lac	2.2	180	1.1	30	0.34	
Lagenorhynchus albirostris	White-beaked D	Lal	2.6	230	1.2	40		
Lagenorhynchus obscurus	Dusky D	Lad	1.9	70	0.9	10		7
Lipotes vexillifer	Baiji	Lv	2.3	120	1.1	5		
Monodon monoceros	Narwhal	Mm	4.1	1,000	1.6	80		7.5
Neophocoena phocaenoides	Finless P	Np	1.6	35	0.8			5.5
Orcaella brevirostris	Irrawaddy D	Ob	2.1	85	0.9	12.3		
Orcinus orca	Killer W	Oo	6	3,000	2.4			15
Phocoena phocoena	Harbor P	Pp	1.6	65	0.7	5	0.55	4.5
Phocoenoides dalli	Dall's P	Pd	1.9	123	1	25		5.5
Physeter macrocephalus	Sperm W	Pm	11	20,000	4	1016	2.1	11.2
Platanista gangetica	Susu	Pg	2.2	69	0.9			
Pontoporia blainvillei	Franciscana	Pb	1.5	41	0.7	7.9		3.6
Pseudorca crassidens	False killer W	Pc	4.5	700	1.7	80		12
Sotalia fluviatilis	Tucuxi	Sfl	1.7	36	0.7		0.36	
Sousa spp.	Humpbacked D	So	2.1	85	1			
Stenella coeruleoalba	Striped D	Sco	2.2	131	1		0.66	10.7
Stenella clymene	Clymene D	Scl	1.8	62				
Stenella attenuata	Spotted D	Sa	2.0	100	0.8		0.57	11
Stenella frontalis	Atlantic spotted D	Sfr	2.0	110	1			
Stenella longirostris	Spinner D	Slo	1.7	60	0.7		0.56	6.5
Steno bredanensis	Rough-toothed D	Sbr	2.3	122	1			11
Tursiops spp.	Bottlenose D	Tu	3.0	200	1.1	32		10
Ziphius cavirostris	Cuvier's beaked W	Zc	5.8	3,000	2.7			

Sources (in order of preference):

1. Evans (1987) for group size and peak calving season
2. Burns et al. (1993) for bowheads; Horwood (1987) for minke whales; Horwood (1990) for sei whales
3. *Handbook of Marine Mammals*, vols. 3, 4, 5, and 6 (Ridgway and Harrison 1985, 1989, 1994, 1999)
4. *Reproduction in whales, dolphins and porpoises: Proceedings of the conference, Cetacean Reproduction, Estimating Parameters for Stock Assessment and Management, La Jolla, CA, 28 Nov.–7 Dec. 1981*, ed. W. F. Perrin, R. L. Brownell, and D. P. DeMaster (1984). Reports of the International Whaling Commission, special issue 6.
5. Evans (1987) for measures other than group size and peak calving season
6. Kasuya (1995)
7. Watson (1981): distribution charts used for mean latitude, masses, and lengths—no life history measures were taken from this source (an outlying and unlikely birth mass for *Sousa* from this source was omitted)

Note: Taxonomy as in *Handbook of Marine Mammals* (Ridgway and Harrison 1985, 1989, 1994, 1999). Data are presented for those species for which the most complete data exist. Data are presented for genera when taxonomy is very uncertain and/or data are few (for example, *Sousa, Balaena, Tursiops*). Species with few good data are excluded (e.g., most beaked whales; Fraser's dolphin, *Lagenodelphis hosei*). When data were available for more than one population of a species, information for the population with the largest-sized animals was used.

Length at maturity[i]	Gestation[j]	Lactation[k]	Interbirth interval[l]	Mortality[m]	Longevity[n]	Peak calving season[o]	Group size[p]	Mean latitude[q]	Habitat[r]
13.2	13.5	10.5	3.5			2	5	70	5
8.1	10	6	1.2	0.10	45	2	2	50	5
14	14.5	6	2	0.07	60	2	2	40	5
12	12	6	2			12	2	20	5
24	10.5	7	2.5		85	1	1	40	5
19.9	11.3	6.5	2	0.05	95	2	5	50	5
	12	5.5					4	42	5
11.7	13.5	7	2	0.06		2	9	45	4
14.5	10	6.5	3			3	6	50	5
12.1	11.3	5	2			2	1	35	5
10.5	17		3		71	2	11	48	7
							10	44	3
1.6	12				18	3	6	50	3
1.6								27	3
				0.15		3	4	43	3
3	14.2	22	3	0.13	32	2	12	65	4
2	9.5	6	2		20	4	255	35	5
3.6	15	24	3.3	0.07	59	2	30	45	6
3.9	15	24	5	0.10	63	2	27	20	6
2.8						3	15	25	7
6	12		2		27	2	5	65	7
1.7	10.5	10	1.25		28	3	8	5	1
3	9	12			28		3	30	7
2.1							5	32	6
1.8	11					3	55	45	7
2.1	11	18	2.5		27	4	55	55	5
	10					4	10	60	5
1.7	12	12	2.5			3	10	45	5
	10					3	2	30	1
3.4	15.3	18	3	0.11	52	2	11	70	4
1.2	11	10	2		23	2	6	5	2
					30	3	6	12	2
5	16	18	5	0.02	85	4	20	45	5
1.4	10.5	10	1		20	1	5	55	3
1.9	11	24	2		35	12	10	55	6
8.7	14.5	24	5	0.06	65	3	25	20	7
	10.5		2		18		5	28	1
1.4	10.8	9	2		13	2	2	30	3
3.9	15.5	21		0.06	63	2	30	20	6
1.6	10.2					2	3	10	2
						3	5	20	3
2.2	12	13	4		57.5	5	152	35	7
							102	20	7
1.8	11.4	18	3	0.10	46	5	252	20	6
1.9							252	28	5
1.7	10.7	18	3		20	4	125	15	5
2.2					32		25	15	7
2.3	12	19	1.4	0.02	45	3	12	30	4
5.3					30		5	35	7

[a]W, whale; P, porpoise; D, dolphin
[b]Abbreviation used in figures in this chapter
[c]Median length of adult female (m)
[d]Median mass of adult female (kg)
[e]Median birth length (m)
[f]Median birth mass (kg)
[g]Growth (m) during first year of life
[h]Mean age at first birth for female in unexploited population (years) (where possible; sometimes age at first conception plus gestation)
[i]Mean length at female sexual maturity (m)
[j]Mean gestation period (months)
[k]Mean lactation period (months)
[l]Modal interbirth interval (years)
[m]Mean adult female mortality per year
[n]Estimate of maximum female longevity (years)
[o]Estimate of duration of peak calving season in months (from Evans 1987: 208–211)
[p]Midpoint (rounded down) of the ranges of "typical group sizes" (from Evans 1987: 161, 163–166); generally overestimates modal group size
[q]Approximate mean absolute latitude in degrees
[r]1, river; 2, river and inshore; 3, inshore; 4, inshore and shelf; 5, shelf; 6, shelf and deep water; 7, deep water. Where a species spans a range of these habitats (e.g., spinner dolphins, humpback whales), the median habitat is selected.

Box 9.1

Methods of Obtaining Life History Data from Female Cetaceans

There are two general sources of information on cetacean life history measures: studies of carcasses provided by the whaling industry (important for baleen whales, sperm whales, and pilot whales), incidental capture (important for many dolphin species), stranding (especially important for beaked whales and other rare species), or scientific kills (minor significance for sperm and minke whales); and long-term studies of individually identified living animals (important for bottlenose dolphins, killer whales, humpback whales, gray whales, and right whales) in the wild or in captivity. The principal methods used to obtain the measures given in table 9.1 include the following.

Mass

Most mass estimates come from dead animals, after either intentional or incidental capture or stranding. Weighing large whales is logistically difficult, and is rarely done. For these animals, there may be biases caused by selection of particularly large or small specimens, as well as by blood loss during dismemberment before weighing. Many species, but especially the baleen whales, show pronounced seasonal patterns of mass loss and gain.

Length

Length is probably the most accurate measure of those given in table 9.1. It is usually taken from dead animals, but can also be derived from aerial or shipboard photography (e.g., Whitehead and Payne 1981; Gordon 1990; Withrow and Angliss 1992). In earlier times, it was not clear precisely what was meant by "length," but this is now universally standardized as the snout tip to fluke notch measurement.

Aging

A method of aging animals is needed to estimate age at sexual maturity and longevity, as well as for some methods of calculating mortality rates and interbirth intervals. In a few studies of individually identified animals, the birth dates of many members of the population are known or can be estimated with some accuracy (e.g.,

Olesiuk et al. 1990 for killer whales), allowing precise age determination.

In studies of carcasses, age is usually determined from the number of layers in some body part. In baleen whales it has been customary to use laminations in the waxy ear plug (Purves and Mountford 1959), whereas among odontocetes, teeth are sectioned (Perrin and Myrick 1980); this is also sometimes possible with living wild animals from whom a tooth is removed during temporary restraint (Hohn 1980a). However, there are still uncertainties for some species (e.g., belugas: Brodie 1989) concerning the rates at which the layers are laid down.

Size at Birth

Length and mass at birth is usually determined by comparing the sizes of the largest fetuses and smallest calves. For some species accurate sizes of captive newborns are available, although captivity itself could bias these measures.

Reproductive Rates

Information on the reproductive status of females is important in a number of ways. It allows an estimate of the age and length at sexual maturity, the interbirth interval, and the gestation period. It also can indicate adolescent sterility and reproductive senescence. These data can be either longitudinal (the reproductive histories of individual females) or cross-sectional (the distribution of reproductive states among a population of females, perhaps indexed by other measures such as age or length).

Accurate longitudinal data are the most revealing and least biased type of information, but this has been achieved in only a few remarkable long-term studies of individually identified animals, principally killer whales off Vancouver Island (Olesiuk et al. 1990), bottlenose dolphins near Sarasota, Florida, and in Shark Bay, Australia (Connor et al., chap. 4, this volume), humpback whales of the Gulf of Maine (Clapham, chap. 7, this volume), right whales off Peninsula Valdés, Argentina (Payne 1986), and gray whales of the eastern North Pacific (Jones 1990). In other studies of living animals, cross-sectional techniques have been used to estimate reproductive rates (e.g., Whitehead 1990a for sperm whales; Miller et al. 1992 for bowheads).

In almost all studies of living cetaceans in the wild, female reproduction has been measured by the presence of a living calf—miscarriage and neonatal mortality are not recorded. Exceptions are the studies of bottlenose dolphins near Sarasota, Florida, where pregnancy is detected using ultrasound on temporarily restrained animals (Wells et al. 1987; R. S. Wells, unpublished data), and in Shark Bay, Australia, where advanced pregnancy can be observed in free-living animals (J. Mann, personal observation; fig. 9.18).

During whaling operations, females were examined for the presence of a fetus, which, unfortunately, was sometimes missed if small (Best 1989). The ovaries of an individual animal also give a useful condensed reproductive history through the presence and number of the different corpora (International Whaling Commission 1984b). Thus, studies of harvested animals often included both cross-sectional and longitudinal data. However, because of uncertainties in aging methods (see above) and in interpreting information on corpora (International Whaling Commission 1984b), these longitudinal data are not of the same quality as those from the most extensive studies of identified living animals (e.g., Olesiuk et al. 1990 for killer whales).

It is important to emphasize that the methods based on internal examination of dead females examine pregnancies, while those using observation of living individuals or populations count successful live births, and in some cases, only those that survive to a few months of age. As perinatal mortality is often high in cetaceans, these can be very different rates.

Pronounced seasonal availability is characteristic of mysticete prey, either because the prey are short-lived, so that their usable biomass is closely tied to seasonal primary productivity (e.g., copepods), or because their schooling or migratory behavior varies seasonally, as with capelin (*Mallotus villosus*: Department of Fisheries and Oceans, Canada 1991). The seasonality of birth (reason 1 above) and weaning (reason 3), as well as the migrations (reason 4), of most female baleen whales, are closely related to this seasonal availability of prey (Evans 1987; Kasuya 1995).

With about half a year of plenty followed by six months of starvation, female mysticetes obtain the requirements of

Table 9.2. Pearson correlation coefficients between life history variables of female cetaceans (using species as units; pairwise deletion)

		L	BL/L	BM/M	LA/IBI	LONG	LA	MA	GE	IBI	PCS	LM/L
Adult length	L											
Birth length/ adult length	BL/L	−0.57										
Birth mass/adult mass	BM/M	−0.60	0.55									
Lactation/ interbirth interval	LA/IBI	−0.56	0.29	0.58								
Longevity	LONG	0.73	−0.28	−0.54	−0.40							
Lactation	LA	−0.48	0.55	0.52	0.56	0.09						
Maturity	MA	0.34	−0.28	−0.27	−0.31	0.68	0.13					
Gestation	GE	0.14	0.05	−0.16	−0.06	0.57	0.53	0.39				
Interbirth interval	IBI	0.02	0.25	0.00	−0.31	0.49	0.57	0.53	0.62			
Peak calving season	PCS	−0.14	0.04	0.26	0.19	−0.01	0.25	0.16	0.03	−0.02		
Length maturity/ adult length	LM/L	−0.13	0.27	0.08	−0.13	−0.06	0.01	−0.12	−0.14	0.13	0.14	
First-year growth/adult length	NG/L	0.74	−0.28	−0.57	−0.59	0.56	−0.74	−0.02	−0.46	−0.36	−0.40	0.14

pregnancy and early lactation in one feeding season, seem to minimize the energy demands of the intervening period (by migrating to less energy-demanding waters for birth and migrating back to take full advantage of the start of productivity as soon as possible: see Clapham, chap. 7, this volume), and wean their offspring during the summer of the first year, when food is most available for the offspring. Thus strong environmental cyclicity appears to constrain the female mysticete's reproductive schedule. However, the females of the tropical Bryde's whale *(Balaenoptera edeni),* which live in a much less seasonal environment than most other mysticetes, have gestation and lactation periods similar to those of other mysticetes, but much less of a seasonal breeding period.

In humpback and gray whales *(Eschrichtius robustus),* whose seasonal calving migrations take them to very specific places, conceptions also occur at such times, suggesting seasonal conspecific aggregation as a function for breeding synchrony (reason 5). However, as gestation in these species is about one year and most interbirth intervals are two or more years, the primary aggregation is of nonconceptive calving females. So there is a puzzle as to why males and nonpregnant females should migrate to these specific calving grounds. This question is discussed further by Clapham in chapter 7 of this volume.

In contrast to that of mysticetes, odontocete food is very varied. In some cases there is pronounced seasonal variability in the availability of prey: seasonally breeding pinnipeds or migrating salmon for killer whales (Baird, chap. 5, this volume); annual species of squid such as *Illex illecebrosus* for pilot whales (Mercer 1975); fish whose abundance is determined by river levels, which vary seasonally, for botos (Best and da Silva 1989). In these species reproduction may be related to the seasonal availability of these foods, either for the mother (reason 1) or the weaning calf (reason 3). However, many odontocetes eat animals that live for several years (such as flying fish or herring, *Clupea harengus*) and are not obviously more or less available or nutritious at different seasons. In these species we must look at alternative explanations for seasonal reproduction. For instance, small to medium-sized species at high latitudes, such as belugas and harbor porpoises, usually give birth in midsummer, when the waters are warmest and the small calves will lose the least energy to thermal regulation (reason 4).

Synchrony of Estrus

Females of several species of Cetacea seem to have estrous periods more closely synchronized than would be expected simply because of seasonal variation in environmental conditions. Among humpback (Clapham, chap. 7, this

volume) and gray whales (Rice and Wolman 1971; Jones et al. 1984) most females show estrus within an interval of just a few weeks. In sperm whales *(Physeter macrocephalus),* there is synchrony within social groups (Best and Butterworth 1980). However, in other social species, such as long-finned pilot whales, there is no evidence of estrus synchrony (Martin and Rothery 1993).

Why female cetaceans would synchronize their ovulations within a season is not well understood. Estrus synchrony might have several possible functions:

1. The "dilution effect": Synchronized ovulations would result in synchronized births. A calf may be relatively safer from predators if there are other, similar-sized calves nearby.
2. Reduction in male harassment (another form of the dilution effect), since there are fewer males vying for each female. We would expect synchronization to occur if male harassment is a problem for breeding females, especially if they have a postpartum estrus—as humpbacks sometimes do (Clapham, chap. 7, this volume). Reduction of harassment by males would improve the ability of a female humpback with a dependent calf to control pregnancy and care for her offspring during this vulnerable period.
3. Allomaternal care of calves, in which non-mothers behave in ways that benefit mothers or their calves (with possible benefits including vigilance for predators, defense against predators, allosuckling, and availability of playmates; see below), may be more efficient if calves within a group are of the same age. This is most likely to be important in species with permanent groups, such as pilot whales.
4. Reproductive synchrony may allow females to manipulate competitive asymmetries among immatures to the benefit of their own offspring (e.g., primates: Pereira 1995).
5. Alternatively, apparent synchrony of estrus within groups could be caused by their shared environment, and have no adaptive basis (Best and Butterworth 1980).

Gestation Period

Most cetacean species have gestation periods of ten to twelve months, fitting in with an annual seasonal cycle, which allows birth and conception at the same time of year, and the possibility of a postpartum estrus in highly seasonal populations. In the baleen whales and small odontocetes, gestation period is not obviously related to adult length (fig. 9.1). Extraordinary fetal growth rates (27 mm/day in the blue whale, *Balaenoptera musculus:* Kasuya 1995) allow the larger baleen whales to give birth to substantial calves

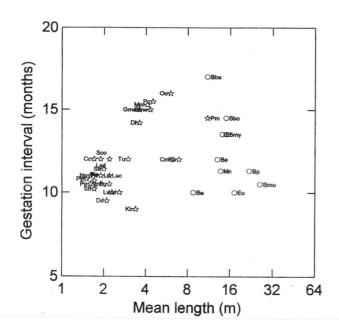

Figure 9.1. Gestation interval plotted against log of body length for odontocete (*) and mysticete (O) species. Species abbreviations are from table 9.1.

within a year of conception. However, as Kasuya (1995) notes, there are a few medium-sized to large toothed whale species, including the sperm whale, killer whale, and pilot whale, with gestation periods of over a year and relatively slow fetal growth rates. Most of the odontocetes lie close to a curve of increasing gestation period with size.

The larger odontocetes may be restricted in the rate at which they can transfer energy to their offspring, and so exceed the convenient annual gestation period. Alternatively, or perhaps additionally, these long gestations could be related to the larger brains found in these species. In mammals, gestation length is predicted better by neonatal brain size than by metabolic factors (Sacher and Staffeldt 1974), although this does not seem to be the case within the odontocetes (Marino 1997).

Three odontocete species have conspicuously short gestations for their sizes (see fig. 9.1). These are the two sperm whale species for which there are gestation data and the northern bottlenose whale *(Hyperoodon ampullatus)*, the only large (>4 m) odontocete with an estimated gestation period of less than fourteen months. These odontocete species with relatively short gestation periods do not have particularly small calves (fig. 9.2), but are deep divers, a potentially interesting pattern. However, some, or all, of these apparently short gestations may result from poor data.

Duration of Lactation

Duration of lactation is extremely variable among cetacean species, with reported weaning ages ranging from about six

months in the baleen whales to three to six years in bottlenose dolphins (Connor et al., chap. 4, this volume). Weaning age is also very variable among individuals within populations (e.g., short-finned pilot whales: Kasuya and Marsh 1984; long-finned pilot whales: Martin and Rothery 1993). Most of the methods used to determine the duration of lactation are problematic (see box 9.1; Oftedal 1997). Thus the estimates of duration of lactation in table 9.1 are likely to be underestimates and should be viewed especially cautiously.

However, there are some very pronounced trends. The two cetacean suborders are clearly separated (Oftedal 1997). In baleen whales, lactation generally lasts about six or seven months—calves are weaned during the summer months of high food abundance—and there is no clear relationship of lactation duration to body size (fig. 9.3). In contrast, odontocetes generally wean their calves after eight to twenty-four months, with larger species suckling longer and all the larger odontocetes nursing for at least a year (fig. 9.4). Over and above the estimated species means, there are some records of odontocetes apparently suckling for up to thirteen years (sperm whales: Best et al. 1984, although see Oftedal 1997 for cautions about the methods employed in this study).

There are two weaning strategies used by mammalian offspring and mothers. With gradual weaning, infants can acquire necessary foraging skills before they are completely dependent on solid foods. With sudden weaning, as in the

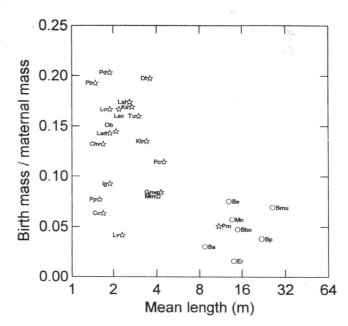

Figure 9.2. Birth mass as a proportion of maternal mass, plotted against log of body length for odontocete (*) and mysticete (O) species.

Figure 9.3. Gray whale mother and calf in Bahia Magdalena, Mexico. Baleen whale calves frequently swim alongside their mothers. Delphinid calves tend to swim under the mother. (Photograph by Robin Baird.)

phocid seals (Bonner 1984), mothers must sufficiently fatten their offspring so that they can survive the sudden transition from total dependence on milk to total dependence on solid food. The breeding-feeding cycles of most mysticetes and the high fat content of their milk suggest that ingestion of solid food overlaps with nursing only briefly. This sudden weaning strategy is appropriate when calving and feeding areas are spatially separated (Oftedal 1997). In contrast, like the otariid seals—fur seals and sea lions—odontocetes wean their offspring gradually (Oftedal 1997). For example, bottlenose dolphin calves begin successfully catching and consuming small fish by four to six months of age, even though they are likely to continue nursing for at least another three years (Mann and Smuts 1999; fig. 9.5). Such a strategy may be necessary if prey are hard to catch.

Cetacean milk has more fat than does milk from all other mammalian taxa except the pinnipeds (Oftedal and Iverson 1995; see table 9.3). In general, the higher the fat content of the milk, the earlier weaning tends to take place (Bonner 1984). Mysticete milk is higher in fat (30–53% at about mid-lactation) than odontocete milk (10–30% at about mid–lactation: Oftedal 1997). The costs of lactation can be measured indirectly by correlating female blubber thickness with reproductive state. Reliably, blubber layers are much thinner in lactating than in pregnant or "resting" (nonpregnant, nonlactating) baleen females (Lockyer

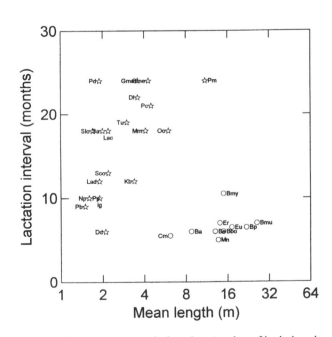

Figure 9.4. Lactation interval plotted against log of body length for odontocete (*) and mysticete (○) species.

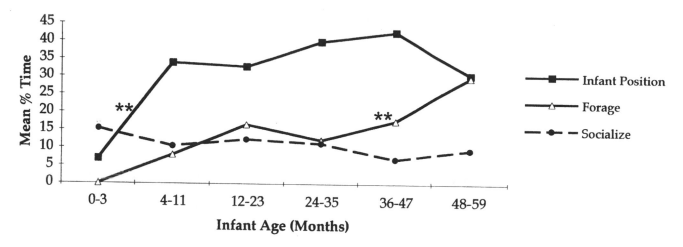

Figure 9.5. Developmental changes in infant activity (proportion of time spent socializing, foraging, and in "infant position"—a measure of infant dependence since all nursing occurs from infant position) from birth to weaning for twenty-eight infant bottlenose dolphins in Shark Bay, Australia. Infants were observed for a total of 962 hours. Sample sizes ranged from four to twenty-one infants per age class. Standard deviations were always less than half the mean value. Statistically significant changes in activity budgets between age intervals are marked with **. Infants significantly increased infant position swimming during the newborn period (Mann-Whitney U Test, $p < .01$). There were no developmental changes in time spent socializing. Weaned infants were excluded. Calves significantly increased foraging at all ages.

1981b; Rice and Wolman 1971). A pregnant blue whale at 119,000 kg is thought to store 45,000 kg of blubber or fat, most of which will be converted into milk for a calf that gains 17,000 kg in mass before weaning (Lockyer 1981b). Lactation output can be estimated by mammary gland mass, but the paucity of samples limits such calculations (Oftedal 1997). However, based on comparisons with phocids, otariids, and terrestrial mammals, Oftedal (1997) estimates that cetacean mothers probably contribute 2–4 kg of milk per kilogram of offspring weight gain.

Four factors influence the duration of lactation. First, age at weaning will be greater than the age at which an offspring is able to feed itself (as cetaceans, unlike phocids, show no evidence of a post-weaning fast). This age will be related to the methods used to find and catch food (Brodie 1969). Killer whales may take longer to learn how to safely and efficiently catch pinnipeds from beaches (Baird, chap. 5, this volume) than right whales to sieve copepod schools. Long-finned pilot whale calves who were stranded with pod members had an average of 10 prey species in their stomachs, compared with the adults, who had an average of 153 (Gannon et al. 1997), suggesting that the young cannot process, handle, or capture the diversity of prey that adults can. In general, animals that have an abundance of food available, such as the mysticetes, will be able to feed themselves at younger ages than those for which food is more difficult to catch, such as bottlenose dolphins (see figs. 9.5, 9.6, and 9.7).

Second, nursing duration is determined by its relationship with offspring fitness. Specific nutrients present in

milk may still be important for a young animal that can feed itself, improving growth or disease resistance. The additional food intake of juveniles that nurse while feeding themselves may increase growth rates, making the juvenile less vulnerable to predators, decrease the age at first repro-

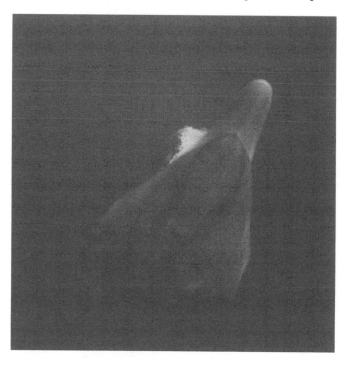

Figure 9.6. Bottlenose dolphin calf mangling a blowfish. Calves occasionally catch and apparently "play with" toxic blowfish. They don't consume the fish, but the neurotoxin in its skin may provide a tingling sensation. This calf mouthed and tossed the fish for over twenty-five minutes. (Photograph by Janet Mann.)

Figure 9.7. Bottlenose dolphin calf catching garfish. In Shark Bay, Australia, calves regularly trap small fish at the surface, a behavior called "snacking." (Photograph by Janet Mann.)

duction, or improve future reproduction in other ways. Once again, these effects are likely to be more important in more food-limited and social species such as sperm whales than in the mysticetes.

Third, the duration of lactation is influenced by the effect of lactation on the female's ability to produce future offspring. Lactation is costly, especially during peak lactation when fat content is highest (Oftedal 1984). Female odontocetes increase caloric intake by roughly 50% during lactation compared with pregnancy and cycling (e.g., Cheal and Gales 1992; Amundin 1986; Recchia and Read 1989). The costs of lactation may be double those of pregnancy in mysticetes (Lockyer 1984). Therefore lactation is a constraint on future reproduction. In all dolphin and whale species that have been intensively studied, ovulation can take place during lactation. Mysticete females may ovulate and conceive during the earliest stages of lactation, the postpartum estrus (e.g., gray whales: Rice and Wolman 1971; humpbacks: Chittleborough 1958; fin whales, *Balaenoptera physalus:* Laws 1961; sei whales, *Balaenoptera borealis:* Gambell 1968b). In contrast, when ovulation overlaps with lactation in odontocetes, it is toward the end of lactation: many females become pregnant before the current offspring is fully weaned (e.g., long-finned pilot whales: Mar-

tin and Rothery 1993). In bottlenose dolphins, calves are typically weaned during the first ten months of the mother's next twelve-month pregnancy (Mann et al. 2000). By becoming pregnant during the latter stage of lactation, female cetaceans may hedge their bets against early fetal loss. For example, if a female loses her fetus, she may continue nursing her current offspring for another year. One Shark Bay bottlenose dolphin female (with a nursing three-year-old calf) was herded by adult males, became visibly pregnant, but apparently lost the fetus within seven or eight months after conception. She continued to nurse her daughter for another year, weaning her at four and a half years of age, when she was, again, six months pregnant (J. Mann et al., unpublished data). This strategy may partially explain the variability in weaning age among odontocetes. Despite these possibilities, the postpartum estrus is quite rare in mysticetes (Lockyer 1984), and many odontocetes wean offspring years before their next pregnancy or successful birth (e.g., short-finned pilot whales: Kasuya and Marsh 1984). Thus the effects of nursing current offspring on future reproductive success seem to be very variable, probably varying with the health and energy stores of the female. For older females, in which the prospects of future pregnancies may not be good, lactation is often prolonged, as predicted by parental investment theory (Pianka and Parker 1975). In sperm whales, newly mature, ten-year-old females lactated for about twenty months, whereas fifty-year-old females lactated for forty-two months (Best et al. 1984). Similar trends of longer lactation with greater maternal age have been found in some delphinids (e.g., Kasuya and Marsh 1984) and other mammals (e.g., Paul et al. 1993).

Fourth, the degree of weaning conflict may influence age at weaning (an interaction between the second and third points discussed above). Parents and offspring weigh the benefits of current offspring fitness and future offspring fitness differently, with offspring favoring themselves and parents on average favoring all offspring equally (Trivers 1974). This difference produces the potential for mother-offspring conflict regarding the duration of lactation, with offspring being selected to nurse for longer than is optimal for the mother. How this conflict is resolved can affect the duration of lactation. Although weaning conflict is not always apparent in studies of parent-offspring behavior (Bateson 1994), there is some evidence of it among cetaceans. Following Trivers (1974), a mother is expected to decrease her role in maintaining proximity to the calf as a function of calf age as calf survival becomes less dependent on her nursing and care. The calf is expected to increase its role in maintaining proximity to the mother, particularly as maternal and calf interests diverge. Both of the focal animal studies that examined proximity maintenance (approach-leave rates) in wild cetaceans show this pattern (southern right whales, *Balaena australis:* Taber and Thomas 1982; bottlenose dolphins: Mann and Smuts 1999). Right whale females have been observed planting their mammaries on the seafloor or swimming belly-up at the surface to prevent a calf's nursing attempts (Payne 1995). Similarly, the provisioned females at Monkey Mia in Shark Bay sometimes remain in shallow water despite the bumping and constant whistling of their offspring, who are futilely attempting to nurse or assume infant position (J. Mann, personal observation; Mann and Smuts 1998).

In general, duration of lactation in cetaceans seems most related to food supply. The mysticetes, which generally seem to have seasonally abundant, easily eaten food, suckle for a few months and wean their offspring suddenly, whereas the larger odontocetes (e.g., killer whales and sperm whales), whose food is probably scarcer and more challenging to locate or capture, lactate for periods of years as their offspring gradually develop hunting skills.

Interbirth Intervals

With a pronounced birth season, most female cetaceans are constrained to have an integer number of years between births. The median interbirth interval following "surviving" and/or nonsurviving offspring varies between species (from about one year in minke whales, *Balaenoptera acutorostrata,* to five years in unexploited sperm whales) and between populations of the same species (e.g., Best et al. 1984). Interbirth intervals have been estimated using at least four quite different techniques (see box 9.1), whose results are not easily comparable (Perrin and Reilly 1984). Therefore, only the most obvious trends should be considered from a general survey, such as that represented in table 9.1.

Interspecifically, as might be expected, interbirth interval correlates with gestational time ($r = 0.62$) and the duration of lactation ($r = 0.57$). Most mysticetes have two or three years between births, whereas in the odontocetes interbirth interval is much more variable. In both suborders, interbirth interval generally increases with body size (fig. 9.8). The two beaked whale species for which there are estimates, Baird's beaked whale *(Berardius bairdii)* and the northern bottlenose whale, have interbirth intervals more similar to mysticetes of their size than to other odontocetes.

Intraspecifically, interbirth interval seems related to environmental conditions. For sperm whales, it appears shorter in the highly productive waters of the Humboldt Current west of South America than in other study areas (Best et al. 1984). In a number of cetacean species—both

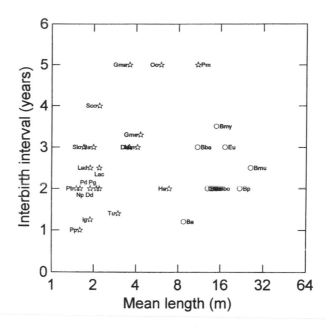

Figure 9.8. Interbirth interval plotted against log of body length for odontocete (*) and mysticete (○) species.

odontocetes and mysticetes—interbirth interval is estimated to have declined as populations were reduced during exploitation (Fowler 1984), although Mizroch and York (1984) have questioned whether trends of increasing pregnancy rate with depletion really exist in the mysticetes.

Reproductive Senescence

Although most mammals show an age-specific fertility decline, few species show clear evidence of menopause, defined here as termination of reproductive function well before expected age of death (e.g., Dunbar 1987; Fowler and Smith 1981; Austad 1994; Hill and Hurtado 1997). Physiological data unequivocally indicate menopause for short-finned pilot whales (Marsh and Kasuya 1984; figs. 9.9, 9.10), and demographic data strongly suggest the same pattern for killer whales (Olesiuk et al. 1990). Female short-finned pilot whales have a significant postreproductive stage in their life histories: they cease to produce calves at about thirty-six years of age, when their life expectancy is still at least fourteen years (Marsh and Kasuya 1984; fig. 9.11). In *resident* killer whales, most females give birth to their last calf by thirty-nine years of age, but have mortalities of less than 0.05 per year for the next twenty years (Olesiuk et al. 1990). Females of a number of other odontocete species (including sperm whales, long-finned pilot whales, false killer whales, *Pseudorca crassidens,* and bottlenose dolphins) show reduced fertility with age (Marsh and Kasuya 1986; Martin and Rothery 1993). In some of these species there are indications of reproductive senescence, but the

Figure 9.9. Short-finned pilot whales in the Bay of La Paz, Mexico. They are one of two mammals besides humans to definitively show menopause. (Photograph by Robin Baird.)

data and analyses are not conclusive (Marsh and Kasuya 1986). In contrast, there is little evidence for an age-related decline in pregnancy rates in mysticete species (Marsh and Kasuya 1986).

The function of menopause is not known. Menopause could have no adaptive significance per se, either because human menopause is an artifact of our recently extended life spans (reviewed by Austad 1994), or because, as Williams (1957) proposed, it is a result (by-product) of pleiotropic effects. Traits that favor reproduction early in life could be favored by natural selection, even if such traits are costly to late reproduction (selection weakens with advancing age). Packer et al. (1998) recently boosted this argument with an analysis of lion *(Panthera leo)* and baboon *(Papio anubis)* data. However, 20–25% of wild adult short-

Figure 9.10. Short-finned pilot whales socializing in the Bay of La Paz, Mexico. Good observation conditions in the bay make long-term study possible. (Photograph by Robin Baird.)

finned pilot whale and killer whale females (Marsh and Kasuya 1984; Olesiuk et al. 1990) are postreproductive, suggesting that menopause is indeed functional (see fig. 9.11). Both the proportion of postreproductive females and the length of the postreproductive period highlight the qualitative differences between "true menopause" and reproductive senescence in general, which typically involves degeneration or cessation of reproduction close to the age of death—a pattern found in many mammals.

Several adaptive functions have been proposed for menopause. The "stop-early" hypothesis suggests that females with extensive maternal investment may have greater reproductive success by ceasing reproduction and successfully rearing their last and other extant offspring than by continuing to give birth until the end of their lives (Mayer 1982; Hill and Hurtado 1996). This would be particularly true for species with stable family groups with overlapping generations (such as those found in humans, killer whales, sperm whales, and pilot whales). Reliable weaning and behavioral data are not available to test this hypothesis.

The "grandmother" hypothesis (Hawkes et al. 1998) proposes that a long postmenopausal life span can be favored through the extension of mother-child food-sharing, which allows older females to enhance their daughters' fertility by investing directly in grandchildren and other close kin. Several sources of evidence would not support this hypothesis for either killer whales or pilot whales. First, although mammal-eating *transient* killer whales share prey, daughters emigrate at maturity (Baird, chap. 5, this volume). Second, postreproductive *resident* (fish-eating) killer

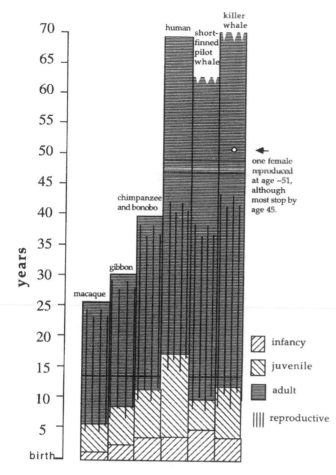

Figure 9.11. Typical female life history patterns of several primate taxa compared with short-finned pilot whales and killer whales. As with humans, the reproductive period appears to be conserved in pilot whales and killer whales, with an extended postreproductive span. Notably, whale females begin reproduction earlier than humans, but may have later weaning ages. Infancy is defined as the period of nursing. The juvenile period includes post-weaning to age at first birth. Expected and maximum life span for both whale species are uncertain, but pilot whale females can live into their sixties and killer whale females into their eighties. (Adapted from Hawkes et al. 1998 and Napier and Napier 1967; data from Harvey et al. 1987 [macaques, averaged across several species], Leighton 1987 [gibbons], Goodall 1986 and Nishida 1990 [chimpanzees], Kano 1992 [bonobos] Hawkes et al. 1998 [humans], Marsh and Kasuya 1984; Kasuya and Marsh 1984 [short-finned pilot whales], Olesiuk et al. 1990 [killer whales].)

whale females do not preferentially associate with grandoffspring or other calves, and cases of fish sharing have not been observed (J. Mann and R. W. Baird, unpublished data). Third, short-finned pilot whales generally consume small prey that would not easily be shared. Fourth, pilot whales and killer whales do not appear to show a relatively long prereproductive growth period for their body size compared with other, nonmenopausal odontocetes. Hawkes et al. (1998) argue that as human postreproductive

lives lengthen, age at maturity should also be delayed because a long prereproductive growth period would be paid off by a long adulthood (see fig. 9.11). Fifth, although human female mortality during parturition may increase with advancing age, such costs are less likely for whales.

It is notable that the cetacean species in which menopause has been found, pilot and killer whales, possess matrilineally based social systems. There is some evidence that cultural processes are very important within the matrilineal groups of these animals (Whitehead 1998; see below), suggesting another explanation for menopause. If an older female's role as a source of information significantly increases her descendants' fitness, and reproduction toward the end of her life decreases survival, menopause could be adaptive (L. Rendell and H. Whitehead, unpublished data), even in groups from which there is some dispersal.

In sum, few data are available to explain menopause in cetacean species. Three approaches may prove useful. The relationship between weaning age and maternal age needs to be quantified. Offspring and grandoffspring fitness measures for whale families with and without surviving postreproductive females might provide some clues. Behavioral observations of menopausal females are necessary to determine what currency of benefits they might provide to kin.

The decline in pregnancy rate with age in many odontocetes is more easily explained. Life history theory predicts that as females age, they should allocate a greater proportion of their resources to each offspring by increasing interbirth intervals and nursing each offspring longer (Clutton-Brock 1984). Mysticetes, which have much faster growing young (fig. 9.12), wean their offspring at a mean of 25% of the interbirth interval, as compared with 54% for odontocetes. Thus, an increase in maternal investment in an offspring is less likely to affect the interval until the next offspring is born in mysticetes.

Longevity and Mortality

Data on longevity and mortality are few and not especially reliable. The smallest cetaceans seem to live only into their teens, and have natural adult mortality rates of about 10% per year. In contrast, females of the larger species, both mysticetes and odontocetes, can reach sixty or more years of age, and may have mortalities of 5% per year or less. The extreme seems to be bowhead whales *(Balaena mysticetus)*, which may live well over a hundred years—possibly the world's oldest animals (George et al. 1998). Interspecifically, the longevity and mortality of females are negatively correlated ($r = -0.59$), as expected, and both are closely related to body size ($r = 0.56$ for longevity, $r = -0.45$ for mortality).

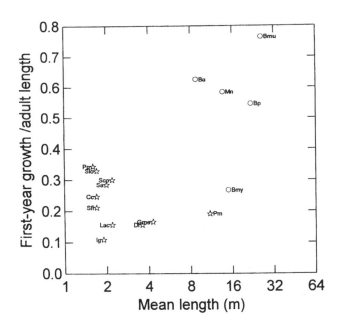

Figure 9.12. Change in length during first year of life as a proportion of maternal length, plotted against log of body length for odontocete (*) and mysticete (O) species.

In only one instance are reliable age-specific mortality rates available for wild cetaceans. Female *resident* killer whales off British Columbia have strongly "U-shaped" mortality curves, with immatures and fifty- to sixty-year-olds having about 2–4% annual mortality, and animals in their late teens and early twenties rarely dying (Olesiuk et al. 1990). Data from long-finned pilot whales also suggest a "U-shaped" mortality curve (Bloch et al. 1993). It is probable that this typical mammalian pattern of decreased mortality in midlife is present in other species, and that females of all large odontocete species (such as sperm whales and bottlenose whales) lead very safe lives during their early reproductive years.

Adolescence

Adolescent Sterility

Sexual maturity for cetaceans is usually defined as either first ovulation or first conception (see International Whaling Commission 1984a). However, females may begin cycling without ovulating (nonovulatory or nonfecundable cycles) or ovulate without conceiving (nonconceptive cycles). Females of several long-lived terrestrial mammals, especially primates (e.g., Short 1976), have several nonovulatory (infertile) cycles before their first ovulation, or nonconceptive cycles before their first pregnancy. This phenomenon is known as adolescent sterility.

Evidence for female adolescent sterility has been found

in killer whales (Perrin 1982), common dolphins (*Delphinus delphis:* Collet and Harrison 1980; Kirby and Ridgway 1984), and possibly Pacific white-sided dolphins (*Lagenorhynchus obliquidens:* Harrison et al. 1972). Nonconceptive cycles, whether ovulatory or not, are likely to be common among odontocetes before the first pregnancy because the costs of poor mate choice may be particularly severe in mammals with prolonged and intensive parental care.

The adaptive significance of this pattern is thought to be that the females are physiologically attractive to males and can thus "practice" their mating strategies before incurring the costs of poor mate choice (Short 1976). For adolescent females to gain "adult-like" experience with mature males, they must be sexually "attractive" to males, exhibiting the chemical or physiological cues (including secondary sex characteristics) associated with fertility without actually being fertile. Nonovulatory cycling is not restricted to adolescence; it may increase the overall duration of attractiveness for mature females if either confusing paternity or enhancing mate choice options is functional (e.g., Connor et al., chap. 4, this volume).

There is no evidence for nonovulatory cycling in mysticetes. Some data indicate that first ovulations do result in conceptions for baleen whales (Gambell 1968b). Adolescent sterility might not be expected to occur in baleen whales because of the limitations of strict breeding and feeding cycles (i.e., juvenile/adolescent female baleen whales should stay in productive waters to increase their mass).

Learning to Parent

In studies of mammalian allomaternal care (care of the young by animals other than the mother), female immatures, compared with all other age and sex classes, appear to be the most attracted to young and are most likely to attempt to carry, feed, or care for the offspring of other females (Hrdy 1976; Reidman 1982; Gittleman 1985). The age and sex class bias of infant attraction generally supports the "learning to parent" hypothesis for allomaternal care. Young female mammals have the most to gain from practice parenting when first-born mortality is high (due to primipara inexperience) and offspring investment is high, a pattern demonstrated for some primates (e.g., see Fairbanks 1990, 1993).

One study demonstrated that inexperienced bottlenose dolphin females (immature females and those who had never had a surviving calf) were much more likely to associate with newborns (zero to three months) who were away from their mothers than adult (experienced) females, supporting the "learning to parent" hypothesis for allomaternal care (Mann and Smuts 1998). Adult and immature females were equally likely to associate with mothers and newborn calves when mother and calf were together (closer than 2 m), but inexperienced females preferentially associated with newborn calves away from their mothers, sometimes accompanying newborns more than 100 m away. Mothers did not appear to benefit from these associations by foraging more when their calves had escorts than when the calves were alone. Young females may benefit from associating with newborn calves, but they do not appear to provide respite to mothers from caregiving duties.

Mother and Calf

Singleton Births

Although twin fetuses are sometimes found, and twin births occasionally observed (Olesiuk et al. 1990), both twins rarely survive (International Whaling Commission 1984a). We know of only one report of surviving twins for any cetacean: twin killer whales off Vancouver Island that apparently survived to at least seven and a half years of age (Olesiuk et al. 1990). Two factors are likely to favor singleton births (fig. 9.13) in Cetacea: the need for large neonates, and the importance of fast growth in calves.

Size at Birth

Birth mass in cetaceans is usually about 3–8% of maternal mass in larger cetaceans, although in the small odontocetes it can range up to 20% (see fig. 9.2). Cetacean newborns seem to have a minimum size of 0.6 m and 5 kg. This lower limit might be due to the problems of maintaining temperature at small body sizes. However, there is no obvious tendency for small cetaceans of cold waters (such as harbor porpoises and Commerson's dolphins, *Cephalorhynchus commersonii*) to have larger calves than similar-sized females from tropical waters (such as the river dolphins) (see fig. 9.2; table 9.1). Instead, small cold-water cetaceans have other means of maintaining body heat in themselves and their calves. Worthy and Edwards (1990) found that, compared with a similar-sized spotted dolphin that lives in tropical waters, a harbor porpoise has less surface area, and blubber that is twice as thick and insulates twice as well. These heat conservation mechanisms of small cold-water cetaceans permit the production of quite small calves. Thus the convergence on a minimum calf size of 0.6 m and 5 kg over a wide range of water temperatures may be a coincidence, or may have another functional or proximal explanation.

Sex Ratio and Sex-Biased Investment

Cetacean populations invariably show 1:1 birth sex ratios (e.g., Best et al. 1984; Kasuya and Marsh 1984; Lockyer 1984). This pattern is common among mammals (Clutton-Brock and Iason 1986) and is theoretically expected, assuming a similar rate of return per unit investment in the two sexes (Fisher 1930; Maynard Smith 1980).

However, some females in a population may be more likely to produce one sex over the other, or to preferentially

Figure 9.13. Long-beaked common dolphin mother and calf off the coast of Mexico. (Photograph by Robin Baird.)

invest in one sex over the other, if the payoffs benefit maternal inclusive fitness. For instance, Trivers and Willard (1973) predict that females in good condition, and so more able to invest in offspring, should improve their inclusive fitness by producing more offspring of the sex with greater reproductive variance, usually males (although see Leimar 1996; Hewison and Gaillard 1999). We discuss four potential examples of such phenomena, for humpback, sperm, and short-finned pilot whales and bottlenose dolphins.

Wiley and Clapham (1993) found that the sex distribution of young-of-the-year humpback whales in the Gulf of Maine was related to the interval since the birth of the previous surviving calf—males were disproportionately born after long interbirth intervals—and suggest that this observation may support the Trivers and Willard (1973) hypothesis (see Clapham, chap. 7, this volume, for details).

There are two intriguing possible observations of sex-biased investment. Data from 806 short-finned pilot whales caught during whaling operations off the Pacific coast of Japan indicated that males aged up to fifteen years occasionally had milk in their stomachs, but no females over the age of seven did (Kasuya and Marsh 1984; Marsh and Kasuya 1984). Males grow faster than females: by two or three years of age, males are consistently longer than females (Kasuya and Matsui 1984), and by adulthood, they are approximately twice the mass of females (Kasuya and Marsh 1984). Similarly, Best and his colleagues found lactose in the stomachs of thirteen-year-old sperm whale males, but the oldest female with lactose was seven years old (Best 1979; Best et al. 1984). Both of these species are highly sexually dimorphic, and males reach reproductive maturity later than females (Connor et al., chap. 10, this volume). Prolonged lactation by mothers of males could be a proximal result of increased solicitation by sons (Redondo et al. 1992), and have no benefit for the mother. Alternatively, mothers may show a net benefit through the resultant increase in the son's reproductive success (theoretical arguments in Wilson and Pianka 1963; Trivers and Willard 1973; Maynard Smith 1980).

Sex-biased investment in daughters is rare among mammals, but has recently been documented in Shark Bay bottlenose dolphins (J. Mann et al., unpublished data). The majority of sons were weaned at age three, but the majority of daughters were weaned at age four or older. The frequency of nursing is not known. Shark Bay bottlenose dolphins show little sexual dimorphism in body size (Connor et al., chap. 4, this volume), thus increased lactational investment in sons is unlikely to afford them increased competitive ability. Because male-male alliances are critical for reproductive competition, weaned males may be able to

accelerate the development of such relationships. Furthermore, because most sons mate with their mothers during dependency, but never do so after weaning, mothers may wean their sons to avoid conceptive incestuous matings. The minimum age of sperm viability is unknown. However, these factors would not explain why daughters are nursed longer. Prolonged nursing of females may enhance daughters' reproductive success if body size is linked to age of first birth, interbirth intervals, or infant mortality. Long-term data will help answer these questions.

Precocial Offspring

Unlike the majority of mammalian species, which produce altricial offspring, cetaceans give birth to precocial young that have well-developed sensory and locomotion abilities (Derrickson 1992; fig. 9.14). Mammals that produce altricial offspring either have a safe den to leave them in or some means of keeping the young animal attached to the mother or another adult (such as the pouches of marsupials, or the hair that some newborn primates hold onto). There are no safe dens for infant cetaceans and no fur or pouches (except the uterus) on streamlined adult cetaceans. The neonatal mortality of wild cetaceans is high. It is estimated to be about 43% during the first six months of life for *resident* killer whales, whose adult females lead very safe lives (Olesiuk et al. 1990), 32% during the first six months for captive bottlenose dolphins (Sweeney 1977), and 35% during the first year for long-finned pilot whales and gray whales (Sergeant 1962; Swartz and Jones 1983; Sumich and Harvey 1986). Much of this mortality is probably due to avoidable natural dangers, including predators. Proximity to the mother, or possibly other adults, usually reduces such dangers, as do speed, maneuverability, and good sensory abilities. These attributes also allow the mother some independence (which may be necessary for foraging) while caring for the offspring.

Neonatal Growth

Growth of the neonate is very different for the two suborders of Cetacea: mysticetes (with the interesting exception of the bowhead whale) add about 60% of maternal length during the first year after birth, over twice the rate for most odontocetes (see fig. 9.12). In neither suborder is there any obvious relationship between first-year growth, as a proportion of maternal length, and the size of the mother.

Nursing Behavior

The different neonatal growth rates of the two suborders of Cetacea are inextricably linked to divergent lactation strategies. Some mysticete mothers (humpback and gray

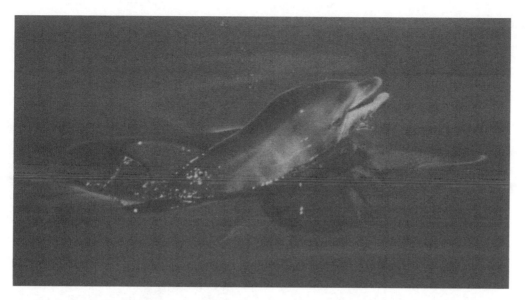

Figure 9.14. Newborn bottlenose dolphin "Mouse" swimming in echelon (alongside the mother). Note fetal folds and surfacing pattern. Newborn calves tend to bring the head out of the water when surfacing, and it takes a month or two before they consistently roll smoothly at the surface to breathe. Mouse nursed from her mother "Mini" for over six and a half years. (Photograph by Janet Mann.)

whales) fast during much of lactation, but this is never the case with odontocetes. Like other animals who fast during parturition and the early stages of lactation (e.g., ursids and phocid seals), baleen whales produce milk very high in fat, low in water, and low in protein (Oftedal and Iverson 1995). It is thought that this composition helps the mother conserve water and maximizes calf growth. In contrast, odontocete milk is considerably less nutritious (Oftedal 1997). Oftedal (1997) estimates energy output through milk (corrected for maternal metabolic body size) to be $0.40-1.06$ MJ/kg$^{0.75}$d for mysticetes and $0.09-0.17$ MJ/kg$^{0.75}$d for odontocetes, leading to the substantial differences in neonatal growth of the two suborders.

Suckling or nursing rates are unknown for wild cetaceans, but have been reported for captive delphinids (e.g., bottlenose dolphins: Reid et al. 1995; Cockcroft et al. 1990; beluga whales: Drinnan and Sadleir 1981). During field observations, nursing can sometimes be inferred based on seeing the calf's rostrum pressed to the mother's mammary slit (e.g., Mann and Smuts 1998; fig. 9.15). However, "nursing" observations do not necessarily mean either that milk is being transferred or that the dyad observed is mother and offspring. The subject of one of J. C. D. Gordon's (1987b) observations of presumed suckling in sperm whales turned out to be an immature male.

Allonursing is rarer in mammals than was originally thought (Packer et al. 1992), but there are several lines of evidence suggesting its occurrence in sperm whales (Whitehead and Weilgart, chap. 6, this volume). Allonurs-

ing provides obvious nutritional benefits to the calf and, if the mother dies, may enhance its chances of adoption. Two captive bottlenose dolphin adult females began spontaneously lactating and successfully adopted two six- and seven-month-old calves within one week of introduction, although the calves required dietary supplements to make the transition (Ridgway et al. 1995). No adoptions have been definitively recorded in the wild.

Several hundred bouts of nursing have been observed in known mother-infant pairs of wild bottlenose dolphins in Shark Bay, Australia (1,679 hours of observation of calves from zero to four years of age: Mann and Smuts 1998). In three instances, calves attempted to nurse from non-mothers. In two of these three cases, the recipients were immature females. In the third case, the adult female had lost her calf six weeks earlier. Although nursing is observed opportunistically, infant position (close, under the mother: fig. 9.16) can always be observed, and this is the position from which all bottlenose dolphin calves nurse (Mann and Smuts 1998). Since Shark Bay bottlenose dolphin calves spend less than 0.01% of their time in infant position with non-mothers, we can conclude that allonursing is quite rare in this population.

Mother-Calf Separation

Among the Cetacea, separations between calf and mother (greater than a few body lengths) are largely the result of the foraging behavior of the mother. Fasting baleen mothers need not separate from their calves at all. Humpback

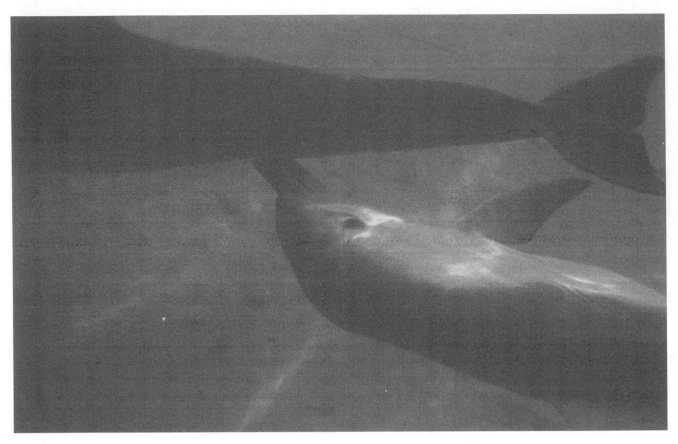

Figure 9.15. Yearling bottlenose dolphin nursing. When nursing, the infant opens its mouth slightly and inserts the tongue in the mammary slit. Nursing bouts typically last four to ten seconds. (Photograph by Janet Mann.)

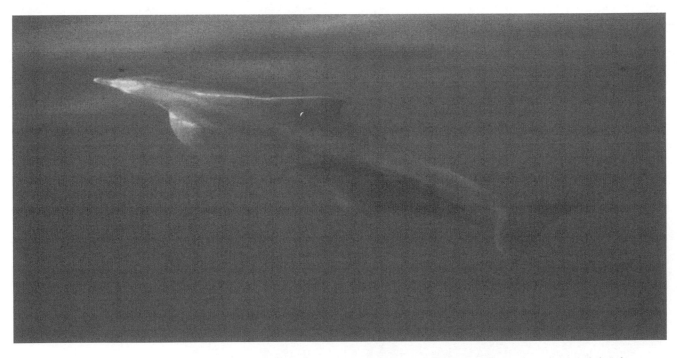

Figure 9.16. Infant position swimming in bottlenose dolphins in Shark Bay, Western Australia. Infant's head and dorsal side lightly touch the mother's abdomen. Infants stop swimming in this position when they are weaned. All nursing occurs from infant position. (Photograph by Janet Mann.)

whale females virtually never separate from their calves on the breeding grounds (Tyack and Whitehead 1983), and southern right whale mothers and calves also stay consistently close (Taber and Thomas 1982). However, postpartum estrus, and the resulting attraction of males, may cause separations or injuries to calves in some mysticetes (e.g., right whales: Payne 1995), although such instances are likely to be rare. In contrast, nursing odontocetes do feed, and so there are greater opportunities for separation. Two aspects of cetacean foraging are particularly likely to cause separations between mothers and infants: rapid accelerations to chase prey and deep diving.

Mann and Smuts (1998, 1999) studied the separation patterns of mother and infant bottlenose dolphins during the first three months of life. Infants often lost physical and visual contact with their mothers during foraging. Although most separations between mothers and infants occurred during foraging, infants separated from their mothers in a variety of other contexts. Infants effectively negotiated these separations, often darting away from their mothers for several minutes. The early development of the signature whistle (Tyack, chap. 11, this volume) in this species is likely to be linked to selection pressures of early separations that result from the mothers' foraging needs.

In Bernard and Hohn's (1989) study of spotted dolphins, pregnant females fed primarily on squid, while lactating females principally ate flying fish. Fish provides more calories and protein than squid, but less fat and water. If the greater energetic requirements of lactating females alone forced them to concentrate on flying fish, why don't all females do the same? Bernard and Hohn propose that lactating females may be constrained from deep diving for squid because this requires them to leave their calves (who are less competent divers) at the surface.

Sperm whale mothers routinely separate from their offspring, even when they are very young (Whitehead and Weilgart, chap. 6, this volume). These separations generally happen when females make deep foraging dives (Whitehead 1996a). Sperm whale infants seem to have poor diving ability and so do not follow their mothers (Best et al. 1984; Whitehead and Weilgart, chap. 6, this volume). To care for calves during these separations, sperm whales have adopted a system of babysitting in which calves accompany other members of their group (Whitehead and Weilgart, chap. 6, this volume).

The northern bottlenose whale seems to make even deeper dives than the sperm whale (Hooker and Baird 1999), and so mother-infant relationships in this species can give an important perspective on the role of deep diving in mother-calf separations. Gowans (1999) has found that,

in contrast to the sperm whale but like the shallow-water bottlenose dolphins, female bottlenose whales have a loose network of associates. Initial observations of mother-calf relationships suggest a mixed pattern, with some bottlenose whale calves showing consistent and long-term proximity to one female, presumably the mother, and others having very varied associations, including being left alone at the surface, or being clustered with just one immature male, even when newborn (Gowans 1999). Thus relationships between deep-diving females and their calves seem varied, both inter- and intraspecifically.

Babysitting and Allomaternal Care

Descriptions of allomaternal behavior in captive and wild odontocetes (e.g., bottlenose dolphins: Brown and Norris 1956; Tavolga and Essapian 1957; Gurevich 1977; Leatherwood 1977; Allen 1977; Wells 1991a; sperm whales: Caldwell and Caldwell 1966; Best 1979; spinner dolphins: Johnson and Norris 1994; killer whales: Haenel 1986; harbor porpoises: Anderson 1969; bottlenose whales: Gowans 1999), although not in mysticetes, are widespread. In the wild, these observations generally consist of other animals accompanying calves while they are separated from their mothers, or "babysitting" (according to the definition in Kleiman and Malcolm 1981). But is this babysitting really allomaternal care, and if so, is it altruistic?

Behavior by an animal that is not the mother of a calf that benefits the calf and its mother and would not be carried out if the calf were absent can be considered allomaternal care (following the definitions of Woodroffe and Vincent 1994 and Mann and Smuts 1998). So it is important to ask, first, whether the behavior of the allomother benefits the calf or the mother; second, whether the allomother is doing anything it would not normally do if the calf were not there; and third, if so, whether such changes in behavior are costly.

In most cases of cetacean allomaternal care, these questions have not been answered or even addressed. However, given the risks that calves face from predators, it is likely that the presence of a nearby adult is of general benefit to the calf, so that the first question can be cautiously answered in the affirmative. There is evidence that sperm whales change their dive schedules to benefit calves, so positively answering the second question and thus suggesting that there is allomaternal care in this species (Whitehead 1996a). However, allomothering sperm whales probably incur little or no cost, so this behavior may not be altruistic (Whitehead 1996a). In other cetacean species it is unclear whether allomothers are changing their normal behavior to benefit calves, let alone whether there are costs involved. In fact, some interactions between newly parturient wild or captive bottlenose dolphin mothers and other conspecifics are agonistic (e.g., McBride and Hebb 1948; Caldwell et al. 1968; Caldwell and Caldwell 1977; Eastcott and Dickinson 1987; Thurman and Williams 1986; Mann and Smuts 1998).

Mating Systems and Mate Choice: The Female Perspective

Mating systems have largely been viewed as the results of male adaptation to maximize individual mating success given the spatial distribution and reproductive physiology of females and the behavior of other males (e.g., Wrangham and Rubenstein 1986; fig. 9.17). But females need to mate too, and mating systems can have important influences on female behavior and reproductive success, just as female behavior toward males is an important aspect of some mating systems, such as leks.

Traditional characterizations of mating systems (e.g., "monogamous," "promiscuous," "polygynous") are confusing and do not always focus on the important differences between mating systems. From a female cetacean's perspective, mating systems may be characterized by asking the following questions:

1. What controls her estrous cycle? Is she monoestrous (one cycle per year) or polyestrous (more than one), and are cycles induced (requiring copulation and/or the presence of a male to trigger ovulation) or spontaneous (independent of copulation)?
2. How many potential mates are available at each estrus?
3. Can a female select mates?
4. How many mates does a female have during an estrus?
5. When a female has multiple mates, are there mechanisms that bias paternity (first mate, last mate, etc.)?

The ovulations of female cetaceans seem to be generally spontaneous (e.g., Benirschke et al. 1980; Kirby and Ridgway 1984). Some baleen whales tend to be monoestrous (humpbacks, gray whales), and others may be polyestrous (Bryde's whale: Best 1977), but all odontocetes appear to be polyestrous (International Whaling Commission 1984a). Polyestrous females may have several opportunities to mate with desirable males, although they may incur the cost of male harassment during each cycle. Monoestrous females may receive less harassment, but have limited chances to choose males. Connor and colleagues (1996; chap. 10, this volume) suggest that polyestrous cycling in bottlenose dolphins allows females to mate with multiple males, confusing paternity and reducing the risk of infanticide, or alternatively, allows females to exercise greater

Figure 9.17. Three male gray whales following a female during the breeding season in Bahia Magdalena, Mexico. (Photograph by Robin Baird.)

choice over who fathers their offspring (Sillén-Tullberg and Møller 1993; van Schaik et al. 1998). Since there is no evidence for infanticide and we know little about female choice, the benefits of polyestrous cycling to females remain unclear.

Cetaceans may include some of the very few mammalian species in which access to males can be an important determinant of female reproductive success. Some of the balaenopterids are so scattered over large tracts of ocean that special loud, far-ranging sounds may have evolved to bring the sexes together (Payne and Webb 1971; Watkins et al. 1987; Payne 1995). However, these sounds (with ranges of thousands of kilometers for blue and fin whales) seem louder than necessary for this purpose, suggesting that they have been amplified by sexual selection (Whitehead 1981). In sperm whales, the breeding members of the two sexes are latitudinally segregated, which, together with intense selection on males by whalers, seems to have limited female breeding success in some areas (Whitehead and Weilgart, chap. 7, this volume).

However, most cetaceans are either gregarious—spec-tacularly so in the case of the oceanic dolphins—or concentrate with conspecifics in particular areas with high availability of resources, as in most inshore odontocetes. Therefore, the majority of female cetaceans should have access to a number of potential mates, although some may be effectively excluded because of dominance relationships among nearby males (as suggested for sperm whales by Watkins et al. 1993).

Evidence for the active choice of mates by females is accruing in a range of taxa, and female choice has the potential to be an important determinant of reproductive success in both sexes (Bateson 1983; Andersson 1994; Eberhard 1996). With the tentative exception of Baird's beaked whales, whose life history characteristics lead Kasuya et al. (1997) to propose the existence of paternal care of weaned offspring, there is no evidence that male cetaceans assist their mates in any way other than by providing sperm. However, females may have been selected to choose males (fig. 9.18), or behave in ways so that they mate with males, who have genes that are useful to their offspring. They may also, at times, avoid mating with any male if mating is not

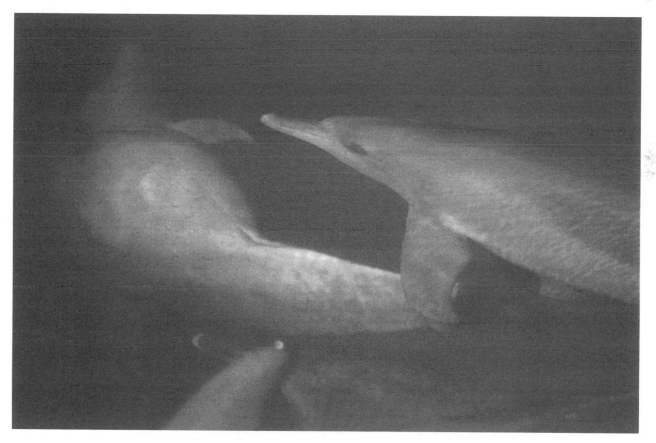

Figure 9.18. Pregnant bottlenose dolphin female "keel-rubbing" on male. Such affiliative interactions are common when males join female groups. (Photograph by Janet Mann.)

in their long-term interest or harassment by ardent males threatens the female or her offspring.

In the three-dimensional fluid structure of the marine environment, female cetaceans probably find it easier to avoid unwanted matings than do many terrestrial female mammals, which can have copulations forced upon them (e.g., Galdikas 1995; Smuts and Smuts 1993). Other aspects of the marine environment, especially excellent long-distance sound transmission and low travel costs, may assist females in recruiting and choosing potential mates. Female choice may be responsible for the evolution of some of the most remarkable cetacean vocalizations, such as the songs of humpback whales or the slow clicks of male sperm whales (Tyack, chap. 11, this volume). There are several suggestions that female cetaceans may be able to avoid mating or induce competition among males. Female right whales may move into shallow waters, roll over, or raise their tails into the air to prevent unwanted matings (Payne 1995). Female humpbacks lead competing males on physically exhausting high-speed chases, which might allow them to select the most physically able mate (Tyack and Whitehead 1983). The only known example of male cetaceans being able to sequester females is the case of the male bottlenose dolphin coalitions in Shark Bay, Australia (Connor et al. 1992b). Herding of females might occur in other populations of this or other species. It would probably be most feasible when waters are very shallow and males form coalitions (as in Shark Bay: Connor et al., chap. 4, this volume).

The number of males that mate with a female during one estrus varies between species. This is indicated both by observations and by variation in testis size, an indication of the degree of sperm competition (see Harcourt et al. 1981; Harvey and Harcourt 1984). Multiple matings during one estrus are likely to be common in right, gray, and bowhead whales and in many delphinids (Connor et al., chap. 10, this volume). In contrast, female balaenopterids and sperm whales are likely to have only one or few mates per estrus (Whitehead and Weilgart, chap. 6; Connor et al., chap. 10, this volume).

When females have multiple mates per estrus, as is likely in right whales (e.g., Brownell and Ralls 1986; Payne 1995), there is the potential for sperm competition. The mechanics of sperm competition among cetaceans are unknown, but it has the potential to affect the optimal behavior of both males (how to behave to maximize the probability of paternity: e.g., by mate guarding) and females (when to be selective about mates) (Gomendio and Roldan 1993).

In summary, as in most other mammals, variation in cetacean mating systems much more obviously influences the behavior and life histories of males than females. However, mating systems may affect the behavior of females (e.g., female right whales avoiding the attentions of ardent males), and the behavior of females toward potential mates may be an important part of some cetacean mating systems (e.g., inducing high-speed chases in humpback whales). Although female choice is usually subtle, it is increasingly seen as a vital factor in the evolution of behavior and life histories through sexual selection (Birkhead and Møller 1993). The limited evidence available suggests that female cetaceans choose their mates in a variety of ways, and that this choice is a major factor in the evolution of male behavior.

Social Behavior of Females

Grouping Behavior

Groups can enhance the reproductive success of their members in a variety of ways. The general costs and benefits of group living among cetaceans are discussed by Connor in chapter 9 of this volume, so here we will only summarize a few issues that principally concern females, as indicated by differences in grouping behavior between males and females.

In all mysticetes and many odontocetes, sex differences in grouping behavior are not pronounced. However, there are differences in some odontocete species, which are probably functionally important. For instance, several offshore dolphin species (including the pantropical spotted dolphin, *Stenella attenuata*, spinner dolphin, and striped dolphin, *Stenella coeruleoalba*) are found in groups that vary enormously in size (tens to thousands), substructure, and sex ratio (Scott and Perryman 1991; Perrin et al. 1994b). In pantropical spotted dolphins, females that are neither pregnant nor lactating mainly live in schools without males (Perrin and Reilly 1984; Scott 1991). Clymene dolphin (*Stenella clymene*) groups, which are smaller than those of the more offshore species, also appear to segregate by sex (Perrin and Mead 1994). The principal functions of grouping in these species are most likely increased foraging efficiency and vigilance for predators (Connor, chap. 9, this volume). Females in different stages of their reproductive cycles may segregate on account of different food requirements (Bernard and Hohn 1989) or other behavioral differences related to the presence or absence of a calf (Mann and Smuts 1999).

Of all cetaceans, grouping behavior differs most between the sexes in the sperm whale: females are highly social, forming complex, partly stable kin-based groups, whereas adult males appear to be much less social (Whitehead and Weilgart, chap. 6, this volume). Calf protection is probably a key function for the stable groups of females (Whitehead and Weilgart, chap. 6, this volume).

Dominance

Dominance has not been systematically studied for any wild cetacean, and there is only one quantitative study of female dominance in captive bottlenose dolphins (Samuels and Gifford 1997). Agonistic interactions involving bottlenose dolphin females are rarely observed, either in captivity (Samuels and Gifford 1997) or in the wild (<0.01 times per hour: Mann and Smuts 1999), making it difficult to assess dominance relationships or whether they are of biological significance. Most resources of importance to female cetaceans cannot be monopolized easily, so that we might perhaps expect enforcement of dominance relationships to be of little consequence.

Reproductive suppression is a form of female-female competition in which dominant females, either through harassment or other forms of social competition, cause subordinate females to suppress or delay cycling. Reproductive suppression has been documented in a variety of mammals, including wolves (*Canis lupus:* e.g., Derix et al. 1993), rodents (e.g., Wolff 1992; Vandenbergh and Coppola 1986), and primates (Abbott 1984; Harcourt 1987; Snowdon 1996). In social animals, the benefits of group living are not equal for all members of the group. If there is an optimal group size (e.g., *transient* killer whales: Baird, chap. 5, this volume), or resource competition favors fewer group members, reproductive suppression is one way dominant females (typically older) can favor their own reproduction at the cost of that of other females in the group. Alternatively, subordinate or young females may gain by helping kin or delaying emigration or reproduction until conditions are right (e.g., resources or mates are available). Reproductive suppression is most likely in species in which females form long-term cooperative groups (Snowdon 1996), so that in cetaceans it might be most expected in species such as sperm whales or killer whales. In neither species has there been any systematic search for evidence of reproductive suppression.

Matrilineal Cultures

Evidence is accumulating that cetacean mothers transmit important information to their daughters or other members of their matriline by instruction or social learning. Examples include the use of sponges as foraging tools by bottlenose dolphins in Shark Bay, Australia (Smolker et al. 1997; Connor et al., chap. 4, this volume), and killer whales intentionally stranding on beaches to catch seals (Guinet and Bouvier 1995; Baird, chap. 5, this volume). The skills needed for these activities are forms of culture, by at least some definitions of this term, and their transmission mechanism makes them particularly interesting.

Probably the most important difference between human and other animal societies lies in the importance of culture: much of the behavior of humans is culturally determined, whereas that of nonhumans seems not to be. Nonhuman animals do possess cultures, learning a range of information, such as foraging methods and song types, from each others' behavior (Laland 1992). However, crucially, while much important human cultural information (including language, caste, and dietary preferences) is transmitted "vertically"—from parent to offspring—nonhuman culture is mostly learned from non-parents, "horizontally" (Laland 1992), and is not particularly stable over generations (e.g., Tomasello 1994). Vertically transmitted cultures, especially when stable over many generations, are much more likely to have profound effects on genetic evolution and to lead to nonadaptive behavior through "runaway" effects (Boyd and Richerson 1985; Laland 1992; Laland et al. 1996).

Transmission of culture within stable matrilineal cetacean groups is effectively vertical. The best example of a stable, vertically transmitted cetacean culture is the vocal dialects of killer whales (Ford 1991; Baird, chap. 5; Tyack, chap. 11, this volume). These dialects are passed down through matrilineal pods and are stable over many generations, which also seems to be the case with sperm whale coda dialects (Whitehead et al. 1998; Whitehead and Weilgart, chap. 6, this volume). These matrilineally transmitted cultures seem to include other important information, such as methods of catching prey in killer whales and defending against predators in sperm whales (Whitehead et al. 1998; Heimlich-Boran and Heimlich-Boran, in press). Heimlich-Boran and Heimlich-Boran (in press) have suggested that feeding techniques culturally acquired from pod members may have initiated the divergence between the "fish-eating" and "mammal-eating" forms of killer whale (see Baird, chap. 5, this volume).

Most marine cetaceans live in a continuous habitat that is not divided into exclusive territories or fragmented by impassable barriers (Connor, chap. 8, this volume). In such circumstances, cultural innovations are potentially more universal in their significance.

Thus, it seems possible that cetacean societies, and especially those with stable matrilineal groups, such as those of killer, sperm, and pilot whales, contain cultures that are qualitatively more similar to human cultures than are those of most terrestrial mammals (Whitehead 1998). Stable, vertically transmitted cultures may therefore explain curious attributes of these species, such as nonadaptive mass strandings, low genetic diversity, and menopause (Whitehead 1998; see above).

Synthesis

Relationships between Life History Variables

Many of the life history variables that we have considered in this review are well correlated (see table 9.2). Some of these correlations, such as those between the duration of lactation and interbirth interval ($r = 0.57$), are structural and to be expected. However, others, including that between adult length and relative neonatal growth ($r = 0.74$), are potentially more interesting.

A principal components analysis of these data showed that much of the variation and correlation could be explained by three orthogonal axes, which were well represented by body length, interbirth interval (as well as life span, age at maturity, and durations of gestation and lactation), and the length of the peak calving season. Therefore, all the life history variables, except peak calving season, which is poorly correlated with all the other variables, are quite well represented by any of the plots in figures 9.1, 9.2, 9.4, or 9.8. We have summarized the general pattern in figure 9.19.

Mysticetes and odontocetes are clearly separated. The mysticetes are much faster-living than the large odontocetes. Within the mysticetes, there is no obvious taxonomic pattern or relationship between body length and the speed of life history processes, with the possible exception of similarities among the rorquals. The bowhead seems to have

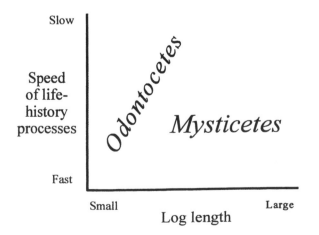

Figure 9.19. Summary of variation in life history variables of female cetaceans.

unusually slow life processes for a mysticete, or indeed, any animal.

The odontocetes lie roughly along a one-dimensional continuum, with larger animals having slower life history processes (indicated by long gestation) and smaller ones living "in the fast lane" (Read and Hohn 1995). There are two linked mechanisms that could produce this correlation. Larger odontocetes seem to be safer from predators; they have lower mortalities (see above) and so greater longevities. They are thus likely to have evolved in populations closer to their carrying capacities, and will have a harder time obtaining nutrients for themselves and their gestating or suckling offspring. If growth is slow and feeding offspring difficult, the large, long-lived female odontocete will have a low reproductive rate and mature late (see fig. 6.13).

Life History and the Environment

As female life history variables seem to be well controlled by natural selection (Stearns 1992), we might expect that there would be strong interspecific correlations between these variables and the environment. Two measures of each cetacean species' environment are shown in table 9.1: the mean absolute latitude of the species distribution (negatively related to sea temperature), and the habitat along an inshore-offshore nominal scale. However, neither of these measures is well correlated with any of the life history variables, or their principal components. The relationships between the life histories of female cetaceans and their environments basically reduce to one unsurprising constraint: large, slow-reproducing animals are not found in inshore or freshwater environments.

Stearns (1992: 307) points out several reasons why habitats do not map directly onto life histories. One of these is that a habitat can appear very different when observed at different spatial and temporal scales. For instance, harbor porpoises, pilot whales, and fin whales all eat herring and squid in the North Atlantic and can be seen in the same areas, such as the entrance of the Bay of Fundy (e.g., Katona et al. 1993). However, these three species are close to the apices of the triangular pattern that emerges from our analyses of life history (see fig. 9.19). From the perspective of a small harbor porpoise, the entrance of the Bay of Fundy is a valuable but dangerous habitat; for a fin whale, it is a seasonally very rich but safe environment; and for a group of pilot whales, it is part of a patchwork of resources covering a large sea area.

Female Cetaceans in the Mammalian Context

In some respects, compared with those of other families and superfamilies of large mammals, cetacean life histories are not exceptional: gestations of nine to fifteen months, weaning at four to thirty-six or more months, and litter sizes of one are common in other large mammals (table 9.3). However, in other ways, the female cetaceans are extraordinary.

The most obvious of these extremes is size. The largest female cetacean (the blue whale) outweighs the largest female noncetacean mammal (the African elephant, *Loxodonta africana*) by a factor of 14, and five of the eight cetacean superfamilies have members larger than elephants. Large size is possible anatomically in Cetacea because of the greatly reduced gravitational load of a nearly neutrally buoyant animal in water. It is also possible ecologically because of the availability of large amounts of food for these huge animals. In the case of the mysticetes, advanced filtering techniques allow the animals to capture whole schools (or parts of them) in one gulp (the balaenopterids) or to sift productive prey patches efficiently and nearly continuously (gray and right whales). The larger odontocetes (sperm whale, Baird's beaked whale, bottlenose whale) are principally mid- or deep-water squid eaters. We know very little of the behavior, abundance, or distribution of such prey (Clarke 1980), but they must be both numerous and reasonably easily caught to support the large populations of sperm and beaked whales that existed before whaling.

With the exception of a few outliers (likely recording errors), few, if any, noncetacean mothers among the large mammals give birth to young that weigh more than 15% of their own mass In contrast, in about half the small cetacean species, the newborn is 15–22% of maternal mass (see fig. 9.2). The 22% of mother's mass for newborn Dall's porpoises is an extreme among large mammals, but some bats have even larger offspring. Females of the smallest cetacean species can have relatively bigger offspring than other large mammals because of the supportive nature of the aquatic habitat: in contrast to terrestrial mammals, the offspring in utero is not an additional gravitational load, so that the structural parts of the mother's anatomy are not overly stressed by carrying the large fetus (Economos 1983).

Table 9.3. The ranges of some life history and other measures for families and superfamilies of large mammals

Superfamily/ Family	Common name	Milk fat	Mass (kg)	Birth mass ratio	Maturity (years)	Gestation (months)	Lactation (months)	Litter size	Longevity (years)
Platanistoidea	River dolphins	13%	41–120	4–19%	4	10–11	9–10	1	13–28
Delphinoidea	Dolphins/porpoises	22–30%	30–3,000	6–29%	4–15	9–15	6–36	1	13–85
Ziphioidea	Beaked whales	—	4,500–11,000	—	10–12	12–17	—	1	27–71
Physeteroidea	Sperm whales	24%	272–20,000	5–17%	11	9–14	12–24	1	28–65
Balaenidae	Right whales	22%	54,000–90,000	—	10–15	10–13	6–10	1	—
Neobalaenidae	Pygmy right whales	—	3,200	—	—	12	5	1	—
Eschrichtiidae	Gray whales	53%	31,500	2	9	13	7	1	—
Balaenopteridae	Rorquals	24–40%	6,600–105,000	3–7	5–12	10–12	4–10	1	41–95
Macropodidae	Kangaroos, etc.	3–7%	1.5–32	≪0.1	0.7–2	1	6–11	1	8–20
Suidae	Pigs	—	6.6–275	1–3	0.5–3	3–6	2–8	1–13	7–24
Cebidae	New World monkeys	1–2%	0.3–15	5–18	2–8	4–7	5–22	1	12–40
Cercopithecidae	Old World monkeys	3–5%	0.7–50	5–13	3–5	5–6	3–12	1	25–45
Hominoidea	Apes[a]	—	30–200	2–5	7	9	24–48	1	30–40
Hylobatidae	Gibbons	—	5–11	—	5–7	7–8	18–24	1	25
Canidae	Dogs	3–13%	1–75	1–4	1–2	2	2–3	2–14	10–18
Ursidae	Bears	17–25%	50–410	0.1–1.5	3–5	3–9	18–30	2	20–30
Felidae	Cats	9–11%	1.5–280	0.7–3	1–6	2–4	1–4	1–3	10–18
Hyaenidae	Hyenas	—	7–65	1–3	1.5–3	3	2–24	2–3	25
Phocidae	True seals	47–61%	50–900	3–16	2–7	11	0.2–2	1	14–46
Otariidae	Eared seals	25–53%	50–300	6–12	3–7	12	4–12	1	17–25
Odobenidae	Walrus	—	1,250	4.4	6–8	12	24	1	40
Proboscidae	Elephants	5–7%	4,000–7,500	2	7–12	22	24	1	70
Equidae	Horses	1–2%	275–350	8–10	2	12	6–8	1	20
Rhinocerotidae	Rhinos	0.2%	800–1,800	4–5	4–6	14–16	18–24	1	35–45
Camelidae	Camels	4%	45–1,000	5–14	1–6	8–14	5–24	1	15–40
Cervidae	Deer	5–11%	7–800	2–12	0.5–3	5–9	2–12	1–2	10–30
Giraffidae	Giraffes	5%	210–1,900	8	2–5	14–15.5	6–17	1	25
Sirenia	Dugong, manatee	—	200–600	5[b]	3–5	12–14	1–30	1	60
Hippopotamidae	Hippos	—	180–3,200	1–2	4–6	6–8	6–8	1	40
Bovidae	Antelope	5–14%	24–1,000	2–15[b]	0.5–3.5	5–10	2–12	1–2	10–30

Sources: Data for cetaceans from table 9.1; milk data from Oftedal and Iverson (1995); other mammal data from Grzimek (1990).

[a] Humans excluded.

[b] Excluded are recorded birth weight ratios of 26% for dugongs and 40% for royal antelope, as both seem likely to be errors.

Some of the larger cetaceans also show mammalian extremes in the age of female sexual maturity. Among noncetacean species, only the female elephant and some primates (at up to twelve years) have a mean age of sexual maturity greater than eight years. In contrast, most of the larger cetaceans become mature at age ten or older. The estimated mean ages of first birth of fifteen years for killer whales and bowheads are (known) extremes among mammals.

The longevities of smaller cetaceans, usually fifteen to thirty years, are in the general range of many large mammals. However, the largest cetaceans may live for sixty to over a hundred years, a span rivaled only by humans, elephants, and manatees (another completely aquatic mammal). Thus, in their life histories, the large cetaceans are extreme among mammals, both in their size and in their longevities and slow rates of maturation. The evolution of these slow life processes is likely linked to adults' relative safety from natural predators and high survival rates. In contrast, the small cetaceans have life histories not obviously different from those of similar-sized terrestrial mammals, with the exception of giving birth to relatively larger young.

Conclusions

This general overview of female reproductive strategies leaves many issues aside, principally because of a lack of good data. For instance, there are apparent outliers in some of the measures, as indicated in the figures and tables of this chapter. Some of these, such as the short duration of lactation in common dolphins, may result from dubious data. Others, such as the long lactation, slow early growth, and great longevity of bowhead whales, seem to be real, and deserve attention.

Within the range of mammalian variation, the reproductive strategies of female cetaceans are characterized by the production of a large, single, and precocial offspring and by prolonged maturation, gestation, lactation, and life spans. Included among the female Cetacea are mammalian extremes in size (31 m blue whales), relative size of newborn for a large mammal (22% of mother's mass for Dall's porpoises), and duration of lactation (up to thirteen years in sperm whales), and the only clear evidence, outside humans, of menopause (in killer whales and pilot whales).

The mysticetes and odontocetes show very different life history patterns. The mysticetes generally have abundant, easily caught food supplies seasonally available. Their life histories are only weakly related to body size, but show strong seasonal cycles. They grow fast and have relatively (for large cetaceans) short periods of gestation (about eleven months) and lactation (about six months). Groups are usually small and impermanent.

The life history of the female odontocete is very dependent on her size. Smaller odontocetes are vulnerable to predators. They mature relatively young (at five to ten years of age) and have relatively short lactation periods (less than one year). Predators are probably less of a problem for larger odontocetes, which have lower mortality. With evolution taking place in populations near carrying capacity, females of these larger species show all the attributes of "K-selected"[1] life histories, including long maturation, gestation, and lactation periods.

Some female cetaceans, and especially the larger odontocetes, live in highly structured, kin-based groups, which are particularly interesting in several ways. For instance, they can foster the development of alloparental care, as seems to be found in sperm whales. Models and results from terrestrial mammals indicate reproductive suppression developing in such groups, but whether they can be generalized to the marine environment is an open question at present. Finally, there is growing evidence that culture is transmitted vertically within matrilineal groups of cetaceans in a manner that is rare outside human society.

---

1. Although the terms "r-selection" and "K-selection" have largely been discarded in recent years for good theoretical and empirical reasons (Stearns 1992), the principal concerns were with "r-selection," and "K-selected" so beautifully describes the attributes of the larger odontocetes that we have retained it here.

10

MALE REPRODUCTIVE STRATEGIES AND SOCIAL BONDS

RICHARD C. CONNOR, ANDREW J. READ, AND RICHARD WRANGHAM

AT OVER 2.5 m in length, the tusk of the male narwhal *(Monodon monoceros)* is one the most impressive instruments of male-male competition among mammals. Direct observations and head scarring indicate that male narwhals "fence" with their tusks (plate 4; Silverman and Dunbar 1980; Best 1981; Gerson and Hickie 1985; Hay and Mansfield 1989). Our knowledge of narwhal mating behavior has advanced little beyond those initial observations, but long-term studies on other species have illuminated male reproductive strategies in a murky world where, relative to terra firma, sound is more important than vision, the cost of locomotion is low, and maneuverability may play a large role in determining the outcome of male-male contests. Here we review our current understanding of male reproductive strategies and social bonds in whales and dolphins, focusing on variation within and between species as well as comparisons with similar terrestrial systems. We hope to show that further studies on cetacean mating systems will broaden significantly our understanding of ecological influences on mating strategies in both the marine and terrestrial arenas.

Most male mammals do not participate in parental care, but instead direct their reproductive effort toward inseminating females. The strategies individual males employ to gain mating opportunities depend primarily on the distribution of females. Because females direct most of their reproductive effort toward parental care, they are the limiting or "choosy" sex, and are concerned primarily with obtaining enough resources to maximize their reproduction (Trivers 1972). Female distribution is thus determined primarily by the distribution of resources and threats to their ability to translate those resources into reproductive suc-

cess; namely, predators, conspecific competitors, and secondarily, the strategies of males (Bradbury and Vehrencamp 1977; Emlen and Oring 1977; Wrangham 1980a).

Although few species have been studied in detail, we will risk a few generalizations about cetacean mating systems as a preface to the more extended discussion to follow. Only about 5% of mammalian species exhibit obligate monogamy, usually in cases in which male parental care occurs (Clutton-Brock 1989b). There are no known cases of male parental care, or obligate monogamy, in cetaceans. In the vast majority of mammals males attempt to mate with more than one female during a given breeding period, and cetaceans appear to be unexceptional in this regard. A common male mating strategy among terrestrial mammals, resource defence, is currently unknown among cetaceans, likely because of the large ranges and mobility of cetaceans and their prey. The best prospect for finding resource defense among male cetaceans might be in the relatively restricted aquatic habitat of river dolphins. Defense of receptive females by roving males is emerging as a common cetacean mating strategy, with interesting variation in whether females are defended by individuals or alliances, and in the composition of those alliances. Testis size indicates that males and females of some species may be highly promiscuous, and we can expect less emphasis on female defense in those species. Finally, at least one species of cetacean, the humpback whale, appears to form leks.

Sexual Selection in Male Cetaceans

Three kinds of sexual selection are currently recognized: intrasexual competition, mate choice, and mate coercion

(Smuts and Smuts 1993; Clutton-Brock and Parker 1994). Males may compete directly by fighting or indirectly by producing more (and perhaps more competitive) sperm (e.g., Harcourt et al. 1981). Mate choice is usually the prerogative of females in polygynous species (but see Berger 1989). Females may choose males on the basis of their fighting ability (e.g., Cox and Le Boeuf 1977), or males may demonstrate their quality with anatomical and behavioral displays. There are several strong candidates for behavioral displays by male cetaceans that might be used by females to discriminate among males. Mate coercion is known to occur in one species. Below, we review evidence for male-male competition, male displays, and coercion of females before turning to a comparative and ecological analysis of male mating strategies in a few well-known cetaceans.

Male-Male Competition

Fighting

Sexual size dimorphism. Patterns of sexual size dimorphism (SSD, measured as the ratio of male standard length to female standard length) vary markedly among cetacean families. In many mammals, there is a general trend toward greater SSD with increasing body size (Ralls 1977). When data from twenty-nine genera of cetaceans are examined (table 10.1), however, there is no clear relationship ($r^2 = 0.01$) between body size and SSD (fig. 10.1). In both the smallest (river dolphins and porpoises) and largest (baleen whales) cetaceans, females are larger than males (Ralls 1976). Due to their habit of seasonal feeding and fasting, baleen whales are subject to strong selective pressures on energy storage capacity and body size. As noted by Brodie (1975), the large body size of baleen whales allows individuals to store massive energy reserves for use during extended periods away from the feeding grounds. Adult female whales have greater energy requirements than males due to the costs of pregnancy and lactation, so larger females may store more energy. Similar arguments may extend to the small porpoises and river dolphins, although the relationship between energetics and body size in these animals remains unclear.

Within odontocetes as a whole (twenty-four genera), there is a weak positive correlation ($r^2 = 0.19$) between body size and SSD (fig. 10.2). The largest odontocete, the sperm whale, exhibits the greatest SSD of any cetacean, with males one and a half times longer and three times heavier than females (Lockyer 1981a). The direction and degree of SSD among beaked whales is particularly interesting. Females are slightly larger than males in the genus

Berardius, but males are larger in *Ziphius* and *Hyperoodon*. Members of the rich and complex genus *Mesoplodon* appear to be monomorphic in size, although our understanding of patterns of growth in these animals is very limited.

The clearest example of increasing SSD with body size occurs within eleven genera of the family Delphinidae (fig. 10.3). Here there is a clear and positive relationship ($r^2 = 0.50$) between SSD and body size. Only in the smallest genera of dolphins (*Cephalorhynchus*) are females larger than males. The greatest SSD among delphinids occurs in pilot whales, where males are one and a third times longer than females (Kasuya and Marsh 1984).

Similar relationships between the degree of sexual size dimorphism and body size have also been reported for other groups of mammals and for some birds, reptiles, and insects (e.g., Berry and Shine 1980; Ralls 1977; Fairbairn 1990). Until recently this result was difficult to reconcile with sexual selection theory, which predicts that males will be larger than females to a degree predicted by the intensity of male-male competition (Webster 1992). Webster found that New World blackbirds (Icterinae) exhibit a positive relationship between SSD and body size because large species tend to be more polygynous than small species. Further, the relationship between body size and SSD disappeared if the effects of mating system were removed (Webster 1992). Similarly, SSD is strongly related to the operational sex ratio in primates, and a weak positive relationship between size dimorphism and female body weight vanished when operational sex ratio was controlled (Mitani et al. 1996). Webster (1992) and Mitani et al. (1996) used more accurate measures of the intensity of male-male competition than previous analyses. For example, Mitani et al. (1996) incorporated variation in interbirth interval, which strongly affects the number of available females, in their measure of primate operational sex ratio (see also Plavcan and van Schaik 1997). Although data for an equivalent study in delphinids, including female reproductive rates, the duration of the breeding season and of estrus, and the number of estrous cycles before conception, are presently unavailable (see Mitani et al. 1996), such an analysis would be well worth pursuing.

The existence of a profound sexual difference in body size provides a starting point for explorations of mating systems and reproductive strategies in cetaceans. If we knew nothing about sperm whales other than the relative size of males and females, for example, we could at least predict that the species is likely to be polygynous and that male-male competition plays an important role in determining the reproductive success of males. The large body size of a male sperm whale is obviously advantageous in contests

Table 10.1. Data used in allometric analyses of sexual size dimorphism in cetaceans

Species	Female SL (cm)	Male SL (cm)	SSD ratio	Data type[a]	Reference
Cephalorhynchus commersoni	134	130	0.97	P	Lockyer et al. 1988
Pontoporia blainvillei	140	131	0.94	S	Brownell 1989
Sotalia fluviatilis	145	146	1.01	S	Best and da Silva 1984
Phocoena phocoena	157	144	0.92	A	Read and Tolley 1997
Neophocaena phocaenoides	158	166	1.05	A	Shirakihara et al. 1993
Phocoenoides dalli	187	204	1.09	A	Newby 1982
Lagenorhynchus obscurus	189	187	0.99	S	Van Waerebeek 1992
Stenella attenuata	190	206	1.08	A	Perrin et al. 1976
Delphinis delphis	196	208	1.06	S	Evans 1994
Inia geoffrensis	199	222	1.11	A	Best and da Silva 1984, 1989
Platanista gangetica	213	191	0.90	S	Kasuya 1972; Reeves and Brownell 1989
Lissodelphis borealis	217	263	1.21	S	Leatherwood and Walker 1979
Lipotes vexillifer	230	174	0.76	S	Peixun 1989
Tursiops truncatus	249	266	1.07	A	Read et al. 1993
Grampus griseus	276	288	1.04	S	Perrin and Reilly 1984
Delphinapterus leucas	328	377	1.15	A	Doidge 1990
Globicephala macrorhynchus	364	474	1.30	A	Kasuya and Marsh 1984
Mondon monoceros	400	470	1.18	A	Hay 1980
Pseudorca crassidens	452	541	1.20	A	Perrin and Reilly 1984
Mesoplodon europaeus	455	450	0.99	S	Mead 1984; J. G. Mead, personal communication
Orcinus orca	580	670	1.16	P	Christensen 1984
Ziphius cavirostris	599	607	1.01	S	Ross 1984
Hyperoodon ampullatus	747	817	1.09	S	Benjaminsen 1972
Physeter macrocephalus	1,052	1,603	1.52	A	Bannister 1969
Berardius bairdii	1,100	1,070	0.97	A	Kasuya 1977
Eschrichtius robustus	1,297	1,243	0.96	A	Rice and Wolman 1971
Megaptera novaeangliae	1,378	1,298	0.94	A	Chittleborough 1965
Balaena mysticetus	1,520	1,460	0.96	S	Johnson et al. 1981
Balaenoptera physalus	2,270	2,140	0.94	P	Panfilov 1978

Note: Only one species per genus was included to avoid taxonomic bias.
[a]Whenever possible, values of male and female asymptotic lengths were derived from growth curves (A). If such data were unavailable, mean lengths of physically mature male and females were used (P). If neither type of data were available, the mean values of sexually mature individuals were used (S). A minimum of four individuals were required from each sex for estimates of mean length at physical or sexual maturity.

of strength and hitting power. However, the three-dimensional underwater habitat should elevate the relative importance of maneuverability, which favors smaller individuals. Stirling (1975) suggested that selection for greater maneuverability may account for the smaller size of male pinnipeds that fight underwater compared with species that fight on land or ice. Given that all of their contests occur underwater, we might expect selection for maneuverability to influence SSD in cetaceans as well.

We suspect that the importance of maneuverability in contests will be relatively greater in smaller cetaceans. The relative importance of size in winning contests might be greater in large mammals than in small, more maneuverable species (e.g., Geist 1966). This idea is worth considering because it presents an alternative to the "increasing strength of sexual selection" explanation for why SSD increases with body size in delphinids. Consider that there are two components to male combat: "quickness" and "hitting power." Quickness affects the rate at which strikes can

be delivered and avoided, which may be combined into a strike ratio of number of hits delivered/received. In addition to strike ratio, contests will be determined by the power of strikes, which will increase with body size. Quickness decreases with increasing body size, and hitting power increases, but the two factors are unlikely to change at the same rate. Particularly for species that fight in three dimensions, a size increase might produce a small gain in hitting power that is offset by a large loss in quickness. In larger species, increases in body size may produce substantial gains in hitting power that are not seriously offset by small losses in quickness.

Weapons. Whales and dolphins, like many mammals, may use many parts of the body as a weapon. The rake marks and scars present on the bodies of many odontocetes are testimony to the use of teeth as weapons. However, it is likely that the most serious fighting in cetaceans includes strikes with the peduncle, flukes, and other parts of the

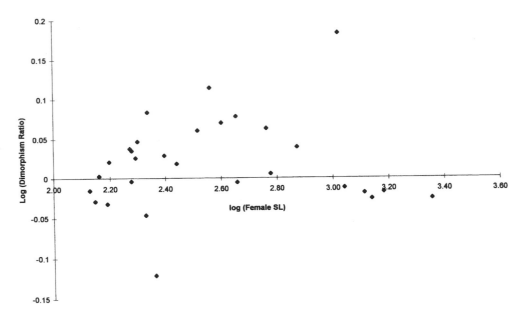

Figure 10.1. Allometry of sexual size dimorphism (SSD) in cetaceans. Data on dimorphism ratios and female standard length (cm) are taken from table 10.1. Only one species per genus is included to avoid taxonomic bias. The correlation between SSD and body size is very weak ($r^2 = 0.01$).

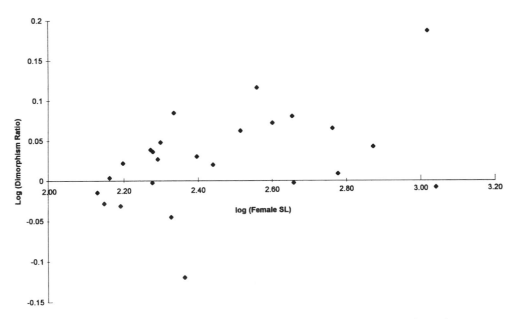

Figure 10.2. Allometry of sexual size dimorphism (SSD) in odontocetes. Data on dimorphism ratios and female standard length (cm) are taken from table 10.1. Only one species per genus is included to avoid taxonomic bias. The correlation between SSD and body size is weak ($r^2 = 0.19$); the regression equation is given by $\log(\mathrm{DR}) = -0.21 = 0.09 \log(\mathrm{SL})$.

body that may not leave obvious external wounds. Forty-two harbor porpoises apparently killed by bottlenose dolphins in the Moray Firth, Scotland, exhibited extensive internal injuries, including fractured ribs and vertebrae, punctured lungs, and ruptured liver capsules (Ross and Wilson 1995). Remarkably, 36% of the battered porpoises exhibited no obvious skin damage. A lethal altercation among captive killer whales, which possess a formidable dental arsenal, was decided by a blow from the peduncle (J. McBain, personal communication). Practically, these observations suggest that reliance upon external scars, often a necessity given the scarcity of direct observations (Norris 1967a), may lead observers to underestimate the frequency and severity of aggression among cetaceans.

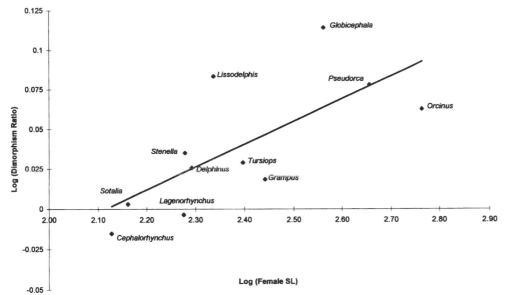

Figure 10.3. Allometry of sexual size dimorphism (SSD) in delphinids. Data on dimorphism ratios and female standard length (cm) are taken from table 10.1. Only one species per genus is included to avoid taxonomic bias. The correlation coefficient is 0.50, and the regression equation is given by $\log(\text{DR}) = -0.30 + 0.14 \log(\text{SL})$.

Odontocetes that feed primarily on cephalopods tend to have teeth that are reduced in number and size. Mac-Leod (1998) suggests that such tooth reduction has allowed selection to co-opt and modify some teeth for combat. The use of modified teeth in aggressive interactions should result in higher levels of scarring, which may have favored scars as advertisements of male quality in squid-eating species. Advertised scars can be produced by slowing the rate at which scar tissue is repigmented (MacLeod 1998), or in the case of Risso's dolphin *(Grampus griseus)*, possibly even by highlighting scars by leaving a border of scar tissue around a center of repigmented skin (Connor 1994). Mac-Leod's model does not explain why scarring is not extensive in species whose teeth may not have been modified for fighting, but are clearly useful as weapons (e.g., killer whales).

Teeth modified for fighting are commonly found in the Ziphiidae (McCann 1974). With the exception of *Tasmacetus shepherdi*, the ziphiids are characterized by the absence of all but one or two pairs of mandibular teeth. These "battle teeth," as they have been called, erupt around the time of sexual maturity in males. Only in Baird's beaked whale *(Berardius bairdii)* are they present in females. From patterns of scarring, it appears that males employ their teeth in tusklike fashion, thrusting at each other with closed mouths, producing single or parallel scars on their victims. Scarring from these encounters is much more prevalent on mature males than on adult females. Additional support for such fights comes from the extremely dense bone that occupies the normally cartilaginous meso-rostral canal in some mesoplodonts (Heyning 1984). The rostrum of the aptly named *Mesoplodon densirostris* contains the densest

bone found in any mammal (Debuffrenil and Casinos 1995).

The battleteeth of northern bottlenose whales are unexceptional and males are not heavily scarred. However, a mass of dense bone supports a forehead in males that is large, square, and flat compared to females (Gray 1882). Gowans and Rendell (1999) observed two males repeatedly head-butting each other. The head-butting was sometimes preceded by tight circling and followed by one male chasing the other.

The most impressive-looking weapon among cetaceans is the tusk of the male narwhal, but are appearances deceiving? Best (1981) concluded that the tusk was likely to be involved in male-male display and assessment, but not combat, because he considered it "fragile" and "brittle." Subsequent investigation revealed that narwhal tusks are not fragile; although they are not capable of withstanding significant longitudinal loads, they are well adapted for withstanding side impacts (Brear et al. 1993). Brear et al. suggested that tusks are used to display to females, and that male fighting is "not to wound" rival males, but to establish that their tusks are honest advertising. As an ornament, the calcium-laden tusk might reflect a male's ability to garner resources. However, the "honest ornament" hypothesis of Brear et al. (1993) explicitly requires that the "honesty" of the ornament designed to impress females be tested, not by the female concerned, but by rival males. This unusual requirement reveals a weakness in their hypothesis. If females require potential mates to test the strength of their tusks in male-male contests, the most likely criterion of female choice would then be simply who wins the contests. We therefore suggest that the function of the tusk is to help males win contests, including the wounding of rival males when contests favor escalated aggression. The lack of observations of escalated fighting with tusks is in keeping with their lethal potential, the fact that narwhals are long-lived iteroparous mammals (Enquist and Leimar 1990), and, just as importantly, that male narwhals have yet to be subjected to systematic behavioral study. The frequent occurrence of broken tusks (34% of 314 individuals: Porsild 1922; 40% of 68: Silverman and Dunbar 1980), a reflection of their fragility to Best (1981), may also result from their use in escalated fighting. Occasionally broken tusk tips have been found embedded in other narwhals, and even in the broken tusks of other narwhals (Best 1981)!

There are a few other morphological modifications that may function as weapons in cetaceans. Many eastern spinner dolphins, *Stenella longirostris,* develop a protruding keel of connective tissue just posterior to the anus, referred to as the postanal hump (Norris et al. 1994). This feature is more prominent in some populations than in others (Perrin 1972). Observations of aggressive encounters between male spinner dolphins in Hawaii reveal that individuals assume a characteristic S-shaped posture prior to fighting, in which the peduncle is bent downward and forward, emphasizing the size of the hump (Norris et al. 1994). Male spinners, like many other dolphins, often use their flukes to strike each other; it is not known what role, if any, the postanal hump may play in such attacks.

Southern right whales *(Balaena australis)* have rough, thickened patches of skin called callosities. Payne and Dorsey (1983) found that callosities were more numerous and larger on known and suspected males than on females, and that scrape marks were also more common on males. They suggested that the scrape marks result from the use of callosities during aggressive encounters.

Sperm Competition

Variation in testis size. In species in which estrous females mate with more than one male, selection should favor males who can produce large amounts of sperm (Parker 1970; Short 1979). To the extent that sperm production capacity is reflected by testis size, the latter may correlate with the type of mating system. In other words, for a given body size, testis size should correlate with female promiscuity (e.g., Harvey and Harcourt 1984).

Kenagy and Trombulak (1986) examined variation in testis size and body mass among 133 species of mammals. The odontocetes in their sample, five delphinids and a phocoenid, had larger testes, relative to body mass, than the other large mammals in the sample. Aguilar and Monzon

(1992) examined testis size, body size, and sexual size dimorphism in 54 species of cetaceans and found that delphinids, phocoenids, kogiids, and balaenids had larger testes than predicted for a mammal of similar body mass, while ziphiids and pontoporids had relatively small testes.

Selection for fighting ability and sperm competition intersects as follows: where males maintain exclusive mating access to multiple females during the entire estrous period, selection for fighting ability will be high and sperm competition low. In monogamous species, selection for both fighting ability and sperm competition will be low. In species in which males compete for access to estrous females but females mate with multiple males, selection for both fighting ability and sperm competition will be high. In species in which males do not compete for access to females and females mate with multiple males, selection for fighting ability will be low and sperm competition will be high. There is empirical support for this type of a classification scheme. Harvey and Harcourt (1984) found that, in primates, species with multiple-male breeding systems tend to have large testes for their body size, and single-male polygynous species have relatively small testes, reflecting a mating pattern in which females mate only with the resident male. Sexual size dimorphism and relative canine size are greatest among polygynous species that live in single-male or multiple-male systems (see also Plavcan and van Schaik 1992).

Do delphinid cetaceans reflect these correlates of sexual selection? To complement our description of sexual size dimorphism in delphinids, we conducted an allometric analysis of testis size in this family (fig. 10.4). We found reliable data on maximum testis mass for ten genera of delphinids; data sets that include both ontogenetic and seasonal variation in testis mass are sorely lacking in all cetaceans, as noted below. There was a significant positive regression between male body size and testis mass in delphinids ($r^2 = 0.69$). Unfortunately, reliable data on body mass are lacking for many of these genera, so it was not possible to conduct a true allometric analysis (regressing testis mass on body mass). Delphinid genera with relatively large testes include *Lagenorhynchus, Delphinus, Grampus,* and *Orcinus;* those with relatively small testes include *Cephalorhynchus* and *Tursiops.* It is worth emphasizing that all of the delphinid testes in our sample are large compared with mammals as a whole (Kenagy and Trombulak 1986). These data suggest that sperm competition is important in all delphinids, unless some aspect of mating underwater favors larger testes independently of sperm competition. The large range of testis mass in this family suggests that, although monogamy seems an unlikely mating system for any delphinid, the degree of sperm competition may vary widely, and that it would be informative to explore further the relationship between testis size and SSD in delphinids.

To examine this relationship, we used the residuals (deviations of each point from the expected value) of the regression equations in figures 10.3 and 10.4. In allometric equations, the residual variation of each point derives from two sources: measurement error in the dependent variable and real deviation of each taxon from the expected relationship (see Relsr 1985). If we plot the residuals of SSD and testis mass against each other, we can divide the genera into four quadrants (fig. 10.5). In quadrant A, we find genera with relatively large testes and low relative values of SSD; these delphinids, such as *Lagenorhynchus*, should exhibit multiple-male breeding systems that involve extensive sperm competition (van Waerebeek and Read 1994). In quadrant B, we find genera, such as *Delphinus*, with relatively large testes and high relative values of SSD, suggesting multiple-male breeding systems in which both fighting and sperm competition play a role in male reproductive success. We note that *Lagenorhynchus* and *Delphinus* are both small, similar-sized delphinids, so sized-based

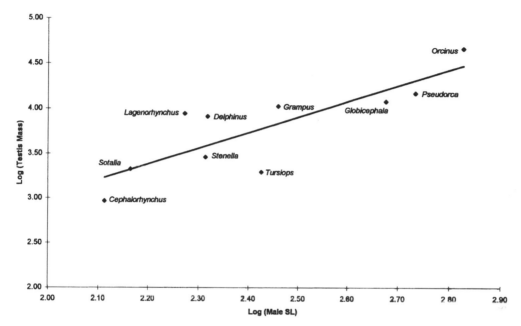

Figure 10.4. Allometry of testis size in delphinids. Data on male standard length (cm) and maximum combined testis mass values (g, without epididymes) are taken from table 10.2. The correlation coefficient is 0.69, and the regression equation is given by: logTM = −0.44 + 1.74 log(SL).

Table 10.2. Data used in allometric analyses of testis mass in delphinid cetaceans

Species	Male SL (cm)[a]	SSD ratio	Testis mass (g)[b]	Reference
Sotalia fluviatilis	146	1.01	2,120	Best and da Silva 1984
Lagenorhynchus obscurus	187	0.99	8,815	van Waerebeek and Read 1994
Grampus griseus	288	1.04	10,600	Perrin and Reilly 1984
Tursiops truncatus	266	1.07	1,966	Perrin and Reilly 1984
Stenella attenuata	206	1.08	2,896	Hohn et al. 1985
Delphinus delphis	208	1.06	8,170	Ross 1979
Cephalorhynchus commersoni	130	0.97	930	Goodall et al. 1988
Pseudorca crassidens	541	1.20	14,800	Perrin and Reilly 1984
Orcinus orca	670	1.16	46,200	Perrin and Reilly 1984
Globicephala melas	474	1.30	12,000	Desportes et al. 1993

Note: Only one species per genus was used to avoid taxonomic bias.
[a]From table 10.1.
[b]Maximal combined testis mass (without epididymes).

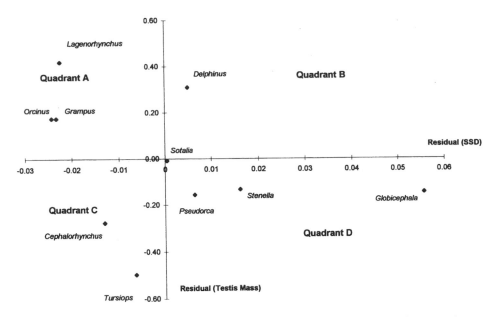

Figure 10.5. Residual analysis of SSD and testis mass in delphinids. Residual values are taken from the allometric equations of SSD and testis mass in figures 10.3 and 10.4, respectively.

differences in selection for SSD (e.g., maneuverability) should not be a factor here.

Taxa with relatively high values of SSD but relatively small testes fall into quadrant C—genera such as *Globicephala,* in which we might expect considerable fighting ability and a reduced role for sperm competition. Finally, quadrant D includes genera such as the small *Cephalorhynchus,* with relatively low values of SSD and relatively small testes. Genera in this quadrant are expected to exhibit less sperm competition and fighting ability, but again, the small size of *Cephalorhynchus* may favor maneuverability over size in fighting.

The genus for which we have the most behavioral information, *Tursiops,* exhibits relatively small testes and a moderate degree of dimorphism. From this analysis, we would predict that sperm competition is relatively less important for bottlenose dolphins than for most other delphinids. This genus is sexually dimorphic in size, but not strikingly so (see Tolley et al. 1995). We caution that the striking variation in male alliance formation among *Tursiops* populations (Connor et al., chap. 4, this volume) may be accompanied by differences in sperm competition and testis size that are not represented in our limited sample. Nonetheless, the difference in testis size between *Tursiops* and *Lagenorhynchus* is striking. Behavioral observations suggest that female bottlenose dolphins mate with multiple males, but that males, cooperating in alliances of two or three individuals, attempt to monopolize females for periods of up to several weeks (Connor et al., chap. 4, this volume). Similar

behavioral data from *Lagenorhynchus* are lacking, but we can predict from their much larger testes that attempts to monopolize females should play a much smaller role in male *Lagenorhynchus* mating success and may be absent altogether.

Other strongly dimorphic odontocetes, such as *Physeter,* also have large testes (Aguilar and Monzon 1992), suggesting that males fight for access to females but that females nonetheless mate with more than one male. Male-male combat is thought to be important in sperm whale mating (Whitehead and Weilgart, chap. 6, this volume), but the mating habits of females are unknown. In other cetaceans, strong male-male competition may be present in the absence of striking sexual size dimorphism. In all eleven species of baleen whales, females are larger than males, even though humpback males engage in vigorous combat for access to females (Clapham, chap. 7, this volume).

Baleen whales exhibit marked variation in maximum testis size (Brownell and Ralls 1986). Right whales have the largest testes of any living animal. One individual with an estimated mass of 74 metric tons had a combined testis mass of 972 kg (Omura et al. 1969). By comparison, a 25.9 m blue whale weighing an estimated 107 metric tons had a combined testis mass of only 70 kg. In general, the balaenids and the gray whale have relatively large testes, while the balaenopterids and the pygmy right whale (the sole member of the family Neobalaenidae) have smaller testes. Brownell and Ralls (1986) caution that many of the testis size values in their analysis may be negatively biased because they were not measured from males on or near the breeding grounds. Like toothed whales, many baleen whales exhibit seasonal changes in testis size. The average testis size in minke whales is 40% greater during the breeding season than during the feeding season (Best 1982). Gray whales migrating toward the breeding grounds have testes that are 70% larger than those of whales migrating away from the breeding grounds (Rice and Wolman 1971).

Differences in the degree of seasonal breeding may confound attempts to evaluate the importance of sperm competition (Kenagy and Trombulak 1986), particularly where maximum testis size is used as the measure of male investment in sperm production (e.g., Brownell and Ralls 1986 and in our analysis here). If males concentrate their sperm production into a short season, then they may have a larger maximum testis size than males that mate with the same number of females over a more extended season (Dunbar and Cowlishaw 1992). Refinement of these interspecific comparisons must await better data sets on seasonal and ontogenetic variation in testis mass and function for all cetacean species.

Penis length appears to follow the trend for testis size in baleen whales. Right and bowhead whales have relatively long penises, roughly 14% of their body length, compared with blue and fin whales, which have penises only 7–9% of their length (Brownell and Ralls 1986). This observation accords with the prediction that species with multiple-male mating should have longer penises, allowing them to place sperm closer to the egg (Parker 1984; for a similar result in primates, see Dixon 1987; Harcourt and Gardiner 1994). Alternatively, penis morphology may be a product of "cryptic" female choice (Eberhard 1996).

Maturation, male fighting, and sperm competition. Unfortunately, with current data, it is not possible to assess patterns of interspecific variation in the attainment of sexual maturity in male cetaceans in any reliable way. The process of sexual maturation in male cetaceans is gradual and unlike that in females, in which the first ovulation defines the attainment of sexual maturity. To date, only a handful of studies have addressed both seasonal and ontogenetic variation in testis size, structure, and function. Without a better understanding of this variation and a better definition of what exactly constitutes sexual maturity in male cetaceans, it is not possible to draw meaningful comparisons between species.

The gradual attainment of sexual maturity in male cetaceans has encouraged a distinction between males that are physically mature (capable of producing sperm) and males that are socially mature (capable of effectively competing for females). We suggest that this dichotomy, insofar as it implies that physically mature but "socially immature" males are not reproducing, is not useful and may even mislead. Why produce sperm at all if there is no prospect of mating? If socially immature males are capable of sperm production, it follows that they must have at least an occasional opportunity to mate, and that such opportunities might arise, not from head-to-head competition with larger males, but via alternative tactics. One prediction we can make at this point is that physically mature males that are not yet large enough to compete directly with larger males may invest more heavily in sperm production, and thus have relatively larger testes (Parker 1990a,b; Stockley and Purvis 1993).

Female Choice

What little we know of female choice in cetaceans is reviewed by Whitehead and Mann in chapter 9 of this volume; here we will be concerned only with male displays. Determining whether male displays are directed at choosy females, rival males, or both is always problematic, and systematic observations do not exist to allow us to make such distinctions in cetaceans. With the exception of the scars on some species and the narwhal's tusk, discussed above, we are unaware of any candidates for male ornaments in cetaceans. We will limit ourselves to a brief mention of a few candidate physical and acoustic displays by males that may increase their attractiveness to females.

Male sperm whales on breeding grounds produce distinctive "metallic" clicks at slow repetition rates that might honestly reflect male size (Whitehead and Weigart, chap. 6; Tyack, chap. 11, this volume). Male humpback whales produce elaborate songs (Clapham, chap. 7, this volume). Although humpback song has not been demonstrated to attract receptive females, its length, complexity, and parallels with bird song are difficult to explain in any other way (Tyack 1981, chap. 11, this volume). Two other balaenopterids, the blue and fin whales, and one balaenid, the bowhead whale, produce vocalizations that may be songs, albeit not nearly as complex as humpback whale songs (Clark 1990). Sufficient recordings have not been made from other balaenopterids to rule out the possibility that they produce simple songs, but the gray and right whales have been recorded extensively and clearly do not (Clark 1990).

Male bottlenose dolphins perform a variety of displays around female consorts, sometimes singly and sometimes synchronously with another male (Connor et al. 1992a). Such displays often occur during consortships maintained by coercion, suggesting that female choice may still operate in coercively maintained associations (see below).

Mate Coercion

In Shark Bay, Western Australia, male bottlenose dolphins cooperate in pairs and trios to form aggressively maintained consortships with individual females that may last from a few minutes to over a month (Connor et al. 1992b, 1996, chap. 5, this volume). Exactly how herding increases the males' chances of mating with the female is unknown, but there are three possibilities (Connor et al. in 1996): (1) herding males could force copulation; (2) they might win a "war of attrition" with a female who has other preferences but a limited receptive period; (3) herding might be a form of mate guarding. If females prefer to mate with multiple males, then the function of herding by males could be as much to prevent a female from mating with other males as to prevent other males from having access to her (Connor et al. 1996).

Connor et al. (1996) also proposed that the risk of infanticide, a form of delayed sexual coercion, might underlie the multiple-cycling reproductive strategy of female bottlenose dolphins in Shark Bay. The first evidence of infanticide among wild cetaceans was uncovered recently in bottlenose dolphins in the Moray Firth, Scotland (Patterson et al. 1998), and off the coast of Virginia (Dunn et al. 1998). In general, many odontocetes exhibit characteristics that are thought to favor infanticide by males in terrestrial mammals, including (1) breeding that is often seasonal but broadly so, allowing time for females that have lost infants to conceive again in the same season (see Hrdy 1979); (2) a lactation period that exceeds the duration of gestation, so that postpartum mating cannot be employed to eliminate the advantage of infanticide (van Schaik and Kappeler 1997); and (3) year-round association between males and females, which van Schaik and Kappeler (1997) have argued is a strategy in primates to reduce the risk of infanticide by "strange males." In odontocetes, predation risk offers a strong alternative explanation for this last characteristic (Connor, chap. 8, this volume).

Mating Strategies and Social Bonds in Male Cetaceans

Are Any Cetaceans Monogamous?

The relatively small testis size and lack of sexual size dimorphism in the boto (*Inia geoffrensis*) led Best and da Silva (1984) to suggest that it might possess a monogamous mating system. Our analysis indicates that, although its testes are relatively small (at least compared with delphinids of similar size), the boto exhibits moderate sexual size dimorphism (dimorphism ratio = 1.11). Brownell (1989) noted that another river dolphin, the franciscana (*Pontoporia blainvillei*), exhibits reversed sexual dimorphism (dimorphism ratio = 0.94) and relatively small testes, characteristics suggesting a single-male mating system (quadrant D in fig. 10.5). Brownell also suggested that, due to the rarity of paternal care in mammals, it is unlikely that the franciscana is monogamous. We also consider river dolphins unlikely prospects for monogamy, but believe that studies of the social behavior of remaining river dolphin populations will provide fascinating contrasts with the mating systems of delphinids. Particularly interesting comparisons could be made with smaller delphinid species that share similar features of dimorphism and testis size, such as *Cephalorhynchus*.

Aggregating versus Solitary Baleen Whales

Of the eleven recognized species of baleen whales, three are known to aggregate on calving grounds, where extensive mating takes place. One of the aggregating species is the gray whale *(Eschrichtiidae)*; the other two, the right and humpback whale, are balaenids. While similar aggregations might occur but remain undetected in some cases, it is unlikely that aggregations have been missed in many balaenopterids. Behavioral biologists might not have thoroughly scoured the seas for blue and fin whale breeding grounds, but the economically motivated whalers did, without success (Payne 1995). Why some species aggregate and others do not is unclear. Mating strategies of the dispersed balaenopterids are virtually unknown, but the aggregating species have been reasonably well studied.

The differences between the aggregating baleen whales are striking. Two of the aggregating species, the gray and right whales, have very large testes for their body size. Although some male-male fighting may occur in right whales (Payne and Dorsey 1983; see above), it is not pronounced, and behavioral observations of multiple-male mating groups, including simultaneous erections, support the hypothesis that females mate with multiple males in both species (Swartz 1986; Payne and Dorsey 1983; Payne 1995). Even simultaneous intromission has been reported in right whales (Brownell and Ralls 1986). In contrast, the humpback whale has smaller testes and a smaller penis, and males engage in intense fighting over females (Clapham, chap. 7, this volume).

The bowhead whale appears to share some characteristics with all three of these species. The relative size of bowhead testes is similar to that of gray whales; they are not nearly as large as those of right whales, but much larger than those of humpback whales (Brownell and Ralls 1986). Conceptions in the North Pacific population are thought to occur during the late winter and early spring (Koski et al. 1993), and mating behavior observed during the spring migration resembles that of right whales (Würsig and Clark 1993). On the other hand, bowheads produce vocalizations that appear to be simple songs; while not nearly as complex as humpback whale songs, they are significant in this comparison because right and gray whales do not sing at all (Würsig and Clark 1993). Further studies of the bowhead whale would clearly be rewarding, especially during the late winter, when many conceptions are thought to occur.

Few explanations have been offered for the variation in baleen whale mating systems. Swartz (1986) suggests that a key feature might be the length of the mating season, pointing out that the gray whale mating season is much shorter than that of humpback whales. With more females present at a given time, males could find females more difficult to monopolize. Any hypothesis is necessarily tentative at this juncture, given that the patterns of female receptivity and the percentage of conceptions that occur on calving grounds for the aggregating species are unknown (Brownell and Ralls 1986).

Do Humpback Whales Lek?

Several authors (e.g., Mobley and Herman 1985; Clapham 1996, chap. 7, this volume) have noted that the humpback whale mating system does not appear to violate Bradbury's (1985) restrictive definition of lekking: (1) no male parental care, (2) an arena where males gather and females come to mate, (3) resource-free display sites, and (4) opportunities for females to exercise mate choice (Clapham 1996, chap. 7, this volume). Höglund and Alatalo (1995) define a lek more broadly as an "aggregated male display that females attend primarily for the purpose of fertilization." There are several points that are worth further discussion here. First, male-male dominance interactions and territoriality are common in leks (and in some definitions of lekking: e.g., Alcock 1998), but are not always used as distinguishing criteria and are not part of the above definitions (see Höglund and Alatalo 1995). Thus, in contrast to Clapham (1996), we do not feel that data on male-male interactions will be critical for deciding whether humpback whales lek. Second, lekking behavior is sometimes found in association with other male mating strategies (see Höglund and Alatalo 1995). Thus, if male-male combat in "rowdy" groups is an alternative strategy to displaying, then such behavior does not bear on the question of whether dis-

playing males are lekking. Neither is the mobility of some singing humpbacks of particular relevance given that males shift display sites in some terrestrial species as well (Höglund and Alatalo 1995).

One possible challenge to the lek hypothesis for humpback whales is found in the second of Bradbury's criteria and in the definition of Höglund and Alatalo: are female humpback whales gathering in breeding areas to mate with males, or for some other purpose? Females may seek breeding sites on protected banks because they are safer for newborns (Whitehead and Moore 1982), but this hypothesis does not account for the presence of nonlactating females (Clapham 1996). At any rate, the issue is not whether these sites are a resource females require for safe breeding, but whether males control access to breeding sites in order to obtain matings (Bradbury 1985)—and they do not. The island breeding sites of humpback whales may be "landmarks," similar to the hill or treetop breeding sites of some insects (Thornhill and Alcock 1983). However, some hypotheses for the evolution of landmark breeding aggregations, and leks in general (the "hotspot" model: Bradbury et al. 1986), begin with females concentrating for one reason or another (see Thornhill and Alcock 1983). We therefore suggest that, in accordance with the definitions of Bradbury (1985) and Höglund and Alatalo (1995), male humpback whales that sing on island arenas are lekking, and that this conclusion will not be altered by future findings on male-male interactions, male movements, alternative male mating strategies, and the reasons why females began to visit islands in the first place.

Roving Odontocetes

Where females range widely and live in small groups or solitarily, males may rove in search of estrous females, whom they attempt to consort with or guard (Clutton-Brock 1989b). Although it makes intuitive sense that males should rove between dispersed, solitary females, what if females live in groups? Whitehead (1990c) modeled the options of roving versus residency for males seeking mates in non-territorial species with group-living females. In general, males are expected to rove if the duration of female estrus is greater than the time it takes males to travel between groups. Surprisingly, this result was independent of female group size (Whitehead 1990c).

With our present, albeit limited, knowledge of cetacean mating systems, there are no obvious candidates for residency by breeding males, and it appears likely that many male cetaceans are rovers. The more solitary, non-aggregating baleen whales are likely rovers, and males may rove among females in aggregating species as well. In some odontocetes, such as the bottlenose dolphin, females do not live in stable groups, and males rove in search of estrous females, whom they attempt to herd or guard (Connor et al., chap. 4, this volume). Some, like the sperm whales observed by Whitehead (1993), do live in stable groups. The typical roving male mammal is a solitary figure, but many male cetaceans rove in groups. Here we focus on two problems. First, given a roving strategy, when should individuals rove alone, and when should they rove in groups? Second, given that individuals rove in groups, with whom should they rove?

Roving Alone

The largest toothed whale, the sperm whale, also presents one of the most extreme cases of sexual size dimorphism among mammals. Male sperm whales may reach over three times the mass of mature females (Whitehead and Weilgart, chap. 6, this volume). Sperm whales were long suspected to be polygynous, although ideas about the exact nature of sperm whale polygyny varied (Best 1979). It was often assumed that female schools were attended by a male for the duration of the mating season in a kind of harem polygyny (e.g., Berzin 1972; Caldwell et al. 1966; see Whitehead 1993). By following sperm whale groups for days at a time, however, Whitehead (1993) found that mature male sperm whales were transient features in female groups, staying for variable periods, but typically about eight or nine hours, with a particular group. Male sperm whales move between female groups, probably in search of estrous females (Whitehead 1993; Whitehead and Weilgart, chap. 6, this volume). Roving is expected in species such as sperm whales and elephants in which males continue to grow long after attaining sexual maturity, resulting in large differences in male fighting ability (Whitehead 1994).

The feeding requirements of attaining large size may explain why male sperm whales disperse from their natal group and why they cease to associate with other males as they grow (Best 1979; Whitehead and Weilgart, chap. 6, this volume). Male sperm whales range into higher latitudes than females (Whitehead and Weilgart, chap. 6, this volume), and it is likely that the resources (larger squid) they obtain in these regions enable them to achieve a larger size than they would if they remained in tropical waters (Best 1979). Prey dispersion may also increase with prey size, which would increase feeding competition among males and limit their ability to associate as they mature. Feeding competition would remain a factor even for males that have attained large size, given their need to rebuild fat stores between visits to the tropical breeding grounds (Weilgart et al. 1996).

Roving Alone versus Roving in Alliances

Given the pattern of bond formation among some smaller, less dimorphic odontocetes (see below), and the fact that smaller male sperm whales do travel together, it seems fair to ask why sperm whales have not opted for an alternative strategy of growing to a smaller size (at which feeding competition might be less severe) and forming male-male bonds. Here we outline two conceptual models that might explain why some male cetaceans rove in groups and some do not.

Encounter rates. If males form alliances because of conflicts with other males, alliance formation will be favored in habitats where males encounter each other frequently. If a male has a low probability of encountering a rival male while attending a female, he should be better off traveling alone. At higher encounter rates, it should be better to travel with an alliance partner if the cost of sharing copulations is less than the benefit of gaining access to more females. Other factors being equal, males should encounter each other more often in areas of higher population density. Gamboa (1978) found that *Polistes* wasp co-foundresses occurred more often in areas of greater nest density. He suggested that single wasps were more likely to have their nests usurped in high-density areas, so that the cost of nest sharing was outweighed by the benefit of greater nest defense. A priori, sperm whales, and other large cetaceans, should be found at lower population densities than smaller cetaceans. Thus, unless female estrous periods increase in length proportionally with increasing size, a male cetacean of a larger species tending a female might expect to encounter rivals less frequently than a male of a smaller species. However, "other factors" will often not be equal, and these may significantly affect male encounter rates; such factors include temporary concentrations of food during the breeding season, individual variation in ranging patterns, and the distance at which males can detect females in estrus. Variation within and between species in the ecological factors that affect male encounter rates will allow future testing of this hypothesis.

Relative body size and the utility of alliances. Is there something about being very large that favors roving alone versus roving with others? Alliances may give two smaller individuals an advantage over a larger or more dominant one (e.g., Noë 1994), but the degree to which this holds true may not be constant across a range of body sizes. The costs of becoming larger (e.g., the costs of traveling to higher latitudes, or the energetic costs of growing additional tissue) and the benefits of an additional increase in body size may change as body size increases. For example, if hitting power increases exponentially with size, alliances in very large species may be relatively ineffective against larger individuals compared with alliances in smaller species.

Roving with Other Males

Female alliances usually occur in contests over resources that the victors can share (Wrangham 1980a; van Hooff and van Schaik 1992). With a litter size of one, a female cetacean in estrus does not present a divisible resource; mating can be shared, but fertilizations cannot. Thus, the conditions for male alliance formation should be more restrictive than those for female alliance formation (van Hooff and van Schaik 1992, 1994). The only well-documented example of alliance formation in cetaceans is found in bottlenose dolphins, particularly among male dolphins in Shark Bay, Western Australia (Connor et al. 1992a,b, chap. 4, this volume). However, reports of strong associations between males in several species suggest that male alliance formation may be present in other cetaceans, including baleen whales. In Hawaii, strongly bonded groups of three to six spinner dolphins (*Stenella longirostris*) were composed mostly, but not entirely, of males (Östman 1994). In the eastern tropical Pacific, subgroups of three to eight adult male spotted dolphins (*S. attenuata*) were observed swimming in unison while their schools were trapped in tuna nets (Pryor and Shallenberger 1991). Strong male-male bonds have also been reported in Atlantic spotted dolphins (*S. frontalis*: Herzing 1996). In *resident* killer whales of British Columbia, some males exhibit their strongest bond with other males. These other males are likely to be relatives, as male *residents* do not disperse from their natal pods (Baird, chap. 5, this volume). The first behavioral study of a ziphiid, the northern bottlenose whale (*Hyperoodon ampullatus*), revealed consistent associations among males (Gowans 1999). Mature male sperm whales compete for females on a solitary basis (Whitehead 1993), but mass strandings by all-male groups (see Whitehead and Weilgart, chap. 6, this volume) and loose aggregations of males at sea (Caldwell and Caldwell 1966; Christal and Whitehead 1997) raise the possibility of male-male bonds in this species.

Male humpback whales usually compete singly over estrous females, but several researchers report the occasional occurrence of male pairs that are observed together repeatedly in competitive groups, but do not behave agonistically toward each other (Clapham et al. 1992; Brown and Cork-

eron 1995; Clapham, chap. 7, this volume). In gray whales, Samaras (1974) reported trios of two males and one female, which he interpreted as a mating couple and an additional helper male. Brownell and Ralls (1986) suggested that the males might be in direct competition to mate with the female gray whale, but a third possibility is that the males are a coalition or alliance, cooperating to guard the female.

As noted earlier, the very large testes of right whales suggest a highly promiscuous mating system. Intense male-male fighting of the sort observed in humpback whales (Clapham, chap. 7, this volume) is not apparent in right whales; rather, several right whales may attempt to mate with a single female simultaneously. Payne (1995) suggests that males may even help each other to mate with reluctant females (who roll belly-up to avoid their suitors) by pushing the female underwater, where another male may more easily perform intromission. Alternatively, the mating male might simply be parasitizing the other males' efforts to push the female down for mating. Whether these male groups are stable over time or composed of relatives is unknown.

Roving with Females

Mother-son alliances in killer whales. The overall community structure of *resident* killer whales is well established, and patterns of grouping and philopatry are known (Baird, chap. 5, this volume), but our knowledge of social relationships among these animals is poor. Nevertheless, there are some intriguing observations that are not easily explained by current hypotheses about the social and ecological basis for killer whale relationships.

To summarize, researchers have identified two communities of *resident* killer whales in the coastal waters of British Columbia and Washington State. Individuals within communities associate in pods, and pods associate only with other pods from their community, although the ranges of the two communities overlap (Baird, chap. 5, this volume). Multiple-pod associations occur regularly within killer whale communities, and a "greeting ceremony" often occurs when pods from the same community meet, including rubbing and socio-sexual behavior (Osborne 1986).

The most stable associations within pods are matrilineal units, which range from two to nine individuals and include up to four generations (Baird, chap. 5, this volume). A typical matrilineal unit contains a grandmother, her adult son, her adult daughter, and the offspring of her daughter. Individuals rarely spend more than a few hours apart from other members of their matrilineal unit. One to eleven matrilineal units associate in subpods, which temporarily travel apart from other members of the pod. Pods contain one to three subpods. A process of gradual fissioning along maternal lines is thought to be responsible for the formation of new matrilineal groups, subpods, and pods.

Two kinds of observations suggest that females may bond more closely with their adult sons than with their adult daughters and other pod members. First, mothers maintain closer associations with their adult sons than with their adult daughters. This is reflected in association patterns and in swimming formation: next to juveniles, adult sons swim closest to the mother, in a position either beside or slightly behind her (Bigg et al. 1990b). Bigg et al. suggested that when a daughter has a calf, her bond with her mother weakens. Second, matrilineal groups containing adult males tend to travel more independently of other pod members than groups without them. For example, one matrilineal group with two adult males often left the rest of the pod to travel alone or with another pod.

How can we explain strong mother-son bonds in killer whales? Previous discussions of male philopatry have focused on the benefits provided by adult males to female relatives and their offspring. However, of the six matrilines with adult males that Bigg et al. (1990b) described as traveling independently of other pod members, three contained only the mother and her one or two adult sons, with no juveniles that the adult male might be able to assist. Further, some groups contain only postreproductive females and their adult sons, so these sons cannot be investing in their mothers' future offspring either. We suggest that mother-son associations are based on benefits provided by the mother to her son. Old females may aid their sons' foraging efforts, or they may form effective alliance partners for their sons in agonistic encounters with other males. Such alliances have not been described in killer whales, but this may reflect both poor observation conditions and a true rarity of escalated aggression in this species (Baird, chap. 5, this volume). Scarring suggests that agonistic encounters do occur (Heimlich-Boran 1988).

There are two other "foraging" hypotheses worth considering here. First, the weakening of the mother-daughter association with the birth of a grandoffspring may result from changes in the ability of individuals to participate in cooperative foraging. Cooperative foraging in *residents* involves herding or concentrating schools of fish (Heimlich-Boran 1988). Mothers may prefer not to forage with daughters burdened by a calf, unless the mother has a calf as well. Another possibility is that mothers with calves may need to forage separately from others, either because of a shift in the mother's dietary needs or for calf protection (e.g., Bernard and Hohn 1989 for spotted dolphins, *Sten-*

ella attenuata). Both of these hypotheses predict that mothers with an adult daughter and son should preferentially accompany the son when the daughter has a calf, but they should accompany the daughter when both have young calves.

Bain (1989) suggested that larger males may dive deeper to feed than females, thus dispersing "ecologically" while maintaining social philopatry, which allows the males to assist offspring of related females. However, the available evidence suggests that males do not dive deeper than females (Bisther and Vongraven 1995; see also Baird, chap. 5, this volume). At any rate, this hypothesis does not explain why males travel more with postreproductive females than with other female relatives with dependent offspring.

Mother-son bonds in other cetaceans. Natal philopatry by both sexes and strong mother-son bonds may be widespread in the subfamily Globicephalinae, which, in addition to pilot whales, includes pygmy killer whales *(Feresa attenuata)*, false killer whales *(Pseudorca crassidens)*, and melon-headed whales *(Peponocephala electra)*. (Le Duc et al. [1999] omit killer whales from the Globicephalinae and include Risso's dolphin, *Grampus griseus*.) Natal philopatry would help explain otherwise puzzling behavioral observations such as those of Porter (1979), who observed a pod of twenty-nine false killer whales remaining in tight formation around a large dying male for three days.

Apart from killer whales, the best data on natal philopatry are found in the long-finned pilot whale *(Globicephala melas)*. Using microsatellite DNA analysis on complete schools of 90 and 103 long-finned pilot whales captured in drive fisheries in the Faeroe Islands, Amos et al. (1993) found that mature males neither mate within nor disperse from their natal pods. They suggested that male pilot whales may encounter potential mates either by temporarily leaving their natal group to rove in search of mates or by encountering mates while traveling with their natal group. The schools analyzed by Amos et al. were likely aggregations of several pilot whale pods. Sergeant (1962) reported that the average size of schools captured in drive fisheries of long-finned pilot whales is several times larger (mean = 85) than reported for pelagic schools of the same species (mean = 20). Similarly, in short-finned pilot whales *(G. macrorhyncus)*, larger aggregations tend to be captured in drive fisheries than is typical of offshore sightings (Kasuya and Marsh 1984).

Heimlich-Boran (1993) studied the association patterns of 559 short-finned pilot whales in the Canary Islands. Using terminology from killer whale studies, he tentatively identified thirty-one *resident* pods, in which most individuals were observed on multiple occasions and were occasionally found in mixed-pod associations. Fifteen *transient* pods were sighted only once, and never with pods of *resident* whales. Although sample sizes did not permit analysis of intrapod association patterns, the strong sexual dimorphism in this species allowed identification of mature males within *resident* pods. Pod size ranged from two to thirty-three individuals, and the number of mature males per pod ranged from zero to six. The pod sizes reported by Heimlich-Boran are similar to those in other reports of pilot whales at sea, and strikingly similar to those of the *resident* killer whales studied by Bigg et al. (1990b).

These studies raise two issues. First, the large schools of pilot whales captured in drive fisheries and analyzed genetically might be aggregations of smaller, more stable "pods" similar in size to pods found in *resident* killer whales. In this case, the genetic analysis of Amos et al. (1993) does not rule out the possibility that males may move between closely associating pods. Second, the large schools may be the equivalent of *resident* killer whale clans. Killer whale clans were discovered based on shared calls. The degree of call sharing is thought to reflect a history of pod fission, but because call sharing does not correlate well with current interpod association patterns, clans should not be considered a level of killer whale social structure. The drive fishery data suggest that the "clan" may be an important unit of social structure in pilot whales. Both issues could be addressed with a combined study of association patterns and genetic sampling. Clearly, results to date indicate that further studies on the social systems and ecology of globicephalids may reveal fascinating variation on a theme (natal philopatry by both sexes) that is rare among terrestrial mammals.

The Unusual Case of Baird's Beaked Whale

As in many beaked whales, female Baird's beaked whales (*Berardius bairdii*) are slightly larger than males, but this species is the only one in which both males and females have paired "battle teeth." Kasuya et al. (1997) presented data from a Japanese *Berardius* fishery showing that males attain sexual maturity four years before females (at six to eleven vs. ten to fifteen years of age); males live thirty years longer than females; the adult sex ratio is strongly biased toward males; and females exhibit high annual ovulation (0.47) and pregnancy (0.30) rates. The ratio of sexually mature females to males in their sample was 1:3.3. Twenty-two of the males, but no females, were between fifty-five and eighty years old. Kasuya (1995) and Kasuya et al. (1997) speculated that male Baird's beaked whales

care for weaned calves, allowing females to shorten the interval between births.

How could such a system evolve? Kasuya (1995) suggested that either females move between groups or, as in *resident* killer whales, male and female relatives remain together for life. As noted above, a suggested benefit of social philopatry by male killer whales is the opportunity to help care for dependent relatives in their pod (Baird, chap. 5, this volume). It is possible that Baird's beaked whales have traveled even further down this path. Since male confidence of paternity is not likely to be high, Kasuya et al. (1997) favor the natal philopatry hypothesis, in which males are caring for the offspring of close relatives (e.g., their mother or sisters). It is unclear, however, that the male parental care hypothesis can account for the shorter female life span, which should be associated with females taking greater reproductive risks. The typical mammalian pattern of higher mortality and shorter life spans for males is associated with male-male competition for mates. In polyandrous species, females may compete for mates who engage in parental care, but this hypothesis fails if care-giving male Baird's beaked whales are related to the females. Alternatively, a shorter female life span might be obtained if females suffer the additional burden of participating in mate competition on behalf of their male relatives or protecting related calves in the group from infanticidal males. This hypothesis, in which alliances are composed of related males and females, could explain the presence of battle teeth in females.

Could sampling biases account for the Baird's whaling data? Kasuya et al. (1997) maintain that because males and females are similar in size, it is unlikely that the data would be biased by gunners making size-based selections. They further argue that geographic sexual segregation is unlikely to explain the "missing females" because previous whaling efforts in the same region, but covering an area encompassing three populations of *Berardius*, found a similar male-biased sex ratio (Omura et al. 1955). A third possible bias is behavior: older females might be more solitary (and less visible) or avoid boats more. Kasuya et al. (1997) cannot exclude this possibility, but consider it unlikely to explain the total lack of females over fifty-four years old. The significance of the data on interbirth intervals also remains unclear. Without knowing infant mortality rates (which could be high for a number of reasons, including hunting pressure), it is impossible to know whether the high ovulation and pregnancy rates of Baird's beaked whales reflect interbirth intervals that are shorter than those of similar-sized odontocetes. Clearly, the social structure of this intriguing species merits further scrutiny.

Variation in Male Bonds and Reproductive Strategies

Variation Within Populations: Alternative Mating Strategies

It is not uncommon to find males in the same population employing different mating strategies, or individual males switching between strategies (e.g., Arak 1988; Caro and Bateson 1986; Gross 1996). Several mechanisms have been offered to explain how selection can sustain more than one mating strategy in a population. Genetically based alternative or mixed strategies are rare (Gross 1996). More commonly, different strategies (or "tactics": Gross 1996) have unequal payoffs and are conditional upon environmental cues, possibly in combination with genetic influences, that indicate their likelihood of success (e.g., Caro and Bateson 1986; Arak 1988). Individuals may use their relative fighting ability or dominance rank in strategy selection (e.g., Hogg 1984).

In species in which females choose mates on the basis of male displays, males often differ in attractiveness (Arak 1988). The optimal strategy for an unattractive male in such circumstances may be to employ a different strategy, such as searching actively for females, attempting to intercept females by remaining near a displaying male as a "satellite," or "sneaking" copulations. However, the alternative strategies of males may also change the cost-benefit equation of choice for females to the males' advantage by, for example, making it too costly to continue searching for higher-quality males.

One of the most promising candidates for alternative mating strategies in cetaceans comes from observations of male humpback whales on their low-latitude breeding grounds (Tyack and Whitehead 1983; Clapham, chap. 7, this volume). Although mating has not been observed in humpbacks, several kinds of male behaviors have been described that are thought to reflect different mating tactics. Single males produce "song," an acoustic display that may attract potential mates and repel other males (Clapham, chap. 7, this volume). Frankel et al. (1995) found that singers were spaced farther apart than non-singers, and suggested that this finding supports an intrasexual display role for singing. However, given that some proportion of non-singers likely were females, their result is questionable. Further, a spacing function for singing does not automatically imply competition. Arak et al. (1990) found that female bushcrickets *(Tettigonia viridissima)* choose the nearest male and do not travel among a number of males, comparing them. As a result, singing male bushcrickets that were regularly spaced attracted more females than males that

were clumped. Because each male benefits from dispersion, the spacing among singing crickets should not be considered an outcome of male-male competition. The possibility remains that male humpback singers might compete for a limited number of singing locations or for particular areas containing more females, but this has not been demonstrated.

On the other hand, as Frankel et al. (1995) and others have argued, the complexity of humpback song implies a role for female choice. Arak et al. (1990) suggested that choosing the closest male may be adaptive if there are significant movement costs to exercising further choice (Searcy and Anderson 1986). Female humpback whales may exercise more choice than female bushcrickets, but there may also be costs to choice for female humpback whales that limit their movements on the breeding grounds and maintain the advantage of a regular spacing system among males (see below).

In "competitive" or "rowdy" groups, more than one male travels with a female and competes with the others for the "principal escort" position next to her. Such groups are characterized by male-male aggression, often resulting in bloody wounds. Females in competitive groups may seek to associate with fighting males to ensure that they mate with a male of high quality. If this is the case, we should expect females to travel extensively through the breeding grounds to attract a cohort of competing males, much like estrous red kangaroos (Jarman 1991).

That individual males may switch from one behavior to another is indicated by observations of males interrupting song bouts to join a competitive group (Tyack 1981). Single male escorts sometimes accompany females with calves, which has been interpreted as a strategy in which males anticipate the possibility that a female will come into estrus (Clapham, chap. 7, this volume). Alternatively, these groups might represent associations between estrous females and single males that have not been discovered by other males.

No data are yet available to determine whether individual males employ one of these behaviors more than another, or, if they do, whether their tactics are age-related (see Caro and Bateson 1986). Clapham (1996, chap. 7, this volume) suggested that singing is a secondary strategy available to males that are unable to compete successfully in competitive groups. This hypothesis runs counter to the more common finding that secondary tactics reflect "low-quality males attempting to circumvent female choice" (Sullivan 1989). If females use singing as a choice criterion, then males that are unable to attract females by singing may become satellites or rovers that attempt to intercept females moving toward singers. Tyack (1981, 1983) found that singers and other humpback whales commonly charged toward playbacks of sounds from competitive groups and that most groups moved away from playbacks of song (but see Mobley et al. 1988; Clapham, chap. 7, this volume, for interesting exceptions). Both of these results are consistent with the hypothesis that singing is the primary strategy among male humpback whales, but are inconsistent with the hypothesis that direct competition is primary. If singers are unable to compete in competitive groups, they should not approach them, and groups should not avoid singers.

A neglected possibility is that alternative male mating strategies are maintained by alternative female mating strategies. The usual notion of alternative strategies being used by low-quality males attempting to thwart female choice puts the entire focus on females attempting to maximize benefits by choosing high-quality males, rather than on females attempting to minimize mating costs. If females differ significantly in mating costs (we will assume that the benefits of mating with high-quality males remain equal), then alternative female choices can maintain alternative male strategies. Some females on breeding grounds are accompanied by calves, while others are not. The cost of participation in competitive groups could be higher for estrous females accompanied by calves, who might prefer single male escorts or to choose from among several solitary singers. Unencumbered females might enjoy the benefits of mating with a high-quality male in a competitive group at less cost.

Alternative male strategies have been described in one other cetacean, the bottlenose dolphin. Wells et al. (1987) described two patterns of behavior in male bottlenose dolphins in Sarasota Bay, Florida. Most males in Sarasota form strong bonds with other males and travel widely. A few males, however, are solitary and tend to exhibit smaller ranges. Preliminary genetic analyses indicate that both single males and males in pairs father offspring, but it is not yet clear which strategy is more successful (Duffield et al. 1994; Connor et al., chap. 4, this volume). In Shark Bay, Australia, only older males who have lost their alliance partners are solitary, suggesting that solitary males might simply be making the best of a bad job (see Connor et al., chap. 4, this volume). Bottlenose dolphins in Sarasota differ in whether they form alliances or not, but in Shark Bay, two different patterns of alliance formation exist among males, whose ranges overlap extensively (Connor et al. 1999). In addition to the previously described stable alliances of two or three individuals that associate with one or two other alliances (Connor et al. 1992a,b), Connor

et al. (1999) recently found highly labile trio formation within a group of fourteen males. Males in this "super-alliance" formed coalitions of three individuals that herded individual females, but the membership of the trios changed often as individuals in the super-alliance switched alliance partners. Members of the super-alliance also attacked teams of stable alliances (Connor et al. 1999). The basis for this variation, which has no parallel in other mammals, is not known.

There is one other mammal, the cheetah *(Acinonyx jubatus)*, in which males form alliances of two or three individuals and herd individual females (Caro 1994a). Male cheetahs exhibit two strategies: non-territorial males are solitary and travel widely and furtively, appearing to avoid resident males, who cooperatively defend territories with one or two other males (Caro 1994a). Physiological differences support the hypothesis that non-territorial male cheetahs are unable to compete with territorial males (Caro et al. 1989; Caro 1994a). Non-territorial males exhibit higher cortisol levels (suggesting higher stress levels), higher white blood cell counts, and are in poorer condition overall than territorial males (Caro et al. 1989; Caro 1994a). Caro (1994a) argues that the poor condition of non-territorial males is a result of a stressful solitary tactic, perhaps forced on them by unfavorable demographic circumstances. Like bottlenose dolphins, cheetah males sometimes became solitary after loosing an alliance partner. In other respects the two systems differ. Unlike solitary cheetahs, solitary dolphins travel less extensively than alliances, and there is no evidence that either solitary dolphins or dolphins in alliances are territorial. Is it not yet known whether or not solitary male bottlenose dolphins are in poorer condition than males in alliances.

Variation Between Populations

Sympatric populations of killer whales. Transient killer whales live sympatrically with *resident* killer whales, but the two exhibit striking behavioral differences (Baird, chap. 5, this volume). *Transient* killer whales specialize on marine mammal prey, whereas *resident* killer whales feed mostly on fish. Baird (chap. 5, this volume) makes a convincing case that the optimal group size of *transient* killer whales hunting seals is three individuals, which places a strong constraint on their group size that is not present in *resident* killer whales. The smaller group sizes of *transient* killer whales strongly affect the numbers and kinds of social bonds between adults. In particular, it is becoming increasingly evident that some male *transient* killer whales, apparently second-born sons, disperse from their maternal group (Baird, chap. 5, this volume). Observations suggest that dispersing males continue to travel alone rather than joining other groups. Thus, some male *transient* killer whales may be forced into a solitary, roving strategy if they leave their maternal group.

Allopatric populations of bottlenose dolphins. Alliance formation is clearly an important mating strategy in male dolphins in Shark Bay, Western Australia, but alliances, and especially second-order alliances, may not be a constant in all bottlenose dolphin populations (Connor et al., chap. 4, this volume). In Sarasota Bay, Florida, most adult males form pairs, but there appear to be more solitary males and fewer trios than in Shark Bay (Connor et al., chap. 4, this volume). Consistent associations between alliances have not been reported elsewhere, nor has anything like the super-alliance in Shark Bay been observed elsewhere. The strongest contrast to Shark Bay is found in an isolated population in the Moray Firth, Scotland, where several years of observation have failed to reveal any consistent associations between adult-sized dolphins similar to those of alliance members in Shark Bay or male pairs in Sarasota (Wilson 1995).

Long-term studies of bottlenose dolphins in different habitats should provide an opportunity to elucidate important ecological influences on social bonds (Connor et al., chap. 4, this volume). The preliminary indications of differences in male association patterns between the Sarasota, Shark Bay, and Moray Firth sites are especially exciting in this regard (see also Smolker et al. 1992). In chapter 4 of this volume, differences in the rates at which potential rivals encounter each other and in the distributions of predators and resources were considered as possible factors favoring alliance formation. Population differences in sexual size dimorphism were related to male use of alliances for coercing females (Connor et al., chap. 4, this volume), but the degree of sexual size dimorphism may also determine whether alliances are favored in relation to male-male contests. Plavcan and van Schaik (1997) found that the relative size of canines in primates, an index of selection for male fighting ability, was inversely related to participation in coalitions to resolve disputes. A similar argument might be made for the apparent inverse relationship between sexual size dimorphism and prevalence of alliance formation among bottlenose dolphin populations. Sexual size dimorphism is minimal in Shark Bay, where alliance formation is most pronounced, and is greater in Sarasota, where alliance formation is relatively reduced (Cockroft and Ross 1990a; Cheal and Gales 1992; Connor et al., chap. 4, this volume). Observations of extensive alliance formation in a population with large SSD would eliminate this hypothesis.

The hypothesis that higher encounter rates between rivals favor alliance formation (Connor et al., chap. 4, this volume; see above) can be extended to account for male coercion of females (herding) (Connor et al. 1996). Consider an alliance that herds females and another that merely follows them, and assume that both defend females against rival alliances. The herding alliance pays the cost of herding, but the following alliance may pay a higher cost in conflict with other males if they follow the female into areas of her range that they do not normally use because of the presence of rival alliances. The herding alliance does not suffer this cost because they keep the female in their normal range, or possibly even in a restricted portion of their range. Approximately half of the seventy females recorded in consortships with focal alliances in the core study area of 60 km² in Shark Bay (Connor et al. 1992a; 1996) were previously unknown to observers, suggesting the these females were escorted outside of their normal range during herding. Subsequent sightings of some of these females in a larger area confirmed that they normally ranged in areas heavily used by other male alliances.

We do not yet know enough about bottlenose dolphins to generate an exhaustive list of hypotheses to explain interpopulation differences in male alliance formation. For example, the hypotheses presented here do not consider the potentially confounding influence on alliance formation of interpopulation variation in patterns of female receptivity. However, the hypotheses presented here are testable and provide a framework within which testing and generation of further hypotheses may proceed.

Dolphins and Apes: Convergent Evolution of Male Mating Strategies?

Convergence offers a special opportunity for understanding the evolutionary origins of social bonds, so it is fortunate that the odontocetes provide several interesting cases. Weilgart et al. (1996; Whitehead and Weilgart, chap. 6, this volume) explored the remarkable convergence between sperm whale and elephant social organization. In this section, we examine two cases of apparent social convergence with apes (see boxes 10.1 and 10.2). Our purpose is to understand how similar the dolphin and ape patterns are, and to assess the significance of their convergence for understanding the roles of ecological pressures and cognitive abilities as influences on social evolution.

First, bottlenose dolphins and common chimpanzees are similar in having a dispersed fission-fusion social organization in which males form their strongest bonds with other males, and in which males temporarily guard females. This combination of social patterns is rare elsewhere, though known in spider monkeys (*Ateles* sp.: Strier 1994) and chee-tahs (Caro 1994a). The second system has not yet been found in other mammals. Bonobos *(Pan paniscus)* and killer whales both live in relatively cohesive groups, in which males bond more strongly with their mothers than with other males, and in which mate guarding has not yet been seen.

Chimpanzees and Bottlenose Dolphins

Female bottlenose dolphins and common chimpanzees are both seasonally polyestrous, and appear to have similar interbirth intervals and ages of sexual maturity (Goodall 1986; Schroeder 1990). Associations between males and females intensify in both species when females are receptive (Wells et al. 1987; Smolker et al. 1992; Wrangham 1993). Male-female association patterns indicate a promiscuous mating system, confirmed in chimpanzees by observations of mating and genetic paternity analysis (Goodall 1986; Gagneux et al. 1997; Wells et al. 1987; Connor et al. 1996; chap. 4, this volume). These similarities are accompanied, in both species, by frequent use of aggression directed at rival males. However, although the distribution of receptive females is similar in the two species, and both species have strong male-male bonds (box 10.1), the male bonds are used in different ways. Chimpanzees use them more often to determine social rank and in intercommunity aggression, whereas dolphins use them more directly in mating competition.

Male chimpanzees, however, do use aggression to obtain copulations. In the early receptive phase of her sexual cycle, a female chimpanzee mates opportunistically without experiencing male-male competition. Several days later, about the time when her peri-ovulatory period begins (within three or four days of ovulation), males initiate mate-guarding. As is true of bottlenose dolphins, patterns of alliance formation differ among populations of chimpanzees. In Gombe and Mahale, "possessive" males compete on their own against rivals approaching the female. In Kibale, by contrast, aggression in this context can be coalitionary, with pairs or more males jointly defending access to the female (Ngogo community, Watts 1998; Kanyawara community, Wrangham pers. obs.). As an alternative to mate-guarding, a male may form a consortship, i.e., an exclusive association between a single female and a single male, either through friendly association or by aggressive herding. Consort pairs typically travel in peripheral areas of the community range. Possessive strategies are more frequent, but from the male's perspective they are often less effective because rival males invariably obtain a proportion of matings (partly because females try to escape from their guards).

Consortships that typically last for a few days but may stretch to three months appear to have a higher probability of conception. Once initiated, they involve little aggression (Tutin 1979; Nishida and Hiraiwa-Hasegawa 1986; Goodall 1986).

In Shark Bay, consortships are the most common form of mating relationship, lasting from a few minutes to over a month (Connor et al. 1992a,b, 1996, chap. 4, this volume). Nearly all dolphin consortships involve a single female being herded by an alliance of two or three males. The duration of consortships suggests that, like chimpanzees, male dolphins may begin herding females before they are maximally attractive. Dolphin consortships encompass the "possessive" category, in the sense that herded females are generally escorted in the males' normal range rather than in "peripheral areas." Thus, compared with chimpanzees, male dolphins may have a higher risk of encountering rivals during consortships, possibly explaining why aggression toward herded females may occur throughout a dolphin consortship.

In summary, the distribution of receptive females and the pattern of aggressively established consortships is rather similar for chimpanzees and bottlenose dolphins; however, although male-male bonds in both species ultimately enhance mating success, the bonds take different forms and are used in different ways. Accordingly, the convergence is not quite as close as it first appears.

Bonobos and Killer Whales

Bonobos and killer whales share a tendency for strongly developed relationships between mothers and their adult sons. In both cases, adult males remain in their mother's social group whether or not she is alive. We ask here how the mother-son bond is related to male mating strategies.

Like chimpanzees, male bonobos sometimes interfere aggressively with the copulations of rivals (5% of copulations: Kano 1996), but in contrast to chimpanzees, neither aggressive mate guarding nor consortships have been recorded (Kano 1992). A possible factor reducing the aggressiveness of mating males is that they appear unable to detect the timing of ovulation (Wrangham and Peterson 1996); this implies that all copulations are equally valuable to them, and are too numerous to be worth contesting individually. Nevertheless, male bonobos of higher dominance rank mate more often than lower-ranking males (Kano 1996). It is not yet clear exactly how high-ranking males achieve more matings, because the relative importance of female choice and male rank has not been studied.

Mothers are important allies for male bonobos competing with each other for high rank (Box 10.2). Males appear

Box 10.1

Chimpanzees and Bottlenose Dolphins: Social Bonds

Both bottlenose dolphins and chimpanzees live in classic fission-fusion societies, in which individuals associate in labile parties averaging fewer than ten individuals (Goodall 1986; Smolker et al. 1992). Chimpanzees live in semi-closed communities of up to a hundred or more individuals, whereas bottlenose dolphin social networks are apparently unbounded (Smolker et al. 1992). In both chimpanzees and bottlenose dolphins, the strongest bonds are found among adult males. Male philopatry is the norm among chimpanzees, and may occur in bottlenose dolphins (Connor et al., chap. 4, this volume).

Among male chimpanzees there are two levels of alliances: between specific partners in aggressive interactions within communities, and between community members in aggression against neighboring communities (Goodall 1986; Nishida 1990). Within communities, alliances are formed principally by pairs of high-ranking males in contests with opposing high-ranking males, but these can extend to include lower-ranking males as part of a "clique." Allied males associate closely, frequently groom each other, and support each other in contests (Nishida and Hiraiwa-Hasegawa 1987; Wrangham et al. 1992). Coalitional aggression occurs principally in the context of achieving or protecting dominance status (Nishida and Hiraiwa-Hasegawa 1987), though cooperative mate guarding is also known from one site (Watts 1998; Wrangham unpublished data). Among bottlenose dolphins, by contrast, both levels of alliance formation are seen principally in the context of gaining access to sexually attractive females, a function similar to that of male alliances in baboons (Connor et al. 1992a).

While alliances may last for years among chimpanzees (e.g., over six years in Kibale Forest: R. W. Wrangham, personal observation), they are more often short-lived, with junior allies shifting partners over a period of months in a highly strategic manner (Nishida 1983; Nishida and Hiraiwa-Hasegawa 1987). Not surprisingly, therefore, allies vary in their genetic relationships, from matrilineal brothers to unrelated males (Nishida and Hiraiwa-Hasegawa 1987; Goldberg and Wrangham 1997). Many dolphin alliances are more durable, lasting, for example, up to twenty years in Sarasota (Wells 1991c; Connor et al., chap. 4, this volume). However,

alliances in the super-alliance in Shark Bay (Connor et al. 1999; chap. 4, this volume) were much more labile than those of chimpanzees; in this case, the frequent partner changes may function to maintain cooperative relationships within the group (Connor et al. 1999). In general, association coefficients within male chimpanzee alliance pairs (e.g., 50–90%, Wrangham et al. 1992) are usually less than within the stable bottlenose dolphin pairs and trios in Shark Bay and Sarasota (70–100%; chapter 4). They are more similar in strength to association coefficients among Shark Bay pairs and trios (Smolker et al. 1992) and among most members of the super-alliance (R. C. Connor et al., unpublished data). Similarly, the opportunistic nature of alliances among individual chimpanzees seems closer to the "second-order" alliances formed among the more stable dolphin pairs and trios (Connor et al. 1992a,b).

The sociability of females varies widely within populations of bottlenose dolphins, probably more than within chimpanzee communities. Some female dolphins are relatively solitary, whereas others spend as much time in groups as males (Wells et al. 1987; Smolker et al. 1992; Connor et al., chap. 4, this volume), and may engage in affiliative contact behavior and even form temporary coalitions against harassing males (Connor et al. 1992b). Among chimpanzees, parity is a major determinant of gregariousness: females without young can be as gregarious as males, unlike mothers who are more solitary (Wrangham 1999).

Chimpanzee female behavior varies more between than within populations. Particularly in East African populations, mothers spend much more time alone than males, rarely groom or form agonistic coalitions with each other, and show little preference in their associations for particular females (Nishida and Hiraiwa-Hasegawa 1987; Nishida 1990; Wrangham 1993); thus, they behave like the most solitary female bottlenose dolphins. In some West African and captive populations, however, female chimpanzee grouping and alliance patterns resemble those of males (Baker and Smuts 1994). Population differences in the distribution of food resources, and thus the potential for intrasexual competition, may explain the differences in affiliative patterns among females, both among chimpanzees and between chimpanzees and bottlenose dolphins.

to benefit in two ways, both through higher rank, and because the presence of his mother enables a male to spend more time with her female associates (Kano 1996). Male investment in matrilineal kin, by contrast, appears to be barely more frequent than in chimpanzees, suggesting little importance to this aspect of the mother-son bond. Among bonobos, therefore, the influence of a mother on her son's fitness appears to be closely linked to his mating success.

The details of mating relationships in killer whales remain obscure, although genetic data indicate that all mating occurs within the community (Hoelzel and Dover 1991a). Behavioral data suggest that mating often involves individuals of different pods: most bouts of heterosexual behavior observed in the southern *resident* community involved whales from more than one pod (fourteen of eighteen: Heimlich-Boran 1993). The fact that killer whales have relatively large testes (see fig. 10.5) suggests that females mate with multiple males.

Heimlich-Boran (1993) argued that a low confidence of paternity in killer whales favors males investing in matrilineal kin rather than attempting to monopolize mates (e.g., assisting juveniles in the capture of sea lions: Lopez and Lopez 1985). The idea that the mother-son bond is a consequence of low confidence of paternity has precedence in the human "avunculate" system (Alexander 1979), although the human example involves a shift from one kind of nepotism to another rather than a shift from mating effort to nepotism.

Earlier we suggested that mother-son alliances in killer whales represent a shift in mating strategies (alliance partners) rather than a shift from mating effort to nepotism. More data are needed to establish how much male killer whales invest in kin and whether mother-son bonds are a cause or a consequence of male nepotism.

Another issue is whether killer whales conceal ovulation (Heimlich-Boran 1993). We suspect that concealed ovulation in bonobos is a consequence rather than a cause of the more continuous association between potential mating partners—sons and their mother's unrelated female associates. Concealment of ovulation by extending the period of attractiveness will reduce male harassment (fighting over ovulating females) only if males are already present in the group. Otherwise, extending attractiveness would increase the males' presence in the group, resulting in more rather than less harassment. Similar selection for concealed ovulation may be absent in killer whales because, in contrast to bonobos, all individuals in killer whale pods are relatives, and thus unlikely mating partners.

In sum, it is not yet clear how similar the mother-son relationships of bonobos and killer whales are, and whether they have a common functional basis. Present indications, however, are that they serve more of a mate acquisition role in bonobos, and either a mate acquisition or a kin investment role in killer whales.

The relationship between the chimpanzee mating system (male bonding, with less gregarious females) and the bonobo mating system (mother-son bonding, with more social females) has been argued to represent a transition from a primitive (chimpanzee) to a derived (bonobo) form, and to be caused by specific ecological changes (Wrangham and Peterson 1996). Although bottlenose dolphins and killer whales are not as closely related as the two species of chimpanzees, and the polarity of the change is unknown, the social parallels prompt us to ask whether similar processes might underlie the shift between male-male and mother-son bonds in both taxa.

Ecological and Cognitive Constraints

In the apes, the key to understanding the transition from male-male to mother-son bonds is thought to be the cost of grouping (Wrangham 1986; Wrangham and Peterson 1996). Female chimpanzees, especially mothers burdened by offspring, experience such a high cost of grouping that for much of the time they are forced to live solitarily (Wrangham 1999). Such mothers cannot form alliances with each other, and can indeed be highly competitive (Pusey et al. 1997). Bonobos, by contrast, have a food regime that ameliorates feeding competition compared with chimpanzees. As a result, bonobo mothers can travel together at low cost. It pays them to do so, as it does for chimpanzees during the occasions when they also experience low feeding competition, because they are then able to form mutually supportive bonds (albeit with nonrelatives). The development of female-female bonds sets off a chain of consequences, including female-female support against male aggressors, reduction of male power, masking of ovulation, and maternal support for males (Wrangham and Peterson 1996). In this highly abbreviated and schematic account, the key point is that the bonobo system emerges when females in a male-bonded community can tolerate the ecological costs of sociality.

A parallel relation between feeding competition and social bonds in delphinids has been nicely illustrated in the contrast between *resident* and *transient* killer whales (Baird, chap. 5, this volume). Baird and Dill (1996) showed that, in *transient* killer whales, the optimal group size for hunting seals is three individuals. Per capita success decreases in larger groups, which may explain why second-born sons of *transient* females are forced to travel separately from the group (Baird, chap. 5, this volume) and why daughters split

Box 10.2

Bonobos and Killer Whales: Social Bonds

Killer whales and bonobos live in relatively stable groups (pods and communities, respectively), which may have affiliative interactions with neighbors. In both cases, mother-son associations are striking.

Bonobo communities generally contain 50–120 individuals (Nishida and Hiraiwa-Hasegawa 1987). Like chimpanzees, male bonobos remain in their natal community and females emigrate. Bonobos form temporary foraging parties like chimpanzees, but unlike chimpanzees, these parties often travel in the same area, maintaining contact with vocalizations before reaggregating in the evening (Furuichi 1989; Fruth and Hohmann 1994). Intercommunity interactions cover a range from aggressive to relatively peaceful. At the peaceful extreme, members of different bonobo communities have been observed to intermingle and engage in sexual and sociosexual behaviors for many hours a day, for several days at a time (Idani 1991).

In killer whales, interpod associations may not reflect genealogy (Ford 1991). Affiliative, but not aggressive, interactions have been observed between killer whale pods of the same community. Pods from different communities have overlapping ranges but do not associate (Baird, chap. 5, this volume), suggesting that pods may defend the community range.

Mother-son relationships among bonobos are visible through adult males maintaining a lifelong close association with their mothers (Ihobe 1992). Mothers support their sons in agonistic interactions against other males, and are therefore important for male rank acquisition (Kano 1992). Importantly, mothers can recruit female allies (who are normally unrelated) to participate in an interaction (Furuichi 1989; Kano 1992). Sons may also defend the mother on the rare occasions that she is attacked (Kano 1992). Although male killer whales maintain similarly strong associations with their mothers, direct observations of sons being aided by mothers or other female relatives are lacking.

Male-male bonds, and aggression between males in general, are less prominent in bonobos than in chimpanzees, not only between but also within communities (Kano 1992; Ihobe 1992). The frequency of intragroup male coalitionary activity in bonobos is low. Though there is some evidence for matrilineal brothers supporting each other, support interactions appear relatively unsophisticated (e.g., one male chases another and a third joins in), lacking the recruitment and coordinated displays found among male chimpanzees (Ihobe 1992). There are striking cases in which maternal brothers have failed to form defensive coalitions when one was attacked, the victim instead seeking the support of his mother (Kano 1992). Males who form close associations appear to be those whose mothers have died (Furuichi 1989).

In sum, the mother-son bond appears to be the most important alliance in determining a male bonobo's rank. The success of mother-son alliances probably depends on the mother's rank, and on her ability to recruit other females in support. Male-male alliances may be a strategy of older orphaned males, with success, again, depending on numbers.

One group of killer whales observed by Bigg et al. (1990b) had no mother, and consisted only of two adult males. Bigg et al. speculated that these males' mother might have died. If true, as in bonobos, tight male-male associations in killer whales might become more important for males who have lost their mothers. Males may also separate from their matrilineal groups to travel and socialize temporarily with other males (Rose 1992; Heimlich-Boran 1993).

Although female bonobos associate with each other much more frequently than female chimpanzees do (Kano 1992), grooming among females occurs at relatively low rates (Furuichi 1989). Female bonobos (but not female chimpanzees) often engage in mutual genital rubbing, "g-g rubbing," at food sites, apparently to reduce tension and to establish bonds (Kano 1992). Female bonding may also be important in intercommunity contests over feeding areas, because females are actively involved in intergroup aggression in escalated encounters (T. Kano, personal communication).

Female killer whales, unlike female bonobos, remain in their natal groups. As adults, they do not associate as strongly with their mothers as males do, possibly because they are with their own offspring (Bigg et al. 1990b). Some matrilineal groups are linked by a com-

Richard C. Connor, Andrew J. Read, and Richard Wrangham

mon mother, with one daughter being in a separate matrilineal group. Bonds between sisters may weaken after their mother dies (Bigg et al. 1990b), beginning a process that is thought eventually to lead to pod fission.

The two proposed functions of female-female alliances in bonobos are defense of food and cooperation against males (Wrangham 1993), with defense against males appearing to be more important. There is no evidence that female killer whales cooperate against males, but a few observations suggest that killer whale pods may function as alliances in defense of feeding areas (see Baird, chap. 5, this volume).

off to form their own matrilineal group once they begin reproducing. Baird and Dill suggest that the larger and more variable size of *resident* pods reflects reduced feeding competition in these individuals. Their analysis opens the way for discovering whether differences in the costs of grouping can contribute to an explanation of the social differences between bottlenose dolphins and killer whales.

Can ecology alone explain the evolution of dolphin and ape social bonds? Dolphins and apes share relatively large brains, which several authors have linked with social bonding in general (Barton 1996) and complex alliance formation in particular (Harcourt 1992; Connor et al. 1992a). This observation suggests a simple explanation for the puzzling combination of convergence in bonding patterns on the one hand with differences in bond benefits on the other hand. We propose that species with higher cognitive ability can benefit relatively easily from use of complex alliances, regardless of the specific advantages they bring. By this argument, convergence occurs as a result of two factors. First, there are parallel ecological constraints that allow certain combinations of individuals the opportunity to form bonds. Second, there are parallel cognitive abilities to take advantage of these opportunities, even though the nature of the advantages might differ by species. The implication

is that the flexibility of behavioral strategies that is conferred by advanced cognition creates new social opportunities. Although multilevel alliances, for example, might be ecologically possible among a smaller-brained fusion-fission species, cognitive constraints would prevent their evolution.

The development of conceptual schemes for explaining odontocete social bonds is, of course, at an early stage. Already it appears, however, that comparison with terrestrial mammals will be worthwhile. As such comparisons proceed, they will be able to take into account a variety of factors that we have not had room to consider here. For example, there are clearly substantial differences in the costs of locomotion between marine and terrestrial forms, which will influence ecological constraints on grouping (see Connor, chap. 8, this volume). There is also evidence of intertaxon differences in the degree of intrapopulation and interpopulation variation in social strategies, with odontocetes showing more variation than terrestrial mammals at both levels. These problems offer rich opportunities for the development of an odontocete socioecology that ultimately promises to inform the analysis of social evolution in terrestrial mammals as well (Connor et al. 1998).

11

FUNCTIONAL ASPECTS OF CETACEAN COMMUNICATION

Peter L. Tyack

MY GOAL for this chapter is to present studies of cetacean communication from the functional perspective of behavioral ecology and ethology. This may involve a view of communication that is not familiar to readers from other disciplines. For example, the mathematical theory of communication usually frames communication between one signaler and one recipient (fig. 11.1A). The signaler is presumed to know something that the recipient may not know, and it is assumed to produce a signal to communicate information to the recipient. The information in the signal is defined by the ability of the signal to reduce uncertainty in the recipient (Shannon and Weaver 1949; Bradbury and Vehrencamp 1998). This classic idea of communication involving reduction of uncertainty is superficially similar to our typical usage regarding human communication transmitting knowledge by language.

By contrast, the functional view I follow in this chapter attempts to look at communication from a much broader perspective (fig. 11.1B–F). The following list describes several non-mutually exclusive features of communication that may not be captured in the classic view:

Advertisement. A signaler may produce an "advertisement" signal more to influence a decision of the recipient than to exchange information. This view emphasizes the role of communication in manipulating the receiver, possibly to its detriment (Dawkins and Krebs 1978; Krebs and Dawkins 1984). Receivers that are predictably bombarded by advertisements may develop sales resistance.

Tonic communication. The classic view of communication uses a model in which one signal transmits information to a receiver, often evoking an immediate response (see fig. 11.1A). In some situations, though, signalers produce long strings of signals without an obvious response from a receiver (fig. 11.1B). Receivers may monitor signals for a long time before making a choice, as when a female chooses a male based upon advertisement displays. There are other situations in which changes in the rate of signaling may convey information (Schleidt 1973).

Deception. A signaler may signal to mislead or to increase uncertainty in the receiver. A variety of theoretical analyses of situations in which the interests of animals conflict suggest that rather than signaling to allow an opponent to predict future behavior, animals might produce signals that either minimize predictability (Maynard Smith 1974) or that actively deceive the opponent (Cheney and Seyfarth 1990; Whitehead and Weilgart, chap. 6, this volume). The classic view of fighting assessment emphasizes the role of threats in predicting future attacks or retreats, while the deception view emphasizes the possibility of bluffs or feints (fig. 11.1C).

Environment. Most signals are modified as they pass through the environment from sender to receiver. This modification need not represent only degradation of the signal, but may provide information to the receiver. For example, songbirds may estimate the range of a singer by assessing degradation of the signal (McGregor et al. 1983; McGregor and Krebs 1984b; Morton 1982, 1986; Naguib 1998). Animals such as bats and dolphins are well known to learn about their

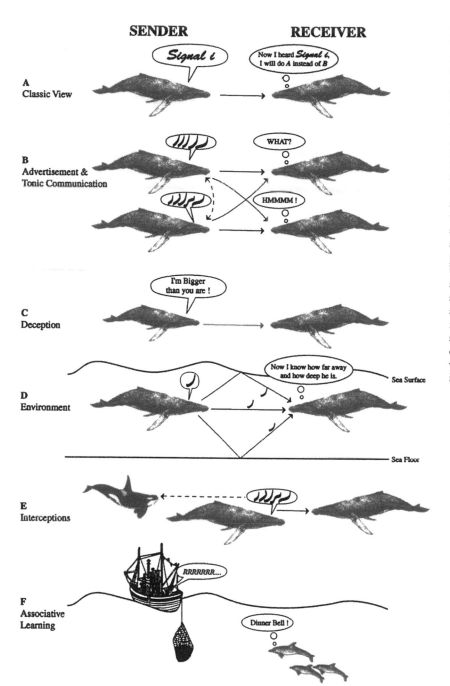

SENDER **RECEIVER**

A
Classic View

B
Advertisement &
Tonic Communication

C
Deception

D
Environment

E
Interceptions

F
Associative
Learning

Figure 11.1. Comparison of classical "reduction of uncertainty" view of communication (A) with the ecological view of communication presented in this chapter. This view includes the concepts of advertising and tonic communication (B): viewing communication as a potentially manipulative process, potentially involving many parties and many signals. It does not assume that all signals are "honest," but entertains the possibility of deception (C). It acknowledges that signals change as they pass through the environment, and that this can inform the receiver about both the signaler and the physical environment (D). There may be multiple intended recipients of a signal, and others may intercept the signal, using information to their own advantage (E). Unintended recipients may even learn beneficial responses to signals from other species (F), as when dolphins suddenly appear at a fishing vessel when a winch that releases fish is turned on.

environment by listening to echoes of sounds they themselves have made. Animals may also learn about their environment from sounds they use for communication and from the sounds of other animals as well (Tyack 1997).

Interception. A signaler may produce a signal to communicate with one specific class of receivers, but other receivers may detect the signal and respond to it, to their advantage and to the detriment of the signaler (Myrberg 1981). These interceptors may be predators (Ryan et al. 1982) or parasites (Cade 1975) of the signaler, or they may be competitors, such as satellite males that may seek out females attracted by the advertisement displays of a territorial male (Cade 1979). Risks of interception may influence the active range of a signal: a signaler may limit the intensity of a signal so it cannot be detected beyond the expected range of the intended recipient. Risks of interception may also influence the frequency range or sensory mode of a signal, or influence the evolution of displays that can be directed toward the intended recipient in such a manner that they are difficult to detect from different directions.

Associative learning and interspecific communication. Akin to interception is the situation in which a receiver may learn the contexts in which a signaler may produce a signal intended for some other recipient. For example, vervet monkeys may respond appropriately to the alarm calls of a bird and may even recognize that the mooing of cows is associated with the presence of a human predator, the Masai (Cheney and Seyfarth 1985).

I will develop an analogy to illustrate the difference between the classic and functional ecological views of communication. During the past few decades, while behavioral ecologists and ethologists have developed the functional view of communication in animals, a human communication system has actually changed from a simple classic mode of operation to a more complex form of communication involving many aspects of the functional perspective I have described above. Starting in the 1960s, the U.S. Department of Defense funded the development of an early version of the Internet, called the ARPANET, to link computers by phone lines to allow more efficient use of limited computer resources (Hafner and Lyon 1996). It quickly became obvious that electronic mail using this network could speed up communication of commands and questions and answers between the military, defense contractors, and academics. While each message typically had one originator and one intended receiver, a complex distributed network for sending these messages was developed on the ARPANET. Some of the design features of this distributed network were developed by a consultant concerned with the vulnerability of centralized communication systems in case of a nuclear attack (Hafner and Lyon 1996). Each computer was linked to several others, so that if one node of the network stopped functioning, the message could immediately be rerouted along the best new path. Now, in the 1990s, this network has opened up to a much broader world. While the initial Internet was designed for classic models of information exchange between two individuals for mutual benefit, cyberspace is now a completely different landscape. One person may send a message to thousands of recipients with ease. Internet users may be attacked by spam—mass advertisements indiscriminately sent to unwilling receivers. Malicious computer programmers devote enormous ingenuity to designing computer viruses—sequences of code not designed to transfer information to the receiving computer, but rather to take over some of its resources or to interfere with its operation. A message may be sent through one computer not to communicate with it, but rather to obscure the actual identity of the signaler.

The world in which animals communicate is more like cyberspace than like the original ARPANET. Over the long course of evolutionary time, animals have evolved adaptations to take advantage of any opportunity provided by communication systems, and this exploitation has stimulated countermeasures. The fullest view of communication is not limited to educational messages between cooperating partners, but also includes competition, advertisement, parasitism, and predation. What we observe today is the product of complex balances of the costs and benefits of many of these relationships. In this chapter, I present a functional view of cetacean communication in this broad ecological context.

Many recent reviews of animal communication emphasize elements of this functional perspective (e.g., Alcock 1998; Bradbury and Vehrencamp 1998; Krebs and Davies 1993), but it is less common in previous reviews of cetacean communication (e.g., Herman 1980) or of vocal behavior and hearing in cetaceans (e.g., Richardson et al. 1995b). One reason for this contrast is that it is often easier for a human to study the functions of communication in animals that share our own terrestrial environment than to work in a foreign environment, which often requires special methods. When humans study communication in terrestrial animals, we naturally can rely upon our own senses, which are adapted for the terrestrial environment, to detect communicative displays. We also draw upon our familiarity with the terrestrial environment to infer the intended recipient of a display. For example, when a cricket calls or a bird sings, we usually have an intuitive sense of how far away an animal might sense the display. The senses of marine animals are adapted to a very different environment, one that is more difficult for biologists to sense directly and to understand intuitively.

When we monitor terrestrial animals, we also can integrate information from all of our senses as we follow an interaction, often tapping skills from our own species-specific communication system. For example, when a non-human primate calls, it may also direct its gaze toward the intended recipient of the call. Since the human primate also uses gaze in this way, it is relatively simple for a biologist to interpret the interaction. Direction of gaze has even proved a powerful response measure for playback experiments with wild primates (Cheney and Seyfarth 1980). All of these factors make it easier for us to understand the communicative context of displays by terrestrial animals than by marine animals such as cetaceans.

Phylogenetic similarities between human observers and their nonhuman primate subjects have costs and benefits for the observer. To the extent that displays such as facial expressions are homologous, the observer may more easily be able to understand and interpret the display. On the other hand, this similarity raises the potential for unquestioned or unwarranted anthropomorphism. When humans observe a chimpanzee, they may be more susceptible to anthropomorphic interpretation than when watching an animal as foreign as a dolphin. One can seldom study wild cetaceans simply by watching and listening to them and inferring what is happening, as is possible with some terrestrial animals. Biologists who study cetaceans in the wild have needed to develop a suite of novel methods and study designs (see Whitehead et al., chap. 3, this volume). Difficulties in identifying which cetacean produces a particular vocalization have been an obstacle to teasing apart the patterns of signal and response that make up a system of communication. In order to overcome these difficulties, studies of communication in cetaceans have used an unusually diverse array of methods. While this has taken extra time and effort, the results ultimately may be less subjective and more valid than those of studies relying upon easier but more subjective methods. Some of the methods initially developed for cetacean research, such as acoustic localization, are now being used with other taxonomic groups to answer questions where the simpler methods are inadequate (McGregor et al. 1997).

However, the evolutionary distance between cetaceans and primates has not prevented anthropomorphism from influencing our understanding of cetacean communication. John Lilly was one of the first scientists to discover the remarkable imitative abilities of dolphins. Marine mammals do stand out as radically different from most nonhuman terrestrial mammals in this ability. Since humans also use vocal imitation, Lilly jumped to the conclusion that if dolphins have large brains and can imitate, they must possess a communication system very similar to human language:

> All of dolphin culture may be transmitted in somewhat the way primitive human tribes transmit knowledge from one generation to the next with long folk tales and legends. . . . (Lilly 1975: 17)

I believe that rather than leaping to analogies with human language, it is a more profitable research strategy to study how cetaceans actually use communication to solve specific problems in their natural environment (Tyack 1993). I will present cetacean results in a broad comparative perspective, including examples from many different animal groups. My presentation of these results in the functional framework of behavioral ecology is designed to highlight the many areas of commonality with studies of communication in terrestrial animals. The basic theories of behavioral ecology are equally applicable in the sea and on land. I will also highlight areas in which cetaceans differ from terrestrial animals, areas of particular interest from a comparative perspective. Some of these differences stem from the physical properties of seawater versus land (Tyack 1998); others, such as the effect of diving upon vocalization, depend upon the interaction of biology with the physical properties of the environment (Janik and Slater 1997).

The phylogenetic distance from cetaceans to many other mammalian groups makes convergent characters particularly interesting. For an example harking back to Lilly's comparisons with humans, while vocal imitation is critical for human language and music, most nonhuman terrestrial animals appear unable to modify their vocal repertoire based upon what they hear (Janik and Slater 1997). Several groups of marine mammals, including seals, whales, and dolphins, have highly developed skills for vocal learning. If their terrestrial ancestors did not have these skills, then this means that vocal learning evolved independently in at least two mammalian taxa that independently entered the sea: seals and cetaceans. This provides two independent groups for analyzing the selective pressures leading to the evolution of vocal learning. By comparison, if such a skill were found in chimpanzees, it would be difficult to determine whether this similarity with humans arose from a shared ancestor with the trait or through independent evolutionary events.

The basic questions I will address include "What are the functions for which cetaceans evolved particular signals?" and "What are the factors that cause these signals to have particular design features?" The structure of the problems and the features of communication described above are very important for deciphering what a communication signal is designed for and why it has particular physical features. Here are some examples of new results using a variety of methods to study cetacean communication, followed by some of the kinds of questions on communicative functions that will be addressed in this chapter:

Mechanisms of sound production. The sperm whale *(Physeter macrocephalus)* devotes about a quarter of its body length to the spermaceti organ, which appears to function primarily to produce loud clicks (see Whitehead and Weilgart, chap. 6, this volume). The vocal repertoire of sperm whales involves variations in these clicks.

- What justifies such a massive investment in such a unique sound production organ?
- What are the critical features of sperm whale clicks driving the evolution of this organ?

Correlations between structure and function of songs. Humpback whales sing complex series of sounds that may last for tens of minutes before repeating (Payne and McVay 1971). These songs sound so musical to our human ears that recordings of them have become commercial best-sellers.

- Why do humpbacks sing these songs?
- Why does humpback song have such a complicated acoustic structure?

Vocal learning. Humans are the only terrestrial mammal with well-developed abilities of vocal learning, yet many marine mammals, including whales, dolphins, and seals, are exceptional vocal mimics. These unusual imitative abilities form the basis of the popular stories that dolphins have a "language."

- What are the functions of vocal learning in the natural communication systems of marine mammals?

Echolocation and communication. Dolphins have a remarkable system for echolocation using high-frequency clicks (Au 1993), and they also produce a diverse array of sounds used in social communication. Biologists often appear to assume that echolocation and vocal communication are independent abilities.

- Might some echolocation signals also transmit information to other animals?
- Might the evolution of skills for echolocation influence the evolution of vocal communication, or vice versa?

Long-range communication. Biologists using technology initially designed to track submarines have been able to hear loud low-frequency vocalizations of fin and blue whales from hundreds or even thousands of kilometers away (Costa 1993; Clark 1994, Clark 1995).

- Do the whales themselves communicate over such long ranges?
- What can a fin whale near Iceland need to know from a fin whale near Bermuda?
- The sounds of fin and blue whales produce echoes

from the seafloor and distant seamounts. Do whales use these echoes to orient or navigate? Might these sounds function both for communication and for echolocation?

This chapter attempts to put what we know about cetacean communication into its ecological and social contexts. The marine environment presented ancestral cetaceans with a new set of selection pressures for each modality of sending and receiving signals. I will briefly discuss how the marine environment influences communication in each sensory modality: chemical, tactile, electrical, visual, and acoustic. The rest of the chapter focuses on acoustic communication in terms of the different kinds of communication problems faced by whales and dolphins. This analysis asks how communication works as a system: for example, who are the intended recipients of a signal? Are there unintended recipients who intercept the signal to their own advantage? This question includes the potential for interspecific communication, particularly in predator-prey relations. I close the chapter by comparing echolocation and communication, by discussing the role of vocal learning in mammals, and by relating patterns of communication in different species of cetaceans to differing problems posed by the social systems of each species.

Sensory Modalities for Communication in Cetaceans

Tens of millions of years ago, the ancestors of cetaceans gradually evolved from a terrestrial existence to living in the sea for their entire lives. These terrestrial mammals, most closely related to modern ungulates (see Eisenberg 1981 for a general review), underwent enormous changes as they adapted to the marine environment. Not only did their bodies have to change to allow them to swim more efficiently, but their sensory systems and vocal apparatus had to adapt to functioning in the dense water medium. The physical characteristics of the ocean environment also greatly modified the usefulness of different sensory modalities for achieving particular communicative ends. For example, vision is important among terrestrial animals for sensing objects at long ranges. Of all the ways to transmit information through the sea, however, sound is the best for communicating over a distance. Whales may hear one another at ranges of up to hundreds of kilometers, but they see one another underwater at ranges of no more than tens of meters. This means that hearing may be more important than vision as a distance sense for marine animals, and that vocal communication may be used preferentially for rapid long-distance communication.

Chemical Communication

Chemical communication is common among terrestrial mammals and many marine organisms. Chemical communication was almost certainly an important mode of communication among the terrestrial ancestors of cetaceans, but it appears to be limited among cetaceans. The olfactory bulbs and nerves, which function in terrestrial mammals for sensing airborne odors, are reduced in mysticetes and absent in odontocetes (Breathnach 1960; Morgane and Jacobs 1972). Little is known about how whales and dolphins may sense waterborne chemicals (Kuznetsov 1979). Most experiments on the chemical senses of cetaceans have tested responsiveness to basic tastes such as sweet, sour, salty, or bitter (Friedl et al. 1990), and the sensitivity of dolphins to taste appears to be about an order of magnitude less than that of humans. There have been some suggestions of use of pheromones among cetaceans (e.g., Norris and Dohl 1980b). M. C. Caldwell and D. K. Caldwell (1972b, 1977) point out the presence of pores from anal glands in bottlenose dolphins, and they suggest that these might release a pheromone. Norris (1991b) speculates that male spinner dolphins *(Stenella longirostris)* use chemical cues to sense the reproductive state of adult females. Little is known about how cetaceans might detect pheromones, and more experimental research is needed to test for pheromones and for specialized abilities to sense them. If cetaceans have only limited use of chemical communication, this may in part stem from the limited ranges of diffusion in water compared with the mobility of these animals.

Tactile Communication

Touch is important for communication at short range in most cetacean species (e.g., Chapters 4–7, this volume). As they evolved a streamlined, hydrodynamically efficient body, most cetacean species lost external sensory hairs or vibrissae. However, most cetaceans retain a few residual hair follicles on the rostrum or upper jaw, and well-innervated hair follicles are present in the Amazon River dolphin, *Inia geoffrensis* (Simpson and Gardner 1972). The skin of cetaceans is well innervated and is very sensitive to touch. Nerve endings are particularly dense in the dermis near the eyes, blowhole, jaw, flukes, vulva, and perineum (Simpson and Gardner 1972; Kolchin and Bel'kovich 1973). Ridgway and Carder (1990) used somatosensory evoked potentials to assess the sensitivity of bottlenose dolphins to vibratory skin stimulation. They suggested that the most sensitive areas are at the angle of the gape of the jaw and around the eyes, snout, melon, and blowhole. These areas are about as sensitive as human skin on the lips or fingers.

Dolphins and whales may rub or caress one another with their flippers or other parts of the body. Up to one-third of the members of an active school of wild spinner dolphins have been estimated to engage in caressing at any one time (Johnson and Norris 1994). Gentle rubbing seems to play an important role in maintaining affiliative relationships in some dolphin species, perhaps analogous to that of social grooming in primates (Norris 1991b; Samuels et al. 1989).

For many cetacean species, sexual contact appears to have a variety of social and communicative functions in addition to procreation. Sexual activity is often reported for all-male groups (e.g., Newman 1976 for gray whales), and copulation is commonly observed between animals that are not sexually mature (Connor et al., chap. 4, this volume). D. K. Caldwell and M. C. Caldwell (1972a) report that infant male bottlenose dolphins in captivity attempt to mate with their mothers within a few weeks of birth. Brodie (1969) has suggested that nursing in toothed whales may not only function in nutrition, but may also take on an affiliative communicative role, reinforcing the mother-calf bond.

Cetaceans engage in a variety of contact behaviors in aggressive interactions, but few studies have isolated a signal role as opposed to the physical displacement, pain, or harm the contact causes. This finding raises an important distinction that illustrates the usefulness of the concept of information for communication researchers: If an animal is said to be responding to a communicative signal, then it must be responding to the information sent, rather than to the physical effects of the signaling action.

Electrical Communication

Many groups of freshwater and marine fish are able to sense electrical signals that can propagate over limited distances in water (Hopkins 1977). Some species, such as the electric eel, are also well known to be capable of producing electrical discharges powerful enough to stun predators or prey. Other groups of freshwater fish, such as the gymnotid fishes of South America and the mormyriform fishes of Africa, produce complex electrical signals that are used for social communication. Electrical communication is characterized by signals that travel very rapidly, fade out quickly, and have ranges typically limited to less than a meter or so in water. Since so few aquatic organisms are highly sensitive to electrical signals, they also form a relatively private channel for communication, reducing the risk that predators or prey will detect a signaling animal. Cetaceans are like most mammals in not being particularly sensitive to electric fields. They are unlikely to use electrical signals themselves, and would be relatively insensitive to such signals from prey organisms.

Even though cetaceans do not use electrical signaling, there is one area where comparisons with electric fishes are useful. This involves potential interactions between using signals to communicate and using signals to learn about the environment. Many fish that have both electric organs and electroreceptors can detect objects in their environment by sensing distortions in the electric field emitted by their own bodies. Some fish can also detect the electric fields generated by other organisms, such as prey. The same abilities to send and receive electrical stimuli may be used by these fish either for communication or to learn about their environment. Some cetacean species have a similar ability to use sound either to communicate or to learn about their environment through echolocation, and this ability creates the potential for evolutionary interactions between these two functions.

Visual Communication

Most marine cetaceans have a well-developed sense of vision, but several species of dolphins living in turbid riverine environments have eyes with reduced capabilities for forming optical images. The platanistid river dolphins, *Platanista minor* and *Platanista gangetica*, have no lens in the eye at all, and the transparency of the cornea is limited because it is vascularized (Dawson 1980). While visual acuity is reduced in these species, they may be able to form crude images using the narrow aperture of the pupil in a manner analogous to a pinhole camera. Most other cetaceans have transparent and colorless lenses and corneas, but the Amazon river dolphin, *Inia geoffrensis,* has a lens that is deep yellow in color (Dawson 1980). It has been suggested that in the brown waters typical of most of the Amazon, this yellow filter might function as well as a clear one.

The eyes of many cetaceans have specific adaptations for underwater vision. Terrestrial eyes rely upon refraction as airborne light enters the aqueous cornea, but this effect is greatly reduced in aquatic animals, and the optics of cetacean eyes have been modified as a result. Deep-diving cetaceans enter a cold and dark environment in which blue-green light penetrates the best. The photopigments of cetaceans are shifted toward the blue end of the spectrum compared with those of terrestrial mammals, and they also have a tapetum lucidum that reflects light and increases the sensitivity of the retina at low light levels (Mobley and Helweg 1990). Cetacean eyes also must be protected against the strong fluctuations in temperature and pressure that occur during dives. A few hundred meters below the

sea surface, the water temperature hovers only a few degrees above 0°C.

Anyone who has observed a captive bottlenose dolphin leaping 5 m into the air to catch a small object would not be surprised to hear that these animals have in-air vision that is nearly as acute as their vision underwater (Herman et al. 1975). Many species of toothed whales clearly can be seen to observe objects in the air when their heads are out of the water. Baleen whales and some odontocetes engage in a specific behavior, called spyhopping, that is thought to be used for inspection of objects in air (Madsen and Herman 1980). It has been suggested that in-air vision may play a role in foraging, surveillance, and orientation among cetaceans (Mobley and Helweg 1990).

Mobley and Helweg (1990) discussed evidence that cetaceans have adaptations favoring motion detection, which might select for moving versus static visual stimuli. They also presented evidence that bottlenose dolphins are excellent at estimating the distance to an object. The visual skills that dolphins display when catching a ball in a show are likely to be equally useful for detecting, tracking, and capturing prey. These visual skills may complement the echolocation skills of odontocetes, and are likely to be particularly useful during daylight at close range. For example, some bottlenose dolphins in the wild hit fish so hard with their flukes that they knock the fish into the air (Wells et al. 1987). After hitting a fish, the dolphin can track the fish in the air and catch the fish in its mouth as the fish hits the water.

Most cetacean species are reported to use some visual signals as communicative displays. Both aggressive and sexual interactions often involve visual signals at close range. Many aggressive visual signals in cetaceans follow patterns that are common among other mammals, including vigorous movements of the head toward another animal, jerking the head, opening the mouth, or even making threats that resemble biting actions (Overstrom 1983; Samuels and Gifford 1997). Some behaviors appear to function to increase the apparent size of a male, and these behaviors may function as visual displays (see Clapham, chap. 7, this volume). For example, male humpback whales *(Megaptera novaeangliae)* competing for access to females may lunge with their jaws open, expanding the pleated area under the lower jaw with water. Several observers have suggested that this may function to increase the apparent size of a competitor (Tyack and Whitehead 1983; Baker and Herman 1984). Other male cetaceans may have thickened areas of callused skin. While these secondary sexual features may function as weapons or armor during fights between males, they may also function as a visual signal to either potential competitors or potential mates (Payne and Dorsey 1983). Some cetaceans retain white pigment where they have tooth rakes from fights (sperm whales: Best 1979; *Grampus griseus* and *Mesoplodon carlhubbsi:* Leatherwood et al. 1983). This might function in a similar way as a visual badge of fighting experience or fighting ability. Visual signals that have been identified in submissive interactions among dolphins include flinching, looking away, and orienting the body away from another animal (Samuels and Gifford 1997). Thrusting or presenting the genital region toward another animal may function as sexual visual signals. Direction of gaze is an important visual cue among primates, and Pryor (1991) suggests that gaze cues may be important for cetaceans in clear water as well as in air.

Many cetacean species have distinctive pigmentation patterns (e.g., Yablokov 1963; Mitchell 1970; Perrin 1997). Most biologists have emphasized the role of pigment patterns as camouflage or disruptive coloration against visually hunting predators (e.g., Madsen and Herman 1980; Würsig et al. 1990). Biologists find these pigmentation patterns to be very useful for identifying species and even individual animals, but little is known about whether cetaceans use them as signals in their own social interactions. Variation in pigmentation and morphology is correlated with age-sex classes among dolphins of the genus *Stenella* (Perrin et al. 1991). The pigmentation patterns among species such as the humpback whale are highly individually distinctive (Katona et al. 1979), so animals that swim within visual range would certainly have sufficient cues to discriminate a large number of individuals.

Exhaling to produce underwater bubbles creates a set of visual displays that are unique to aquatic animals. Some dolphins occasionally blow streams of bubbles that are highly synchronized with the production of whistle vocalizations (D. K. Caldwell and M. C. Caldwell 1972a). These bubble streams are a highly visible marker identifying who vocalized, but it is not known whether dolphins respond to this visual accompaniment of the acoustic signal. Humpback whales produce bubble streams in aggressive interactions during the breeding season (see Clapham, chap. 7, this volume). In competitive groups, they emit streams of bubbles typically in a line as long as 30 m (Tyack and Whitehead 1983). A male escorting the female in such a group may place a bubble stream between a challenging male and the female, perhaps as a visual screen. Pryor (1991) reports that a young male spotted dolphin *(Stenella attenuata)* used a cloud of bubbles created by a breach as a visual screen to facilitate escape. Humpback whales are also reported to use bubbles in an unusual form of interspecific communication. In certain circumstances, feeding humpbacks produce a series of large bubbles. Their prey seem to respond to these bubbles, and the humpbacks appear to use these bubble "nets" to increase the concentration of prey before they engulf a mouthful (see Clapham, chap. 7, this volume). As was discussed with aggressive contact behaviors, the concept of communication is usually limited to signals that evoke a response through information transfer, rather than through the physical effects of the signal. It is not known whether bubble nets act as a signal to fish or whether the bubbles act as an obstacle or physically concentrate the fish prey.

With the exception of bubbles, which are unique to aquatic animals, the visual signaling of marine mammals seems similar to that of their terrestrial relatives. Terrestrial animals also have male secondary sexual characteristics such as ornaments or weapons, they have visual agonistic displays that appear to be ritualized from fighting behavior, they use gaze cues, and they use pigmentation for camouflage. However, the range of vision is much more limited in water than in air, with daytime vision usually limited to a few meters in the sea.

Acoustic Communication and Echolocation

Communication using the auditory modality is emphasized throughout the rest of this chapter. In this section, auditory processing and echolocation will be discussed. An understanding of auditory processing is important for evaluating which acoustic features of a vocal communication signal may be most salient to the recipient of the signal. Echolocation is an important vocal and auditory adaptation for many dolphins and toothed whales.

Adaptations of cetaceans for sound production and hearing underwater. The basic mechanisms for sound production are similar in cetaceans and terrestrial mammals: both make sound by passing air under pressure past membranes that vibrate. However, the differing densities of air and water make for some differences, as do the special needs of animals that dive while vocalizing. When a terrestrial animal vocalizes, it usually must open its mouth to propagate sound into the surrounding air. When cetaceans vocalize underwater, sound vibrations in their soft tissues, which are about the same density as seawater, transfer well to the surrounding medium. This means that cetaceans do not need to open their mouths or blowholes when they vocalize underwater. When terrestrial animals vocalize, they usually simply do so while exhaling, and fill their lungs on the next inhalation. However, many cetaceans vocalize underwater when they cannot breathe again for tens of minutes. For example, humpback whales may sing for 10–20 minutes without surfacing to breathe and without emitting air bubbles. This observation suggests that if sound production involves the flow of air in the vocal tract, this air may need to be stored in the upper respiratory tract and recycled in between vocalizations.

Cetaceans have an auditory anatomy that follows the basic mammalian pattern, with some modifications to adapt to the demands of hearing in the sea. The typical mammalian ear is divided into three sections: the outer ear, the middle ear, and the inner ear or cochlea. The outer ear is separated from the inner ear by the tympanic membrane, or eardrum. In terrestrial mammals, the outer ear, eardrum, and middle ear function to transmit airborne sound to the inner ear, where the sound is detected in a fluid. Since cetaceans already live in a fluid medium, they do not require this matching, and cetaceans do not have an air-filled external ear canal.

The inner ear of cetaceans shares the same basic design with that of terrestrial mammals. The inner ear is where sound energy is converted into neural signals that are transmitted to the central nervous system via the auditory nerve. Sound enters the inner ear, or cochlea, via the oval window. This acoustic energy causes a membrane, called the basilar membrane, to vibrate. The motion of the membrane creates a shear force on hair cells, generating acoustically stimulated neural signals. The key to how the mammalian inner ear operates involves the mechanical tuning of the basilar membrane. This membrane is stiff and narrow at the basal end, near the oval window, causing it to vibrate when excited by high frequencies. Farther into the cochlea, at the apical end, the basilar membrane becomes wider and floppier, making it more sensitive to lower frequencies. Sensory cells at different positions along the basilar membrane are excited by different frequencies, and their rate of firing is proportional to the amount of sound energy in the frequency band to which they are sensitive. Thus the mammalian ear basically measures sound energy in a series of frequency bands.

The ability of mammals to discriminate different frequencies appears to be related to the density of neurons receiving input from sensory cells on the basilar membrane. The density of these ganglion cells is expressed as the number of cells per millimeter along the basilar membrane. Humans and the greater horseshoe bat *(Rhinolophus ferrumequinum)* are mammals with excellent hearing. They average about 1,000 ganglion cells per millimeter (humans: Schuknecht and Gulya 1986; bat: Bruns and Schmieszek 1980; reviewed in Ketten 1990). The greater horseshoe bat has an area along the basilar membrane that is particularly sensitive, corresponding to the frequency of its echoloca-

tion calls. This "acoustic fovea" has a density of 1,750 ganglion cells/mm (Bruns and Schmieszek 1980). As might be expected for animals that rely so heavily upon hearing, cetaceans have an unusually high density of these ganglion cells (Ketten 1990). Pacific white-sided dolphins, *Lagenorhynchus obliquidens,* and spotted dolphins, *Stenella attenuata,* have about 2,000 ganglion cells/mm. The bottlenose dolphin has about 2,500 ganglion cells/mm. The harbor porpoise averages 2,750 cells/mm, and Ketten (1997) suggests that porpoises may also have an acoustic fovea at the frequency of their echolocation calls. Among mammals, dolphins have extraordinary abilities of discriminating different frequencies, and can detect a change of as little as 0.2% in frequency (Thompson and Herman 1975). This is roughly equivalent to the abilities of humans revealed in similar psychoacoustic tests.

Hearing ranges in cetaceans. The hearing ranges of animals can be estimated anatomically. The resonant frequencies of the basilar membrane can be estimated by measuring its dimensions. Ketten (1994) has analyzed inner ears from twelve cetacean species and suggests that there are three basic patterns. The baleen whales have inner ears that appear to be specialized for low-frequency hearing. The apical end is unusually wide and floppy, probably allowing these whales to hear sounds best in the range of approximately 20–200 Hz. All of the toothed whales studied by Ketten (1994) had inner ears that were specialized for high frequencies rather than very low frequencies. All of the smaller odontocetes had measurements from the basal end of the basilar membrane that were consistent with an upper limit of hearing well above 100 kHz. There appeared to be two basic odontocete inner ear types. The inner ears of animals such as *Tursiops* appeared to be consistent with a generalist adaptation for high-frequency hearing with best frequencies ranging from 40–70 kHz, while the inner ears of porpoises and the platanistid dolphins appeared to be particularly specialized for hearing above 100 kHz.

For cetaceans that are held in captivity, hearing ranges can be measured directly by training these animals to respond to tones. Audiograms plot the level of sound that is just detectable by a subject as a function of frequency. The lower a sound level a subject can detect, the more sensitive its hearing. Audiograms thus typically have a U-shaped curve, with low thresholds (best hearing) in a central frequency range and higher thresholds (less sensitive hearing) at higher or lower frequencies. Figure 11.2 presents audiograms for a range of toothed whales that have been tested in this way. All of these odontocetes have hearing that is specialized for sensitivity at frequencies well above what

Box 11.1

Definitions of Acoustic Terms

Frequency (Hz, kHz), wavelength, and bandwidth. A sound that we perceive as a pure tone has a sinusoidal pattern of pressure fluctuations. The *frequency* of these pressure fluctuations is measured in cycles per second. The modern name for the unit of frequency is the *Hertz,* and just as 1,000 meters are called a kilometer, 1,000 Hertz are called a *kiloHertz,* abbreviated kHz. The *wavelength* of this tonal sound is the distance from one measurement of the maximum pressure to the next maximum. Sound passes through a homogeneous medium with a constant speed, *c.* The speed of sound in water is approximately 1,500 m per second, or roughly five times the value in air, which is 340 m/second. The speed of sound *c* relates the frequency *f* to the wavelength λ by the following formula: $c = \lambda f$. Not all sounds are tonal. Sounds that have energy in a range of frequencies, say in the frequency range between 200 and 300 Hz, would be described as having a *bandwidth* of 100 Hz.

Sound intensity, sound pressure; decibel, dB; micro-Pascal, μPa. Sound *intensity* is the amount of energy per unit of time (power) flowing through a unit of area. The intensity of a sound equals the acoustic pressure squared, divided by a proportionality factor that is specific for each medium. This factor is called the specific acoustic resistance of the medium and equals the density of the medium, ρ, times the speed of sound, *c.*

$$Intensity = \frac{Pressure^2}{\rho c} \qquad (11.1)$$

If *I* and *Iref* are two intensities, then their difference in decibels (dB) is calculated as follows:

$$Intensity\ difference\ (dB) = 10 \log\frac{I}{I_{ref}} \qquad (11.2)$$

For the intensity levels and pressure levels to be comparable in dB, the difference in sound pressure is defined as follows:

$$Pressure\ difference\ (dB) = 20 \log\frac{P}{P_{ref}} \qquad (11.3)$$

This maintains the appropriate proportionality of intensity and pressure (if $I \propto P^2$ then $\log I \propto 2 \log P$) for sounds in the same medium. As an example, take a sound measured to be 10 times the pressure reference. This would be a pressure difference of 20 dB re P_{ref} by equation (11.3). Since intensity is proportional to pressure squared, the intensity of this sound would be 10^2 or 100 times the intensity of the reference. This would still be 20 dB re the reference intensity, by the definition of intensity in equation (11.2).

The primary definition of the decibel is as a ratio of intensities. The decibel always compares a pressure or intensity with a reference unit. The standard underwater reference *Iref* is the intensity of a sound having a pressure level of 1 μPa (Urick 1983). The microPascal is a unit of pressure: $1\ \mu Pa = 10^{-6}$ Pascal $= 10^{-6}$ Newtons/m². Both intensities and pressures are referred to a unit of pressure, 1 μPa, in a form of shorthand. When an *intensity* is referred to *pressure* of 1 μPa, it really means referred to the intensity of a continuous sound of pressure equal to 1 μPa. Confusion about decibel references and pressure versus intensity can have significant practical consequences (Chapman and Ellis 1998).

Doppler. When echoes from a moving source bounce off a moving target, the frequency of the echo is shifted by the Doppler effect. The change in frequency $\Delta f = (2vf)/c$, where *v* is the difference in velocity between source and target and *c* is the speed of sound.

Hydrophone. A hydrophone is an underwater microphone that converts the pressure fluctuations of a sound into voltage fluctuations.

Received Level and Source Level. If you measure the pressure of a sound at some location with a calibrated hydrophone, you can report the received level of that sound in decibels, simply as 20 log $P_{measured}/P_{ref}$. If $P_{ref} = 1\mu$Pa, then this would be expressed as Received Level = XX dB re 1μPa. However, if this was measuring a specific sound source, and you want to report how loud it was, it is conventional to report the source level as measured one meter from the sound source. Source levels are expressed as Source Level = XX dB re 1μPa at 1 m. If you measure the sound farther away, and you know how sound energy was lost as it propagated from 1 m to where you measured it, you can correct the received level to give an estimated source level.

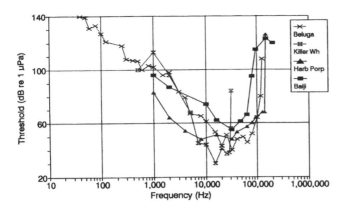

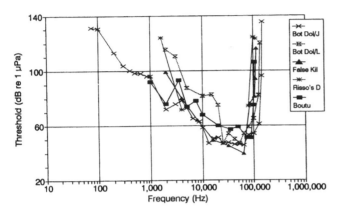

Figure 11.2. Audiograms of a variety of toothed whales tested in captivity. (From Richardson et al. 1995b.)

humans hear (reviews in Au 1993 and Richardson et al. 1995b). Their best frequencies extend from about 20 kHz, the upper limit of human hearing, to about 70 kHz. The upper limit of hearing in most toothed whales that have been tested extends above 100 kHz, or more than five times the upper limit of human hearing. Ambient noise levels are usually low in this high-frequency region (Urick 1983). In such a low-noise environment, the ability to detect a faint signal may depend upon having hearing that is sensitive compared with the typical noise levels. Noise levels tend to be much higher at lower frequencies, lowering the benefit of such sensitive hearing in this frequency range. The sensitivity of hearing in most of these toothed whales drops off at frequencies below about 10 kHz. Hearing has not been well studied below 1 kHz, but these animals appear to be relatively insensitive to low-frequency sounds (Richardson et al. 1995b). Hearing is still important at these low frequencies, however; many odontocetes produce vocal signals with dominant energy below several kHz (e.g., Connor and Smolker 1996; Overstrom 1983).

The frequency range of hearing has never been tested in baleen whales. Hearing is usually tested with trained ani-

mals, and baleen whales are so big that only a few have been kept for short periods in captivity. However, as mentioned above, the frequency tuning of the inner ear suggests they are specialized for low-frequency hearing (Ketten 1994). As we shall see, the vocalizations of baleen whales concentrate sound energy well below 1 kHz, and some species, such as fin and blue whales, produce sounds with dominant energy below 20 Hz, well below the frequencies humans can hear well. There tends to be a rough correlation between the frequencies at which most animal species can hear best and the frequencies typical of their vocalizations, so these low-frequency vocalizations of baleen whales also suggest the importance of low-frequency hearing.

Echolocation. Echolocation, defined as the ability to produce high-frequency clicks and to detect echoes that bounce off distant objects, is highly specialized in some odontocete species. Echolocation has been studied most in the smaller toothed whales that can be kept in captivity (Au 1993). Several distinct kinds of echolocation signals have been reported for different species. For example, figure 11.3 shows two examples of odontocete echolocation pulses, one from the bottlenose dolphin *(Tursiops truncatus)* and one from the harbor porpoise *(Phocoena phocoena).* The *Tursiops* pulse has a sudden onset, increasing to its peak level in tens of microseconds. The maximum peak-to-peak sound source level is quite high, >220 dB re 1 μPa at 1 m (see Box 11.1 for an explanation of how acousticians measure intensity of sound in decibels, or dB). These louder levels tend to be recorded from animals in open waters echolocating on distant objects. Peak-to-peak source levels as low as 150–160 dB re 1 μPa at 1 m have also been reported, typically from captive animals in tanks (Evans 1973). The *Tursiops* pulse in figure 11.3A was recorded from a trained dolphin as it was echolocating in open waters on a distant artificial target. When wild bottlenose dolphins are recorded as they echolocate, their pulses are typically reported to have energy in a broad range of frequencies from about 100 to 130 kHz (Au 1993). The *Tursiops* pulse in figure 11.3A has a peak frequency of 117 kHz and a bandwidth of 37 kHz. The echolocation pulses of harbor porpoise (fig. 11.3B) are quite different from those of bottlenose dolphins. They have a high-frequency component in the frequency band of 120–150 kHz, with sound source levels around 150–160 dB re 1 μPa at 1 m (Goodson et al. 1995; Møhl and Anderson 1973; Kamminga and Wiersma 1981). These porpoise pulses have a longer duration (hundreds vs. tens of microseconds), lower level, and narrower bandwidth (10–15 kHz vs. 30–60 kHz) than pulses from bottlenose dolphins (Au 1993). The

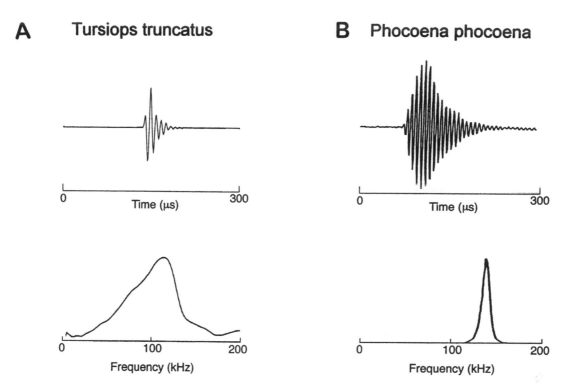

A Tursiops truncatus   **B** Phocoena phocoena

Figure 11.3. Waveforms (top) and spectra (bottom) of clicks from the bottlenose dolphin, *Tursiops truncatus* (A) and the harbor porpoise, *Phocoena phocoena* (B). The bottlenose dolphin figure is an average from an entire click train (adapted from Au 1980). The harbor porpoise figure is from a single click from a young animal (adapted from Kamminga 1988).

high-frequency hearing abilities of these small toothed whales are typically related to the need for these animals to detect faint echoes from their high-frequency echolocation clicks (Au 1993).

Odontocetes can vary their echolocation clicks depending on background noise. For example, the echolocation clicks of a beluga whale, *Delphinapterus leucas,* tested in San Diego had peak energy in the frequency band 40–60 kHz (Au et al. 1985). When the beluga was moved to Kaneohe Bay, Hawaii, where there is a high level of noise in these frequencies, it shifted the frequency of its clicks upward to mostly above 100 kHz. Moore and Pawloski (1991) investigated whether dolphins could be trained to modify the loudness or peak frequencies of their clicks. They successfully trained a bottlenose dolphin to produce clicks that were either high or low source level (>205 vs. <195 dB re 1 μPa at 1 m) or that had high or low peaks in frequency (>105 kHz vs. <60 kHz). Echolocating dolphins can discriminate the shape or composition of targets (e.g., Kamminga and van der Ree 1976), but there is little evidence of dolphins modifying their echolocation signals for different kinds of targets.

Many echolocation tasks involve ranges of less than several meters. Dolphins often inspect objects with echoloca-

tion at ranges as close as a few centimeters away. Bottlenose dolphins usually wait to hear the echo from a target before they produce the next click, and as these dolphins close in on the target, the interval between pulses usually decreases (Au 1993). This sounds to our ears like individual clicks blending into a buzz sound, but the dolphins are capable of much better temporal resolution in their hearing, and can resolve the individual clicks.

Dolphins do not use echolocation only at short ranges, but can detect objects at much greater distances than they can typically see them. An echolocating dolphin can detect a 2.5 cm metal target about 72 m away (Murchison 1980). The greater potential range of echolocation compared with vision may make it particularly useful for detecting obstacles or prey at a distance. If a cetacean were swimming rapidly in murky water, at depth, or at night, it seldom could see an obstacle rapidly enough to avoid it, but echolocation could be used to detect an obstacle far enough away to give even fast-swimming animals plenty of time to respond. Many cetaceans also feed in turbid water, at depth, or at night when there is little downwelling light from the sky. Some of these animals may visually detect luminescent prey nearby, but there are many circumstances in which vision has a more limited range than echolocation for detecting prey underwater. Most studies of dolphin echolocation have taken place under carefully controlled conditions with captive animals and artificial targets. Little is known about how wild dolphins use echolocation to perform tasks such as obstacle avoidance or prey detection, prey selection, or prey capture.

The problems of studying how cetaceans use echolocation to forage are particularly difficult for deep-diving animals such as sperm whales, because it is so difficult to observe them feeding at depths of many hundreds of meters. When sperm whales dive and forage, they tend to produce series of click sounds with relatively stable interpulse intervals of 0.5–2.0 seconds (Whitehead and Weilgart 1990, 1991). This association of what Weilgart and Whitehead (1988) call "usual clicks" with diving and feeding has led most biologists to hypothesize that these regular click trains function for echolocation (Backus and Schevill 1966; J. C. D. Gordon 1987a; Whitehead and Weilgart 1990). Watkins (1980), on the other hand, argued that sperm whale clicks were not suited to echolocation of prey. For example, the clicks of sperm whales are lower in frequency, longer in duration, and much less directional than the high-frequency clicks of dolphins (Au 1993). These observations led Watkins to argue that usual clicks are social signals used by diving whales to maintain contact with one another.

The echolocation hypothesis for the usual clicks of sperm whales has not been definitively tested. However, Goold and Jones (1995) have used acoustic models to evaluate the potential range at which the clicks of sperm whales might be used to detect their squid prey. These calculations necessarily involve assumptions and rough estimates of some parameters, but they suggest that sperm whales might detect squid at a range of 200–680 m. This range is consonant with the low end of the interpulse intervals observed in usual clicks, assuming the whale waits to make a new click until it hears an echo return. The speed of sound in seawater is near 1,500 m/sec, so the round-trip travel time to a target 750 m away would be about one second. If sperm whales producing usual clicks are waiting for potential echo returns before producing the next click, then the 0.5–2.0-second intervals between usual clicks would suggest maximum working ranges between about 375–1,500 m. While the shortest intervals correspond to the estimated detection range, the longer intervals involve much longer round-trip travel times.

Unfortunately, Goold and Jones (1995) did not correct for the bandwidth of hearing in their application of the sonar equation, and this inflated their estimated range of detection. They defined the threshold at which sperm

whales are just able to detect echoes from their clicks as occurring when the echo level equals the ambient noise level. To make a biologically realistic estimate, it is critical to match the noise and echo levels in equivalent bands, optimally ones that are appropriate for the hearing of the animal. Goold and Jones did not do this, but rather compared a broadband estimate of the source level of sperm whale clicks across a frequency range of thousands of Hz to a spectral level of noise in 1 Hertz band. Most mammals integrate sound energy over frequency bands roughly about one-third octave in breadth. If the spectral noise estimates used by Goold and Jones are corrected to third-octave band levels, leaving all other aspects of the calculation the same, the estimated detection range drops from 200 m to 68 m at 2 kHz and from 680 m to 108 m at 10 kHz. These reduced ranges match the round-trip travel time much less well than those resulting from the unrealistic echo:noise comparisons.

The only way to find out whether sperm whales do use their clicks for echolocation is to conduct empirical tests. Sperm whales have seldom been held in captivity for long enough to conduct the kind of tests that have been so successful with dolphins. Tags that can record sounds at the whale are important new tools for conducting this kind of research with free-ranging animals (e.g., see Fletcher et al. 1996 and Burgess et al. 1998 for acoustic recording tags for deep-diving elephant seals). If such a tag had sufficient dynamic range, it ought to be able to record clicks of individual sperm whales throughout the dive cycle, as well as detect the hypothesized echo returns at the whale. If echoes were detected, and if they were followed by specific responses such as pursuit and capture of prey or avoidance of obstacles, then this would provide much stronger evidence for echolocation.

The only echolocation system that has been demonstrated in cetaceans involves the use of high-frequency clicks by small odontocetes. These animals clearly have evolved a highly specialized system for echolocation. However, sound may be used to explore the environment even among cetaceans that are not specialized for high-frequency echolocation. Several biologists have suggested that whales might be able to sense echoes of low-frequency vocalizations from distant bathymetric features to orient or navigate (Norris 1967b, 1969; Payne and Webb 1971; Thompson et al. 1979). Some baleen whales produce loud low-frequency sounds that are particularly well suited for long-range propagation. For example, finback (*Balaenoptera physalus*) and blue whales (*B. musculus*) produce series of loud sounds well below the lowest frequencies humans can hear, centered around 10–30 Hz. The sounds of blue whales last several tens of seconds (Cummings and Thompson 1971a; Edds 1982), while the most common sounds reported from finbacks comprise series of one-second pulses (Watkins et al. 1987). Biologists have recently used bottom-mounted hydrophones to locate and track whales over long ranges, including one blue whale tracked for >1,700 km over 43 days (Costa 1993). These results have confirmed the predictions of Payne and Webb (1971) and Spiesberger and Fristrup (1990), who used acoustic models to estimate that these low-frequency whale calls could function for communication or orientation over very long ranges. Payne and Webb (1971), Watkins et al. (1987), and McDonald et al. (1995) conclude that these signals have better design features for communication than for listening for echoes from features on the seafloor. On the other hand, Clark (1993) has suggested that the low-frequency calls of finback or blue whales would produce easily identifiable echoes from seamounts hundreds of kilometers away. Even if these sounds are used primarily for communication, echoes from these signals could be very useful for orientation of migrating animals. If these uses of sound to explore the environment met an important biological need, they could influence the further evolution of cetacean signals.

None of these suggestions that whales may use low-frequency sounds to echolocate on bathymetric features have been tested, but there is suggestive evidence that migrating bowhead whales, *Balaena mysticetus*, use echoes from their calls to detect ice obstacles. Vocalizing bowhead whales avoid floes of deep ice at ranges much farther than the limit of underwater visibility (Clark 1989; George et al. 1989). Ellison et al. (1987) used acoustic models to show that deep-keeled ice may produce strong echoes from the low-frequency calls of migrating bowhead whales, and they suggested that bowhead whales may use these echoes to sense and avoid deep ice.

Functional Categories of Acoustic Communication Signals in Cetaceans

A signaling animal incurs costs such as the energy and time required to produce the signal and the risks of attracting predators or alerting prey. Some animals that are not the intended recipients of the signal may overhear it and respond to it to their own benefit, and often to the detriment of the signaler. Myrberg (1981) calls these unintended recipients "interceptors" of the signal. Interspecific communication often includes interactions in which predators or prey may intercept signals. This risk of detection by an interceptor may represent a significant component of the

cost of signaling in many settings. If production of the signal is to confer a selective advantage, then these costs must be offset by a benefit to the signaler induced by changes in the behavior of a recipient of the signal. There are a variety of benefits associated with different potential recipients. Common forms of intraspecific communication involve signaling to potential mates or potential competitors. In species with parental care, communication is often required between parent and offspring. In social species, animals often communicate to maintain contact with other members of the group. In species with individual-specific social relationships, signals are required for individual recognition and for maintaining these relationships.

Interspecific Communication and Interception

Interception by prey. One potential problem with producing echolocation sounds for detecting prey is that an echolocating predator runs a risk of the prey intercepting the echolocation signal. This advance warning may help the prey escape or avoid detection. Odontocete predators may be able to alter their echolocation strategies to reduce the probability that acoustically sensitive prey (including seals and other cetacean species) will intercept their echolocation signals. For example, in the Puget Sound area of the Pacific Northwest, there are two populations of killer whales, *Orcinus orca* (see Baird, chap. 5, this volume). One population feeds on marine mammals, a prey that is sensitive to the frequencies of killer whale clicks; the other population feeds on salmon, a prey that is likely to be much less sensitive. Barrett-Leonard et al. (1996a) report that mammal-eating killer whales vary the intensity, repetition rate, and spectral composition within their click trains, apparently making these clicks more difficult for their acoustically sensitive prey to identify than the regular click series of fish-eating killer whales.

There is suggestive evidence that even some species of fish may have special abilities to intercept the high-frequency echolocation sounds of their odontocete predators (Mann et al. 1997). Fish such as American shad *(Alosa sapidissima)*, alewives *(Alosa pseudoharengus)*, herring *(Alosa aestivalis)*, and cod *(Gadus morhua)* are able to detect intense sounds of frequencies much higher than is typical of their own vocalizations (Astrup and Møhl 1993; Dunning et al. 1992; Mann et al. 1997; Nestler et al. 1992). Some of these species are prey for echolocating odontocetes. The only known natural sources of sounds with the intensity and frequency of these ultrasonic stimuli are the clicks of echolocating toothed whales. Clupeid fishes have an unusual specialization for hearing: an air-filled chamber that abuts the inner ear and is thought to increase the sensitivity of hearing (Blaxter et al. 1981). This auditory specialization is shared by all living clupeids, suggesting an origin in early clupeids, which date well before the origin of echolocating cetaceans. If this auditory specialization not only enhances sensitivity at low frequencies, but also enables ultrasonic hearing in clupeids, then this ultrasonic sensitivity may be a preadaptation (Mann et al. 1997), rather than having evolved specifically to enable interception of odontocete predators. Several clupeids respond with escape behavior when they hear ultrasonic pulses, and this response may represent an adaptation for escaping odontocete predators.

Responses of cetaceans to their own predators. Animals may produce signals to confuse or startle an approaching predator. For a visual example, dwarf sperm whales of the genus *Kogia* release a cloud of opaque reddish anal fluid when they are disturbed (Scott and Cordaro 1987). As with squid ink, this release of opaque fluid is thought to act as a visual screen, giving the whale better odds of concealment or escape. Whales and dolphins may also intercept the signals of their predators and respond to avoid detection or capture. One of the most common predators of many cetaceans is another cetacean, the killer whale (Jefferson et al. 1991). Killer whales often vocalize while foraging (Ford 1989), providing their cetacean prey the opportunity to intercept these signals and avoid capture. When potential prey intercept the calls of a dangerous predator, they often show strong responses. These strong, easy-to-observe responses make this a good phenomenon to study using playback experiments. For example, when Fish and Vania (1971) played killer whale calls to beluga whales, *Delphinapterus leucas,* that were feeding on salmon in an Alaskan river, the beluga whales showed a strong avoidance response. Gray whales migrating along the coast of California responded to playback of killer whale calls by swimming rapidly inshore into beds of kelp (Cummings and Thompson 1971b; Malme et al. 1983).

Intraspecific Communication and Interception

This section is organized around the potential intraspecific recipients of a signal, along with the problem(s) that the signaling is designed to resolve. These categories include the following signaling functions:

- increase one's chances of mating
- agonistic relationships or fighting assessment
- parent-offspring recognition

- maintaining the coordination and cohesion of individuals within a group
- maintaining individual-specific social relationships

The "communication-as-manipulation" view discussed at the beginning of this chapter treats signals as advertisements designed to manipulate the choices of recipients, and leads one to expect that the recipients may be selected to develop a certain amount of sales resistance. Just as the signaler balances costs and benefits when deciding whether to signal, the receiver may balance the costs and benefits of responding at all to a signal or of selecting any of several potential responses. The structure of these displays cannot be understood unless one considers how the signal is designed to manipulate the choices of animals that receive the display. For example, when we think of an animal responding to a signal, we usually think of an immediate response, such as the avoidance responses of animals hearing the sounds of a predator. However, one animal may produce a long series of signals to modify the outcome of one choice by a receiving animal. The receiver may make its choice only after hearing hundreds or thousands of advertisements from many different signalers. If our model of communication was limited to expecting immediate responses based upon transfer of information from a single signal, we would have a hard time understanding these kinds of advertisements.

Signals for Improving Chances of Mating

A common form of animal advertisement is the reproductive advertisement display made by an animal that is seeking to improve its chances of mating. Charles Darwin (1871) coined the term sexual selection to describe the evolutionary selection pressures for traits that are concerned with increasing mating success. Darwin described two basic modes of sexual selection: those that increase the likelihood that an animal will outcompete a conspecific of the same sex for fertilization of or by a member of the opposite sex (intrasexual selection), and those that increase the likelihood that an animal will be chosen by a potential mate (intersexual selection). More recent reviews have included a third mode, in which a male may attempt to limit the choice of a female by coercing her to mate with him and not with other males (Smuts and Smuts 1993; Clutton-Brock and Parker 1994).

Sexual selection can create different selection pressures for reproductive advertisement displays than for signals designed for simpler exchanges of information. Selection is likely to favor these other signals being quieter and more difficult to detect to avoid the costs of interception. Reproductive advertisement displays are typically designed to be attention-getting. This involves potential risks, such as attracting a predator, but the intended receiver may use the advertisement in part to evaluate how well the displaying animal can deal with these risks. Since the biologist is more likely to observe advertisement signals than other, more-difficult-to-intercept signals, advertisements are better known than many other kinds of animal signals, such as the quiet interchanges of mother and young when they are close together, predator alarm calls, or the nighttime flight calls of migrating birds.

Examples of reproductive advertisement displays include the songs of birds and whales. Songs are usually defined as acoustic displays in which sequences of discrete sounds are repeated in a predictable pattern.

Songs of humpback whales. The songs of humpback whales are the best-known advertisement display in the cetaceans. Humpback songs have a hierarchical structure that was described by Payne and McVay (1971). Each song is made up of three to nine themes that tend to be sung in a particular order, and it often takes about ten to fifteen minutes before a singer returns to the initial theme. Each theme is made up of phrases that repeat a variable number of times before a new theme is heard. Each phrase lasts about fifteen seconds and contains a series of sounds. Figure 11.4 shows a spectrogram of a humpback song recorded in waters near the Hawaiian Islands and made during a period when songs contained up to nine themes.

Humpback song is so loud and distinctive that it is relatively easy to record, and biologists now have recordings of song from many humpback breeding areas over periods of many years. There appears to be a strong force for vocal convergence in the songs of humpback whales on a breeding area at any one time, coupled with progressive change in all aspects of the song over time (K. B. Payne et al. 1983; Payne and Payne 1985). Recordings of humpback song made from different whales in the same breeding area at the same time are quite similar. At any one time, all of the singers within a breeding area sing the same themes in the same order, and the individual sounds that make up the song are quite similar. However, the song changes dramatically from month to month and from year to year. For example, theme 3 from figure 11.4 was on its way out in March of 1977. In hundreds of recordings made later that spring and for years afterward, it was never heard after the end of March 1977. There is no indication that these changes in the song reflect changes in the message, for the

whales appear to be engaged in similar interactions even as the song changes.

Humpback song is usually recorded during the winter in the tropical breeding grounds of these whales (see Clapham, chap. 7, this volume). This observation led Schevill (1964: 310) to suggest that the song "may be an audible manifestation of more fundamental urges." The primary reason that we can discuss the social functions of humpback song at all rests on the ability of biologists to identify and follow singing whales at sea, and to track their interactions with other whales. Singing humpback whales often surface to breathe once per song during a particular theme, "blowing" or breathing during the silent intervals between notes (Tyack 1981). Biologists can deploy a hydrophone from a small boat to listen for singers. If they can get close enough to one singer that it is louder than the others in the background, they can often locate the singer by careful visual scanning during this part of the song. Once a singer has been identified in this way, its sex can be determined by underwater photography of the genital slit (Glockner 1983) or by taking a small tissue sample for genetic determination of sex (Baker et al. 1991; Palsbøll et al. 1992; see Whitehead et al., chap. 3, this volume). The continuous vocalizations and predictable surfacing behavior of singers allow biologists to follow them for hours. These observations have shown that almost all singing humpbacks are lone adult males, and that their songs are repeated in bouts that can last for many hours.

Observation from elevated shore stations has been important for studying species, such as humpbacks, that concentrate in coastal waters. Observers on land can use a surveyor's theodolite to pinpoint the location of each whale surfacing, and can communicate by radio with vessels that can obtain more detailed behavioral and acoustic data by following whale groups (Tyack 1981). This combination of shore- and ship-based observations has revealed interactions of singing whales with groups nearly 10 km away (Tyack and Whitehead 1983), and is well suited to observing whales for periods up to ten hours. These visual locations can also be linked to acoustic locations from arrays of hydrophones (Frankel et al. 1995). This acoustic location technique provides some evidence suggesting that singers tend to be evenly spaced. The interactions of whales are so much slower than our own pace that it is often necessary to make a plot of movement patterns and behavioral displays to make sense out of an interaction that took many hours to unfold.

A variety of results have encouraged biologists to suggest that humpback song plays a role in male-male competition:

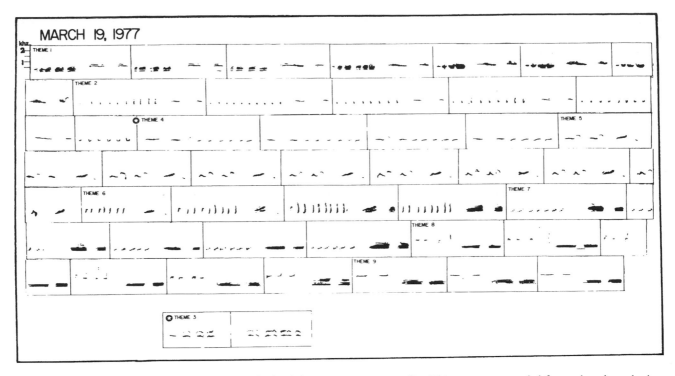

MARCH 19, 1977

Figure 11.4. Spectrogram of the song of a humpback whale, *Megaptera novaeangliae*. This song was recorded from a lone humpback on 19 March 1977 in the Hawaiian Islands. Each line represents 120 seconds; this song took seven lines, or fourteen minutes, before repeating the initial theme. The phrase boundaries are marked by vertical lines. The *x*-axis indicates time; the *y*-axis indicates frequency in kHz. The third theme was not included in this song, but was in the next song of this whale, as indicated by the star in the spectrogram. (From K. B. Payne et al. 1983; courtesy Roger Payne.)

• Song appears to maintain distance between singers (Tyack 1981; Frankel et al. 1995; Helweg et al. 1992).

• No known females were attracted to playbacks of song (Tyack 1983; Mobley et al. 1988).

• Aggressive interactions (particularly between singers and known males) are much more commonly observed than sexual interactions (particularly between singers and known females) (Tyack 1981, 1982; Darling 1983).

These behavioral observations are clearly consistent with the idea that song plays a role in mediating male-male interactions.

However, just because humpback song appears to be used in male-male interactions does not mean that it is not also used by females to select a mate. Both intra- and intersexual selection often operate at the same time on the same display (see Catchpole 1982 for a discussion regarding bird song). The use of songs to mediate spacing among singers says nothing about whether females are also an important audience. Females are often more discriminating than males in responding to an advertisement display such as song. None of the song playbacks conducted with humpback whales duplicated all of the potentially relevant fea-

tures of song, and this may account for some of the lack of response of females to playbacks. Furthermore, as Catchpole (1982) points out for songbirds, aggressive male-male interactions are much more common and obvious in many species than male-female interactions. Just because the responses of male humpbacks to song are seen more frequently than those of females does not mean that the subtler and rarer responses of females to singers are not biologically significant. The critical question here is whether females choose a mate based upon his song. Copulation has never been observed in humpback whales, so little evidence has been collected to address this question. However, genetic analysis of paternity is a much more direct indicator of male reproductive success. Extensive efforts to biopsy humpback whales now make genetic analysis of paternity a realistic option for this species (Clapham and Palsbøll 1997).

In the absence of direct data on the relative function of song in male-male competition versus female choice, there may be some indirect evidence from the acoustic structure of the song. Much more is known about relationships between acoustic structure and social function in the songs of birds than in whales. In the next few paragraphs, I will apply some of these songbird results to humpback song. Several ornithologists have attempted to relate specific acoustic features of the songs of birds to the relative importance of intra- versus intersexual selection. Catchpole and Slater (1995) suggest that songs used to attract females may be selected to be long, complex, and continuous, while songs used for intrasexual competition are likely to be shorter, simpler, and sung with gaps, so that a singer can listen for rival males. Catchpole (1982) suggests that continuous singing, lack of matched countersinging between males, and lack of a singing response to song playback are also diagnostic of a female attraction role for song. Humpback whales often sing continuously for hours, and they do not respond to the song of another male by immediately matching the sounds produced by that male, nor do they respond to song playbacks by singing themselves. All of these features are consistent with the hypothesis that humpback males sing in part to influence the mating choices of females.

There are also some acoustic features of humpback song that have been associated with intrasexual selection. Baker and Cunningham (1985) suggest that intrasexual selection tends to select for rapid song change and convergence in the songs of neighboring males. The basic idea is that young males may benefit from mimicking the songs of territorial neighbors with greater attractiveness to females, and the more attractive males may then have to change their song to avoid this competitive mimicry. Humpback males on the breeding grounds are not thought to be territorial, and there is no evidence that this kind of arms race occurs with humpback whales, but both rapid change and convergence between individuals are striking features of humpback song.

There are problems with extrapolating from models of bird song to singing whales, because whales live in an environment with significant differences from those of most songbirds. In particular, while birds must lay eggs in a nest, whales are not bound to a fixed site, but meander at will. During the breeding season, females with young may seek out clear, protected waters, but humpbacks do not stay in one site for long, nor does either sex appear to defend territories on the breeding ground. Tyack (1981, 1982) followed interactions between singers and other whales within 5–10 km of a shore station, and the same individual whales were seldom sighted across days within this site. If males do not have small territories and consistent neighbors, then they would be unlikely to form the subdialects described by Baker and Cunningham (1985) for birds. Also, sound propagates so far in the ocean that the songs of whales have a much longer range than those of most birds. These differences may help account for the unusual pattern of vocal convergence and continual change in the songs of entire populations of humpback whales, a pattern that differs from that of most songbirds.

In summary, there is evidence suggesting that humpback song plays a role in both male-male competition and female choice. Observations of interactions between humpbacks emphasize the role of song in mediating interactions between males, but females may also monitor, approach, and join with singers. The acoustic structure of humpback song clearly has attributes associated with both intra- and intersexual selection, and is more complex than would be expected for a signal used only in male-male interactions (Helweg et al. 1992).

Songs of bowhead whales. In addition to humpback whales, several other species of baleen whales are reported to produce songs that are thought to be reproductive advertisement displays. However, the behavioral contexts of song production are not as well understood in these species as in humpback whales. Bowhead whales spend their winter breeding season in Arctic waters, where biologists have few opportunities to observe them. The songs of bowhead whales have been recorded in the spring as they migrate past Point Barrow, Alaska (Ljungblad et al. 1982). Bowhead songs are simpler than those of humpbacks, consisting of a few sounds that repeat in the same order for many song repetitions. Like humpback songs, bowhead songs appear to change year after year (Clark and Johnson 1984; Clark 1990; Würsig and Clark 1993). However, little is known about behavior concurrent with singing, and there are few reports of bowheads observed during their winter breeding season in Arctic waters.

Long-range communication in finback whales. In the section on echolocation, I described low-frequency vocalizations of finback and blue whales that appear to be adapted for long-range propagation in the ocean. As described in that section, there has been speculation that these signals may function for long-range echolocation of large-scale geographic targets such as seamounts, which might be used for orientation and navigation. However, it has also been suggested that the long series of low-frequency pulses produced by finback whales may also be a reproductive advertisement display produced by males (Watkins et al. 1987). Finback whales produce pulses with energy in a range roughly between 15 and 30 Hz (Watkins et al. 1987), near the lowest frequencies that humans can hear. Each pulse lasts on the order of one second and contains twenty cycles. The source level of the pulses ranges from about 170 to 180 dB re 1 μPa at 1 m (Patterson and Hamilton 1964). Particularly during the breeding season, finbacks produce series of pulses in a regularly repeating pattern. These bouts of pulsing may last for longer than one day (Watkins et al. 1987). The seasonal distribution of these pulse series has been measured near Bermuda, and it matches the breeding season quite closely (Watkins et al. 1987). This correlation has been used to suggest that the pulse series may function as reproductive advertisement displays (Watkins et al. 1987). However, finback whales are common in waters near the latitude of Bermuda only during the winter breeding season, so few would be likely to be heard there at other seasons even if they did vocalize at that time. Similar recordings in more polar waters will be required to test how frequently these whales produce these series of pulses outside of the breeding season.

Finback whales disperse into tropical oceans during the mating season, unlike other species such as humpback and gray whales, which are thought to congregate in well-defined breeding grounds. The functional importance of a signal adapted to long-range communication is obvious for animals that disperse over ocean basins for breeding. However, even though biologists can hear finback pulses at ranges of hundreds of kilometers by listening at appropriate depths, the effective range of these signals for finbacks themselves is not known. Until very recently, marine bioacousticians have had few methods that would allow us to detect whether a whale was responding to the calls of another whale more than several kilometers away. The longest range over which whales have been observed to respond to the sounds of other whales is approximately 10 km (Tyack and Whitehead 1983; Watkins and Schevill 1979).

The U.S. Navy has recently made networks of bottom-mounted hydrophone arrays available to biologists (Costa 1993). These arrays will allow biologists to track vocalizing whales hundreds of kilometers away, and may help us to ascertain whether whales actually communicate in vocal exchanges over these long ranges. Just because biologists can detect these signals over such long ranges, however, does not mean that the intended recipients are that far away. If the signals are used in competitive interactions, for example, and if the louder vocalizer has an advantage, then selection could favor the evolution of signals much louder than required to be just detectable at the typical range of the intended recipient.

Mate guarding in bottlenose dolphins. The discussion above of reproductive advertisement displays emphasizes female choice and competition between males. There are also species in which males may adopt a strategy of attempting to preempt female choice by guarding a receptive female and preventing her from mating with other males (Smuts and Smuts 1993; Clutton-Brock and Parker 1994). As described by Connor et al. (chap. 4, this volume), groups of two or three adult male bottlenose dolphins in Shark Bay, Western Australia, may form consortships with an adult female. A coalition of males may start such a consortship by chasing and herding a female away from the group in which they initially find her. Some of these consortships appear to be attempts by the males to limit choice of mate by the female, who may try to escape from the males.

Males in these coercive consortships produce distinctive trains of low-frequency clicks, called "pops" (Connor and Smolker 1996). These pops have peak frequencies of 300–3,000 Hz, far below the clicks typically reported for echolocation. They occur at rates of about 6–12 per second in trains of about 3–30 pops. When a herded female hears these pops, she shows a strong tendency to turn toward the popping male and to approach him within seconds. A male may produce aggressive displays toward a female who fails to approach after he pops. These may include head jerks, chases, or physical contact. Much of this aggression is also typically associated with the production of pops by males. Connor and Smolker (1996) argue that males that are consorting with and guarding a female produce these pops to induce her to remain close.

Signals to Competitors: Predictive Signaling or Fighting Assessment

Animals often find themselves in a situation in which they must compete with a conspecific for access to a critical resource. Competitors often confront one another in a contest in which there will be a winner and a loser. In most of these contests, it will be worth a competitor's while to gather some information on the willingness and ability of its opponent to fight. Both competitors often have a shared interest in gaining this information, and many contests start with a ritualized phase of exchanging displays (see chapter 21 of Bradbury and Vehrencamp 1998 for a review of threat display contests). While competitive contests sometimes involve physical contact and injury, they may be settled prior to such escalation through a series of displays. If one contestant concludes from this initial assessment phase that it would be likely to lose a fight, or that the potential cost of injury outweighs the potential benefit of winning, then it may gain by breaking off and leaving the contest (Krebs and Davies 1993). In this case, it may benefit from producing a submissive display to signal an end to the contest to the opponent.

The communicative signals used in agonistic interactions have several functions. Some features of signals may correlate with attributes that are good predictors of fighting ability, such as age, physical condition, or size. These basic attributes cannot be changed without considerable cost. If a contestant is using a signal to estimate the fighting ability of an opponent, then the signal need not necessarily predict an immediate behavioral response by the signaler, but it is critical that the signal involve features that are reliable indicators of fighting ability. Other agonistic signals may be used to assess the willingness of an opponent to escalate or to predict its next actions. Some theoretical analyses of fighting emphasize the costs of producing signals that help an opponent to predict one's behavior (e.g., Maynard Smith 1974). These analyses highlight the potential benefit of bluffing or of tailoring signals so they do not allow an opponent to predict immediate responses. However, there is evidence of predictive signaling from a diverse array of species. The potentially high risks of making a mistake in assessing an opponent in an all-out fight may encourage a more or less "honest" assessment phase (Hauser and Nelson 1991). For example, male African elephants *(Loxodonta africana)* produce vocal, chemical, and postural cues about their willingness to fight other males during the breeding season (Poole 1987). Poole (1987, 1989) argues that these cues provide accurate signals that males frequently use in deciding whom and when to fight. If the competitors are closely matched, they may escalate to more costly displays, which may serve the functions of more accurate assessment of the competitor or increasing the likelihood that the competitor will stand down.

The best description of agonistic interactions and displays in cetaceans stems from observation of captive bottlenose dolphins, *Tursiops truncatus.* Samuels and Gifford (1997) conducted a quantitative assessment of agonistic interactions among bottlenose dolphins. Their analysis focused on a specific set of aggressive or submissive behaviors. Following earlier work with primates (Hausfater 1975), a dolphin was said to have lost an interaction if and only if it produced a submissive response to either a neutral or an aggressive behavior from its opponent. This use of submissive behaviors to define which animal won or lost the interaction differed from most previous work on dominance in dolphins, which focused more on aggressive than on submissive behaviors. The submissive behaviors involved either fleeing from an opponent and leaving the contest or "flinching." This "flinching" behavior, which was previously little noted, involves an immediate recoil in response to an action by the opponent.

Most of the aggressive behaviors defined by Samuels and Gifford (1997) were behaviors that are directly involved in

a fight; these included chasing, pinning, ramming, hitting, or biting an opponent. However, they also included threat displays. A graded series of threat displays has been defined by Overstrom (1983) for captive bottlenose dolphins (fig. 11.5). Overstrom suggests that the earliest stages of a threat may involve one dolphin simply directing pulsed sounds toward another. The dolphin may escalate the threat by producing an open mouth threat display while emitting distinctive bursts of pulses. Overstrom suggests that the longer in duration or the louder in sound intensity these burst-pulses are, the stronger the threat. As another step in the escalation, the animal may accentuate this display with abrupt vertical head movements. This is similar to the head jerk display that Connor and Smolker (1996) associate with the pop vocalizations produced by wild male dolphins when they are aggressively herding an adult female. The jaw clap is a commonly cited aggressive display in dolphins, and is interpreted by Overstrom as one of the most intense threat displays. A dolphin starts the jaw clap display with an open mouth. The jaw clap consists of a rapid and abrupt closure of the gaping jaw, accompanied by an intense pulsed sound (fig. 11.6). Overstrom seldom recorded a dolphin producing a jaw clap when it had not already produced some burst-pulsed sounds, and the probability of an animal producing a jaw clap increased as the duration of the burst-pulsed sounds increased. Overstrom's framework for describing a graded series of threats in bottlenose dolphins should be viewed more as a hypothesis than as a definitive result, pending further analysis of how these displays function in agonistic interactions.

Many of the agonistic visual displays used by bottlenose dolphins are related to movements used to inflict injury. For example, the open mouth display looks like the first step in preparing to bite. Norris and Møhl (1983) suggested that some click sounds used by dolphins and toothed whales may be so intense that they could be used to stun or threaten conspecifics. Overstrom (1983) suggested that the loudest burst-pulsed sounds or the jaw clap may be intense enough to produce tactile or auditory irritation. If these suggestions were true, then lower-intensity pulsed calls might be used as a direct threat, similar to the way an open mouth display may threaten a bite. However, there is no evidence that these calls do function as weapons, so these interpretations remain speculative.

Most of the examples of agonistic signals described above involve "predictive signaling," in which a signal may reliably indicate future behavior, such as whether a competitor is ready to escalate (Hauser and Nelson 1991). The escalating series of aggressive displays described by Overstrom (1983) fit this "predictive signaling" model. The

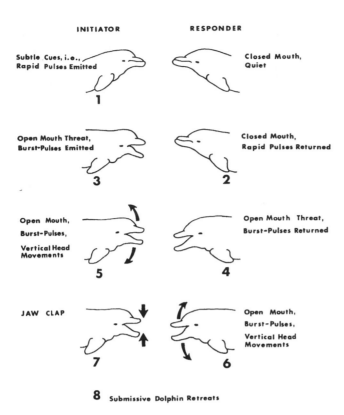

Figure 11.5. Visual and acoustic threat displays of the bottlenose dolphin, *Tursiops truncatus*. (From Overstrom 1983.)

pops described by Connor and Smolker (1996) also may be interpreted as early stages of a threat to escalate aggression if the consorting female does not approach the popping male. The pop vocalization may reliably indicate the readiness of the male to attempt to herd the female more aggressively.

Let us now switch focus from predictive signaling to signals that allow animals to assess the fighting ability of opponents. Body size is a good predictor of the outcome of fights in many animal species (Krebs and Davies 1993). A variety of animal species assess the lowest frequency of vocalization as an indicator of body size (e.g., Davies and Halliday 1978 for anurans; Clutton-Brock and Albon 1979 for a terrestrial mammal). There is clear evidence for such an effect in some marine fishes. Myrberg et al. (1993) demonstrated a strong correlation between the body size of the bicolor damselfish, *Pomacentrus partitus*, and the fundamental frequency of a chirp sound produced by males. They suggested that differences in the peak frequency of chirps result from differences in the volume of the swim bladder. The basic principle here is that if a sound is made by a resonant cavity, then the frequency of the sound will be lower for an animal with a larger resonator. If males judge the size of opponents by listening to the lowest peak frequency of their calls, then this will create selection pressure for males to make the lowest-frequency sounds they

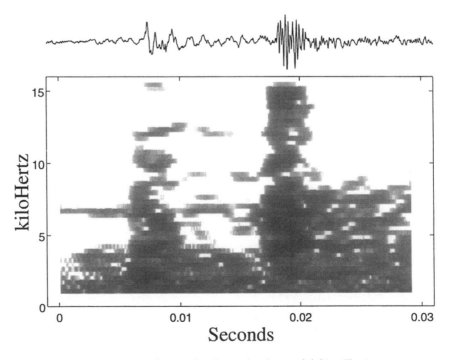

Figure 11.6 Spectrogram of a jaw clap from a bottlenose dolphin, *Tursiops truncatus*.

PETER L. TYACK

can. However, if the minimum peak frequency a male can produce is constrained by the volume of a resonating cavity, and if the resonator volume correlates with body size, then males may be constrained to produce an honest advertisement of their body size.

However, the link between frequency of vocalizations and body size depends upon how the sounds are produced. In fishes such as the toadfish, *Opsanus tau,* the fundamental frequency of the call is affected more by the contraction rate of sound-producing muscles than by the volume of the swim bladder (Fine 1978). In this case, the frequency of the vocalizations may not reliably indicate the size of the displaying fish, as was suggested for the bicolor damselfish. A small toadfish might be able to slow the rate of contraction of its sound-producing muscles to make a call lower in frequency than that of a larger competitor. This could be interpreted as the weaker contestant making a bluff by faking a display that inflates its apparent size and abilities. If an animal can use a signal to bluff about its motivation or ability, then it will pay the opponent to call the bluff. In many situations, it may even pay the opponent to ignore this kind of signal. This kind of logic has led biologists to emphasize the importance of signals that cannot be faked because they are inherently linked to an attribute associated with fighting ability.

While the frequency of a resonator is inversely proportional to its size, the comparison between damselfish and toadfish shows that the association between frequency of vocalizations and body size depends upon how the sounds are produced. The same issue is important in mammals. In some terrestrial mammals, there is little correlation between body size and the fundamental frequency of vocalizations (Fitch 1997). It appears for some primates that a different acoustic feature, the separation between formants, is a better indicator of size. Unfortunately, there are few models of sound production in cetaceans sufficiently detailed to predict which acoustic features of a signal are inherently related to attributes of the animal that may be relevant for mate choice or fighting assessment. The sperm whale is the cetacean species for which we have a model of vocal production that is relevant to this issue. As mentioned by Whitehead and Weilgart (chap. 6, this volume), an enormous fraction of the volume of the sperm whale body is devoted to an unusual organ called the spermaceti organ. The spermaceti organ lies dorsal and anterior to the skull and can have a length up to 40% of the length of the whale. Norris and Harvey (1972) argue that the spermaceti organ may function to generate the clicks that are the dominant vocalization of sperm whales. The clicks usually recorded from sperm whales are made up of a burst of pulses with

very regular interpulse spacing (Backus and Schevill 1966; Goold and Jones 1995). Norris and Harvey (1972) suggest that this regular spacing may result from reverberation within the spermaceti organ. They suggest that the spermaceti organ has an efficient reflector of sound at the posterior end and a partial reflector of sound at the anterior end. They propose that the source of the sound energy in the click comes from a strong valve in the nasal passage, called the *museau du singe,* at the anterior end of the spermaceti organ. They suggest that the first pulse within the click is produced as the initial sound made by this valve is transmitted directly into the water, while the remaining pulses within the click result from reverberation of the posteriorly directed component of the initial sound between the posterior and anterior reflectors within the spermaceti organ. Each time the sound arrives at the anterior reflector, some of the sound energy passes out into the ocean medium, while some reflects back within the spermaceti organ. If this hypothesis is correct, then the interpulse interval (IPI) could represent an accurate indicator of the length of the spermaceti organ. Gordon (1991) measured the length of sperm whales in the wild, along with the IPI of their clicks, and found a clear relation between IPI and estimated size of the spermaceti organ.

Mature male sperm whales produce a distinctive click vocalization, called slow clicks because of the long interclick interval (Weilgart and Whitehead 1988). Males can produce both slow clicks and usual clicks, and one solitary maturing male switched between slow clicks and usual clicks with no intermediate forms (Weilgart and Whitehead 1988). Slow clicks differ from usual clicks not only in their longer interclick interval, but also in their increased loudness, longer duration, and lower-frequency intensity peaks, near 2 kHz (fig. 11.7). Slow clicks tend to have a "ringing" sound near 2 kHz, perhaps indicating a structure in the sound production apparatus that resonates at this frequency. Sperm whales are sexually dimorphic, with adult males growing to 16 m in length, while females grow to about 12 m (Best 1979). The spermaceti organ also takes up a greater proportion of body length in adult males than in females or smaller males (Nishiwaki et al. 1963). If some of the acoustic features of slow clicks necessarily scale to the size and condition of adult sperm whales, then these may qualify as honest advertisements that may be used by males for fighting assessment or by females in mate choice.

Other acoustic features may be directly related to the condition of a male in other cetacean species. For example, Chu and Harcourt (1986) suggest that humpback females may select singing males based upon how long they can stay underwater and hold their breath. As described above,

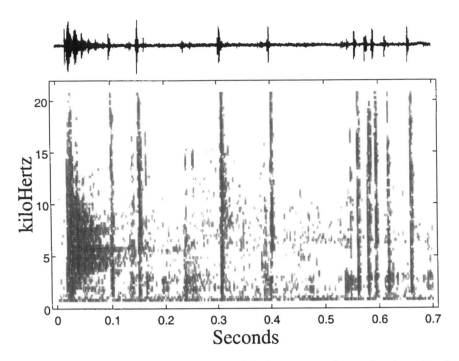

Figure 11.7. Spectrogram comparing one slow click (between 0 and 0.1 sec) and a series of usual clicks (to right of slow click) of sperm whales. The *x*-axis indicates time; the *y*-axis indicates frequency in kHz. Recordings are from a sperm whale group near Dominica, West Indies. (Data from William Watkins, Woods Hole Oceanographic Institution.)

most singers surface once per song cycle, and when they do, there are distinctive changes in the song. As a singing whale nears the surface, the lowest frequencies of the song drop out. This is due to the physics of underwater acoustics: as a sound source comes within a fraction of a wavelength of the surface, the sound does not propagate as well as it would at greater depth. This makes it difficult for a singing whale to come to the surface to breathe without broadcasting a reliable cue of his surfacing. Chu and Harcourt (1986) argue that breath-holding ability may be a good indicator of a male's stamina and physical condition. Breath holding may be particularly important for diving mammals. The problem with this argument is that, as described above, song duration changes as humpbacks slowly evolve every feature of their song. In the beginning of one year, the song may average seven minutes, whereas four months later it may have doubled in length (K. B. Payne et al. 1983). The next season, it may start long and decrease in length. Each individual whale is more likely to sing songs of the current length than what he was singing a few months earlier or later (Guinee et al. 1983). If humpbacks were using song to advertise their breath-holding ability, then each individual would be expected either to always sing as long as he was able, or to sing longest at that part of the breeding season when his chances of mating were highest. These predictions do not match the observations

that whales at any one time sing songs of similar duration and that the songs change over time with no repeated seasonal pattern.

Recognition systems and contact calls. When one animal interacts with another, the appropriate behavior often depends upon the identity of the other animal. Almost every species faces this problem when selecting a mate: an individual must find a mate from the other sex of the same species. The reproductive isolating mechanisms used to reduce inappropriate mating can be thought of as a recognition system. Other socioecological settings require other kinds of recognition systems. If a mother hears a begging call from an infant, the proper response may depend upon whether it is her own infant or not. If an animal becomes separated from its group and hears sounds of conspecifics, it may be critical to determine whether the sounds are coming from its own group or from a different one.

All animals have evolved recognition systems to discriminate conspecifics from animals of other species. Depending upon the social system typical of a species, they may evolve systems for discriminating kin, mates, neighbors, strangers, and individual companions. As Beecher (1989) points out, the resolution of these recognition systems depends upon the within-class and between-class variability of the signals used and upon the level of detail that can be sensed by the perceptual system. We do not yet know much about variation in the perceptual abilities used by cetaceans for resolving recognition signals, but there are a variety of cetacean acoustic signals that appear well designed to allow recognition between parents and young, between different social groups, and between specific individuals.

The need for recognition systems is particularly acute when animals meet after a separation. For example, if a human mother gives birth to an infant in a room and stays with the infant in that room, there is little need for a recognition system. However, in many hospitals, the infant is taken away from the parents into a nursery with many other infants. As soon as this happens, there is a serious danger of mixups unless the infant is carefully tagged and identified. A variety of features of cetacean life in the marine environment may require more frequent exercise of recognition mechanisms. Cetaceans have no nests or central places where specific individuals or groups can reliably be found. Cetaceans are highly mobile, and show great variability in their movement patterns. For many cetacean species, location cues are much less useful than they are for many terrestrial animals. Cetaceans within a group frequently swim out of sight of one another, causing increasing likelihood of separation unless the animals exchange signals using the acoustic modality, which is the only modality capable of rapidly transmitting information over the ranges at which these animals routinely separate.

The term "contact call" is used for sounds that function to help an animal keep in touch with its group. "Isolation calls" are a kind of contact call produced when animals are losing or have lost contact with one another, and they typically elicit immediate approach. Contact calls need not necessarily elicit an immediate reaction. When members of a group are separated but within acoustic range of one another, they may regularly call in order to keep in touch with at least one other member of the group, without otherwise altering their activity or movement pattern. The terms "contact" and "keep in touch" are unfortunate because they imply close physical proximity. This is not the case for contact calls; rather, these calls are used to keep an open communication channel between animals that are staying within the effective range of the communication signals. In the ocean, this range can extend over many kilometers. As might be expected for calls of mutual benefit to sender and receiver that pose a risk of interception, contact calls are often relatively short in duration and faint if the animals are not separated by great distances. For example, the flight calls of migrating passerine birds are shorter and fainter than their songs. Isolation calls of primates are often longer and louder than contact calls of the same species (Robinson 1982).

If animals can use contact calls to track one another continuously from the time they separate to the time they rejoin, then these calls may reduce the need for recognition at reunion. Much of the time, however, contact calls may function in a recognition system by containing enough information to allow the listener to recognize the caller. The relevant information may vary depending upon the social setting. In the parent-offspring context, the parent may need to distinguish its own offspring from other young in the area. In species with stable groups, an animal may simply need to maintain contact with other members of its group, while discriminating them from members of other groups. In species in which individuals share strong bonds within fluid groupings, it may be necessary for each individual to recognize and track the contact calls of specific individuals within a group. In the next three subsections, I will describe vocal signals that appear to play a role in mediating recognition between parents and their offspring, between members of a group, and between specific individuals.

For all three of these recognition problems, there is a clear pattern of correlation in which those species with the most obvious examples of each kind of signal live in socioecological settings with high demands for that signal function. While the discussion will separate these three different kinds of recognition, they all share a similar framework in which the communication signals function to make or regain contact between animals who must recognize and discriminate between different individuals or classes of individuals.

Signals for parent-offspring recognition. In many mammalian and avian species, parents invest heavily in their own offspring, spending significant amounts of time and energy to feed their young and to maintain an appropriate physical environment for them. All mammalian young are born dependent upon the mother. Most need to suckle frequently, and many species depend upon the mother for thermoregulation and for protection from parasites and predators. Mammals have sophisticated systems for chemical communication during pregnancy so that a fetus obtains the appropriate nutrients and physical environment. Once the young are born and the direct umbilical communications are broken, they still need to signal the mother to maintain homeostasis and to receive appropriate levels of nutrients. When the young are in contact with the mother, they may produce a "distress" call if they require immediate care, or a "begging" call to request food.

If dependent young are mobile and might be separated from a parent, they need to be able to signal their location to the parent so they can rapidly be reunited. In most mammalian species, when mother and offspring are separated, the young produce "isolation" calls that are used for regaining contact. These isolation calls are produced by infants within days of birth and are particularly elicited by separation from the mother. Most mammalian isolation calls are frequency-modulated tonal calls, and are longer and louder than other infant calls. These calls appear to represent a widespread and basic mammalian adaptation. Examples come from a variety of terrestrial taxa, including primates (Newman 1985), felids (Buchwald and Shipley 1985), bats (Balcombe 1990), and ungulates (Nowak 1991).

The investment that a parent puts into its offspring presents an opportunity for other animals of the same or different species to parasitize the parent. Parasitism of parental care is best documented for birds, in which a parasitic bird may lay an egg in the nest of a conspecific (Brown and Brown 1998; Lyon 1998; Petrie and Møller 1991) or of another species (Payne 1998). Young birds that are receiving inadequate parental care may parasitize the parental care of unrelated adults. They may seek care from unrelated adults temporarily, or they may actually be adopted by these adults until they are independent (Pierotti and Murphy 1987; K. M. Brown et al. 1995). This kind of parasitism, in which the young parasitize adults, has also been reported for some marine mammals. For example, Trillmich (1981) reports that young Galápagos fur seals, *Arctocephalus galapagoensis,* may attempt to suckle from females other than their mother.

Among some primate species, infants run a risk of being taken by an animal other than the mother. While these other animals may be highly attracted to infants, unrelated animals seldom provide as much care as the mother. The chances of survival for such an infant may be low unless it is reunited with its biological mother. Similar cases of animals other than the mother associating with a young infant have been reported for several species of odontocetes (see Whitehead and Mann, chap. 9, this volume), both in the wild and in captivity.

The risks described in the preceding paragraphs may create selection pressures either for parents to isolate their offspring from conspecifics or for a system for mothers and young to recognize one another. Among some mammals, there appears to be a two-part process by which a mother finds and then recognizes her offspring. For example, a ewe uses visual and auditory cues at a distance to learn that a lamb may need care (Alexander 1977; Alexander and Shil-

lito 1977). However, once she has approached a lamb, she relies upon olfactory cues before letting it suckle (Alexander 1978). Similarly, when a female Galápagos fur seal arrives at the beach after feeding, both she and her pup may exchange calls to find one another (Trillmich 1981). Before actually allowing the pup to suckle, however, she sniffs the pup, apparently using olfactory cues for final recognition. While olfaction was probably important for individual recognition among the terrestrial ancestors of cetaceans, the sense of olfaction has been either lost or greatly reduced among modern cetaceans (see the section above on chemical communication). Vision functions over much shorter ranges underwater than in air. This leaves acoustic signals as the primary ones for individual recognition in cetaceans.

Most birds also have a poorly developed sense of olfaction and rely upon acoustic signals for parent-offspring recognition. In many cases, the calls used by the young to beg for food or by parent and young to reunite contain sufficient information for individual recognition. The ability of different bird species to recognize individuals using these signals appears to correlate with the likelihood of misallocation of parental care. For example, barn swallows (*Hirundo rustica*) raise their young in nests that are far from other broods, so location is a good predictor of kinship throughout the period of parental care. Young cliff swallows (*Petrochelidon pyrrhonota*), on the other hand, intermingle within a colony while still being fed by their parents. There is evidence from these closely related species that animals evolve systems to recognize their own young if the ecological setting involves a sufficient risk of providing care to the wrong offspring or of withholding care from the correct one. The chicks of barn swallows make a begging call, but their parents do not distinguish between the calls of their own and unrelated chicks (Medvin and Beecher 1986). Cliff swallow parents can discriminate the begging calls of their own offspring from those of other young (Stoddard and Beecher 1983). Cliff swallows have evolved a more distinctive begging call in the young and also a more rapid discrimination of begging calls by adults (Loesche et al. 1991). Similar results suggest that colonial birds switch from location cues to identifying the calls of their own offspring at the time when the young from different broods intermix (Beer 1970; Miller and Emlen 1975). These results suggest that evolution favors investment in parent-offspring recognition if the risk of misallocation of care outweighs the cost of the recognition system.

The young of many dolphin and other odontocete species are born into groups made up of many adult females with their young, and they rely upon an unusually extended mother-young bond. Bottlenose dolphin calves typically remain with their mothers for three to six years (Wells et al. 1987). Sperm whales and short-finned pilot whales *(Globicephala macrorhynchus)* suckle their young for up to fifteen years (Best 1979; Kasuya and Marsh 1984). Dolphin calves are precocious in sensory and locomotor skills, and they swim out of sight of the mother within the first few weeks of life (Mann and Smuts 1993). Sperm whale calves may not be able to stay with their mother on deep foraging dives, and may remain separated from her nearer the surface for tens of minutes (Weilgart and Whitehead 1986; Whitehead and Weilgart, chap. 6, this volume). Young calves often associate with animals other than the mother during these separations. This combination of early calf mobility and prolonged dependence would appear to select for early development of a mother-offspring recognition system in these species. The prevalence of alloparental care in these species (see Connor et al., chap. 4; Whitehead and Weilgart, chap. 6, this volume) may favor a more generalized caregiver-calf recognition system.

Some baleen whales show a pattern different from this prolonged and highly social period of dependence in the young. In the seasonally migratory baleen whales, a young calf must migrate thousands of kilometers within months of birth, and the young of most species are weaned within one-half to one year (Tyack 1986a). When an adult female humpback whale has a calf, she seems to avoid other mother-calf pairs (Tyack 1982). Very few groups of humpbacks with more than one calf are ever sighted on the breeding grounds. Some baleen whales may have an ecological setting more like that of barn swallows than cliff swallows, in which a mother and calf may stay away from other mothers with young during most of the period of dependency. This reduces the potential for misallocation of parental care. In these cases, there may be reduced selection for parent-offspring recognition. One caveat about these conclusions concerns our ignorance of how mothers and calves maintain contact on the migration, and the social setting for migration. This is an important setting for further behavioral research.

In the beginning of this section, I mentioned that in many terrestrial mammals, infants produce frequency-modulated tonal "isolation" or "distress" calls. A very similar call is produced by dolphin species in which the young are raised within a social group or school. In most of these species, dolphin infants produce frequency-modulated tonal calls, called whistles, within the first few days of life. Not only the acoustic structure of dolphin whistles, but also the context in which they are produced, is very similar to the isolation or distress calls of many terrestrial mammals. Dolphins of all ages often whistle when alarmed or distressed, leading to some early descriptions of alarm or distress whistles (Lilly 1963b; Busnel and Dziedzic 1968). While dolphins do tend to whistle in these contexts, there is little evidence for a species-specific alarm or distress whistle with a particular acoustic structure that differs from that of whistles produced in other contexts (Caldwell et al. 1990).

Observations of captive dolphins suggest that whistles function to maintain contact between mothers and young calves (McBride and Kritzler 1951). When a mother and her young calf are forcibly separated in the wild, both whistle at high rates (Sayigh et al. 1990). During voluntary separations between wild mothers and calves in Shark Bay, Western Australia, whistling can be heard after the calf turns toward the mother (Smolker et al. 1993). This observation suggests that mothers and calves can keep track of each other's general location during normal separations, and that they use whistles to reunite as they approach each other.

The ecological and social settings of sperm whales would suggest a premium on early development of a communication system for mothers to find and recognize their calves after separations. Newborn calves appear unable to dive as deep as their mothers must dive to feed. In one of the few published observations of a sperm whale birth, Weilgart and Whitehead (1986) report that a calf was left on the surface for twenty minutes as the mother dove. Yet sperm whales have not been observed to produce whistles, and they primarily produce click sounds. Newborn sperm whales have been recorded making relatively unstereotyped sequences of clicks (Watkins et al. 1988). Tonal sounds have also been recorded in settings where a calf was present, and this is the only setting in which tonal sounds have been reported for sperm whales (J. C. D. Gordon 1987a; Watkins et al. 1988). Further research is required to test whether these tonal calls are used in a fashion similar to the isolation calls of other mammals, or whether sperm whale calves might use clicks to meet the same functional need to reunite after separations.

There are some hints among cetaceans of the same kind of matching between social setting and parent-offspring recognition system as has been shown for some bird species (Beer 1970; Miller and Emlen 1975; Stoddard and Beecher 1983; Medvin and Beecher 1986; Loesche et al. 1991). I mentioned above that humpback whale mother-calf pairs remain close together in clear tropical waters. They also appear to isolate themselves from other mothers and calves, and this may reduce the importance of a system to find and recognize each other (Whitehead and Mann, chap. 9, this volume). The vocalizations of humpback whales have been well studied on the breeding grounds. Biologists have followed and recorded thousands of social groups. While vocal patterns have been well documented from lone males (song) and from groups of males competing for access to a female (so-called social vocalizations: Silber 1986), there are no reports of vocalizations from mothers and calves. I have spent tens of hours listening within range of mother-calf pairs and have never heard a vocalization from them. If their use of vocal signals to maintain contact was as frequent as that of bottlenose dolphin mother-calf pairs, then this level of monitoring effort would have recorded such signals. Herman and Tavolga (1980) have argued that among the odontocetes, the species not known to produce whistles are more solitary than those that do whistle. Mothers with young of non-whistling species may be solitary enough to have reduced needs for a vocal system to recognize their own offspring. However, further studies of communication between mothers and infants, and of the social settings of mothers and infants, will be needed to test whether cetaceans show the same kind of matching between social setting and individual recognition systems that has been shown for many bird species. In particular, studies need to be designed for unbiased estimates of call rates and the behavioral contexts of vocalizations in different social groups (see Mann, chap. 2; Whitehead et al., chap. 3, this volume). This is particularly important for the poorly studied platanistid river dolphins and phocoenid porpoises, which have been described as solitary based upon fragmentary data. Herman and Tavolga (1980) categorized platanistid river dolphins as "non-whistling," but whistles have been reported for several platanistid species (Mizue et al. 1971; Jing et al. 1981; Wang et al. 1995). Further research on the vocalizations of this group is also urgently needed.

Signals for maintaining coordination and cohesion of groups. Individual animals that live in social groups may need to remain with a particular group to obtain the fullest benefits of group living. If these animals are mobile, they may need a communication system allowing them to maintain or regain contact with members of their group at the distances over which they separate. There are two cetacean species for which there is strong evidence for stable groups. These two species, killer whales and sperm whales, are described by review chapters in this volume. As Baird (chap. 5, this volume) points out, the most stable groups documented among mammals occur among fish-eating killer whales studied in the inshore waters of the Pacific Northwest. Neither sex disperses from its natal group; the only way group composition changes is by birth, death, or rare fissions of

very large groups (Bigg et al. 1990b). Adult female sperm whales live in matrilineal family units that are stable for years (Whitehead and Weilgart, chap. 6, this volume).

Both sperm whales and killer whales tend to aggregate while they are socializing, but individuals often disperse out of sight of other group members while feeding. Groups of sperm whales near the Galápagos Islands tend to spend about three-quarters of each day diving and foraging in small dispersed clusters and about one-quarter of the day aggregated in large social groups (Whitehead and Weilgart 1991). While killer whales in the inshore waters of British Columbia tend to be sighted in stable pods, these pods break up during foraging into small subgroups that disperse over areas of several square kilometers (Ford 1989). In both species, individuals would appear to need some mode of communication to allow animals separated by several kilometers to maintain contact or to reunite on a daily basis.

This problem would easily be solved by vocal contact calls. Loud calls of sperm and killer whales can be heard over ranges of 5 km or more under normal circumstances. However, a single species-specific contact call could cause problems for maintaining group cohesion. In areas of high whale density, several groups might swim within acoustic range of one another, and individuals that are dispersed while foraging might be confused by calls from several groups coming from different directions. Individual sperm and killer whales also face a problem in that their pod or family unit may join with any of a large number of other groups, and when these larger groups split up, each individual must find its own pod or family unit. This problem would seem to call for a group-distinctive contact call to act as a badge of group membership. In fact, killer and sperm whales are the only cetacean species for which group-specific vocal repertoires have been reported.

Killer whales produce a variety of vocalizations, including echolocation clicks, whistles, and pulsed calls. Ford (1989) separates the pulsed calls into two general categories of discrete and variable calls. The discrete calls can easily be further broken down into stable call types, while the variable calls cannot be sorted into such well-defined categories (Ford 1989, 1991). Figure 11.8 shows spectrograms of several stereotyped discrete calls from killer whales recorded in the waters off British Columbia. These spectrograms show the slight variations that define different subtypes of two common discrete calls from this community, N7 and N8. Individual pulses can be seen as vertical lines in the beginning of each of the N8 calls. In other parts of these spectrograms, the interval between pulses is usually less than the window size for spectral analysis. This creates a pattern that may look like the multiple harmonics of a

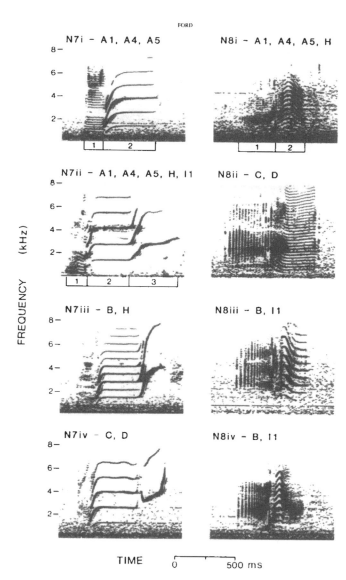

N7i – A1, A4, A5   N8i – A1, A4, A5, H

N7ii – A1, A4, A5, H, I1   N8ii – C, D

N7iii – B, H   N8iii – B, I1

N7iv – C, D   N8iv – B, I1

FREQUENCY (kHz)

TIME   0   500 ms

Figure 11.8. Spectrograms of discrete calls N7 (left column) and N8 (right column) from killer whales of the northern *resident* community in waters off British Columbia. Each of the spectrograms represents a repeated subtype of the basic N7 or N8 call. This is labeled above each spectrogram, along with the pods that have been recorded producing this call. The x-axis indicates time; the y-axis indicates frequency in kHz. (From Ford 1991.)

tonal signal, but the apparent harmonics actually represent the frequency proportional to the period of the interpulse interval (Watkins 1967). As long as one uses the same settings for spectrographic analysis, the apparent harmonics cause no problems for categorizing these discrete calls. Ford (1989) points out that these signals are relatively broadband and contain abrupt frequency shifts, which enhance their potential for detection in noise and for localization.

While all of the pulsed calls of killer whales are thought to function primarily in social communication, the discrete calls are most associated with foraging and traveling, and

the more variable pulsed calls are associated with socializing (Ford 1989). When killer whales are foraging, each pod usually breaks up into subgroups, and often several different pods will feed in the same area (Ford 1989). A foraging killer whale will often produce a series of the same discrete call, or two or more whales will exchange series of the same call. Discrete calls are also produced at high rates when two different pods meet and temporarily join.

Each pod of killer whales has a group-specific repertoire of discrete calls that is stable for many years (Ford 1991; Strager 1995). Figure 11.9 illustrates the presence or absence of different discrete calls within the northern *resident* community of killer whales. Each pod has been recorded producing between eight and fourteen of these calls. No two pods have the same repertoire of these calls. These group-specific repertoires are thought to indicate pod affiliation, maintain pod cohesion, and coordinate activities of pod members. Ford (1989) argues that these repertoires function to maintain the spacing and coordination of foraging and traveling whales. Production of discrete calls would allow dispersed subgroups of foraging whales to keep track of each other's location. After several pods have finished foraging together, they may use their group-specific repertoires to regroup to form the original pods. Ford (1989: 743) argues that these "repertoires of multiple discrete calls have evolved in killer whales to increase the reliability and efficiency of intrapod communication and to maintain the integrity of the pod."

The only other cetacean species in which group-specific dialects have been reported is the sperm whale (Weilgart and Whitehead 1997). Sperm whales make a variety of click sounds. So far, I have discussed the usual clicks produced during foraging dives and the slow clicks produced by adult males. When sperm whales are socializing, they tend to repeat series of clicks, lasting 0.5–1.5 seconds, that follow a precise rhythm. Each of these distinct rhythmic click patterns is called a coda (Watkins and Schevill 1977c). As with the discrete calls of killer whales, codas can be easily classified into discrete categories based on the number of clicks they contain and the timing of the intervals between clicks (Moore et al. 1993; Weilgart and Whitehead 1993). Figure 11.10 illustrates examples of codas from sperm whales recorded in waters near Dominica, West Indies.

Individual sperm whales often repeat a particular coda pattern several times in a row (Backus and Schevill 1966; Watkins and Schevill 1977c). Watkins and Schevill (1977c) used an array of hydrophones to locate where each coda was produced. They reported that over periods of tens of minutes, each coda pattern appeared to come from one direction, and different coda patterns came from different

Call	A1	A4	A5	B	C	D	H	I1
N1								
i	X							
ii				X				X
iii					X	X		
iv							X	X
v		X						
N2	X	X	X					
N3	X	X	X	X	X	X	X	X
N4	X	X	X					
N5								
i	X	X	X	X			X	X
ii				X			X	X
N7								
i	X	X	X					
ii	X	X	X				X	X
iii				X				X
iv					X	X		
N8								
i	X	X	X				X	
ii					X	X		
iii				X				X
iv				X				X
N9								
i	X							
ii		X						
iii			X					
N10	X	X	X					
N11								
i	X	X	X	X				
ii				X	X	X	X	
N12	X	X	X	X	X	X	X	X
N13		X	X					
N16								
i				X				
ii					X	X		
iii							X	X
iv								X
N17				X				
N18				X	X			
N19		X						
N20				X	X	X		X
N21				X				
N27	X							
N47	X							
Total	14	14	13	14	9	8	9	13

Figure 11.9. Patterns of production of discrete calls by eight pods of killer whales in the northern *resident* community. An X in the column of a pod means that the call has been recorded from that pod; a blank space means that the call has not been reported. (From Ford 1991.)

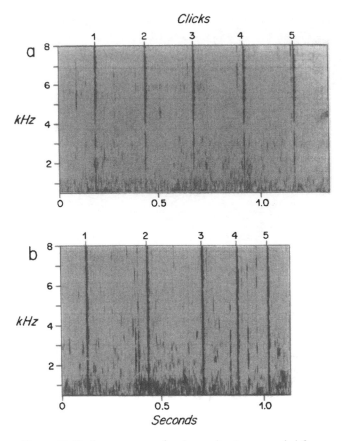

Figure 11.10. Spectrograms of coda vocalizations recorded from sperm whales in waters near Dominica, West Indies. The timing of intervals from Dominican sperm whales can usually be categorized as regular (A) or by a series of long intervals followed by a series of short intervals (B). (From Moore et al. 1993.)

directions. This observation led them to suggest that each sperm whale produces an individually distinctive coda. Whales that are close together—say, within a few hundred meters—often exchange codas (Watkins and Schevill 1977c). Sperm whales that are exchanging individually distinctive codas may match the coda typical of the other whale (Watkins and Schevill 1977c).

Later research on sperm whale codas has revealed that in each geographic area, the most common codas are shared by different individuals in different groups. The two most common codas from Dominican sperm whales involve five equally spaced or "regular" clicks fig. 11.10A) and a series of two long intervals followed by a series of two short intervals (fig. 11.10B). Codas involving equally spaced click intervals are similar in different oceans. For example, the regular codas from Dominican sperm whales are similar to those reported from sperm whales in the Galápagos (Weilgart and Whitehead 1993). By contrast, irregular codas from sperm whales in the Galápagos differ from those from

Dominica, tending to start with short intervals followed by long intervals (Weilgart and Whitehead 1993). Weilgart and Whitehead (1997) have analyzed coda repertoires by recording session, date, group, place, area, and ocean. They find strong similarities in coda repertoires recorded from the same group of whales over periods of several years, along with a weaker pattern of geographic variation. This finding leads them to describe sperm whale codas as group-specific dialects. As with the discrete calls of killer whales, the weakest link in this argument is the lack of data on the vocal repertoires of individual sperm whales within these groups.

Weilgart and Whitehead (1997) suggest that codas function primarily for intragroup communication. The clicks that make up codas appear to be less intense than usual or slow clicks. Weilgart and Whitehead (1997) argue that codas may have an effective range that does not extend beyond 600 m, limiting communication to within groups. However, this is almost certainly an underestimate of the effective range of codas, and sperm whales have been seen to react to codas of other groups. Sperm whales sometimes abruptly silence their own vocalizations after distant codas from some other group of sperm whales are heard (Watkins et al. 1985). The impression this gives to a listener is that the whales cease vocalizing to listen for the faint codas of distant whales. Adult female sperm whales with young tend to form temporary groupings of two matrilineal units (Whitehead and Weilgart, chap. 6, this volume). This means that sperm whales within a unit frequently join or split with other units. When they detect a distant unit, they face a problem of deciding whether to join or avoid that unit. Weilgart and Whitehead (1997) show that groups within one area tend to have more similar coda repertoires than more distant groups. Sperm whales may listen to the similarity of call repertoires to assess a group within acoustic range; for example, to decide whether or not to challenge strangers, to affiliate with a neighboring group, or to avoid a group altogether. While these intergroup interactions may not be as common as intragroup responses to codas, more research is needed to define the effective range of codas and to evaluate whether interception of codas from other groups influences interactions between groups.

Whitehead and Weilgart (1991) systematically recorded vocal and visually observable behaviors as they followed groups of sperm whales in the Galápagos. They found that usual clicks were strongly associated with periods when the whales were diving and foraging in dispersed subgroups, and that codas were strongly associated with periods when the whales were socializing in tight aggregations at the surface. This pattern differs from the usage of discrete calls in killer whales, which predominate during foraging, but Weilgart and Whitehead (1993) have an interpretation of the function of codas similar to that proposed by Ford (1989) for discrete calls in killer whales. Weilgart and Whitehead (1993) hypothesize that codas function to maintain the cohesion of groups of sperm whales as they aggregate following dispersed foraging. It is also possible that the different coda repertoires of different groups could result from cultural or genetic drift, with no adaptive function. Testing these hypotheses about the possible functions of codas will require more detailed observations of call usage by individual killer and sperm whales as they come together after dispersal. There is an urgent need to study the fine-grained patterns of communication using methods that can link data on acoustic location of calls with follows of individual whales.

If an engineer were designing a system to identify group membership, she might assign one distinctive call as a badge for each group. Sperm and killer whales have a more complex system in which individual calls may be shared between groups, but the entire call repertoire of each group is distinctive. One reason for this apparent complexity may involve difficulties in evolving a system where each group develops a distinctive call. In addition, if the number of distinctive features in one call is limited, then a larger repertoire might allow for a larger number of groups to be distinguished. However, this system has the potential disadvantage that an individual may need to monitor a series of calls for some time before being able to distinguish between some groups.

Current research on calls for maintaining cohesion in cetacean groups suggests a correlation between the presence of and the need for such calls. However, there is a paucity of data describing how sperm or killer whales use these calls as members of a group reunite, or as members of one group detect a distant group and make a decision about whether or not to join with it. The main reason for this lack of data stems from a methodological difficulty. It is relatively easy to follow a group of killer or sperm whales in the wild and to record sounds in their presence. However, it is extremely difficult to identify which animal produces a call. This problem forces one to analyze the data by group and prevents one from studying patterns of signal and response of individuals within a group. Miller and Tyack (1998) report the development of an array of hydrophones that can be towed behind a boat and used to locate where killer whale sounds are coming from. This kind of system can be integrated with individual-focused follows of visual behavior to identify which individual produces a call. This technique may allow observers to better define the contexts in which an individual whale produces a call as well as the vocal and behavioral responses of other individuals. Such an approach holds great promise for better defining functional usage of these calls.

Signals for maintaining individual-specific social relationships. A common theme runs through many chapters of this book, suggesting that individual-specific social relationships may be important elements of the social behavior of many dolphins and toothed whales. If animals are to maintain these individual-specific relationships, they must be able to recognize and differentiate between different individuals. Another mammalian taxon in which individual-specific social relationships are important is the primates, and the mechanisms of individual recognition by primates have been well studied. Some primates can recognize individuals by voice cues (e.g., Cheney and Seyfarth 1980), but visual recognition of the face is of primary importance for individual recognition in many primate species. Primates appear to have evolved a system for recognizing and differentiating between many different individuals using special mechanisms for processing individually distinctive features of the face (Gross 1992; Tovee and Cohen-Tovee 1993).

If some cetaceans also rely upon individual-specific social relationships, then they will require mechanisms for individual recognition. Many cetacean species have enough individual variation in visual features, such as pigmentation or scars and notches, that human biologists can discriminate between many individuals (Hammond et al. 1990). However, as described above, vision is much less valuable as a distance sense under the sea than in many terrestrial environments. If cetaceans are to broadcast a signal of their identity or recognize individuals at ranges of greater than 10 m or so, they must rely upon acoustic signals.

The whistle vocalizations described for dolphins in the context of mother-infant recognition provide one of the most striking cases of individually distinctive vocalizations. Most early papers on dolphin whistles tried to associate specific whistle patterns with a particular behavioral context, such as alarm or distress, and they appeared to assume that all or most of these whistle patterns were shared by all members of a species (Dreher and Evans 1964; Lilly 1963b). However, Caldwell and Caldwell (1965) recorded whistles from five recently caught wild bottlenose dolphins in a variety of captive contexts, and they reported that each individual dolphin tended to produce its own individually distinctive whistle, which they called a signature whistle. Signature whistles have also been reported for the common

dolphin, *Delphinus delphis* (Caldwell and Caldwell 1968), the Pacific white-sided dolphin, *Lagenorhynchus obliquidens* (Caldwell and Caldwell 1971), and the spotted dolphin, *Stenella plagiodon* (Caldwell et al. 1973). Caldwell et al. (1990) reviewed whistle repertoires from 126 captive bottlenose dolphins of both sexes and a wide range of ages. The primary method of identifying which dolphin produced a whistle was recording dolphins when they were isolated, and signature whistles made up about 94% of each individual's whistle repertoire in this context. These signature whistles were distinctive between individuals and stable over many years.

Caldwell et al. (1990) also reported that bottlenose dolphins produce an extremely variable array of whistles that are not individually distinctive. These non-signature whistles are called variant whistles. While variant whistles made up only about 6% of the whistles in the Caldwell et al. (1990) data set, which emphasized isolated animals, they are more common in other contexts. For example, Tyack (1986b) reported that variant whistles made up 23% of the repertoire of two dolphins that were interacting socially, and Janik et al. (1994) reported that variant whistles were much more common when a dolphin was being trained than when it was isolated. The increased tendency of dolphins to produce signature whistles in isolation supports the hypothesis that dolphins use signature whistles to maintain contact with individuals from which they have been separated (Janik and Slater 1998). Variant whistles can be very diverse, but there can also be considerable overlap in the repertoires of variant whistles from different individuals. Tyack (1986b) and Janik et al. (1994) identified several classes of variant whistles, such as rise, flat, or down whistles, that were repeated by each individual and appeared to be shared across different individuals in different studies.

Additional evidence for signature whistles comes from a study of free-ranging bottlenose dolphins in inshore waters near Sarasota, Florida. Approximately a hundred free-ranging dolphins have been the focus of this long-term field study of population biology and behavior, which has involved extensive observations and censuses (Scott et al. 1990a; Wells et al. 1987; Wells 1991). This study also includes, on a more or less annual basis, a temporary capture-release component, in which dolphins are briefly held in a net corral and are then released. During the period when they are restrained and isolated from other dolphins in the net corral, the vocalizations of individual dolphins can be recorded with suction cup hydrophones placed directly on the head of each animal. A library of 398 recording sessions, most containing hundreds of whistles from an identified individual, has been obtained from 134 known individuals. Many of these dolphins were first recorded at one or two years of age and have been recorded over spans of ten to twenty years. All but a very few individuals have shown a stable and distinctive signature whistle over the entire time span during which they have been recorded (Sayigh et al. 1990), similar to the pattern reported by Caldwell et al. (1990) for captive dolphins. Figure 11.11 shows spectrograms of the signature whistle of an adult female over a period of eleven years and of one of her calves at one and three years of age. These whistles were recorded in an unusual setting, when the dolphin was restrained and separated from other dolphins. However, there is evidence that these whistles are similar to whistles recorded under more normal circumstances. Sayigh (1992) followed groups of free-ranging Sarasota dolphins (average group size of three to seven animals) and found that the signature whistles recorded during these follows matched those recorded previously from the same individuals using suction cup hydrophones.

There are several details of analyzing the acoustic structure of whistles that are crucial to the categorization of signature whistles. When biologists analyze the song of a bird, what is defined as a song is not limited to signals with continuous energy. If these continuous signals, called syllables, are seldom produced in isolation and usually produced in a regular and predictable series, then it is the series, called the song, that becomes the primary unit of analysis. Biologists who have had the opportunity to study large repertoires of whistles from identified individual dolphins have tended to define the acoustic structure of signature whistles in a manner similar to that of bird song, as opposed to syllables. For example, the signature whistles of mother 16 and calf 140 shown in figure 11.11 are each made up of three continuous elements, which are called "loops" by Caldwell et al. (1990). Since these are typically produced as a unit with regular interloop spacing and highly variable spacing between the three-loop units, they are analyzed as one three-loop whistle instead of as three separate whistles. Even if a dolphin varies the number of loops in a signature whistle, as long as the variation involves repetition of the same loop with regular spacing, it is categorized as one signature whistle (Tyack and Sayigh 1997). Dolphins sometimes appear to produce incomplete sections of a signature whistle. If such a whistle precisely matches a portion of a signature whistle, Tyack (1986b) would categorize it as a segment of a signature whistle. Inspection of the whistles of mother 16 in figure 11.11 reveals how precisely dolphins can repeat the frequency-time contour of a loop over more than a decade.

McCowan (1995) proposed an alternative way to ana-

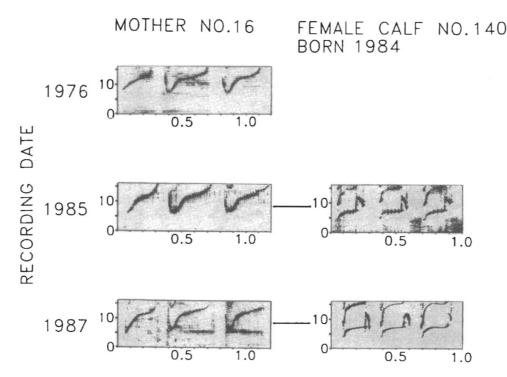

Figure 11.11. Spectrograms of signature whistles from one wild adult female bottlenose dolphin recorded over a period of eleven years, and from her daughter at one and three years of age. Note the stability of both signature whistles. The x-axis indicates time in seconds; the y-axis indicates frequency in kHz. (From Sayigh et al. 1990.)

FUNCTIONAL ASPECTS OF CETACEAN COMMUNICATION

lyze dolphin whistles. The basic unit of her analysis is a continuous trace of whistle energy in a spectrogram. This would split many signature whistles into their component loops, a modification that would affect later analysis as much as if one compared an analysis of bird song with an analysis of the same data broken up into each song syllable. Furthermore, most analyses of signature whistles compare the detailed changes in frequency with time that make up the loop and whistle. McCowan instead used a technique that assigns little weight to the actual durations and frequencies of the whistle, but rather compares the relative changes in frequency across twenty evenly spaced segments of each whistle. These are then fed into a cluster analysis, which, as might be expected, yields quite different whistle categories from those derived by the typical approach (Janik 1999). McCowan and Reiss (1997) used this method to categorize whistles of captive dolphins and reported results similar to those of "studies published before the advent of the 'signature whistle hypothesis'—a large whistle repertoire within social groups, sharing of whistle types across social groups, and a predominant but not individualized whistle type" (180).

These differing results from different methods of categorizing whistles can be resolved only by testing how dolphins themselves perceive whistles. The hypothesis that dolphins use signature whistles to broadcast individual identity assumes that dolphins can use acoustic features of whistles to recognize different individuals. McCowan and Reiss (1995, 1997) appear to be arguing that most whistles do not contain individually distinctive features. Some studies have been conducted that are germane to this issue. Captive dolphins have been trained to categorize a sample of whistles from the same individual as similar and samples of whistles from different individuals as different, even if these are novel whistles from dolphins with which they have not interacted. A captive male bottlenose dolphin was able to discriminate signature whistles from two different males, in a design using a large sample of whistles from both individuals (Caldwell et al. 1969). Recall of this discrimination remained high for as long as the animal was retested, up to twenty-two days. Experimental playbacks with wild bottlenose dolphins have also demonstrated that mothers and offspring can recognize each others' signature whistles even after calves became independent from their mothers (Sayigh 1992; Sayigh et al. 1999).

Quantitative signal processing also indicates that signature whistles provide sufficient information to distinguish individuals. The question of whether the interindividual variability of signature whistles is greater than their intraindividual variability was tested analytically by Buck and

Tyack (1993), who developed a computer algorithm to compare the similarity of pairs of whistle contours. This algorithm measures differences in absolute frequency while allowing different segments of the whistle to vary their time axis to maximize fit in the fundamental frequency of the whistle. It was used to sort three randomly chosen signature whistles recorded from each of ten wild bottlenose dolphins of the Sarasota population during temporary captures. Five dolphins produced signature whistles without repetitive loops, and the other five produced multi-loop whistles. The algorithm correctly matched fifteen of fifteen of the whistles without repetitive loops and fourteen of fifteen central loops from multi-loop whistles.

Bottlenose dolphins appear to develop a highly stable signature whistle within the first year or so of life, yet observations of dolphins that are interacting socially and/or acoustically suggest that imitation may play an important role in the natural communication system of dolphins. Janik (1998) found whistle matching in 24% of 188 whistle interactions among wild bottlenose dolphins in the Moray Firth, Scotland. In one study of two captive adult dolphins, Tyack (1986b) found that each dolphin imitated the signature whistle of the other at rates of about 25% (i.e. 25% of all occurrences of each signature whistle were imitations produced by the other dolphin). Other studies have reported rates of signature whistle imitation near 1% among captive dolphins that were in acoustic but not physical contact (Burdin et al. 1975; Gish 1979; Janik and Slater 1998). These imitated signature whistles are not just produced immediately after the partner makes its signature whistle, but can become incorporated into a dolphin's whistle repertoire. For example, after a period of silence, one dolphin might produce a copy of another dolphin's signature whistle. The two animals in the Tyack (1986b) study were first housed together at about five years of age, well after signature whistles are developed. This ability of adult dolphins to add new whistles to their vocal repertoire through imitation of auditory models has been well established in experimental studies in which dolphins were trained to imitate novel synthetic whistle-like sounds (e.g., Richards et al. 1984).

If one dolphin can produce precise imitations of another dolphin's signature whistle, then this raises questions about whether a dolphin hearing a signature whistle can reliably predict which individual made the whistle. We do not know whether dolphins can discriminate imitated whistles from the original signature whistle, but many imitated whistles have acoustic features that allow biologists to discriminate them from the original (Tyack 1991b). Little is known about the precise functions of whistle imitation. If the intended recipient of an imitated whistle is the animal that typically produces that whistle, then that animal is unlikely to be confused about whether it is an imitated whistle or not. In this context, imitated whistles may function to initiate an interaction with the animal whose signature whistle has been imitated (Tyack 1993). Among songbirds, call matching appears to provide a competitive advantage for young males (R. B. Payne 1982, 1983). The deceptive mimicry hypothesis suggests that call matching might allow a young bird to deceive a mate or competitor into accepting it as an experienced resident (McGregor and Krebs 1984a). While dolphins do not have the same pattern of territoriality, a dolphin could use an imitated whistle to attempt to deceive another dolphin about the identity of the whistler. In order to succeed, such deception would have to be rare enough not to interfere with the usual reliability of individual recognition by signature whistles. All of these considerations suggest that imitation of signature whistles is not likely to interfere with their role in individual recognition.

McCowan and Reiss (1997) raised a different set of questions about the proportion of whistles produced by captive dolphins that are unique to individuals. They recorded whistles from small groups of captive dolphins held together in small pools. Their methods differed from earlier studies of signature whistles, both in the categorization scheme described above and because they limited their whistle sample to occasions when the individual producing the whistle could be identified because the whistle was accompanied by a stream of bubbles from the blowhole. Emission of bubbles allows one to identify whistles from dolphins that are interacting closely, but is so rare that Caldwell et al. (1990) raised questions about its reliability and effectiveness.

McCowan and Reiss (1997) categorized 185 bubble-stream whistles from ten adult dolphins in three social groups. Their categorization method identified twenty-eight whistle types, 29% of which were shared across groups, 25% of which were shared only within groups, and 46% of which were unique to individuals. Differences in the methods used to categorize whistles make it difficult to compare the McCowan and Reiss (1995, 1997) data to other analyses of signature whistles. The percentage of whistles unique to individuals reported by McCowan and Reiss (1997) was lower than the 70–90% typically reported for signature whistles, but if they recorded imitated signature whistles, those would probably have been scored as whistles shared within a group. The most common whistle was a simple upsweep produced by all of the adults. This is similar to the rise whistle reported by Tyack (1986b), who identified whistles from two captive dolphins using

small telemetry devices, called vocalights, attached to each dolphin's head. However, the upsweep accounted for 97 of 185 bubble-stream whistles, but only 17 of 284 vocalight whistles. Janik et al. (1994) found that rise whistles were the second most common whistle type (305/1743), after the signature whistle (1098/1743) of the subject.

The primary results of McCowan and Reiss (1997) are that they observed fewer whistles unique to individuals, and many more upsweeps, than has been reported in other whistle samples. Janik and Slater (1998) studied whistles of four captive bottlenose dolphins and found that these differences may relate to the social context in which the whistles were recorded. They reported that each dolphin produced its signature whistle when in a separate pool from the other three, but that rise whistles were most common when all the dolphins were swimming together in one group. When these captive dolphins were undisturbed and together in one pool, only 2.4% of their whistles were signature whistles. These results make sense in terms of the signature whistle hypothesis, which suggests that dolphins use signature whistles to maintain contact with specific individuals when they are separated or isolated.

The Janik and Slater (1998) results emphasize how important it is to study variation in whistle repertoires within a behavioral context. Most work on signature whistles has emphasized recordings from the isolation context, so it might be expected to overestimate the percentage of signature whistles compared with other contexts. On the other hand, bubble-stream whistles may also represent a biased sample of whistles. As mentioned in the section on visual displays, bubble streams may be used as a display by cetaceans. If a dolphin chooses to add a visual display to a whistle, the intended recipient may be likely to be within view, so this context may be particularly different from the isolation setting, where other dolphins are out of view. In addition, the highly unusual addition of bubble streams to a whistle does not necessarily occur at random, but may be more likely for some whistle types, such as upsweeps. Their rarity, unreliability, and potential for bias led Caldwell et al. (1990) to consider bubble streams a method of last resort for identifying whistles from captive bottlenose dolphins.

Resolution of these alternative interpretations of dolphin whistles will depend upon several directions for future research. Experiments on how dolphins perceive whistles are urgently needed to test which methods of categorizing whistles most closely match how dolphins categorize whistles themselves. In addition, new methods are required to identify whistles unbiased by the activity or context in which the dolphins are engaged. A promising start in this direction was provided by Janik (1997), who located individual wild dolphins from a shore station and used an array of hydrophones to locate the whistles of wild dolphins. He was able to link the acoustic locations of whistles to the sightings of specific individual dolphins being followed from shore, and neither observation method affected the behavior of the dolphins in any way.

Another cetacean species reported to have individually distinctive signals is the sperm whale. In the first paper to define the codas of sperm whales, Watkins and Schevill (1977c) suggested that each individual sperm whale may produce an individually distinctive coda. Since that time, both Moore et al. (1993) and Weilgart and Whitehead (1993) have reported that many sperm whales within a geographic area may share several common coda types, such as those illustrated in figure 11.10. Weilgart and Whitehead (1993) reject the hypothesis that codas are individual identifiers, except over short time periods, because many individuals produce more than one coda type, and there are many fewer coda types (as classified by Moore et al. 1993 and Weilgart and Whitehead 1993) than individuals in the population. Weilgart and Whitehead (1997) interpret codas as group-specific dialects. However, the existence of shared codas does not rule out the existence of individually distinctive codas as well. Weilgart and Whitehead (1997) partitioned variance in codas from the large scale of ocean and area down to the group, and found that codas within a group were similar over several years. However, they were not able to compare coda repertoires of individual whales within a group. Their results are consistent with either group-specific dialects or a repertoire combining shared codas and individually distinctive codas. The small number of coda types does not rule out the possibility of an individual identification function among subsets of the population. Codas are categorized by humans rather broadly by timing and number of clicks. It is likely that sperm whales can achieve finer distinctions among codas. Bradbury and Vehrencamp (1998) point out that it is common for signals to have hierarchical levels of variation, in which a mean pattern might indicate species or group identity, and small deviations from the mean may indicate individual identity.

A broad range of communication and echolocation functions has been hypothesized for sperm whale clicks. These different functions are by no means mutually exclusive, especially for different click patterns such as slow clicks, regular clicks, and codas. I cannot help but feel that our understanding of sperm whale codas is like that of the blind men feeling small parts of the elephant—each individual study has had such a limited view that different studies of codas can easily be expected to have come to different conclusions. I feel that our most important task is to determine the coda repertoires of individual sperm whales within a variety of behavioral contexts. Further research using acoustic recording tags or acoustic location to identify codas from known individuals over days, weeks, and years may provide a new perspective on this problem.

Echolocation May Influence Communication in Cetaceans

Many cetaceans use their vocal and auditory systems for echolocation as well as communication. Many treatments of vocal behavior in cetaceans suggest that high-frequency clicks are used exclusively for echolocation and that all other sounds are used exclusively for social communication. While this assumption of a rigid dichotomy between echolocation and communication is often implicit, it may overly restrict our hypotheses about the functions of cetacean vocalizations. Some cetaceans may use low-frequency sounds for biosonar, sounds that are typically thought of as serving a communicative function. For example, while humpback song appears to function as a reproductive advertisement display (Tyack 1981), whales may also learn about their environment from listening to echoes of bottom reverberation from sounds used in song (Tyack 1997). I have discussed how different biologists have argued that the 20 Hz pulses of finback whales may function either as a low-frequency echolocation system (e.g., Norris 1967, 1969; Payne and Webb 1971; Thompson et al. 1979) or for social communication (Payne and Webb 1971; Watkins et al. (1987); McDonald et al. 1995), perhaps as a reproductive advertisement display (Watkins et al. 1987). On the other side of this issue, there are several dolphin species that appear to use "echolocation" clicks for social communication. Several of the species that specialize in high-frequency hearing, including the phocoenid porpoises and dolphins of the genus *Cephalorhynchus*, are not known to produce any of the sounds typically associated with social communication in other dolphins (Amundin 1991; Dawson and Thorpe 1990). The clicks they use in echolocation may also function in social communication (Amundin 1991; Dawson 1991; Dawson and Thorpe 1990). Amundin (1991) associated relatively stereotyped patterns of repetition rate of "echolocation" clicks with specific social contexts in the harbor porpoise. Dawson (1991) also found in Hector's dolphin, *Cephalorhynchus hectori*, that specific kinds of complex clicks were associated with large groups and specific group activities. High repetition rates of Hector's dolphin clicks were associated with aerial and aggressive behavioral contexts rather than feeding. If predators of these species, such as killer whales, cannot hear the high frequencies of these clicks, then it may be advantageous to use them for communication as well as echolocation. Sperm whales also have a vocal repertoire limited primarily to clicks, and biologists have argued that these clicks are used for echolocation (Backus and Schevill 1966; Goold and Jones 1995; J. C. D. Gordon 1987a; Whitehead and Weilgart 1990) as well as communication (Watkins and Schevill 1977c; Whitehead and Weilgart 1991).

Tyack (1997) has argued that research on the evolution of echolocation in cetaceans suffers from a dearth of studies of ecological function and from a lack of broad comparative reviews. If studies of cetacean sonar included more analysis of the problems for which sonar may have evolved, we might have a much broader view of echolocation in cetaceans. For example, low-frequency sound is better suited than high-frequency sound for long-range sonar in the sea, and many targets of great importance to cetaceans, such as large bathymetric features, are also well suited to low-frequency sonar. Clark (1993) suggests that the low-frequency calls of baleen whales that are so well suited for long-range propagation in the sea, such as those described for blue and finback whales, may function for long-range orientation through detecting echoes from large distant features such as seamounts. Even if these signals have a primarily communicative function, these whales may also detect and respond to reverberation from bathymetric features. This possibility has been little studied, but there is suggestive evidence that bowhead whales migrating through the ice may use low-frequency calls to detect large areas of ice where they may not be able to surface to breathe (Ellison et al. 1987; Clark 1989 and George et al. 1989).

These examples blend features typically associated with the separate domains of sonar and communication. I would like to suggest that auditory and vocal skills evolved to function in one of these domains may preadapt animals for developing abilities in the other domain.

Vocal Learning

Most of this review has suggested parallels between communication in cetaceans and in other taxa. There is one area, however, in which cetaceans show pronounced differences from their terrestrial mammalian relatives. Janik and Slater (1997) review evidence for vocal learning among mammals. Their definition of vocal learning concentrates on "production learning": evidence that the acoustic morphology of an animal's signals is modified by auditory exposure. Very few nonhuman terrestrial mammals have been shown to have an ability for production learning. Janik and

Slater (1997) further divide evidence for vocal learning into modification of the timing of vocalizations, which may involve simple modification of the timing of exhalation, versus modification of frequency parameters, which is more likely to involve complex coordination of the entire vocal apparatus. The only nonhuman mammals for which they find convincing evidence of vocal learning of frequency parameters are the following:

- Greater horseshoe bat, *Rhinolophus ferrumequinum*
- Harbor seal, *Phoca vitulina*
- Humpback whale, *Megaptera novaeangliae*
- Beluga whale, *Delphinapterus leucas*
- Bottlenose dolphin, *Tursiops truncatus*

Some of the strongest evidence for vocal learning comes from species that have been reported to imitate man-made sounds in captivity. A few individual harbor seals (Ralls et al. 1985) and beluga whales (Eaton 1979) have been reported to imitate the sounds of human speech. Many bottlenose dolphins have been shown to imitate man-made whistle-like sounds (Caldwell and Caldwell 1972c; Herman 1980). Dolphins can also be trained using food and social reinforcement to imitate man-made whistle-like sounds (Evans 1967; Richards et al. 1984; Sigurdson 1993). Evidence of vocal learning in the other two species stems from observations of natural behavior. The echolocation call of horseshoe bats rises in frequency over the first year or two of life and then has a decreasing frequency with increasing age. The echolocation calls of young with older mothers are lower in frequency than the calls of young with younger mothers, suggesting that young horseshoe bats match the call of their mother (Jones and Ransome 1993). The vocal convergence at any one time within a population of singing humpback whales, coupled with the rapid changes in the song over time, provides evidence for vocal learning in these animals (K. B. Payne et al. 1983).

There is a striking lack of evidence for these kinds of vocal imitation among nonhuman terrestrial mammals. Since vocal learning is so important for human communication, the lack of evidence for vocal learning is particularly striking in nonhuman primates. This observation highlights the importance of investigations into the evolutionary origins of vocal learning in other mammals. The question can be addressed by studying the current utility of learned displays and then comparing the phylogenetic relationships of species with and without skills for vocal learning.

There is some evidence for a link between echolocation and vocal learning among some bats and dolphins. Several species of bats have hearing that is particularly sensitive in a narrow frequency band. This frequency band is so narrow that the Doppler shift in frequency of their outgoing pulse, induced by the relative motion of bat and target, can shift the echo outside of the bat's best hearing band. These bats modify their outgoing signal to maintain the appropriate frequency of the Doppler-shifted echo, a process called Doppler compensation. This frequency shift is required for effective operation of their sonar system. While this frequency shift takes place in milliseconds, and involves changing one's outgoing pulse depending upon the echo characteristics of one's own earlier vocal production, it is otherwise similar to the frequency matching between mother and infant that is reported as evidence for vocal learning in the horseshoe bat. Beluga whales and bottlenose dolphins have been shown to imitate a variety of pulsed and whistle-like sounds. Both of these species are able to echolocate, and, as in the case of Doppler compensation in bats, vocal learning may play a role in the effective operation of their sonar. Both species have been shown to be able to shift the peak frequency of their echolocation clicks, either as a result of differing ambient noise (belugas: Au et al. 1985) or through training (bottlenose dolphins: Moore and Pawloski 1991). These findings suggest that the requirements of echolocation may have selected for a simple form of vocal learning in these species.

Bottlenose dolphins are skilled at imitating whistle-like sounds, and whistle matching appears to play a role in their natural communication system. Their use of whistles in individual recognition, coupled with the problems associated with voice cues for individual recognition in diving animals, suggests that vocal learning may play a critical role in individual recognition and in maintaining individual-specific social relationships in whistling odontocetes (Tyack 1991a; Janik and Slater 1997). The diving habit of cetaceans may create significant differences in how they perform vocal individual recognition compared with terrestrial mammals. Slight variations in the vocal tracts of terrestrial animals cause predictable differences in the voices of individuals. Many of the features that distinguish the calls of individual terrestrial animals appear to be subtle cues that result from these variations in the vocal tracts of different individuals. These involuntary characteristics of voice are not likely to be reliable cues for diving animals, however. The vocal tract is an air-filled cavity, and gases halve in volume for every doubling of pressure as an animal dives. Since different parts of the vocal tract are more or less elastic, changes in volume will lead to changes in shape. These depth-induced changes in the vocal tract are likely to outweigh the subtle developmental differences that lead to voice differences. For example, the whistles of a beluga

whale recorded at different depths show strong differences in their frequency spectra (Ridgway 1997). If diving animals rely upon individually distinctive calls, they may be unable to use voice cues and may need to create distinctive calls by learning to modify acoustic features under voluntary control, such as the frequency modulation of whistles.

Selective pressures for the evolution of complex advertisements appear to have played a role in the evolution of vocal learning in songbirds and baleen whales. As in the case of most oscine songbirds, vocal learning in humpback whales has been described only for their song, which is a reproductive advertisement display that has evolved by sexual selection. Indirect evidence also suggests that sexual selection was a significant factor in the evolution of vocal learning in some seals. Evidence for vocal learning among seals (Ralls et al. 1985) is particularly interesting from an evolutionary perspective, because the pinnipeds evolved from a different terrestrial ancestor than the cetaceans. This suggests that there were at least two independent origins of vocal learning among marine mammals. Vocal imitation in harbor seals has been reported only for adult males (Ralls et al. 1985), and adult male harbor seals have been reported to produce repetitive acoustic displays during the breeding season (Hanggi and Schusterman 1994). Many seals produce songlike advertisement displays during the breeding season (Ray et al. 1969; Sjare and Stirling 1993; Stirling 1973; Thomas et al. 1983). Further research is urgently needed on the potential role of vocal learning in these seal songs. Among the cetaceans, there appear to be a variety of current functions for vocal learning, including echolocation, individual identification, and producing advertisement displays. This variety makes it difficult to determine whether vocal learning has arisen several times independently in this taxon, and if not, which was the original function for which the skill evolved. Janik (in press) suggests that vocal learning initially evolved in cetaceans for individual recognition, and only later played a role in evolution of songs in baleen whales. Whatever its origin, cetaceans use vocal learning for developing a remarkable array of signals with different social and sonar functions.

Comparison of Social Signals with Social Organization

This volume has described a great diversity of social systems, life histories, and mobility among cetaceans. Data on acoustic communication and social behavior also suggest diverse patterns of variation in communicative signals, from individually distinctive signals, to group-specific vocal repertoires, to signals shared among groups over broad geographic areas, to vocal dialects in different areas. Socioecological comparisons between species suggest diverse functions for these signals among cetaceans.

There is a clear correlation between the types of social bonds and the types of communication signals seen in different cetacean groups (Tyack 1986a). Individual-specific signals have been reported for species, such as the bottlenose dolphin, with strong individual social bonds; group-specific vocal repertoires have been reported for species, such as killer whales, with stable groups; and population-specific advertisement displays have been reported among species, such as humpback whales, in which adults appear to have neither stable bonds nor stable groups. Vocal learning appears to be involved in the development of many of these signals. It is possible that vocal learning evolved de novo in these different taxa as independent solutions to the different problems posed by their differing social organizations. However, once a flexible system of vocal development evolved to solve one problem, it may have allowed the flexibility to solve different problems. As the descendants of animals that had evolved abilities of vocal learning branched into other niches, they may have used vocal learning to other ends.

The rarity with which vocal learning has evolved in animals suggests that the evolution of a system of vocal development with this kind of flexibility may be a slow and unlikely process. If a system for vocal communication cannot be modified through learning, but changes via the evolution of genetic predispositions, then modification of the system may take evolutionary time scales. Yet animal populations may face new socioecological opportunities on ecological time scales that are shorter than the evolutionary scales. For example, bottlenose dolphins that reside in inshore local habitats may benefit from different group sizes and structures for feeding and predator avoidance than migratory offshore populations. The recognition systems required to maintain these different societies are also likely to be quite different. While I separated the functional problems of parent-offspring recognition from those of group and individual recognition, it appears that dolphins may use whistles, and sperm whales may use clicks, for all of these recognition problems. We cannot assume that cetaceans have evolved specific signals independently for each of these functions. It is also possible that cetaceans have evolved an open system of vocal development in which individual animals develop vocal repertoires through learning that match their locally adapted social groupings. Suppose that cetaceans develop a signal distinctive enough for parent-offspring recognition and then imitate common signals in their environment. Even if they followed the same system of vocal development, animals living in stable social groups might then develop group-distinctive repertoires, while animals living in fluid societies might develop repertoires characterized by more individual-distinctive signals and a diverse set of other calls. Further research will be required to test whether these kinds of patterns reflect an adaptive system in which vocal learning allows fine-tuning of a communication system to local demands, or whether vocal learning may create variation in signals with little adaptive significance, as has been suggested for some songbird dialects (Andrew 1962; Bitterbaum and Baptista 1979; Wiens 1982).

One of the most promising areas for future research on communication among cetaceans involves detailed behavioral studies of how individuals of different age and sex classes in a variety of different species use communication signals. New techniques for identifying which individual is vocalizing in the wild, and for following details of social interaction, will better integrate studies of cetacean communication with social interaction. Longitudinal studies of individuals will be important for studying vocal development. The current status of knowledge of communication in cetaceans highlights the importance of cetaceans for studying relationships between social signals and social structure, interactions between echolocation and communication, social influences on vocal development, and the evolutionary origins of vocal learning.

12

SCIENCE AND THE CONSERVATION, PROTECTION, AND MANAGEMENT OF WILD CETACEANS

HAL WHITEHEAD, RANDALL R. REEVES, AND PETER L. TYACK

CETACEANS HAVE become, with good reason, important icons of the conservation movement. Most of the large whales, and many of the smaller cetaceans, that have lived in the twentieth century have been hunted with methods that, in addition to killing them, often cause suffering and pain. Many of the animals that have not been killed outright have been injured following encounters with fishing gear or ship strikes, disturbed by human-generated underwater noise, or contaminated by pollutants. As humans increase their exploitation of marine and aquatic resources, individual dolphins, porpoises, and whales suffer, their societies are disrupted, and their populations decline.

Although most kinds of marine organisms are adversely affected by human activities, many people, including scientists, have become particularly concerned about cetaceans. The reasons for focusing on these animals include their ecological significance, their economic value (both dead and alive), and their potential as indicators of marine and aquatic pollution. In the case of the great whales, the history of their exploitation (fig. 12.1) provides a particularly clear picture of the devastating effects of human greed. Apart from their symbolic significance, cetaceans are especially intriguing because of their unparalleled size, sound producing and sound processing abilities, and diving proficiency. For many humans, including the authors of this book, a principal attraction of cetaceans is that they have cognitive abilities and social structures of a complexity and flexibility seen only among the most intelligent terrestrial mammals—the apes and elephants. For all these reasons, and others discussed at the end of this chapter, the view that cetacean populations should be conserved prevails throughout much of human society. Moreover, there is a growing belief that cetaceans, as individuals, should be protected from harm caused by human activity.

During the twentieth century science took on an increasingly important and complex role in the interactions between humans and cetaceans (Samuels and Tyack, chap. 1, this volume). Often, scientists have directed their efforts at improving the survival prospects of cetacean populations; less often, at enhancing the welfare of individual animals. Their motives have varied. In the first three-quarters of the twentieth century, a common goal of cetacean research was to optimize the use of whales and dolphins by humans (Samuels and Tyack, chap. 1, this volume). This utilitarian objective has long dominated mainstream thinking about conservation (e.g., Holt and Talbot 1978). The general failure to achieve the goal of sustainable use (particularly for the large whales), combined with changes in attitude toward wild animals (especially those perceived to be more cognitively and socially developed: see Lavigne et al. 1999), has led to shifts in the range and distribution of attitudes among scientists (Samuels and Tyack, chap. 1, this volume). Instead of focusing on ways to optimize human benefits, many cetologists now direct their efforts toward minimizing the impacts of human activities on cetaceans. Others remain committed to the goal of sustainable exploitation.

But how much can science contribute to conservation and protection? What information is obtainable? How should we act, given what we do and do not know, and

Figure 12.1. A finback whale being butchered in Williamsport, Newfoundland, in 1971, in the final phase of Canadian commercial whaling. Products of the hunt were exported to Japan. (Photograph by R. Olsen.)

equally important, what we can and cannot know? And what can we say about the long-term prospects for the whales, dolphins, and porpoises? This chapter is principally about those areas in which the study and protection of individual cetaceans and their populations interact. We first review the anthropogenic threats that cetaceans face and summarize the status of species. We then examine the ability of science to evaluate the status of cetaceans in regard to those threats, and to try to ameliorate them.

The Threats

Past Exploitation

The populations of most species of large whales are well below their pre-exploitation levels (Clapham et al. 1999). Some populations, such as those of the northern right whale *(Balaena glacialis),* number only in the hundreds (National Marine Fisheries Service 1991; Best 1993). Very low population size is of concern for a number of reasons,

especially in species that existed, and presumably evolved, at much higher population densities. Small populations are at greater risk from stochastic catastrophic events (Shaffer 1981). It is conceivable, for example, that one large ship passing through a courtship group of a dozen or so right whales could have a substantial impact on the North Atlantic population of about three hundred individuals (M. W. Brown et al. 1995).

Per capita reproduction may decline at low population sizes (the Allee effect or depensation) if individuals fail to find suitable mates or do not reproduce for other density-dependent reasons. At low post-whaling population densities, widely scattered individuals, such as the few hundred adult female blue whales *(Balaenoptera musculus)* in the Antarctic Ocean (International Whaling Commission, in press), may fail to encounter appropriate breeding partners. This may be especially likely for species that, unlike humpback *(Megaptera novaeangliae)* and gray *(Eschrichtius robustus)* whales, do not migrate seasonally to geographically well-defined breeding grounds. Simply encountering a member of the opposite sex might not be sufficient. For example, if female mating behavior evolved in a context in which females consistently had opportunities to choose among multiple males, they may have been selected to ignore solitary males and to mate only in situations in which they could exercise some choice, or to mate only with males that surpass some threshold (as seems to be the case in red jungle fowl, *Gallus gallus:* Zuk et al. 1990). This would lead to apparently maladaptive behavior in situations of artificially low population density (e.g., sperm whales: Whitehead and Weilgart, chap. 6, this volume).

Low population size can also decrease genetic diversity and increase homozygosity (Nunney and Campbell 1993). For example, North Atlantic right whales have lower mitochondrial genetic diversity than southern right whales, *Eubalaena australis* (Schaeff et al. 1997). The total population of right whales in the North Atlantic is only a few hundred, with little evidence of increase, while at least two populations of southern right whales are over a thousand and increasing steadily (Best 1993). It is not always clear, however, whether, or how, genetic diversity affects population fitness (Nunney and Campbell 1993). Amos (1996) points out that cetacean populations probably have not been depleted to low enough numbers for long enough to show significant reductions in nuclear genetic diversity. He cautions, however, that deleterious effects of inbreeding can appear long before the nuclear DNA molecule exhibits appreciable loss of variability.

In addition to reducing population size, exploitation can affect populations in more subtle ways (e.g., Geist 1971). If the growth rate of an individual or the size at physical maturity is heritable, profit-induced or legally mandated selection of larger whales by whalers may have left a population of slower-growing or physically smaller genotypes. Just as trophy hunting of ungulates or killing of elephants for ivory may lead to populations with smaller horns, antlers, or tusks (Geist 1971), highly selective hunting of narwhals *(Monodon monoceros)* might result in those morphs with long, massive tusks being reduced or eliminated from the population (see fig. 12.2).

In species for which reproduction or survival is particularly dependent on social factors, exploitation can reduce the fitness of those animals that are not deliberately removed from the population (e.g., African elephants, *Loxodonta africana:* Poole and Thomsen 1989). There is some evidence for such effects in cetaceans. In *resident* killer whales *(Orcinus orca)* off Vancouver Island (see Baird, chap. 5, this volume), the southern community, which was reduced by live capture operations between 1962 and 1977, has had a lower population growth rate than the northern community, which was less affected by live capture (Olesiuk et al. 1990). This difference may be due to the reduced proportion of males in the southern community (Olesiuk et al. 1990). The pregnancy rate in some sperm whale *(Physeter macrocephalus)* populations fell as the proportion of mature males was greatly reduced by selective whaling (Whitehead and Weilgart, chap. 6, this volume). The disruption of stable, kin-based social structures (Baird, chap. 5; Whitehead and Weilgart, chap. 6, this volume) by whaling could also reduce the reproductive rates or survival of unexploited females and young animals. Such effects can persist for many years after the end of exploitation (Whitehead and Weilgart, chap. 6, this volume).

Some populations of large whales that may have been reduced to a few hundred or fewer have shown encouraging increases at rates of a few percent per year since protection. These include the eastern Pacific gray whale, several populations of southern right whales, and most humpback whale populations (Best 1993). However, others, including northern right whales, some bowhead *(Balaena mysticetus)* populations, western Pacific gray whales, and Antarctic blue whales, remain precariously small despite decades of protection from whaling (Clapham et al. 1999). Illegal catches of these species following official protection has almost certainly been a factor in their slowness of recovery (Clapham et al. 1999; Tormosov et al. 1998).

Present Exploitation

Massive, direct exploitation of cetaceans is generally viewed as a problem of the past (Hofman 1995). As indicated in table 12.1, however, many thousands of whales, dolphins, and porpoises still die each year in directed hunts. In most cases, the method of capture (usually harpoons or nets) causes pain and suffering to the animal (Mitchell et al. 1986). Members of the majority of cetacean species are killed intentionally in at least part of their range, and for about 30% of species, the hunting takes more than just an occasional individual. Direct exploitation comes in several general forms:

Aboriginal hunts. For the most part, aboriginal hunting, aimed principally at providing food and other products for local consumption, is not perceived as representing a threat to the survival of cetacean species or populations. For example, the whaling for sperm whales and smaller odontocetes at Lamalera, Indonesia, involves the use of relatively simple technology and occurs on a small geographic scale. The annual take, which does not exceed about forty sperm whales and at most a few tens of individuals from other species (Barnes 1991; Rudolph et al. 1997), is probably small relative to the sizes of the affected cetacean populations. Whaling for bowhead whales by Alaskan Eskimos was regarded during the 1970s as a threat to the species (Mitchell and Reeves 1980), but this hunt, which kills a few tens of animals each year, is now well regulated and considered sustainable (International Whaling Commission 1992).

However, aboriginal subsistence whaling is not always sustainable. Several populations of belugas (*Delphinapterus leucas*) in Canada and Greenland have been severely depleted by overhunting during the last few decades (Reeves and Mitchell 1989; Heide-Jørgensen and Reeves 1996). The sustainability of narwhal hunting by indigenous people in Canada and Greenland (fig. 12.2) is uncertain, and the cash value of both the skins and tusks of these animals warrants close attention to this hunt to ensure against overexploitation (Reeves 1993).

To date, the humaneness of aboriginal hunting has received much less attention than has that of commercial hunting by nonaboriginal whalers. However, since aboriginal hunters may take hours to kill their prey, and frequently wound whales but fail to secure them, their hunting should be of at least as much concern, from an animal welfare perspective, as that by Norwegian, Japanese, and other technologically advanced whalers (see Mitchell et al. 1986).

Hunts in developing countries. With increasing human populations, decreasing abundance of natural food resources, and commercialization of many aspects of their lives, coastal residents of developing countries are looking for new opportunities to obtain food and make money. In some developing countries, such as Sri Lanka, the hunting of cetaceans may be increasing (Mulvaney 1996; fig. 12.3). In most cases the animals are eaten, although off southern Chile large numbers of dolphins and porpoises have been caught for use as crab bait (Lescrauwaet and Gibbons 1994). Rudimentary information on the numbers and species of animals killed is available from only a few countries, such as Sri Lanka and Peru (Read et al. 1988; Leatherwood 1994; Van Waerebeek and Reyes 1994). Reports from these places suggest that for some cetacean populations, such as the dusky dolphin (*Lagenorhynchus obscurus*) off Peru (Van Waerebeek 1994), the exploitation is not sustainable (International Whaling Commission 1995b). There are substantial catches in other parts of the world as well (in Asia, Africa, Latin America, and the West Indies), but documentation is generally poor or outdated (e.g., Caldwell and Caldwell 1975). Attempts at legal regulation of such hunts have not always been successful, as during the earlier phases of the crab-bait fishery in southern Chile (International Whaling Commission 1995b).

Hunts in or by developed countries. Cetacean hunts continue in some developed countries, such as the pilot whale (*Globicephala melas*) drives in the Faeroe Islands, minke (*Balaenoptera acutorostrata*) whaling off Norway, and kills of small whales and dolphins by coastal communities in Japan (fig. 12.4). There is generally good information on these hunts, but some, such as the drive fishery for striped dolphins (*Stenella coeruleoalba*) in Japan, do not appear to be sustainable (Kasuya 1999a).

Whaling for scientific purposes. The practice of issuing national permits that allow otherwise protected species to be killed for scientific research has long been entrenched in the International Whaling Commission (IWC). Following the implementation of the IWC moratorium on commercial whaling, both Japan and Norway began taking hundreds of minke whales each year, ostensibly for scientific purposes. Although it is unlikely to endanger populations on its own, scientific whaling has been plagued by questions as to whether the research is bona fide or necessary, or whether the primary motive is to permit whaling operations to continue.

Culls. Fishermen are sometimes convinced that cetaceans are harming their livelihood, usually by eating fish that have been caught, or have the potential of being caught, in fishing gear (Mulvaney 1996; Northridge and Hofman 1999). The result may be culls, which can be conducted informally by shooting at cetaceans while fishing, or formally as a fishery enhancement program. Part of the justification for Norway's hunt of minke whales is that they are consuming commercially important fish (Earle 1996). Except when cetaceans take fish that have already been caught, or when they physically damage fishing gear, there is no clear evidence that their presence in the ecosystem is harmful to fisheries (Earle 1996). Nor is there any evidence that a cull of cetaceans has actually resulted in increased yield from a commercially valuable fish population. Marine ecosystems are dynamic, complex, and hard to study, so the likelihood of obtaining unequivocal proof, one way or the other, of the avowed benefits of a cull is very low.

Live capture. Cetaceans are sometimes captured alive for display, research, or conservation purposes (Reeves and Mead 1999). This practice is generally diminishing because of concerns about the welfare and rights of cetaceans in captivity, reasonable success in captive breeding (especially

A

B

Figure 12.2. Aboriginal hunting on Baffin Island, Canada. *(Top)* Removing the valuable tusk from a male narwhal. *(Bottom)* Preparing to flense a pregnant female narwhal. (Photographs by R. R. Reeves.)

of bottlenose dolphins, *Tursiops* spp.), and new techniques allowing research that previously required captive animals to be conducted instead using wild, provisioned, or temporarily restrained animals. The effects of live capture operations on wild populations would generally be expected to be small because of the few animals involved. However, the removal of sixty-eight killer whales from coastal waters off British Columbia and Washington State between 1962 and 1977 included an unsustainable take of about forty-eight individuals from the southern *resident* community

HAL WHITEHEAD, RANDALL R. REEVES, AND PETER L. TYACK

Figure 12.3. A by-catch of dolphins, such as these Risso's (*Grampus griseus,* top) and spinner (*Stenella longirostris,* bottom) dolphins, in Sri Lanka has led to the development of a marketing system and directed hunts. (Photographs from the Steve Leatherwood collection; courtesy R. R. Reeves.)

(Olesiuk et al. 1990). Also, local populations of other coastal species, such as the Irrawaddy dolphin (*Orcaella brevirostris*), finless porpoise (*Neophocaena phocaenoides*), hump-backed dolphin (*Sousa chinensis*), and bottlenose dolphin, are susceptible to depletion by intensive live capture operations, particularly when the populations are already under pressure from fishery by-catch, competition with fisheries, pollution, and other types of habitat degradation (Smith 1991; Reeves and Leatherwood 1994). Live capture and maintenance in semi-natural reserves is being attempted as a last-ditch means of preventing the extinction of China's Yangtze River dolphin, the baiji (*Lipotes vexillifer:* Leatherwood and Reeves 1994; Perrin 1999; see fig. 12.8).

By-catch and Other Incidental Effects

By-catch refers to the incidental capture of non-target species in fisheries. By-catch affects nearly every cetacean species, whether odontocete or mysticete, large or small, riverine, inshore, or offshore (fig. 12.5; International Whaling Commission 1994a). In the cases of the vaquita (the Gulf of California porpoise, *Phocoena sinus*) and probably the baiji, by-catch is the primary threat to species survival. In 1992, an international workshop on cetacean mortality in passive nets and traps considered by-catches of populations of harbor porpoise (*Phocoena phocoena*), hump-backed dolphin, striped dolphin, and two populations of bottlenose dolphin to be unsustainable (International Whaling Commission 1994a). In addition to its population consequences, death from entanglement in fishing gear is often likely to be slow and painful.

By-catch can occur with almost any fishery, but gill nets (including drift nets) and tuna seines (especially when directed at dolphin schools) cause the greatest mortality—many thousands of animals per year in some fisheries (International Whaling Commission 1994a; Wade 1995; Read 1996). (Dolphins caught in purse seines set around and under them do not, technically, constitute by-catch, even though they are non-target species and their killing is incidental.) During the last few years, it has become apparent that midwater trawls take substantial numbers of cetaceans in several parts of the world (e.g., Fertl and Leatherwood 1997; Couperus 1997; Dans et al. 1997). In some developing countries, such as Sri Lanka and Peru, the salvage and use of by-caught carcasses has led to the development of directed cetacean hunts (Reeves and Leatherwood 1994; Read 1996; see fig. 12.3).

Cetaceans are incidentally killed or injured by other human activities as well, including shipping—a particular concern for the northern right whale (National Marine

Figure 12.4. Schools of rough-toothed dolphins *(Steno bredanensis)* occasionally come close enough to shore to draw the attention of Japan's dolphin hunters. Although these are rare events, many tens of animals can be killed in one drive. There is no international regulation of these catches, although the Sub-Committee on Small Cetaceans of the International Whaling Commission's Scientific Committee does subject the Japanese catches to rigorous scrutiny. (Photograph by R. White-Hubbs Marine Research Institute.)

Figure 12.5. By-catch: Dall's porpoise on the deck of a salmon vessel. (Photograph by W. Everett.)

Fisheries Service 1991; M. W. Brown et al. 1995; Katona and Kraus 1999)—and explosions, which can be caused either by military exercises or by industrial blasting (Richardson et al. 1995b).

Habitat Loss and Degradation

The destruction of habitat is a serious threat to cetacean biodiversity, particularly for inshore and freshwater species. River-dwelling species are especially vulnerable. As the world's major rivers are used and transformed, they become less and less capable of supporting dolphins and porpoises (Perrin et al. 1989; Leatherwood and Reeves 1994). The Yangtze, the Ganges, the Indus, and other dolphin-inhabited rivers are being radically altered by dams, embankments, deforestation, pollution, and overfishing (Brownell et al. 1989; fig. 12.6). The baiji is very close to extinction, and the bhulan, or Indus river dolphin *(Platanista minor)*, may not be far behind (Reeves and Chaudhry 1998). The Amazon and Orinoco, with much lower surrounding human population densities than the large Asian rivers, are less altered, and the South American river dolphins have significantly better prospects in the next few decades (Leatherwood and Reeves 1994; Vidal et al. 1997).

Pressures from growing human populations are also changing some coastal environments that are important for inshore cetaceans (Kemp 1996) (fig. 12.7). Harmful developments include sewage outfalls, agricultural runoff, moorings, dredging, blasting, dumping, port construction, and hydroelectric projects. Large hydroelectric projects certainly have major downstream ecological effects, but these are poorly understood. Concern has been raised about the potential effects of dam construction on belugas because of their habit of occupying estuaries in summer. Most attention, to date, has focused on the dammed rivers flowing into Hudson and James bays in the eastern Canadian Subarctic (Bunch and Reeves 1992; Lawrence et al. 1992).

Although the riverine and inshore odontocetes may be most threatened by habitat destruction, baleen whales that depend on particular coastal areas for important parts of their life cycle are also at risk. Gray whales use traditional calving lagoons, whose development may be harmful: gray whales left Laguna Guerrero Negro, in Baja California, for several years while there was considerable disturbance (including dredging) associated with a salt works (Bryant et al. 1984). Another proposed salt works, at Laguna San Ignacio, is causing concern (Dedina and Young 1995). A possible contributory factor to the lack of recovery of the northern right whale is that some of its traditional calving grounds (such as Delaware Bay) are no longer suitable because of human development (International Whaling Commission 1986; National Marine Fisheries Service 1991).

Even deep-water cetaceans may be vulnerable to the effects of habitat degradation. A population of about 230 northern bottlenose whales *(Hyperoodon ampullatus)* makes

Figure 12.6. Taunsa Barrage, Pakistan, which fragments the population of the Indus river dolphin. (Photograph by R. R. Reeves.)

SCIENCE AND THE CONSERVATION, PROTECTION, AND MANAGEMENT OF WILD CETACEANS

Figure 12.7. Hump-backed dolphins in Hong Kong harbor. Hong Kong waters are among the most polluted anywhere on Earth. During the last decade, conservationists have called attention to the plight of hump-backed dolphins and finless porpoises in Hong Kong and elsewhere along the coast of southern China. Both species survive despite competition with humans for ever scarcer fish stocks, vessel noise, filth from toxic waste dumping and raw sewage disposal, and the risks of entanglement and collisions. It seems unlikely that these animals can persist over the long term in the face of rampant development. (Photograph from the Steve Leatherwood collection; courtesy R. R. Reeves.)

heavy use of a deep submarine canyon on the edge of the Scotian Shelf called the Gully. Petrochemical development around the Gully threatens this population (Whitehead et al. 1997b).

Competition with Fisheries

Just as it is very difficult to assess whether cetaceans are harming fisheries, the possible trophic effects of fishing on whale, dolphin, and porpoise populations are largely speculation. However, as fishing industries are making radical changes to almost all freshwater and marine ecosystems, it is likely that cetaceans are more threatened by fisheries than the reverse (Earle 1996).

Pollution

Until quite recently, oceans, and even large rivers, were perceived as almost infinite sinks that could absorb human waste with no impact on humans and little harm to the environment. But this is not so: even the deepest and most offshore waters show signs of our pollution. For many pollutants, these signs are particularly clear in cetaceans because they are long-lived and exist near the top of the food chain. Additionally, they store energy (and so pollutants) in their blubber, and contaminants are transferred to infants in maternal milk (e.g., Aguilar and Borrell 1994; Ridgway and Reddy 1995). Cetaceans also have a lower capacity to metabolize some PCB isomers than most other mammals (Tanabe and Tatsukawa 1992; Tanabe et al.

1994). One example illustrating the degree to which foreign, and often toxic, substances are biomagnified in cetaceans is the striped dolphin in the western North Pacific. Individuals taken near Japan had PCB and DDT levels 10^7 times higher than the surrounding seawater (Tanabe and Tatsukawa 1992). Tanabe et al. (1994) point out that cetaceans inhabiting pristine areas far from industrial centers sometimes contain higher concentrations of PCBs than coastal and terrestrial mammals living near pollution sources. It is not unusual for PCB levels in cetacean blubber to reach 200–800 ppm, well over the 50 ppm level normally requiring goods to be labeled as toxic (Cummins 1988).

A wide range of contaminants are found in whales, dolphins, and porpoises, including heavy metals (such as mercury, lead, and cadmium), organochlorines (such as PCBs, DDT, dioxins, etc.), and polycyclic aromatic hydrocarbons (Johnston et al. 1996; O'Shea 1999). These chemicals and other agents, including oil, plastic debris, and sewage (O'Shea 1999; Reijnders et al. 1999), have the potential to affect cetaceans. In a few instances, such as stranded animals whose digestive tracts are blocked by plastic bags (e.g., Viale et al. 1992), mortality can be linked directly to pollution. There are also suggestive correlations, such as the very high pollutant levels (including metals, organochlorines, and polycyclic aromatic hydrocarbons) and extreme incidence of tumors in beluga whales in the St. Lawrence River (Martineau et al. 1994). However, controlled experimental studies of the effects of pollution on cetaceans are logistically difficult and expensive, and they raise ethical concerns. Thus, assessment of the impacts of pollution on individual cetaceans and their populations is largely circumstantial and inferential (Tanabe and Tatsukawa 1992; O'Shea et al. 1999).

Pollution raises two general areas of concern. First, pollution may lower the ability of the immune system to respond to naturally occurring diseases (e.g., Lahvis et al. 1995), resulting in increased mortality. This factor may explain some of the large die-offs of cetaceans in polluted marine habitats during the last fifteen years (e.g., Kuehl et al. 1991). Second, pollution may impair the reproductive system, as found by experimental work with seals (Reijnders 1986) and as has been suggested for St. Lawrence belugas (Martineau et al. 1994).

Pollutant levels are generally higher in inshore or coastal animals of northern temperate latitudes, such as the harbor porpoise, than in offshore animals, such as the oceanic dolphins, although the long-range transport of pollutants ensures that they find their way into all marine ecosystems (Tanabe and Tatsukawa 1992; Reijnders 1996). They are found in the tissues of polar cetaceans (Muir et al. 1992) as well as deep-water whales (Law et al. 1996). Since pollutants bioaccumulate along the food chain, the baleen whales, which tend to feed at lower trophic levels, generally have lower concentrations of most contaminants than odontocetes (O'Shea and Brownell 1994).

Noise and Disturbance

The physical characteristics of the underwater environment favor the use of sound, rather than sight or some other sense, for sensing over distances of more than a few meters. As discussed by Tyack in chapter 11 of this volume, cetaceans have evolved a sophisticated auditory sense, and they use a broad range of vocal signals to communicate and to investigate their underwater environment. Humans have learned to use sounds in the same wide range of frequencies to explore the ocean. Noise is also incidentally produced by most marine activities. As a result, seafaring humans have the potential to interfere acoustically with the lives of cetaceans.

A fine and detailed review of the state of knowledge about the impacts of noise on marine mammals is provided by Richardson et al. (1995b). Acoustics is a technically sophisticated subject with its own terminology (see Urick 1983; Richardson et al. 1995b; box 11.1). In addition to their frequency (usually given in Hz or cycles per second) and intensity (usually given in decibels [dB] re 1 μPa), sounds can also be categorized as those that are continuous, over periods of seconds or more, and those that are transient.

Extreme among transient sounds is the pressure wave produced by an explosion. Explosives are used underwater for a variety of purposes, including military activities, construction, and oceanographic or geophysical research. Very little is known about the effects of blasts on marine mammals. Estimates have been based on extrapolation from experiments in which terrestrial mammals were submerged and exposed to underwater explosions at close range (Richardson et al. 1995b). These animals showed injuries such as hemorrhaging in the lungs and contusion and ulceration of the gastrointestinal tract. The auditory system is designed to transfer sound energy efficiently into the ear, so loss of hearing typically occurs at much lower noise levels than those needed to damage other tissues. Damage to the inner ear has been reported in Weddell seals (Leptonychotes weddellii), collected from McMurdo Sound, Antarctica, shortly after a series of underwater explosions (Bohne et al. 1985, 1986). In Newfoundland, two humpback whales died following injury to their inner ears, apparently caused by exposure to powerful underwater blasts used for construction, even though every attempt was made in advance to ensure that no whales were near the blast site (Ketten et al. 1993; Todd et al. 1996).

Although explosions produce the most obvious impacts on marine mammals, only a few individuals are affected at a time. In contrast, continuous noise is part of the environment of huge numbers of cetaceans at any given time. Since the advent of motorized ships about a century ago, commercial shipping has changed the acoustic environment in which cetaceans live. The noise from ships puts so much acoustic energy into the ocean that it dominates the average ambient deep-water conditions all over the planet in the frequency region from 20–200 Hz (Urick 1983). This problem is growing: shipping is reported to have increased ambient noise levels in the oceans tenfold from 1950 to 1975 (Urick 1986). Other more recent human marine activities, such as those associated with the oil and gas industry, the commercial, research, and military use of sonar, and oceanographic acoustic experimentation (especially when many fixed sources are operated for long periods, as in the proposed full implementation of the Acoustic Tomography of Ocean Climate or ATOC project), also increase noise levels in the sea.

Exposure to continuous noise can affect hearing either temporarily or permanently. Unfortunately, we do not know what levels of exposure are safe for cetaceans, nor do we know how serious the potential effects of existing noise sources might be. Noise, both continuous and transient, can also disrupt the behavior of cetaceans. When whales, dolphins, and porpoises are disturbed or confronted with novel stimuli, they often stop vocalizing. This silence presumably makes it easier for the animals to listen and renders them more difficult to detect (Herman and Tavolga 1980). The suspension of vocalizations from otherwise active groups of whales is often the first clue that they are reacting to a disturbance. For example, sperm whales off St. Lucia responded to sonar signals and ship propeller noise (often at considerable distance) associated with the U.S. military invasion of Grenada in 1983 by becoming silent (Watkins et al. 1985). These whales also showed longer-term changes in behavior: they were much quieter and much more wary of the research vessel than during other research cruises in the same area. In the southern Indian Ocean, sperm whales appeared to react to the sounds of a seismic vessel over 300 km away, as well as to the pre-ATOC Heard Island Feasibility Test sounds, by ceasing vocalizations (Bowles et al. 1994).

Noise from ships may cause marine mammals to change their behavior and avoid a noise source at surprisingly long ranges. Beluga whales and narwhals in the Canadian High

Arctic have been observed to avoid icebreakers approaching them at ranges of 45–60 km (Finley et al. 1990). Belugas were displaced by as much as 80 km, their groups became dispersed, and their diving became asynchronous. It took as long as 48 hours after the ship had passed for these belugas to resume normal activities near the ice edge. A mass stranding of Cuvier's beaked whales *(Ziphius cavirostris)* in Greece probably resulted from an unusually strong behavioral reaction to a loud military sonar being tested in the area (Frantzis 1998).

If a noise is sufficiently loud or annoying, it can cause animals to avoid certain areas, effectively reducing or degrading their habitat. Chronic noise and human activity can cause animals to abandon areas permanently (or at least until these stimuli are reduced), as in the case of the abandonment of Laguna Guerrero Negro by gray whales (see above). The offshore oil and gas industry is an important source of noise. In the Beaufort Sea, bowhead whales reacted to and avoided noises of seismic vessels, dredging, and drill ships at ranges of kilometers (Richardson et al. 1990). Migrating gray whales responded to playback of oil industry noise by slowing down and altering course well away from the sound source, at ranges of 1–3 km (Richardson et al. 1995b).

These responses of bowhead and gray whales to continuous sounds have been documented at relatively low levels of exposure, 110–120 dB. Many modern sound sources, from ships to sonars, can ensonify large areas of ocean with levels louder than this, out to ranges of hundreds of kilometers. Most playback experiments have used fainter sources and have documented avoidance responses at ranges of hundreds to thousands of meters. Little is known about the potential effects of these louder sources at greater ranges. Even relatively minor effects could be biologically significant if the affected area included a large fraction of a species' habitat.

Special concerns are raised by the sounds of vessels and aircraft that are used to approach whales and dolphins for science or tourism. Although the sounds produced by these vessels are not as loud as those from commercial shipping, their much closer range makes them a cause for concern. They may also lead to nonacoustic forms of disturbance (including swimmers and boaters attempting to touch or feed animals, and biopsy darting and tagging by scientists). In most areas, both scientists and whale-watchers are sufficiently scarce and conscientious that the disturbance they cause is unlikely to result in either long-term effects on individual animals or population-level consequences. However, in special circumstances—for instance, where unregulated whale-watching takes place on a breeding ground—there is need for caution.

Chapter 11 of this volume describes how cetaceans use sound to communicate, to orient in the ocean environment, to find their prey, and to detect predators. If human activities increase the level of ambient noise, then cetaceans may be less effective in performing these critical activities.

In general, the disruption of behavior or the masking of sound is likely to occur at much lower exposure levels, or signal-to-noise ratios, than those that might cause physical injury. Increases in noise of 20–30 dB above background have been shown to reduce growth and reproduction in a variety of marine organisms, including fish (Banner and Hyatt 1973) and shrimp (Lagardere 1982). If similarly modest increases in exposure were to affect the growth and reproduction of cetaceans, then it would be reasonable to expect loud, persistent sounds to affect populations over large fractions of their range. While these kinds of effects may be difficult to detect and measure in the wild, they could pose a greater risk to most cetacean populations than the possibility that a small number of animals near a source might be injured.

Global Climate Change

It is generally agreed that the Earth's climate is changing systematically in response to human activities (Intergovernmental Panel on Climate Change 1995). These changes are likely to affect virtually all life, including cetaceans (International Whaling Commission 1997; Tynan and DeMaster 1997). In the long term, they may be the decisive anthropogenic threats to some whales and dolphins (MacGarvin and Simmonds 1996). The two principal changes are increases in ultraviolet radiation caused by ozone depletion in the upper atmosphere (principally a result of transport of anthropogenic chlorofluorocarbons into the upper atmosphere) and planetary warming resulting from increased carbon dioxide levels in the atmosphere.

Currently, ozone depletion mainly affects the Antarctic and other areas of the Southern Hemisphere. Increased ultraviolet radiation decreases rates of photosynthesis in phytoplankton, leading to reduced productivity and thus potentially harming all parts of the food chain (MacGarvin and Simmonds 1996). Increased ultraviolet radiation could also affect cetaceans directly, especially those with low skin pigmentation living in areas subject to ozone holes, but indirect effects—resulting from the effects of increased radiation on other organisms—are likely to be more important (MacGarvin and Simmonds 1996; International Whaling Commission 1997).

HAL WHITEHEAD, RANDALL R. REEVES, AND PETER L. TYACK

Global warming will melt polar icecaps, raising sea level and sea temperature and altering currents (Intergovernmental Panel on Climate Change 1995). This effect is likely to change the distribution of ocean production, the community composition of marine ecosystems, and thus the availability of food for cetaceans (MacGarvin and Simmonds 1996). It may also change the suitability of traditional calving grounds or migration routes. These changes may have positive or negative effects on particular populations and individuals. In general, species with flexible diets and lifestyles, such as bottlenose dolphins, are likely to be more resilient to changes in the environment than stenophagous species, such as the blue whale, and species with special habitat requirements, such as the gray whale.

While it is increasingly accepted that both global warming and ozone depletion are being caused or enhanced by human activities, predictions regarding their severity are varied and uncertain (Intergovernmental Panel on Climate Change 1995). There are also great difficulties in assessing the effects of any climate change on cetaceans (International Whaling Commission 1997), so the implications for whale, dolphin, and porpoise populations of the systematic changes in our environment that are taking place right now will remain almost entirely conjectural.

Interactions Between and Among Threats

In this summary we have mentioned several ways in which threats may interact with one another and with natural stressors. For example, by-catch can evolve into a directed hunt, pollution may increase susceptibility to naturally occurring pathogens, and noise could decrease the ability of dispersed members of a highly depleted population to find mates. There are many other possible kinds of interactions, or synergies, between or among threats. The cumulative or synergistic effects of human activities can substantially increase the danger to cetaceans. The effects of obvious, direct threats to cetaceans are difficult enough to study, but definitive studies of interactive effects from multiple threats are almost impossible. It is important to bear in mind the principle that interacting threats often have a greater impact than would be predicted by simply summing the results of each threat considered independently.

Status of Whale, Dolphin, and Porpoise Species

IUCN Red List Classifications

If there is a general consensus on the conservation status of most of the world's whales, dolphins, and porpoises, it

should be reflected in the work of the Cetacean Specialist Group (CSG) of the Species Survival Commission of the World Conservation Union (IUCN). The CSG is a voluntary network of about sixty to seventy cetologists from around the world. The views of the CSG are expressed in two recent publications: *Dolphins, Porpoises, and Whales of the World: The IUCN Red Data Book* (Klinowska 1991) and *Dolphins, Porpoises, and Whales: 1994–1998 Action Plan for the Conservation of Cetaceans* (Reeves and Leatherwood 1994). The first of these describes the conservation status of each species based on information current through about 1990. The second is less detailed but more recent (current through 1994) and prescriptive (see also Leatherwood and Reeves 1994 and Reeves and Leatherwood 1996 for brief updates).

Among the responsibilities of the CSG is to provide advice to the IUCN concerning the Red List status of all cetacean species. In 1994 the IUCN adopted a new system for assessing status and assigning species to categories of threat. The aim of this new system was to introduce greater objectivity into the process. The first Red List to be developed under the new system was published in 1996 (Baillie and Groombridge 1996; table 12.1). Although everyone involved in producing this ambitious document was aware of the difficulty, and inappropriateness, of assigning a global status to species with disjunct geographic populations, it was agreed that this should be done as a first step in a longer-term process. Consequently, all species of cetaceans were given a global designation. In a few instances, when particular geographic populations had been relatively well studied and there was a strong basis for giving them a separate status, this was also done. The result is far from perfect, and the list will require ongoing attention to make sure that it reflects the best information available.

In this recent Red List (table 12.1), 74% of the thirty-eight cetacean species for which sufficient data were available were either considered threatened (fourteen species) or would be so if current conservation measures were stopped (fourteen species). The two critically endangered species, the baiji (fig. 12.8) and vaquita, are both small animals, but also listed as threatened are some of the largest animals on Earth: the blue, fin, and northern right whales. The nearly 50% of cetacean species listed as "data deficient" seems to imply that lack of scientific knowledge is a major barrier to the conservation of cetacean populations.

How Species Are Threatened

Nearly all cetacean species are known to face anthropogenic threats of some kind. The few exceptions are several beaked

Table 12.1. Conservation status of and threats to whales, dolphins, and porpoises

Species[a]	Common name[b]	Status[c]	Recent directed catch[d]	Low numbers from previous directed catch[e]	Recent by-catch[d]	Competition or culling (historical or recent)[d]	Habitat degradation[d,h]	Chemical pollution[d,h]	Acoustic pollution[d,h]	Other[f,h]
Platanista gangetica	Ganges D, Susu	E	***	#	***		***	?	?	s
Platanista minor	Indus D, Bhulan	E	*	#	***		***	?		
Lipotes vexillifer	Yangtze D, Baiji	C		#	****	?	****	?	?	s,b,c
Pontoporia blainvillei	La Plata D, Franciscana	I			***					
Inia geoffrensis	Amazon D, Boto	V	*		**	?	***	?		
Delphinapterus leucas	White W, Beluga	V	***	#		*	?	*	?	
Monodon monoceros	Narwhal	I	**	?					?	
Phocoena phocoena	Harbor P	V	**	#	***	?		?	?	
Phocoena spinipinnis	Burmeister's P	I	?		***					
Phocoena sinus	Vaquita	C			****		?			
Neophocaena phocaenoides	Finless P	I		?	***	?	***	?		s
Phocoena dioptrica	Spectacled P	I	?		**					
Phocoenoides dalli	Dall's P	D	***		***					
Steno bredanensis	Rough-toothed D	I	?		**					
Sousa chinensis	Indo-Pacific Hump-backed D	I	?		**	?	***	?		
Sousa teuszii	Atlantic Hump-backed D	I	?		?	?	?			
Sotalia fluviatilis	Tucuxi	I	*		**	?	?	?	?	
Lagenorhynchus albirostris	White-beaked D	U	**		*					
Lagenorhynchus acutus	Atlantic white-sided D	U	**		*					
Lagenorhynchus obscurus	Dusky D	I	***		***					
Lagenorhynchus obliquidens	Pacific white-sided D	U	*		**	**				
Lagenorhynchus cruciger	Hourglass D	U								
Lagenorhynchus australis	Peale's D	I	**		**					
Grampus griseus	Risso's D	I	**		**	**				
Tursiops spp.	Bottlenose D	I	**		**	**	?			
Stenella frontalis	Atlantic spotted D	I	**		**					
Stenella attenuata	Pantropical Spotted D	D	**		***	*				
Stenella longirostris	Spinner D	D	**		**		?			
Stenella clymene	Clymene D	I	*		**					
Stenella coeruleoalba	Striped D	D	***	#	**			?		
Delphinus spp.	Common D	U	**		***	?				
Lagenodelphis hosei	Fraser's D	I	**		**					
Lissodelphis borealis	Northern right whale D	U			***					
Lissodelphis peronii	Southern right whale D	I			*					
Orcaella brevirostris	Irrawaddy D	I	*	?	***		?			
Cephalorhynchus commersonii	Commerson's D	I	**		**					
Cephalorhynchus eutropia	Black D	I	**		**					
Cephalorhynchus heavisidii	Heavisides's D	I			**					
Cephalorhynchus hectori	Hector's D	V			***				?	
Peponocephala electra	Melon-headed W	U	*		*					
Feresa attenuata	Pygmy killer W	I	*		**					
Pseudorca crassidens	False killer W	U	**		**	**				
Orcinus orca	Killer W	D	*	?	*	**				
Globicephala melas	Long-finned pilot W	U	**	?	**					
Globicephala macrorhynchus	Short-finned pilot W	D	**	?	*	?				
Tasmacetus shepherdi	Shepherd's beaked W	I								
Berardius bairdii	Baird's beaked W	D	**	?	*					

Species[a]	Common name[b]	Status[c]	Recent directed catch[d]	Low numbers from previous directed catch[e]	Recent by-catch[d]	Competition or culling (historical or recent)[d]	Habitat degradation[d,h]	Chemical pollution[d,h]	Acoustic pollution[d,h]	Other[f,h]
Berardius arnuxii	Arnoux's beaked W	D								
Mesoplodon spp. (13 spp.)	Mesoplodonts	I	?		**					
Ziphius cavirostris	Cuvier's beaked W	I	?		*				*	
Hyperoodon ampullatus	N. bottlenose W	D	*	#					?	
Hyperoodon planifrons	S. bottlenose W	D		*						
Physeter macrocephalus	Sperm W	V	**	##	**	*			?	♂/♀, p, c
Kogia breviceps	Pygmy sperm W	U		*						p
Kogia sima	Dwarf sperm W	U		*						p
Balaena mysticetus	Bowhead W	D	**	###					?	
Eubalaena australis	Southern right W	D		###	*		*			c
Eubalaena glacialis	Northern right W	E		###	***		?		?	c
Caperea marginata	Pygmy right W	U		*						
Eschrichtius robustus	Gray W	D	**	###	*		?		*	
Balaenoptera acutorostrata	Minke W	N/D[g]	**	#	**	**			?	c
Balaenoptera borealis	Sei W	E		##					?	c
Balaenoptera edeni	Bryde's W	I	*	#					?	
Balaenoptera musculus	Blue W	E		###					?	c
Balaenoptera physalus	Fin W	E	*	##	*	?			?	
Megaptera novaeangliae	Humpback W	V	*	##	**				?	b

Source: Conservation status from Baillie and Groombridge (1996); threats summarized from Klinowska (1991) and Reeves and Leatherwood (1994), augmented by more recent information available to the authors.

[a]Classification as in Reeves and Leatherwood (1994), with changes for consistency with appendix 2 of this volume.

[b]W, whale; D, dolphin; P, porpoise.

[c]U, lower risk—least concern; N, lower risk—near threatened (close to qualifying for Vulnerable); D, lower risk—conservation-dependent (focus of taxon-specific or habitat-specific conservation program, the cessation of which would result in the taxon qualifying for one of the threatened categories below within a period of five years); V, vulnerable (facing a high risk of extinction in the wild in the medium-term future); E, endangered (facing a very high risk of extinction in the wild in the near future); C, critically endangered (facing an extremely high risk of extinction in the wild in the immediate future); I, data deficient (inadequate information).

[d]Threats: *, known effect on members of species; **, kills tens or more of animals per year; ***, serious threat to one or more populations of species; ****, likely to cause extinction of species; ?, suggested threat.

[e]#, population reduced by pre-1980 catches; ##, population substantially reduced by pre-1980 catches; ###, population almost extirpated by pre-1980 catches.

[f]s, ship/boat traffic; c, collisions with ships; p, ingestion of plastic; b, blasting; skewed adult sex ratio (♂/♀).

[g]In the Red List (Baillie and Groombridge 1996), the two species of minke whale are listed: *B. acutorostrata* (status N) and *B. bonaerensis* (status D).

[h]For these headings, the evidence for biologically significant effects is usually not conclusive, and we have been cautious in our interpretations. Exposure to chemical and acoustic pollution, for example, is universal, but there are only a few examples in which evidence for effects has been marshaled. These headings are illustrative, not definitive.

whale species about which almost nothing is known, and which may or may not be threatened by humans (other beaked whale species are known to be threatened: see table 12.1), and the hourglass dolphin (*Lagenorhynchus cruciger:* fig. 12.9).

The apparently unthreatened status of the hourglass dolphin is illustrative. Why does this species seem to be safe? Three factors help explain its status (Goodall et al. 1997): first, hourglass dolphins are truly pelagic, almost never coming close to land, where most anthropogenic threats to cetaceans originate; second, they are found only in Antarc-

tic and sub-Antarctic waters, where by-catch in fisheries is currently negligible; and third, they are small and thus do not warrant the attention of the pelagic whalers who seriously depleted all populations of large cetaceans in the Antarctic.

The hourglass dolphin may not be quite as safe from the effects of human activities as our table implies. It may soon be taken as by-catch in expanding fisheries, or it may suffer from the effects of ozone depletion on the Antarctic marine food web. What is nevertheless instructive about the hourglass dolphin's unique combination of attributes

Figure 12.8. The Yangtze River dolphin, or baiji, is the world's most endangered cetacean. In the most recent survey of almost the entire habitat of this species, only about twenty-five animals were sighted. This is an adult male, Qi Qi, who, having been found "severely injured," was brought to the Wuhan Institute of Hydrobiology in January 1980. In mid-1999 he was alive, and in reasonably good health, in his pool at Wuhan. (Photograph by S. Leatherwood.)

is that their inverse defines three principal classes of threat to other cetaceans: proximity to shore, vulnerability to by-catch, and consequences of size.

Inshore and riverine cetaceans. Animals found close to shore are particularly subject to the destruction and modification of habitat, by-catch, directed hunts, chemical pollution, and noise. The relative combination of these different threats varies from population to population. In some cases, as with the vaquita, one particular threat, by-catch, threatens the entire species population. In other cases, as

Figure 12.9. Hourglass dolphins in offshore Antarctic waters: possibly the least threatened of cetaceans. (Photograph by S. Hooker.)

with river dolphins, no single threat predominates, but populations are endangered by several acting together.

Offshore cetaceans. The major current threat to offshore cetaceans is by-catch in fisheries. The extent of pelagic fishing is huge (Read 1996), creating a gauntlet of gill nets and other life-threatening gear within the home ranges of most temperate zone pelagic animals.

Large whales. The fresh carcass of a large whale is economically very valuable (ca. $20,000 for a minke whale, according to Kuronuma and Tisdell 1994), and this makes it a tempting commercial target. The elimination of most large whales from the oceans between 1900 and 1975 is a testament to the willingness of profit-driven humans to cause massive changes at a global scale. The large whales remain vulnerable for two principal reasons: first, their populations are artificially small, and so susceptible to threats other than whaling (see above); and second, there is plenty of economic incentive to resume widespread commercial whaling. The low potential rates of increase of populations of large whales (Whitehead and Mann, chap. 9, this volume) make it difficult to manage large commercial hunts on a sustainable basis, although the Scientific Committee of the International Whaling Commission has made considerable progress at designing a risk-averse approach to doing so (Kirkwood 1992; Cooke 1994).

In addition, large cetaceans are likely to be particularly dependent on the ability to hear long-range and thus low-frequency sounds (for food finding, mate finding, and navigation: Tyack, chap. 11, this volume), and so these animals may be especially vulnerable to the effects of anthropogenic noise, which is particularly pervasive at lower frequencies (Committee on Low Frequency Sound and Marine Mammals 1994). The large odontocetes, such as sperm and killer whales, that eat high on the food web accumulate toxins, and all large whales suffer from by-catch in fisheries. Even though large whales are more frequently able to break free of fishing gear than are the smaller dolphins and porpoises, they often end up towing nets and lines, and this in turn can lead to a slow, painful process of injury and debilitation, and possibly death (fig. 12.10). Injurious encounters with fishing gear can have a decisive effect on small populations of large whales (e.g., right whales, National Marine Fisheries Service 1991).

Conservation, Protection, and Science

We can evaluate the status of cetaceans in relation to human threats in two principal ways, focusing either on the threats themselves or on the animals (and their populations). These two approaches are compared in figure 12.11.

Threat-Based Approach

A threat-based approach involves an effort to estimate the significance of a given threat. It is necessary to evaluate the extent and magnitude of the threat within the geographic area of concern, as well as to develop an understanding of its effects on individual animals (both short-term and long-term) and populations.

With some threats, such as by-catch and directed hunts, especially those in developing countries, it is often hard to estimate with any precision how many animals and what species are being affected. Even in sophisticated directed hunts, the data provided to scientists and managers are sometimes falsified or unreliable in other ways (Rørvik 1987; Best 1989; Kasuya 1991, 1999b; Yablokov 1994; Tormosov et al. 1998).

For other threats, including chemical pollution and noise, evaluating the effects on individual animals is logistically difficult, especially in the long term. Useful quantitative extrapolations to the level of the population are usually impossible. In social animals, like most cetaceans, effects on individuals cannot simply be summed to estimate population impacts; one animal may be affected indirectly by harm done to other members of its community (Whitehead and Weilgart, chap. 6, this volume).

Figure 12.10. Gray whale on a California beach entangled in fishing gear.
(Photograph courtesy of S. Leatherwood).

SCIENCE AND THE CONSERVATION, PROTECTION, AND MANAGEMENT OF WILD CETACEANS

323

Threat-based approach	Animal/population-based approach
Identify threat	Examine population
Assess its occurrence in cetacean habitat	Are animals suffering? Is population at risk (small/declining)?

Evaluate its effect on individual animals
    short-term
    long-term
	If yes, then which threat(s) are responsible?

Evaluate its effect on populations	How can threats(s) be ameliorated?
If effect of threat on individuals/populations is serious, how can it be ameliorated?	

Figure 12.11. Approaches toward the protection of cetaceans from anthropogenic threats.

The threat-based approach, even if it worked well for individual threats considered in isolation, is inadequate for situations in which several threats operate cumulatively or synergistically. Its results can be misleading when complex interactions are involved.

These and other shortcomings mean that the threat-based approach will rarely, if ever, operate in the ideal manner outlined in figure 12.11. It is not useless, however. Examining the nature of threats may suggest useful general priorities for action; for instance, such research might indicate which noises or which chemical pollutants are most likely to have serious consequences (Box 12.1). It is important that such analyses be undertaken before a new potential threat is introduced into cetacean habitat.

Individual- or Population-Based Approach

The first step in an individual-based approach is to determine whether there is a problem for individual animals. It is sometimes obvious that wild animals are suffering (if, for instance, they have been caught in a gill net), but in most other cases (for instance, if the stressor is noise) it may be hard to determine whether they are suffering. Certain wild cetaceans may behave differently from most other members of their species (e.g., dive for shorter periods), but is this difference the result of an anthropogenic threat (e.g., disturbance from whale-watching boats), and is it important to the lives of the animals (Tyack 1989; Gordon et al. 1992)?

At the level of the population, there are similar difficulties. A range of methods can be used to estimate abundance. These include aerial and shipboard surveys, mark-recapture analyses using artificial or natural marks, and

analyses of data from hunting (such as catch per unit effort). These methods can give reasonable order-of-magnitude estimates for a cetacean population, indicating which populations number in the hundreds and which in the thousands. However, with the exception of mark-recapture estimates (or simple counts) from populations in which nearly all animals are individually known (e.g., the *resident* killer whales near Vancouver Island: Olesiuk et al. 1990; the right whales of the western North Atlantic: Knowlton et al. 1994), few of these estimates are especially precise. A c.v. (coefficient of variation = standard error/mean) of 0.2 (indicating a 95% confidence interval of about $\pm 40\%$) is unusually good for an estimate of cetacean abundance. For instance, following well-planned shipboard surveys of the cetaceans off California, Barlow (1995) made estimates with c.v.s for twenty-four species or groups of species. The lowest c.v. was 0.279, and eighteen of the twenty-four estimates had c.v.s greater than 0.5, indicating 95% confidence intervals spanning the range from approximately zero to more than twice the estimated abundance. For aerial surveys off California, the lowest c.v. was 0.24, and seven of nineteen estimates had c.v.s greater than 0.5 (Forney et al. 1995).

This imprecision means that detecting trends in population size is difficult (Taylor and Gerrodette 1993). With their low reproductive rates (Whitehead and Mann, chap. 9, this volume), most cetacean populations cannot grow at more than a few percent per year, so a declining trend of any magnitude can quickly become serious. Calculations using formulae provided by Gerrodette (1987) show that using annual estimates with c.v.s of 0.2, ten years of surveys are needed, at the very least, to detect changes

Box 12.1

Protecting Cetaceans against Habitat Degradation

As we have seen in this chapter, habitat degradation is a major threat to cetaceans. At the same time, cetaceans, in their roles as large, charismatic, and relatively visible upper-level predators, can provide a focus for the protection of marine and freshwater habitats. Thus, the protection of cetacean habitats is not only vital for the animals themselves, but may also have beneficial consequences for other parts of their ecosystems.

An important characteristic of cetacean habitat is its large spatial scale. National boundaries mean nothing to whales, dolphins, and porpoises, so, in their attempts to exploit concentrations of whales, the commercial whalers covered the globe. This made it obvious that whaling could be regulated effectively only by an international body, leading to the formation of the International Whaling Commission (IWC) in 1949 (Holt 1985). Thanks in part to the moratorium on commercial whaling and other management measures instituted by the IWC, whaling currently may not pose as significant a threat to cetaceans as the more insidious, unintentional threat of habitat degradation, particularly from chemical and noise pollution. This degradation also operates over large scales of both time (in the case of persistent chemicals) and space.

Aquatic pollution pays as little heed to national boundaries as cetaceans do. Chemical pollutants released into the air, into rivers, and onto the land all find their way to the most remote parts of the planet. The physics of sound in water also means that ocean noise pollution is a global problem. The sounds of a supertanker, an explosion, or a low-frequency sonar can be heard all over an ocean basin. Therefore, the need for international regulation of human activities that degrade ocean habitat may be as urgent now as the need for international regulation of whaling was fifty years ago.

The last decades have seen some success in international agreements to regulate ocean pollution. For example, the International Maritime Organization has administered a Marine Pollution, or MARPOL, convention "concerned with the prevention and control of marine pollution from ships" (Wallace 1994: 1649). The MARPOL treaty was stimulated by several disastrous oil spills. Prior to the treaty, most oil that polluted the ocean was dumped intentionally when ships washed their tanks with seawater and dumped used oil from machinery into the sea (National Research Council 1985). This treaty, which, among other provisions, sets standards for oil tankers and requires inspections of ships, has had an important role in making intentional pollution unacceptable in principle. It is thought to have helped to reduce oil pollution from tankers substantially (National Research Council 1985). While it initially focused on oil, the treaty covers all pollution, and its later annexes focus on other noxious fluids, packaged hazardous materials, sewage, garbage, and plastics (Wallace 1994). Although enforcement can be a problem, MARPOL forces companies to develop plans for managing waste at sea, and it raises the cost of polluting.

Some cetaceans live in partially enclosed seas where water circulation is limited. For example, the Baltic, Black, and Mediterranean seas are primarily affected by pollution originating in the nations that surround them. In such situations, regional agreements to protect habitats may be effective. For example, in 1976 many of the nations bordering the Mediterranean signed the Convention for the Protection of the Mediterranean Sea against Pollution (Wallace 1994). This convention defines "pollution" as follows: "the introduction by man directly or indirectly, of substances or *energy* [italics ours] into the marine environment resulting in such deleterious effects as harm to living resources" (Wallace 1994: 2146). This definition clearly includes noise pollution, which involves the introduction of acoustic energy into the marine environment. The administrative structure is therefore in place to regulate both chemical and noise pollution in the Mediterranean, a sea badly degraded by human development in many ways.

These examples indicate that regional and international agreements are both feasible and realistic mechanisms for helping to protect cetacean habitat. Such agreements, as well as other approaches to habitat protection, need a scientific basis. First, it is important to define the nature of the threats and what species and habitats are at risk (from either the individual-based or threat-based approaches described in the chapter: see fig. 12.11). Second, standards must be set for levels of exposure to chemical or noise pollution that are unacceptable. Once management systems are in place, we need to continue to monitor both the levels of pollution in critical habitats and the responses of cetaceans individually and as populations, so that insufficient regulations can be reformulated and unnecessary ones discarded.

The history of international regulation of whaling suggests that there will be mistakes along the way to developing effective international institutions to protect marine ecosystems, but if we do not start now, even with our far from adequate knowledge, there will be even less room for error.

---

of a few percent per year in population size. If the estimates are less precise, with c.v.s of 0.5, then this figure becomes twenty years.

Even when a trend in abundance is detected, it is often impossible to distinguish it from a change in the population's distribution (Forney 1999). Such changes are frequently caused by a change in the distribution or abundance of prey species (Whitehead and Carscadden 1985). Surveys rarely cover the entire potential range of members of a population. There is a particular problem if the geographic distribution of survey effort varies between surveys (e.g., Hammond 1990).

An additional or alternative approach to the monitoring of abundance is to observe trends in population parameters, such as per capita birth and death rates. A population with few young animals or a high mortality rate may be in trouble. For instance, the low calving rate of sperm whales in the eastern tropical Pacific is a cause for concern (Clarke et al. 1980; Whitehead et al. 1997a; Whitehead and Weilgart, chap. 7, this volume).

For some cetacean populations that are decreasing, or small and not increasing, a threat or combination of threats causing the problem can be identified. For instance, by-catch is clearly the principal immediate problem for the vaquita, habitat degradation and by-catch for the baiji. In other cases, however, such as the apparent failure of Antarctic blue whales to recover (Clapham et al. 1999), the problems are not so clear.

Sei *(Balaenoptera borealis)* and fin whale populations were seriously depleted by commercial whalers in the 1950s and 1960s. In this instance, there appeared to be excellent information on the status of the populations, and the threat was severe and well documented. Nevertheless, some scientists continued to argue that the data were insufficiently conclusive to take action, and effective management decisions were long delayed (McVay 1974).

In summary, the individual- or population-based approach to the conservation of cetaceans is hampered in

several ways, but especially by our inability to determine whether individuals are experiencing problems and the difficulty of obtaining the data required to clearly demonstrate population trends.

It is increasingly recognized that our understanding of cetacean biology and population dynamics is going to remain inadequate in the foreseeable future. Thus, following the precautionary principle, we need to be prepared to act before the scientific evidence is conclusive (Taylor and Gerrodette 1993; Mayer and Simmonds 1996; Katona and Kraus 1999). For example, Barlow et al. (1997) argue that evidence for declining vaquita abundance should be taken seriously, even though the trend, as documented from survey data, is not statistically significant (at $P = .05$). The precautionary principle is espoused by some politicians and encoded in certain agreements and legislation (e.g., the Food and Agriculture Organization of the United Nations *Code of Conduct for Responsible Fisheries* 1995; Northridge and Hofman 1999). However, generally speaking, governments act only if those concerned for cetaceans make a strong case that a development is harming or will harm the animals, whereas developers and polluters rarely are required to give legitimate proof that their operations will have minimal or no impact (Mayer and Simmonds 1996 and experience of authors).

Management and Mitigation

In addition to examining threats and monitoring effects on individuals and populations, scientists may have another role once a threat has been identified: Can the threat be ameliorated, or kept at an acceptable level?

The regulation of whaling by the International Whaling Commission (IWC) represents one major example of an attempt to do this. Although the IWC is highly politicized, it has a Scientific Committee whose mandate is to provide objective advice on the status of whale populations and on the ability of stocks to withstand different levels of exploitation. The IWC Scientific Committee is itself politicized, as evidenced by long-running statistical battles between those scientists working for whaling nations and those representing countries or organizations with a more protectionist stance. Despite these internal tensions, the Committee has made substantial and important progress in supporting conservation principles in recent years.

The IWC failed to manage whaling on anything remotely resembling a sustainable basis during the 1950s and the first half of the 1960s (McHugh 1974). This was at least partially due to a lack of clear advice from the Scientific Committee, which had little expertise in population biology (McVay 1974). The situation changed with the appointment in 1960 of the Committee of Three (later four) experts on the scientific management of fish populations, whose advice to radically cut quotas was finally heeded in 1965 (Allen 1980).

The IWC Scientific Committee then concentrated their efforts on rational management of whaling and on ensuring that whale populations could sustain removals. The New Management Procedure, adopted in 1974, used information on the biology of the species to set quotas, stock by stock. Its aim was to keep populations near the level at which sustainable yields could be maximized (Allen 1980). However, despite the massive quantity of data potentially available from all the whales being killed, knowledge of the biology of the animals remained insufficient. Except in a very few cases, the IWC Scientific Committee continued to have difficulty agreeing on unambiguous advice (Cooke 1991).

The failure of the New Management Procedure was one incentive for the moratorium on commercial whaling instituted in 1986. During the moratorium, the IWC Scientific Committee has developed a Revised Management Procedure (RMP: Kirkwood 1992; Cooke 1994; International Whaling Commission 1994b). This system of managing the exploitation of whale populations uses only minimal biological information, instead regulating catch levels principally on the basis of trends in abundance from independent sighting surveys. Computer simulations suggest that the procedure safeguards against the substantial depletion of any population under all likely scenarios, but such safety is achieved by keeping takes very low compared with population sizes.

The RMP (which was designed for baleen whale populations) has yet to be used to manage any cetacean hunt, although Norwegian authorities use a process that largely follows the RMP to manage their harvest of minke whales, which continues in defiance of the IWC moratorium. Un-

fortunately, the RMP is impractical for most of the directed catches in the world today, which take place in developing countries and principally target odontocetes (see table 12.1).

Much more than science is needed to address most of the threats facing cetaceans. Even in those rare instances in which the nature and magnitude of a threat are clearly presented in scientific terms, the challenge of finding solutions remains. The realms of education, sociology, law, and management are often at least as relevant as science in finding solutions: Devise a strategy for providing fishermen with alternative employment. Establish a protected area. Stop the building of a dam. Enforce regulations. These imperatives always require perspectives and expertise from outside a purely scientific framework.

One of the best examples of integrating scientific and nonscientific dimensions is the work of Jon Lien in Newfoundland. There, fishermen have experienced a substantial, and unwelcome, problem with cetacean by-catch. Lien has worked with the fishermen in setting up a program for releasing entrapped whales (Lien 1994). This same partnership has led to the invention of clangers and pingers to warn cetaceans of the presence of fishing gear and thus prevent entanglement in the first place.

What Is the Future of Whales, Dolphins, and Porpoises?

In table 12.2, we attempt to evaluate the status of some of the major threats facing cetaceans, in terms of their geographic and taxonomic scope, how many animals or populations are affected, whether the threat is sustainable at its current level, temporal trends in the level of the threat, and its inertia—how long the effect on cetaceans would continue if the source of the threat were removed.

At current levels, by-catch, habitat degradation, and directed hunting seem to be the most serious threats to cetaceans. Recent national laws and international agreements may lead to a general decrease in levels of by-catch over the next decade, especially in pelagic waters and developed countries (Donovan 1994). Technical fixes, such as clangers, pingers, and other modifications to fishing gear, may help reduce rates of by-catch in some contexts (Kraus et al. 1997, but see Dawson et al. 1998). These positive developments are counterbalanced, however, by the likelihood that by-catch, directed catch, and habitat degradation will all increase in developing countries. The result is that riverine and coastal species, especially those that are endemic

Table 12.2. The principal threats facing cetaceans

Threat	Geographic scope	Phylogenetic scope	Scale of threat	Currently sustainable?	Trend in level of threat	Inertia[a]
Effects of past exploitation	All oceans and some rivers	Large whales, and some smaller species	Serious for several populations	—	Lessening	Decades
Aboriginal subsistence hunting	Local or regional	>10 species	Serious for some populations	Some sustainable, some not, others uncertain	Stable	Instantaneous (by definition)
Commercial hunts in developing countries	Widespread inshore	Many species	Serious for some populations	Some cases probably not sustainable	Generally increasing	Instantaneous
Commercial hunts in or by developed countries	Local or regional	A few species	Serious for a few populations	Some sustainable, some probably not	Fairly stable, or declining	Instantaneous
Scientific whaling	A few areas	Minke whale (at present)	Hundreds of individuals	Sustainable	Fairly stable	Instantaneous
Culls	A few areas	A few species	Tens to hundreds of individuals	Uncertain	Fairly stable	Instantaneous
By-catch	Very widespread	Most species	Serious threat, many populations	Often unsustainable	Lessening in some areas, increasing in others	Instantaneous for actual fishing, years for ghost-fishing
Habitat degradation	Mainly coastal and riverine	Coastal marine and freshwater species	Very serious threat to a few populations	No	Increasing	Decades to permanent
Noise	All habitats (Northern Hemisphere worse)	All species	??	—	Increasing	Instantaneous
Competition with fisheries	Almost all cetacean habitat	Almost all species	?	—	Increasing	Years to decades
Heavy metals	More inshore	Odontocetes	?	—	Increasing?	At least decades
Organochlorines	More inshore, but all oceans	Principally odontocetes	Potential threat to all cetacean populations	—	Increasing?	At least decades
Polycyclic aromatic hydrocarbons	Local, inshore	A few species	Local threat to individuals	—	Increasing?	
Plastic debris	All habitat	Many species (especially sperm whales?)	Individuals, at least	—	Increasing?	Years
Ozone depletion	Especially Southern Ocean at present	??	??	—	Increasing	Decades?
Global warming	All habitat	Many species?	??	—	Increasing	Decades?

[a]Approximate time scale for duration of threat once anthropogenic causes of threat have been stopped

or have disjunct local populations, will continue to lose ground.

There is little chance for a significant reduction in anthropogenic ocean noise. In fact, the continuing high rate of economic and technical development suggests that noise levels will increase. However, noise pollution has one major advantage over most other indirect threats to cetaceans: it

can be abated almost instantaneously simply by reducing the source level.

We are very uncertain about the impacts of several of the threats (listed at the bottom of table 12.2): competition with fisheries, chemical pollution, plastic pollution, and global climate change. These threats are geographically widespread, increasing, and have considerable inertia—un-

HAL WHITEHEAD, RANDALL R. REEVES, AND PETER L. TYACK

like noise, their effects cannot suddenly be stopped if we change our behavior. For instance, there is no feasible way of controlling the dispersal into the ocean of the large stock of PCBs that has accumulated in dump sites and in the sediments of lakes, estuaries, and coastal zones (Reijnders 1996).

Despite a lack of clear knowledge of the effects of many of these threats, it is vital that they be minimized.

What Can Be Done?

The Role of the Scientist

Scientists have an important role to play in the conservation of cetaceans. They must help identify and call attention to conservation problems. Their expertise is critical for helping decision makers understand the nature and scale of the problems, the options available for addressing problems, and what the consequences of not addressing problems in a timely and appropriate manner will be (see, for example, Mangel et al. 1996).

Scientists have a responsibility to use their special training to advance knowledge. But they are also citizens who have opinions, make value judgments, and act on behalf of causes in which they believe. In this respect, they are like other people, although they usually have a better knowledge of certain subjects than most nonspecialists. As people who often spend the best years of their lives acquiring the needed analytical skills, raising the necessary funds, and carrying out expeditions to study the animals, scientists are themselves, in today's parlance, stakeholders in the conservation of cetaceans.

Even within their areas of specialization, scientists are much more than cogs in a machine that converts experiments or observations into rational management. The claim that science is, or even should be, value-free and totally objective is fraudulent (Glass 1965). All scientists make choices about what to study and how to study it. When interpreting findings from research in nature, it is inevitable that the experiences, beliefs, and preferences of the researcher will affect his or her results. What science provides is a rigorous set of boundaries that, when respected, ensure openness and honesty (= transparency) in what is said and written (Glass 1965), allowing other scientists, managers, and the general public to make informed use of the results. In turn, these other scientists, managers, and the general public have an important role in guiding the scientific process. They share the responsibility of identifying the types of research that will be most useful and ensuring that research results are used appropriately. In the next section, we use the information summarized earlier in this chapter to suggest how science can facilitate cetacean conservation and protection.

How Can Science Best Contribute to the Conservation and Protection of Cetaceans?

Long-term monitoring. For less accessible (especially offshore) populations, standard surveys every five years or so might indicate major trends in population size. The work of the Inter-American Tropical Tuna Commission and the National Marine Fisheries Service in the eastern tropical Pacific is a good example of long-term monitoring of trends in offshore populations (e.g., Buckland and Anganuzzi 1988).

Acoustic eavesdropping using hydrophone arrays is potentially a very efficient way to monitor trends in abundance and changes in distribution of many cetaceans, whose vocalizations are frequent, allow identification to species, and are audible over very long ranges (Tyack, chap. 11, this volume). Some use is already being made of military hydrophone networks in this way (Clark 1995).

For accessible populations, long-term studies of identified individuals provide valuable information (Mann, chap. 2, this volume). Even if the population seems healthy and relatively unthreatened (as with the killer whales around Vancouver Island), the insights gained (e.g., on natural rates of mortality, social organization, or temporal variability in population processes, such as reproductive parameters) are of great benefit when assessing threats to other populations and individuals (Thompson and Mayer 1996).

Long-term monitoring of trends in levels of pollutants and other threats (such as noise), both in animals and in their environment, can warn of potential dangers and provide a basis for assessing progress in abatement (O'Shea et al. 1999; see Box 12.1).

Basic research on little-known cetaceans. Scientists must develop a basic knowledge of the social and population structure of those cetaceans that are little known at present. These would include all species of beaked whales, some riverine and inshore dolphins and porpoises, and many pelagic dolphins (see Tyack et al., Epilogue, this volume).

Strengthening the scientific basis for cetacean conservation. The population biology behind the principle of sustainable use of cetacean populations has been carefully examined by the Scientific Committee of the International Whaling Commission. In the last few years, application of the precautionary principle to cetacean populations has also begun to receive scientific scrutiny (e.g., Taylor and Gerrodette

1993). Extensive efforts, overseen by the International Whaling Commission's Technical Committee, have been made to account for the socioeconomic impacts of whaling regulations on whaling communities and to ensure that the killing of whales is carried out humanely. The work of Clark (1975) and others has addressed the market dynamics of commercial whaling. Comparatively little attention has been given to other areas of concern. For example, many campaigns against catching cetaceans in fishing gear are accompanied by gruesome images suggesting implicitly that this by-catch involves terrible pain and suffering. If this is important for decision making, we need to know how, when, and how much cetaceans do suffer pain when caught in fishing gear, or when living in tanks, or swimming near underwater blasts. Some organizations and governments assign special rights and protections to certain animals (cetaceans are a prime example) based upon implicit notions that these animals are more advanced than others. Can we objectively compare the intelligence, self-awareness, ability to suffer, social organization, or culture of cetaceans to those of terrestrial mammals, including humans, and use this information to assign rights to members of different species? Scientists should examine these exceedingly challenging issues with open minds.

Choice of research methods. The rationale for, and alternatives to, research methods that disturb whales should be carefully considered. When two or more alternative methods can be used to obtain the data needed to answer a particular question, the less intrusive should be preferred. For instance, collection of sloughed skin can replace biopsy darting as a source of tissue for some genetic analyses of some species (Amos et al. 1992). Particularly intrusive methods, which may cause suffering or death to animals, should not be used unless the potential gains to other animals are very large and no alternative methods are available.

Study of human-cetacean interactions. Studies of cetaceans should be integrated with studies of humans. The human side of a human-cetacean interaction is often most amenable to study (Thompson and Mayer 1996), and virtually all solutions to conservation problems require some kind of change in human behavior (Mangel et al. 1996). Human behavior is usually resistant to change. However, with knowledge of how humans act and what is important to them, conservation actions can be tailored so as to least affect, and be most consistent with, pre-existent human behavior. Such conservation actions will almost always be more effective than those planned without regard for the people involved. To do this well, we need good bridges

between the animal- (or population-) centered views of cetologists and the human-centered perspectives of anthropologists and sociologists.

Communicating with the public. Scientists should convey the results of their research, and other concerns, to the public. One of Mangel et al.'s (1996) principles for the conservation of wild living resources is that effective conservation requires communication that is interactive, reciprocal, and continuous. Almost invariably, governments do not act to protect animals against threats, especially those from powerful organizations, unless there is public pressure to do so. Scientists should not only ensure that their findings on the effects of particular threats (e.g., how gray whales respond to the sounds of oil exploration) are communicated to the public, but they should also convey what they have learned about the lives of the animals themselves. If the public understands something about the social organization of a community of bottlenose dolphins (Connor et al., chap. 5, this volume), they can much better appreciate the potential effects of a proposed causeway that would divide their habitat. Some prominent cetologists, such as Roger Payne and the late Kenneth Norris, have been eloquent communicators about the lives of the animals that they have studied (see Norris 1974; Payne 1995). It is important that scientists, who know whales, dolphins, and porpoises better than anyone else, articulate why it is important to conserve them (Naess 1986; Ehrlich 1996).

Conservation, Protection, and Behavior

The applied field of conservation biology is seen as being most closely linked to the academic discipline of population biology. However, both the behavior and social structure of species are highly relevant to the types of problems faced by individual animals and populations, as well as to their solutions (Curio 1996; Strier 1997). This is particularly the case with cetaceans, as has been recognized for some time by the IWC (International Whaling Commission 1986). There is a behavioral component in all six of the areas listed above in which we think that science can most help cetacean conservation.

Management of Threats

Actions that will help cetaceans and many other forms of aquatic life include:

- Reduce pollution, including chemical contaminants, noise, and debris in aquatic habitats
- Reduce or eliminate gill netting and other unselective fishing methods

- Make sure that any exploitation of marine and riverine resources is biologically sustainable, without resort to artificial enhancement programs
- Reduce or eliminate the production of atmospheric pollutants that cause global climate change or ozone depletion
- Support, or if necessary create, international bodies responsible for managing human activities (see box 12.1) and conserving cetacean populations (e.g., the IWC) and marine habitat, and develop international structures to regulate more recently recognized threats, such as noise pollution

To achieve these and other conservation objectives, it is almost always necessary to change how we humans behave. Such change is best achieved through education, so that those who harm cetaceans or their habitat will voluntarily alter their ways, and understand the rationale for doing so.

When the threat is critical and immediate, however, or when the persons responsible for the threatening activities are either uneducable or unresponsive (as in commercial operations governed by immediate profit, such as whaling or shipping companies), then legislation backed by enforcement is vital. To be effective, such legislation must use available scientific information in a rational manner. For instance, most effort at regulating noise and disturbance has been directed at individual acts of acoustic harassment (Tyack 1989). This strategy is unlikely to provide adequate protection from the more diffuse, and possibly cumulative, threats of exposure to chronically elevated noise levels and habitat degradation. The sometimes conflicting goals of cetacean conservation and human access to the sea would both be served by research aimed at defining safe sound exposure levels and by the development of regulations designed to prevent or mitigate impacts on cetacean populations.

Sometimes neither education nor legislation will be effective. When very poor people directly or indirectly exploit an endangered population, satisfactory solutions may be hard to find (Perrin 1999).

An obvious way to protect cetaceans is by establishing and managing marine protected areas. The general trend is toward the establishment of marine protected areas in which natural ecological systems of some integrity can be protected (Agardy 1994). However, for cetaceans, such initiatives are of less immediate importance than affording protection from one or more outstanding threats over a substantial part of a population's range. This often requires that human activities be regulated in some way over areas larger than most ecologically based marine protected areas.

In some instances, cetaceans function as flagship, or umbrella, species, providing the impetus for protecting large water areas and thus conferring benefits on other species and processes that occur there. An example is the 1,600 km² Stellwagen Bank U.S. National Marine Sanctuary, established in 1993 largely for the benefit of humpback whales (Phillips 1996).

A number of protected areas have been specifically designed for cetaceans. Two of them, both created by the IWC—the Indian Ocean Sanctuary established in 1979 (Leatherwood and Donovan 1991) and the Southern Ocean Sanctuary established in 1994 (International Whaling Commission 1995a)—are very large and are intended to protect cetaceans from commercial whaling. Smaller protected areas are designed to address particular conservation problems. These include the 1,170 km² Banks Peninsula Marine Mammal Sanctuary in New Zealand, established in 1989 to protect endemic Hector's dolphins (Cephalorhynchus hectori) from by-catch in commercial and recreational gill net fisheries (Dawson and Slooten 1993).

Protected areas are of little worth unless they actually provide the intended protection. The Hector's dolphin sanctuary changed human practices by reducing fishing effort, and thus by-catch, almost certainly benefiting the animals (Donoghue 1996). However, this is an exceptional case. For almost all other marine protected areas, their establishment has caused little or no change in the level of threat faced by cetaceans using the area. Instead, their principal utility has been to forestall future adverse developments: commercial whaling in the case of the Southern Ocean Sanctuary, and a variety of threats, including ocean dumping of sewage and fishing for prey species, in or near the Stellwagen Bank U.S. National Marine Sanctuary.

Why We Should Care

As populations of whales, dolphins, and porpoises decline and disappear, as individuals suffer and die, humans lose. Our loss takes several forms, which vary in importance among different individuals and organizations. Each type of loss is valid and significant for substantial numbers of people. The reasons for conserving and protecting cetaceans include the following:

Preservation of biodiversity. Cetaceans are the only mammals that live their entire lives in the sea far from land. They are unusual in other ways, including diving behavior, intelligence, size, and social organization. Without the cetaceans, the Earth's faunal variety and richness would be diminished. Biodiversity has intrinsic value (Wilson 1984;

Soulé 1986). It also has direct benefits in the lives of humans (some of which are covered below).

Maintenance of ecological role. Cetaceans play significant roles in some marine ecosystems, and their absence would affect other species (Katona and Whitehead 1988; Bowen 1997). Most obvious are the parasites and commensals that live on or in cetaceans, but there are others, such as the benthic fauna that use cetacean carcasses (Smith et al. 1989).

Nutritional and economic value. Cetacean carcasses can provide food and other products that have economic as well as nutritional value. Alternatively, some cetaceans have considerable economic potential when alive: the value of whale-watching around the world was estimated to be about U.S.$500 million in 1994 (Hoyt 1995), roughly equivalent to 25,000 minke whale carcasses.

Potential scientific knowledge. We have much to learn from cetaceans, especially in areas in which they are unusual or particularly well adapted. These areas include anatomy, diving physiology, migration, social organization, cognition, acoustic communication, and echolocation.

Cultural and spiritual significance. For some humans, especially aboriginal people who continue to depend on wildlife for many of life's necessities, the hunting of cetaceans has cultural and spiritual significance. It is important to these people that such hunting be sustained. Conversely, many other humans are attracted to cetaceans as fellow cognitively and socially advanced mammals. They are stirred by the size of whales, the grace and beauty of dolphins, and the juxtaposition of familiar behavior such as suckling and vocal communication with long, deep dives into what is, for us, an alien world. Such people feel that their lives are enriched by the presence of free-living cetaceans, and that therefore the animals should not be harmed.

Animal welfare. Many human societies share the precept that animals should be spared unnecessary pain and suffering: that we should behave humanely toward them. Many cetaceans suffer pain from human activities. For example, death from entanglement in fishing nets, whether the animal was caught intentionally or unintentionally, is likely preceded by fear and suffering. Even if the individual survives the encounter with fishing gear, it may lose a limb or have to live with debilitating infection and pain.

Animal rights. The idea that nonhuman animals have rights is controversial in the West, but some Eastern and traditional cultures accept it without question. The animal rights movement has growing support in postindustrial society (Singer 1990). Clearly all animals cannot have the same rights, so some explicit or implicit system of priorities is required. Such systems are usually based on cognitive capability, ability to suffer pain, social complexity, or phylogenetic closeness to humans. Cetaceans rank high among animals in all such schemes (except phylogenetic proximity to humans), and there is a growing belief that whales, dolphins, and porpoises have a right to life independent of their current or future utility to humans (D'Amato and Chopra 1991; Stroud 1996). World Wildlife Fund-U.S., responding to the feelings of its membership, considers the large whales to be exempt from its general policy that encourages sustainable use of nonhuman animals (Fuller 1995, as quoted in Blichfeldt 1996).

Symbolic value. To many, though not all (see Kalland 1993), the whales and dolphins have become symbolically special (Barstow 1992), representing the successes and failures of the conservation movement. This is partly because of their size, cognitive abilities, and other special biological attributes, but also because of the magnitude of the holocaust they have experienced. As the great conservationist Peter Scott told the IWC in 1972 (Phillips 1996): "The feeling is now abroad that if we can't save the largest animals in the world we have little chance of saving the biosphere itself and therefore of saving our own species."

EPILOGUE

The Future of Behavioral Research on Cetaceans in the Wild

Peter L. Tyack, Richard C. Connor,
Janet Mann, and Hal Whitehead

THIS BOOK was initially developed with conscious reference to a book called *Primate Societies* (Smuts et al. 1987), which reviewed the status of primate field research. The editors of *Primate Societies* compared their volume with the first major collection of papers reviewing primate behavioral field studies, *Primate Behavior: Field Studies of Monkeys and Apes* (DeVore 1965). The editors noted that with the passing of two decades, the breadth and depth of studies had markedly increased from about a dozen species described in DeVore (1965) to over one hundred in Smuts et al. (1987). They also noted a corresponding increase in the depth of understanding of the best-studied species. In comparison to primates, our understanding of cetacean social structure is limited to fewer than a dozen cetacean species. However, this review has convinced us not only that cetaceans are prime candidates for studying a variety of important issues in behavioral biology, but also that a combination of new techniques and better exploitation of old opportunities offer tremendous potential for rapid growth in studying cetacean societies. Thus, in some ways, we stand close to the level of knowledge of primate societies in 1965, but we expect, as happened to primatology, that there will be a rapid acceleration toward an understanding of the social lives of whales and dolphins in the next twenty years.

Long-Term Studies of Identified Individual Cetaceans

Most field studies of animal behavior rely on familiarity with a population of identified individual animals that can be repeatedly sampled. It can take many years before this infrastructure is sufficiently established to enable many of the most interesting behavioral studies. One of the most striking features of cetaceans is the number of species and sites for which this infrastructure is already available. Hammond et al. (1990) list over a dozen such species with well-established long-term studies of identified individuals. Many of these studies initially focused on population biology because of its immediate relevance for protecting threatened and endangered species. Whatever their initial purpose, these long-term studies of identified individual cetaceans provide an extraordinary opportunity to jump-start behavioral field studies. Any of these studies that recorded which identified individuals were sighted together in the same group can be used directly to study the structure of social organization and the persistence of individual associations using methods like those described by Whitehead (1995a, 1997). In some of these populations, such as humpback whales in the North Atlantic and sperm whales in waters near the Galápagos Islands, genetic data from biopsy samples are linked to the individual identifications. These data dramatically increase the richness of behavioral studies and open the door for new studies, such as research on mating systems using paternity analysis or on kinship relations within social groups (e.g., Amos et al. 1993; Clapham and Palsbøll 1997; Richard et al. 1996a).

Methods for Behavioral Studies

Many studies of identified individual cetaceans maximize the number of individuals identified by using survey methods in which a research vessel or aircraft stays with a group only until all of the animals have been identified. We cannot overemphasize the value of following individuals or groups and recording behavior systematically when this is feasible and can be done in an unbiased manner. Such follows are extremely useful for understanding social behavior and social relationships, as is outlined in chapter 2 of this volume. The development and application of systematic quantitative methods of sampling behavior from cetaceans in the wild will raise the mark for behavioral studies and will rapidly provide—for appropriate species—a much more detailed view of social relationships than has previously been attained. Where possible, the kinds of methods applied by Samuels and Gifford (1997) to studying dominance relations in captive bottlenose dolphins need to be taken into the field to provide even more fine-grained views of social interaction and social relationships in wild cetaceans.

Many cetaceans live in circumstances in which such rigorous behavioral sampling is currently impossible, and even when it is possible, such methods cover only a small range of the relevant temporal and spatial scales. Most cetaceans cannot be followed and observed after they dive; it is usually difficult to identify which individual is vocalizing in the wild; and field biologists seldom can integrate information on vocal and visually observable behavior. As chapter 3 points out, cetologists have adapted and devised a set of methods for circumventing these limitations. Technological advances are opening new windows for studying cetaceans in the wild. With sophisticated tags and acoustic arrays, or with more flexible viewing platforms such as small remotely operated airships or underwater vehicles (Nowacek 1999), we will soon be able to follow and systematically observe animals and groups in circumstances that used to be noted as "start of dive," "very large group," or "dusk, return to shore." Sophisticated tags are now able to record visual, acoustic, and other data continuously from a particular individual throughout the dive cycle (Davis et al. 1992, 1999a, 1999b; Fletcher et al. 1996). In addition, new video and bioacoustic techniques are improving our abilities to follow individuals and to integrate visual and acoustic data on the individuals being followed.

Most cetacean field-workers have worked in study areas with a size that is limited more by the mobility of the researchers than of their subjects. Many cetaceans migrate thousands of kilometers, and there has always been a mismatch between the scale over which they range and the scale over which biologists could study them. It has been difficult to study their movement patterns over ranges beyond tens of kilometers and periods of months or more.

Tags that transmit data to satellites have started to remedy this gap (e.g., Mate 1989; Watkins et al. 1996). However, current satellite tags have limited longevity of attachment (a month or so), and data telemetry is very limited. New satellite systems with more frequent and better location data and the ability to telemeter more information will be available soon, but improved methods for long-term attachment to cetaceans are needed to minimize any detrimental effect on the animal while increasing the longevity of the tag attachment. Another promising development for large-scale observations involves the use of bottom-mounted hydrophone arrays. Originally developed by the U.S. Navy to track submarines, bottom-mounted arrays have the potential to expand our ability to locate vocalizing whales at ranges of hundreds of kilometers and to study the scale over which whales communicate and coordinate movements and activities (Costa 1993; McDonald et al. 1995; Stafford et al. 1998).

Cetacean Systematics

The past decade has seen tremendous development in comparative methods for testing hypotheses about the evolution of characters related to social evolution (e.g., sexual size dimorphism in primates: Mitani et al. 1996). These methods would be extremely useful for examining similar questions in cetacean biology, but they require a robust phylogeny (Harvey and Pagel 1991). Unfortunately, uncertainties abound at most levels in cetacean systematics. We are even uncertain as to whether the studies described in chapter 4 on the bottlenose dolphin are actually comparisons of one species, several species, or even members of the same genus. These problems are likely to be resolved by the recent developments in molecular and morphological studies of cetacean systematics. Better phylogenies will allow cetacean biologists to participate more fully in modern comparative methods to study the evolution of characters, including those related to social behavior and communication.

Critical Issues for Further Study

Spatial Scale

One must be particularly conscious of issues of spatial scale when studying social interactions between cetaceans. Cetaceans live their lives, and especially their social lives, over a wider range of spatial scales than most terrestrial mammals. Many cetaceans routinely swim 100 km per day, and many whales have seasonal migrations of many thousands of kilometers. Interactions involving the potential for physical contact may differ from those that are more remote; this suggests a focus on close-range interactions. On the other hand, whales may be able to hear one another's vocalizations at ranges of tens or even hundreds of kilometers. Whales may be able to interact acoustically and coordinate movements over ranges much larger than biologists have normally considered (Connor, chap. 8; Tyack, chap. 11, this volume). The maximum range over which whales conduct different kinds of interactions is an important question needing more research. Thus, scale is of primary importance when examining cetacean social structure, with temporal scales of association from minutes to decades, and spatial scales of interaction from centimeters to tens or even hundreds of kilometers.

How Do Ecological Factors Affect Social Behavior in Cetaceans?

When the ancestors of modern cetaceans entered the marine environment, they were confronted with a new set of constraints and opportunities. A promising area for comparative research involves analyzing how features of the marine environment might have affected the evolution of cetacean social systems. Mammals on terra firma live on a complex substrate on which they can find resources, territory that can be defended, and refuges for protection, particularly for the young. By contrast, cetaceans live in a fluid environment with no refuges and in which most resources are highly patchy in space and time and difficult to defend (Connor, chap. 8, this volume). When a terrestrial mammal stops moving, it stays put. If a cetacean were to stop swimming, it would be carried by currents. The weight of a cetacean is supported by the water, and the costs of locomotion appear to be significantly lower for cetaceans than for most terrestrial mammals (Connor, chap. 8, this volume). Sound carries so well underwater that cetaceans can also communicate over longer ranges than most terrestrial mammals, often more than 10 km. There are opportunities for more research exploring how these factors may influence the evolution of mammalian social systems in marine versus terrestrial environments. For example, the definitions of lek-like mating systems have primarily derived from studies of terrestrial animals. Imagine how all of the factors just listed would influence the evolution of lek-like mating systems for cetaceans, as is discussed for humpback whales in chapter 7.

A basic problem for socioecology concerns the relationship between the grouping patterns of a species and the distribution of prey and predators. There are hints of fascinating socioecological interactions in cetaceans. For example, Whitehead (1996a; Whitehead and Weilgart, chap. 6, this volume) suggests that adult female sperm whales within a family group may desynchronize their foraging dives when young calves are in the group so that an adult remains at the surface near calves that cannot dive deep. This strategy may provide protection against predators. Some studies of foraging in cetaceans suggest that they may coordinate or cooperate in order to improve foraging success, but more tightly focused studies are required to define the mechanisms and outcomes of social foraging (Connor, chap. 8, this volume). For example, what are the sensory mechanisms by which animals find their prey? If animals use sound to find prey, does this broadcast the location of prey to conspecifics?

We need studies that actually map prey distributions as well as prey behavior during cetacean feeding. For example, do groups of cetaceans actually concentrate or corral prey? More detailed behavioral observations are also needed on cetacean defense strategies as well as the distribution and movement of cetacean predators. Many of the same advances in technology that allow us to track cetaceans should also allow us to track the movements of their predators and prey. Studying the distribution of cetacean prey has been notoriously difficult, but new and powerful methods are becoming available. Fisheries biologists and biological oceanographers have developed powerful tools to map the distribution of animals that are important prey for cetaceans. High-frequency active sonars can map prey in three dimensions (chap. 7 in Clay and Medwin 1977; Greene et al. 1994; Wiebe 1995). These sonars are particularly useful when combined with simultaneous net or video sampling to verify the species composition of the sonar targets (Benfield et al. 1996). Marine mammal biologists are beginning to use these instruments as part of an integrated approach to study foraging ecology in enough detail to be useful for socioecology (Croll et al. 1998).

Parental Care

All cetaceans have a strong bond between mother and young. Cetacean females give birth to single offspring; with interbirth intervals of several years, a female is unlikely to have more than a dozen young during her lifetime. While adult cetaceans generally have low mortality rates, their infants (even of species with presumably low predation risk, such as killer whales) have a much higher mortality. It follows that the fitness of a female will be largely determined by her success in protecting her offspring through the dangerous early months of life.

While they are the largest of all cetaceans, baleen whales have the shortest duration of parental care, usually less than one year. Mysticete mothers undergo an enormous energetic demand during pregnancy and lactation as they grow a calf large enough to become independent after its first or second seasonal migration. On the other hand, delphinids and the larger toothed whales have much slower growth and development, and mothers may provide parental care for a decade or more (Whitehead and Mann, chap. 9, this volume).

There is a fascinating contrast in cetacean young between their prolonged dependence and their extremely precocial locomotor and sensory abilities. Cetaceans must be able to swim and surface to breathe at birth, so cetacean infants exhibit a degree of locomotor precociality resembling that of "following" ungulates. Even though many infant terrestrial mammals can signal at an early age, they cannot move far, and are dependent on their caregiver to approach them. Cetacean infants, on the other hand, can leave the mother to explore conspecifics, prey, or objects, and have more control over the maintenance of proximity to the mother. The sensory abilities of cetaceans appear to be well developed at birth, as might be expected for infants that must orient as they swim and must be able to reunite with their caregivers. More research is needed to understand the functions of the prolonged period of dependency in cetacean young, particularly since they are so precocious in locomotor and sensory skills.

Philopatry and Dispersal in Cetaceans

Neither male nor female *resident* killer whales disperse from their natal pod, a pattern that may extend to species in the Globicephalinae, such as the long-finned pilot whales (Amos et al. 1993; Baird, chap. 5, this volume). More generally, a large number of cetaceans, including baleen whales, are not highly territorial and have extensive ranges. The high degree of mobility in cetaceans gives them the opportunity to maintain social bonds with their mother or other members of their natal group even after dispersal (Connor, chap. 8, this volume). Studies that provide ecological and social explanations for patterns of philopatry and dispersal in cetaceans will find a large audience outside of cetology.

Individual-Specific Social Relationships and Vocal Learning

As is true for many primates, individual-specific social relationships play an important role in the social organization of some, and probably most, cetacean species. This is made particularly obvious by the strong social bonds between individuals within the highly fluid groups of species with fission-fusion societies, such as bottlenose dolphins (Connor et al., chap. 4, this volume). Bottlenose dolphins appear to maintain these individual-specific relationships by keeping in contact using individually distinctive signature whistles. Dolphins use vocal learning in the development of signature whistles and show remarkable skills of vocal imitation (Tyack and Sayigh 1997). Terrestrial mammals are more likely to use visual cues, such as face recognition, or voice cues that result from slight differences in the vocal tract. Visual cues have limited range underwater, and voice cues are unlikely to be useful for diving mammals, in which increasing pressure will modify the shape of air-filled vocal tracts. The need for individually distinctive vocal signals and the problems of voice cues in diving animals may have been factors in the evolution of vocal learning in cetaceans (Janik and Slater 1997; Tyack, chap. 11, this volume). There is little evidence for this remarkable ability in nonhuman terrestrial mammals, but it has evolved in a variety of marine mammal species. Further research on the role of vocal learning in acoustic communication by cetaceans will shed light on the current utility and evolutionary origin of this important cognitive ability. More research is also needed on the mechanisms by which cetaceans produce sounds and whether auditory feedback is required to maintain a stable signal as an animal dives.

Comparative Studies of Brain Evolution

Many odontocete cetacean species invest heavily in brains. The sperm whale has the largest brain on the planet (Whitehead and Weilgart, chap. 6, this volume), and some odontocetes, such as bottlenose dolphins, are exceeded only by humans in relative brain size (Ridgway and Brownson 1984). These facts have stimulated wild speculations in the popular literature, but most serious studies of the evolution of large brains are limited to primates, and few consider other taxa such as cetaceans. Even behavioral biologists with interests limited to the evolution of large brains in humans should be interested in "outgroups" that have independently evolved large brains (Marino 1998). Brain tissue is metabolically expensive, taking up to 20% of basal metabolic rate in large-brained animals (Sokoloff 1977). This observation raises questions of whether large brains may be "cheaper" for delphinid odontocetes than for terrestrial mammals or, alternatively, raises questions regarding the benefits that pay for their upkeep. Further work on the energy budgets, diet, and foraging ecology of cetaceans is required to address the cost side of this equation.

Many models of the evolution of large brains in primates and other mammals implicate factors suggesting that cetaceans would be particularly unlikely candidates to evolve large brains. For example, many such models implicate tool use and fine motor skills involving the hands (Wilson 1998). Cetaceans do not have hands and build much less sophisticated tools than many insects, birds, and mammals. However, over the past few decades, some primatologists have suggested that large brains evolved in primates because they improved the ability of individuals to model their social environment (Jolly 1966; Humphrey 1976; Kummer 1982; Dunbar 1992). Cetaceans are particularly interesting subjects for comparative studies on the social functions of intelligence because they represent another taxon in which large brains are associated with complex individual-specific social relationships. Many of these relationships, such as the alliances between adult male dolphins (Connor et al., chap. 4, this volume), include the tension between cooperation and competition identified by Humphrey (1976) as so critical for the evolution of social intelligence.

Critical Species for Further Study

There is every reason for studies on the best-known cetaceans to continue apace, but the review represented in this volume has highlighted some critical data gaps in species that urgently need further study. We present several suggestions below.

Platanistid River Dolphins

Several of the platanistid river dolphins are endemic to disturbed riverine habitats, and all are threatened. Described in sketchy reports as being relatively solitary and without complex social organization, the platanistid river dolphins have morphological features similar to those of some early fossil cetaceans and have brains one-half to one-third the size of a delphinid of similar body size (Ridgway 1986; Connor et al. 1992a). Comparisons of platanistids to other odontocetes provide an outstanding opportunity to examine the social and ecological correlates of the evolution of large brains in cetaceans.

Cephalorhynchus and Phocoenid Porpoises

There is a fascinating case of apparent convergence in morphology, vocalizations, and ecological niche between a genus in the Delphinidae, called *Cephalorhynchus,* and the phocoenid porpoises (Watkins et al. 1977; Dudok van Heel 1981). For example, while most delphinids (Tyack, chap. 11, this volume) and river dolphins (Wang et al. 1995) use frequency-modulated whistles in social communication, neither dolphins of the genus *Cephalorhynchus* nor phocoenid porpoises are known to produce whistles (Tyack, chap. 11, this volume). These last two non-whistling groups are reported to use sequences of clicks for social

communication (Amundin 1991; Dawson 1991). The clicks they use for echolocation also are similar, being higher in frequency and narrower in bandwidth than those of most delphinids. Detailed studies of identified individual animals are urgently needed to define the patterns of social behavior, acoustic communication, echolocation, foraging, and social organization in these taxa.

Beaked Whales

The beaked whales (Ziphiidae) are the least known of the cetaceans. The twenty or so species in this family are all inhabitants of the deep ocean, but they vary considerably. Some species have highly unusual anatomical structures, such as the teeth of the male strap-toothed whale *(Mesoplodon layardii),* which grow over the upper jaw and prevent the jaw from opening more than a centimeter or so. There are also considerable differences in sexual dimorphism. In the bottlenose whales *(Hyperoodon),* the males are considerably larger than females and have unusually developed cranial structures, while in Baird's and Arnoux's beaked whales *(Berardius)* it is the females that are larger (Mead 1989b; Balcomb 1989). Studies of living beaked whales are in their infancy, although it is possible to photographically identify individuals of at least some species (e.g., Whitehead et al. 1997c for the northern bottlenose whale). Baird's beaked whale appears particularly interesting because of the suggestion, based on demographic studies, that there is the potential for extended male care of the young (Kasuya and Brownell 1989; Kasuya et al. 1997; Connor et al., chap. 10, this volume).

Killer Whales and Globicephalinae

In addition to their unusual patterns of philopatry (see above), members of the Globicephalinae exhibit remarkable patterns of life history and parental investment. Short-finned pilot whales *(Globicephala macrorhynchus)* may suckle their young for up to fifteen years, and up to 25% of adult females may be postreproductive, ceasing to ovulate past the age of forty even though the maximum life span exceeds sixty years (Kasuya and Marsh 1984). Short-finned pilot whales and killer whales may be the only nonhuman species in which females are known to undergo menopause (Whitehead and Mann, chap. 9, this volume). Intriguingly, in the closely related long-finned pilot whale *(G. melas),* ovulation continues throughout life, even though pregnancy is rare past the age of forty (Martin and Rothery 1993). Longitudinal studies of identified individuals are needed to track the social organization of schools with such extended periods of parental care and to investigate the role of postreproductive females in killer whales and in the Globicephalinae.

Critical Habitats for Further Study

Even among the best-studied cetacean species, there are several reasons to develop new field studies in specific habitats. For example, killer whales have been well studied in temperate inshore coastal waters, but the overwhelming majority of killer whales are pelagic, with many living in the polar regions (Baird, chap. 5, this volume). While these animals are easiest to study in coastal waters, offshore and polar sites must also be studied if we are to obtain a representative view of killer whales in the primary habitats to which this species is adapted.

Testing hypotheses about the ecological functions of social behavior often requires careful selection of study sites in habitats with particular features. For example, most dolphin species living in open pelagic habitats are sighted in larger groups than those living in coastal areas (Norris and Dohl 1980b). Is there a greater need for social defense from predation in habitats with no refuge? Answering this question requires further study of offshore populations. Even coastal populations of bottlenose dolphins may experience different levels of predation (e.g., Moray Firth versus Sarasota: Connor et al., chap. 4, this volume), the examination of which will greatly facilitate testing of the hypothesis relating group size to social defense from predation.

Conservation and Research

Chapter 12 highlights the major conservation issues for cetaceans. Here we would like to point out the very close connections between conservation and cetacean behavioral research. Some of the most endangered cetaceans are river dolphins. Little is known of their social behavior, yet they fill a critical phylogenetic space and ecological niche. Our basic understanding of the evolution of cetacean behavior will be permanently impoverished if any of these species goes extinct. It is hard to imagine a more dramatic example of how conservation can affect comparative studies of cetacean behavior.

Studies of behavior can also address important conservation problems by monitoring the health of populations and illuminating unrecognized risks to cetaceans. Many of the most important threats to cetaceans today involve habitat degradation. These threats may have subtle initial effects and may be slow to develop, but they can be much more difficult to reverse than a directed hunt. This makes it all the more important to monitor the health of wild populations more closely. For example, some chemical contaminants can disrupt endocrine functions; one way their effects can be monitored is by studying reproductive behavior. Scientists who have studied acoustic communication in cetaceans have slowly come to understand how critical the acoustic environment is for these animals. This research, which was initially motivated as pure research, has highlighted the need to better understand the potential impacts of risks such as noise pollution in the ocean (Richardson et al. 1995b).

The Social Cetacean

Despite the difficulties of studying cetaceans, and the precarious conservation status of some species, we have learned something of the social lives of these animals. We have found richness, both in the diversity of cetacean social systems and in the social relationships of individual animals. For a cetacean, and perhaps especially for a large odontocete, conspecifics are much more than just competitors for resources, sources of milk for the young, and sources of mating opportunities for adults. The degrees of cooperation observed among female sperm whales (chap. 6), among male bottlenose dolphins (chap. 4), and among relatives of both sexes in killer whales (chap. 5) are remarkable. The social structures of some cetaceans fit models and classifications developed for socially advanced terrestrial mammals such as elephants and primates. Previously undocumented types of social organization, such as the philopatry of both sexes found in killer whales, may have evolved and may persist because of the very different constraints and opportunities afforded by the marine environment.

As we learn to look more deeply at cetaceans, new layers of complexity become apparent. Our appreciation of the complexity of a cetacean's social system is largely determined by the effort spent on studying it. The most detailed studies, those of bottlenose dolphins in Sarasota, Florida, and Shark Bay, Australia, are beginning to show the importance of individual-specific social relationships in cetacean societies. On the other hand, most species described as being relatively asocial, such as platanistid dolphins and porpoises, have not been subjects of intensive studies of social behavior in the wild.

In 1982 David Gaskin published an important book on the ecology of whales and dolphins. Toward the beginning of the chapter on social behavior, he posed two questions (Gaskin 1982: 115):

(1) Is there any real social structure in cetacean populations?
(2) Do cetaceans have highly developed social behavior?

After reviewing what was then known of the social structure of cetaceans, he concluded, "Statements that assume the existence of a high order of social evolution in the Cetacea are, frankly, not supported to any extent by the observations of free-ranging populations" (151). Gaskin's views were controversial (Reeves 1983), but not without some justification at that time, given the state of knowledge. Fifteen years later, the results summarized in this book justify a radically contrary conclusion: cetaceans are among the most socially diverse and complex orders of mammals. And research on cetacean societies is in its infancy compared with that on primates, carnivores, or ungulates. We are confident that as research progresses, cetacean societies will reveal unexpected and diverse forms of social behavior that will not only be fascinating in their own right, but will also provide important comparisons of general interest to behavioral biologists.

PETER L. TYACK, RICHARD C. CONNOR, JANET MANN, AND HAL WHITEHEAD

APPENDIX 1

Cetacean Phylogeny and Evolution

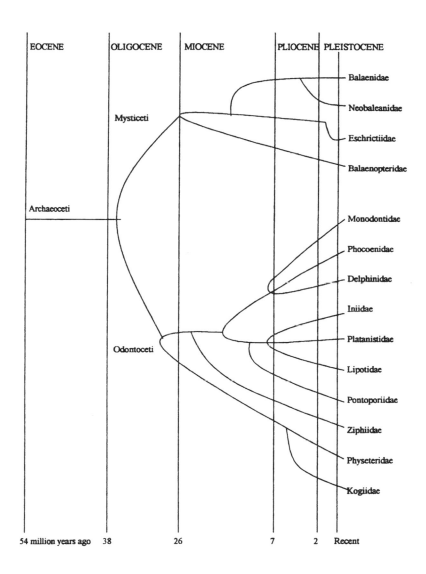

APPENDIX 2

Cetacean Taxonomy

Species name	Common name	Other names	Distribution	Source
CETACEA				
Mysticeti				
Balaenidae				
Balaena australis[a]	Southern right whale	*B. antarctica* *B. antipodarum* *B. temminckii*	Southern Hemisphere: antarctic to temperate waters	Desmoulins 1822 Rice 1998
Balaena glacialis[a]	Northern right whale, black right whale, nordcaper	*B. biscayensis* *B. japonica* *B. nordcaper* *B. seiboldii* *B. cisarctica*	Northern Hemisphere: temperate to tropical waters	Müller 1776 Rice (1998) suggests collapsing *B. australis* and *B. glacialis* and just calling them *B. glacialis*
Balaena mysticetus	Bowhead or Greenland right whale; arctic right whale; ahvik		Northern Hemisphere: arctic waters	Linnaeus 1758
Neobalaenidae				
Caperea marginata	Pygmy right whale	*C. antipodarum*	Southern Hemisphere: cold temperate waters	Gray 1870
Eschrichtiidae				
Eschrichtius robustus	Gray whale	*E. gibbosus* *E. glaucus*	North Pacific: warm temperate to arctic waters	Lilljeborg 1861
Balaenopteridae				
Balaenoptera acutorostrata	Minke whale, lesser rorqual, piked whale Dwarf minke whale	*B. bonaerensis* *B. davidsoni* *B. huttoni* *B. minimus* *B. rostrata*	Worldwide: arctic to tropical waters	Lacépède 1804
Balaenoptera bonaerensis	Antarctic minke whale	—	Southern Hemisphere	Burmeister 1865 Rice (1998) separates this species but other sources include it with *B. acutorostrata*
Balaenoptera borealis	Sei whale	*B. rostrata* *B. schlegellii*	Worldwide: cold temperate to tropical waters	Lesson 1828
Balaenoptera edeni	Bryde's whale, Eden's whale, Sittang whale	*B. brydei*	Worldwide: warm temperate to tropical waters	Anderson 1879
Balaenoptera brydei	—	—	—	Olsen 1913 Rice (1998) separates this species but other sources include it with *B. edeni*
Balaenoptera musculus	Blue whale	*B. brevicauda* *B. gigas* *B. indica* *B. intermedia* *B. major* *B. sibbaldii* *B. sibbaldius* *B. sulfureus*	Worldwide: arctic to tropical waters	Linnaeus 1758

Species name	Common name	Other names	Distribution	Source
Balaenopteridae				
Balaenoptera physalus	Fin whale, finback whale, common rorqual	*B. antiquorum* *B. boops* *B. gibbar* *B. patachonica* *B. velifera*	Worldwide: arctic to tropical waters	Linnaeus 1758
Megaptera novaeangliae	Humpback whale	*M. braziliensis* *M. burmeisteri* *M. lalandii* *M. longimana* *M. longipinna* *M. nodosa* *M. versabilis*	Worldwide: cold temperate to tropical waters	Borowski 1781
Odontoceti				
Physeteroidea				
Physeteridae				
Physeter macrocephalus	Sperm whale, cachalot	*P. australasianus* *P. australis* *P. catodon*	Worldwide: antarctic, cold temperate waters, tropical waters	Linnaeus 1758 (Mead and Brownell 1993 call this species *P. catodon* and group in family with *Kogia*)
Kogiidae				
Kogia breviceps	Pygmy sperm whale, lesser cachalot	*K. floweri* *K. goodei* *K. grayii*	Worldwide: temperate to tropical waters	de Blainville 1838
Kogia sima	Dwarf sperm whale		Worldwide: warm temperate to tropical waters	Owen 1866
Ziphioidea				
Ziphiidae				
Berardius arnuxii	Arnoux's beaked whale		Southern Hemisphere: circumpolar, temperate waters	Duvernoy 1851
Berardius bairdii	Baird's beaked whale	*B. vegae*	North Pacific: temperate waters	Stejneger 1883
Hyperoodon ampullatus	Northern bottlenose whale	*H. butskoph* *H. latifrons* *H. rostratus*	North Atlantic: arctic to cold temperate waters	Forster 1770
Hyperoodon planifrons	Southern bottlenose whale, flatheaded bottlenose whale	*H. burmeisterei*	Southern Hemisphere: circumpolar, antarctic to temperate waters	Flower 1882
Indopacetus pacificus	Indo-Pacific beaked whale, Longman's beaked whale	*M. pacificus*	Indian Ocean and W South Pacific: tropical waters	Longman 1926
Mesoplodon bidens	Sowerby's beaked whale, North Sea beaked whale	*M. dalei* *M. micropterus* *M. sowerbyensis* *M. sowerbyi*	North Atlantic and Baltic Sea: temperate waters	Sowerby 1804
Mesoplodon bowdoini	Andrews' beaked whale, deepcrest beaked whale		Southern Hemisphere, South Pacific and Indian Oceans: cold temperate waters	Andrews 1908
Mesoplodon bahamondi	Bahamonde's beaked whale	—	—	Reyes et al. 1996 Rice (1998) separates this species, but other sources do not recognize it
Mesoplodon carlhubbsi	Hubbs' beaked whale, arch beaked whale		North Pacific: temperate waters	Moore 1963
Mesoplodon densirostris	Blainville's beaked whale, densebeak whale	*M. sechellensis*	Worldwide: temperate to tropical waters	de Blainville 1817
Mesoplodon europaeus	Gervais' beaked whale, Antillean beaked whale, Gulf Stream beaked whale	*M. gervaisi*	North Atlantic: temperate to tropical waters	Gervais 1855
Mesoplodon ginkgodens	Ginkgo-toothed beaked whale, Japanese beaked whale	*M. hotaula*	North Pacific and Indian Oceans: warm temperate to tropical waters	Nishiwaki and Kamiya 1958
Mesoplodon grayi	Gray's beaked whale, scamperdown whale, Haast's beaked whale, small-toothed beaked whale	*M. australis* *M. haasti*	Southern Hemisphere: cold temperate waters	von Haast 1876

Species name	Common name	Other names	Distribution	Source
Mesoplodon hectori	Hector's beaked whale	*M. knoxi*	Southern Hemisphere, North Pacific: temperate waters	Gray 1871
Mesoplodon layardii	Strap-toothed beaked whale, Layard's beaked whale, long-toothed beaked whale	*M. floweri* *M. guntheri* *M. longirostris* *M. thomsoni* *M. traversii*	Southern Hemisphere: temperate waters	Gray 1865
Mesoplodon mirus	True's beaked whale		North Atlantic, South Atlantic coast of South Africa, Australia: temperate waters	True 1913
Mesoplodon peruvianus	Pygmy beaked whale, Peruvian beaked whale, lesser beaked whale		E South and E North Pacific: cold temperate to tropical waters	Reyes, Mead, and Van Waerebeek 1991
Mesoplodon stejnegeri	Stejneger's beaked whale, Bering Sea beaked whale, saber-toothed whale		North Pacific: cold temperate waters	True 1885
Tasmacetus shepherdi	Shepherd's beaked whale, Tasman beaked whale		Southern Hemisphere: cold temperate waters (N.Z.)	Oliver 1937
Ziphius cavirostris	Cuvier's beaked whale, goosebeaked whale	*Z. australis* *Z. capensis* *Z. chathamensis* *Z. indicus*	Worldwide: cold temperate to tropical waters	Cuvier 1823

Platanistoidea
Platanistidae

Species name	Common name	Other names	Distribution	Source
Platanista gangetica	Ganges river dolphin, Ganges susu, bhulan, Indian river dolphin, blind river dolphin		India, Nepal, Bhutan, and Bangladesh: freshwater only	Roxburgh 1801 Rice (1998) does not separate *P. gangetica* from *P. minor* but considers them subspecies: *P. g. gangetica, P. g. minor*
Platanista minor	Indus river dolphin, Indus susu, bhulan	*P. indi*	Pakistan: freshwater only	Owen 1853

Iniidae

Species name	Common name	Other names	Distribution	Source
Inia geoffrensis	Boto, bouto, Amazon river dolphin	*I. boliviensis*	Freshwater only: Peru, Ecuador, Brazil, Bolivia, Venezuela, Columbia	de Blainville 1817

Lipotidae

Species name	Common name	Other names	Distribution	Source
Lipotes vexillifer	Baiji, Chinese river dolphin, Yangtze river dolphin, whitefin dolphin, pei c'hi, whiteflag dolphin		China: freshwater only	Miller 1918

Pontoporiidae

Species name	Common name	Other names	Distribution	Source
Pontoporia blainvillei	La Plata dolphin, franciscana	*P. tenuirostris*	Brazil to Argentina: coastal waters from Doce River	Gervais and d'Orbigny 1844

Delphinoidea
Monodontidae

Species name	Common name	Other names	Distribution	Source
Delphinapterus leucas	Beluga, white whale, belukha	*D. albicans* *D. beluga* *D. catodon* *D. dorofeevi* *D. marisalbi*	Circumpolar in arctic seas: arctic to cold temperate waters	Pallas 1776
Monodon monoceros	Narwhal	*M. microcephalus* *M. monodon* *M. narhval* *M. vulgaris*	Arctic Ocean	Linnaeus 1758

Delphinidae

Species name	Common name	Other names	Distribution	Source
Cephalorhynchus commersonii	Commerson's dolphin, piebald dolphin, Jacobite	*C. floweri*	Argentina to Chile	Lacépède 1804
Cephalorhynchus eutropia	Black dolphin, white-bellied dolphin, Chilean dolphin, eutropia dolphin	*C. albiventris* *C. obtusata*	Chile	Gray 1846
Cephalorhynchus heavisidii	Heaviside's dolphin, hastate dolphin	*C. hastatus*	South Africa to perhaps S. Angola	Gray 1828

Species name	Common name	Other names	Distribution	Source
Delphinidae				
Cephalorhynchus hectori	Hector's dolphin, pied dolphin, whitefronted dolphin	*C. albifrons*	New Zealand: coastal waters	van Bénéden 1881
Delphinus delphis	Common dolphin, shortbeaked common dolphin, saddleback dolphin, offshore common dolphin, whitebelly dolphin	*D. bairdii* *D. capensis* *D. tropicalis*	Worldwide: temperate and tropical waters	Linnaeus 1758
Delphinus capensis	Longbeaked common dolphin, saddleback dolphin, neritic common dolphin, Cape dolphin, Baird's dolphin	—	—	Rice (1998) and LeDuc et al. (1999) consider this a separate species, while sources include it with *D. delphis*
Delphinus tropicalis	Arabian common dolphin, saddleback dolphin, Malabar common dolphin	—	—	Rice (1998) considers this a separate species. LeDuc et al. (1999) consider its status unresolved
Feresa attenuata	Pygmy killer whale, slender blackfish	*F. intermedius* *F. occulta*	Worldwide: tropical to warm temperate waters	Gray 1874
Globicephala macrorhynchus	Short-finned pilot whale, shortfinned blackfish	*G. brachypterus* *G. scammonii* *G. seiboldii*	Worldwide: tropical and temperate waters	Gray 1846
Globicephala melas	Long-finned pilot whale, longfinned blackfish, caa'ing whale	*G. edwardii* *G. globiceps* *G. leucosagmaphora* *G. svineval*	North Atlantic and southern Oceans: cold–temperate waters	Traill 1809
Grampus griseus	Risso's dolphin, grampus, whiteheaded grampus, gray grampus, mottled grampus	*G. rissoanus* *G. stearnsii*	Worldwide: temperate to tropical waters	Cuvier 1812
Lagenodelphis hosei	Fraser's dolphin, shortsnouted whitebelly dolphin, Hose's dolphin, Sarawak dolphin.		Worldwide: warm temperate to tropical waters	Fraser 1956
Lagenorhynchus acutus	Atlantic white-sided dolphin	*L. gubernator* *L. leucopleurus* *L. perspicillatus*	North Atlantic: cold temperate waters	Gray 1828
Lagenorhynchus albirostris	White-beaked dolphin	*L. pseudotursio*	North Atlantic: cold temperate waters	Gray 1846
Lagenorhynchus australis[b]	Peale's dolphin, blackchinned dolphin	*L. amblodon* *L. chilöensis*	Chile to Argentina	Peale 1848
Lagenorhynchus cruciger[b]	Hourglass dolphin	*L. albigena* *L. bivattus* *L. clanculus* *L. wilsoni*	Southern Hemisphere: antarctic and cold–temperate waters	Quoy and Gaimard 1824
Lagenorhynchus obliquidens[b]	Pacific white-sided dolphin, Pacific striped dolphin	*L. longidens* *L. ognevi*	North Pacific: cold temperate waters	Gill 1865
Lagenorhynchus obscurus[b]	Dusky dolphin	*L. breviceps* *L. fitzroyi* *L. similis* *L. superciliosus*	Southern Hemisphere: cold temperate continental waters	Gray 1828
Lissodelphis borealis	Northern right whale dolphin		North Pacific: cold temperate waters	Peale 1848
Lissodelphis peronii	Southern right whale dolphin	*L. leucorhamphus*	Southern Hemisphere: cold temperate waters	Lacépède 1804
Orcaella brevirostris	Irrawaddy dolphin, pesut	*O. fluminalis*	SE Asia, N Australia and Papua New Guinea: tropical coastal waters and large rivers	Owen 1866/1869 in Gray 1866
Orcinus orca	Killer whale, orca	*O. ater* *O. capensis* *O. gladiator* *O. rectipinna*	Worldwide	Linnaeus 1758
Peponocephala electra	Melon-headed whale, many-toothed blackfish, little blackfish, Electra dolphin	*P. asia* *P. fusiformis* *P. pectoralis*	Worldwide: tropical to warm temperate waters	Gray 1846

Species name	Common name	Other names	Distribution	Source
Pseudorca crassidens	False killer whale	*P. destructor* *P. meridionalis*	Worldwide: temperate to tropical waters	Owen 1846
Sotalia fluviatilis	Tucuxi, gray river dolphin	*S. guianensis* *S. pallida* *S. tucuxi*	W South Atlantic: coastal waters, estuaries, and rivers	Gervais and Deville 1853
Sousa chinensis	Indo-Pacific hump-backed dolphin, Chinese white dolphin, Bornean white dolphin	*S. borneensis* *S. lentiginosa* *S. plumbea* *S. zambezicus* *S. sinensis*	Indian Ocean: coastal waters and rivers	Osbeck 1765
Sousa plumbea	Indian humpback dolphin, plumbeous dolphin, speckled dolphin, freckled dolphin	—	—	Cuvier 1829 in Rice 1998 *Rice (1998) considers this a separate species while other sources refer to it as a synonym for S. chinensis*
Sousa teuszi	Atlantic hump-backed dolphin		E South Atlantic: coastal waters in river mouths	Kükenthal 1892
Stenella attenuata	Pantropical spotted dolphin or bridled dolphin	*S. albirostratus* *S. brevimanus* *S. capensis* *S. consimilis* *S. dubia* *S. graffmani* *S. malayanus* *S. pseudodelphis* *S. punctata* *S. velox*	Worldwide: temperate to tropical waters	Gray 1846
Stenella clymene	Clymene dolphin, Atlantic spinner dolphin or short-snouted spinner dolphin	*S. metis* *S. normalis*	Atlantic Ocean including Gulf of Mexico: warm temperate to tropical waters	Gray 1846
Stenella coeruleoalba	Striped dolphin, euphrosyne dolphin, blue-white dolphin, Meyen's dolphin	*S. asthenops* *S. crotaphiscus* *S. euphrosyne* *S. styx* *S. tethyos*	Worldwide: cold temperate to tropical waters	Meyen 1833
Stenella frontalis	Atlantic spotted dolphin, bridled dolphin	*S. doris* *S. froenatus* *S. plagiodon*	Atlantic Ocean, including the Gulf of Mexico	Cuvier 1829
Stenella longirostris	Spinner dolphin, long-snouted spinner dolphin	*S. alope* *S. centro-americana* *S. microps* *S. orientalis* *S. roseiventris*	Worldwide: warm temperate to tropical waters	Gray 1828
Steno bredanensis	Rough-toothed dolphin	*S. compressus* *S. frontatus* *S. perspicillatus* *S. rostratus*	Worldwide: warm temperate to tropical waters	G. Cuvier in Lesson 1828
Tursiops truncatus	Bottlenose dolphin, bottle-nosed dolphin	*T. aduncus* *T. gephyreus* *T. gillii* *T. nesarnack* *T. nuuanu*	Worldwide: temperate to tropical waters	Montagu 1821
Tursiops aduncus	Indian Ocean bottlenose dolphin, Indo-Pacific bottlenose dolphin, gadamu	*T. truncatus*	Indian and Pacific Ocean, coastal waters	Ehrenberg 1833
Phocoenidae				
Neophocaena phocaenoides	Finless porpoise, little Indian porpoise	*N. asiaeorientalis* *N. melas* *N. sunameri*	Indo-Pacific: warm temperate to tropical waters	Cuvier 1829
Phocoena dioptrica	Spectacled porpoise	*P. stornii*	Southern Hemisphere: cold temperate waters	Lahille 1912

Species name	Common name	Other names	Distribution	Source
Phocoenidae				
Phocoena phocoena	Harbor porpoise or common porpoise	*P. americana* *P. communis* *P. lineata* *P. relicta* *P. vomerina*	N Pacific and N Atlantic: arctic to cold temperate waters	Linnaeus 1758
Phocoena sinus	Vaquita, Gulf porpoise, cochito, pygmy porpoise		Gulf of California, warm temperate waters	Norris and McFarland 1958
Phocoena spinipinnis	Burmeister's porpoise, black porpoise	*P. philippii*	Southern Hemisphere: coastal temperate waters of South America	Burmeister 1865
Phocoenoides dalli	Dall's porpoise, True's porpoise, whitesided porpoise, whitefin porpoise	*P. truei*	North Pacific: cold temperate waters	True 1885

Sources: Species names (left column) are from Rice (1998). The other primary source used for "other" and "common" names was Mead and Brownell (1993). If there were discrepancies, we used Rice's version. The taxonomic status of many cetacean species remains uncertain. The status of several delphinic species is being revised (see below). We recommend Rice (1998) for a complete review of cetacean systematics and distribution.

[a]Rice (1998) uses *Balaena* rather than *Eubalaena* because genetic and morphological evidence favor grouping the right whales with the same genus as bowheads.

[b]LeDuc et al. (1999) have classified four species of *Lagenorhynchus* as *Sagmatias* because of closer genetic and morphological relationships to *Lissodelphis* and *Cephalorhynchus*. See cladogram below.

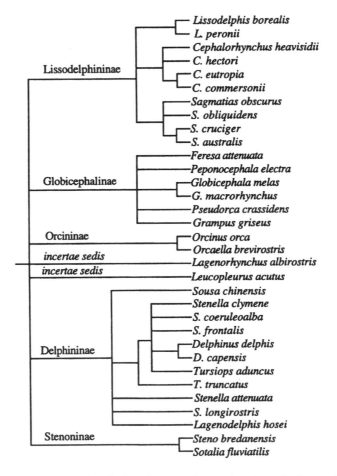

Systematic revision of delphinidae based on cytochrome *b* analyses (LeDuc et al. 1999).

REFERENCES

Abbott, D. 1984. Behavioral and physiological suppression of fertility in subordinate marmoset monkeys. *Am. J. Primatol.* 6:169–86.

Abernethy, R. B., Baker, C. S., and Cawthorn, M. W. 1992. Abundance and genetic identity of humpback whales *(Megaptera novaeangliae)* in the Southwest Pacific. Final report to the International Whaling Commission.

Abrahams, M. V., and Colgan, P. W. 1985. Risk of predation, hydrodynamic efficiency, and their influence on school structure. *Environ. Biol. Fish.* 13:195–202.

Acevedo-Gutiérrez, A., Brennan, B., Rodriques, P., and Thomas, M. 1997. Resightings and behavior of false killer whales *(Pseudorca crassidens)* in Costa Rica. *Mar. Mamm. Sci.* 13:307–14.

Agardy, M. T. 1994. Advances in marine conservation: The role of marine protected areas. *Trends Ecol. Evol.* 9:267–70.

Agler, B. A., Robertson, K. A., Dendanto, D., Katona, S. K., Allen, J. M., Frohock, S. E., Seipt, I. E., and Bowman, R. S. 1992. The use of photographic identification for studying individual fin whales *(Balaenoptera physalus)* in the Gulf of Maine. *Rep. Int. Whal. Commn.* 42:711–22.

Aguilar, A. 1983. Organochlorine pollution in sperm whales, *Physeter macrocephalus,* from the temperate waters of the eastern North Atlantic. *Mar. Pollut. Bull.* 9:349–52.

Aguilar, A., and Borrell, A. 1994. Reproductive transfer and variation of body load of organochlorine pollutants with age in fin whales *(Balaenoptera physalus)*. *Arch. Environ. Contam. Toxicol.* 27:546–54.

Aguilar, A., and Lockyer, C. H. 1987. Growth, physical maturity and mortality of fin whales, *Balaenoptera physalus,* inhabiting the temperate waters of the northeast Atlantic. *Can. J. Zool.* 65:253–64.

Aguilar, A., and Monzon, F. 1992. Interspecific variation in testes size in cetaceans: A clue to reproductive behaviour? In *European research on cetaceans: Proceedings of the sixth annual conference of the European Cetacean Society,* ed. P. G. H. Evans. Cambridge: European Cetacean Society.

Alcock, J. 1975. *Animal behavior: An evolutionary approach.* 1st edition. Sunderland, MA: Sinauer Associates.

———. 1993. *Animal behavior: An evolutionary approach.* 5th edition. Sunderland, MA: Sinauer Associates.

———. 1998. *Animal behavior: An evolutionary approach.* 6th edition. Sunderland, MA: Sinauer Associates.

Alexander, G. 1977. Role of auditory and visual cues in mutual recognition between ewes and lambs in merino sheep. *Appl. Anim. Ethol.* 3:65–81.

———. 1978. Odour, and the recognition of lambs by merino ewes. *Appl. Anim. Ethol.* 4:153–58.

Alexander, G., and Shillito, E. E. 1977. The importance of odour, appearance, and voice in maternal recognition of the young in the merino ewe. *Appl. Anim. Ethol.* 3:127–35.

Alexander, R. D. 1974. The evolution of social behavior. *Annu. Rev. Ecol. Syst.* 5:325–83.

———. 1979. *Darwinism and human affairs.* Seattle: University of Washington Press.

Allee, W. C. 1933. Review: *Gorillas in a native habitat. Ecology* 14:319–20.

Allen, A. R. 1984. Report of the sub-committee on Southern Hemisphere minke whales, appendix 14. Effect of the value of Z used in constructing "ideal" age length keys on the values of r_{II} calculated using the "ideal" keys. *Rep. Int. Whal. Commn.* 34:99–100.

Allen, G. M. 1916. The whalebone whales of New England. *Mem. Boston Soc. Nat. Hist.* 8:107–322.

Allen, J. A. 1874. Review: Scammon's *Marine Mammals of the Northwestern Coast and American Whale-Fishery. Am. Nat.* 8: 632–35.

Allen, J. F. 1977. Dolphin reproduction in an ocean area in Australia and Indonesia. In *Breeding dolphins: Present status, suggestions for the future,* ed. S. H. Ridgway and K. Benirschke. U.S. Marine Mammal Commission Report MMC-76/07, Washington, DC.

Allen, K. R. 1980. *Conservation and management of whales.* Seattle: University of Washington Press.

Altmann, J. 1974. Observational study of behavior: Sampling methods. *Behaviour* 49:227–67.

Altmann, S. A. 1991. Diets of yearling female primates *(Papio cynocephalus)* predict lifetime fitness. *Proc. Nat. Acad. Sci. USA* 88:420–23.

Altmann, S. A., and Altmann, J. 1970. *Baboon ecology.* Chicago: University of Chicago Press.

Amos, B. 1993. Use of molecular probes to analyse pilot whale pod structure: Two novel analytical approaches. *Symp. Zool. Soc. Lond.* 66:33–48.

———. 1996. Levels of variability in cetacean populations have probably changed little as a result of human activities. *Rep. Int. Whal. Commn.* 46:657–58.

Amos, B., and Hoelzel, A. R. 1991. Long-term preservation of whale skin for DNA analysis. In *Genetic ecology of whales and dolphins: Incorporating the proceedings of the workshop on the genetic analysis of cetacean populations,* ed. A. R. Hoelzel, 99–

103. Reports of the International Whaling Commission, special issue 13. Cambridge: International Whaling Commission.

Amos, B., Barrett, J., and Dover, G. A. 1991a. Breeding behaviour of pilot whales revealed by DNA fingerprinting. *Heredity* 67:49–55.

———. 1991b. Breeding system and social structure in the Faroese pilot whale as revealed by DNA fingerprinting. In *Genetic ecology of whales and dolphins: Incorporating the proceedings of the workshop on the genetic analysis of cetacean populations,* ed. A. R. Hoelzel, 255–68. Reports of the International Whaling Commission, special issue 13. Cambridge: International Whaling Commission.

Amos, B., Schloetterer, C., and Tautz, D. 1993. Social structure of pilot whales revealed by analytical DNA profiling. *Science* 260:670–72.

Amos, B., Twiss, S., Pomeroy, P., and Anderson, S. 1995. Evidence for mate fidelity in the gray seal. *Science* 268:1897–1900.

Amos, W., and Dover, G. A. 1990. DNA fingerprinting and the uniqueness of whales. *Mammal Rev.* 20:23–30.

Amos, W., and Hoelzel, A. R. 1990. DNA fingerprinting cetacean biopsy samples for individual identification. In *Individual recognition of cetaceans: Use of photo-identification and other techniques to estimate population parameters,* ed. P. S. Hammond, S. A. Mizroch, and G. P. Donovan, 79–85. Reports of the International Whaling Commission, special issue 12. Cambridge: International Whaling Commission.

Amos, W., Whitehead, H., Ferrari, M. J., Glockner-Ferrari, D. A., Payne, R., and Gordon, J. 1992. Restrictable DNA from sloughed cetacean skin: Its potential for use in population analysis. *Mar. Mamm. Sci.* 8:275–83.

Amundin, M. 1986. Breeding in bottle-nosed dolphin, *Tursiops truncatus,* at the Kalmarden Dolphinarium. *Int. Zoo Yrbk.* 24/25:263–71.

———. 1991. Click repetition rate patterns in communicative sounds from the harbour porpoise, *Phocoena phocoena.* Chapter in Sound production in odontocetes with emphasis on the harbour porpoise, *Phocoena phocoena.* Ph.D. dissertation, University of Stockholm.

Andersen, S., and Dziedzic, A. 1964. Behavior patterns of captive harbour porpoise *Phocaena phocaena* (L.). *Bulletin de l'Institut Oceanographique de Monaco* 63(1316):1–20.

Anderson, J. 1878. *Anatomical and zoological researches: Comprising an account of the zoological results of two expeditions to western Yunnan in 1868 and 1875; and a monograph of the two cetacean genera,* Platanista *and* Orcaella. London: Bernard Quaritch.

Anderson, S. 1969. Epimeletic behavior in a captive harbour porpoise, *Phocaena phocaena.* In *Investigations on Cetacea,* ed. G. Pilleri, vol. 1. Berne: Braine Anatomy Institute.

Andersson, M. 1994. *Sexual selection.* Princeton, NJ: Princeton University Press.

André, M. 1997. Distribution and conservation of the sperm whale *(Physeter macrocephalus)* in the Canary Islands. Ph.D. dissertation, University of Las Palmas de Gran Canaria, Spain.

Andrew, R. J. 1962. Evolution of intelligence and vocal mimicking. *Science* 137:585–89.

Andrews, R. C. 1908. Notes upon the external and internal anatomy of *Balaena glacialis* Bonn. *Bull. Am. Mus. Nat. Hist.* 24(10):171–82.

———. 1909. Observations on the habits of the finback and humpback whales of the eastern North Pacific. *Bull. Am. Mus. Nat. Hist.* 26:213–27.

Appel, F. C. 1964. The intellectual mammal. *Saturday Evening Post,* Jan. 4–11, 30–32.

Arak, A. 1988. Callers and satellites in the natterjack toad: Evolutionary decision rules. *Anim. Behav.* 36:416–32.

Arak, A., Eriksson, T., and Radester, T. 1990. The adaptive significance of acoustic spacing in male bushcrickets *Tettigonia viridissima:* A perturbation experiment. *Behav. Ecol. Sociobiol.* 26:1–7.

Arnbom, T. 1987. Individual identification of sperm whales. *Rep. Int. Whal. Commn.* 37:201–4.

Arnbom, T., Papastavrou, V., Weilgart, L. S., and Whitehead, H. 1987. Sperm whales react to an attack by killer whales. *J. Mammal.* 68:450–53.

Arntz, W. E. 1986. The two faces of El Niño 1982–83. *Meeresforschung* 31:1–46.

Astrup, J., and Møhl, B. 1993. Detection of intense ultrasound by the cod *Gadus morhua. J. Exp. Biol.* 182:71–80.

Au, W. W. L. 1980. Echolocation signals of the Atlantic bottlenose dolphin *(Tursiops truncatus)* in open waters. In *Animal sonar systems,* ed. R.-G. Busnel and J. F. Fish, 251–82. New York: Plenum Press.

———. 1993. *The sonar of dolphins.* New York: Springer-Verlag.

Au, W. W. L., Carder, D. A., Penner, R. H., and Scronce, B. L. 1985. Demonstration of adaptation in beluga whale echolocation signals. *J. Acoust. Soc. Am.* 77:726–30.

Austad, S. N. 1984. A classification of alternative reproductive behaviors and methods of field-testing ESS models. *Am. Zool.* 24:309–19.

———. 1994. Menopause: An evolutionary perspective. *Exp. Gerontol.* 29:255–63.

Avise, J. C., and Vrijenhoek, R. C. 1987. Mode of inheritance and variation of mitochondrial DNA in hybridogenetic fishes of the genus *Poeciliopsis. Mol. Biol. Evol.* 4:415–525.

Axelrod, R., and Hamilton, W. D. 1981. The evolution of cooperation. *Science* 211:1390–96.

Backus, R. H. 1961. Stranded killer whale in the Bahamas. *J. Mammal.* 42:418–19.

Backus, R., and Schevill, W. E. 1962. *Physeter* clicks. In *Whales, dolphins, and porpoises,* ed. K. S. Norris. Berkeley: University of California Press.

Baillie, J., and Groombridge, B. (eds.). 1996. *1996 IUCN Red List of Threatened Animals.* Gland, Switzerland: IUCN.

Bain, D. E. 1989. An evaluation of evolutionary processes: Studies of natural selection, dispersal, and cultural evolution in killer whales *(Orcinus orca).* Ph.D. thesis, University of California, Santa Cruz.

———. 1990. Examining the validity of inferences drawn from photo-identification data, with special reference to studies of the killer whale *(Orcinus orca)* in British Columbia. In *Individual recognition of cetaceans: Use of photo-identification and other techniques to estimate population parameters,* ed. P. S. Hammond, S. A. Mizroch, and G. P. Donovan, 93–100. Reports of the International Whaling Commission, special issue 12. Cambridge: International Whaling Commission.

Baird, R. W. 1994. Foraging behaviour and ecology of *transient* killer whales. Ph.D. thesis, Simon Fraser University, Burnaby, B.C.

———. 1995. Should I stay or should I go?: Costs and benefits of natal philopatry in two sympatric populations of killer whales. Abstracts of the Eleventh Biennial Conference on the Biology of Marine Mammals, Orlando, FL.

———. 1998. Studying diving behavior of whales and dolphins using suction-cup attached tags. *Whalewatcher* 32(1):3–7

———. 1999. Status of killer whales in Canada. Contract report to the Committee on the Status of Endangered Wildlife in Canada, Ottawa.

Baird, R. W., and Dill, L. M. 1995. Occurrence and behaviour of *transient* killer whales: Seasonal and pod-specific variability, foraging behaviour and prey handing. *Can. J. Zool.* 73:1300–1311.

———. 1996. Ecological and social determinants of group size in *transient* killer whales. *Behav. Ecol.* 7:408–16.

Baird, R. W., and Stacey, P. J. 1988. Variation in saddle patch pigmentation in populations of killer whales *(Orcinus orca)* from British Columbia, Alaska, and Washington State. *Can. J. Zool.* 66:2582–85.

Baird, R. W., and Whitehead, H. 1999. Association patterns of mammal-eating killer whales. Manuscript.

Baird, R. W., Abrams, P. A., and Dill, L. M. 1992. Possible indirect interactions between *transient* and *resident* killer whales: Implications for the evolution of foraging specializations in the genus *Orcinus. Oecologia* 89:125–32.

Baird, R. W., Dill, L. M., and Hanson, M. B. 1998. Diving behaviour of killer whales. Abstract submitted to the Twelfth Biennial Conference on the Biology of Marine Mammals, Monaco.

Bakeman, R., and Gottman, J. M. 1986. *Observing interaction: An introduction to sequential analysis.* Cambridge: Cambridge University Press.

Baker, C. S., and Herman, L. M. 1981. Migration and local movement of humpback whales *(Megaptera novaeangliae)* through Hawaiian waters. *Can. J. Zool.* 59:460–69.

———. 1984a. Aggressive behaviour between humpback whales *(Megaptera novaeangliae)* wintering in Hawaiian waters. *Can. J. Zool.* 62:1922–37.

———. 1984b. Seasonal contrasts in the behavior of the humpback whale. *Cetus* 5:14–16.

Baker, C. S., Herman, L. M., Perry, A., Lawton, W. S., Straley, J. M., and Straley, J. H. 1985. Population characteristics and migration of summer and late season humpback whales *(Megaptera novaeangliae)* in southeastern Alaska. *Mar. Mamm. Sci.* 1:304–23.

Baker, C. S., Herman, L. M., Perry, A., Lawton, W. S., Straley, J. M., Wolman, A. A., Kaufman, G. D., Winn, H. E., Hall, J. D., Reinke, J. M., and Östman, J. 1986. Migratory movement and population structure of humpback whales *(Megaptera novaeangliae)* in the central and eastern North Pacific. *Mar. Ecol. Prog. Ser.* 31:105–19.

Baker, C. S., Perry, A., and Herman, L. M. 1987. Reproductive histories of female humpback whales, *Megaptera novaeangliae,* in the North Pacific. *Mar. Ecol. Prog. Ser.* 41:103–14.

Baker, C. S., Lambertsen, R. H., Weinrich, M. T., Calambokidis, J., Early, G., and O'Brien, S. J. 1991. Molecular genetic iden-

tification of the sex of humpback whales *(Megaptera novaeangliae).* In *Genetic ecology of whales and dolphins: Incorporating the proceedings of the workshop on the genetic analysis of cetacean populations,* ed. A. R. Hoelzel, 105–11. Reports of the International Whaling Commission, special issue 13. Cambridge: International Whaling Commission.

Baker, C. S., Straley, J. M., and Perry, A. 1992. Population characteristics of individually identified humpback whales in southeastern Alaska: Summer and fall 1986. *Fish. Bull.* 90: 429–37.

Baker, C. S., Weinrich, M. T., Early, G., and Palumbi, S. R. 1993. Genetic impact of an unusual group mortality among humpback whales. *J. Hered.* 84:281–90.

Baker, C. S., Slade, R. W., Bannister, J. L., Abernethy, R. B., Weinrich, M. T., Lien, J., Urban, J., Corkeron, P., Calambokidis, J., Vasquez, O., and Palumbi, S. R. 1994. Hierarchical structure of mitochondrial DNA gene flow among humpback whales *Megaptera novaeangliae* world-wide. *Mol. Ecol.* 3: 313–27.

Baker, K. C., and Smuts, B. B. 1994. Social relationships of female chimpanzees: Diversity between captive social groups. In *Chimpanzee cultures,* ed. R. W. Wrangham, W. C. McGrew, F. B. M. de Waal, and P. G. Heltne. Cambridge, MA: Harvard University Press.

Baker, M. C., and Cunningham, M. A. 1985. The biology of bird song dialects. *Behav. Brain Sci.* 8:85–133.

Balcomb, K. C. 1989. Bairds beaked whale *(Berardius bairdii* Stejneger, 1883) and Arnouxs beaked whale *(Berardius arnuxii* Duvernoy, 1851). In *Handbook of marine mammals,* vol. 4, *River dolphins and the larger toothed whales,* ed. S. H. Ridgway and R. Harrison. London: Academic Press.

Balcomb, K. C., and Nichols, G. 1982. Humpback whale censuses in the West Indies. *Rep. Int. Whal. Commn.* 32:401–6.

Balcomb, K. C., Boran, J. R., and Heimlich, S. L. 1982. Killer whales in greater Puget Sound. *Rep. Int. Whal. Commn.* 32: 681–85.

Balcombe, J. P. 1990. Vocal recognition of pups by mother Mexican free-tailed bats *Tadarida brasiliensis mexicana. Anim. Behav.* 39:960–66.

Baldridge, A. 1972. Killer whales attack and eat a gray whale. *J. Mammal.* 53:898–900.

Ballance, L. T. 1990. Residence patterns, group organization, and surfacing associations of bottlenose dolphins in Kino Bay, Gulf of California, Mexico. In *The bottlenose dolphin,* ed. S. Leatherwood and R. R. Reeves. San Diego: Academic Press.

Banner, A., and Hyatt, M. 1973. Effects of noise on eggs and larvae of two estuarine fish. *Trans. Am. Fish. Soc.* 108.

Bannister, J. L. 1969. The biology and status of the sperm whale off western Australia—an extended summary of results of recent work. *Rep. Int. Whal. Commn.* 19:70–76.

———. 1990. Southern right whales off western Australia. In *Individual recognition of cetaceans: Use of photo-identification and other techniques to estimate population parameters,* ed. P. S. Hammond, S. A. Mizroch, and G. P. Donovan, 279–88. Reports of the International Whaling Commission, special issue 12. Cambridge: International Whaling Commission.

Baraff, L., and Weinrich, M. T. 1993. Separation of humpback whale mothers and calves on feeding ground in early autumn. *Mar. Mamm. Sci.* 9(4):431–34.

REFERENCES

Baraff, L. S., Clapham, P. J., Mattila, D. K., and Bowman, R. 1991. Feeding behaviour of a humpback whale in low-latitude waters. *Mar. Mamm. Sci.* 7:197–202.

Barlow, J. 1995. The abundance of cetaceans in California waters. Part I: Ship surveys in summer and fall of 1991. *Fish. Bull.* 93:1–14.

Barlow, J., and Clapham, P. 1997. A new birth-interval approach to estimating demographic parameters of humpback whales. *Ecology* 78:535–46.

Barlow, J., Baird, R. W., Heyning, J. E., Wynne, K., Manville, A. M., Lowry, L. F., Hanan, D., Sease, J., and Burkanov, V. N. 1994. A review of cetacean and pinniped mortality in coastal fisheries along the west coast of the USA and Canada and the east coast of the Russian Federation. In *Gillnets and cetaceans: Incorporating the proceedings of the symposium and workshop on the mortality of cetaceans in passive fishing nets and traps,* ed. W. F. Perrin, G. Donovan, and J. Barlow, 405–26. Reports of the International Whaling Commission, special issue 15. Cambridge: International Whaling Commission.

Barlow, J., Gerrodette, T., and Silber, G. 1997. First estimates of vaquita abundance. *Mar. Mamm. Sci.* 13:44–58.

Barnard, C. J., and Sibly, R. M. 1981. Producers and scroungers: a general model and its application to captive flocks of house sparrows. *Anim. Behav.* 29:543–50.

Barnes, R. F. W. 1982. Mate searching behaviour of elephant bulls in a semi-arid environment. *Anim. Behav.* 30:1217–23.

Barnes, R. H. 1991. Indigenous whaling and porpoise hunting in Indonesia. In *Cetaceans and cetacean research in the Indian Ocean Sanctuary,* ed. S. Leatherwood and G. P. Donovan. Marine Mammal Technical Report no. 3. Nairobi: UNEP.

Barrett-Lennard, L. G., Ford, J. F. B., and Heise, K. A. 1996a. The mixed blessing of echolocation: Differences in sonar use by fish-eating and mammal-eating killer whales. *Anim. Behav.* 51:553–65.

Barrett-Lennard, L. G., Smith, T. G., and Ellis, G. M. 1996b. A cetacean biopsy system using lightweight pneumatic darts, and its effect on the behavior of killer whales. *Mar. Mamm. Sci.* 12:14–27.

Barros, N. B., and Odell, D. K. 1990. Food habit of bottlenose dolphins in the southeastern United States. In *The bottlenose dolphin,* ed. S. Leatherwood and R. R. Reeves. San Diego: Academic Press.

Barstow, R. 1992. Whales are uniquely special. In *Why whales?* ed. N. Davies, A. M. Smith, S. R. Whyte, and V. Williams. Bath: Whale and Dolphin Conservation Society.

Bartholomew, G. A. 1942. The fishing activities of double-crested cormorants on San Francisco Bay. *Condor* 44:13–21.

———. 1974. The relation of the natural history of whales to their management. In *The whale problem: A status report,* ed. W. E. Schevill, G. C. Ray, and K. S. Norris. Cambridge, MA: Harvard University Press.

Barton, R. A. 1996. Neocortex size and behavioual ecology in primates. *Proc. R. Soc. Lond.* B 263:173–77.

Bastian, J. 1967. The transmission of arbitrary environmental information between bottlenose dolphins. In *Animal sonar systems: Laboratoire de Physiologie Acoustique,* ed. R.-G. Busnel. France: Jouy-en-Josas.

Bateson, G. 1974. Observations of a cetacean community. In *Mind in the waters,* ed. J. McIntyre. New York: Charles Scribner's Sons.

Bateson, P. P. G. 1983. *Mate choice.* Cambridge: Cambridge University Press.

———. 1991. Are there principles of behavioural development? In *The development and integration of behaviour: Essays in honour of Robert Hinde,* ed. P. P. G. Bateson. Cambridge: Cambridge University Press.

———. 1994. The dynamics of parent-offspring relationships in mammals. *Trends Ecol. Evol.* 9:399–403.

Bateson, P. P. G., and Hinde, R. A. 1976. Editorial. In *Growing points in ethology,* ed. P. P. G. Bateson and R. A. Hinde. Cambridge: Cambridge University Press.

Baylis, H. A. 1920. Observations on the genus *Crassicauda. Am. Mag. Nat. Hist.* 9:410–19.

Beale, T. 1835. *A few observations on the Nat. Hist. of the sperm whales, with an account of the rise and progress of the fishery, and of the modes of pursuing, killing, and "cutting in" that animal, with a list of its favorite places of resort.* London: Effingham Wilson, Royal Exchange.

Bearzi, G., Notarbartolo-di-Sciara, G., and Politi, E. 1997. Social ecology of bottlenose dolphins in the Kvarneri (Northern Adriatic Sea). *Mar. Mamm. Sci.* 13:650–68.

Beddington, J. R., and Kirkwood, G. P. 1980. On the mathematical structure of possible sperm whale models. In *Sperm whales,* 57–58. Reports of the International Whaling Commission, special issue 2. Cambridge: International Whaling Commission.

Bednarz, J. C. 1988. Cooperative hunting in Harris hawks *(Parabuteo unicinctus). Science* 239:1525–27.

Bednekoff, P. A. 1997. Mutualism among safe, selfish sentinels: a dynamic game. *Am. Nat.* 150:373–92.

Beecher, M. D. 1989. Signalling systems for individual recognition: An information theory approach. *Anim. Behav.* 38:248–61.

Beer, C. G. 1970. Individual recognition of voice in the social behavior of birds. In *Adv. Study Behav.,* vol. 3, ed. J. S. Rosenblatt, C. G. Beer, and R. A. Hinde. New York: Academic Press.

Bejder, L., Fletcher, D., and Bräger, S. 1998. A method for testing association patterns of social animals. *Anim. Behav.* 56:719–25.

Bejder, L., Dawson, S. M., and Harraway, J. A. 1999. Responses by Hector's dolphins to boats and swimmers in Porpoise Bay, New Zealand. *Mar. Mamm. Sci.* 15:738–50.

Bel'kovich, V. M. 1991. Herd structure, hunting, and play: Bottlenose dolphins in the Black Sea. In *Dolphin societies: Discoveries and puzzles,* ed. K. Pryor and K. S. Norris. Berkeley: University of California Press.

Bel'kovich, V. M., Krushinskaya, N. L., and Gurevich, V. S. 1970. *The behavior of dolphins in captivity.* No. 50701. Washington, DC: Joint Publications Research Service.

Bel'kovich, V. M., Ivanova, E. E., Yefremenkova, O. V., Kozarovitsky, I. B., and Kharitonov, S. P. 1991. Searching and hunting behavior in the bottlenose dolphin *(Tursiops truncatus)* in the Black Sea. In *Dolphin societies: Discoveries and puzzles,* ed. K. Pryor and K. S. Norris. Berkeley: University of California Press.

van Bénéden, P. J. 1881. Notice sur un nouveau dauphin de la

Nouvelle-Zélande. *Academie Royale de Belgique, Bulletin* ser. 3, 4:877.

Benfield, M. C., Davis, C. S., Wiebe, P. H., Gallager, S. M., Lough, R. G., and Copley, N. J. 1996. Video plankton recorder estimates of copepod, pteropod and larvacean distributions from a stratified region of Georges Bank with comparative measurements from a MOCNESS sampler. *Deep-Sea Res.* (Part 2) 43:1925–45.

Benirschke, K., Johnson, M. L., and Benirschke, R. J. 1980. Is ovulation in dolphins, *Stenella longirostris* and *Stenella attenuata,* always copulation-induced? *Fish. Bull.* 78(20):507–28.

Benjaminsen, T. 1972. On the biology of the bottlenose whale, *Hyperoodon ampullatus* (Forster). *Norw. J. Zool.* 20:233–41.

Bennett, F. D. 1840. *Narrative of a whaling voyage around the globe from the year 1833 to 1836.* London: Richard Bentley.

Bercovich, F. B. 1988. Coalitions, cooperation, and reproductive tactics among adult male baboons. *Anim. Behav.* 36:1198–1209.

Berger, J. 1989. Female reproductive potential and its apparent evaluation by male mammals. *J. Mammal.* 70:347–58.

———. 1992. Facilitation of reproductive synchrony by gestation adjustment in gregarious mammals: A new hypothesis. *Ecology.* 73(1):323–29.

Berggren, P. 1995. Foraging behaviour by bottlenose dolphins (*Tursiops* sp.) in Shark Bay, Western Australia. Abstract, Eleventh Biennial Conference on the Biology of Marine Mammals, Orlando, FL.

Berglund, A., Magnhagen, C., Bisazza, A., Konig, B., and Huntingford, F. 1993. Female-female competition over reproduction. *Behav. Ecol.* 4:184–87.

Bernard, H. J., and Hohn, A. A. 1989. Differences in feeding habits between pregnant and lactating spotted dolphins (*Stenella attenuata*). *J. Mammal.* 70:211–15.

Berry, J. F., and Shine, R. 1980. Sexual size dimorphism in turtles (Order Testudines). *Oecologia* 44:185–91.

Berubé, M., and Pallsbøll, P. 1996. Identification of sex in cetaceans by multiplexing with three ZFX and ZFY specific primers. *Mol. Ecol.* 5:283–87.

Berzin, A. A. 1972. *The sperm whale.* Jerusalem: Israel Program for Scientific Translations.

Berzin, A. A., and Vladimirov, V. L. 1983. A new species of killer whale (Cetacea, Delphinidae) from Antarctic waters. *Zool. Zhurnal* 62:287–95.

Best, P. B. 1967. The sperm whale (*Physeter catodon*) off the west coast of South Africa. 1. Ovarian changes and their significance. *Div. Sea Fish. Invest. Rep.* 61:1–27.

———. 1968. The sperm whale (*Physeter catodon*) off the west coast of South Africa. 2. Reproduction in the female. *Div. Sea Fish. Inv. Rep.* 66:1–32.

———. 1969a. The sperm whale (*Physeter catodon*) off the west coast of South Africa. 3. Reproduction in the male. *Div. Sea Fish. Invest. Rep.* 72:1–20.

———. 1969b. The sperm whale (*Physeter catodon*) off the west coast of South Africa. 4. Distribution and movements. *Div. Sea Fish. Invest. Rep.* 78:1–12.

———. 1970. The sperm whale (*Physeter catodon*) off the west coast of South Africa. 5. Age, growth and mortality. *Div. Sea Fish. Invest. Rep.* 79:1–27.

———. 1977. Two allopatric forms of Bryde's whale off South Africa. In *Report of the Special Meeting of the Scientific Committee on Sei and Bryde's Whales, International Whaling Commission, La Jolla, CA, Dec. 1974,* 10–35. Reports of the International Whaling Commission, special issue 1. Cambridge: International Whaling Commission.

———. 1979. Social organization in sperm whales, *Physeter macrocephalus.* In *Behavior of marine mammals: Current perspectives in research,* vol. 3, *Cetaceans,* ed. H. E. Winn and B. L. Olla. New York: Plenum Press.

———. 1982. Seasonal abundance, feeding, reproduction, age and growth in minke whales off Durban (with incidental observations from the Antarctic). *Rep. Int. Whal. Commn.* 32:759–86.

———. 1983. Sperm whale stock assessments and the relevance of historical whaling records. In *Historical whaling records: Including the proceedings of the International Workshop on Historical Whaling Records, Sharon, MA, Sept. 12–16, 1977,* ed. M. F. Tillman and G. P. Donovan, 41–55. Reports of the International Whaling Commission, special issue 5. Cambridge: International Whaling Commission.

———. 1989. Some comments on the BIWS catch record data base. *Rep. Int. Whal. Commn.* 39:363–69.

———. 1993. Increase rates in severely depleted stocks of baleen whales. *ICES J. Mar. Sci.* 50:169–86.

Best, P. B., and Butterworth, D. S. 1980. Timing of oestrus within sperm whale schools. In *Sperm whales,* 137–40. Reports of the International Whaling Commission, special issue 2. Cambridge: International Whaling Commission.

Best, P. B., and Underhill, L. G. 1990. Estimating population size in southern right whales *(Eubalaena australis)* using naturally marked animals. In *Individual recognition of cetaceans: Use of photo-identification and other techniques to estimate population parameters,* ed. P. S. Hammond, S. A. Mizroch, and G. P. Donovan, 183–89. Reports of the International Whaling Commission, special issue 12. Cambridge: International Whaling Commission.

Best, P. B., Canham, P. A. S., and Macleod, N. 1984. Patterns of reproduction in sperm whales, *Physeter macrocephalus.* In *Reproduction in whales, dolphins and porpoises: Proceedings of the conference, Cetacean Reproduction, Estimating Parameters for Stock Assessment and Management, La Jolla, CA, 28 Nov.– 7 Dec. 1981,* ed. W. F. Perrin, R. L. Brownell Jr., and D. P. DeMaster, 51–79. Reports of the International Whaling Commission, special issue 6. Cambridge: International Whaling Commission.

Best, R. C. 1981. The tusk of the narwhal (*Monodon monoceros* L.): Interpretation of its function (Mammalia: Cetacea). *Can. J. Zool.* 59:2386–93.

Best, R. C., and da Silva, V. M. F. 1984. Preliminary analysis of reproductive parameters of the boutu, *Inia geoffrensis,* and the tucuxi, *Sotalia fluviatilis,* in the Amazon River system. In *Reproduction in whales, dolphins and porpoises: Proceedings of the conference, Cetacean Reproduction, Estimating Parameters for Stock Assessment and Management, La Jolla, CA, 28 Nov.– 7 Dec. 1981,* ed. W. F. Perrin, R. L. Brownell Jr., and D. P. DeMaster, 361–69. Reports of the International Whaling Commission, special issue 6. Cambridge: International Whaling Commission.

———. 1989. Amazon river dolphin, boto. In *Handbook of ma-*

rine mammals, vol. 4: *River dolphins and the larger toothed whales,* ed. S. H. Ridgway and R. J. Harrison. New York: Academic Press.

Bigg, M. A. 1979. Interaction between pods of killer whale off British Columbia and Washington. Abstracts of the Third Biennial Conference on the Biology of Marine Mammals, Seattle, WA.

———. 1982. An assessment of killer whale *(Orcinus orca)* stocks off Vancouver Island, British Columbia. *Rep. Int. Whal. Commn.* 32:655–66.

———. 1994. The development of our study. In *Killer whales,* ed. J. K. B. Ford, G. M. Ellis, and K. C. Balcomb. Vancouver: UBC Press; Seattle: University of Washington Press.

Bigg, M. A., and Wolman, A. A. 1975. Live-capture killer whale *(Orcinus orca)* fishery, British Columbia and Washington, 1962–73. *J. Fish. Res. Bd. Can.* 32:1213–21.

Bigg, M. A., MacAskie, I. B., and Ellis, G. 1976. Abundance and movements of killer whales off eastern and southern Vancouver Island with comments on management. Unpublished report, Arctic Biological Station, Ste. Anne de Bellevue, Quebec.

Bigg, M. A., Ellis, G. M., Ford, J. K. B., and Balcomb, K. C. 1987. *Killer whales: A study of their identification, genealogy, and natural history in British Columbia and Washington State.* Nanaimo, B.C.: Phantom Press.

Bigg, M. A., Ellis, G. M., Ford, J. K. B., and Balcomb, K. C. 1990a. Feeding habits of the *resident* and *transient* forms of killer whale in British Columbia and Washington State. Abstracts of the Third International Orca Symposium, Victoria, B.C., March 1990.

Bigg, M. A., Olesiuk, P. F., Ellis, G. M., Ford, J. K. B., and Balcomb, K. C. 1990b. Social organization and genealogy of resident killer whales *(Orcinus orca)* in the coastal waters of British Columbia and Washington State. In *Individual recognition of cetaceans: Use of photo-identification and other techniques to estimate population parameters,* ed. P. S. Hammond, S. A. Mizroch, and G. P. Donovan, 383–405. Reports of the International Whaling Commission, special issue 12. Cambridge: International Whaling Commission.

Bingham, H. C. 1932. *Gorillas in a native habitat.* Publication no. 426. Washington, DC: Carnegie Institute.

Birkhead, T., and Møller, A. 1993. Female control of paternity. *Trends Ecol. Evol.* 8:100–104.

Bisther, A., and Vongraven, D. 1995. Studies of the social ecology of Norwegian killer whales *(Orcinus orca).* In *Whales, seals, fish and man: Proceedings of the International Symposium on the Biology of Marine Mammals in the North East Atlantic, Tromsø, Norway, 29 November–1 December 1994,* ed. A. S. Blix, L. Walløe, and Ø. Ulltang. Amsterdam: Elsevier.

Bitterbaum, E., and Baptista, L. F. 1979. Geographical variation in songs of California house finches *(Carpodacus mexicanus). Auk* 96:462–74.

Black, N. A., Schulman-Janiger, A., Ternullo, R. L., and Guerrero-Ruiz, M. 1997. *Killer whales of California and western Mexico: A catalog of photo-identified individuals.* NOAA Technical Memo NOAA-TM-NMFS-SWFSC-247. La Jolla, CA: Southwest Fisheries Science Center, National Marine Fisheries Service.

de Blainville, H. M. D. 1817. Dauphins. In *Nouveau Dictionaire d'Histoire Naturelle.* Paris: Chez Deterville.

———. 1838. Sur les cacholots. *Annales Françaises et Étrangers d'Anatomie et de Physiologie* 2:335–37.

Blaxter, J. H. S., Denton, E. J., and Gray, J. A. B. 1981. Acoustico-lateralis system in clupeid fishes. In *Hearing and sound communication in fishes,* ed. W. N. Tavolga, A. N. Popper, and R. R. Fay. New York: Springer.

Blaylock, R. A., Hain, J. W., Hansen, L. J., Palka, D. L., and Waring, G. T. 1995. *U.S. Atlantic and Gulf of Mexico marine mammal stock assessments.* NOAA Technical Memo NMFS-SEFSC-363.

Blichfeldt, G. 1996. In the name of conservation. *Oryx* 30:84–87.

Bloch, D., and Lockyer, C. 1988. Killer whales *(Orcinus orca)* in Faroese waters. *Rit Fiskideildar* 11:55–64.

Bloch, D., Lockyer, C. H., and Zachariassen, M. 1993. Age and growth parameters of the long-finned pilot whale off the Faroe Islands. In *Biology of Northern Hemisphere pilot whales: A collection of papers,* ed. G. P. Donovan, C. H. Lockyer, and A. R. Martin, 163–207. Reports of the International Whaling Commission, special issue 14. Cambridge: International Whaling Commission.

Boehm, C. 1992. Segmentary warfare and the management of conflict: Comparison of East African chimpanzees and patrilineal-patrilocal humans. In *Coalitions and alliances in humans and other animals,* ed. A. H. Harcourt and F. B. M. de Waal. Oxford: Oxford University Press.

Bogoslovskaya, L. S., Votorogov, L. M., and Semenova, T. N. 1982. Distribution and feeding of gray whales off Chukotka in the summer and autumn of 1980. *Rep. Int. Whal. Commn.* 32:385–89.

Bohne, B. A., Thomas, J. A., Yohe, E. R., and Stone, S. H. 1985. Examination of potential hearing damage in Weddell seals *(Leptonychotes weddelli)* in McMurdo Sound, Antarctica. *Antarctic J.* 20:174–76.

Bohne, B. A., Bozzay, D. G., and Thomas, J. A. 1986. Evaluation of inner ear pathology in Weddell seals. *Antarctic J.* 21:208.

Boice, R. 1981. Captivity and feralization. *Psychol. Bull.* 89:407–21.

Bond, J. 1999. Genetic analysis of the sperm whale *(Physeter macrocephalus)* using microsatellites. Ph.D. dissertation, Cambridge University, Cambridge, England.

Bonner, W. N. 1984. Lactation strategies in pinnipeds: Problems for a marine mammalian group. *Symp. Zool. Soc. Lond.* 51:253–72.

Borowski, G. H. 1781. *Gemeinnüzzige Naturgeschichte des Thierreichs.* Berlin: Gottllieb August Lange.

Borsani, J. F., Pavan, G., Gordon, J. C. D., and Notarbartolo-di-Sciara, G. 1997. Regional vocalizations of the sperm whale: Mediterranean codas. *Eur. Res. Cetaceans* [Abstracts] 10:78–81.

Bose, N., and Lien, J. 1989. Propulsion of a fin whale *(Balaenoptera physalus):* Why the fin whale is a fast swimmer. *Proc. R. Soc. Lond.* B 237:175–200.

Bose, N., Lien, J., and Ahia, J. 1992. Measurements of the bodies and flukes of several cetacean species. *Proc. R. Soc. Lond.* B 242:163–73.

Bowen, W. D. 1997. Role of marine mammals in aquatic ecosystems. *Mar. Ecol. Prog. Ser.* 158:267–74.

REFERENCES

Bowers, C. A., and Henderson, R. S. 1972. *Project deep ops: Deep object recovery with pilot and killer whales.* Naval Undersea Center Technical Publication 306.

Bowles, A. E., Smultea, M., Würsig, B., DeMaster, D. P., and Palka, D. 1994. Relative abundance and behavior of marine mammals exposed to transmissions from the Heard Island Feasibility Test. *J. Acoust. Soc. Am.* 96:2469–84.

Boyd, R., and Richerson, P. 1985. *Culture and the evolutionary process.* Chicago: University of Chicago Press.

Bradbury, J. W. 1985. Contrasts between insects and vertebrates in the evolution of male display, female choice and lek mating. *Fort. Zool.* 31:273–89.

———. 1986. Social complexity and cooperative behavior in delphinids. In *Dolphin cognition and behavior: A comparative approach,* ed. R. J. Schusterman, J. A. Thomas, and F. G. Wood. Hillsdale, NJ: Lawrence Erlbaum Associates.

Bradbury, J. W., and Gibson, R. M. 1983. Leks and mate choice. In *Mate choice,* ed. P. P. G. Bateson. Cambridge: Cambridge University Press.

Bradbury, J. W., and Vehrencamp, S. L. 1977. Social organization and foraging in emballonurid bats. III. Mating systems. *Behav. Ecol. Sociobiol.* 2:1–17.

———. 1998. *Principles of animal communication.* Sunderland, MA: Sinauer Associates.

Bradbury, J. W., Gibson, R. M., and Tsai, I. M. 1986. Hotspots and the dispersion of leks. *Anim. Behav.* 34:1649–1709.

Bräger, S., Würsig, B., Acevedo, A., and Henningsen, T. 1994. Association patterns of bottlenose dolphins *(Tursiops truncatus)* in Galveston Bay, Texas. *J. Mammal.* 75:431–37.

Braslau-Schneck, S. 1994. Innovative behaviors and synchronization in bottlenose dolphins. MA thesis, University of Hawaii.

Brault, S., and Caswell, H. 1993. Pod-specific demography of killer whales *(Orcinus orca). Ecology* 74:1444–54.

Brear, K., Currey, J. D., Kingsley, M. C. S., and Ramsay, M. 1993. The mechanical design of the tusk of the narwhal *(Monodon monoceros:* Cetacea). *J. Zool.* 230:411–23.

Breathnach, A. S. 1960. The cetacean central nervous system. *Biol. Rev.* 35:187–230.

Brennan, B., and Rodriguez, P. 1994. Report of two orca attacks on cetaceans in Galápagos. *Notic. Galápagos* 54:28–29.

Brodie, P. F. 1969. Duration of lactation in Cetacea: An indicator of required learning? *Am. Midl. Nat.* 82:312–14.

———. 1975. Cetacean energetics: An overview of intraspecific size variation. *Ecology* 56:152–61.

———. 1977. Form, function and energetics of Cetacea: A discussion. In *Functional anatomy of marine mammals,* vol. 3, ed. R. J. Harrison. New York: Academic Press.

———. 1989. The white whale—*Delphinapterus leucas* (Pallas 1776). In *Handbook of marine mammals,* vol. 4, *River dolphins and the larger toothed whales,* ed. S. H. Ridgway and R. Harrison. London: Academic Press.

Brown, B. E., and Halliday, R. G. 1983. Fisheries resources of the Northwest Atlantic: Some responses to extreme fishing perturbations. In *Proceedings of the expert consultation to examine changes in abundance and species composition of neritic fish resources,* ed. E. Por. FAO Fisheries Report no. 291.

Brown, C. R. 1986. Cliff swallow colonies as information centers. *Science* 234:83–85.

Brown, C. R., and Brown, M. B. 1998. Fitness components associated with alternative reproductive tactics in cliff swallows. *Behav. Ecol.* 9(2):158–71.

Brown, C. R., Brown, M. B., and Shaffer, M. L. 1991. Food-sharing signals among socially foraging cliff-swallows. *Anim. Behav.* 42:551–64.

Brown, D. H., and Norris, K. S. 1956. Observations of captive and wild cetaceans. *J. Mammal.* 37:311–26.

Brown, K. M., Woulfe, M., and Morris, R. D. 1995. Patterns of adoption in ring-billed gulls: Who is really winning the inter-generational conflict? *Anim. Behav.* 49:321–31.

Brown, M. R., and Corkeron, P. J. 1995. Pod characteristics of migrating humpback whales *(Megaptera novaeangliae)* off the east Australian coast. *Behaviour* 132:163–79.

Brown, M. R., Corkeron, P. J., Hale, P. T., Schultz, K. W., and Bryden, M. M. 1995. Evidence for a sex-segregated migration in the humpback whale *(Megaptera novaeangliae). Proc. R. Soc. Lond.* B 259:229–34.

Brown, S. G. 1975. Marking of small cetaceans using "Discovery" type whale marks. *J. Fish. Res. Bd. Can.* 32:1237–40.

———. 1977. Whale marking: A short review. In *A voyage of discovery,* ed. M. Angel. Oxford: Pergamon Press.

———. 1978. Whale marking techniques. In *Animal marking: Recognition marking of animals in research,* ed. B. Stonehouse. Baltimore: University Park Press.

Brown, M. W., George, M. Jr., and Wilson, A. C. 1979. Rapid evolution of animal mitochondrial DNA. *Proc. Natl. Acad. Sci. USA* 76:1967–71.

Brown, M. W., Helbig, R., Boag, P. T., Gaskin, D. E., and White, B. N. 1991a. Sexing beluga whales *(Delphinapterus leucas)* by means of DNA markers. *Can. J. Zool.* 69:1971–76.

Brown, M. W., Kraus, S. D., and Gaskin, D. E. 1991b. Reaction of North Atlantic right whales *(Eubalaena glacialis)* to skin biopsy sampling for genetic and pollutant analysis. In *Genetic ecology of whales and dolphins: Incorporating the proceedings of the workshop on the genetic analysis of cetacean populations,* ed. A. R. Hoelzel, 81–89. Reports of the International Whaling Commission, special issue 13. Cambridge: International Whaling Commission.

Brown, W. M., Allen, J. M., and Kraus, S. D. 1995. The designation of seasonal right whale conservation areas in the waters of Atlantic Canada. In *Marine protected areas and sustainable fisheries,* ed. N. L. Shackell and J. H. M. Willison. Wolfville, Nova Scotia: Science and Management of Marine Protected Areas Association.

Brownell, R. L. Jr. 1989. Franciscana. In *Handbook of marine mammals,* vol. 4, *River dolphins and the larger toothed whales,* ed. S. H. Ridgway and R. J. Harrison. New York: Academic Press.

Brownell, R. L. Jr., and Ralls, K. 1986. Potential for sperm competition in baleen whales. In *Behaviour of whales in relation to management: Incorporating the proceedings of a workshop of the same name held in Seattle, Washington, 19–23 April 1982,* ed. G. P. Donovan, 97–112. Reports of the International Whaling Commission, special issue 8. Cambridge: International Whaling Commission.

Brownell, R. L. Jr., Zhou, K., and Liu, J. 1989. *Biology and con-*

servation of the river dolphins. Gland, Switzerland: IUCN Species Survival Commission.

Bruns, V., and Schmieszek, E. T. 1980. Cochlear innervation in the greater horseshoe bat: Demonstration of an acoustic fovea. *Hearing Res.* 3:27–43.

Bryant, P. J., Lafferty, C. M., and Lafferty, S. K. 1984. Reoccupation of Laguna Guerrero Negro, Baja California, Mexico, by gray whales. In *The gray whale, Eschrichtius robustus,* ed. M. L. Jones, S. L. Swartz, and S. Leatherwood. Orlando, FL: Academic Press.

Buchwald, J. S., and Shipley, C. 1985. A comparative model of infant cry. In *Infant crying,* ed. B. M. Lester and C. F. Z. Boukydis. New York: Plenum.

Buck, J., and Tyack, P. L. 1993. A quantitative measure of similarity for *Tursiops truncatus* signature whistles. *J. Acoust. Soc. Am.* 94:2497–2506.

Buckingham, M. J., Potter, J. R., and Epifanio, C. L. 1996. Seeing underwater with background noise. *Sci. Am.* 274(2): 86–90.

Buckland, S. T., and Anganuzzi, A. A. 1988. Estimated trends in abundance of dolphins associated with tuna in the eastern tropical Pacific. *Rep. Int. Whal. Commn.* 38:411–37.

Budker, P. 1953. Les campagnes baleinières 1949–52 au Gabon. *Mammalia* 17:129–48.

Bullen, F. T. 1902. *The cruise of the Cachalot round the world after sperm whales.* New York: D. Appleton.

Bullock, T. H., and Gurevitch, V. S. 1979. Soviet literature on the nervous system and psychobiology of Cetacea. *Int. Rev. Neurobiol.* 21:47–127.

Bunch, J. N., and Reeves, R. R. 1992. Proceedings of a workshop on the potential cumulative impacts of development in the region of Hudson and James bays, 17–19 June 1992. *Can. Tech. Rep. Fish. Aquat. Sci.* 1838:1–39.

Burdin, V. I., Reznik, A. M., Skornyakov, V. M., and Chupakov, A. G. 1975. Communication signals of the Black Sea bottlenose dolphin. *Sov. Phys. Acoust.* 20:314–18.

Burgess, W. C., Tyack, P. L., Le Boeuf, B. J., and Costa, D. P. 1998. A programmable acoustic recording tag and first results from free-ranging northern elephant seals. *Deep-Sea Res.* 45: 1327–51.

Burke, T. 1989. DNA fingerprinting and other methods for the study of mating success. *Trends Ecol. Evol.* 4:139–44.

Burke, T., and Bruford, M. W. 1987. DNA fingerprinting in birds. *Nature* 327:149–52.

Burley, N. 1981. Sex ratio manipulation and selection for attractiveness. *Science* 211 (13):721–22.

Burmeister, H. 1865. Description of a new species of porpoise in the museum of Buenos Aires. *Proc. Zool. Soc. Lond.* 1865(1):228–31.

Burns, J. J., Montague, J. J., and Cowles, C. J. 1993. *The bowhead whale.* Lawrence, KS: Society for Marine Mammalogy.

Busnel, R.-G. 1973. Symbiotic relationship between man and dolphins. *NY Acad. Sci. Trans.* 35:112–31.

Busnel, R.-G., and Dziedzic, A. 1968. Etude des signaux acoustiques associé à des situations détresse chez certain cétacés odontocètes. *Ann. Inst. Océanogr. Monaco* 46:109–44.

Buss, L. W. 1981. Group living, competition, and the evolution of cooperation in a sessile invertebrate. *Science* 213:1012–14.

Butler, R. W., and Jennings, J. G. 1980. Radio tracking of dolphins in the eastern tropical Pacific using VHF and HF equipment. In *A handbook on biotelemetry and radio tracking,* ed. C. J. Amlaner and D. W. MacDonald. Oxford: Pergamon Press.

Butterworth, D. S., Borchers, D. L., Chalis, S., De Decker, J. B., and Kasamatsu, F. In press. Estimates of abundance for Southern Hemisphere blue, fin, sei, humpback, sperm, killer and pilot whales from the 1978/79 to 1990/91 International Whaling Commission/IDCR sighting survey cruises, with extrapolations to the area south of 30 S for the first five species based on Japanese scouting vessel data. *Rep. Int. Whal. Commn.*

Byers, J. A., and Moodie, J. D. 1990. Sex-specific maternal investment in pronghorn and the question of a limit on differential provisioning in ungulates. *Behav. Ecol. Sociobiol.* 26:157–64.

Byrne, R., and Whiten, A. (eds.). 1988. *Machiavellian intelligence: Social expertise and the evolution of intellect in monkeys, apes, and humans.* Oxford: Oxford University Press.

Cade, W. 1975. Acoustically orienting parasitoids: Fly phonotaxis to cricket song. *Science* 190:1312–13.

———. 1979. The evolution of alternative male reproductive strategies in field crickets. In *Sexual selection and reproductive competition in insects,* ed. M. Blum and A. Blum, 343–79. London: Academic Press.

Cairns, S. J., and Schwager, S. 1987. A comparison of association indices. *Anim. Behav.* 3:1454–69.

Calambokidis, J., and Baird, R. W. 1994. Status of marine mammals in the Strait of Georgia, Puget Sound and the Juan de Fuca Strait and potential human impacts. *Can. Tech. Rep. Fish. Aquat. Sci.* 1948:282–300.

Calambokidis, J., Cubbage, J. C., Steiger, G. H., Balcomb, K. C., and Bloedel, P. 1990a. Population estimates of humpback whales in the Gulf of the Farallones, California. In *Individual recognition of cetaceans: Use of photo-identification and other techniques to estimate population parameters,* ed. P. S. Hammond, S. A. Mizroch, and G. P. Donovan, 325–33. Reports of the International Whaling Commission, special issue 12. Cambridge: International Whaling Commission.

Calambokidis, J., Langelier, K. M., Stacey, P. J., and Baird, R. W. 1990b. Environmental contaminants in killer whales from Washington, British Columbia and Alaska. Abstracts of the Third International Orca Symposium, Victoria, B.C., March 1990.

Calambokidis, J., Steiger, G. H., Evenson, J. R., Flynn, K. R., Balcomb, K. C., Claridge, D., Bloedel, P., Straley, J. M., Baker, C. S., von Ziegesar, O., Dalheim, M., Waite, J. M., Darling, J. D., Ellis, G., and Green, G. A. 1996. Interchange and isolation of humpback whales off California and other North Pacific feeding grounds. *Mar. Mamm. Sci.* 12:215–26.

Caldwell, D. K. 1955. Evidence of home range of an Atlantic bottlenose dolphin. *J. Mammal.* 36:304–5.

Caldwell, D. K., and Brown, D. H. 1964. Tooth wear as a correlate of described feeding behavior by the killer whale. *Bull. S. Calif. Acad. Sci.* 63:128–40.

Caldwell, D. K., and Caldwell, M. C. 1972a. *The world of the bottlenose dolphin.* Philadelphia: Lippincott.

———. 1975. Dolphin and small whale fisheries of the Caribbean and West Indies: Occurrence, history, and catch statistics—with special reference to the Lesser Antillean island of St. Vincent. *J. Fish. Res. Bd. Can.* 32:1105–10.

———. 1977. Cetaceans. In *How animals communicate,* ed. T. A. Sebeok. Bloomington: Indiana University Press.

Caldwell, D. K., Caldwell, M. C., and Rice, D. W. 1966. Behavior of the sperm whale *Physeter catodon* L. In *Whales, dolphins and porpoises,* ed. K. S. Norris. Berkeley: University of California Press.

Caldwell, M. C., and Caldwell, D. K. 1964. Experimental studies on factors involved in care-giving behavior in three species of the cetacean family Delphinidae. *Bull. S. Calif. Acad. Sci.* 63: 1–20.

———. 1965. Individualized whistle contours in bottlenosed dolphins *(Tursiops truncatus). Nature* 207:434–35.

———. 1966. Epimeletic (care-giving) behavior in Cetacea. In *Whales, dolphins, and porpoises,* ed. K. S. Norris. Berkeley: University of California Press.

———. 1967. Dolphin community life. *Los Angeles County Mus. Nat. Hist. Q.* 5(4):12–15.

———. 1968. Vocalizations of naïve captive dolphins in small groups. *Science* 159:1121–23.

———. 1971. Statistical evidence for individual signature whistles in the Pacific whitesided dolphin *(Lagenorhynchus obliquidens). Cetology* 16:1–21.

———. 1972b. Behavior of marine mammals. In *Mammals of the sea,* ed. S. H. Ridgway. Springfield, IL: Charles C. Thomas.

———. 1972c. Vocal mimicry in the whistle mode by an Atlantic bottlenosed dolphin. *Cetology* 9:1–8.

Caldwell, M. C., Brown, D. H., and Caldwell, D. K. 1963. Intergeneric behavior by a captive Pacific pilot whale. *Los Angeles County Museum Contributions in Science* 70:1–12.

Caldwell, M. C., Caldwell, D. K., and Townsend, B. G. Jr. 1968. Social behavior as a husbandry factor. Symposium of Disease and Husbandry of Aquatic Mammals (mimeograph).

Caldwell, M. C., Caldwell, D. K., and Hall, N. R. 1969. An experimental demonstration of the ability of an Atlantic bottlenosed dolphin to discriminate between whistles of other individuals of the same species. Los Angeles County Museum of Natural History Foundation Technical Report no. 6.

Caldwell, M. C., Caldwell, D. K., and Turner, R. H. 1970. Statistical analysis of the signature whistle of an Atlantic bottlenosed dolphin with correlation between vocal changes and level of arousal. Los Angeles County Museum of Natural History Foundation Technical Report no. 8.

Caldwell, M. C., Caldwell, D. K., and Miller, J. F. 1973. Statistical evidence for individual signature whistles in the spotted dolphin, *Stenella plagiodon. Cetology* 16:1–21.

Caldwell, M. C., Caldwell, D. K., and Tyack, P. L. 1990. A review of the signature whistle hypothesis for the Atlantic bottlenose dolphin, *Tursiops truncatus.* In *The bottlenose dolphin,* ed. S. Leatherwood and R. R. Reeves. San Diego: Academic Press.

Cameron, E. Z. 1998. Is suckling behaviour a useful predictor of milk intake? A review. *Animal Behaviour* 56:521–32.

Campagna, C., Le Boeuf, B. J., and Cappozzo, L. 1988. Group raids: A mating strategy of male southern sea lions. *Behaviour* 105:224–49.

Campagna, C., Bisioli, C., Quintana, F., Perez, F., and Vila, A. 1992. Group breeding in sea lions: Pups survive better in colonies. *Anim. Behav.* 43:541–48.

Carl, G. C. 1946. A school of killer whales stranded at Estevan Point. *B.C. Prov. Mus. Nat. Hist. Anth. Rep.* 1945:21–28.

Carlson, C. A. 1992. Variation in the behavior of humpback whales: A study of individuals. Ph.D. dissertation, Dalhousie University, Halifax, Nova Scotia.

Carlson, C. A., Mayo, C. A., and Whitehead, H. 1990. Changes in the ventral fluke pattern of the humpback whale *(Megaptera novaeangliae),* and its effects on matching: Evaluation of its significance to photo-identification research. In *Individual recognition of cetaceans: Use of photo-identification and other techniques to estimate population parameters,* ed. P. S. Hammond, S. A. Mizroch, and G. P. Donovan, 105–11. Reports of the International Whaling Commission, special issue 12. Cambridge: International Whaling Commission.

Caro, T. M. 1994a. *Cheetahs of the Serengeti plains.* Chicago: University of Chicago Press.

———. 1994b. Ungulate antipredator behaviour: Preliminary and comparative data from African bovids. *Behaviour* 128: 189–228.

Caro, T. M., and Bateson, P. 1986. Organization and ontogeny of alternative tactics. *Anim. Behav.* 34:1483–99.

Caro, T. M., and Hauser, M. D. 1992. Is there teaching in nonhuman animals? *Q. Rev. Biol.* 67:151–74.

Caro, T. M., Fitzgibbon, C. D., and Holt, M. E. 1989. Physiological cost of behavioural strategies for male cheetahs. *Anim. Behav.* 38:309–17.

Carpenter, C. R. 1964. Social behaviour of non-human primates. In *Naturalistic behaviour of non-human primates,* ed. C. R. Carpenter. University Park: Pennsylvania State University Press.

Catchpole, C. K. 1982. The evolution of bird sounds in relation to mating and spacing behavior. In *Acoustic communication in birds,* vol. 1: *Production, perception and design features of sounds,* ed. D. E. Kroodsma and E. H. Miller. New York: Academic Press.

Catchpole, C. K., and Slater, P. J. B. 1995. *Bird song.* Cambridge: Cambridge University Press.

Cato, D. H. 1991. Songs of humpback whales: The Australian perspective. *Mem. Queensland Mus.* 30:277–90.

CeTAP. 1982. A characterization of marine mammals and turtles in the mid- and north Atlantic areas of the U.S. outer continental shelf. Bureau of Land Management contract no. AA551-CTB-48. Cetacean and Turtle Assessment Program, University of Rhode Island.

Chapman, C. A., Wrangham, R. W., and Chapman, L. J. 1994. Party size in chimpanzees and bonobos. In *Chimpanzee cultures,* ed. R. W. Wrangham, W. C. McGrew, F. B. M. de Waal, and P. G. Heltne. Cambridge, MA: Harvard University Press.

Chapman, D. M. F., and Ellis, D. D. 1998. The elusive decibel: Thoughts on sonars and marine mammals. *Can. Acoust.* 26: 29–31.

Charnov, E. L., and Berrigan, D. 1990. Dimensionless numbers and life history evolution: Age at maturity versus the adult lifespan. *Evol. Ecol.* 4:273–75.

Charnov, E. L., and Krebs, J. R. 1975. The evolution of alarm calls: Altruism or manipulation. *Am. Nat.* 109:107–12.

Cheal, A. J., and Gales, N. J. 1992. Growth, sexual maturity and food intake of Australian Indian Ocean bottlenose dolphins, *Tursiops truncatus*, in captivity. *Aust. J. Zool.* 40:215–23.

Cheney, D. L. 1987. Interactions and relationships between groups. In *Primate societies*, ed. B. B. Smuts, D. L. Cheney, R. M. Seyfarth, R. W. Wrangham, and T. T. Struhsaker. Chicago: University of Chicago Press.

Cheney, D. L., and Seyfarth, R. M. 1980. Vocal recognition in free-ranging vervet monkeys. *Anim. Behav.* 28:362–67.

———. 1985. Social and non-social knowledge in vervet monkeys. *Phil. Trans. R. Soc. Lond.* B 308:187–201.

———. 1990. How monkeys see the world. Chicago: University of Chicago Press.

Cheney, D. L., and Wrangham, R. W. 1987. Predation. In *Primate societies*, ed. B. B. Smuts, D. L. Cheney, R. M. Seyfarth, R. W. Wrangham, and T. T. Struhsaker. Chicago: University of Chicago Press.

Cheney, D. L., Seyfarth, R. M., Smuts, B. B., and Wrangham, R. W. 1987. The study of primate societies. In *Primate societies*, ed. B. B. Smuts, D. L. Cheney, R. M. Seyfarth, R. W. Wrangham, and T. T. Struhsaker. Chicago: University of Chicago Press.

Childerhouse, S. J., and Dawson, S. M. 1996. Stability of fluke marks used in individual photoidentification of sperm whales at Kaikoura, New Zealand. *Mar. Mamm. Sci.* 12:447–51.

Childerhouse, S. J., Dawson, S. M., and Slooten, E. 1995. Abundance and seasonal residence of sperm whales at Kaikoura, New Zealand. *Can. J. Zool.* 73:723–31.

Chittleborough, R. G. 1953. Aerial observations on the humpback whale, *Megaptera nodosa* (Bonnaterre), with notes on other species. *Aust. J. Mar. Freshw. Res.* 4:219–26.

———. 1954. Studies on the ovaries of the humpback whale, *Megaptera nodosa* (Bonnaterre), on the eastern Australian coast. *Aust. J. Mar. Freshw. Res.* 5:35–63.

———. 1955a. Aspects of reproduction in the male humpback whale, *Megaptera nodosa* (Bonnaterre). *Aust. J. Mar. Freshw. Res.* 6:1–29.

———. 1955b. Puberty, physical maturity, and relative growth of the female humpback whale, *Megaptera nodosa* (Bonnaterre), on the western Australian coast. *Aust. J. Mar. Freshw. Res.* 6:315–27.

———. 1958. The breeding cycle of the female humpback whale, *Megaptera nodosa* (Bonnaterre). *Aust. J. Mar. Freshw. Res.* 9:1–18.

———. 1959a. Australian marking of humpback whales. *Norsk Hvalfangst-Tidende* 48:47–55.

———. 1959b. Determination of age in the humpback whale, *Megaptera nodosa* (Bonnaterre). *Aust. J. Mar. Freshw. Res.* 10: 125–43.

———. 1960. Apparent variations in the mean length of female humpback whales at puberty. *Norsk Hvalfangst-Tidende* 49: 120–24.

———. 1965. Dynamics of two populations of the humpback whale, *Megaptera novaeangliae* (Borowski). *Aust. J. Mar. Freshw. Res.* 16:33–128.

Christal, J. 1998. An analysis of sperm whale social structure: Patterns of association and genetic relatedness. Ph.D. dissertation, Dalhousie University, Halifax, Nova Scotia.

Christal, J., and Whitehead, H. 1997. Aggregations of mature male sperm whales on the Galápagos Islands breeding ground. *Mar. Mamm. Sci.* 13:59–69.

Christal, J., Whitehead, H., and Lettevall, E. 1998. Sperm whale social units: Variation and change. *Can. J. Zool.* 76:1431–40.

Christensen, I. 1984. Growth and reproduction of killer whales, *Orcinus orca*, in Norwegian coastal waters. In *Reproduction in whales, dolphins and porpoises: Proceedings of the conference, Cetacean Reproduction, Estimating Parameters for Stock Assessment and Management, La Jolla, CA, 28 Nov.–7 Dec. 1981*, ed. W. F. Perrin, R. L. Brownell Jr., and D. P. DeMaster, 253–58. Reports of the International Whaling Commission, special issue 6. Cambridge: International Whaling Commission.

———. 1990. A note on recent strandings of sperm whales *(Physeter macrocephalus)* and other cetaceans in Norwegian waters. *Rep. Int. Whal. Commn.* 40:513–15.

Christensen, I., Haug, T., and Oien, N. 1992. Seasonal distribution, exploitation and present abundance of stocks of large baleen whales (Mysticeti) and sperm whales *(Physeter macrocephalus)* in Norwegian and adjacent waters. *ICES J. Mar. Sci.* 49:341–55.

Chu, K. C. 1988. Dive times and ventilation patterns of singing humpback whales *(Megaptera novaeangliae)*. *Can. J. Zool.* 66: 1322–27.

Chu, K. C., and Harcourt, P. 1986. Behavioral correlations with aberrant patterns in humpback whale songs. *Behav. Ecol. Sociobiol.* 19:309–12.

Chu, K. C., and Nieukirk, S. 1988. Dorsal fin scars as indicators of sex, age and social status in humpback whales *(Megaptera novaeangliae)*. *Can. J. Zool.* 66:416–20.

Ciampi, E. 1964. The status pet. *Saturday Evening Post*, Jan. 4–11, 22–29.

Clapham, P. J. 1992. The attainment of sexual maturity in humpback whales. *Can. J. Zool.* 70:1470–72.

———. 1993a. Social and reproductive biology of North Atlantic humpback whales. Ph.D. dissertation, University of Aberdeen.

———. 1993b. Social organization of humpback whales on a North Atlantic feeding ground. *Symp. Zool. Soc. Lond.* 66: 131–45.

———. 1994. Maturational changes in patterns of association in male and female humpback whales, *Megaptera novaeangliae*. *J. Zool.* 234:265–74.

———. 1996. The social and reproductive biology of humpback whales: An ecological perspective. *Mammal. Rev.* 26:27–49.

Clapham, P. J., and Brownell, R. L. Jr. 1996. Potential for interspecific competition in baleen whales. *Rep. Int. Whal. Commn.* 46:361–67.

Clapham, P. J., and Mattila, D. K. 1990. Humpback whale songs as indicators of migration routes. *Mar. Mamm. Sci.* 6:155–60.

Clapham, P. J., and Mayo, C. A. 1987. Reproduction and recruitment of individually identified humpback whales, *Megaptera novaeangliae*, observed in Massachusetts Bay, 1979–1985. *Can. J. Zool.* 65:2853–63.

———. 1990. Reproduction of humpback whales, *Megaptera novaeangliae*, observed in the Gulf of Maine. In *Individual recognition of cetaceans: Use of photo-identification and other techniques to estimate population parameters,* ed. P. S. Hammond, S. A. Mizroch, and G. P. Donovan, 171–75. Reports of the International Whaling Commission, special issue 12. Cambridge: International Whaling Commission.

Clapham, P. J., and Mead, J. G. 1999. *Megaptera novaeangliae. Mammalian Species.* 604:1–9.

Clapham, P. J., and Palsbøll, P. J. 1997. Molecular analysis of paternity shows promiscuous mating in female humpback whales (*Megaptera novaeangliae,* Borowski). *Proc. R. Soc. Lond.* B 264:95–98.

Clapham, P. J., Palsbøll, P. J., Mattila, D. K., and Vásquez, O. 1992. Composition and dynamics of humpback whale competitive groups in the West Indies. *Behaviour* 122:182–94.

Clapham, P. J., Baraff, L. S., Carlson, C. A., Christian, M. A., Mattila, D. K., Mayo, C. A., Murphy, M. A., and Pittman, S. 1993a. Seasonal occurrence and annual return of humpback whales in the southern Gulf of Maine. *Can. J. Zool.* 71:440–43.

Clapham, P. J., Mattila, D. K., and Palsbøll, P. J. 1993b. High-latitude-area composition of humpback whale competitive groups in Samana Bay: Further evidence for panmixis in the North Atlantic population. *Can. J. Zool.* 71:1065–66.

Clapham, P. J., Palsbøll, P. J., and Matilla, D. K. 1993c. High-energy behaviors in humpback whales as a source of sloughed skin for molecular analysis. *Mar. Mamm. Sci.* 9:213–20.

Clapham, P. J., Bérubé, M. C., and Mattila, D. K. 1995. Sex ratio of the Gulf of Maine humpback whale population. *Mar. Mamm. Sci.* 11:227–31.

Clapham, P. J., Webmore, S. E., Smith, T. D., and Mead, J. G. 1999. Length at birth and at independence in humpback whales. *Journal of Cetacean Research and Management* 1 (in press).

Clapham, P. J., Young, S. B., and Brownell, R. L. Jr. 1999. Baleen whales: Conservation issues and the status of the most endangered populations. *Mammal Rev.* 29:35–60.

Clark, A. B. 1978. Sex ratio and local resource competition in a prosimian primate. *Science* 201:163–65.

Clark, C. W. 1975. *Mathematical bioeconomics: The optimal management of renewable resources.* New York: Wiley.

———. 1980. A real-time direction finding device for determining the bearing to the underwater sounds of southern right whales, *Eubalaena australis. J. Acoust. Soc. Am.* 68:508–11.

———. 1982. The acoustic repertoire of the southern right whale, a quantitative analysis. *Anim. Behav.* 30:1060–71.

———. 1989. Call tracks of bowhead whales based on call characteristics as an independent means of determining tracking parameters. *Rep. Int. Whal. Commn.* 39:111–12.

———. 1990. Acoustic behavior of mysticete whales. In *Sensory abilities of cetaceans,* ed. J. Thomas and R. Kastelein. New York: Plenum Press.

———. 1991. Moving with the herd. *Nat. Hist.,* March, 38.

———. 1993. Bioacoustics of baleen whales: From infrasonics to complex songs. Abstract. *J. Acoust. Soc. Am.* 94:1830.

———. 1994. Blue deep voices: Insights from the Navy's Whales '93 program. *Whalewatcher* 28:6–11.

———. 1995. Application of US Navy underwater hydrophone arrays for scientific research on whales. *Rep. Int. Whal. Commn.* 45:210–13.

Clark, C. W., and Clark, J. M. 1980. Sound playback experiments with southern right whales *(Eubalaena australis). Science* 207:663–64.

Clark, C. W., and Ellison, W. T. 1988. Numbers and distributions of bowhead whales, *Balaena mysticetus,* based on the 1985 acoustic study off Pt. Barrow, Alaska. *Rep. Int. Whal. Commn.* 38:365–70.

Clark, C. W., and Johnson, J. H. 1984. The sounds of the bowhead whale, *Balaena mysticetus,* during the spring migrations of 1979 and 1980. *Can. J. Zool.* 62:1436–41.

Clarke, M. R. 1980. Cephalopoda in the diet of sperm whales of the Southern Hemisphere and their bearing on sperm whale biology. *Discovery Rep.* 37:1–324.

———. 1986. Cephalopods in the diet of odontocetes. In *Research on dolphins,* ed. M. M. Bryden and R. Harrison. Oxford: Clarendon Press.

Clarke, M. R., Martins, H. R., and Pascoe, P. 1993. The diet of sperm whales (*Physeter macrocephalus* Linnaeus 1758) off the Azores. *Phil. Trans. R. Soc. Lond.* B 339:67–82.

Clarke, R., and Paliza, O. 1988. Intraspecific fighting in sperm whales. *Rep. Int. Whal. Commn.* 38:235–41.

Clarke, R., Aguayo, A., and Paliza, O. 1980. Pregnancy rates of sperm whales in the southeast Pacific between 1959 and 1962 and a comparison with those from Paita, Peru, between 1975 and 1977. In *Sperm whales,* 151–58. Reports of the International Whaling Commission, special issue 2. Cambridge: International Whaling Commission.

Clay, C. S., and Medwin, H. 1977. *Acoustical oceanography.* New York: Wiley.

Clemmons, J. R., and Buchholz, R. 1997. Preface. In *Behavioral approaches to conservation in the wild,* ed. J. R. Clemmons and R. Buchholz. Cambridge: Cambridge University Press.

Clutton-Brock, T. H. 1984. Reproductive effort and terminal investment in iteroparous animals. *Am. Nat.* 123:212–29.

———. 1985. Size, sexual dimorphism, and polygyny in primates. In *Size and scaling in primate biology,* ed. W. L. Junger. New York: Plenum Press.

———. 1988. *Reproductive success.* Chicago: University of Chicago Press.

———. 1989a. Female transfer, male tenure and inbreeding avoidance in social mammals. *Nature* 337:70–71.

———. 1989b. Mammalian mating systems. *Proc. R. Soc. Lond.* B 235:339–72.

———. 1998. Reproductive concessions and skew in vertebrates. *Trends Ecol. Evol.* 7:288–92.

Clutton-Brock, T. H., and Albon, S. D. 1979. The roaring of red deer and the evolution of honest advertisement. *Behaviour* 69:145–70.

Clutton-Brock, T. H., and Harvey, P. H. 1977. Primate ecology and social organization. *J. Zool.* 183:1–39.

————. 1978. Introduction. In *Readings in sociobiology,* ed. T. H. Clutton-Brock and P. H. Harvey. San Francisco: W. H. Freeman.

Clutton-Brock, T. H., and Iason, G. R. 1986. Sex ratio variation in mammals. *Q. Rev. Biol.* 61:339–74.

Clutton-Brock, T. H., and Parker, G. A. 1994. Sexual coercion in animal societies. *Anim. Behav.* 49:1345–65.

————. 1995. Punishment in animal societies. *Nature* 373:209–16.

Clutton-Brock, T. H., Guinness, F. E., and Albon, S. D. 1982. *Red deer: The behaviour and ecology of two sexes.* Chicago: University of Chicago Press.

Clutton-Brock, T. H., O'Riain, M. J., Brotherton, P. N. M., Gaynor, D., Kansky, R., Griffin, A. S., and Manser, M. 1999. Selfish sentinels in cooperative mammals. *Science* 284:1640–44.

Cockcroft, V. G., and Ross, G. J. B. 1990a. Age, growth and reproduction of bottlenose dolphins, *Tursiops truncatus,* from the east coast of southern Africa. *Fish. Bull.* 88:289–302.

————. 1990b. Food and feeding of the Indian Ocean bottlenose dolphin off Southern Natal, South Africa. In *The bottlenose dolphin,* ed. S. Leatherwood and R. R. Reeves. San Diego: Academic Press.

————. 1990c. Observations on the early development of a captive bottlenose dolphin calf. In *The bottlenose dolphin,* ed. S. Leatherwood and R. R. Reeves. New York: Academic Press.

Cockcroft, V. G., Cliff, G., and Ross, G. J. B. 1989. Shark predation on Indian Ocean bottlenose dolphins *Tursiops truncatus* off Natal, South Africa. *S. Afr. J. Zool.* 24:305–10.

————. 1990. Observations on the early development of a captive bottlenose dolphin calf. In *The bottlenose dolphin,* ed. S. Leatherwood and R. R. Reeves. New York: Academic Press.

Colgan, P. W. 1978. *Quantitative ethology.* New York: John Wiley & Sons.

Collet, A., and Harrison, R. J. 1980. Ovarian characteristics, corpora lutea and corpora albicantia in *Delphinus delphis* stranded on the Atlantic coast of France. *Aquat. Mamm.* 8:69–76.

Committee on Low-Frequency Sound and Marine Mammals. 1994. *Low-frequency sound and marine mammals: Current knowledge and research needs.* Washington, DC: National Academy Press.

Condy, P. R., van Ararde, R. J., and Bester, M. N. 1978. The seasonal occurrence and behavior of killer whales, *Orcinus orca,* at Marion Island. *J. Zool.* 184:449–64.

Connor, R. C. 1986. Pseudo-reciprocity: Investing in mutualism. *Anim. Behav.* 34:1562–66.

————. 1995a. Altruism among non-relatives: Alternatives to the Prisoners Dilemma. *Trends Ecol. Evol.* 10:84–86.

————. 1995b. The benefits of mutualism: A conceptual framework. *Biol. Rev.* 70:427–57.

————. 1996. Partner preferences in by-product mutualisms and the case of predator inspection in fish. *Anim. Behav.* 51:451–454.

Connor, R. C., and Heithaus, M. R. 1996. Approach by great white shark elicits flight response in bottlenose dolphins. *Mar. Mamm. Sci.* 12:602–6.

Connor, R. C., and Norris, K. S. 1982. Are dolphins reciprocal altruists? *Am. Nat.* 19(3):358–74.

Connor, R. C., and Smolker, R. A. 1985. Habituated dolphins (*Tursiops* sp.) in Western Australia. *J. Mammal.* 66:398–400.

————. 1990. Quantitative description of a rare behavioral event: A bottlenose dolphins behavior toward her deceased offspring. In *The bottlenose dolphin,* ed. S. Leatherwood and R. R. Reeves. San Diego: Academic Press.

————. 1995. Seasonal changes in the stability of male-male bonds in Indian Ocean bottlenose dolphins (*Tursiops* sp.). *Aquat. Mamm.* 21:213–16.

————. 1996. "Pop" goes the dolphin: A vocalization male bottlenose dolphins produce during consortships. *Behaviour* 133:643–62.

Connor, R. C., Smolker, R. A., and Richards, A. F. 1992a. Dolphin alliances and coalitions. In *Coalitions and alliances in humans and other animals,* ed. A. H. Harcourt and F. B. M. de Waal. Oxford: Oxford University Press.

————. 1992b. Two levels of alliance formation among bottlenose dolphins (*Tursiops* sp.). *Proc. Natl. Acad. Sci. USA* 89:987–90.

Connor, R. C., Richards, A. F., Smolker, R. A., and Mann, J. 1996. Patterns of female attractiveness in Indian Ocean bottlenose dolphins. *Behaviour* 133:37–69.

Connor, R. C., Mann, J., Tyack, P. L., and Whitehead, H. 1998. Social evolution in toothed whales. *Trends Ecol. Evol.* 13:228–32.

Connor, R. C., Heithaus, M. R., and Barre, L. M. 1999. Superalliance of bottlenose dolphins. *Nature* 397:571–72.

Constantine, R., Visser, I., Buurman, D., Buurman, R., and McFadden, B. 1998. Killer whale (*Orcinus orca*) predation on dusky dolphins (*Lagenorynhcus obscurus*) in Kaikoura, New Zealand. *Mar. Mamm. Sci.* 14:324–30.

Cooke, J. G. 1986. A review of some problems relating to the assessment of sperm whale stocks, with reference to the western North Pacific. *Rep. Int. Whal. Commn.* 36:187–90.

————. 1991. Introduction and overview. In *Dolphins, porpoises and whales of the world: The IUCN Red Data Book,* ed. M. Klinowska. Gland, Switzerland: IUCN.

————. 1994. The management of whaling. *Aquat. Mamm.* 20:129–35.

Corbet, G. B., and Hill, J. E. 1991. *A world list of mammalian species.* 3d edition. Natural History Museum publication. Oxford: Oxford University Press.

Cords, M. 1987. Forest guenons and patas monkeys: Male-male competition in one-male groups. In *Primate societies,* ed. B. B. Smuts, D. L. Cheney, R. M. Seyfarth, R. W. Wrangham and T. T. Struhsaker, 98–111. Chicago: University of Chicago Press.

Corkeron, P. J. 1990. Aspects of the behavioral ecology of inshore dolphins *Tursiops truncatus* and *Sousa chinensis* in Moreton Bay, Australia. In *The bottlenose dolphin,* ed. S. Leatherwood and R. R. Reeves. San Diego: Academic Press.

Corkeron, P. J., and Connor, R. S. 1999. Why do baleen whales migrate? *Mar. Mamm. Sci.* In press.

Corkeron, P. J., Morris, R. J., and Bryden, M. M. 1987. Interactions between bottlenose dolphins and sharks in Moreton Bay, Queensland. *Aquat. Mamm.* 13:109–13.

Cornell, L. H., Asper, E. D., Antrim, J. E., Searles, S., Young, W. G., and Goff, T. 1987. Progress report: Results of a long-range captive breeding program for the bottle-nose dolphin

Tursiops truncatus and *Tursiops truncatus gilli. Zool. Biol.* 6: 41–54.

Costa, D. P. 1993. The secret life of marine mammals. *Oceanography* 6:120–28.

Costa, D. P., Worthy, G. A. J., Wells, R. S., Read, A. J., Scott, M. D., Irvine, A. B., and Waples, D. M. 1993. Seasonal changes in the field metabolic rate of bottlenose dolphins, *Tursiops truncatus*. Abstract, Tenth Biennial Conference on the Biology of Marine Mammals, Galveston, TX.

Cote, I. M., and Poulin, R. 1995. Parasitism and group size in social animals: A meta-analysis. *Behav. Ecol.* 6:159–65.

Couperus, A. S. 1997. Interactions between Dutch midwater trawl and Atlantic white-sided dolphins (*Lagenorhynchus acutus*) southwest of Ireland. *J. N.W. Atl. Fish. Sci.* 22:209–18.

Cox, C. R., and Le Boeuf, B. J. 1977. Female incitation of male competition: A mechanism of mate selection. *Am. Nat.* 111: 317–35.

Craig, J. L., and Jamieson, I. G. 1990. Pukeko: Different approaches and some different answers. In *Cooperative breeding in birds*, ed. P. B. Stacey and W. D. Koenig. Cambridge: Cambridge University Press.

Cranford, T. W., Amundin, M., and Norris, K. S. 1996. Functional morphology and homology in the odontocete nasal complex: Implications for sound generation. *J. Morphol.* 228: 1–64.

Croll, D. A., Tershy, B. R., Hewitt, R. P., Demer, D. A., Fiedler, P. C., Smith, S. E., Armstrong, W., Popp, J. M., Kiekhefer, T., Lopez, V. R., Urban, J., and D. Gendron. 1998. An integrated approach to the foraging ecology of marine birds and mammals. *Deep-Sea Res. II* 45:1353–71.

Cubbage, J. C., and Rugh, D. J. 1982. Bowhead whale length estimates and calf counts in the eastern Beaufort Sea, 1980. *Rep. Int. Whal. Commn.* 32:371–73.

Cummings, W. C., and Thompson, P. O. 1971a. Underwater sounds from the blue whale, *Balaenoptera musculus. J. Acoust. Soc. Am.* 50:1193–98.

Cummings, W. C., and Thompson, P. O. 1971b. Gray whales, *Eschrichtius robustus*, avoid the underwater sounds of killer whales, *Orcinus orca. Fish. Bull.* 69:525–30.

Cummins, J. E. 1988. Extinction: The PCB threat to marine mammals. *Ecologist* 18:193–95.

Curio, E. 1978. The adaptive significance of avian mobbing. *Z. Tierpsychol.* 48:175–83.

———. 1996. Conservation needs ethology. *Trends Ecol. Evol.* 11:260–63.

Curio, E., and Regelmann, K. 1986. Predator harassment implies a real deadly risk: A reply to Hennessy. *Ethology* 72:75–78.

Curren, K., Bose, N., and Lien, J. 1994. Swimming kinematics of a harbor porpoise (*Phocoena phocoena*) and an Atlantic white-sided dolphin (*Lagenorhynchus acutus*). *Mar. Mamm. Sci.* 10:485–92.

Curry, B. E. 1997. Phylogenetic relationships among bottlenose dolphins (genus *Tursiops*) in a worldwide context. Ph.D. dissertation, Texas A&M University, College Station.

Curry, B. E., and Smith, J. 1997. Phylogenetic structure of the bottlenose dolphin (*Tursiops truncatus*): Stock identification and implications for management. *Mar. Mamm. Sci.* Spec. Publ. no. 3:227–47.

Curry, B. E., Milinkovitch, M., Smith, J., and Dizon, A. E. 1995.

Stock structure of bottlenose dolphins, *Tursiops truncatus*. Abstract, Eleventh Biennial Conference on the Biology of Marine Mammals, Orlando, FL.

Cuvier, G. 1812. Rapport fait a la classe des sciences mathematiques et physiques, sur divers cétacés pris sur les cotes de France, principalement sur ceux qui sont echoues pres des Paimpol, le 7 janvier 1812. *Ann. du Mus. d'Hist. nat.*, xix, 1812, 1–16, pl. i.

———. 1823. Recherchees sur les ossemens fossiles, ou l'on retablit les caracteres de plusiers animaux dont les revolutions du globes detruit les especes. In *1re partie, contenant les rongeurs, les edentes, et les mammifères marins*, ed. G. Dufour and E. d'Ocagne. Paris.

———. 1829. Le règne animal distribué d'après son organisation, pour servir de base à l'histoire naturelle des animaux et d'introduction à l'anatomie comparée. Règne Anim., Nouv. Ed. 1:288.

Dahlheim, M. E. 1988. Killer whale (*Orcinus orca*) depredation on longline catches of sablefish (*Anoplopoma fimbria*) in Alaskan waters. National Marine Fisheries Service, Northwest and Alaska Fisheries Center Processed Report 88-14.

Dahlheim, M. E., and Matkin, C. O. 1994. Assessment of injuries to Prince William Sound killer whales. In *Marine mammals and the* Exxon Valdez, ed. T. R. Loughlin. San Diego: Academic Press.

Dahlheim, M. E., and Towell, R. G. 1994. Occurrence and distribution of Pacific white-sided dolphins (*Lagenorhynchus obliquidens*) in southeastern Alaska, with notes on an attack by killer whales (*Orcinus orca*). *Mar. Mamm. Sci.* 10:458–64.

Dahlheim, M. E., Ellifrit, D. K., and Swenson, J. D. 1997. *Killer whales of southeast Alaska: A catalogue of photo-identified individuals*. Seattle: Day Moon Press.

D'Amato, A., and Chopra, S. K. 1991. Whales: Their emerging right to life. *Am. J. Int. Law* 85:21–62.

Dans, S. L., Crespo, E. A., García, N. A., Reyes, L. M., Pedraza, S. N., and Alonso, M. K. 1997. Incidental mortality of Patagonian dusky dolphins in mid-water trawling: Retrospective effects from the early 1980s. *Rep. Int. Whal. Commn.* 47:699–703.

Darling, J. D. 1983. Migrations, abundance and behavior of Hawaiian humpback whales, *Megaptera novaeangliae* (Borowski). Ph.D. dissertation, University of California at Santa Cruz.

———. 1988. Whales: An era of discovery. *Nat. Geogr.* 174(6): 872–909.

Darling, J. D., and Cerchio, S. 1993. Movement of a humpback whale (*Megaptera novaeangliae*) between Japan and Hawaii. *Mar. Mamm. Sci.* 9:84–89.

Darling, J. D., and McSweeney, D. J. 1985. Observations on the migrations of North Pacific humpback whales (*Megaptera novaeangliae*). *Can. J. Zool.* 63:308–14.

Darling, J. D., and Mori, K. 1993. Recent observations of humpback whales (*Megaptera novaeangliae*) in Japanese waters off Ogasawara and Okinawa. *Can. J. Zool.* 71:325–33.

Darwin, C. R. 1871. *The descent of man, and selection in relation to sex.* New York: Appleton.

Davies, N. B., and Halliday, T. R. 1978. Deep croaks and fighting assessment in toads, *Bufo bufo. Nature* 391:56–58.

Davis, R. W., Collier, S. O., Hagey, W., Williams, T. M., and Le Boeuf, B. J. 1999a. A video system and three-dimensional recorder for marine mammals: Using virtual reality to study diving behavior. In *Wildlife Telemetry*, ed. Y. LeMaho. In press.

Davis, R. W., Fuiman, L. A., Williams, T. M., Collier, S. O., Hagey, W. P., Katanous, S. B., Kohin, S., and Horning, M. 1999b. Hunting behavior of a marine mammal beneath the arctic fast ice. *Science* 283:993–96.

Davis, R. W., Wartzok, D., Elsner, R., and Stone, H. 1992. Attaching a small video camera to a Weddell seal: A new way to observe diving behavior. In *Sensory Abilities of Aquatic Mammals*, ed. J. Thomas and R. Kastelein. New York: Plenum Publishing Corp., 631–42.

Dawbin, W. H. 1966. The seasonal migratory cycle of humpback whales. In *Whales, dolphins and porpoises*, ed. K. S. Norris. Berkeley: University of California Press.

Dawbin, W. H., and Eyre, E. J. 1991. Humpback whale songs along the coast of western Australia and some comparison with east coast songs. *Mem. Queensland Mus.* 30:249–54.

Dawbin, W. H., and Gill, P. C. 1991. Humpback whale survey along the west coast of Australia: A comparison of visual and acoustic observations. *Mem. Queensland Mus.* 30:255–57.

Dawkins, R., and Krebs, J. R. 1978. Animal signals: Information or manipulation. In *Behavioural ecology: An evolutionary approach*, ed. J. R. Krebs and N. B. Davies. Oxford: Blackwell Scientific Publications.

Dawson, S. M. 1991. Clicks and communication: The behavioural and social contexts of Hectors dolphin vocalizations. *Ethology* 88:265–76.

Dawson, S. M., and Slooten, E. 1993. Conservation of Hector's dolphins: The case and process which led to establishment of the Banks Peninsula Marine Mammal Sanctuary. *Aquat. Conserv. Mar. Freshw. Ecosystems* 3:207–21.

Dawson, S. M., and Thorpe, C. W. 1990. A quantitative analysis of the sounds of Hectors dolphin. *Ethology* 86:131–45.

Dawson, S. M., Chessum, C. J., Hunt, P. J., and Slooten, E. 1995. An inexpensive, stereographic technique to measure sperm whales from small boats. *Rep. Int. Whal. Commn.* 45: 431–36.

Dawson, S. M., Read, A., and Slooten, E. 1998. Pingers, porpoises and power: Uncertainties with using pingers to reduce bycatch of small cetaceans. *Biol. Conserv.* 84(2):141–46.

Dawson, W. W. 1980. The cetacean eye. In *Cetacean behavior: Mechanisms and functions*, ed. L. M. Herman. New York: Wiley-Interscience.

Day, D. 1987. *The whale war.* San Francisco: Sierra Club Books.

Debuffrenil, N., and Casinos, A. 1995. Observations on the microstructure of the rostrum of *Mesoplodon densirostris* (Mammalia, Cetacea, Ziphiidae)—the highest density bone known. *Ann. Sci. Natur-Zool. Biol. Anim.* 16:21–32.

Dedina, S., and Young, E. 1995. Conservation and development in the gray whale lagoons of Baja California Sur, Mexico. Final Report for Marine Mammal Commission Contract T10155592. NTIS PB96-113154.

Defran, R. H., and Pryor, K. 1980. The behavior and training of cetaceans in captivity. In *Cetacean behavior: Mechanisms and functions*, ed. L. M. Herman. New York: John Wiley & Sons.

Defran, R. H., and Weller, D. W. 1999. The occurrence, distribution, and site fidelity of bottlenose dolphins *(Tursiops truncatus)* in San Diego, California. *Mar. Mamm. Sci.* 15:366–80.

Defran, R. H., Shultz, G. M., and Weller, D. W. 1990. A technique for the photographic identification and cataloging of dorsal fins of the bottlenose dolphin *(Tursiops truncatus)*. In *Individual recognition of cetaceans: Use of photo-identification and other techniques to estimate population parameters*, ed. P. S. Hammond, S. A. Mizroch, and G. P. Donovan, 53–55. Reports of the International Whaling Commission, special issue 12. Cambridge: International Whaling Commission.

Defran, R. H., Weller, D. W., Kelly, D. L., and Espinoza, M. A. 1999. Range characteristics of Pacific bottlenose dolphins within the Southern California Bight. *Mar Mammal Sci.* 15: 381–93.

Delany, M. J. 1978. Introduction: Marking animals for research. In *Animal marking: Recognition marking of animals in research*, ed. B. Stonehouse. Baltimore: University Park Press.

DeMaster, D. P. 1984. Review of techniques used to estimate the average age at attainment of sexual maturity in marine mammals. In *Reproduction in whales, dolphins and porpoises: Proceedings of the conference, Cetacean Reproduction, Estimating Parameters for Stock Assessment and Management, La Jolla, CA, 28 Nov.–7 Dec. 1981*, ed. W. F. Perrin, R. L. Brownell Jr., and D. P. DeMaster, 175–80. Reports of the International Whaling Commission, special issue 6. Cambridge: International Whaling Commission.

Department of Fisheries and Oceans, Canada. 1991. The science of capelin—a variable resource. St John's, Newfoundland, Canada: Communications Branch, Department of Fisheries and Oceans.

Derix, R., van Hooff, J., De Vriess, H., and Wensing, J. 1993. Male and female mating competition in wolves: Female suppression vs. male intervention. *Behaviour* 127(1–2):141–74.

Derrickson, E. M. 1992. Comparative reproductive strategies of altricial and precocial eutherian mammals. *Funct. Ecol.* 6:57–65.

Desmoulins, A. 1822. Nord-Caper-Austral, *Balaena australis.* In *Dictionaire classique d'histoire naturelle*, ed. Bory de Saint-Vincent. Paris: Ray et Gravier.

Desportes, G., Saboureau, M., and Lacroix, A. 1993. Reproductive maturity and seasonality of male long-finned pilot whales, off the Faroe Islands. In *Biology of Northern Hemisphere pilot whales: A collection of papers*, ed. G. P. Donovan, C. H. Lockyer, and A. R. Martin, 233–62. Reports of the International Whaling Commission, special issue 14. Cambridge: International Whaling Commission.

DeVore, I. (ed.). 1965. *Primate behavior: Field studies of monkeys and apes.* New York: Holt, Rinehart and Winston.

DeVore, I., and Hall, K. R. L. 1965. Baboon ecology. In *Primate behavior: Field studies of monkeys and apes*, ed. I. DeVore. New York: Holt, Rinehart, and Winston.

Dewsbury, D. A. 1973. Introduction. In *Comparative psychology: A modern survey*, ed. D. A. Dewsbury and D. A. Rethlingshafer. New York: McGraw-Hill.

———. 1984. *Comparative psychology in the twentieth century.* Stroudsburg, PA: Hutchinson Ross.

Dewsbury, D. A., and Rethlingshafer, D. A. 1973. *Comparative psychology: A modern survey.* New York: McGraw-Hill.

Dillon, M. C. 1996. Genetic structure of sperm whale populations assessed by mitochondrial DNA sequence variation. Ph.D. dissertation, Dalhousie University, Halifax, Nova Scotia.

di Natale, A., and Mangano, A. 1985. Mating and calving of the Sperm Whale in the central Mediterranean Sea. *Aquat. Mamm.* 1:7–9.

di Natale, A., and di Sciara, G. N. 1994. A review of the passive fishing nets and traps used in the Mediterranean Sea and of their cetacean by-catch. In *Gillnets and cetaceans: Incorporating the proceedings of the symposium and workshop on the mortality of cetaceans in passive fishing nets and traps,* ed. W. F. Perrin, G. Donovan, and J. Barlow, 189–202. Reports of the International Whaling Commission, special issue 15. Cambridge: International Whaling Commission.

Dixon, A. F. 1987. Observations on the evolution of the genitalia and copulatory behaviour in male primates. *J. Zool.* 213:423–43.

Dizon, A. E., Southern, S. O., and Perrin, W. F. 1991. Molecular analysis of mtDNA types in exploited populations of spinner dolphins *(Stenella longirostris).* In *Genetic ecology of whales and dolphins: Incorporating the proceedings of the workshop on the genetic analysis of cetacean populations,* ed. A. R. Hoelzel, 183–202. Reports of the International Whaling Commission, special issue 13. Cambridge: International Whaling Commission.

Doidge, D. W. 1990. Age-length and age-weigh comparisons in the beluga, *Delphinapterus leucas. Can. Bull. Fish. Aquat. Sci.* 224:59–68.

Dolphin, W. F. 1987a. Dive behavior and estimated energy expenditure of foraging humpback whales in southeast Alaska. *Can. J. Zool.* 65:354–62.

———. 1987b. Observations of humpback whale, *Megaptera novaeangliae*—killer whale, *Orcinus orca,* interactions in Alaska: Comparison with terrestrial predator-prey relationships. *Can. Field Nat.* 101:70–75.

———. 1987c. Prey densities and foraging humpback whales, *Megaptera novaeangliae. Experientia* 43:468–71.

Donoghue, M. 1996. The New Zealand experience—one country's response to cetacean conservation. In *The conservation of whales and dolphins: Science and practice,* ed. M. P. Simmonds and J. Hutchinson. Chichester: Wiley.

Donovan, G. P. 1994. Developments on issues relating to the incidental catches of cetaceans since 1992 and the UNCED Conference. In *Gillnets and cetaceans: Incorporating the proceedings of the symposium and workshop on the mortality of cetaceans in passive fishing nets and traps,* ed. W. F. Perrin, G. Donovan, and J. Barlow, 609–13. Reports of the International Whaling Commission, special issue 15. Cambridge: International Whaling Commission.

Dorsey, E. M. 1983. Exclusive adjoining ranges in individually identified minke whales *(Balaenoptera acutorostrata)* in Washington state. *Can. J. Zool.* 61:174–81.

Dorsey, E. M., Richardson, W. J., and Würsig, B. 1989. Factors affecting surfacing, respiration, and dive behavior of bowhead whales, *Balaena mysticetus,* summering in the Beaufort Sea. *Can. J. Zool.* 67(7):1801–15.

Dorsey, E. M., Stern, S. J., Hoelzel, A. R., and Jacobsen, J. 1990. Minke whales *(Balaenoptera acutorostrata)* from the West Coast of North America: Individual recognition and small-scale site fidelity. In *Individual recognition of cetaceans: Use of photo-identification and other techniques to estimate population parameters,* ed. P. S. Hammond, S. A. Mizroch, and G. P. Donovan, 357–68. Reports of the International Whaling Commission, special issue 12. Cambridge: International Whaling Commission.

dos Santos, M. E., and Lacerda, M. 1987. Preliminary observations of the bottlenose dolphin *(Tursiops truncatus)* in the Sado Esturay (Portugal). *Aquat. Mamm.* 13:65–80.

dos Santos, M. E., Ferreira, A. J., and Harzen, S. 1995. Rhythmic sound sequences emitted by aroused bottlenose dolphins in the Sado estuary, Portugal. In *Sensory systems of aquatic mammals,* ed. R. A. Kasteliein, J. A. Thomas, and P. E. Nachtigall. Woerden, The Netherlands: De Spil Publishers.

Douglas-Hamilton, I., and Douglas-Hamilton, O. 1975. *Among the elephants.* New York: Viking Press.

Dreher, J., and Evans, W. E. 1964. Cetacean communication. In *Marine bioacoustics,* vol. 1, ed. W. N. Tavolga. Oxford: Pergamon Press.

Drinnan, R. L., and Sadleir, R. M. F. S. 1981. The suckling behavior of a captive beluga *(Delphinapterus leucas)* calf. *Appl. Anim. Ethol.* 7:179–85.

Dudok van Heel, W. H. 1981. Investigations on cetacean sonar. III. A proposal for an ecological classification of cetaceans in relation to sonar. *Aquat. Mamm.* 8:65–68.

Dudzinski, K. M., Clark, C. W., and Würsig, B. 1995. A mobile video/acoustic system for simultaneous underwater recording of dolphin interactions. *Aquat. Mamm.* 21:187–93.

Dufault, S., and Whitehead, H. 1993. Assessing the stock identity of sperm whales in the eastern equatorial Pacific. *Rep. Int. Whal. Commn.* 43:469–75.

———. 1995a. An assessment of changes with time in the marking patterns used for photo-identification of individual sperm whales, *Physeter macrocephalus. Mar. Mamm. Sci.* 11:335–43.

———. 1995b. An encounter with recently wounded sperm whales *(Physeter macrocephalus). Mar. Mamm. Sci.* 11:560–63.

———. 1995c. The geographic stock structure of female and immature sperm whales in the South Pacific. *Rep. Int. Whal. Commn.* 45:401–5.

———. 1998. Regional and group-level differences in fluke markings and fluke notch type of sperm whales of the South Pacific. *J. Mammal.* 79:514–20.

Dufault, S., Whitehead, H., and Dillon, M. 1999. An examination of the current knowledge on the stock structure of sperm whales *(Physeter macrocephalus)* worldwide. *J. Cetacean Res. Mgmt.* 1:1–10.

Duffield, D. A., and Miller, K. W. 1988. Demographic features of killer whales in oceanaria in the United States and Canada, 1965–1987. *Rit Fiskideildar* 11:297–306.

Duffield, D. A., and Wells, R. S. 1986. Population structure of bottlenose dolphins: Genetic studies of bottlenose dolphins along the central west coast of Florida. Contract Report to National Marine Fisheries Service, Southeast Fisheries Center. Contract no. 45-WCNF-5-00366.

———. 1991. The combined application of chromosome protein and molecular data for the investigation of social unit structure and dynamics in *Tursiops truncatus.* In *Genetic ecology of whales and dolphins: Incorporating the proceedings of the workshop on the genetic analysis of cetacean populations,* ed. A. R. Hoelzel, 155–70. Reports of the International Whaling Commission, special issue 13. Cambridge: International Whaling Commission.

Duffield, D. A., Wells, R. S., Scott, M. D., Chamberlin-Lea, J., and Sheehy, R. R. 1991. Paternity in a free-ranging bottlenose dolphin society. Abstract, Ninth Biennial Conference on the Biology of Marine Mammals, Chicago.

Duffield, D. A., Wells, R. S., Lenox, J. S., and Moors, T. 1994. Analysis of paternity in a free-ranging bottlenose dolphin society by DNA fingerprinting and behavioral coefficients of association. Abstract, International Symposium on Marine Mammal Genetics, La Jolla, CA.

Duffield, D. A., Odell, D. K., McBain, J. F., and Andrews, B. 1995. Killer whale *(Orcinus orca)* reproduction at Sea World. *Zoo Biol.* 14:417–30.

Duffus, D. A. 1993. Recreational use, valuation, and management of killer whales *(Orcinus orca)* on Canada's Pacific coast. *Environ. Cons.* 20:149–56.

Duffus, D. A., and Baird, R. W. 1995. Killer whales, whalewatching and management: A status report. *Whalewatcher* 29(2): 14–17.

Duffus, D. A., and Dearden, P. 1992. Whales, science, and protected area management in British Columbia, Canada. *George Wright Forum* 9:79–87.

Dunbar, R. I. M. 1976. Some aspects of research design and their implications in the observational study of behaviour. *Behaviour* 58:79–98.

———. 1987. Demography and reproduction. In *Primate societies,* ed. B. B. Smuts, D. L. Cheney, R. M. Seyfarth, R. W. Wrangham, and T. T Struhsaker. Chicago: University of Chicago Press.

———. 1992. Neocortex size as a constraint on group size in primates. *J. Hum. Evol.* 20:469–93.

Dunbar, R. I. M., and Cowlishaw, G. 1992. Mating success in male primates. Dominance rank, sperm competition and alternative strategies. *Anim. Behav.* 44:1171–73.

Duncan, P., and Vigne, N. 1979. The effect of group size in horses on the rate of attacks by blood-sucking flies. *Anim. Behav.* 27:623–25.

Dunn, D. G., Barco, S., McLellan, W. A., and Pabst, D. A. 1998. Virginia Atlantic bottlenose dolphin *(Tursiops truncatus)* strandings: Gross pathological findings in ten traumatic deaths. Abstract, Atlantic Coast Dolphin Conference, Sarasota, FL.

Dunning, D. J., Ross, Q. E., Geoghegan, P., Reichle, J. J., Menezes, J. K., and Watson, J. K. 1992. Alewives avoid high-frequency sound. *N. Am. J. Fish. Mgmt.* 12:407–16.

Duvernoy, G. L. 1851. Mémoire sur les charact. *Annales des Sciences Naturelles* 3, 15:41.

D'Vincent, C. G., Nilson, R. M., and Hanna, R. E. 1985. Vocalization and coordinated feeding behavior of the humpback whale in southeastern Alaska. *Sci. Rep. Whales Res. Inst.* 36: 41–47.

Earle, M. 1996. Incidental catches of small cetaceans. In *The conservation of whales and dolphins: Science and practice,* ed. M. P. Simmonds and J. Hutchinson. Chichester: Wiley.

Eastcott, A., and Dickinson, T. 1987. Underwater observations of the suckling and social behaviour of a new-born bottlenosed dolphin *(Tursiops truncatus). Aquat. Mamm.* 13(2):51–56.

Eaton, R. L. 1979. A beluga whale imitates human speech. *Carnivore* 2:22–23.

Eberhard, W. G. 1996. *Female control: Sexual selection by cryptic female choice.* Princeton, NJ: Princeton University Press.

Economos, A. C. 1983. Elastic and/or geometric similarity in mammalian design? *J. Theor. Biol.* 103:167–82.

Edds, P. L. 1982. Vocalizations of the blue whale, *Balaenoptera musculus,* in the St. Lawrence River. *J. Mammal.* 63:345–47.

Eggert, L. S., Lux, C. A., O'Corry-Crowe, G. M., and Dizon, A. E. 1998. Dried dolphin blood on fishery observer records provides DNA for genetic analyses. *Mar. Mamm. Sci.* 14: 136–43.

Ehrenberg, C. G. 1828, 1833. *Mammalia,* Decades I and II. Vol. 1 in *Symbolae Physicae, seu Icones et descriptiones Corporum Naturalium novorum aut mins cogitorum, quae ex itineribus per Libyam, Aegyptum, Nubiam, Dongalam, Syriam, Arabiam et Habessiniam publico institutis sumptu Friderici Guilelemi Hemprich et Christiani Godofredi Ehrenberg studio annis MDCCCXX-MDCCCXXV redierunt, Pars Zoologica.* Officina Academica, Berolini (Berlin).

Ehrlich, P. R. 1996. Environmental anti-science. *Trends Ecol. Evol.* 11:393.

Einhorn, R. N. 1967. Dolphins challenge the designer. *Electronic Design* 25: 49–64.

Eisenberg, J. F. 1981. *The mammalian radiations.* Chicago: University of Chicago Press.

Ellis, R. 1991. *Men and Whales.* New York: Knopf.

Ellison, W. T., Clark, C. W., and Bishop, G. C. 1987. Potential use of surface reverberation by bowhead whales, *Balaena mysticetus,* in under-ice navigation. *Rep. Int. Whal. Commn.* 37: 329–32.

Emlen, S. T. 1984. Cooperative breeding in birds and mammals. In *Behavioural ecology: An evolutionary approach,* 2d ed., ed. J. R. Krebs and N. B. Davies. Oxford: Blackwell Scientific Publications.

———. 1997. Predicting family dynamics in social vertebrates. In *Behavioural ecology: An evolutionary approach,* 4th ed., ed. J. R. Krebs and N. B. Davies. Oxford: Blackwell Scientific Publications.

Emlen, S. T., and Oring, L. W. 1977. Ecology, sexual selection, and the evolution of mating systems. *Science* 197:215–23.

Enquist, M., and Leimar, O. 1990. The evolution of fatal fighting. *Anim. Behav.* 39:1–9.

Epple, G. 1978. Reproductive and social behavior of marmosets with special reference to captive breeding. *Primate Med.* 10: 50–62.

Essapian, F. S. 1962. Courtship in the saddle-backed porpoises, *Delphinus delphis,* L. 1758. *Z. Saugertierkunde* 27:211–17.

———. 1963. Observations on abnormalities of parturition in captive bottle-nosed dolphins, *Tursiops truncatus,* and concurrent behavior of other porpoises. *J. Mammal.* 44:405–14.

Evans, P. G. H. 1987. *The natural history of whales and dolphins.* New York: Facts on File Publications.

Evans, W. E. 1967. Vocalization among marine mammals. In

Marine bioacoustics, vol. 2, ed. W. N. Tavolga. Oxford: Pergamon Press.

———. 1973. Echolocation by marine delphinids and one species of fresh-water dolphin. *J. Acoust. Soc. Am.* 54:191–99.

———. 1974. Radio telemetric studies of two species of small odontocete cetaceans. In *The whale problem: A status report*, ed. W. E. Schevill, G. C. Ray, and K. S. Norris. Cambridge, MA: Harvard University Press.

———. 1994. Common dolphin, white-bellied porpoise. In *Handbook of marine mammals*, vol. 5: *The first book of dolphins*, ed. S. H. Ridgway and R. J. Harrison. New York: Academic Press.

Evans, W. E., and Bastian, J. 1969. Marine mammal communication: Social and ecological factors. In *The biology of marine mammals*, ed. H. T. Andersen. New York: Academic Press.

Evans, W. E., and Dreher, J. J. 1962. Observations on scouting behavior and associated sound production by the Pacific bottlenose porpoise (*Tursiops gilli* Dall). *Bull. S. Calif. Acad. Sci.* 61:217–26.

Fagen, R. 1993. Primate juveniles and primate play. In *Juvenile primates: Life-history, development, and behavior*, ed. M. E. Pereira and L. A. Fairbanks. New York: Oxford University Press.

Fairbairn, D. J. 1990. Factors influencing sexual size dimorphism in temperate waterstriders. *Am. Nat.* 136:61–86.

Fairbanks, L. A. 1988. Vervet monkey grandmothers: Effects on mother-infant relationships. *Behaviour* 104:176–88.

———. 1990. Reciprocal benefits of allomothering for female vervet monkeys. *Anim. Behav.* 40:553–62.

———. 1993. Juvenile vervet monkeys: Establishing relationships and practicing skills for the future. In *Juvenile primates: Life-history, development and behavior*, ed. M. E. Pereira and L. A. Fairbanks, 211–27. New York: Oxford University Press.

Fairbanks, L. A., and McGuire, M. T. 1986. Age, reproductive value and dominance-related behaviour in vervet monkey females: Cross-generational influences on social relationships and reproduction. *Anim. Behav.* 34:1710–21.

Fairfield, C. P., Waring, C. T., and Sano, M. H. 1993. Pilot whales incidentally taken during the distant water fleet Atlantic mackerel fishery in the mid-Atlantic Bight, 1984–88. In *Biology of Northern Hemisphere pilot whales: A collection of papers*, ed. G. P. Donovan, C. H. Lockyer, and A. R. Martin, 107–16. Reports of the International Whaling Commission, special issue 14. Cambridge: International Whaling Commission.

Falls, J. B. 1992. Playback: A historical perspective. In *Playback and studies of animal communication*, ed. P. K. McGregor. New York: Plenum Press.

Faucher, A., and Whitehead, H. 1991. Population biology, social structure and movements of northern bottlenose whales (*Hyperoodon ampullatus*) in The Gully, Nova Scotia. Abstracts, Ninth Biennial Conference on the Biology of Marine Mammals, Chicago.

Feinholz, D. M. 1995. Northern range extension, abundance, and distribution of Pacific coastal bottlenose dolphins (*Tursiops truncatus gilli*) in Monterey Bay, California. Abstract, Eleventh Biennial Conference on the Biology of Marine Mammals, Orlando, FL.

Félix, F. 1994. Ecology of the coastal bottlenose dolphin *Tursiops truncatus* in the Gulf of Guayaquil, Ecuador. *Invest. Cetacea* 25:235–56.

Felleman, F. L., Heimlich-Boran, J. R., and Osborne, R. W. 1991. The feeding ecology of killer whales (*Orcinus orca*) in the Pacific Northwest. In *Dolphin societies: Discoveries and puzzles*, ed. K. Pryor and K. S. Norris. Berkeley: University of California Press.

Fertl, D. 1994. Occurrence patterns and behavior of bottlenose dolphins (*Tursiops truncatus*) in the Galveston ship channel, Texas. *Tex. J. Sci.* 46:299–317.

Fertl, D., and Leatherwood, S. 1997. Cetacean interactions with trawls: A preliminary review. *J. N. W. Atl. Fish. Sci.* 22:219–48.

Fertl, D., and Würsig, B. 1995. Coordinated feeding by Atlantic spotted dolphins (*Stenella frontalis*) in the Gulf of Mexico. *Aquat. Mamm.* 21:3–5.

Fichtelius, K., and Sjolander, S. 1972. *Smarter than man? Intelligence in whales, dolphins, and humans.* New York: Random House.

Findlay, K. P., Best, P. B., Peddemors, V. M., and Gove, D. 1994. The distribution and abundance of humpback whales on their southern Mozambique breeding grounds. *Rep. Int. Whal. Commn.* 44:311–20.

Fine, M. L. 1978. Seasonal and geographical variation of the mating call of the oyster toadfish *Opsanus tau. Oecologia* 36:45–47.

Finley, K. J., Miller, G. W., Davis, R. A., and Greene, C. R. 1990. Reactions of belugas, *Delphinapterus leucas*, and narwhals, *Monodon monoceros*, to ice-breaking ships in the Canadian high arctic. *Can. Bull. Fish. Aquat. Sci.* 224:97–117.

Fish, J. F., and Vania, J. S. 1971. Killer whale, *Orcinus orca*, sounds repel white whales, *Delphinapterus leucas. Fish. Bull.* 69:531–35.

Fisher, R. A. 1930. *The genetical theory of natural selection.* Clarendon Press, Oxford.

———. 1958. *The genetical theory of natural selection.* 2d edition. New York: Dover Publications.

Fitch, W. T. 1997. Vocal tract length and formant frequency dispersion correlate with body size in rhesus macaques. *J. Acoust. Soc. Am.* 102(2):1213–22.

FitzGibbon, C. D. 1994. The costs and benefits of predator inspection behavior in Thomson's gazelles. *Behav. Ecol. Sociobiol.* 34:139–48.

FitzGibbon, C. D., and Fanshawe, J. H. 1988. Stotting in Thomson's gazelles: An honest signal of condition. *Behav. Ecol. Sociobiol.* 23:69–74.

Fleiss, J. L. 1981. *Statistical methods for rates and proportions.* New York: John Wiley & Sons.

Fletcher, S., Le Boeuf, B. J., Costa, D. P., Tyack, P. L., and Blackwell, S. B. 1996. Onboard acoustic recording from diving northern elephant seals. *J. Acoust. Soc. Am.* 100:2531–39.

Flórez-González, L. 1991. Humpback whales *Megaptera novaeangliae* in the Gorgona Island, Colombian Pacific breeding waters: Population and pod characteristics. *Mem. Queensland Mus.* 30:291–95.

Flórez-González, L., Capella, J. J., and Rosenbaum, H. C. 1994. Attack of killer whales (*Orcinus orca*) on humpback whales

(*Megaptera novaeangliae*) on a South American Pacific breeding ground. *Mar. Mamm. Sci.* 10:218–22.

Flower, W. H. 1882. On the cranium of the new species of *Hyperoodon* from the Australian seas. *Proc. Zool. Soc. Lond.* 1882: 392–96.

Food and Agriculture Organization of the United Nations. 1995. *Code of conduct for responsible fisheries.* Rome: Food and Agriculture Organization of the United Nations.

Ford, J. K. B. 1989. Acoustic behaviour of *resident* killer whales (*Orcinus orca*) off Vancouver Island, British Columbia. *Can. J. Zool.* 67(3):727–45.

———. 1991. Vocal traditions among *resident* killer whales (*Orcinus orca*) in coastal waters of British Columbia. *Can. J. Zool.* 69:1454–83.

Ford, J. K. B., and Fisher, H. D. 1982. Killer whale (*Orcinus orca*) dialects as an indicator of stocks in British Columbia. *Rep. Int. Whal. Commn.* 32:671–79.

———. 1983. Group-specific dialects of killer whales (*Orcinus orca*) in British Columbia. *AAAS Sel. Symp.* 76:129–61.

Ford, J. K. B., and Hubbard-Morton, A. B. 1990. Vocal behavior and dialects of *transient* killer whales in coastal waters of British Columbia, California and southeast Alaska. In Abstracts of the Third International Orca Symposium, Victoria, BC, March 1990: 6.

Ford, J. K. B., Ellis, G. M., and Nichol, L. M. 1992. Killer whales of the Queen Charlotte Islands—a preliminary study of the abundance, distribution and population identity of *Orcinus orca* in the waters of Haida Gwaii. Report to South Moresby/ Gwaii Haanas National Park Reserve, Canadian Parks Service.

Ford, J. K. B., Ellis, G. M., and Balcomb, K. C. 1994. *Killer whales.* Vancouver: UBC Press; Seattle: University of Washington Press.

Forney, K. A. 1999. Trends in harbor porpoise abundance off central California, 1986–95: Evidence for interannual changes in distribution? *J. Cetacean Res. Mgmt.* 1:73-80.

Forney, K. A., Barlow, J., and Carretta, J. V. 1995. The abundance of cetaceans in California waters. Part II: Aerial surveys in winter and spring of 1991 and 1992. *Fish. Bull.* 93:15–26.

Forster, J. R. 1770. In *1770–71 Travels into North America,* ed. P. Kalm. London.

Foster, M. S. 1981. Cooperative behavior and social organization of the Swallow-tailed Manakin (*Chiroxiphia caudata*). *Behav. Ecol. Sociobiol.* 9:167–77.

Foster, W. A., and Treherne, J. E. 1981. Evidence for the dilution effect in the selfish herd from fish predation on a marine insect. *Nature* 293:466–67.

Fowler, C. W. 1984. Density dependence in cetacean populations. In *Reproduction in whales, dolphins and porpoises: Proceedings of the conference, Cetacean Reproduction, Estimating Parameters for Stock Assessment and Management, La Jolla, CA, 28 Nov.–7 Dec. 1981,* ed. W. F. Perrin, R. L. Brownell Jr., and D. P. DeMaster, 373–79. Reports of the International Whaling Commission, special issue 6. Cambridge: International Whaling Commission.

Fowler, C. W., and Smith, T. D. (eds.). 1981. *Dynamics of large mammal populations.* New York: Wiley.

Fragaszy, D. M., Boinski, S., and Whipple, J. 1992. Behavioral sampling in the field: Comparison of individual and group sampling methods. *Am. J. Primatol.* 26:259–75.

Frank, L. G. 1986. Social organization of the spotted hyaena (*Crocuta crocuta*). II. Dominance and reproduction. *Anim. Behav.* 34:1510–27.

Frank, S. A. 1995. Mutual policing and repression of competition in the evolution of cooperative groups. *Nature* 377:520–22.

———. 1996. Policing and group cohesion when resources vary. *Anim. Behav.* 52:1163–69.

Frankel, A. S., Clark, C. W., Herman, L. M., and Gabriele, C. M. 1995. Spatial distribution, habitat utilization, and social interactions of humpback whales, *Megaptera novaeangliae,* off Hawai'i, determined using acoustic and visual techniques. *Can. J. Zool.* 73:1134–46.

Frantzis, A. 1998. Does acoustic testing strand whales? *Nature* 392:29.

Fraser, F. C. 1956. A new Sarawak dolphin. *Sarawak Mus. J.* 7(8):478–503.

———. 1974. Report on Cetacea stranded on the British coasts from 1948–1966. *Publ. Brit. Mus. Nat. Hist.* no. 14.

Frazer, J. F. D., and Huggett, A. St. G. 1973. Specific foetal growth rates of cetaceans. *J. Zool.* 169:111–26.

Friedl, W. A., Nachtigall, P. E., Moore, P. W. B., Chun, N. K. W., Haun, J. E., Hall, R. W., and Richards, J. L. 1990. Taste reception in the Pacific bottlenose dolphin (*Tursiops truncatus gilli*) and the California sea lion (*Zalophus californianus*). In *Sensory abilities of cetaceans,* ed. J. Thomas and R. Kastelein. New York: Plenum.

Fristrup, K. M., and Harbison, G. R. 1993. Vision and sperm whale foraging. Abstract, Tenth Biennial Conference of the Biology of Marine Mammals, Galveston, TX.

Fruth, B., and Hohman, G. 1994. Comparative analyses of nest-building behavior in bonobos and chimpanzees. In *Chimpanzee cultures,* ed. R. W. Wrangham, W. C. McGrew, F. B. M. de Waal, and P. G. Heltne. Cambridge, MA: Harvard University Press.

Fuller, K. 1995. President's note. *Conservation Issues (WWF),* 2.

Furuichi, T. 1989. Social interactions and the life history of female *Pan paniscus* in Wamba, Zaire. *Int. J. Primatol.* 10:173–97.

Gadgil, M. 1982. Changes with age in the strategy of social behavior. In *Perspectives in ethology,* vol. 5, ed. P. P. G. Bateson and P. H. Klopfer. New York: Plenum Press.

Gagneux, P., Woodruff, D. S., and Boesch, C. 1997. Furtive mating in female chimpanzees. *Nature* 387:358–59.

Galdikas, B. M. F. 1995. *Reflections of Eden: My years with the orangutans of Borneo.* Boston: Little Brown.

Galef, B. G. Jr. 1996. Historical origins: The making of a science. In *Foundations of animal behavior: Classic papers with commentaries,* ed. L. D. Houck and L. C. Drickamer. Chicago: University of Chicago Press.

Gambell, R. 1968a. Aerial observations of sperm whale behaviour based on observations, notes and comments by K. J. Pinkerton. *Norsk Hvalfangst-Tidende* 57:126–38.

———. 1968b. Seasonal cycles and reproduction in sei whales in the Southern Hemisphere. *Discovery Rep.* 35:31–134.

———. 1973. Some effects of exploitation on reproduction in whales. *J. Reprod. Fertil. Suppl.* 19:531–51.

Gambell, R., Lockyer, C., and Ross, G. J. B. 1973. Observations on the birth of a sperm whale calf. *South Afr. J. Sci.* 69:147–48.

Gamboa, G. J. 1978. Intraspecific defense: Advantage of social cooperation among paper wasp foundresses. *Science* 199: 1463–65.

Gannon, D. P., Read, A. J., Craddock, J. E., and Mead, J. G. 1997. Stomach contents of long-finned pilot whales *(Globicephala melas)* stranded on the U.S. mid-Atlantic coast. *Mar. Mamm. Sci.* 13(3):405–18.

Gans, C. 1978. All animals are interesting! *Am. Zool.* 18:3–9.

Gaskin, D. E. 1967. Luminescence in a squid *Moroteuthis* sp. (probably *ingens* Smith) and a possible feeding mechanism in the sperm whale *Physeter catodon* L. *Tuatara* 15:86–88.

———. 1982. *The ecology of whales and dolphins.* London: Heinemann.

———. 1987. Updated status of the right whale, *Eubalaena glacialis*, in Canada. *Can. Field Nat.* 101:295–309.

Gaskin, D. E., Smith, G. J. D., Watson, A. P., Yasui, W. Y., and Yurick, D. B. 1984. Reproduction in the porpoises (Phocoenidae): Implications for management. In *Reproduction in whales, dolphins and porpoises: Proceedings of the conference, Cetacean Reproduction, Estimating Parameters for Stock Assessment and Management, La Jolla, CA, 28 Nov.–7 Dec. 1981*, ed. W. F. Perrin, R. L. Brownell Jr., and D. P. DeMaster, 135–48. Reports of the International Whaling Commission, special issue 6. Cambridge: International Whaling Commission.

Geist, V. 1966. The evolution of horn-like organs. *Behaviour* 27: 175–214.

———. 1971. A behavioural approach to the management of wild ungulates. In *The scientific management of animal and plant communities for conservation: The eleventh symposium of the British Ecological Society, University of East Anglia, Norwich, 7–9 July 1970*, ed. E. Duffey and A. S. Watt. Oxford: Blackwell.

George, J. C., Clark, C. W., Carrol, G. M., and Ellison, W. T. 1989. Observations on the ice-breaking and ice navigation behavior of migrating bowhead whales *(Balaena mysticetus)* near Point Barrow, Alaska, Spring 1985. *Arctic* 42:24–30.

George, J. C., Bada, J., Zeh, J., Scott, L., Brown, S. E., and O'Hara, T. 1998. Preliminary age estimates of bowhead whales via aspartic acid racemization. Document SC/50/AS10 presented to the 50th Annual Meeting of the Scientific Committee of the International Whaling Commission, Muscat, Oman.

Geraci, J. R., Anderson, D. M., Timperi, R. J., St. Aubin, D. J., Early, G. A., Prescott, J. H., and Mayo, C. A. 1989. Humpback whales *(Megaptera novaeangliae)* fatally poisoned by dinoflagellate toxins. *Can. J. Fish. Aquat. Sci.* 46:1895–98.

Gerrodette, T. 1987. A power analysis for detecting trends. *Ecology* 68:1364–72.

Gerson, H. B., and Hickie, J. P. 1985. Head scarring on male narwhals *(Monodon monoceros)*: Evidence for aggressive tusk use. *Can. J. Zool.* 63:2083–87.

Gervais, F. L. P. 1855. *Histoire naturelle des mammifères.* Paris: L. Curmer.

Gervais, F. L. P., and Deville, E. 1853. Sur les mammifères marins qui frequentent les co. *Bulletin de la Société Centrale d'Agriculture et des Comices Agricoles du département de l'Héault*, 40: 140–55.

Gervais, F. L. P., and d'Orbigny, A. C. V. D. 1844. Séance du 27 avril 1844. Mammalogie. *Bulletin de la Société Philomathique de Paris*, 1844:39.

Giard, J., and Michaud, R. 1997. L'observation des rorquals communs sous surveillance par tilimirie VHF. *Naturaliste Canadien* 121:25–30.

Gill, P. C., and Burton, C. L. K. 1995. Photographic resight of a humpback whale between Western Australia and Antarctic Area IV. *Mar. Mamm. Sci.* 11:96–100.

Gill, P. C., and Thiele, D. 1997. A winter sighting of killer whales *(Orcinus orca)* in Antarctic sea ice. *Polar Biol.* 17: 401–4.

Gill, T. 1865. On two species of Delphinidae, from California, in the Smithsonian Institution. *Proc. Acad. Nat. Sci. Phila.* 1865:177–78.

Ginsberg, J. R., and Young, T. P. 1992. Measuring association between individuals or groups in behavioural studies. *Anim. Behav.* 44(2):377–79.

Giraldeau, L.-A., and Caraco, T. 1993. Genetic relatedness and group size in an aggregation economy. *Evol. Ecol.* 7:429–38.

Gish, S. L. 1979. A quantitative description of two-way acoustic communication between captive Atlantic bottlenosed dolphins (*Tursiops truncatus* Montagu). Ph.D. dissertation, University of California at Santa Cruz.

Gittleman, J. L. 1985. Functions of communal care in mammals. In *Evolution: Essays in honour of John Maynard Smith*, ed. P. J. Greenwood, P. H. Harvey, and M. Slatkin. Cambridge: Cambridge University Press.

Glass, B. 1965. The ethical basis of science. *Science* 150:1254–61.

Glickman, S. E., Frank, L. G., Davidson, J. M., Smith, E. R., and Siiteri, P. K. 1987. Androstenedione may organize or activate sex-reversed characteristics in female spotted hyenas. *Proc. Natl. Acad. Sci. USA* 84:3444–47.

Glockner, D. A. 1983. Determining the sex of humpback whales *(Megaptera novaeangliae)* in their natural environment. In *Communication and behavior of whales*, ed. R. Payne. Boulder, CO: Westview Press.

Glockner D. A., and Venus, S. C. 1983. Identification, growth rate and behavior of humpback whale, *Megaptera novaeangliae*, cows and calves, in the waters off Maui, Hawaii, 1977–79. In *Communication and behavior of whales*, ed R. S. Payne. Boulder, CO: Westview Press.

Glockner-Ferrari, D. A., and Ferrari, M. J. 1985. Individual identification, behavior, reproduction and distribution of humpback whales, *Megaptera novaeangliae*, in Hawaii. National Technical Information Service Report no. MMC-83/06, Springfield, VA.

———. 1990. Reproduction in the humpback whale *(Megaptera novaeangliae)* in Hawaiian waters, 1975–1988: The life history, reproductive rates, and behaviour of known individuals identified through surface and underwater photography. In *Individual recognition of cetaceans: Use of photo-identification and other techniques to estimate population parameters*, ed. P. S. Hammond, S. A. Mizroch, and G. P. Donovan, 161–69. Reports of the International Whaling Commission, special issue 12. Cambridge: International Whaling Commission.

Godin, J. G. J., and Davis, S. A. 1995. Who dares, benefits: Predator approach behaviour in the guppy *(Poecilia reticulata)* deters predator pursuit. *Proc. R. Soc. Lond.* B 259:193–200.

Goldberg, T. L., and Wrangham, R. W. 1997. Genetic correlates of social behavior in wild chimpanzees: Evidence from mitochondrial DNA. *Anim. Behav.* 54:559–70.

Goley, P. D., and Straley, J. M. 1994. Attack on gray whales *(Eschrichtius robustus)* in Montery Bay, California, by killer whales *(Orcinus orca)* previously identified with Glacier Bay, Alaska. *Can. J. Zool.* 72:1528–30.

Gomendio, M., and Roldan, E. R. S. 1993. Mechanisms of sperm competition: Linking physiology and behavioural ecology. *Trends Ecol. Evol.* 8:95–100.

Gompper, M. E. 1996. Sociality and asociality in white-nosed coatis *(Nasua narica):* Foraging costs and benefits. *Behav. Ecol.* 7:251–63.

Gonzalez, F. T. 1994. The use of photoidentification to study the Amazon river dolphin, *Inia geoffrensis,* in the Colombian Amazon. *Mar. Mamm. Sci.* 10(3):348–53.

Goodall, J. 1965. Chimpanzees of the Gombe Stream Reserve. In *Primate behavior,* ed. I. DeVore. New York: Holt, Rinehart and Winston.

———. 1967. *My friends, the wild chimpanzees.* Washington, DC: National Geographic Society.

———. 1971. *In the shadow of man.* Boston: Houghton Mifflin.

———. 1986. *The chimpanzees of Gombe: Patterns of behavior.* Cambridge, MA: Harvard University Press.

Goodall, R. N. P., Galeazzi, A. R., Leatherwood, S., Miller, K. W., Cameron, I. S., Kastelein, R. K., and Sobral, A. P. 1988. Studies of Commerson's dolphins, *Cephalorhynchus commersonii,* off Tierra del Fuego, 1976–1984, with a review of information on the species in the South Atlantic. In *Biology of the genus* Cephalorhynchus, ed. R. L. Brownell Jr. and G. P. Donovan, 3–70. Reports of the International Whaling Commission, special issue 9. Cambridge: International Whaling Commission.

Goodall, R. N. P., Baker, A. N., Best, P. B., Meyer, N., and Miyazaki, N. 1997. On the biology of the hourglass dolphin, *Lagenorhynchus cruciger* (Quoy and Gaimard, 1824). *Rep. Int. Whal. Commn.* 47:485–99.

Goodman, D. 1984. Statistics of reproductive rate estimates, and their implication for population protection. In *Reproduction in whales, dolphins and porpoises: Proceedings of the conference, Cetacean Reproduction, Estimating Parameters for Stock Assessment and Management, La Jolla, CA, 28 Nov.–7 Dec. 1981,* ed. W. F. Perrin, R. L. Brownell Jr., and D. P. DeMaster, 161–74. Reports of the International Whaling Commission, special issue 6. Cambridge: International Whaling Commission.

Goodson, A. D., Kastelein, R. A., and Sturtivant, C. R. 1995. Source levels and echolocation signal characteristics of juvenile harbour porpoises *(Phocoena phocoena)* in a pool. In *Harbour porpoises: Laboratory studies to reduce bycatch,* ed. P. E. Nachtigall, J. Lien, W. W. L. Au, and A. J. Read. Woerden: De Spil.

Goodyear, J. D. 1993. A sonic/radio tag for monitoring dive depths and underwater movements of whales. *J. Wildl. Mgmt.* 57:503–13.

Goold, J. C. 1996. Signal processing techniques for acoustic measurement of sperm whale body lengths. *J. Acoust. Soc. Am.* 100:3431–41.

———. 1999. Behavioural and acoustic observations of sperm whales in Scapa Flow, Orkney Islands. *J. Mar. Biol. Assoc. UK* 79:541–50.

Goold, J. C., and Jones, S. E. 1995. Time and frequency domain characteristics of sperm whale clicks. *J. Acoust. Soc. Am.* 98:1279–91.

Gordon, D. M. 1987. The dynamics of group behavior. In *Perspectives in ethology,* vol. 7, ed. P. P. G. Bateson and P. H. Klopfer. New York: Plenum Press.

Gordon, J. C. D. 1987a. Behaviour and ecology of sperm whales off Sri Lanka. Ph.D. dissertation, Cambridge University, Cambridge, England.

———. 1987b. Sperm whale groups and social behaviour observed off Sri Lanka. *Rep. Int. Whal. Commn.* 37:205–17.

———. 1990. A simple photographic technique for measuring the length of whales from boats at sea. *Rep. Int. Whal. Commn.* 40:581–88.

———. 1991. Evaluation of a method for determining the length of sperm whales *(Physeter macrocephalus)* from their vocalizations. *J. Zool.* 224:301–14.

Gordon, J. C. D., and Moscrop, A. 1996. Underwater noise pollution and its significance for whales and dolphins. In *The conservation of whales and dolphins: Science and practice,* ed. M. P. Simmonds and J. Hutchinson. Chichester: Wiley.

Gordon, J. C. D., and Steiner, L. 1992. Ventilation and dive patterns in sperm whales, *Physeter macrocephalus,* in the Azores. *Rep. Int. Whal. Commn.* 42:561–65.

Gordon, J. C. D., Leaper, R., Hartley, F. G., and Chappell, O. 1992. *Effects of whale-watching vessels on the surface and underwater acoustic behaviour of sperm whales off Kaikoura, New Zealand.* Science and Conservation Series, no. 52. Wellington, New Zealand: Department of Conservation.

Gordon, J. C. D., Moscrop, A., Carlson, C., Ingram, S., Leaper, R., and Young, K. In press. Distribution, movements and residency of sperm whales off Dominica, Eastern Caribbean: Implications for the development and regulation of the local whale watching industry. *Rep. Int. Whal. Commn.*

Gosling, L. M., and Petrie, M. 1981. The economics of social organization. In *Physiological ecology,* ed. C. R. Townsend and P. Calow. Oxford: Blackwell Scientific Publications.

Gould, J. L. 1982. *Ethology: The mechanisms and evolution of behavior.* New York: W. W. Norton.

Gould, S. J. 1981. *The mismeasure of man.* New York: W. W. Norton.

———. 1997. This view of life: Seeing eye to eye. *Nat. Hist.* 106:14–18, 60–62.

Gould, S. J., and Lewontin, R. C. 1979. The spandrels of San Marco and the Panglossian paradigm: A critique of the adaptationist programme. *Proc. R. Soc. Lond.* B 205:581–98.

Gowans, S. 1999. Social organization and population structure of northern bottlenose whales in the Gully. Ph.D. dissertation, Dalhousie University, Halifax, Nova Scotia.

Gowans, S., and Rendell, L. 1999. Head-butting in Northern bottlenose whales *(Hyperoodon ampullatus):* a possible function for big heads? *Mar. Mamm. Sci.* 15: in press.

Gowans, S., and Whitehead, H. 1995. Distribution and habitat partitioning by small odontocetes in the Gully, a submarine canyon on the Scotian Shelf. *Can. J. Zool.* 73:1599–1608.

Gray, D. 1882. Notes on the characters and habits of the bottlenose whale (*Hyperoodon rostratus*). *Proc. Zool. Soc. Lond.* 1882: 726–31.

Gray, J. A. B., and Denton, E. J. 1991. Fast pressure pulses and communication between fish. *J. Mar. Biol. Assn. UK* 71:83–106.

Gray, J. E. 1828. *Spicilegia Zoologica,* or original figures and short systematic descriptions of new or unfigured animals. London.

———. 1846a. Description of Cetaceous Animals, with figures of the new species and their skulls. In *The zoology of the voyage of H.M.S Erebus and Terror under the command of Captain Sir James Clark Ross, R.N.F.R.S., during the years 1839 to 1843,* vol. 1, *Mammalia, birds,* 13–53. London: E. W. Janson.

———. 1846b. On the British Cetacea. *Ann. Nat. Hist.* (1846): 82–85.

———. 1865. Descriptions of three species of Dolphins. *Proc. Zool. Soc. Lond.* (1865):735.

———. 1866. Notes of the skulls of dolphins, or bottlenose whales, in the British Museum. *Proc. Zool. Soc. Lond.* 1866(2): 211–16.

———. 1874. *Feresa attenuata.* Journal des Museum Godeffroy, 8:184.

———. 1871. Notes of the Berardius of New Zealand. *Ann. Mag. Nat. Hist.,* 4th ser., 8:115–17.

Greene, C. H., Wiebe, P. H., and Zamon, J. E. 1994. Acoustic visualization of patch dynamics in oceanic ecosystems. *Oceanography* 7:4–12.

Greenwood, P. J. 1980. Mating systems, philopatry and dispersal in birds and mammals. *Anim. Behav.* 28:1140–62.

Gross, C. G. 1992. Representation of visual stimuli in inferior temporal cortex. *Phil. Trans. R. Soc. Lond.* 335:3–10.

Gross, M. R. 1996. Alternative reproductive strategies and tactics: Diversity within sexes. *Trends Ecol. Evol.* 11:92–98.

Grzimek, B. 1990. *Grzimek's encyclopedia of mammals,* vol. 1–5. New York: McGraw-Hill.

Guinee, L., Chu, K., and Dorsey, E. M. 1983. Changes over time in the songs of known individual humpback whales (*Megaptera novaeangliae*). In *Communication and behavior of whales,* ed. R. Payne. Boulder, CO: Westview Press.

Guinet, C. 1990. Sympatrie des deux categories d'orques dans le detroit de Johnstone, Columbie Britannique. *Rev. Ecol. (Terre Vie)* 45:25–34.

———. 1991a. Intentional stranding apprenticeship and social play in killer whales (*Orcinus orca*). *Can. J. Zool.* 69:2712–16.

———. 1991b. L'orque (*Orcinus orca*) autour de l'archipel Crozet—comparaison avec d'autres localites. *Rev. Ecol. (Terre Vie)* 46:321–37.

———. 1992. Comportement de chasse des orques (*Orcinus orca*) autour des iles Crozet. *Can. J. Zool.* 70:1656–67.

Guinet, C., and Bouvier, J. 1995. Development of intentional stranding hunting techniques in killer whale (*Orcinus orca*) calves at Crozet Archipelago. *Can. J. Zool.* 73:27–33.

Guinet, C., and Jouventin, P. 1990. La vie sociale des "baleines tueuses." *La Recherche* 21:508–10.

Gulland, J. A. 1974. Distribution and abundance of whales in relation to basic productivity. In *The whale problem: A status report,* ed. W. E. Schevill, G. C. Ray, and K. S. Norris. Cambridge, MA: Harvard University Press.

Gunter, G. 1942. Contributions to the natural history of the bottlenose dolphin, *Tursiops truncatus* (Montague), on the Texas coast, with particular reference to food habits. *J. Mammal.* 23:267–76.

———. 1951. Consumption of shrimp by the bottlenose dolphin. *J. Mammal* 32:465–66.

Gunther, E. R. 1949. The habits of fin whales. *Discovery Rep.* 25:113–42.

Gurevich, V. S. 1977. Post natal behaviour of an Atlantic bottlenose dolphin calf. In *Breeding dolphins: Present status, suggestions for the future,* ed. S. H. Ridgway and K. Benirschke. Washington, DC: U.S. Marine Mammal Commission Report no. MMC-76/07, Washington, DC.

Gyllensten, U., Wharton, D., and Wilson, A. C. 1985. Maternal inheritance of mitochondrial DNA during backcrossing of two species of mice. *J. Hered.* 76:321–24.

Gyllensten, U., Wharton, D., Josefsson, A., and Wilson, A. C. 1991. Paternal inheritance of mitochondrial DNA in mice. *Nature* 352:255–57.

Haase, B., and Félix, F. 1994. A note on the incidental mortality of sperm whales (*Physeter macrocephalus*) in Ecuador. In *Gillnets and cetaceans: Incorporating the proceedings of the symposium and workshop on the mortality of cetaceans in passive fishing nets and traps,* ed. W. F. Perrin, G. Donovan, and J. Barlow, 481–83. Reports of the International Whaling Commission, special issue 15. Cambridge: International Whaling Commission.

von Haast, J. 1876. Further notes on *Oulodon,* a new genus of ziphioid whales from the New-Zealand seas. *Proc. Zool. Soc. Lond.* 1876:457–58.

Haenel, N. J. 1986. General notes on the behavioral ontogeny of Puget Sound killer whales and the occurrence of allomaternal behavior. In *Behavioral biology of killer whales,* ed. B. Kirkevold and J. S. Lockard. New York: Alan R. Liss.

Hafner, K., and Lyon, M. 1996. *Where wizards stay up late: The origins of the Internet.* New York: Simon and Schuster.

Hain, J. H. W. 1991. Airships for marine mammal research: Evaluation and recommendations. NTIS publication PB92-128271.

Hain, J. H. W., Carter, G. R., Kraus, S. D., Mayo, C. A., and Winn, H. E. 1982. Feeding behaviour of the humpback whale, *Megaptera novaeangliae,* in the western North Atlantic. *Fish. Bull.* 80:259–68.

Hain, J. H. W., Ellis, S. L., Kenny, R. D., Clapham, P. J., Gray, B. K., Weinrich, M. T., and Babb, I. G. 1995. Apparent bottom feeding by humpback whales on Stellwagen Bank. *Mar. Mamm. Sci.* 11:464–79.

Hainsworth, F. R. 1987. Precision and dynamics of positioning by Canada geese flying in formation. *J. Exp. Biol.* 128:445–62.

Hall, H. J. 1985. Fishing with dolphins? Affirming a traditional aboriginal fishing story in Moreton Bay SE. Queensland. In *Focus on Stradbroke: New information on North Stradbroke Island and surrounding areas, 1974–1984.*

Hamburg, D. A. 1987. Foreword. In *Primate societies,* ed. B. B. Smuts, D. L. Cheney, R. M. Seyfarth, R. W. Wrangham, and T. T. Struhsaker. Chicago: University of Chicago Press.

Hamilton, P. K., and Mayo, C. A. 1990. Population characteristics of right whales *(Eubalaena glacialis)* observed in Cape Cod and Massachusetts Bays, 1978–1986. In *Individual recognition of cetaceans: Use of photo-identification and other techniques to estimate population parameters,* ed. P. S. Hammond, S. A. Mizroch, and G. P. Donovan, 203–8. Reports of the International Whaling Commission, special issue 12. Cambridge: International Whaling Commission.

Hamilton, P. V., and Nishimoto, R. T. 1977. Dolphin predation on mullet. *Fla. Sci.* 40:251–52.

Hamilton, W. D. 1964. The genetical evolution of social behavior, I and II. *J. Theor. Biol.* 7:1–52.

———. 1971. The geometry of the selfish herd. *J. Theor. Biol.* 31:295–311.

Hammond, P. S. 1986. Estimating the size of naturally marked whale populations using capture-recapture techniques. In *Behaviour of whales in relation to management: Incorporating the proceedings of a workshop of the same name held in Seattle, Washington, 19–23 April 1982,* ed. G. P. Donovan, 253–82. Reports of the International Whaling Commission, special issue 8. Cambridge: International Whaling Commission.

———. 1990. Heterogeneity in the Gulf of Maine? Estimating humpback whale population size when capture probabilities are not equal. In *Individual recognition of cetaceans: Use of photo-identification and other techniques to estimate population parameters,* ed. P. S. Hammond, S. A. Mizroch, and G. P. Donovan, 135–39. Reports of the International Whaling Commission, special issue 12. Cambridge: International Whaling Commission.

Hammond, P. S., Mizroch, S. A., and Donovan, G. P. 1990. (eds.). *Individual recognition of cetaceans: Use of photo-identification and other techniques to estimate population parameters.* Reports of the International Whaling Commission, special issue 12. Cambridge: International Whaling Commission.

Hancock, D. 1965. Killer whales attack and eat a minke whale. *J. Mammal.* 46:341–42.

Hanggi, E. B., and Schusterman, R. J. 1994. Underwater acoustic displays and individual variation in male harbor seals, *Phoca vitulina. Anim. Behav.* 48:1275–83.

Hansen, L. J. 1990. California coastal bottlenose dolphins. In *The bottlenose dolphin,* ed. S. Leatherwood and R. R. Reeves. San Diego: Academic Press.

Haraway, D. 1989. *Primate visions: Gender, race, and nature in the world of modern science.* New York: Routledge.

Harcourt, A. H. 1987. Behaviour of wild gorillas *(Gorilla gorilla)* and their management in captivity. *Int. Zoo Yrbk.* 26:248–55.

———. 1992. Coalitions and alliances: Are primates more complex than non-primates? In *Coalitions and alliances in humans and other animals,* ed. A. H. Harcourt and F. B. M. de Waal. Oxford: Oxford University Press.

Harcourt, A. H., and de Waal, F. B. M. (eds.). 1992. *Coalitions and alliances in humans and other animals.* Oxford: Oxford University Press.

Harcourt, A. H., and Gardner, J. 1994. Sexual selection and genital anatomy of male primates. *Proc. Roy. Soc. Lond.* B 255:47–53.

Harcourt, A. H., Harvey, P. H., Larson, S. G., and Short, R. V. 1981. Testis weight, body weight and breeding system in primates. *Nature* 293:55–57.

Harlin, A. D., Würsig, B., Baker, C. S., and Markowitz, T. M. 1999. Skin swabbing for genetic analysis: Application to dusky dolphins *(Lagenorhynchus obscurus). Mar. Mamm. Sci.* 15:409–25.

Harnad, S. 1987. *Categorical perception.* Cambridge: Cambridge University Press.

Harrison, R. J., and Ridgway, S. H. 1971. Gonadal activity of some bottlenose dolphins *(Tursiops truncatus). J. Zool.* 165:355–66.

Harrison, R. J., Brownell, R. L. Jr., and Boice, R. C. 1972. Reproduction and gonadal appearances in some odontocetes. In *Functional anatomy of marine mammals,* vol. 1, ed. R. J. Harrison. New York: Academic Press.

Harvey, P. H., and Harcourt, A. H. 1984. Sperm competition, testes size, and breeding systems in primates. In *Sperm competition and the evolution of animal mating systems,* ed. R. J. Smith. Orlando, FL: Academic Press.

Harvey, P. H., and Pagel, M. D. 1991. *The comparative method in evolutionary biology.* Oxford: Oxford University Press.

Harvey, P. H., Martin, R. D., and Clutton-Brock, T. H. 1987. Life histories in comparative perspective. In *Primate societies,* ed. B. Smuts, D. L. Cheney, R. M. Seyfarth, R. W. Wrangham, and T. T. Struhsaker, 181–96. Chicago: University of Chicago Press.

Haury, L. R., McGowan, J. A., and Wiebe. P. H. 1978. Patterns and processes in the space-time scales of plankton distributions. In *Spatial pattern in plankton communities,* ed. J. H. Steel. New York: Plenum.

Hauser, M. D. 1992. Costs of deception: Cheaters are punished in rhesus monkeys *(Macaca mulatta). Proc. Natl. Acad. Sci. USA* 89:12137–39.

Hauser, M. D., and Nelson, D. A. 1991. Intentional signaling in animal communication. *Trends Ecol. Evol.* 6:186–89.

Hauser, N., Peckham, H., and Clapham, P. J. 1999. Humpback whales in the Southern Cook Islands, South Pacific. Report SC/51/CAWS16 submitted to the International Whaling Commission.

Hausfater, G. 1975. Dominance and reproduction in baboons *(Papio cynocephalus):* A quantitative analysis. In *Contributions to primatology,* vol. 7. Basel: Karger.

Hausfater, G., and Hrdy, S. B. 1984. *Infanticide: A comparative and evolutionary perspective.* New York: Aldine.

Hawkes, K., O'Connell, J. F., Blurton Jones, N. G., Alvarez, H., and Charnov, E. L. 1998. Grandmothering, menopause, and the evolution of human life histories. *Proc. Natl. Acad. Sci. USA* 95:1336–39.

Hawkins, A. D., and Myrberg, A. A. Jr. 1983. Hearing and sound communication under water. In *Bioacoustics: A comparative approach,* ed. B. Lewis. New York: Academic.

Hay, K. A. 1980. Age determination in the narwhal, *Monodon monoceros* L. In *Age determination of toothed whales and sirenians: Proceedings of the International Conference on Determining Age of Odontocete Cetaceans and Sirenians, La Jolla, CA, Sept. 5–19, 1978,* ed. W. F. Perrin and A. C. Myrick Jr., 119–32. Reports of the International Whaling Commission, special issue 3. Cambridge: International Whaling Commission.

Hay, K. A., and Mansfield, A. W. 1989. Narwhal. In *Handbook of marine mammals*, vol. 4, *River dolphins and the larger toothed whales*, ed. S. H. Ridgway and R. J. Harrison. New York: Academic Press.

Heezen, B. C., and Johnson, G. L. 1969. Alaskan submarine cables: A struggle with a harsh environment. *Arctic* 22:413–24.

Heide-Jørgensen, M. P., and Reeves, R. R. 1996. Evidence of a decline in beluga, *Delphinapterus leucas,* abundance off West Greenland. *ICES J. Mar. Sci.* 53:61–72.

Heimlich-Boran, J. R. 1986. Fishery correlations with the occurrence of killer whales in greater Puget Sound. In *Behavioral biology of killer whales,* ed. B. Kirkevold and J. S. Lockard. New York: Alan R. Liss.

———. 1988. Behavioral ecology of killer whales *(Orcinus orca)* in the Pacific northwest. *Can. J. Zool.* 66:565–78.

———. 1993. Social organization of the short-finned pilot whale, *Globicephala macrorhyncus,* with special reference to the comparative social ecology of delphinids. Ph.D. thesis, Cambridge University.

Heimlich-Boran, J. R., and Heimlich-Boran, S. L. In press. Social learning in cetaceans: Hunting, hearing and hierarchies. *Symp. Zool. Soc. Lond.* 73.

Heimlich-Boran, S. L. 1986. Cohesive relationships among killer whales. In *Behavioral biology of killer whales,* ed. B. Kirkevold and J. S. Lockard. New York: Alan R. Liss.

Heise, K., Ellis, G., and Matkin, C. 1991. *A catalogue of Prince William Sound killer whales.* Homer, AK: North Gulf Oceanic Society.

Helweg, D. A., Frankel, A. S., Mobley, J. R. Jr., and Herman, L. M. 1992. Humpback whale song: Our current understanding. In *Marine mammal sensory systems,* ed. J. A. Thomas, R. Kastelein, and A. Ya. Supin. New York: Plenum.

Herman, L. M. 1980. Cognitive characteristics of dolphins. In *Cetacean behavior: Mechanisms and functions,* ed. L. M. Herman. New York: Wiley-Interscience.

Herman, L. M., and Antinoja, R. C. 1977. Humpback whales in the Hawaiian breeding waters: Population and pod characteristics. *Sci. Rep. Whales Res. Inst.* 29:59–85.

Herman, L. M., and Tavolga, W. N. 1980. The communication systems of cetaceans. In *Cetacean behavior: Mechanisms and functions,* ed. L. M. Herman, 149–209. New York: Wiley-Interscience.

Herman, L. M., Peacock, M. F., Yunker, M. P., and Madsen, C. J. 1975. Bottlenosed dolphin: Double-slit pupil yields equivalent aerial and underwater diurnal acuity. *Science* 189:650–52.

Hersh, S. L., and Duffield, D. A. 1990. Distinction between northwest Atlantic offshore and coastal bottlenose dolphins based on hemoglobin profile and morphometry. In *The bottlenose dolphin,* ed. S. Leatherwood and R. R. Reeves. San Diego: Academic Press.

Herzing, D. L. 1996. Vocalizations and associated underwater behavior of free-ranging Atlantic spotted dolphins, *Stenella frontalis* and bottlenose dolphins, *Tursiops truncatus. Aquat. Mamm.* 22:61–79.

Herzing, D. L., and Johnson, C. M. 1997. Interspecific interactions between Atlantic spotted dolphins *(Stenella frontalis)* and bottlenose dolphins *(Tursiops truncatus)* in the Bahamas, 1985–1995. *Aquat. Mamm.* 23:85–99.

Hester, F. J. 1984. Possible biases in estimates of reproductive rates of the spotted dolphin, *Stenella attenuata,* in the Eastern Pacific. In *Reproduction in whales, dolphins and porpoises: Proceedings of the conference, Cetacean Reproduction, Estimating Parameters for Stock Assessment and Management, La Jolla, CA, 28 Nov.–7 Dec. 1981,* ed. W. F. Perrin, R. L. Brownell Jr., and D. P. DeMaster, 337–41. Reports of the International Whaling Commission, special issue 6. Cambridge: International Whaling Commission.

Hester, F. J., Hunter, J. R., and Whitney, R. R. 1963. Jumping and spinning behavior in the spinner porpoise. *J. Mammal.* 44:586–88.

Hewison, A. J. M., and Gaillard, J. M. 1999. Successful sons or advantaged daughters? The Trivers-Willard model and sex-biased maternal investment in ungulates. *TREE* 14:229–34.

Heyland, J. D., and Hay, K. 1976. An attack by a polar bear on a juvenile beluga. *Arctic* 29:56–57.

Heyning, J. E. 1984. Functional morphology involved in intraspecific fighting of the beaked whale, *Mesoplodon carlhubbsi. Can. J. Zool.* 62:1645–54.

———. 1988. Presence of solid food in a young calf killer whale *(Orcinus orca). Mar. Mamm. Sci.* 4:68–71.

———. 1997. Sperm whale phylogeny revisited: Analysis of the morphological evidence. *Mar. Mamm. Sci.* 13:596–613.

Heyning, J. E., and Brownell, R. L. Jr. 1990. Variation in external morphology of killer whales. Abstracts of the Third International Orca Symposium, Victoria, B.C., March 1990.

Heyning, J. E., and Dahlheim, M. E. 1988. *Orcinus orca. Mammalian Species* 304:1–9.

Heyning, J. E., and Mead, J. G. 1996. Suction feeding in beaked whales: Morphological and observational evidence. *Contr. Sci.* 464:1–12.

Hill, K., and Hurtado, A. M. 1996. *Ache life history: The ecology and demography of a foraging people.* Hawthorne, NY: Aldine de Gruyter.

———. 1997. The evolution of premature reproductive senescence and menopause in human females: An evaluation of the Grandmother Hypothesis. In *Human nature: A critical reader,* ed. L. Betzig. New York: Oxford University Press.

Hill, R. 1994. Theory of geolocation by light levels. In *Elephant seals: Population ecology, behavior, and physiology,* ed. B. Le Boeuf and R. M. Laws. Berkeley: University of California Press.

Hill, R. N. 1956. *Window in the sea.* New York: Rinehart and Company.

Hinde, R. A. 1966. *Animal behaviour: A synthesis of ethology and comparative psychology.* New York: McGraw-Hill.

———. 1976. Interactions, relationships and social structure. *Man* 11:1–17.

———. 1982. *Ethology: Its nature and relations with other sciences.* Oxford: Oxford University Press.

Hinde, R. A., and Atkinson, S. 1970. Assessing the roles of social partners in maintaining mutual proximity, as exemplified by mother-infant relations in rhesus monkeys. *Anim. Behav.* 18:169–76.

Hiraiwa-Hasegawa, M. 1987. Infanticide in primates and a possible case of male-based infanticide in chimpanzees. In *Animal societies: Theories and facts,* ed. Y. Ito, J. L. Brown, and J. Kikkawa. Tokyo: Japan Scientific Societies Press.

Hobson, E. S. 1978. Aggregating as a defense against predators in aquatic and terrestrial environments. In *Contrasts in behavior,* ed. E. S Reese and F. J. Lighter. New York: Wiley.

Hoelzel, A. R. 1991a. Killer whale predation on marine mammals at Punta Norte, Argentina: Food sharing, provisioning and foraging strategy. *Behav. Ecol. Sociobiol.* 29:197–204.

———— (ed.). 1991b. *Genetic ecology of whales and dolphins: Incorporating the proceedings of the workshop on the genetic analysis of cetacean populations.* Reports of the International Whaling Commission, special issue 13. Cambridge: International Whaling Commission.

————. 1993. Foraging behaviour and social group dynamics in Puget Sound killer whales. *Anim. Behav.* 45:581–91.

Hoelzel, A. R., and Dover, G. A. 1989. Molecular techniques for examining genetic variation and stock identity in cetacean species. In *The comprehensive assessment of whale stocks: The early years,* ed. G. P. Donovan, 81–120. Reports of the International Whaling Commission, special issue 11. Cambridge: International Whaling Commission.

————. 1991a. Genetic differentiation between sympatric killer whale populations. *Heredity* 66:191–95.

————. 1991b. Mitochondrial D-loop DNA variation within and between populations of the minke whale *(Balaenoptera acutorostrata).* In *Genetic ecology of whales and dolphins: Incorporating the proceedings of the workshop on the genetic analysis of cetacean populations,* ed. A. R. Hoelzel, 171–81. Reports of the International Whaling Commission, special issue 13. Cambridge: International Whaling Commission.

————. 1991c. *Molecular genetic ecology.* Oxford: Oxford University Press.

Hoelzel, A. R., and Osborne, R. W. 1986. Killer whale call characteristics: Implications for cooperative foraging strategies. In *Behavioral biology of killer whales,* ed. B. Kirkevold and J. S. Lockard. New York: Alan R. Liss.

Hoelzel, A. R., Ford, J. K. B., and Dover, G. A. 1991. A paternity test case for the killer whale *(Orcinus orca)* by DNA fingerprinting. *Mar. Mamm. Sci.* 7:35–43.

Hoelzel, A. R., Dahlheim, M., and Stern, S. J. 1998a. Low genetic variation among killer whales *(Orcinus orca)* in the eastern North Pacific, and genetic differentiation between foraging specialists. *J. Hered.* 89:121–28.

Hoelzel, A. R., Potter, C. W., and Best, P. B. 1998b. Genetic differentiation between parapatric "nearshore" and "offshore" populations of the bottlenose dolphin. *Proc. R. Soc. Lond.* B 265:1177–83.

Hoese, H. D. 1971. Dolphin feeding out of water on a salt marsh. *J. Mammal.* 52:222–23.

Hofman, R. 1995. The changing focus of marine mammal conservation. *Trends Ecol. Evol.* 10:462–64.

Hofman, R. J., and Bonner, W. N. 1985. Conservation and protection of marine mammals: Past, present and future. *Mar. Mamm. Sci.* 1:109–27.

Hogg, J. T. 1984. Mating in bighorn sheep: Multiple creative male strategies. *Science* 225:526–29.

Höglund, J., and Alatalo, R. V. 1995. *Leks.* Princeton, NJ: Princeton University Press.

Hohn, A. A. 1980a. Age determination and age related factors in the teeth of western North Atlantic bottlenose dolphins. *Sci. Rep. Whales Res. Inst.* 32:39–66.

————. 1980b. Analysis of growth layers in the teeth of *Tursiops truncatus* using light microscopy, microradiography, and SEM. In *Age determination of toothed whales and sirenians: Proceedings of the International Conference on Determining Age of Odontocete Cetaceans and Sirenians, La Jolla, CA, Sept. 5–19, 1978,* ed. W. F. Perrin and A. C. Myrick Jr., 155–60. Reports of the International Whaling Commission, special issue 3. Cambridge: International Whaling Commission.

————. 1997. Design for a multiple-method approach to determine stock structure of bottlenose dolphins in the Mid-Atlantic. NOAA Technical Memo NMFS-SEFSC-401.

Hohn, A. A., Chivers, S. J., and Barlow, J. 1985. Reproductive maturity and seasonality of male spotted dolphins, *Stenella attenuata,* in the eastern tropical Pacific. *Mar. Mamm. Sci.* 1: 273–93.

Hohn, A. A., Scott, M. D., Wells, R. S., Sweeney, J. C., and Irvine, A. B. 1989. Growth layers in teeth from known-age, free-ranging bottlenose dolphins. *Mar. Mamm. Sci.* 5:315–42.

Hölldobler, B., and Wilson, E. O. 1994. *Journey to the ants.* Cambridge, MA: Harvard University Press.

Holt, S. J. 1977. Whale management policy. *Rep. Int. Whal. Commn.* 27:133–35.

————. 1995. Whale mining, whale saving. *Mar. Pol.,* July, 194–213.

Holt, S. J., and Talbot, L. M. 1978. New principles for the conservation of wild living resources. *Wildl. Monogr.* 59:1–33.

Holt, S. J., and Young, N. 1990. *Guide to review the management of whaling.* Washington, DC: Center for Marine Conservation.

van Hooff, J. A. R. A. M., and van Schaik, C. P. 1992. Cooperation in competition: The ecology of primate bonds. In *Coalitions and alliances in humans and other animals,* ed. A. H. Harcourt and F. B. M. de Waal. Oxford: Oxford University Press.

————. 1994. Male bonds: Affiliative relationships among nonhuman primate males. *Behaviour* 130:309–37.

Hooker, S. K., and Baird, R. W. 1999. Deep-diving behaviour of the northern bottlenose whale, *Hyperoodon ampullatus* (Cetacea: Ziphiidae). *Proc. R. Soc. Lond.* B 266:671–76.

Hope, P. L., and Whitehead, H. 1991. Sperm whales off the Galápagos Islands from 1830–50 and comparisons with modern studies. *Rep. Int. Whal. Commn.* 41:273–86.

Hopkins, C. D. 1977. Electrical communication. In *How animals communicate,* ed. T. A. Sebeok. Bloomington: Indiana University Press.

Hopkins, W. J. 1922. *She blows! And sparm at that!* New York: Houghton Mifflin.

Horn, H. S., and Bradbury, J. W. 1984. Behavioural adaptations and life history. In *Behavioural ecology,* 2d ed., ed. J. R. Krebs and N. B. Davies. Oxford: Blackwell Scientific Publications.

Horner, J. R. 1982. Evidence of colony nesting and "site fidelity" among ornithischian dinosaurs. *Nature* 297:675–76.

Horner, J. R., and Makela, R. 1979. Nest of juveniles provides evidence of family structure among dinosaurs. *Nature* 282: 296–98.

Horrocks, J. A., and Hunte, W. 1986. Sentinel behaviour in vervet monkeys: Who sees whom first? *Anim. Behav.* 34: 1566–67.

Horwood, J. 1987. *The sei whale: Population biology, ecology and management*. New York: Croom Helm.

———. 1990. *Biology and exploitation of the minke whale*. Boca Raton, FL: CRC Press.

Hoyt, E. 1984. The whales called "killer." *Natl. Geogr.* 166:221–37.

———. 1990. *Orca, the whale called killer*. 3d edition. Canada: Camden House.

———. 1992. *The performing orca—why the show must stop*. Bath, England: Whale and Dolphin Conservation Society.

———. 1995. *The worldwide value and extent of whale watching 1995*. Bath: Whale and Dolphin Conservation Society.

Hrdy, S. B. 1976. Care and exploitation of nonhuman primate infants by conspecifics other than the mother. *Adv. Stud. Behav.* 6:101–58.

———. 1977. Infanticide as a primate reproductive strategy. *Am. Sci.* 65:40–49.

———. 1979. Infanticide among mammals: A review, classification, and examination of the implications for the reproductive strategies of females. *Ethol. Sociobiol.* 1:13–40.

———. 1981. "Nepotists" and "altruists": The behavior of old females among macaques and langur monkeys. In *Other ways of growing old*, ed. P. T. Amoss and S. Harrell. Stanford, CA: Stanford University Press.

Hubbs, C. L. 1953. Dolphin protecting dead young. *J. Mammal.* 34:498

Hui, C. A. 1987. Power and speed of swimming dolphins. *J. Mammal.* 68:126–32.

Humphrey, N. K. 1976. The social function of intellect. In *Growing points in ethology*, ed. P. P. G. Bateson and R. A. Hinde. Cambridge: Cambridge University Press.

Hussain, F. 1973. Whatever happened to dolphins? *New Sci.*, 25 Jan., 182–84.

Hutt, S. J., and Hutt, C. 1970. *Direct observation and measurement of behavior*. Springfield, IL: Charles C. Thomas.

Idani, G. 1991. Social relationships between immigrant and resident bonobo *(Pan paniscus)* females at Wamba. *Folia Primatol.* 57:83–95.

Ihobe, H. 1992. Male-male relationships among wild bonobos *(Pan paniscus)* at Wamba, Republic of Zaire. *Primates* 33: 163–79.

Ingebrigtsen, A. 1929. Whales caught in the North Atlantic and other seas. *Rapp. Proc. Verb. Cons. Int. Explor. Mer.* 56: 1–26.

Inman, A. J., and Krebs, J. 1987. Predation and group living. *Trends Ecol. Evol.* 2:31–32.

Intergovernmental Panel on Climate Change. 1995. The IPCC assessment of knowledge relevant to the interpretation of Article 2 of the UN Convention on climate change: A synthesis report. Geneva.

International Whaling Commission. 1974. Report 1972–73. *Rep. Int. Whal. Commn.* 24:6–11.

———. 1980. *Sperm whales*. Reports of the International Whaling Commission, special issue 2. Cambridge: International Whaling Commission.

———. 1982. Reports of the sub-committee on sperm whales. *Rep. Int. Whal. Commn.* 32:68–86.

———. 1983. Chairman's report. *Rep. Int. Whal. Commn.* 33: 20–42.

———. 1984a. Report of the Workshop on reproduction in whales, dolphins and porpoises, La Jolla, CA, Nov. 1981. In *Reproduction in whales, dolphins and porpoises: Proceedings of the conference, Cetacean Reproduction, Estimating Parameters for Stock Assessment and Management, La Jolla, CA, 28 Nov.–7 Dec. 1981*, ed. W. F. Perrin, R. L. Brownell Jr., and D. P. DeMaster, 1–24. Reports of the International Whaling Commission, special issue 6. Cambridge: International Whaling Commission.

———. 1984b. Report of the Workshop on Reproduction in Whales, Dolphins and Porpoises, La Jolla, December 1981. Appendix B. Terminology of female reproductive morphology and physiology. In *Reproduction in whales, dolphins and porpoises: Proceedings of the conference, Cetacean Reproduction, Estimating Parameters for Stock Assessment and Management, La Jolla, CA, 28 Nov.–7 Dec. 1981*, ed. W. F. Perrin, R. L. Brownell Jr., and D. P. DeMaster, 18–22. Reports of the International Whaling Commission, special issue 6. Cambridge: International Whaling Commission.

———. 1986. Report of the Workshop on the Behaviour of Whales in Relation to Management, Seattle, Washington, 19–23 April 1982. In *Behaviour of whales in relation to management: Incorporating the proceedings of a workshop of the same name held in Seattle, Washington, 19–23 April 1982*, ed. G. P. Donovan, 1–56. Reports of the International Whaling Commission, special issue 8. Cambridge: International Whaling Commission.

———. 1990. Report of the Workshop on Individual Recognition and the Estimation of Cetacean Population Parameters, La Jolla, 1–4 May 1988. In *Individual recognition of cetaceans: Use of photo-identification and other techniques to estimate population parameters*, ed. P. S. Hammond, S. A. Mizroch, and G. P. Donovan, 3–40. Reports of the International Whaling Commission, special issue 12. Cambridge: International Whaling Commission.

———. 1991. Report of the Workshop on the Genetic Analysis of Cetacean Populations, La Jolla, 27–29 Sept. 1989. In *Genetic ecology of whales and dolphins: Incorporating the proceedings of the workshop on the genetic analysis of cetacean populations*, ed. A. R. Hoelzel, 3–21. Reports of the International Whaling Commission, special issue 13. Cambridge: International Whaling Commission.

———. 1992. Report of the Scientific Committee. *Rep. Int. Whal. Commn.* 42:51–270.

———. 1993. Report of the sub-committee on small cetaceans. *Rep. Int. Whal. Commn.* 43:130–45.

———. 1994a. Report of the workshop on mortality of cetaceans in passive fishing nets and traps. In *Gillnets and cetaceans: Incorporating the proceedings of the symposium and workshop on the mortality of cetaceans in passive fishing nets and traps*, ed. W. F. Perrin, G. Donovan, and J. Barlow, 6–71. Reports of the International Whaling Commission, special issue 15. Cambridge: International Whaling Commission.

———. 1994b. The revised management procedure (RMP) for baleen whales. *Rep. Int. Whal. Commn.* 44:145–52.

———. 1995a. Chairman's Report of the Forty-Sixth Annual Meeting. *Rep. Int. Whal. Commn.* 45:15–40.

———. 1995b. Report of the sub-committee on small cetaceans. *Rep. Int. Whal. Commn.* 45:165–81.

———. 1997. Report of the International Whaling Commission workshop on climate change and cetaceans. *Rep. Int. Whal. Commn.* 47:291–319.

———. In press. Report of the Scientific Committee. *J. of Cetacean Research and Management* 2 (Supplement).

Irvine, A. B., and Wells, R. S. 1972. Results of attempts to tag Atlantic bottlenosed dolphins *(Tursiops truncatus)*. *Cetology* 13:1–5.

Irvine, A. B., Wells, R. S., and Gilbert, P. W. 1973. Conditioning an Atlantic bottle-nosed dolphin, *Tursiops truncatus*, to repel various species of sharks. *J. Mammal.* 54:503–5.

Irvine, A. B., Scott, M. D., Wells, R. S., and Kaufmann, J. H. 1981. Movements and activities of the Atlantic bottlenose dolphin, *Tursiops truncatus*, near Sarasota, Florida. *Fish. Bull.* 79:671–88.

Irvine, A. B., Wells, R. S., and Scott, M. D. 1982. An evaluation of techniques for tagging small odontocete cetaceans. *Fish. Bull.* 80:135–43.

Isbell, L. A. 1991. Contest and scramble competition: Patterns of female aggression and ranging behavior among primates. *Behav. Ecol.* 2:143–55.

Isbell, L. A., and van Vuren, D. 1996. Differential costs of locational and social dispersal and their consequences for female group-living primates. *Behaviour* 133:1–36.

Jacobsen, J. K. 1990. Associations and social behaviors among killer whales *(Orcinus orca)* in the Johnstone Strait, British Columbia, 1979–1986. M.A. thesis, Humboldt State University, Arcata, California.

Jakobsen, P. J., and Johnsen, G. H. 1988. Size-specific protection against predation by fish in swarming waterfleas, *Bosmina longispina*. *Anim. Behav.* 36:986–90.

Jamieson, I. G. 1997. Testing reproductive skew models in a communally breeding bird, the pukeko, *Porphyrio porphyrio*. *Proc. R. Soc. Lond.* B 264:335–40.

Janik, V. M. 1997. Whistle matching in wild bottlenose dolphins. *J. Acoust. Soc. Am.* 101:3136. Abstract.

———. 1998. Functional and organizational aspects of vocal repertoires in bottlenose dolphins *(Tursiops truncatus)*. Ph.D. dissertation, University of St. Andrews, St. Andrews, Scotland.

———. 1999. Pitfalls in the categorization of behavior: A comparison of dolphin whistle categorization methods. *Anim. Behav.* 57:133–43.

———. In press. Origins and implications of vocal learning in bottlenose dolphins. In *Mammalian social learning: Comparative and ecological perspectives*, ed. H. O. Box and K. R. Gibson. Cambridge: Cambridge University Press.

Janik, V. M., and Slater, P. J. B. 1997. Vocal learning in mammals. *Adv. Stud. Behav.* 26:59–99.

———. 1998. Context-specific use suggests that bottlenose dolphin signature whistles are cohesion calls. *Anim. Behav.* 56: 829–38.

Janik, V. M., and Thompson, P. 1996. Changes in surfacing patterns of bottlenose dolphin in response to boat traffic. *Mar. Mamm. Sci.* 12(4):597–602.

Janik, V. M., Denhardt, G., and Todt, D. 1994. Signature whistle variations in a bottlenosed dolphin, *Tursiops truncatus*. *Behav. Ecol. Sociobiol.* 35:243–48.

Janson, C. H., and Goldsmith, M. L. 1995. Predicting group size in primates: Foraging costs and predation risks. *Behav. Ecol.* 3:326–36.

Janzen, D. H. 1985. The natural history of mutualisms. In *The biology of mutualism*, ed. D. Boucher. London: Croom Helm.

Jaquet, N. 1996. Distribution and spatial organization of groups of sperm whales in relation to biological and environmental factors in the South Pacific. Ph.D. dissertation, Dalhousie University, Halifax, Nova Scotia.

Jaquet, N., and Whitehead, H. 1996. Scale-dependent correlation of sperm whale distribution with environmental features and productivity in the South Pacific. *Mar. Ecol. Prog. Ser.* 135:1–9.

———. 1999. Movements, distribution and feeding success of sperm whales in the Pacific Ocean, over scales of days and tens of kilometers. *Aquat. Mamm.* 25:1–13.

Jaquet, N., Whitehead, H., and Lewis, M. 1996. Coherence between 19th century sperm whale distributions and satellite-derived pigments in the tropical Pacific. *Mar. Ecol. Prog. Ser.* 145:1–10.

Jarman, P. J. 1989. Sexual dimorphism in Macropodoidea. In *Kangaroos, wallabies and rat-kangaroos*, ed. G. Grigg, P. Jarman, and I. Hume. New South Wales, Australia: Surrey, Beatty and Sons.

———. 1991. Social behavior and organization in the Macropodoidea. *Adv. Stud. Behav.* 20:1–50.

Jefferson, T. A., Stacey, P. J., and Baird, R. W. 1991. A review of killer whale interactions with other marine mammals: Predation to co-existence. *Mammal Rev.* 21:151–80.

Jeffreys, A. J., Wilson, V., and Thein, S. L. 1985a. Hypervariable "minisatellite" regions in human DNA. *Nature* 314:67–73.

———. 1985b. Individual specific "fingerprints" of human DNA. *Nature* 316:76–79.

Jepson, P. D., Bennett, P. M., Kirkwood, J. K., Kuiken, T., Simpson, V. R., and Baker, J. R. 1997. Bycatch and other causes of mortality in cetaceans stranded on the coasts of England and Wales 1990–1996. Abstracts, Eleventh Annual Conference of the European Cetacean Society, Stralsund, Germany.

Jerison, H. J. 1986. The perceptual worlds of dolphins. In *Dolphin cognition and behavior: A comparative approach*, ed. R. J. Schusterman, J. A. Thomas, and F. G. Wood. Hillsdale, NJ: Lawrence Erlbaum Associates.

Jing, X., Xiao, Y., and Jing, R. 1981. Acoustic signals and acoustic behavior of Chinese river dolphin *(Lipotes vexillifer)*. *Sci. Sin.* 2:233–39.

Johnson, C. M., and Norris, K. S. 1986. Delphinid social organization and social behavior. In *Dolphin cognition and behavior: A comparative approach*, ed. R. J. Schusterman, J. A.

Thomas, and F. G. Wood. Hillsdale, NJ: Lawrence Erlbaum Associates.

———. 1994. Social behavior. In *The Hawaiian spinner dolphin,* ed. K. S. Norris, B. Würsig, R. S. Wells, and M. Würsig, 243–86. Berkeley: University of California Press.

Johnson, C. N. 1986. Sex-biased philopatry and dispersal in mammals. *Oecologia* 69:626–27.

Johnson, D. H., and Dudgeon, D. E. 1993. *Array signal processing: Concepts and techniques.* Englewood Cliffs, NJ: Prentice-Hall.

Johnson, J. H., and Wolman, A. A. 1985. The humpback whale. *Mar. Fish. Rev.* 46:30–37.

Johnson, J. H., Braham, H. W., Krogman, B. D., Marquette, W. M., Sonntag, R. M., and Rugh, D. J. 1981. Bowhead whale research: June 1979 to June 1980. *Rep. Int. Whal. Commn.* 31:461–75.

Johnston, P. A., Stringer, R. L., and Santillo, D. 1996. Cetaceans and environmental pollution: The global concerns. In *The conservation of whales and dolphins: Science and practice,* ed. M. P. Simmonds and J. Hutchinson. Chichester: Wiley.

Jolly, A. 1966. Lemur social behavior and primate intelligence. *Science* 153:510–16.

Jones, E. C. 1971. *Isistius brasiliensis,* a squaloid shark, the probable cause of crater wounds on fishes and cetaceans. *Fish. Bull.* 69:791–98.

Jones, G., and Ransome, R. D. 1993. Echolocation calls of bats are influenced by maternal effects and change over a lifetime. *Proc. R. Soc. Lond.* B 252:125–28.

Jones, M. L. 1990. The reproductive cycle in gray whales based on photographic resightings of females in the breeding grounds from 1977–1982. In *Individual recognition of cetaceans: Use of photo-identification and other techniques to estimate population parameters,* ed. P. S. Hammond, S. A. Mizroch, and G. P. Donovan, 177–82. Reports of the International Whaling Commission, special issue 12. Cambridge: International Whaling Commission.

Jones, M. L., Swartz, S. L., and Leatherwood, S. 1984. *The gray whale, Eschrichtius robustus.* New York: Academic Press.

Jonsgård, A. 1980. On the sex proportion in Norwegian minke whaling. *Rep. Int. Whal. Commn.* 30:389.

Joseph, B. E., Antrom, J. E., and Cornell, L. H. 1987. Commerson's dolphin *Cephalorhyncus commersonii:* A discussion of the first live birth within a marine zoological park. *Zoo Biol.* 6:69–77.

Jurasz, C. M., and Jurasz, V. P. 1979. Feeding modes of the humpback whale, *Megaptera novaeangliae,* in southeast Alaska. *Sci. Rep. Whales Res. Inst. Tokyo* 31:69–83.

Kahn, B. 1991. The population biology and social organization of sperm whales *(Physeter macrocephalus)* off the Seychelles: Indications of recent exploitation. M.Sc. thesis, Dalhousie University, Halifax, Nova Scotia.

Kahn, B., Whitehead, H., and Dillon, M. 1993. Indications of density-dependent effects from comparisons of sperm whale populations. *Mar. Ecol. Prog. Ser.* 93:1–7.

Kalland, A. 1993. Management by totemization: Whale symbolism and the anti-whaling campaign. *Arctic* 46:124–33.

Kamminga, C. 1988. Echolocation signal types of odontocetes. In *Animal sonar: Processes and performance,* ed. P. E. Nachtigall and P. W. B. Moore. New York: Plenum.

Kamminga, C., and van der Ree, A. F. 1976. Discrimination of solid and hollow spheres by *Tursiops truncatus* (Montagu). *Aquat. Mamm.* 4:1–9.

Kamminga, C., and Wiersma, H. 1981. Investigations on cetacean sonar. II. Acoustical similarities and differences in odontocete sonar signals. *Aquat. Mamm.* 8:41–62.

Kano, T. 1992. *The last ape: Pygmy chimpanzee behavior and ecology.* Stanford, CA: Stanford University Press.

———. 1996. Male rank order and copulation rate in a unit-group of bonobos at Wamba, Zaire. In *Great ape societies,* ed. W. C. McGrew, L. F. Marchant, and T. Nishida. Cambridge: Cambridge University Press.

Kanwisher, J. W., and Ridgway, S. H. 1983. The physiological ecology of whales and porpoises. *Sci. Am.* 248:111–21.

Karczmarski, L. 1996. Ecological studies of humpback dolphins *Sousa chinensis* in the Algoa Bay region, Eastern Cape, South Africa. Ph.D. dissertation, University of Port Elizabeth, South Africa.

Kasuya, T. 1972. Some information on the growth of the Ganges river dolphin with a comment on the Indus dolphin. *Sci. Rep. Whales Res. Inst., Tokyo* 25:1–103.

———. 1977. Age determination and growth of Baird's beaked whale with a comment on the foetal growth rate. *Sci. Rep. Whales Res. Inst., Tokyo* 29:1–20.

———. 1991. Density-dependent growth in North Pacific sperm whales. *Mar. Mamm. Sci.* 7:230–57.

———. 1995. Overview of cetacean life histories. In *Whales, seals, fish and man: Proceedings of the International Symposium on the Biology of Marine Mammals in the North East Atlantic, Tromsø, Norway, 29 November–1 December 1994,* ed. A. S. Blix, L. Walløe, and Ø. Ulltang. Amsterdam: Elsevier.

Kasuya, T. 1999a. Review of the biology and exploitation of striped dolphins in Japan. *J. of Cetacean Research and Management* 1:81–100.

Kasuya, T. 1999b. Examination of the reliability of catch statistics in the Japanese coastal sperm whale fishery. *J. of Cetacean Research and Management* 1:109–22.

Kasuya, T., and Brownell, R. L. Jr. 1989. Male parental investment in Baird's beaked whales, an interpretation of the age data. Abstract, Fifth International Theriological Congress, Rome, 523–24.

Kasuya, T., and Marsh, H. 1984. Life history and reproductive biology of the short-finned pilot whale, *Globicephala macrorhynchus,* off the Pacific Coast of Japan. In *Reproduction in whales, dolphins and porpoises: Proceedings of the conference, Cetacean Reproduction, Estimating Parameters for Stock Assessment and Management, La Jolla, CA, 28 Nov.–7 Dec. 1981,* ed. W. F. Perrin, R. L. Brownell Jr., and D. P. DeMaster, 259–310. Reports of the International Whaling Commission, special issue 6. Cambridge: International Whaling Commission.

Kasuya, T., and Matsui, S. 1984. Age determination and growth of the short-finned pilot whale off the Pacific coast of Japan. *Sci. Rept. Whales Res. Inst. Tokyo* 35:57–91.

Kasuya, T., and Miyashita, T. 1988. Distribution of sperm whale stocks in the North Pacific. *Sci. Rep. Whales Res. Inst.* 39:31–75.

Kasuya, T., and Tai, S. 1993. Life history of short-finned pilot whale stocks off Japan and a description of the fishery. In *Biology of Northern Hemisphere pilot whales: A collection of papers,* ed. G. P. Donovan, C. H. Lockyer, and A. R. Martin, 439–73. Reports of the International Whaling Commission, special issue 14. Cambridge: International Whaling Commission.

Kasuya, T., Balcomb K., and Brownell, R. L. Jr. 1997. Life history of Baird's beaked whales off the Pacific coast of Japan. *Rep. Int. Whal. Commn.* 47:969–79.

Kato, H. 1984. Observation of tooth scars on the head of male sperm whale, as an indication of intra-sexual fightings. *Sci. Rep. Whales Res. Inst.* 35:39–46.

Katona, S. K. 1991. Large-scale planning for assessment and recovery of humpback whale populations. *Mem. Queensland Mus.* 30:297–305.

Katona, S. K., and Beard, J. A. 1990. Population size, migrations and feeding aggregations of the humpback whale *(Megaptera novaeangliae)* in the western North Atlantic Ocean. In *Individual recognition of cetaceans: Use of photo-identification and other techniques to estimate population parameters,* ed. P. S. Hammond, S. A. Mizroch, and G. P. Donovan, 295–305. Reports of the International Whaling Commission, special issue 12. Cambridge: International Whaling Commission.

Katona, S. K., and Kraus, S. 1999. The North Atlantic right whale. In *Conservation and management of marine mammals,* ed. J. R. Twiss Jr. and R. R. Reeves, 311–31. Washington, DC: Smithsonian Institution Press.

Katona, S. K., and Whitehead, H. 1981. Identifying humpback whales using their natural markings. *Polar Rec.* 20(128):439–44.

———. 1988. Are Cetacea ecologically important? *Oceanogr. Mar. Biol. Annu. Rev.* 26:553–68.

Katona, S. K., Baxter, B., Brazier, O., Kraus, S., Perkins, J., and Whitehead, H. 1979. Identification of humpback whales by fluke photographs. In *Behavior of marine mammals: Current perspectives in research,* vol. 3, *Cetaceans,* ed. H. E. Winn and B. L. Olla. New York: Plenum Press.

Katona, S. K., Harcourt, P. M., Perkins, J. S., and Kraus, S. D. 1980. *Humpback whales: A catalogue of individuals identified in the western North Atlantic Ocean by means of fluke photographs.* Bar Harbor, ME: College of the Atlantic.

Katona, S. K., Rough, V., and Richardson, D. T. 1993. *A field guide to whales, porpoises and seals from Cape Cod to Newfoundland.* 4th edition. Washington, DC: Smithsonian Institution Press.

Kawakami, T. 1980. A review of sperm whale food. *Sci. Rep. Whales Res. Inst.* 32:199–218.

Keane, B., Waser, P. M., Creel, S. R., Creel, N. M., Elliot, L. F., and Minchella, D. J. 1994. Subordinate reproduction in dwarf mongooses. *Anim. Behav.* 47:65–75.

Keen, S. 1971. A conversation with John Lilly. *Psychology Today,* Dec. 1971, 75–77, 92–93.

Keller, L. 1997. Indiscriminate altruism: Unduly nice parents and siblings. *Trends Ecol. Evol.* 12:99–103.

Keller, L., and Reeve, H. K. 1994. Partitioning of reproduction in animal societies. *Trends Ecol. Evol.* 9:98–102.

Kellogg, A. R. 1928. The history of whales—their adaptation to life in the water. *Q. Rev. Biol.* 3:29–76, 174–208.

———. 1929. What is known of the migrations of some of the whalebone whales. *Annu. Rep. Smithsonian Inst.* 467–94.

Kellogg, W. N., Kohler, R., and Morris, H. N. 1953. Porpoise sounds as sonar signals. *Science* 117:239–43.

Kemp, N. J. 1996. Incidental catches of small cetaceans. In *The conservation of whales and dolphins: Science and practice,* ed. M. P. Simmonds and J. Hutchinson. Chichester: Wiley.

Kenagy, G. J., and Trombulak, S. C. 1986. Size and function of mammalian testes in relation to body size. *J. Mammal.* 67:1–22.

Kenney, R. D. 1990. Bottlenose dolphins off the northeastern United States. In *The bottlenose dolphin,* ed. S. Leatherwood and R. R. Reeves. San Diego: Academic Press.

Kenward, R. E. 1978. Hawks and doves: Factors affecting success and selection in goshawk attacks on wild pigeons. *J. Anim. Ecol.* 47:449–60.

Ketten, D. R. 1990. Three-dimensional reconstructions of the dolphin ear. In *Sensory abilities of cetaceans,* ed. J. Thomas and R. Kastelein. New York: Plenum.

———. 1994. Functional analyses of whale ears: Adaptations for underwater hearing. *IEEE Proceedings in Underwater Acoustics* 1:264–70.

———. 1997. Structure and function in whale ears. *Bioacoustics* 8:103–35.

Ketten, D. R., Lien, J., and Todd, S. 1993. Blast injury in humpback whale ears: Evidence and implications. *J. Acoust. Soc. Am.* 94:1849–50.

Kimura, T., Ozawa, T., and Pastene, L. A. 1997. Sample preparation and analysis of mitochondrial DNA from baleen plates. *Mar. Mamm. Sci.* 13:495–98.

Kirby, V. L., and Ridgway, S. H. 1984. Hormonal evidence of spontaneous ovulation in captive dolphins, *Tursiops truncatus* and *Delphinus delphis.* In *Reproduction in whales, dolphins and porpoises: Proceedings of the conference, Cetacean Reproduction, Estimating Parameters for Stock Assessment and Management, La Jolla, CA, 28 Nov.–7 Dec. 1981,* ed. W. F. Perrin, R. L. Brownell Jr., and D. P. DeMaster. Reports of the International Whaling Commission, special issue 6. Cambridge: International Whaling Commission.

Kirkwood, G. P. 1992. Report of the Scientific Committee, Annex I. Background to the development of revised management procedures. *Rep. Int. Whal. Commn.* 42:236–43.

Kleiman, D. G. 1992. Behavior research in zoos: Past, present, and future. *Zoo Biol.* 11:301–12.

———. 1994. Mammalian sociobiology and zoo breeding programs. *Zoo Biol.* 13:423–32.

Kleiman, D. G., and Malcolm, J. R. 1981. The evolution of male parental investment in mammals. In *Parental care in mammals,* ed. D. J. Gubernick and P. H. Klopfer. New York: Plenum Press.

Klimley, A. P. 1994. The predatory behavior of the white shark. *Am. Sci.* 82:122–33.

Klingel, H. 1965. Notes on the biology of the plains zebra *Equus quagga boehmi* Matschie. *E. Afr. Wildl. J.* 3:86–88.

Klinowska, M. 1988. Are cetaceans especially smart? *New Sci.* 1636:46–47.

———. 1991. *Dolphins, porpoises and whales of the world: The IUCN Red Data Book.* Gland, Switzerland: IUCN.

Knowlton, A. R., Kraus, S. D., and Kenney, R. D. 1994. Reproduction in North Atlantic right whales (*Eubalaena glacialis*). *Can. J. Zool.* 72:1297–1305.

Ko, D., Zeh, J. E., Clark, C. W., Ellison, W. T., Krogman, B. D., and Sonntag, R. M. 1986. Utilisation of acoustic location data in determining a minimum number of spring migrating bowhead whales, unaccounted for by the icebased visual census. *Rep. Int. Whal. Commn.* 36:325–38.

Kohn, M. H., and Wayne R. K. 1997. Facts from feces revisited. *Trends Ecol. Evol.* 12:223–27.

Kolchin, S., and Bel'kovich, V. 1973. Tactile sensitivity in *Delphinus delphis*. *Zool. Zhurnal* 52:620–22.

Kondo, R., Satta, Y., Matsuura, E. T., Ishiwa, H., Takahata, N., and Chigusa, S. I. 1990. Incomplete maternal transmission of mitochondrial DNA in *Drosophila. Genetics* 126: 657–63.

Kooyman, G. L. 1989. *Diverse divers.* Berlin: Springer-Verlag.

Kooyman, G. L., Billups, J. O., and Farwell, W. D. 1983. Two recently developed recorders for monitoring diving activity of marine birds and mammals. In *Experimental biology at sea*, ed. A. G. MacDonald and I. G. Priede, 197–214. New York: Academic Press.

Koski, W. R., Davis, R. A., Miller, G. W., and Withrow, D. E. 1993. Reproduction. In *The bowhead whale*, ed. J. J. Burns, J. J. Montague, and C. J. Cowles. Lawrence, KS: Society for Marine Mammalogy.

Kraus, C., and Gihr, M. 1971. On the presence of *Tursiops truncatus* in schools of *Globicephala melaena* on the Faeroe Islands. *Invest. Cetacea* 3(1):180–82.

Kraus, S. D. 1986. A review of the status of right whales (*Eubalaena glacialis*) in the western North Atlantic with a summary of research and management. National Technical Information Service report no. PB86-154143, Springfield, Virginia.

Kraus, S. D., Prescott, J. H., Knowlton, A. R., and Stone, G. S. 1986. Migration and calving of right whales (*Eubalaena glacialis*) in the western North Atlantic. In *Right whales: Past and present status: proceedings of the Workshop on the Status of Right Whales, New England Aquarium, Boston, MA, 15–23 June 1983*, ed. R. L. Brownell Jr., P. B. Best, and J. H. Prescott, 139–44. Reports of the International Whaling Commission, special issue 10. Cambridge: International Whaling Commission.

Kraus, S. D., Read, A. J., Solow, A., Baldwin, K., Spradlin, T., Anderson, E., and Williamson, J. 1997. Acoustic alarms reduce porpoise mortality. *Nature* 388:525.

Krebs, J. R., and Davies, N. B. 1993. *An introduction to behavioural ecology.* 3d edition. Oxford: Blackwell Scientific Publications.

Krebs, J. R., and Dawkins, R. 1984. Animal signals: Mind reading and manipulation. In *Behavioural ecology: An evolutionary approach*, 2d ed., ed. J. R. Krebs and N. B. Davies. Oxford: Blackwell Scientific Publications.

Kroodsma, D. E. 1989. Suggested experimental designs for song playbacks. *Anim. Behav.* 37:600–609.

Kruse, S. 1991. The interactions between killer whales and boats in Johnstone Strait, B.C. In *Dolphin societies: Discoveries and puzzles*, ed. K. Pryor and K. S. Norris. Berkeley: University of California Press.

Kshatriya, M., and Blake, R. W. 1985. The energetics of whale migration. In *Proceedings of the Sixth Biennial Conference on the Biology of Marine Mammals.* Lawrence, KS: Society for Marine Mammalogy.

Kuiehl, D. W., Haebler, R., and Potter, C. 1991. Chemical residues in dolphins from the US Atlantic coast including Atlantic bottlenose obtained during the 1987/88 mass mortality. *Chemosphere* 22:1071–84.

Kükenthal, W. G. 1892. *Sotalia teuszii* n. sp., ein pflanzenfressender (?) Delphin aus Kamerun. *Zool. Jahrb. Syst.* 6:442.

Kummer, H. 1967. Tripartite relations in hamadryas baboons. In *Social communication among primates*, ed. S. A. Altmann. Chicago: University of Chicago Press.

———. 1968. *Social organisation of Hamadryas baboons.* Chicago: University of Chicago Press.

———. 1982. Social knowledge in free-ranging primates. In *Animal mind—human mind*, ed. D. R. Griffin. Berlin: Springer Verlag.

Kummer, H., and Kurt, F. 1965. A comparison of social behavior in captive and wild hamadryas baboons. In *The baboon in medical research*, ed. H. Vagtborg. Austin: University of Texas Press.

Kuronuma, Y., and Tisdell, C. A. 1994. Economics of minke whale catches: Sustainability and welfare considerations. *Mar. Res. Econ.* 9:141–58.

Kuznetsov, V. B. 1974. A method of studying chemoreception in the Black Sea bottlenose dolphin (*Tursiops truncatus*). In *Morphology, physiology and acoustics of marine mammals*, ed. V. Y. Sokolov. Moscow: Science Press.

———. 1979. Chemoreception in dolphins of the Black Sea: Afalines (*Tursiops truncatus* Montagu), common dolphins (*Delphinus delphis* L.) and porpoises (*Phocoena phocoena* L.). *Doklady Akademii Nauk SSSR* 249:1498–1500.

Lacépède, B. G. E. 1804. *Histoire naturelle des Cétacés.* Paris: Plassan.

Lack, D. 1966. *Population studies of birds.* Oxford: Oxford University Press.

Lagardere, J. P. 1982. Effects of noise on growth and reproduction of crangon-crangon in rearing tanks. *Mar. Biol.* 71:177–86.

Lahille, F. 1912. Nota preliminar sobre una nueva especie de marsopa del Rio de La Plata (*Phocaena dioptrica*). *Anales del Museo nacional de historia natural de Buenos Aires* 23:269–78.

Lahvis, G. P., Wells, R. S., Kuehl, D. W., Stewart, J. L., Rhinehart, H. L., and Via, C. S. 1995. Decreased lymphocyte responses in free-ranging bottlenose dolphins (*Tursiops truncatus*) are associated with increased concentrations of PCBs and DDT in peripheral blood. *Environ. Health Perspect.* 103(4): 67–72.

Laland, K. N. 1992. A theoretical investigation of the role of social transmission in evolution. *Ethol. Sociobiol.* 13:87–113.

Laland, K. N., Richerson, P. J., and Boyd, R. 1996. Developing a theory of animal social learning. In *Social learning in animals: The roots of culture*, ed. C. M. Heyes, and B. G. J. Galef. San Diego: Academic Press.

Lambersten, R. H. 1986. Disease of the common fin whale (*Balaenoptera physalus*): Crassicaudiosis of the urinary system. *J. Mammal.* 67:353–66.

Landeau, L., and Terborgh. J. 1986. Oddity and the confusion effect in predation. *Anim. Behav.* 34:1372–80.

Lansman, R. A., Avise, J. C., and Huettel, M. D. 1983. Critical experimental test of the possibility of "paternal leakage" of mitochondrial DNA. *Proc. Natl. Acad. Sci. USA* 80:1969–71.

Larsen, A. H., Sigurjónsson, J., Óien, N., Vikingsson, G., and Palsbøll, P. J. 1996. Population genetic analysis of mitochondrial and nuclear genetic loci in skin biopsies collected from central and northeastern North Atlantic humpback whales *(Megaptera novaeangliae):* Population identity and migratory destinations. *Proc. R. Soc. Lond.* B 263:1611–18.

Lavigne, D. M., Scheffer, V. B., and Kellert, S. 1999. Changes in North American attitudes toward marine mammals. In *Conservation and management of marine mammals,* ed. J. R. Twiss Jr. and R. R. Reeves, 10–47. Washington, DC: Smithsonian Institution Press.

Law, R. J., Stringer, R. L., Allchin, C. R., and Jones, B. R. 1996. Metals and organochlorines in sperm whales *(Physeter macrocephalus)* stranded around the North Sea during the 1994/1995 winter. *Mar. Pollut. Bull.* 32:72–77.

Lawrence, B., and Schevill, W. E. 1954. *Tursiops* as an experimental subject. *J. Mammal.* 35:225–32.

Lawrence, M. J., Paterson, M., Baker, R. F., and Schmidt, R. 1992. Report of the workshop examining the potential effects of hydroelectric development on beluga of the Nelson River estuary, Winnipeg, Manitoba, Nov. 6–7, 1990. *Can. Tech. Rep. Fish. Aquat. Sci.* 1828:1–39.

Laws, R. M. 1956. Growth and sexual maturity in aquatic mammals. *Nature* 178:193–94.

———. 1959. The foetal growth rates of whales with special reference to the fin whale, *Balaenoptera physalus,* Linn. *Discovery Rep.* 29:281–308.

———. 1961. Reproduction, growth and age of southern fin whales. *Discovery Rep.* 31:327–486.

———. 1970. Elephants as agents of habitat and landscape change in East Africa. *Oikos* 21:1–15.

Laws, R. M., and Parker, I. S. C. 1968. Recent studies on elephant populations in East Africa. *Symp. Zool. Soc. Lond.* 21: 319–59.

Laws, R. M., and Purves, P. E. 1956. The ear plug of the Mysticeti as an indication of age with special reference to the North Atlantic fin whale. *Norsk Hvalfangst-Tidende* 45:413–25.

Laws, R. M., Parker, I. S. C., and Johnstone, R. B. C. 1975. *Elephants and their habitats: The ecology of elephants in North Bunyoro, Uganda.* Oxford: Clarendon.

Leaper, R., Chappell, O., and Gordon, J. 1992. The development of practical techniques for surveying sperm whale populations acoustically. *Rep. Int. Whal. Commn.* 42:549–60.

Leatherwood, S. 1975. Some observations of feeding behavior of bottle-nosed dolphins *(Tursiops truncatus)* in the northern Gulf of Mexico and *(Tursiops* cf. *T. gilli)* off southern California, Baja California, and Nayarit, Mexico. *Mar. Fish. Rev.* 37:10–16.

———. 1977. Some preliminary impressions on the numbers and social behavior of free-swimming bottlenose dolphins calves *(Tursiops truncatus)* in the northern Gulf of Mexico. In *Breeding dolphins: Present status, suggestions for the future,* ed. S. H. Ridgway and K. Benirschke. U.S. Marine Mammal Commission report no. MMC-76/07, Washington, DC.

———. 1991. Memories: David Keller Caldwell. *Mar. Mamm. Sci.* 7:97–99.

———. 1994. Re-estimation of incidental cetacean catches in Sri Lanka. In *Gillnets and cetaceans: Incorporating the proceedings of the symposium and workshop on the mortality of cetaceans in passive fishing nets and traps,* ed. W. F. Perrin, G. Donovan, and J. Barlow, 64–65. Reports of the International Whaling Commission, special issue 15. Cambridge: International Whaling Commission.

Leatherwood, S., and Donovan, G. P. (ed.). 1991. *Cetaceans and cetacean research in the Indian Ocean Sanctuary.* Marine Mammal Technical Report no. 3. Nairobi: UNEP.

Leatherwood, S., and Evans, W. E. 1979. Some recent uses and potentials of radiotelemetry in field studies of cetaceans. In *Behavior of marine mammals: Current perspectives in research,* vol. 3, *Cetaceans,* ed. H. E. Winn and B. L. Olla. New York: Plenum Press.

Leatherwood, S., and Ljungblad, D. K. 1979. Nighttime swimming and diving behavior of a radio-tagged spotted dolphin, *Stenella attenuata. Cetology* 34:1–6.

Leatherwood, S., and Reeves, R. R. 1994. River dolphins: A review of activities and plans of the Cetacean Specialist Group. *Aquat. Mamm.* 20:137–54.

Leatherwood, S., and Walker, W. A. 1979. The northern right whale dolphin *Lissodelphis borealis* Peale in the eastern North Pacific. In *Behavior of marine mammals: Current perspectives in research,* vol. 3: *Cetaceans,* ed. H. E. Winn and B. L. Olla. New York: Plenum Press.

Leatherwood, S., Reeves, R., and Foster, L. 1983. *Sierra Club handbook of whales and dolphins.* San Francisco: Sierra Club Books.

Leatherwood, S., Bowles, A. E., Krygier, E., Hall, J. D., and Ingell, S. 1984. Killer whales *(Orcinus orca)* of Shelikof Strait, Prince William Sound, Alaska and southeast Alaska: A review of available information. *Rep. Int. Whal. Commn.* 34:521–30.

Leatherwood, S., Matkin, C. O., Hall, J. D., and Ellis, G. E. 1990. Killer whales, *Orcinus orca,* photo-identified in Prince William Sound, Alaska, 1976 through 1987. *Can. Field Nat.* 104:362–71.

LeBoeuf, B. J. 1974. Male-male competition and reproductive success in elephant seals. *Am. Zool.* 14:163–76.

LeBoeuf, B. J., and Peterson, R. S. 1969. Social status and mating activity in elephant seals. *Science* 163:91–93.

LeBoeuf, B. J., and Reiter, J. 1988. Lifetime reproductive success in northern elephant seals. In *Reproductive success,* ed. T. H. Clutton-Brock. Chicago: University of Chicago Press.

LeBoeuf, B. J., and Würsig, B. 1985. Beyond bean counting and whale tales. *Mar. Mamm. Sci.* 1:128–48.

LeBoeuf, B. J., Costa, D. P., Huntley, A. C., and Feldkamp, S. D. 1988. Continuous, deep diving in female northern elephant seals, *Mirounga angustirostris. Can. J. Zool.* 66:446–58.

LeDuc, R. G. 1997. Mitochondrial systematics of the delphinidae. Ph.D. dissertation, University of California, San Diego.

LeDuc, R. G., and Curry, B. E. 1996. Mitochondrial DNA sequence analysis indicates need for revision of the genus *Tursiops.* Report of the Scientific Committee of the International Whaling Commission, SC/48/SM27.

LeDuc, R. G., Perrin, W. F., and Dizon, A. E. 1999. Phylogenetic relationships among the delphinid cetaceans based on full cytochrome *b* sequences. *Mar. Mamm. Sci.* 15:619–48.

Lee, P. C. 1987. Nutrition, fertility and maternal investment in primates. *J. Zool.* 213:409–22.

Lee, P. C., and Moss, C. J. 1986. Early maternal investment in male and female African elephant calves. *Behav. Ecol. Sociobiol.* 18:353–61.

Lehner, P. N. 1996. *Handbook of ethological methods.* 2d edition. Cambridge: Cambridge University Press.

Leighton, D. R. 1987. Gibbons: Territoriality and monogamy. In *Primate societies,* ed. B. Smuts, D. L. Cheney, R. M. Seyfarth, R. W. Wrangham, and T. T. Struhsaker, 135–45. Chicago: University of Chicago Press.

Leimar, O. 1996. Life-history analysis of the Trivers and Willard sex-ratio problem. *Behav. Ecol.* 7:316–25.

van Lennep, E. W., and van Utretcht, W. L. 1953. Preliminary report on the study of the mammary glands of whales. *Norsk Hvalangst-Tidende* 42:249–58.

Lescrauwaet, A.-C., and Gibbons, J. 1994. Mortality of small cetaceans and the crab bait fishery in the Magallanes area of Chile since 1980. In *Gillnets and cetaceans: Incorporating the proceedings of the symposium and workshop on the mortality of cetaceans in passive fishing nets and traps,* ed. W. F. Perrin, G. Donovan, and J. Barlow, 485–94. Reports of the International Whaling Commission, special issue 15. Cambridge: International Whaling Commission.

Lesson, R. P. 1828. *Histoire naturelle, generale et particuliere des mammiferes et des oiseaux. Cétacés.* Paris: Baudouin Freres.

Leuthold, W. 1977. *African ungulates.* New York: Springer-Verlag.

Levenson, C., and Leapley, W. T. 1978. Distribution of humpback whales *(Megaptera novaeangliae)* in the Caribbean determined by a rapid acoustic method. *J. Fish. Res. Bd. Can.* 35:1150–52.

Lien, J. 1994. Entanglement of large cetaceans in passive inshore gear in Newfoundland and Labrador (1979–1990). In *Gillnets and cetaceans: Incorporating the proceedings of the symposium and workshop on the mortality of cetaceans in passive fishing nets and traps,* ed. W. F. Perrin, G. Donovan, and J. Barlow, 149–57. Reports of the International Whaling Commission, special issue 15. Cambridge: International Whaling Commission.

Lien, J., and Katona, S. 1990. *A guide to the photographic identification of individual whales based on their natural and acquired markings.* San Pedro, CA: American Cetacean Society.

Lilljeborg, W. 1861. Hvalben, funna i jorden pa Gräsön i Roslagen i Sverige. Forhandlinger ved de Skandinavske Naturforskeres, ottende mode [volume eight], i Kiobenhaven, fra den 8 de til den 14 de Juli 1860, 599–616.

Lilly, J. C. 1958. Electrode and cannulae implantation in the brain by a simple percutaneous method. *Science* 127:1181–82.

———. 1961a. The importance of being in earnest about dialogues of dolphins. *Life* 51(4):68.

———. 1961b. *Man and dolphin.* New York: Doubleday.

———. 1963a. Critical brain size and language. *Persp. Biol. Med.* 6(2):246–55.

———. 1963b. Distress call of the bottlenose dolphin: Stimuli and evoked behavioral responses. *Science* 139:116–18.

———. 1965. Vocal mimicry in *Tursiops:* Ability to match numbers and durations of human vocal bursts. *Science* 147:300–301.

———. 1966. Sonic-ultrasonic emissions of the bottlenose dolphin. In *Whales, dolphins, and porpoises,* ed. K. S. Norris. Berkeley: University of California Press.

———. 1967a. Dolphin-human relationship and LSD-25. In *The use of LSD in psychotherapy and alcoholism,* ed. H. Abramson. New York: Bobbs-Merrill.

———. 1967b. *The mind of the dolphin: A nonhuman intelligence.* New York: Avon.

———. 1975. *Lilly on dolphins.* Garden City, NY: Anchor Press/Doubleday.

———. 1976. The rights of cetaceans under human laws. *Oceans* 2:66–68.

Lilly, J. C. 1978. *Communication between man and dolphin: The possibilities of talking with other species.* New York: Crown.

Lilly, J. C., and Miller, A. M. 1961a. Sounds emitted by the bottlenose dolphin. *Science* 133:1689–93.

———. 1961b. Vocal exchanges between dolphins. *Science* 134:1873–76.

Lima, S. L. 1995a. Back to the basics of anti-predator vigilance: The group-size effect. *Anim. Behav.* 49:11–20.

———. 1995b. Collective detection of predatory attack by social foragers: Fraught with ambiguity? *Anim. Behav.* 50:1097–1108.

Lima, S. L., and Dill, L. M. 1990. Behavioral decisions made under the risk of predation: A review and prospectus. *Can. J. Zool.* 68:619–40.

Lindhard, M., and Strager, H. 1989. Andenes Whale Safari and sperm whale research. Progress report to World Wildlife Fund Sweden, Denmark, Norway.

Linegaugh, R. M. 1976. Soviet dolphin research: An analysis of the literature from 1900 to the present. *Mar. Tech. Soc. J.* 10(3):16–20.

Linnaeus, C. 1758. *Systema naturae per regna tria naturae, secundum classis, ordines, genera, species cum characteribus, differentiis synonymis, locis.* 10th edition. Vol. 1, *Regnum animale,* pt. 1, 1–532.

Lissaman, P. B. S., and Schollenberger, C. A. 1970. Formation flight of birds. *Science* 18:1003–5.

Ljungblad, D. K., Thompson, P. O., and Moore, S. E. 1982. Underwater sounds recorded from migrating bowhead whales, *Balaena mysticetus,* in 1979. *J. Acoust. Soc. Am.* 71:477–82.

Lockyer, C. H. 1976. Body weights of some species of large whales. *J. Cons. Int. Explor. Mer.* 36:259–73.

———. 1977. Observations on diving behaviour of the sperm whale. In *A voyage of discovery,* ed. M. Angel. Oxford: Pergamon.

———. 1981a. Estimates of growth and energy budget for the sperm whale, *Physeter catodon.* In *Mammals in the seas,* 379–487. FAO Fisheries Series, no. 5, vol. 3. Rome: Food and Agriculture Organization of the United Nations.

———. 1981b. Growth and energy budgets of large baleen whales from the Southern Hemisphere. In *Mammals in the seas,* 379–487. FAO Fisheries Series, no. 5, vol. 3. Rome: Food and Agriculture Organization of the United Nations.

———. 1984. Review of baleen whale (Mysticeti) reproduction

and implications for management. In *Reproduction in whales, dolphins and porpoises: Proceedings of the conference, Cetacean Reproduction, Estimating Parameters for Stock Assessment and Management, La Jolla, CA, 28 Nov.–7 Dec. 1981,* ed. W. F. Perrin, R. L. Brownell Jr., and D. P. DeMaster, 27–48. Reports of the International Whaling Commission, special issue 6. Cambridge: International Whaling Commission.

———. 1985. A wild but sociable dolphin off Portreath, north Cornwall. *J. Zool.* 207:605–30.

———. 1987. The relationship between body fat, food resource and reproductive energetic costs in North Atlantic fin whales (*Balaenoptera physalus*). *Symp. Zool. Soc. Lond.* 57:343–61.

———. 1990. Review of incidents involving wild, sociable dolphin, worldwide. In *The bottlenose dolphin,* ed. S. Leatherwood and R. R. Reeves. San Diego: Academic Press.

Lockyer, C. H., and Morris, R. J. 1990. Some observations on wound healing and persistence of scars in *Tursiops truncatus.* In *Individual recognition of cetaceans: Use of photo-identification and other techniques to estimate population parameters,* ed. P. S. Hammond, S. A. Mizroch, and G. P. Donovan, 113–18. Reports of the International Whaling Commission, special issue 12. Cambridge: International Whaling Commission.

Lockyer, C. H., Goodall, R. N. P., and Galeazzi, A. R. 1988. Age and body length characteristics of *Cephalorhynchus commersoni* from incidentally-caught specimens off Tierra del Fuego. In *Biology of the genus* Cephalorhynchus, ed. R. L. Brownell Jr. and G. P. Donovan, 71–83. Reports of the International Whaling Commission, special issue 9. Cambridge: International Whaling Commission.

Loesche, P., Stoddard, P. K., Higgins, B. J., and Beecher, M. D. 1991. Signature versus perceptual adaptations for individual vocal recognition in swallows. *Behaviour* 118:15–25.

Long, D. J., and Jones, R. E. 1996. White shark predation and scavenging on cetaceans in the eastern north Pacific ocean. In *Great white sharks: The biology of* Carcharodon carcharias, ed. A. P. Klimley and D. G. Ainley, 293–307. San Diego: Academic Press.

Longman, H. A. 1926. New records of Cetacea, with a list of Queensland species. Mem. Queensland Mus. 8(3):266–78.

Lopez, J. C., and Lopez, D. 1985. Killer whales (*Orcinus orca*) of Patagonia, and their behavior of intentional stranding while hunting nearshore. *J. Mammal.* 66:181–83.

Lorenz, K. Z. 1937. The companion in the bird's world. *Auk* 54:245–73.

Lott, D. F. 1984. Intraspecific variation in the social systems of wild vertebrates. *Behaviour* 88:266–325.

Lowry, L. F., Burns, J. J., and Nelson, R. R. 1987. Polar bear, *Ursus maritimus,* predation of belugas, *Delphinapterus leucas,* in the Bering and Chukehi Seas. *Can. Field Nat.* 101:141–46.

Lynch, M. 1988. Estimation of relatedness by DNA fingerprinting. *Mol. Biol. Evol.* 5:584–99.

Lynn, S. K. 1995. Movements, site fidelity, and surfacing patterns of bottlenose dolphins on the central Texas coast. M.S. thesis, Texas A&M University.

Lyon, B. E. 1998. Optimal clutch size and conspecific brood parasitism. *Nature* 392(6674):380–83.

Lyrholm, T. 1988. Photoidentification of individual killer whales, *Orcinus orca,* off the coast of Norway, 1983–1986. *Rit Fiskideildar* 11:89–94.

Lyrholm, T., and Gyllensten, U. 1998. Global matrilineal population structure in sperm whales as indicated by mitochondrial DNA sequences. *Proc. R. Soc. Lond.* B 265:1679–84.

Lyrholm, T., Leimar, O., and Gyllensten, U. 1996. Low diversity and biased substitution patterns in the mitochondrial DNA control region of sperm whales: Implications for estimates of time since common ancestry. *Mol. Biol. Evol.* 13:1318–26.

Lyrholm, T., Leimar, O., Johanneson, B., and Gyllensten, U. 1999. Sex-biased dispersal in sperm whales: Contrasting mitochondrial and nuclear genetic structure of global populations. *Proc. R. Soc. Lond.* B 266:347–54.

MacGarvin, M., and Simmonds, M. 1996. Whales and climate change. In *The conservation of whales and dolphins: Science and practice,* ed. M. P. Simmonds and J. Hutchinson. Chichester: Wiley.

Mackintosh, N. A. 1942. The southern stocks of whalebone whales. *Discovery Rep.* 22:197–300.

———. 1965. *The stocks of whales.* London: Fishing News (Books) Ltd.

Mackintosh, N. A., and Wheeler, J. F. G. 1929. Southern blue and fin whales. *Discovery Rep.* 1:257–540.

MacLeod, C. D. 1998. Intraspecific scarring in odontocete cetaceans: An indicator of male "quality" in aggressive social interactions? *J. Zool.* 244:71–77.

Madsen, C. J., and Herman, L. M. 1980. Social and ecological correlates of cetacean vision and visual appearance. In *Cetacean behavior: Mechanisms and functions,* ed. L. M. Herman. New York: Wiley-Interscience.

Major, P. F. 1978. Predator-prey interactions in two schooling fishes, *Caranx ignobilis* and *Stolephorus purpureus. Anim. Behav.* 26:760–77.

Malloné, J. S. 1991. Behaviour of gray whales (*Eschritus robustus*) summering off the northern California coast, from Patrick's Point to Crescent City. *Can. J. Zool.* 69(3):776–82.

Malme, C. I., Miles, P. R., Clark, C. W., Tyack, P., and Bird, J. E. 1983. Investigations of the potential effects of underwater noise from petroleum industry activities on migrating gray whale behavior. Bolt Beranek and Newman Report no. 5366, submitted to Minerals Management Service, U.S. Department of the Interior, NTIS PB86-174174.

Mangel, M., Talbot, J. M. et al. 1996. Principles for the conservation of wild living resources. *Ecol. Appl.* 6:338–62.

Mann, D. A., Zhongmin, L., and Popper, A. N. 1997. A clupeid fish can detect ultrasound. *Nature* 389:341.

Mann, J. 1997. Individual differences in bottlenose dolphin infants. *Family Syst.* 4:35–49.

———. 1998. The relationship between bottlenose dolphin mother-infant behavior, habitat, weaning and infant mortality. Paper presented at the Twelfth Biennial Conference on Biology of Marine Mammals, Monaco, January.

———. 1999. Behavioral sampling methods for cetaceans: A review and critique. *Mar. Mamm. Sci.* 15:102–22.

Mann, J., and Barnett H. 1999. Lethal tiger shark (*Galeocerdo cuvier*) attack on bottlenose dolphin (*Tursiops* sp.) calf: Defense and reactions by the mother. *Mar. Mamm. Sci.* 15:568–74.

Mann, J., and Smuts, B. B. 1993. The behavioral ecology of wild bottlenose dolphins from birth to weaning. Paper presented at the Tenth Biennial Conference on the Biology of Marine Mammals, Galveston, TX.

———. 1998. Natal attraction: Allomaternal care and mother-infant separations in wild bottlenose dolphins. *Anim. Behav.* 55:1097–1113.

———. 1999. Behavioral development of wild bottlenose dolphin newborns. *Behaviour* 136:529–66.

Mann, J., Ten Have, T., Plunkett, J. W., and Meisels, S. J. 1991. Time sampling: A methodological critique. *Child Dev.* 62:227–41.

Mann, J., Smolker, R. A., and Smuts, B. B. 1995. Responses to calf entanglement in free-ranging bottlenose dolphins. *Mar. Mamm. Sci.* 11:168–75.

Mann, J., Connor, R. C., Barre, L. M., and Heithaus, M. R. 2000. Female reproductive success in wild bottlenose dolphins (*Tursiops* sp.).: Life history, habitat, provisioning, and group size effects. *Behav. Ecol.* In press.

Manning, A., and Dawkins, M. S. 1992. *An introduction to animal behaviour.* Cambridge: Cambridge University Press.

Manson, J., and Wrangham, R. W. 1991. Intergroup aggression in chimpanzees and humans. *Curr. Anthropol.* 32(4):369–90.

Marino, L. 1997. The relationship between gestation length, encephalization, and body weight in odontocetes. *Mar. Mamm. Sci.* 13:133–38.

———. 1998. A comparison of encephalization between odontocete cetaceans and anthropoid primates. *Brain Behav. Evol.* 51:230–38.

Marler, P., and Hamilton, W. J. 1966. *Mechanisms of animal behavior.* New York: John Wiley & Sons.

Marsh, H., and Kasuya, T. 1984. Changes in the ovaries of the short-finned pilot whale, *Globicephala macrorhynchus*, with age and reproductive activity. In *Reproduction in whales, dolphins and porpoises: Proceedings of the conference, Cetacean Reproduction, Estimating Parameters for Stock Assessment and Management, La Jolla, CA, 28 Nov.–7 Dec. 1981*, ed. W. F. Perrin, R. L. Brownell Jr., and D. P. DeMaster, 311–35. Reports of the International Whaling Commission, special issue 6. Cambridge: International Whaling Commission.

———. 1986. Evidence for reproductive senescence in female cetaceans. In *Behaviour of whales in relation to management: Incorporating the proceedings of a workshop of the same name held in Seattle, Washington, 19–23 April 1982*, ed. G. P. Donovan, 57–74. Reports of the International Whaling Commission, special issue 8. Cambridge: International Whaling Commission.

———. 1991. An overview of the changes in the role of a female pilot whale with age. In *Dolphin societies: Discoveries and puzzles*, ed. K. Pryor and K. S. Norris. Berkeley: University of California Press.

Marshall, G. J. 1998. Crittercam: An animal-borne imaging and data logging system. *Mar. Tech. Soc. J.* 32:11–17.

Martin, A. R. 1982. A link between the sperm whales occurring off Iceland and in the Azores. *Mammalia* 46:259–60.

Martin, A. R., and Rothery, P. 1993. Reproductive parameters of female long-finned pilot whales (*Globicephala melas*) around the Faroe Islands. In *Biology of Northern Hemisphere pilot whales: A collection of papers*, ed. G. P. Donovan, C. H.

Lockyer, and A. R. Martin, 263–304. Reports of the International Whaling Commission, special issue 14. Cambridge: International Whaling Commission.

Martin, P., and Bateson, P. 1986. *Measuring behaviour: An introductory guide.* Cambridge: Cambridge University Press.

———. 1994. *Measuring behaviour.* 2d edition. Cambridge: Cambridge University Press.

Martin, P., and Kraemer, H. C. 1987. Individual differences in behaviour and their statistical consequences. *Anim. Behav.* 35:1366–75.

Martineau, D., Deguise, S., Girard, C., Lagacé, A., and Bland, P. 1994. Pathology and toxicology of beluga whales from the St. Lawrence estuary, Quebec, Canada. *Sci. Total Environ.* 154:201–15.

Mate, B. 1989. Satellite-monitored radio tracking as a method for studying cetacean movements and behaviour. *Rep. Int. Whal. Commn.* 39:389–91.

Mate, B. R., Harvey, J. T., Hobbs, L., and Maiefski, R. 1983. A new attachment device for radio-tagging large whales. *J. Wildl. Mgmt.* 47:868–72.

Mate, B. R., Prescott, J. H., and Geraci, J. R. 1987. Free ranging movements of a pilot whale from a satellite-monitored radio. Abstract, Seventh Biennial Conference on the Biology of Marine Mammals.

Mate, B. R., Stafford, K. M., and Ljungblad, D. K. 1994a. A change in sperm whale (*Physeter macrocephalus*) distribution correlated to seismic surveys in the Gulf of Mexico. *J. Acoust. Soc. Am.* 965:3268–69.

Mate, B. R., Stafford, K. M., Nawojchik, R., and Dunn, J. L. 1994b. Movements and dive behavior of a satellite-monitored Atlantic white-sided dolphin (*Lagenorhynchus acutus*) in the Gulf of Maine. *Mar. Mamm. Sci.* 10:116–21.

Mate, B. R., Rossbach, K. A., Nieukirk, S. L., Wells, R. S., Irvine, A. B., Scott, M. D., and Read, A. J. 1995. Satellite-monitored movements and dive behavior of a bottlenose dolphin (*Tursiops truncatus*) in Tampa Bay, Florida. *Mar. Mamm. Sci.* 11:452–63.

Mate, B. R., Nieukirk, S. L., and Kraus, S. D. 1997. Satellite-monitored movements of the northern right whale. *J. Wildl. Mgmt.* 61(4):1393–1405.

Matkin, C. O., and Saulitis, E. L. 1994. Killer whale (*Orcinus orca*) biology and management in Alaska. Report no. T75135023, U.S. Marine Mammal Commission, Washington, DC.

Matkin, C. O., Ellis, G. M., Dahlheim, M. E., and Zeh, J. 1994. Status of killer whales in Prince William Sound, 1985–1992. In *Marine mammals and the* Exxon Valdez, ed. T. R. Loughlin. San Diego: Academic Press.

Matkin, C. O., Matkin, D. R., Ellis, G. M., Saulitis, E., and McSweeney, D. 1997. Movements of *resident* killer whales in southeastern Alaska and Prince William Sound, Alaska. *Mar. Mamm. Sci.* 13:469–75.

Matthews, L. H. 1937. The humpback whale, *Megaptera nodosa. Discovery Rep.* 17:7–92.

Mattila, D. K., and Clapham, P. J. 1989. Humpback whales, *Megaptera novaeangliae*, and other cetaceans on Virgin Bank and in the northern Leeward Islands, 1985 and 1986. *Can. J. Zool.* 67:2201–11.

Mattila, D. K., Guinee, L. N., and Mayo, C. A. 1987. Humpback whale songs on a North Atlantic feeding ground. *J. Mammal.* 68:880–83.

Mattila, D. K., Clapham, P. J., Katona, S. K., and Stone, G. S. 1989. Population composition of humpback whales, *Megaptera novaeangliae*, on Silver Bank, 1984. *Can. J. Zool.* 67: 281–85.

Mattila, D. K., Clapham, P. J., Vásquez, O., and Bowman, R. 1994. Occurrence, population composition and habitat use of humpback whales in Samana Bay, Dominican Republic. *Can. J. Zool.* 72:1898–1907.

Mayer, P. J. 1982. Evolutionary advantage of the menopause. *Hum. Ecol.* 10:477–94.

Mayer, S., and Simmonds, M. 1996. Science and precaution in cetacean conservation. In *The conservation of whales and dolphins: Science and practice,* ed. M. P. Simmonds and J. Hutchinson. Chichester: Wiley.

Maynard Smith, J. 1974. The theory of games and the evolution of animal conflicts. *J. Theor. Biol.* 47:209–21.

———. 1976. Group selection. *Q. Rev. Biol.* 51:277–83.

———. 1980. A new theory of sexual investment. *Behav. Ecol. Sociobiol.* 7:247–51.

———. 1983. Game theory and the evolution of cooperation. In *Evolution from molecules to men,* ed. D. S. Bendall. Cambridge: Cambridge University Press.

Mayr, E. 1965. *Animal species and evolution.* Cambridge, MA: Belknap Press.

Mayr, E., and Ashlock, P. D. 1991. *Principles of systematic zoology.* 2d edition. New York: McGraw-Hill.

McBride, A. F. 1940. Meet Mr. Porpoise. *Nat. Hist.* 45: 16–29.

McBride, A. F., and Hebb, D. O. 1948. Behavior of the captive bottlenose dolphin *Tursiops truncatus. J. Comp. Phys. Psychol.* 41:111–23.

McBride, A. F., and Kritzler, H. 1951. Observations on pregnancy, parturition, and post-natal behavior in the bottlenose dolphin. *J. Mammal.* 32:251–66.

McCann, C. 1974. Body scarring on Cetacea—Odontocetes. *Sci. Rep. Whales Res. Inst.* 1974:145–55.

McCowan, B. 1995. A new quantitative technique for categorizing whistles using simulated signals and whistles from captive bottlenose dolphins (Delphinidae, *Tursiops truncatus*). *Ethology* 100:177–93.

McCowan, B., and Reiss, D. 1995. Quantitative comparison of whistle repertoires from captive adult bottlenose dolphins (Delphinidae, *Tursiops truncatus*): A reevaluation of the signature whistle hypothesis. *Ethology* 100:193–209.

———. 1997. Vocal learning in captive bottlenose dolphins: A comparison with humans and nonhuman animals. In *Social influences on vocal development,* ed. C. T. Snowdon and M. Hausberger. Cambridge: Cambridge University Press.

McDonald, D. B., and Potts, W. K. 1994. Cooperative display and relatedness among males in a lek-mating bird. *Science* 266: 1030–32.

McDonald, M. A., Hildebrand, J. A., and Webb, S. C. 1995. Blue and fin whales observed on a seafloor array in the Northeast Pacific. *J. Acoust. Soc. Am.* 98:712–21.

McGregor, P. K., and Krebs, J. R. 1984a. Song learning and deceptive mimicry. *Anim. Behav.* 32:280–87.

———. 1984b. Sound degradation as a distance cue in great tit (*Parus major*) song. *Behav. Ecol. Sociobiol.* 16:49–56.

McGregor, P. K., Krebs, J. R., and Ratcliffe, L. M. 1983. The response of great tits (*Parus major*) to the playback of degraded and undegraded songs: The effect of familiarity with the stimulus song type. *Auk* 100:898–906.

McGregor, P. K., Dabelsteen, T., Clark, C. W., Bower, J. L., Tavares, J. P., and Holland, J. 1997. Accuracy of a passive acoustic location system: Empirical studies in terrestrial habitats. *Ethol. Ecol. Evol.* 9:269–86.

McHugh, J. L. 1974. The role and history of the International Whaling Commission. In *The whale problem: A status report,* ed. W. E. Schevill, G. C. Ray, and K. S. Norris. Cambridge, MA: Harvard University Press.

McIntyre, J. 1974. *Mind in the waters.* New York: Charles Scribner's Sons.

McKaye, K. R., Mughogho, D. E., and Lovullo, T. J. 1992. Formation of the selfish school. *Environ. Biol. Fish.* 35:213–18.

McSweeney, D. J., Chu, K. C., Dolphin, W. F., and Guinee, L. N. 1989. North Pacific humpback whale songs: A comparison of southeast Alaskan feeding ground songs and Hawaiian wintering ground songs. *Mar. Mamm. Sci.* 5:116–38.

McVay, S. 1974. Reflections on the management of whaling. In *The Whale problem: A status report,* ed. W. E. Schevill, G. C. Ray, and K. S. Norris, Cambridge, MA: Harvard University Press.

Mead, J. G. 1975. A preliminary report on the former net fisheries for *Tursiops truncatus* in the western North Atlantic. *J. Fish. Res. Bd. Can.* 32:1155–62.

———. 1984. Survey of reproductive data for the beaked whales (Ziphiidae). In *Reproduction in whales, dolphins and porpoises: Proceedings of the conference, Cetacean Reproduction, Estimating Parameters for Stock Assessment and Management, La Jolla, CA, 28 Nov.–7 Dec. 1981,* ed. W. F. Perrin, R. L. Brownell Jr., and D. P. DeMaster, 91–96. Reports of the International Whaling Commission, special issue 6. Cambridge: International Whaling Commission.

———. 1989a. Beaked whales of the genus *Mesoplodon.* In *Handbook of marine mammals,* vol. 4, *River dolphins and the larger toothed whales,* ed. S. H. Ridgway and R. Harrison. London: Academic Press.

———. 1989b. Bottlenose whales *Hyperoodon ampullatus* (Forster, 1770) and *Hyperoodon planifrons* Flower, 1882. In *Handbook of marine mammals,* vol. 4, *River dolphins and the larger toothed whales,* ed. S. H. Ridgway and R. Harrison. London: Academic Press.

Mead, J. G., and Brownell, R. L. Jr. 1993. Order Cetacea. In *Mammal species of the world,* 2d edition, ed. D. E. Wilson and D. M. Reeder. Washington, DC: Smithsonian Institution Press.

Mead, J. G., and Potter, C. W. 1990. Natural history of bottlenose dolphins along the central Atlantic coast of the United States. In *The bottlenose dolphin,* ed. S. Leatherwood and R. R. Reeves. San Diego: Academic Press.

Medrano, L., Salinas, I., Salas, P., Ladrón de Guevara, P., Aguayo, A., Jacobsen, J., and Baker, C. S. 1994. Sex identification of humpback whales, *Megaptera novaeangliae*, on the wintering grounds of the Mexican Pacific Ocean. *Can. J. Zool.* 72:1771–74.

Medvin, M. B., and Beecher, M. D. 1986. Parent-offspring recognition in the barn swallow *(Hirundo rustica). Anim. Behav.* 34:1627–39.

Mellinger, D. K., and Clark, C. W. 1993. Bioacoustic *transient* detection by image convolution. *J. Acoust. Soc. Am.* 93:2358.

Melville, H. 1851. *Moby-Dick or the whale.* London: Penguin (1972).

Mercer, M. C. 1975. Modified Leslie-DeLury population models of the long-finned pilot whale *(Globicephala melaena)* and annual production of the shore-finned squid *(Illex illecebrosus)* based upon their interaction at Newfoundland. *J. Fish. Res. Bd. Can.* 32:1145–54.

Merle, R. 1969. *The day of the dolphin.* New York: Simon and Schuster.

Meyen, F. J. F. 1833. Beiträge zur zoologie, gesammelt auf einer reise um die erde. *Nova Acta Physico-Medica, Academiae Caesarae Leopoldino-Carolinae,* 16(2):549–610.

Mikhalev, Y. A. 1997. Humpback whales *Megaptera novaeangliae* in the Arabian Sea. *Mar. Ecol. Prog. Ser.* 149:13–21.

Mikhalev, Y. A., Ivashin, M. V., Savusin, V. P., and Zelenaya, F. E. 1981. The distribution and biology of killer whales in the Southern Hemisphere. *Rep. Int. Whal. Commn.* 31:551–66.

Milinkovitch, M. C., Guillermo, O., and Meyer, A. 1993. Revised phylogeny of whales suggested by mitochondrial ribosomal DNA sequences. *Nature* 361:346–48.

Milinski, M. 1977. Do all members of a swarm suffer the same predation? *Z. Tierpsychol.* 45:373–88.

Millar, J. S., and Hickling, G. J. 1990. Fasting endurance and the evolution of mammalian body size. *Funct. Ecol.* 4:5–12.

Miller, D. E., and Emlen, J. T. 1975. Individual chick recognition and family integrity in the ring-billed gull. *Behaviour* 52:124–44.

Miller, G. Sr. 1918. A new river dolphin from China. *Smithsonian Misc. Coll.* 68(9):1–12, 13.

Miller, G. W., Davis, R. A., Koski, W. R., and Crone, M. J. 1992. Calving intervals of bowhead whales: An analysis of photographic data. *Rep. Int. Whal. Commn.* 42:501–6.

Miller, P. J., and Tyack, P. L. 1998. A small towed beamforming array to identify vocalizing *resident* killer whales *(Orcinus orca). Deep-Sea Res.* 45:1389–1405.

Miller, R. C. 1922. The significance of the gregarious habit. *Ecology* 3:122–26.

Mitani, J. C. 1985. Mating behavior of male orangutans in the Kutai Reserve. *Anim. Behav.* 33:392–402.

Mitani, J. C., and Rodman, P. S. 1979. Territoriality: The relation of ranging patterns and home range size to defendability among primate species. *Behav. Ecol. Sociobiol.* 5:241–51.

Mitani, J. C., Gros-Louis, J., and Richards, A. F. 1996. Sexual dimorphism, the operational sex ratio, and the intensity of male competition in polygynous primates. *Am. Nat.* 147:966–80.

Mitchell, C. L., Boinski, S., and van Schaik, C. P. 1991. Competitive regimes and female bonding in two species of squirrel monkeys *(Saimiri oerstedi* and *S. sciureus). Behav. Ecol. Sociobiol.* 28:55–60.

Mitchell, E. D. 1970. Pigmentation pattern evolution in delphinid cetaceans: An essay in adaptive coloration. *Can. J. Zool.* 48:717–40.

———. 1973. Draft report on humpback whales taken under special scientific permit by eastern Canadian land stations, 1969–1971. *Rep. Int. Whal. Commn.* 23:138–54.

———. 1975a. Report of the Scientific Committee, Annex U. Preliminary report on Nova Scotian fishery for sperm whales *(Physeter catodon). Rep. Int. Whal. Commn.* 25:226–35.

———. 1975b. Report on the meeting on smaller cetaceans, Montreal. April 1–11, 1974. *Fish. Res. Bd. Can.* 32:889–983.

———. 1977. Sperm whale maximum length limit: Proposed protection of "harem masters." *Rep. Int. Whal. Commn.* 27:224–27.

———. 1983. Potential of whaling logbook data for studying aspects of social structure in the sperm whale, *Physeter macrocephalus,* with an example—the ship Mariner to the Pacific, 1836–1840. In *Historical whaling records: Including the proceedings of the International Workshop on Historical Whaling Records, Sharon, MA, Sept. 12–16, 1977,* ed. M. F. Tillman and G. P. Donovan, 63–80. Reports of the International Whaling Commission, special issue 5. Cambridge: International Whaling Commission.

Mitchell, E. D., and Reeves, R. R. 1980. The Alaska bowhead problem: A commentary. *Arctic* 33:686–723.

Mitchell, E. D., Reeves, R. R., and Evely, A. (eds.). 1986. Introductory essay. In *Bibliography of whale killing techniques,* 1–12. Reports of the International Whaling Commission, special issue 7. Cambridge: International Whaling Commission.

Mizroch, S. A., and Bigg, M. A. 1990. Shooting whales (photographically) from small boats: An introductory guide. In *Individual recognition of cetaceans: Use of photo-identification and other techniques to estimate population parameters,* ed. P. S. Hammond, S. A. Mizroch, and G. P. Donovan, 39–40. Reports of the International Whaling Commission, special issue 12. Cambridge: International Whaling Commission.

Mizroch, S. A., and York, A. E. 1984. Have pregnancy rates of Southern Hemisphere fin whales, *Balaenoptera physalus,* increased? In *Reproduction in whales, dolphins and porpoises: Proceedings of the conference, Cetacean Reproduction, Estimating Parameters for Stock Assessment and Management, La Jolla, CA, 28 Nov.–7 Dec. 1981,* ed. W. F. Perrin, R. L. Brownell Jr., and D. P. DeMaster, 401–10. Reports of the International Whaling Commission, special issue 6. Cambridge: International Whaling Commission.

Mizroch, S. A., Beard, J., and Lynde, M. 1990. Computer assisted photo-identification of humpback whales. In *Individual recognition of cetaceans: Use of photo-identification and other techniques to estimate population parameters,* ed. P. S. Hammond, S. A. Mizroch, and G. P. Donovan, 63–70. Reports of the International Whaling Commission, special issue 12. Cambridge: International Whaling Commission.

Mizue, K., Nishiwaki, M., and Takemura, A. 1971. The underwater sound of Ganges river dolphin *(Platanista gangetica). Sci. Rep. Whales Res. Inst.* 23:123–28.

Mobley, J. R., and Helweg, D. A. 1990. Visual ecology and cognition in cetaceans. In *Sensory abilities of cetaceans,* ed. J. Thomas and R. Kastelein. New York: Plenum.

Mobley, J. R., and Herman, L. M. 1985. Transience of social affiliations among humpback whales *(Megaptera novaeangliae)* on the Hawaiian wintering grounds. *Can. J. Zool.* 63:762–72.

Mobley, J. R., Herman, L. M., and Frankel, A. S. 1988. Responses of wintering humpback whales *(Megaptera novaeangliae)* to playback recordings of winter and summer vocalizations and of synthetic sound. *Behav. Ecol. Sociobiol.* 23:211–23.

Møhl, B., and Andersen, S. 1973. Echolocation: High frequency component in the click of the harbour porpoise *(Phocoena phocoena* L.). *J. Acoust. Soc. Am.* 54:1368–72.

Montagu, G. 1821. Description of a species of Delphinus which appears to be new. *Mem. Wern. Soc. Nat. Hist.* 3:75–82.

Moore, J. C. 1955. Bottle-nosed dolphins support remains of young. *J. Mammal.* 36:466–67.

———. 1963. Recognizing certain species of beaked whales of the Pacific ocean. *Am. Midl. Nat.* 70(2):396–428.

Moore, K. E., Watkins, W. A., and Tyack, P. L. 1993. Pattern similarity in shared codas from sperm whales *(Physeter catodon). Mar. Mamm. Sci.* 9:1–9.

Moore, P. W. B., and Pawloski, D. 1991. Investigation on the control of echolocation pulses in the dolphin *(Tursiops truncatus).* In *Sensory abilities of cetaceans,* ed. J. Thomas and R. Kastelein. New York: Plenum.

Moore, S. E., and Reeves, R. R. 1993. Distribution and movement. In *The bowhead whale,* ed. J. J. Burns, J. J. Montague, and C. J. Cowles. Lawrence, KS: Society for Marine Mammalogy.

Mooring, M. S., and Hart, B. L. 1992. Animal grouping for protection from parasites: Selfish herd and encounter-dilution effects. *Behaviour* 123:173–93.

Moors, T. M. 1997. Is a "menage a trois" important in dolphin mating systems? Behavioral patterns of breeding female bottlenose dolphins. M.Sc. thesis, University of California, Santa Cruz.

Morgan, M. J., and Godin, J. G. J. 1985. Antipredator benefits of schooling behaviour in a cyprinodontid fish, the banded killifish *(Fundulus diaphanus). Z. Tierpsychol.* 70:236–46.

Morgane, P. J., and Jacobs, J. S. 1972. Comparative anatomy of the cetacean central nervous system. In *Functional anatomy of marine mammals,* vol. 1, ed. R. J. Harrison. New York: Academic Press.

Morozov, D. A. 1970. Dolphins hunting. *Rybnoe Khoziaistvo* 46: 16–17.

Morris, D. 1966. Animal behaviour studies at London Zoo. *Int. Zoo Yrbk.* 6:288–91.

Morton, A. B. 1990. A quantitative comparison of the behaviour of *resident* and *transient* forms of the killer whale off the central British Columbia coast. In *Individual recognition of cetaceans: Use of photo-identification and other techniques to estimate population parameters,* ed. P. S. Hammond, S. A. Mizroch, and G. P. Donovan, 245–48. Reports of the International Whaling Commission, special issue 12. Cambridge: International Whaling Commission.

Morton, E. S. 1982. Grading, discreteness, redundancy, and motivation-structural rules. In *Acoustic communication in birds,* ed. D. E. Kroodsma and E. H. Miller. New York: Academic Press.

———. 1986. Predictions from the ranging hypothesis for the evolution of long distance signals in birds. *Behaviour* 99:65–86.

Moss, C. J. 1975. *Portraits in the wild.* Chicago: University of Chicago Press.

———. 1977. The Amboseli elephants. *Wildl. News* 12(2):9–12.

———. 1983. Oestrous behaviour and female choice in the African elephant. *Behaviour* 86:167–96.

———. 1988. *Elephant memories.* New York: William Morrow.

Moss, C. J., and Poole, J. H. 1983. Relationships and social structure of African elephants. In *Primate social relationships: An integrated approach,* ed. R. A. Hinde. Sunderland, MA: Sinauer Associates.

Muir, D. C. G., Wagemann, R., Hargrave, B. T., Thomas, D. J., Peakall, D. B., and Norstrom, R. J. 1992. Arctic marine ecosystem contamination. *Sci. Total Environ.* 122:75–134.

Müller, O. F. 1776. Zoologiae Danicae Prodromus, seu animalium Daniae et Norvegiae indigenarum characteres, nomina, et synonyma imprimis popularium. *Typis Hallageriis, Havniae* [Copenhagen] 8:1–282.

Mullins, J., Whitehead, H., and Weilgart, L. S. 1988. Behaviour and vocalizations of two single sperm whales, *Physeter macrocephalus,* off Nova Scotia. *Can. J. Fish. Aquat. Sci.* 45:1736–43.

Mullis, K., Faloona, F., Scharf, S., Saiki, R., Horn, G., and Erlich, H. 1986. Specific enzymatic amplification of DNA in vitro: The polymerase chain reaction. *Cold Spring Harbor Symp. Quant. Biol.* 51:263–73.

Mulvaney, K. 1996. Directed kills of small cetaceans worldwide. In *The conservation of whales and dolphins: Science and practice,* ed. M. P. Simmonds and J. Hutchinson. Chichester: Wiley.

Murchison, A. E. 1980. Detection range and range resolution of echolocating porpoise *(Tursiops truncatus).* In *Animal sonar systems,* ed. R.-G. Busnel and J. F. Fish. New York: Plenum Press.

Murphey, R. M., Paranhos da Costa, M. J. R., Gomes da Silva, R., and de Souza, R. C. 1995. Allonursing in river buffalo, *Bubalus bubalis:* Nepotism, incompetence, or thievery? *Anim. Behav.* 49:1611–16.

Myrberg, A. A. Jr. 1981. Sound communication and interception in fishes. In *Hearing and sound communication in fishes,* ed. W. N. Tavolga, A. N. Popper, R. R. Fay. New York: Springer.

Myrberg, A. A. Jr., Ha, S. J., and Shamblott, M. J. 1993. The sounds of bicolor damselfish *(Pomacentrus partitus):* Predictors of body size and a spectral basis for individual recognition and assessment. *J. Acoust. Soc. Am.* 94:3067–70.

Nachtigall, P. E. 1986. Vision, audition, and chemoreception in dolphins and other marine mammals. In *Dolphin cognition and behavior: A comparative approach,* ed. R. J. Schusterman, J. A. Thomas, and F. G. Wood. Hillsdale, NJ: Lawrence Erlbaum Associates.

Naess, A. 1986. Intrinsic value: Will the defenders of wildlife please rise. In *Conservation biology: The science of scarcity and diversity,* ed. M. E. Soulé. Sunderland, MA: Sinauer Associates.

Naguib, M. 1998. Perception of degradation in acoustic signals and its implications for ranging. *Behav. Ecol. Sociobiol.* 42: 139–42.

Napier, J. R., and Napier, P. H. 1967. *A handbook of living primates.* London: Academic Press.

National Marine Fisheries Service. 1990. *Final environmental impact statement on the use of marine mammals in swim-with-the-dolphin programs.* Silver Spring, MD: Office of Protected Resources, National Marine Fisheries Service.

———. 1991. *Recovery plan for the northern right whale* (Eubalaena glacialis). Silver Spring, MD: National Marine Fisheries Service.

———. 1993. *Marine Mammal Protection Act of 1972 Annual Report: Jan. 1, 1990 to Dec. 31, 1990.* Silver Spring, MD: Office of Protected Resources, National Marine Fisheries Service.

National Research Council. 1985. *Oil in the sea: Inputs, fates and effects.* Washington, DC: National Academy Press.

Nemoto, T. 1957. Foods of baleen whales in the northern Pacific. *Sci. Rep. Whales Res. Inst.* 12:33–89.

———. 1964. School of baleen whales in the feeding areas. *Sci. Rep. Whales Res. Inst.* 18:89–110.

Nerini, M. 1984. A review of gray whale feeding ecology. In *The gray whale,* ed. M. J. Jones, S. L. Swartz, and S. Leatherwood. Orlando, FL: Academic Press.

Nestler, J. M., Ploskey, G. R., Pickens, J., Menezes, J., and Schilt, C. 1992. Responses of blueback herring to high-frequency sound and implications for reducing entrainment at hydropower dams. *N. Am. J. Fish. Mgmt.* 12:667–83.

Newby, T. C. 1982. Life history of Dall porpoise (*Phocoenoides dalli,* True 1885) incidentally taken in the Japanese high seas salmon mothership fishery in the Northwestern North Pacific and western Bering Sea, 1978 and 1980. Ph.D. thesis, University of Washington, Seattle.

Newman, J. D. 1985. The infant cry of primates. In *Infant crying,* ed. B. M. Lester and C. F. Z. Boudykis. New York: Plenum.

Newman, J. R. 1976. Observations of sexual behavior in male gray whales, *Eschrichtius robustus. Murrelet* 57:49.

Newman, M. A. 1994. *Life in a fishbowl.* Vancouver: Douglas and McIntyre.

Newman, M. A., and McGeer, P. L. 1966. The capture and care of a killer whale, *Orcinus orca,* in British Columbia. *Zoologica* 51:59–69.

Nichol, L. M., and Shackleton, D. M. 1996. Seasonal movements and foraging behaviour of northern *resident* killer whales *(Orcinus orca)* in relation to the inshore distribution of salmon (*Oncorhynchus* spp.) in British Columbia. *Can. J. Zool.* 74:983–91.

Nishida, T. 1983. Alpha status and agonistic alliance in wild chimpanzees *(Pan troglodytes schweinfurthii). Primates* 24:318–36.

——— (ed.). 1990. *The chimpanzees of the Mahale Mountains: Sexual and life history strategies.* Tokyo: University of Tokyo Press.

Nishida, T., and Hiraiwa-Hasegawa, M. 1987. Chimpanzees and bonobos: Cooperative relationships among males. In *Primate societies,* ed. B. B. Smuts, D. L. Cheney, R. M. Seyfarth, R. W. Wrangham, and T. T Struhsaker. Chicago: University of Chicago Press.

Nishiwaki, M. 1959. Humpback whales in Ryukyuan waters. *Sci. Rep. Whales Res. Inst.* 14:49–87.

———. 1962. Aerial photographs show sperm whales' interesting habits. *Norwegian Whal. Gaz.* 51(10):395–98.

Nishiwaki, M., and Handa, C. 1958. Killer whales caught in the coastal waters off Japan for recent ten years. *Sci. Rep. Whales Res. Inst.* 13:85–96.

Nishiwaki, M., and T. Kamiya. 1958. A beaked whale *Mesoplodon* stranded at Oiso beach, Japan. *Sci. Rep. Whales Res. Inst.* 13:53–83.

Nishiwaki, M., Ohsumi, S., and Maeda, Y. 1963. Change of form in the sperm whale accompanied with growth. *Sci. Rep. Whales Res. Inst. Tokyo* 17:1–4.

Noë, R. 1990. A veto game played by baboons: A challenge to the use of the Prisoners Dilemma as a paradigm for reciprocity and cooperation. *Anim. Behav.* 39:78–90.

———. 1992. Alliance formation among male baboons: Shopping for profitable partners. In *Coalitions and alliances in humans and other animals,* ed. A. H. Harcourt and F. B. M. de Waal. Oxford: Oxford University Press.

———. 1994. A model of coalition formation among male baboons with fighting ability as the crucial parameter. *Anim. Behav.* 47:211–13.

Noë, R., and Hammerstein, P. 1994. Biological markets: Supply and demand determine the effect of partner choice in cooperation, mutualism and mating. *Behav. Ecol. Sociobiol.* 35:1–11.

Noë, R., van Schaik, C. P., and van Hooff, J. A. R. A. M. 1991. The market effect: An explanation for pay-off asymmetries among collaborating animals. *Ethology* 87:97–118.

Noonan, K. M. 1981. Individual strategies of inclusive-fitness-maximizing in *Polistes fuscatus* foundresses. In *Natural selection and social behavior,* ed. R. D. Alexander and D. W. Tinkle. New York: Chiron Press.

Nordhoff, C. 1895. *Whaling and fishing.* New York: Dodd, Mead and Company.

Norman, J. R., and Fraser, F. C. 1937. *Giant fishes, whales and dolphins.* London: Putnam.

Norris, K. S. (ed.). 1966. *Whales, dolphins, and porpoises.* Berkeley: University of California Press.

———. 1967a. Aggressive behavior in Cetacea. In *Aggression and defenses: Neural mechanisms and social patterns,* ed. C. D. Clemente and D. B. Lindsley. Berkeley: University of California Press.

———. 1967b. Some observations on the migration and orientation of marine mammals. In *Animal orientation and navigation,* ed. R. M. Storm. Corvallis: Oregon State University Press.

———. 1969. The echolocation of marine mammals. In *The biology of marine mammals,* ed. H. T. Andersen. New York: Academic Press.

———. 1974. *The porpoise watcher.* New York: Norton.

———. 1985. The use of captive marine mammals in behavior studies. Abstracts of the Sixth Biennial Conference on the Biology of Marine Mammals, Vancouver, BC.

———. 1991a. Looking at wild dolphin schools. In *Dolphin societies: Discoveries and puzzles,* ed. K. Pryor and K. S. Norris. Berkeley: University of California Press.

———. 1991b. *Dolphin days: The life and times of the spinner dolphin.* New York: W. W. Norton.

———. 1991c. Looking at captive dolphins. In *Dolphin societies: Discoveries and puzzles,* ed. K. Pryor and K. S. Norris, 293–303. Berkeley: University of California Press.

———. 1994. Introduction. In *The Hawaiian spinner dolphin*, ed. K. S. Norris, B. Würsig, R. S. Wells, and M. Würsig. Berkeley: University of California Press.

Norris, K. S., and Dohl, T. P. 1980a. Behavior of the Hawaiian spinner dolphin. *Fish. Bull.* 77:821–49.

———. 1980b. The structure and function of cetacean schools. In *Cetacean behavior*, ed. L. M. Herman. New York: John Wiley & Sons.

Norris, K. S., and Harvey, G. W. 1972. A theory for the function of the spermaceti organ of the sperm whale (*Physeter catodon* L.). In *Animal orientation and migration*, ed. S. R. Galler, K. Schmidt-Koenig, G. J. Jacobs, and R. E. Belleville, 397–417. NASA Special Publication 262.

Norris, K. S., and Johnson, C. M. 1994. Social behavior. In *The Hawaiian spinner dolphin*, ed. K. S. Norris, B. Würsig, R. S. Wells, and M. Würsig. Berkeley: University of California Press.

Norris, K. S., and McFarland, N. W. 1958. A new harbour porpoise of the genus *Phocoena* from the Gulf of California. *J. Mammal.* 39:22–39.

Norris, K. S., and Møhl, B. 1983. Can odontocetes debilitate prey with sound? *Am. Nat.* 122:85–104.

Norris, K. S., and Prescott, J. H. 1961. Observations of cetaceans of Californian and Mexican waters. *Univ. Cal. Publ. Zool.* 63: 291–402.

Norris, K. S., and Pryor, K. 1991. Some thoughts on grandmothers. In *Dolphin societies: Discoveries and puzzles*, ed. K. Pryor and K. S. Norris. Berkeley: University of California Press.

Norris, K. S., and Schilt, C. R. 1988. Cooperative societies in three-dimensional space: On the origins of aggregations, flocks, and schools, with special reference to dolphins and fish. *Ethol. Sociobiol.* 9:149–79.

Norris, K. S., and Wells, R. S. 1994. Observing dolphins underwater. In *The Hawaiian spinner dolphin*, ed. K. S. Norris, B. Würsig, R. S. Wells, and M. Würsig. Berkeley: University of California Press.

Norris, K. S., Prescott, J. H., Asa-Dorian, P. V., and Perkins, P. 1961. An experimental demonstration of echolocation behavior in the porpoise, *Tursiops truncatus* (Montagu). *Biol. Bull.* 120:163–76.

Norris, K. S., Evans, W. E., and Ray, G. C. 1974. New tagging and tracking methods for the study of marine mammal biology and migration. In *The whale problem: A status report*, ed. W. E. Schevill, G. C. Ray, and K. S. Norris. Cambridge, MA: Harvard University Press.

Norris, K. S., Stuntz, W. E., and Rogers, W. 1978. *The behaviour of porpoises and tuna in the eastern tropical Pacific yellowfin tuna fishery: Preliminary studies.* National Technical Information Service PB283-970.

Norris, K. S., Würsig, B., Wells, R. S., and Würsig, M. 1994. *The Hawaiian spinner dolphin.* Berkeley: University of California Press.

Northridge, S., and Hofman, R. J. 1999. Marine mammal-fishery interactions. In *Conservation and management of marine mammals*, ed. J. R. Twiss Jr. and R. R. Reeves, 99–119. Washington, DC: Smithsonian Institution Press.

Nowacek, D. P. 1999. Sound use, sequential behavior and ecology of foraging bottlenose dolphins, *Tursiops* truncatus. Ph.D. dissertation, MIT/WHOI Joint Program.

Nowak, R. 1991. Senses involved in discrimination of merino ewes at close contact and from a distance by their newborn lambs. *Anim. Behav.* 42:357–66.

Nunney, L., and Campbell, K. A. 1993. Assessing minimum viable population size: Demography meets population genetics. *Trends Ecol. Evol.* 8:234–39.

O'Barry, R., and Coulbourn, K. 1988. *Behind the dolphin smile.* Chapel Hill, NC: Algonquin Books.

O'Corry-Crowe, G. M., Suydam, R. S., Rosenberg, A., Frost, K. J., and Dizon, A. E. 1997. Phylogeography, population structure and dispersal patterns of the beluga whale *Delphinapterus leucas* in the western Nearctic revealed by mitochondrial DNA. *Mol. Ecol.* 6:955–70.

Oftedal, O. T. 1984. Milk composition, milk yield, and energy output at peak lactation: A comparative review. *Symp. Zool. Soc. Lond.* 51:33–85.

———. 1993. The adaptation of milk secretion to the constraints of fasting in bears, seals, and baleen whales. *J. Dairy Sci.* 76(10):3234–46.

———. 1997. Lactation in whales and dolphins: Evidence of divergence between baleen- and toothed-species. *J. Mammary Gland Biol. Neoplasia* 2:205–30.

Oftedal, O. T., and Iverson, S. J. 1995. Comparative analysis of nonhuman milks. A. Phylogenetic variation in the gross composition of milks. In *Handbook of milk composition*, ed. R. G. Jensen. New York: Academic Press.

Ohno, M., and Fujino, K. 1952. Biological investigations on the whales caught by the Japanese Antarctic whaling fleets, season 1950/51. *Sci. Rep. Whales Res. Inst.* 21:1–78.

Ohsumi, S. 1971. Some investigations on the school structure of sperm whale. *Sci. Rep. Whales Res. Inst., Tokyo* 23:1–25.

Oien, N. 1988. The distribution of killer whales (*Orcinus orca*) in the North Atlantic based on Norwegian catches, 1938–1981, and incidental sightings, 1967–1987. *Rit Fiskideildar* 11:65–78.

Olesiuk, P. F., Bigg, M. A., and Ellis, G. M. 1990. Life history and population dynamics of *resident* killer whales (*Orcinus orca*) in the coastal waters of British Columbia and Washington State. In *Individual recognition of cetaceans: Use of photo-identification and other techniques to estimate population parameters*, ed. P. S. Hammond, S. A. Mizroch, and G. P. Donovan, 209–43. Reports of the International Whaling Commission, special issue 12. Cambridge: International Whaling Commission.

Oliver, W. R. B. 1937. *Tasmacetus shepherdi:* A new genus and species of beaked whale from New Zealand. *Proc. Zool. Soc. Lond.* 3:371–78.

Olsen, Ø. 1913. On the external characters and biology of Bryde's whale (*Balaenoptera grydei*): a new rorqual from the coast of South Africa. *Proc. Zool. Soc. Lond.* 1913:1073–90.

Olson, A. F., and Quinn, T. P. 1993. Vertical and horizontal movements of adult chinook salmon *Oncorhynchus tshawytscha* in the Columbia River. *Fish. Bull.* 91:171–78.

Omura, H. 1953. Biological study on humpback whales in the Antarctic whaling Areas IV and V. *Sci. Rep. Whales Res. Inst.* 8:81–102.

Omura, H., Fujino, K., and Kimura, S. 1955. Beaked whale *Berardius bairdii* off Japan, with a note on *Ziphius cavirostris*. *Sci. Rep. Whales Res. Inst.* 10:89–132.

Omura, H., Ohsumi, S., Nemoto, T., Nasu, K., and Kasuya, T. 1969. Black right whales in the North Pacific. *Sci. Rep. Whales Res. Inst.* 21:1–78.

Osbeck, P. 1765. Reise nach Ostindien und China. In *Rostok,* ed. J. C. Koppe.

Osborne, R. W. 1986. A behavioral budget of Puget Sound killer whales. In *Behavioral biology of killer whales,* ed. B. Kirkevold and J. S. Lockard. New York: Alan R. Liss.

O'Shea, T. J. 1999. Environmental contaminants and marine mammals. In *Biology of marine mammals,* ed. J. E. Reynolds III and S. A. Rommel, 485–563. Washington, DC: Smithsonian Institution Press.

O'Shea, T. J., and Brownell, R. L. Jr. 1994. Organochlorine and metal contaminants in baleen whales: A review and evaluation of conservation implications. *Sci. Total Environ.* 154:179–200.

O'Shea, T. J., Reeves, R. R., and Long, A. K. 1999. Marine mammals and persistent ocean contaminants. Proceedings of the Marine Mammal Commission workshop, Keystone, Colorado, 12–15 October 1998. Bethesda, MD: U.S. Marine Mammal Commission. 150 pp.

Östman, J. 1991. Changes in aggressive and sexual behavior between two male bottlenose dolphins *(Tursiops truncatus)* in a captive colony. In *Dolphin societies: Discoveries and puzzles,* ed. K. Pryor and K. S. Norris. Berkeley: University of California Press.

———. 1994. Social organization and social behavior of Hawaiian spinner dolphins *(Stenella longirostris).* Ph.D. dissertation, University of California, Santa Cruz.

Ostrom, J. H. 1986. Social and unsocial behavior in dinosaurs. In *Evolution of animal behavior: Paleontological and field approaches,* ed. M. H. Nitecki and J. A. Kitchell. New York: Oxford University Press.

Overholtz, W. J., and Nicolas, J. R. 1979. Apparent feeding by the fin whale, *Balaenoptera physalus,* and the humpback whale, *Megaptera novaeangliae,* on the American sandlance, *Ammodytes americanus,* in the Northwest Atlantic. *Fish. Bull.* 77:285–87.

Overstrom, N. 1983. Association between burst-pulse sounds and aggressive behavior in captive Atlantic bottlenosed dolphins *(Tursiops truncatus). Zoo Biol.* 2:93–103.

Owen, R. 1846. *A history of British fossil mammals and birds,* ed. J. van Voorst. London.

———. 1853. Descriptive catalog of the osteological series contained in the museum of the Royal College of Surgeons. *Mus. R. Coll. Surg. Lond.* 2:351–914.

———. 1866/1869. On some Indian cetacea collected by Walter Elliot, Esq. *Trans. Zool. Soc. Lond.* 6(1):17–47.

———. 1990. A retrospective study of captive breeding programs involving *Tursiops truncatus* in South Florida. Paper presented at the International Association for Aquatic Animal Medicine.

Pack, A. A., Salden, D. R., Ferrari, M. J., Glockner-Ferrari, D. A., Herman, L. M., Stubbs, H. A.,, and Straley, J. M. 1998. Male humpback whale dies in competitive group. *Mar. Mamm. Sci.* 14(4):861–73.

Packer, C., and Pusey, A. E. 1982. Cooperation and competition within coalitions of male lions: Kin selection or game theory? *Nature* 296:740–42.

Packer, C., Herbst, L., Pusey, A. E., Bygott, J. D., Hanby, J. P., Cairns, S. J., and Borgerhoff Mulder, M. 1988. Reproductive success in lions. In *Reproductive success,* ed. T. H. Clutton-Brock. Chicago: University of Chicago Press.

Packer, C., Gilbert, D. A., Pusey, A. E., and O'Brien, S. J. 1991. A molecular genetic analysis of kinship and cooperation in African lions. *Nature* 351:562–65.

Packer, C., Lewis, S., and Pusey, A. E. 1992. A comparative analysis of non-offspring nursing. *Anim. Behav.* 43:265–81.

Packer, C., Tatar, M., and Collins, A. 1998. Reproductive cessation in female mammals. *Nature* 392:807–11.

Page, D. C., Mosher, R., Simpson, E. M., Fisher, E. M. C., Mardon, G., Pollack, J., McGillivray, B., de la Chapelle, A., and Brown, L. G. 1987. The sex-determining region of the human Y chromosome encodes a finger protein. *Cell* 51:1091–1104.

Pagel, M. D., and Harvey, P. H. 1993. Evolution of the juvenile period in mammals. In *Juvenile primates: Life-history, development and behavior,* ed. M. E. Pereira and L. A. Fairbanks. Oxford: Oxford University Press.

Palacios, D. M., and Mate, B. R. 1996. Attack by false killer whales *(Pseudorca crassidens)* on sperm whales *(Physeter macrocephalus)* in the Galápagos Islands. *Mar. Mamm. Sci.* 12:582–87.

Pallas, P. S. 1776. *Reise durch verschiedene Provinzen des Russischen Reichs.* St. Petersburg, 3.

Palmeirim, J. M., and Rodrigues, L. 1995. Dispersal and philopatry in colonial animals: The case of *Miniopterus schreibersii. Symp. Zool. Soc. Lond.* 67:219–31.

Palsbøll, P. J., Vader, A., Bakke, I., and El-Gewely, M. R. 1992. Determination of gender in cetaceans by the polymerase chain reaction. *Can. J. Zool.* 70:2166–70.

Palsbøll, P. J., Clapham, P. J., Mattila, D. K., Larsen, F., Sears, R., Siegismund, H. R., Sigurjónsson, J., Vásquez, O., and Arctander, P. 1995. Distribution of mtDNA haplotypes in North Atlantic humpback whales: The influence of behavior on population structure. *Mar. Ecol. Prog. Ser.* 116:1–10.

Palsbøll, P. J., Allen, J., Bérubé, M., Clapham, P. J., Feddersen, T. P., Hammond, P., Jørgensen, H., Katona, S., Larsen, A. H., Larsen, F., Lien, J., Mattila, D. K., Sigurjónsson, J., Sears, R., Smith, T., Sponer, R., Stevick, P., and Øien, N. 1997. Genetic tagging of humpback whales. *Nature* 388:767–69.

Panfilov, B. G. 1978. Postnatal linear growth of fin whales from Indian Ocean waters of the Southern Hemisphere. *Rep. Int. Whal. Commn.* 28:297–300.

Papastavrou, V., Smith, S. C., and Whitehead, H. 1989. Diving behaviour of the sperm whale, *Physeter macrocephalus,* off the Galápagos Islands. *Can. J. Zool.* 67:839–46.

Parfit, M. 1980. Are the dolphins trying to say something, or is it all much ado about nothing? *Smithsonian* 11(7):73–80.

Parker, E. D. 1984. Sperm competition and the evolution of animal mating systems. In *Sperm competition and the evolution of animal mating systems,* ed. R. L. Smith. Orlando, FL: Academic Press.

Parker, G. A. 1970. Sperm competition and its evolutionary consequences in insects. *Biol. Rev.* 45:525–67.

———. 1990a. Sperm competition games: Raffles and roles. *Proc. R. Soc. Lond.* 242:120–26.

————. 1990b. Sperm competition games: Guards and extra pair copulations. *Proc. R. Soc. Lond.* 242:127–33.

Partridge, B. L., Johansson, J., and Kalish, J. 1983. The structure of schools of giant bluefin tuna in Cape Cod Bay. *Environ. Biol. Fish.* 9:253–62.

Paterson, R. A. 1986. An analysis of four large accumulations of sperm whales observed in the modern whaling era. *Sci. Rep. Whales Res. Inst.* 37:167–72.

Paterson, R. A., Quayle, C. J., and van Dyck, S. M. 1993. A humpback whale calf and two subadult Dense-beaked whales recently stranded in southern Queensland. *Mem. Queensland Mus.* 33:291–97.

Patterson, B., and Hamilton, G. R. 1964. Repetitive 20 cycle per second biological hydroacoustic signals at Bermuda. In *Marine bioacoustics*, vol. 1, ed. W. N. Tavolga. Oxford: Pergamon.

Patterson, I. A. P., Reid, R. J., Wilson, B., Grellier, K., Ross, H. M., and Thompson, P. M. 1998. Evidence for infanticide in bottlenose dolphins: An explanation for violent interactions with harbour porpoises? *Proc. R. Soc. Lond. B.* 265:1–4.

Paul, A., Kuester, J., and Podzuweit, D. 1993. Reproductive senescence and terminal investment in female Barbary macaques (*Macaca sylvanus*) at Salem. *Int. J. Primatol.* 14:105–24.

Payne, K. B., and Payne, R. 1985. Large scale changes over 19 years in the songs of humpback whales in Bermuda. *Z. Tierpsychol.* 68:89–114.

Payne, K. B., Tyack, P., and Payne, R. S. 1983. Progressive changes in the songs of humpback whales. In *Communication and behavior of whales*, ed. R. Payne. Boulder, CO: Westview Press.

Payne, P. M., Nicolas, J. R., O'Brien, L., and Powers, K. D. 1986. Distribution of the humpback whale, *Megaptera novaeangliae*, on Georges Bank and in the Gulf of Maine in relation to densities of the sand eel, *Ammodytes americanus. Fish. Bull.* 84:271–77.

Payne, P. M., Wiley, D. N., Young, S. B., Pittman, S., Clapham, P. J., and Jossi, J. W. 1990. Recent fluctuations in the abundance of baleen whales in the southern Gulf of Maine in relation to changes in selected prey. *Fish. Bull.* 88:687–96.

Payne, R. B. 1982. Ecological consequences of song matching: Breeding success and intraspecific song mimicry in indigo buntings. *Ecology* 63:401–11.

————. 1983. The social context of song mimicry: Song matching dialects in indigo buntings. *Anim. Behav.* 31:788–805.

————. 1998. Brood parasitism in birds: Strangers in the nest: Why do birds rear young that are not their own? *Bioscience* 48(5):377–86.

Payne, R. S. 1970. *Songs of the humpback whale.* Phonograph record. Hollywood: Capitol Records.

————. 1976. At home with right whales. *Natl. Geogr.* 149: 322–39.

————. 1980. Research on the behavior of various species of whales. *Natl. Geogr. Res. Rep.* 12:551–64.

———— (ed.). 1983. *Communication and behavior of whales.* American Association for the Advancement of Science Selected Symposia Series, 76. Boulder, CO: Westview Press.

————. 1986. Long-term behavioural studies of the southern right whale (*Eubalaena australis*). In *Right whales: Past and present status: proceedings of the Workshop on the Status of Right Whales, New England Aquarium, Boston, MA, 15–23 June 1983*, ed. R. L. Brownell Jr., P. B. Best, and J. H. Prescott, 161–67. Reports of the International Whaling Commission, special issue 10. Cambridge: International Whaling Commission.

————. 1995. *Among whales.* New York: Simon and Schuster.

Payne, R. S., and Dorsey, E. 1983. Sexual dimorphism and aggressive use of callosities in right whales (*Eubalaena australis*). In *Communication and behavior of whales*, ed. R. S. Payne. Boulder, CO: Westview Press.

Payne, R. S., and Guinee, L. N. 1983. Humpback whale, *Megaptera novaeangliae*, songs as an indicator of "stocks." In *Communication and behavior of whales*, ed. R. S. Payne. Boulder, CO: Westview Press.

Payne, R. S., and McVay, S. 1971. Songs of humpback whales. *Science* 173:585–97.

Payne, R. S., and Webb, D. 1971. Orientation by means of long range acoustic signalling in baleen whales. *Ann. NY Acad. Sci.* 188:110–41.

Payne, R. S., Brazier, O., Dorsey, E. M., Perkins, J. S., Rowntree, V. J., and Titus, A. 1983. External features in Southern Right Whales (*Eubalaena australis*) and their use in identifying individuals. In *Communication and behavior of whales*, ed. R. Payne. American Association for the Advancement of Science Selected Symposia Series 76. Boulder, CO: Westview Press.

Payne, R. S., Rowntree, V., Perkins, J. S., Cooke, J. G., and Lankester, K. 1990. Population size, trends and reproductive parameters of right whales (*Eubalaena australis*) off Peninsula Valdes. In *Individual recognition of cetaceans: Use of photo-identification and other techniques to estimate population parameters*, ed. P. S. Hammond, S. A. Mizroch, and G. Donovan, 271–78. Reports of the International Whaling Commission, special issue 12. Cambridge: International Whaling Commission.

Peale, T. R. 1848. Mammalogy and ornithology. United States Exploring Expedition during the years 1838, 1839, 1840, 1841, 1842. Under the command of Charles Wilkes, USN, 8:33–35.

Peddemors, V. M. 1990. Respiratory development in a captive-born bottlenose dolphin *Tursiops truncatus* calf. *S. Afr. Tydskr. Dierk.* 25(3):178–84.

Pedersen, T. 1952. A note on humpback oil and on the milk and milkfat from this species (*Megaptera nodosa*). *Norsk Hvalfangst-Tidende* 41:375–78.

Peixun, C. 1989. Baiji. In *Handbook of marine mammals*, vol. 4, *River dolphins and the larger toothed whales*, ed. S. H. Ridgway and R. J. Harrison. New York: Academic Press.

Pennycuick, C. J., and Rudnai, J. 1970. A method of identifying individual lions *Panthera leo* with an analysis of the reliability of identification. *J. Zool.* 160:497–508.

Pereira, M. E. 1993. Juvenility in animals. In *Juvenile primates: Life-history, development and behavior*, ed. M. E. Pereira and L. A. Fairbanks. Oxford:Oxford University Press.

————. 1995. Development and social dominance in group-living primates. *Am. J. Primatol.* 37:143–75.

Pereira, M. E., and Fairbanks, L. A. 1993. What are juvenile primates all about? In *Juvenile primates: Life-history, development and behavior,* ed. M. E. Pereira and L. A. Fairbanks. Oxford: Oxford University Press.

Perrin, W. F. 1972. Color patterns of spinner porpoises *(Stenella* cf. *S. longirostris)* of the eastern Pacific and Hawaii, with comments on delphinid pigmentations. *Fish. Bull.* 70:983–1003.

———. 1982. Report of the workshop on identity, structure and vital rates of killer whale populations, Cambridge, England, June 23–25, 1981. *Rep. Int. Whal. Commn.* 32:617–32.

———. 1997. Development and homologies of head stripes in the delphinoid cetaceans. *Mar. Mamm. Sci.* 13:1–43.

———. 1999. Selected examples of small cetaceans at risk. In *Conservation and management of marine mammals,* ed. J. R. Twiss Jr. and R. R. Reeves, 296–310. Washington, DC: Smithsonian Institution Press.

Perrin, W. F., and Donovan, G. P. 1984. Report of the workshop. In *Reproduction in whales, dolphins and porpoises: Proceedings of the conference, Cetacean Reproduction, Estimating Parameters for Stock Assessment and Management, La Jolla, CA, 28 Nov.–7 Dec. 1981,* ed. W. F. Perrin, R. L. Brownell Jr., and D. P. DeMaster, 1–27. Reports of the International Whaling Commission, special issue 6. Cambridge: International Whaling Commission.

Perrin, W. F., and Mead, J. G. 1994. Clymene dolphin—*Stenella clymene* (Gray, 1846). In *Handbook of marine mammals,* vol. 5, ed. S. H. Ridgway and R. Harrison. London: Academic Press.

Perrin, W. F., and Myrick, A. C. Jr. (eds.). 1980. *Age determination of toothed whales and sirenians: Proceedings of the International Conference on Determining Age of Odontocete Cetaceans and Sirenians, La Jolla, CA, Sept. 5–19, 1978.* Reports of the International Whaling Commission, special issue 3. Cambridge: International Whaling Commission.

Perrin, W. F., and Reilly, S. B. 1984. Reproductive parameters of dolphins and small whales of the family *Delphinidae.* In *Reproduction in whales, dolphins and porpoises: Proceedings of the conference, Cetacean Reproduction, Estimating Parameters for Stock Assessment and Management, La Jolla, CA, 28 Nov.–7 Dec. 1981,* ed. W. F. Perrin, R. L. Brownell Jr., and D. P. DeMaster, 97–134. Reports of the International Whaling Commission, special issue 6. Cambridge: International Whaling Commission.

Perrin, W. F., Coe, J. M., and Zweifel, J. R. 1976. Growth and reproduction of the spotted porpoise, *Stenella attenuata,* in the offshore eastern tropical Pacific. *Fish. Bull.* 74:229–69.

Perrin, W. F., Evans, W. E., and Holts, D. B. 1979. Movements of pelagic dolphins *(Stenella* spp.) in the eastern tropical Pacific as indicated by results of tagging, with summary of tagging operations, 1969–1976. NOAA Technical Report NMFS SSRF-737.

Perrin, W. F., Mitchell, E. D., Mead, J. G., Caldwell, D. K., Caldwell, M. C., van Bree, P. J. H., and Dawbin W. H. 1987. Revision of the spotted dolphins *Stenella* spp. *Mar. Mamm. Sci.* 3:99–170.

Perrin, W. F., Brownell, R. L. Jr., Zhou, K., and Liu, J. 1989. *Biology and conservation of the river dolphins.* IUCN Species Survival Commission Occasional Paper 3.

Perrin, W. F., Akin, P. A., and Kashiwada, J. V. 1991. Geographic variation in external morphology of the spinner dolphin *Stenella longirostris* in the eastern Pacific and implications for conservation. *Fish. Bull.* 89:411–28.

Perrin, W. F., Donovan, G., and Barlow, J. (eds.). 1994a. *Gillnets and cetaceans: Incorporating the proceedings of the symposium and workshop on the mortality of cetaceans in passive fishing nets and traps.* Reports of the International Whaling Commission, special issue 15. Cambridge: International Whaling Commission.

Perrin, W. F., Wilson, C. E., and Archer, F. I. 1994b. Striped dolphin *Stenella coeruleoalba* (Meyen, 1833). In *Handbook of marine mammals,* vol. 5, ed. S. H. Ridgway and R. Harrison. London: Academic Press.

Perry, A., Baker, C. S., and Herman, L. M. 1990. Population characteristics of individually identified humpback whales in the central and eastern North Pacific: A summary and critique. In *Individual recognition of cetaceans: Use of photo-identification and other techniques to estimate population parameters,* ed. P. S. Hammond, S. A. Mizroch, and G. P. Donovan, 307–17. Reports of the International Whaling Commission, special issue 12. Cambridge: International Whaling Commission.

Perryman, W. L., and Foster, T. C. 1980. Preliminary report on predation by small whales, mainly the false killer whale, *Pseudorca crassidens,* on dolphins *(Stenella* spp. and *Delphius delphis)* in the eastern tropical Pacific. Southwest Fisheries Center Administrative Report LJ-80-05.

Peterson, M. J. 1992. Whalers, cetologists, environmentalists, and the international management of whaling. *Int. Org.* 46(1): 147–86.

Petricig, R. O. 1993. Diel patterns of strand-feeding behavior by bottlenose dolphins in South Carolina salt marshes. Abstracts, Tenth Biennial Conference on the Biology of Marine Mammals, Galveston, TX.

———. 1995. Bottlenose dolphins *(Tursiops truncatus)* in Bull Creek, South Carolina. Ph.D. dissertation, University of Rhode Island.

Petrie, M., and Møller, A. P. 1991. Laying eggs in others' nests: Intraspecific brood parasitism in birds. *Trends Ecol. Evol.* 6: 315–20.

Phillips, C. 1996. Defining future research needs for cetacean conservation. In *The conservation of whales and dolphins: Science and practice,* ed. M. P. Simmonds and J. Hutchinson. Chichester: Wiley.

Phillips, N. E., and Baird, R. W. 1993. Are killer whales harassed by boats? *Victoria Nat.* 50(3):10–11.

Pianka, E. R., and Parker, W. S. 1975. Age-specific reproductive tactics. *Am. Nat.* 109:453–64.

Pierce, N. E. 1989. Butterfly-ant mutualisms. In *Toward a more exact ecology,* ed. P. J. Grubb and J. B. Whittaker. Oxford: Blackwell Scientific Publications.

Pierce, N. E., Kitching, R. L., Buckley, R. C., Taylor, M. F. J., and Benbow, K. F. 1987. The costs and benefits of cooperation between the Australian lycaenid butterfly, *Jalmenus evagoras,* and its attendant ants. *Behav. Ecol. Sociobiol.* 21:237–48.

Pierotti, R., and Murphy, E. C. 1987. Intergenerational conflict in gulls. *Anim. Behav.* 35:435–44.

Pierroti, R., Swatland, C. A., and Ewald, P. W. 1985. "Brass knuckles" in the sea: The use of barnacles as weapons. In *Proceedings of the Sixth Biennial Conference on the Biology of Marine Mammals* (unpaginated abstract). Lawrence, KS: Society for Marine Mammalogy.

Pitcher, T. J. 1992. Who dares wins: The function and evolution of predator inspection behavior in fish shoals. *Neth. J. Zool.* 42:371–91.

Pitcher, T. J., and Parrish, J. K. 1993. Functions of shoaling behaviour in teleosts. In *Behaviour of teleost fishes,* 2d ed., ed. T. J. Pitcher. London: Chapman and Hall.

Pitcher, T. J., Magurran, A. E., and Winfield, I. 1982. Fish in larger shoals find food faster. *Behav. Ecol. Sociobiol.* 10:149–51.

Pitcher, T. J., Green, D., and Magurran, A. E. 1986. Dicing with death: Predator inspection behaviour in minnow shoals. *J. Fish. Biol.* 28:439–48.

Pitman, R. L., and Chivers, S. J. 1999. Terror in black and white. *Nat. Hist.* 107:26–29.

Plavcan, J. M., and van Schaik, C. P. 1992. Intrasexual competition and canine dimorphism in anthropoid primates. *Am. J. Phys. Anthropol.* 87:461–77.

———. 1997. Intrasexual competition and body weight dimorphism in anthropoid primates. *Am. J. Phys. Anthropol.* 103:37–68.

Poiani, A., and Yorke, M. 1989. Predator harassment: More evidence on the deadly risk. *Ethology* 83:167–69.

Poole, J. H. 1987. Rutting behavior of African elephants: The phenomenon of musth. *Behaviour* 102:283–316.

———. 1989. Mate guarding, reproductive success and female choice in African elephants. *Anim. Behav.* 37:842–49.

———. 1994. Sex differences in the behaviour of African elephants. In *Sex differences in behaviour,* ed. R. V. Short and E. Balaban. Cambridge: Cambridge University Press.

Poole, J. H., and Thomsen, J. 1989. Elephants are not beetles: Implications of the ivory trade for the survival of the African elephant. *Oryx* 23:188–98.

Poole, J. H., Payne, K. B., Langauer, W. R. Jr., and Moss, C. J. 1988. The social contexts of some very low frequency calls of African elephants. *Behav. Ecol. Sociobiol.* 22:385–92.

Porsild, M. P. 1922. Scattered observations of narwhals. *J. Mammal.* 41:8–13.

Porter, J. W. 1979. *Pseudorca* strandings. *Oceans* 10:8–15.

Potter, J. R., Mellinger, D. K., and Clark, C. W. 1994. Marine mammal call discrimination using artificial neural networks. *J. Acoust. Soc. Am.* 96:1255–62.

Prescott, J. H. 1977. Comments on captive births of *Tursiops truncatus* at Marineland of the Pacific (1957–1972). In *Breeding dolphins: Present status, suggestions for the future,* ed. S. H. Ridgway and K. Benirschke. U.S. Marine Mammal Commission Report no. MMC-76/07, Washington, DC.

———. 1981. Clever Hans: Training the trainers, or the potential for misinterpreting the results of dolphin research. *Ann. NY Acad. Sci.* 364:130–36.

Pryor, K. 1975. *Lads before the wind.* New York: Harper and Row.

———. 1991. Non-acoustic communication in small cetaceans: Glance, touch, position, gesture, and bubbling. In *Sensory abilities of cetaceans,* ed. J. A. Thomas and R. A. Kastelein. New York: Plenum.

Pryor, K., and Kang, I. 1980. Social behavior and school structure in pelagic porpoises *(Stenella attenuata* and *S. longirostris)* during purse seining for tuna. NOAA Southwest Fisheries Center Administrative Report LJ-80-11C. La Jolla, CA: Southwest Fisheries Science Center.

Pryor, K., and Kang Shallenberger, I. 1991. Social structure in spotted dolphins *(Stenella attenuata)* in the tuna purse seine fishery in the Eastern Tropical Pacific. In *Dolphin societies: Discoveries and puzzles,* ed. K. Pryor and K. S. Norris, 293–303. Berkeley: University of California Press.

Pryor, K., and Norris, K. S. 1991a. Dolphin politics and dolphin science. In *Dolphin societies: Discoveries and puzzles,* ed. K. Pryor and K. S. Norris. Berkeley: University of California Press.

——— (eds.). 1991b. *Dolphin societies: Discoveries and puzzles.* Berkeley: University of California Press.

———. 1991c. Introduction. In *Dolphin societies: Discoveries and puzzles,* ed. K. Pryor and K. S. Norris. Berkeley: University of California Press.

Pryor, K., and Shallenberger, I. K. 1991. Social structure in spotted dolphins *(Stenella attenuata)* in the tuna purse seine fishery in the eastern tropical Pacific. In *Dolphin societies: Discoveries and puzzles,* ed. K. Pryor and K. J. Norris. Berkeley: University of California Press.

Pryor, K., Haag, R., and O'Reilly, J. 1969. The creative porpoise: Training for novel behavior. *J. Exp. Anal. Behav.* 12:653–61.

Pryor, K., Lindbergh, J., Lindbergh, S., and Milano, R. 1990. A human-dolphin fishing cooperative in Brazil. *Mar. Mamm. Sci.* 6:77–82.

Puente, A. E., and Dewsbury, D. A. 1976. Courtship and copulatory behavior of bottlenosed dolphins *(Tursiops truncatus). Cetology* 21:1–9.

Pulliam, H. R., and Caraco, T. 1984. Living in groups: Is there an optimal group size? In *Behavioral ecology: An evolutionary approach,* 2d ed., ed. J. R. Krebs and N. B. Davies. Oxford: Blackwell Scientific Publications.

Purrington, P. 1955. A whale and her calf. *Nat. Hist.* 64:363.

Purves, P. E., and Mountford, M. D. 1959. Ear-plug laminations in relation to the age composition of a population of fin whales. *Bull. Brit. Mus. Nat. Hist. Zool.* 5:123–61.

Pusey, A. E., and Packer, C. 1998. The ecology of relationships. In *Behavioral ecology: An evolutionary approach,* 4th ed., ed. J. R. Krebs and N. B. Davies. Oxford: Blackwell Scientific Publications.

Pusey, A. E., and Wolf, M. 1996. Inbreeding avoidance in animals. *Trends Ecol. Evol.* 11:201–6.

Pusey, A. E., Williams, J., and Goodall, J. 1997. The influence of dominance rank of the reproductive success of female chimpanzees. *Science* 277:828–31.

Queller, D. C., and Goodnight, K. F. 1989. Estimating relatedness using genetic markers. *Evolution* 43:258–75.

Quinn, T. P., and terHart, B. A. 1987. Movements of adult sockeye salmon *(Oncorhynchus nerka)* in British Columbia coastal waters in relation to temperature and salinity stratification: Ultrasonic telemetry results. *Can. Spec. Publ. Fish. Aquat. Sci.* 96:61–77.

Quinn, T. P., terHart, B. A., and Groot, C. 1989. Migratory orientation and vertical movements of homing adult sockeye salmon, *Oncorhynchus nerka*, in coastal waters. *Anim. Behav.* 37:587–99.

Quoy, J. R. C., and Gaimard, J. P. 1824. Zoologie. In Voyage autour du monde entrepris par ordre du Roi, sous le Ministère et conformément aux instructions de S. Exc. M. Le Vicompte du Bocage, Secrétaire d'état au Départment de la Marine, Exécuté sur les Corvettes de S. M. l'Uranie et la Physicienne, pendant les Années 1817, 1818, 1819 et 1820, ed. L. C. D. de Freycinet, 1–712 Paris: Chez Pillet Aîné.

Rabb, G. B., Woolpy, J. H., and Ginsburg, B. E. 1967. Social relationships in a group of captive wolves. *Am. Zool.* 7:305–11.

Ralls, K. 1976. Mammals in which females are larger than males. *Q. Rev. Biol.* 51:245–70.

———. 1977. Sexual dimorphism in mammals: Avian models and unanswered questions. *Am. Nat.* 111:917–38.

Ralls, K., and Brownell, R. L. Jr. 1991. A whale of a new species. *Nature* 350:560.

Ralls, K., Brownell, R. L. Jr., and Ballou, J. 1977. Differential mortality by sex and age in mammals, with specific reference to the sperm whale. *Rep. Int. Whal. Commn.* (special issue) 1:233–43.

Ralls, K., Fiorelli, P., and Gish, S. 1985. Vocalizations and vocal mimicry in captive harbor seals, *Phoca vitulina. Can. J. Zool.* 63:1050–56.

Ramirez, P. 1988. Comportamiento reproductivo del "cachalote" (*Physeter catodon* L.). *Bol. Lima* 59:29–32.

Ray, C., Watkins, W. A., and Burns, J. J. 1969. The underwater song of *Erignathus* (Bearded Seal). *Zoologica* 54:79–83 + 3 plates.

Ray, G. C., Mitchell, E. D., Wartzok, D., Kozicki, V. M., and Maiefski, R. 1978. Radio tracking of a fin whale *(Balaenoptera physalus). Science* 202:521–24.

Read, A. J. 1996. Incidental catches of small cetaceans. In *The conservation of whales and dolphins: Science and practice,* ed. M. P. Simmonds and J. Hutchinson. Chichester: Wiley.

———. 1998. Possible applications of new technology to marine mammal research and management. Report to the U.S. Marine Mammal Commission, Washington, DC. 36 pp.

Read, A. J., and Gaskin, D. E. 1985. Radio tracking the movements and activities of harbor porpoises, *Phocoena phocoena* (L.), in the Bay of Fundy, Canada. *Fish. Bull.* 83:543–52.

Read, A. J., and Hohn, A. A. 1995. Life in the fast lane: The life history of harbor porpoises from the Gulf of Maine. *Mar. Mamm. Sci.* 11:423–40.

Read, A. J., and Tolley, K. A. 1997. Postnatal growth and allometry of harbour porpoises from the Bay of Fundy. *Can. J. Zool.* 75:122–30.

Read, A. J., van Waerebeek, K., Reyes, J. C., McKinnon, J. S., and Lehman, L. C. 1988. The exploitation of small cetaceans in coastal Peru. *Biol. Conserv.* 46:53–70.

Read, A. J., Wells, R. S., Hohn, A. A., and Scott, M. D. 1993. Patterns of growth in wild bottlenose dolphins, *Tursiops truncatus. J. Zool.* 231:107–23.

Recchia, C., and Read, A. J. 1989. Stomach contents of harbour porpoises, *Phocoena phocoena* L., from the Bay of Fundy. *Can. J. Zool.* 679:2140–46.

Redondo, T., Gomendio, M., and Medina, R. 1992. Sex-biased parent-offspring conflict. *Behaviour* 123(3–4):261–69.

Reeve, H. K. 1991. *Polistes.* In *The social biology of wasps,* ed. K. G. Ross and R. W. Matthews. Ithaca: Cornell University Press.

———. 1998. Game theory, reproductive skew, and nepotism. In *Game theory and animal behavior,* ed. L. A. Dugatkin and H. K. Reeve. Oxford: Oxford University Press.

Reeve, H. K., and Ratnieks, F. L. W. 1993. Queen-queen conflicts in polygynous societies: Mutual tolerance and reproductive skew. In *Queen number and sociality in insects,* ed. L. Keller. Oxford: Oxford University Press.

Reeve, H. K., Emlen, S. T., and Keller, L. 1998. Reproductive sharing in animal societies: Reproductive incentives or incomplete control by dominant breeders? *Behav. Ecol.* 9:267–78.

Reeves, R. R. 1983. Book review: *Cetaceans. Science* 220:709.

———. 1993. Domestic and international trade in narwhal products. *TRAFFIC Bull.* 14:13–20.

Reeves, R. R., and Brownell, R. L. Jr. 1989. Susu. In *Handbook of marine mammals,* vol. 4: *River dolphins and the larger toothed whales,* ed. S. H. Ridgway and R. J. Harrison. New York: Academic Press.

Reeves, R. R., and Chaudhry, A. A. 1998. Status of the Indus river dolphin, *Platanista minor. Oryx* 32:35–44.

Reeves, R. R., and Leatherwood, S. 1994. *Dolphins, porpoises and whales: 1994–1998 action plan for the conservation of cetaceans.* Gland, Switzerland: IUCN.

———. 1996. Cetacean Specialist Group. *Species* 26–27:57–58.

Reeves, R. R., and Mead, J. G. 1999. Marine mammals in captivity. In *Conservation and management of marine mammals,* ed. J. R. Twiss Jr. and R. R. Reeves, 412–36. Washington, DC: Smithsonian Institution Press.

Reeves, R. R., and Mitchell, E. 1988a. Distribution and seasonality of killer whales in the eastern Canadian Arctic. *Rit Fiskideildar* 11:136–60.

———. 1988b. Killer whale sightings and takes by American pelagic whalers in the North Atlantic. *Rit Fiskideildar* 11:7–23.

———. 1989. Status of white whales, *Delphinapterus leucas,* in Ungava Bay and eastern Hudson Bay. *Can. Field Nat.* 103:220–39.

Reeves, R. R., and Whitehead, H. 1997. Status of the sperm whale *(Physeter macrocephalus)* in Canada. *Can. Field Nat.* 111:293–307.

Reeves, R. R., Leatherwood, S., and Papastavrou, V. 1991. Possible stock affinities of humpback whales in the northern Indian Ocean. In *Cetaceans and cetacean research in the Indian Ocean Sanctuary,* ed. S. Leatherwood and G. P. Donovan, 259–69. Marine Mammal Technical Report no. 3. Nairobi: UNEP.

Reid, K., Mann, J., Weiner, J. R., and Hecker, N. 1995. Infant development in two aquarium bottlenose dolphin. *Zoo Biol.* 14(2):135–47.

Reidman, M. L. 1982. The evolution of alloparental care and adoption in mammals and birds. *Q. Rev. Biol.* 57:405–35.

Reijnders, P. J. H. 1986. Reproductive failure of common seals feeding on fish from polluted waters. *Nature* 324:456–57.

———. 1996. Organochlorine and heavy metal contamination in cetaceans: Observed effects, potential impact and future prospects. In *The conservation of whales and dolphins: Science and practice*, ed. M. P. Simmonds and J. Hutchinson. Chichester: Wiley.

Reijnders, P. J. H., Donovan, G. P., Aguilar, A., and Bjørge, A. 1999. Report of the workshop on chemical pollution and cetaceans. *J. Cetacean Res. Mgmt.* (special issue) 1:1–42.

Reist, J. D. 1985. An empirical evaluation of several univariate methods that adjust for size variation in morphometric data. *Can. J. Zool.* 63:1429–39.

Reiter, J., and Le Boeuf, B. J. 1991. Life history consequences of variation in age at primiparity in northern elephant seals. *Behav. Ecol. Sociobiol.* 28:153–60.

Reyes, J. C., Mead, J. G., and van Waerebeek, K. 1991. A new species of beaked whale *Mesoplodon peruvianus* sp. n. (Cetacea: Ziphiidae) from Peru. *Mar. Mamm. Sci.* 7(1):1.

Reyes, J. C., van Waerebeek, K., Cárdenas, J. C., and Yáñez, J. L. 1996. *Mesoplodon bahamondi* sp. n. (Cetacea Ziphiidae), a new living beaked whale from the Juan Fernandez Archipelago, Chile. *Boletin del Museo Nacional de Historia Natural, Chile* 43:31–44.

Reznick, D. N., and Bryga, H. 1987. Life-history evolution in guppies *(Poecilia reticulata):* 1. Phenotypic and genetic changes in an introduction experiment. *Evolution* 41:1370–85.

Reznick, D. N., Bryga, H., and Endler, J. A. 1990. Experimentally-induced life-history evolution in a natural population. *Nature* 346:357–59.

Rice, D. W. 1968. Stomach contents and feeding behavior of killer whales in the eastern North Pacific. *Norsk Hvalfangst-Tidende* 57:35–38.

———. 1989. Sperm whale. *Physeter macrocephalus* Linnaeus, 1758. In *Handbook of marine mammals*, vol. 4, *River dolphins and the larger toothed whales,* ed. S. H. Ridgway, and R. Harrison. London: Academic Press.

———. 1998. *Marine mammals of the world: Systematics and distribution.* Special Publication no. 4. Society for Marine Mammalogy.

Rice, D. W., and Wolman, A. A. 1971. The life history and ecology of the gray whale *(Eschrichtius robustus).* Special Publication no. 3. American Society of Mammalogists.

Rice, M., Carlson, D., Chu, K., Dolphin, W., and Whitehead, H. 1987. Are humpback whale population estimates being biased by sexual differences in fluking behavior? *Rep. Int. Whal. Commn.* 37:333–35.

Richard, A. F. 1987. Malagasy prosimians: Female dominance. In *Primate societies,* ed. B. B. Smuts, D. L. Cheney, R. M. Seyfarth, R. W. Wrangham, and T. T Struhsaker. Chicago: University of Chicago Press.

Richard, K. R., McCarrey, S. W., and Wright, J. M. 1994. DNA sequence from the *SRY* gene of the sperm whale *(Physeter macrocephalus)* for use in molecular sexing. *Can. J. Zool.* 72:873–77.

Richard, K. R., Dillon, M. C., Whitehead, H., and Wright, J. M. 1996a. Patterns of kinship in groups of free-living sperm whales *(Physeter macrocephalus)* revealed by multiple molecular genetic analyses. *Proc. Natl. Acad. Sci. USA* 93:8792–95.

Richard, K. R., Whitehead, H., and Wright, J. M. 1996b. Polymorphic microsatellites from sperm whales and their use in the genetic identification of individuals from naturally sloughed pieces of skin. *Mol. Ecol.* 5:313–15.

Richards, A. F. 1996. Life history and behavior of female dolphins in Shark Bay, Western Australia. Ph.D. dissertation, University of Michigan, Ann Arbor.

Richards, D. G. 1986. Dolphin vocal mimicry and vocal object labeling. In *Dolphin cognition and behavior: A comparative approach,* ed. R. J. Schusterman, J. A. Thomas, and F. G. Wood. Hillsdale, NJ: Lawrence Erlbaum Associates.

Richards, D. G., Wolz, J. P., and Herman, L. M. 1984. Vocal mimicry of computer-generated sounds and vocal labeling of objects by a bottlenosed dolphin, *Tursiops truncatus. J. Comp. Psychol.* 98:10–28.

Richardson, W. J., Fraker, M. A., Würsig, B., and Wells, R. S. 1985. Behaviour of bowhead whales *Balaena mysticetus* summering in the Beaufort Sea: Reactions to industrial activities. *Biol. Conserv.* 32(3):195–230.

Richardson, W. J., Würsig, B., and Greene, C. R. 1990. Reactions of bowhead whales, *Balaena mysticetus,* to drilling and dredging noise in the Canadian Beaufort Sea. *Mar. Environ. Res.* 29:135–60.

Richardson, W. J., Finley, K. J., Miller, G. W., Davis, R. A., and Koski, W. R. 1995a. Feeding, social and migration behavior of bowhead whales, *Balaena mysticetus* in Baffin Bay vs. Beaufort Sea: Regions with different amounts of human activity. *Mar. Mamm. Sci.* 11:1–45.

Richardson, W. J., Greene, C. R. Jr., Malme, C. I., and Thomson, D. H. 1995b. *Marine mammals and noise.* San Diego: Academic Press.

Ridgway, S. H. 1986. Physiological observations on dolphin brains. In *Dolphin cognition and behavior: A comparative approach,* ed. R. J. Schusterman, J. A. Thomas, and F. G. Wood. Hillsdale, NJ: Lawrence Erlbaum Associates.

———. 1987. *The dolphin doctor.* Dublin, NH: Yankee Publishing.

———. 1997. First audiogram for marine mammals in the open ocean and at depth: Hearing and whistling by two white whales down to 30 atmospheres. *J. Acoust. Soc. Am.* 101(5): 3136. Abstract.

Ridgway, S. H., and Brownson, R. H. 1984. Relative brain sizes and cortical surface area in odontocetes. *Acta Zool. Fenn.* 1972:149–52.

Ridgway, S. H., and Carder, D. A. 1990. Tactile sensitivity, somatosensory responses, skin vibrations, and the skin surface ridges of the bottlenose dolphin, *Tursiops truncatus.* In *Sensory abilities of cetaceans,* ed. J. Thomas and R. Kastelein. New York: Plenum.

Ridgway, S. H., and Harrison, R. (eds.). 1985. *Handbook of marine mammals.* Vol. 3. London: Academic Press.

———. 1989. *Handbook of marine mammals.* Vol. 4. *River dolphins and the larger toothed whales.* London: Academic Press.

———. 1994. *Handbook of marine mammals.* Vol. 5. London: Academic Press.

———. 1999. *Handbook of marine mammals.* Vol. 6. London: Academic Press.

Ridgway, S., and Reddy, M. 1995. Residue levels of several organochlorines in *Tursiops truncatus* milk collected at various stages of lactation. *Mar. Pollut. Bull.* 30:609–14.

Ridgway, S., Kamolnick, T., Reddy, M., Curry, C., and Tarpley, R. J. 1995. Orphan-induced lactation in *Tursiops* and analysis of collected milk. *Mar. Mamm. Sci.* 11(2):172.

Ridoux, V., Guinet, C., Liret, C., Creton, P., Steenstrup, R., and Beauplet, G. 1997. A video sonar as a new tool to study marine mammals in the wild: Measurements of dolphin swimming speed. *Mar. Mamm. Sci.* 13:196–206.

Rigley, L. 1983. Dolphins feeding in a South Carolina salt marsh. *Whalewatcher* 7:3–5.

Robeck, T. R., Schneyer, A. L., McBain, J. F., Dalton, L. M., Walsh, M. T., Czekala, N. M., and Kraemer, D. C. 1993. Analysis of urinary immunoreactive steroid metabolites and gonadotropins for characterization of the estrous cycle, breeding period, and seasonal estrous activity of the captive killer whale *(Orcinus orca)*. *Zoo Biol.* 12:173–87.

Robinson, J. G. 1982. Vocal systems regulating within-group spacing. In *Primate communication,* ed. C. T. Snowdon, C. H. Brown, and M. R. Peterson. Cambridge: Cambridge University Press.

Robinson, M. H. 1991. Ethology: Field studies versus zoo studies. *Scientific Proceedings of 1990 International Union of Directors of Zoological Gardens Conference,* 119–64. Copenhagen, Denmark.

Rodman, P. S. 1984. Foraging and social systems of orangutans and chimpanzees. In *Adaptations for foraging in nonhuman primates,* ed. P. S. Rodman and J. G. H. Cant. New York: Columbia University Press.

Rogosa, D., and Ghandour, G. 1991. Statistical models for behavioral observations. *J. Educ. Stat.* 16(3):157–252.

Rood, J. P. 1983. Banded mongoose rescues pack member from eagle. *Anim. Behav.* 31:1261–62.

Rørvik, C. J. 1987. Northeast Atlantic minke whales re-assessed. *Rep. Int. Whal. Commn.* 37:241–52.

Rørvik, C. J., Jonsson, J., Mathisen, O., and Jonsgård, A. 1976. Fin whales, *Balaenoptera physalus,* L. off the west coast of Iceland. Distribution, segregation by length and exploitation. *Rit Fiskideildar* 5:1–30.

Rose, B., and Payne, A. I. L. 1991. Occurrence and behavior of the southern right whale dolphin *Lissodelphis peronii* off Namibia. *Mar. Mamm. Sci.* 7(1):25–34.

Rose, N. A. 1992. The social dynamics of male killer whale, *Orcinus orca,* in Johnstone Strait, British Columbia. Ph.D. thesis, University of California, Santa Cruz.

Ross, G. J. B. 1977. The taxonomy of bottlenosed dolphins *Tursiops* species in South African waters, with notes on their biology. *Ann. Cape Prov. Mus. Nat. Hist.* 11:135–94.

———. 1979. The smaller cetaceans of the south east coast of southern Africa. Ph.D. thesis, University of Port Elizabeth, South Africa.

———. 1984. The smaller cetaceans of the south east coast of southern Africa. *Ann. Cape Prov. Mus. Nat. Hist.* 15:173–410.

Ross, G. J. B., and Cockcroft, V. G. 1990. Comments on Australian bottlenose dolphins and the taxonomic status of *Tursiops aduncus* (Ehrenberg, 1832). In *The bottlenose dolphin,* ed. S. Leatherwood and R. R. Reeves. San Diego: Academic Press.

Ross, H. M., and Wilson, B. 1996. Violent interactions between bottlenose dolphins and harbour porpoises. *Proc. R. Soc. Lond.* B 263:283–86.

Rossbach, K. A., and Herzing, D. L. 1997. Underwater observations of benthic feeding bottlenose dolphins *(Tursiops truncatus)* near Grand Bahama Island, Bahamas. *Mar. Mamm. Sci.* 13(3):498–503.

Rothstein, S. I., and Pierotti, R. 1988. Distinctions among reciprocal altruism, kin selection, and cooperation and a model for the initial evolution of beneficent behavior. *Ethol. Sociobiol.* 9:189–209.

Rowell, T. E. 1994. Not little furry people. Keynote address of the Thirty-first Annual Meeting of the Animal Behavior Society, Seattle, WA.

Rowntree, V. J. 1996. Feeding, distribution and reproductive behavior of cyamids (Crustacea: Amphipoda) living on right and humpback whales. *Can. J. Zool.* 74:103–9.

Roxburgh, W. 1801. An account of a new species of Delphinus, an inhabitant of the Ganges. *Asiatick Researches* (Calcutta edition) 7:170–74.

Rubenstein, D. I., and Wrangham, R. W. 1986. Socioecology: Origins and trends. In *Ecological aspects of social evolution,* ed. D. I. Rubenstein and R. W. Wrangham. Princeton, NJ: Princeton University Press.

Rudolph, P. 1996. New information on the whale fishery at Lamalera, Lembata Island. *SEAMMAM Newsletter* 1:7.

Rudolph, P., Smeenk, C., and Leatherwood, S. 1997. Preliminary checklist of Cetacea in the Indonesian Archipelago and adjacent waters. *Zool. Verhandel* 312:1–48.

Rudwick, M. 1992. *Scenes from deep time.* Chicago: University of Chicago Press.

Ruggerone, G. T., Quinn, T. P., McGregor, I. A., and Wilkinson, T. D. 1990. Horizontal and vertical movements of adult steelhead trout, *Oncorhynchus mykiss,* in the Dean and Fisher channels, British Columbia. *Can. J. Fish. Aquat. Sci.* 47:1963–69.

Rugh, D. J. 1990. Bowhead whales reidentified through aerial photography near Point Barrow, Alaska. In *Individual recognition of cetaceans: Use of photo-identification and other techniques to estimate population parameters,* ed. P. S. Hammond, S. A. Mizroch, and G. P. Donovan, 289–94. Reports of the International Whaling Commission, special issue 12. Cambridge: International Whaling Commission.

Rugh, D. J., Miller, G. W., Withrow, D. E., and Koski, W. R. 1992. Calving intervals of bowhead whales established through photographic identifications. *J. Mammal.* 73(3):487–90.

Rutberg, A. T. 1986. Lactation and fetal sex ratios in American bison. *Am. Nat.* 127:89–94.

Ryan, M. J., Tuttle, M. D., and Rand, A. S. 1982. Bat predation and sexual advertisement in a Neotropical frog. *Am. Nat.* 199:136–39.

Saayman, G. S., and Tayler, C. K. 1973. Social organization of inshore dolphins *(Tursiops aduncus* and *Sousa)* in the Indian ocean. *J. Mammal.* 54:993–96.

———. 1979. The socioecology of humpback dolphins *(Sousa* sp.). In *Behavior of marine mammals: Current perspectives in research,* vol. 3, *Cetaceans,* ed. H. E. Winn and B. L. Olla. New York: Plenum Press.

Saayman, G. S., Tayler, C. K., and Bower, D. 1973. Diurnal activity cycles in captive and free-ranging Indian Ocean bottlenose dolphins (*Tursiops aduncus* Ehrenburg). *Behaviour* 44: 212–33.

Sacher, G. A., and Staffeldt, E. F. 1974. Relationship of gestation time to brain weight for placental mammals: Implications for the theory of vertebrate growth. *Am. Nat.* 108:593–615.

Saiki, R. K., Gelfand, D. H., Stoffel, S., Scharf, S. J., Higuchi, R., Horn, G. T., Mullis, K. B., and Erlich, H. A. 1988. Primer-directed enzymatic amplification of DNA with a thermostable DNA polymerase. *Science* 239:487–94.

Samaras, W. F. 1974. Reproductive behavior of the gray whale, *Eschrichtius robustus*, in Baja California. *Bull. S. Calif. Acad. Sci.* 73:57–64.

Samuels, A., and Gifford, T. 1997. A quantitative assessment of dominance relations among bottlenose dolphins. *Mar. Mamm. Sci.* 13:70–99.

Samuels, A., Sevenich, M., Gifford, T., Sullivan T., and Sustman, J. 1989. Gentle rubbing among bottlenose dolphins. Abstracts, Eighth Biennial Conference on the Biology of Marine Mammals.

Sandell, M., and Liberg, O. 1992. Roamers and stayers: A model on male mating tactics and mating systems. *Am. Nat.* 139: 177–89.

Saulitis, E. L. 1993. The behavior and vocalizations of the "AT" group of killer whales *(Orcinus orca)* in Prince William Sound, Alaska. M.Sc. thesis, University of Alaska, Fairbanks.

Sayigh, L. S. 1992. Development and functions of signature whistles of free-ranging bottlenose dolphins, *Tursiops truncatus*. Doctoral dissertation, Woods Hole Oceanographic Institution.

Sayigh, L. S., Tyack, P. L., Wells, R. S., and Scott, M. D. 1990. Signature whistles of free-ranging bottlenose dolphins, *Tursiops truncatus*: Stability and mother-offspring comparisons. *Behav. Ecol. Sociobiol.* 2:247–60.

Sayigh, L. S., Tyack, P. L., and Wells, R. S. 1993a. Recording underwater sounds of free-ranging dolphins while underway in a small boat. *Mar. Mamm. Sci.* 9:209–13.

Sayigh, L. S., Tyack, P. L., Wells, R. S., Scott, M. D., and Irvine, A. B. 1993b. Individual recognition in free-ranging bottlenose dolphins: A field test using playback experiments. In Abstracts of the Tenth Biennial Conference on the Biology of Marine Mammals, Galveston, TX: 95.

Sayigh, L. S., Tyack, P. L., Wells, R. S., and Samuels, A. 1995a. Signature whistle development in free-ranging bottlenose dolphins. In Abstracts of the Eleventh Biennial Conference on the Biology of Marine Mammals, Orlando, FL: 101.

Sayigh, L. S., Tyack, P. L., Wells, R. S., Scott, M. D., and Irvine, A. B. 1995b. Sex differences in signature whistle production of free-ranging bottlenose dolphins, *Tursiops truncatus*. *Behav. Ecol. Sociobiol.* 36:171–77.

Sayigh, L. S., Tyack, P. L., Wells, R. S., Solow, A., Scott, M. D., and Irvine A. B. 1999. Individual recognition in wild bottlenose dolphins: A field test using playback experiments. *Anim. Behav.* 56:41–50.

Scammon, C. M. 1874. *The Marine Mammals of the Northwestern Coast of North America. Together with an Account of the American Whale-Fishery*. San Francisco: John H. Carmany.

Schaeff, C., Kraus, S., Brown, M., Perkins, J., Payne, R., Gaskin, D., Boag, P., and White, B. 1991. Preliminary analysis of mitochondrial DNA variation within and between the right whale species *Eubalaena glacialis* and *Eubalaena australis*. In *Genetic ecology of whales and dolphins: Incorporating the proceedings of the workshop on the genetic analysis of cetacean populations*, ed. A. R. Hoelzel, 217–23. Reports of the International Whaling Commission, special issue 13. Cambridge: International Whaling Commission.

Schaeff, C. M., Kraus, S. D., Brown, M. W., Perkins, J. S., Payne, R., and White, B. N. 1997. Comparisons of genetic variability of North and South Atlantic right whales (*Eubalaena*), using DNA fingerprinting. *Can. J. Zool.* 75:1073–80.

van Schaik, C. P. 1983. Why are diurnal primates living in groups? *Behaviour* 87:120–44.

———. 1989. The ecology of social relationships amongst female primates. In *Comparative socioecology: The behavioral ecology of humans and other mammals*, ed. V. Standen and R. A. Foley. London: Blackwell.

van Schaik, C. P., and Dunbar, R. I. M. 1990. The evolution of monogamy in large primates: A new hypothesis and some crucial tests. *Behaviour* 115:30–61.

van Schaik, C. P., and Kappeler, D. M. 1997. Infanticide risk and the evolution of male-female association in primates. *Proc. R. Soc. Lond. B* 264:1687–94.

van Schaik, C. P., Noordwijk, M. A., and Nunn, C. L. 1998. Sex and social evolution in primates. In *Comparative primate socioecology*, ed. P. C. Lee. Cambridge: Cambridge University Press.

Schaller, G. B. 1963. *The mountain gorilla: Ecology and behavior*. Chicago: University of Chicago Press.

———. 1965. Field procedures. In *Primate behavior: Field studies of monkeys and apes*, ed. I. DeVore. New York: Holt, Rinehart and Winston.

———. 1972. *The Serengeti lion*. Chicago: University of Chicago Press.

Scheffer, V. B. 1969. Marks on the skin of a killer whale. *J. Mammal.* 50:151.

Schevill, W. E. 1964. Underwater sounds of cetaceans. In *Marine bioacoustics*, vol. 1, ed. W. N. Tavolga. Oxford: Pergamon.

———. 1974. Preface, Glossary. In *The whale problem: A status report*, ed. W. E. Schevill, G. C. Ray, and K. S. Norris. Cambridge, MA: Harvard University Press.

Schevill, W. E., and Backus, R. H. 1960. Daily patrol of a *Megaptera*. *J. Mammal.* 41:279–81.

Schevill, W. E., and Lawrence, B. 1956. Food-finding by a captive porpoise (*Tursiops truncatus*). *Breviora* 53:1–15.

Schevill, W. E., and Watkins, W. A. 1962. *Whale and porpoise voices: A phonograph record*. Woods Hole, MA: Woods Hole Oceanographic Institution.

———. 1966. *Radio-tagging of Whales*. Technical Report. Woods Hole, MA: Woods Hole Oceanographic Institution.

Schevill, W. E., Watkins, W. A., and Backus, R. H. 1964. The 20-cycle signals and *Balaenoptera* (fin whales). In *Marine bioacoustics*, vol. 1, ed. W. N. Tavolga. Oxford: Pergamon Press.

Schleidt, W. M. 1973. Tonic communication: Continual effects of discrete signs in animal communication systems. *J. Theor. Biol.* 42:359–86.

Schmidt-Nielsen, K. 1984. *Scaling: Why is animal size so important*. Cambridge: Cambridge University Press.

————. 1990. *Animal physiology*. Cambridge: Cambridge University Press.

Schmit, R. J., and Strand, S. W. 1982. Cooperative foraging by yellowtail, *Seriola lalandei* (Carangidae), on two species of fish prey. *Copeia* 3:714–17.

Schneider, K., Baird, R. W., Dawson, S., Visser, I., and Childerhouse, S. 1998. Reactions of bottlenose dolphins to tagging attempts using a remotely-deployed suction-cup tag. *Mar. Mamm. Sci.* 14:316–24.

Schroeder, P. J. 1990. Breeding bottlenose dolphins in captivity. In *The bottlenose dolphin*, ed. S. Leatherwood and R. R. Reeves. Orlando, FL: Academic Press.

Schroeder, P. J., and Keller, K. V. 1989. Seasonality of serum testosterone levels and sperm density in *Tursiops truncatus*. *J. Exp. Zool.* 249:316–21.

————. 1990. Artificial insemination of bottlenose dolphins. In *The bottlenose dolphin*, ed. S. Leatherwood and R. R. Reeves. San Diego: Academic Press.

Schuknecht, H. F., and Gulya, A. J. 1986. *Anatomy of the temporal bone with surgical implications*. Philadelphia: Lea and Febiger.

Schulke, F. 1961. Can the dolphin learn to talk? *Life* 51(4):61–66.

Scoresby, W. 1820. *An account of the Arctic regions, with a history and description of the northern whale-fishery*. Vol. 2. *The whale-fishery*. Edinburgh, Scotland: Archibald Constable.

Scott, D. K. 1978. Identification of individual Bewick's swans by bill patterns. In *Animal marking: Recognition marking of animals in research*, ed. B. Stonehouse. Baltimore: University Park Press.

Scott, J., and Fredericson, E. 1951. The causes of fighting in mice and rats. *Physiol. Zool.* 24:273–309.

Scott, M. D. 1991. The size and structure of pelagic dolphin herds. Ph.D. dissertation, University of California, Los Angeles.

Scott, M. D., and Chivers, S. J. 1990. Distribution and herd structure of bottlenose dolphins the eastern tropical Pacific ocean. In *The bottlenose dolphin*, ed. S. Leatherwood and R. R. Reeves. San Diego: Academic Press.

Scott, M. D., and Cordaro, J. G. 1987. Behavioral observations of the dwarf sperm whale, *Kogia simus*. *Mar. Mamm. Sci.* 3: 353–54.

Scott, M. D., and Perryman, W. L. 1991. Using aerial photogrammetry to study dolphin school structure. In *Dolphin societies: Discoveries and puzzles*, ed. K. Pryor and K. S. Norris. Berkeley: University of California Press.

Scott, M. D., Wells, R. S., and Irvine, A. B. 1990a. A long-term study of bottlenose dolphins on the west coast of Florida. In *The bottlenose dolphin*, ed. S. Leatherwood and R. R. Reeves. San Diego: Academic Press.

Scott, M. D., Wells, R. S., Irvine, A. B., and Mate, B. R. 1990b. Tagging and marking studies on small cetaceans. In *The bottlenose dolphin*, ed. S. Leatherwood and R. R. Reeves. San Diego: Academic Press.

Scott, T. M., and Sadove, S. S. 1997. Sperm whale, *Physeter macrocephalus*, sightings in the shallow shelf waters off Long Island, New York. *Mar. Mamm. Sci.* 13:317–21.

Searcy, W. A., and Anderson, M. 1986. Sexual selection and the evolution of song. *Annu. Rev. Ecol. Syst.* 17:507–33.

Selous, E. 1905. *The Bird Watcher in the Shetlands, With Some Notes on Seals—and Digressions*. London: J. M. Dent.

Sergeant, D. E. 1962. The biology of the pilot or pothead whale *Globicephala melaena* (Traill) in Newfoundland waters. *Bull. Fish. Res. Bd. Can.* 132:1–84.

Sergeant, D. E., and Brodie, P. F. 1969. Body size in white whales, *Delphinapterus leucas*. *J. Fish. Res. Bd. Can.* 26:2561–80.

Sergeant, D. E., Caldwell, D. K., and Caldwell, M. C. 1973. Age, growth, and maturity of bottlenose dolphin *(Tursiops truncatus)* from northeast Florida. *J. Fish. Res. Bd. Can.* 30: 1009–11.

Shaffer, M. L. 1981. Minimum population sizes for species conservation. *Bioscience* 31:131–34.

Shane, S. H. 1980. Occurrence, movements, and distribution of bottlenose dolphin, *Tursiops truncatus*, in Southern Texas. *Fish. Bull.* 78:593–601.

————. 1990a. Comparison of bottlenose dolphin behavior in Texas and Florida, with a critique of methods for studying dolphin behavior. In *The bottlenose dolphin*, ed. S. Leatherwood and R. R. Reeves. San Diego: Academic Press.

————. 1990b. Behavior and ecology of the bottlenose dolphin at Sanibel Island, Florida. In *The bottlenose dolphin*, ed. S. Leatherwood and R. R. Reeves. San Diego: Academic Press.

Shane, S. H., and McSweeney, D. 1990. Using photo-identification to study pilot whale social organization. In *Individual recognition of cetaceans: Use of photo-identification and other techniques to estimate population parameters*, ed. P. S. Hammond, S. A. Mizroch, and G. P. Donovan, 259–63. Reports of the International Whaling Commission, special issue 12. Cambridge: International Whaling Commission.

Shane, S. H., and Schmidly, D. J. 1978. The population biology of the Atlantic bottlenose dolphin, *Tursiops truncatus*, in the Aransas Pass area of Texas. U.S. Marine Mammal Commission Report no. MMC-76/11, Washington, DC.

Shane, S. H., Wells, R. S., and Würsig, B. 1986. Ecology, behavior and social organization of the bottlenose dolphin: A review. *Mar. Mamm. Sci.* 2:34–63.

Shannon, C. E., and Wiener N. 1949. *The mathematical theory of communication*. Urbana: University of Illinois Press.

Sharpe, F. A., and Dill, L. M. 1995. The bubble helix: Sonar studies of feeding humpback whales. Abstracts, Eleventh Biennial Conference on the Biology of Marine Mammals, Orlando, FL.

Shavelson, R. J., and Webb, N. M. 1991. *Generalizability theory*. Newbury Park, CA: Sage Publications.

Shea, B. T. 1984. Ontogenetic allometry and scaling: A discussion based on the growth and form of the skull in African apes. In *Size and scaling in primate biology*, ed. W. L. Jungers. New York: Plenum Press.

Shelden, K. E. W., Baldridge, A., and Withrow, D. E. 1995. Observations of Rissos dolphins, *Grampus griseus* with gray whales, *Eschrichtius robustus*. *Mar. Mamm. Sci.* 11(2): 231–40.

Sherman, P. W. 1977. Nepotism and the evolution of alarm calls. *Science* 197:1246–53.

Sherman, P. W., Jarvis, J. U. M., and Alexander, R. D. 1991. *The biology of the naked mole-rat*. Princeton, NJ: Princeton University Press.

Sherman, S. C. 1965. *The voice of the whaleman*. Providence, RI: Providence Public Library.

Shirakihara, M., Takemura, A., and Shirakihara, K. 1993. Age, growth, and reproduction of the finless porpoise, *Neophocaena phocaenoides,* in the coastal waters of western Kyushu, Japan. *Mar. Mamm. Sci.* 9:392–406.

Short, R. V. 1976. The evolution of human reproduction. *Proc. R. Soc. Lond.* B 195:3–24.

———. 1979. Sexual selection and its component parts, somatic and genital selection, as illustrated by man and the great apes. *Adv. Stud. Behav.* 9:131–58.

Siebenaler, J. B., and Caldwell, D. K. 1956. Cooperation among adult dolphins. *J. Mammal.* 37:126–28.

Sigurdson, J. 1993. Whistles as a communication medium. In *Language and communication: Comparative perspectives,* ed. H. L. Roitblat, L. M. Herman, and P. Nachtigall. Hillsdale, NJ: Lawrence Erlbaum Associates.

Sigurjónsson, J., Lyrholm, T., Leatherwood, S., Jonsson, E., and Vikingsson, G. 1988. Photoidentification of killer whales, *Orcinus orca,* off Iceland, 1981 through 1986. *Rit Fiskideildar* 11:99–114.

Silber, G. 1986. The relationship of social vocalizations to surface behavior and aggression in the Hawaiian humpback whale *(Megaptera novaeangliae). Can. J. Zool.* 72:805–11.

Silber, G. K., Newcomer, M. W., and Perez-Cortez, M. H. 1990. Killer whales *(Orcinus orca)* attack and kill a Bryde's whale *(Balaenoptera edeni). Can. J. Zool.* 68:1603–6.

Sillén-Tullberg, B., and Møller, A. P. 1993. The relationship between concealed ovulation and mating systems in anthropoid primates: A phylogenetic analysis. *Am. Nat.* 141:1:25.

Silverman, H. B., and Dunbar, M. J. 1980. Aggressive tusk use by the narwhal *(Monodon monocerus* L.). *Nature* 284:57–58.

Similä, T. 1997. Sonar observations of killer whales *(Orcinus orca)* feeding on herring schools. *Aquat. Mamm.* 23:119–26.

Similä, T., and Ugarte, F. 1993. Surface and underwater observations of cooperatively feeding killer whales in northern Norway. *Can. J. Zool.* 71:1494–99.

Similä, T., Holst, J. C., and Christensen, I. 1996. Occurrence and diet of killer whales in northern Norway: Seasonal patterns relative to the distribution and abundance of Norwegian spring-spawning herring. *Can. J. Fish. Aquat. Sci.* 53: 769–79.

Simpson, J. G., and Gardner, M. B. 1972. Comparative microscopic anatomy of selected marine mammals. In *Mammals of the sea,* ed. S. H. Ridgway. Springfield, IL: Charles C. Thomas.

Sinclair, A. H., Berta, P., Palmer, M. S., Hawkins, J. R., Griffiths, B. L., Smith, M. J., Foster, J. W., Frischauf, A. M., Lovell-Badge, R., and Goodfellow, P. N. 1990. A gene from the human sex-determining region encodes a protein with homology to a conserved DNA-binding motif. *Nature* 346:240–44.

Singer, P. 1990. *Animal liberation*. 2d edition. New York: New York Review.

Sjare, B., and Stirling, I. 1993. The breeding behavior and mating system of walruses. Abstracts, Proceedings of the Tenth Biennial Conference on the Biology of Marine Mammals, Galveston, TX.

Slijper, E. J. 1962. *Whales*. New York: Basic Books.

Slooten, E. 1994. Behavior of Hector's dolphin: Classifying behavior by sequence analysis. *J. Mammal.* 75:956–64.

Slooten, E., Dawson, S. M., and Whitehead, H. 1993. Associations among photographically identified Hector's dolphins. *Can. J. Zool.* 71:2311–18.

Smith, A. 1991. Prisoners in paradise. *Sonar* 6:14–15.

Smith, C. R., Kukert, H., Wheatcroft, R. A., Jumars, P. A., and Deming, J. W. 1989. Vent fauna on whale remains. *Nature* 341:27–28.

Smith, R. I. F. 1986. Evolution of alarm signals: Role of benefits of retaining group members or territorial neighbors. *Am. Nat.* 128:604–10.

Smith, S. C. 1992. Sperm whales and mesopelagic cephalopods in the waters off the Galápagos Islands. M.Sc. thesis, Dalhousie University, Halifax, Nova Scotia.

Smith, S. C., and Whitehead, H. 1993. Variations in the feeding success and behaviour of Galápagos sperm whales *(Physeter macrocephalus)* as they relate to oceanographic conditions. *Can. J. Zool.* 71:1991–96.

Smith, T. G. 1985. Polar bears, *Ursus maritimus,* as predators of belugas, *Delphinapterus leucas. Can. Field Nat.* 99:71–75.

Smith, T. G., Siniff, D. B., Reichle, R., and Stone, S. 1981. Coordinated behavior of killer whales, *Orcinus orca,* hunting a crabeater seal, *Lobodon carcinophagus. Can. J. Zool.* 59: 1185–89.

Smith, W. J. 1977. *The behavior of communicating*. Cambridge, MA: Harvard University Press.

Smolker, R. A., Richards, A. F., Connor, R. C., and Pepper, J. W. 1992. Sex differences in patterns of association among Indian Ocean bottlenose dolphins. *Behaviour* 123:38–69.

Smolker, R. A., Mann, J., and Smuts, B. B. 1993. Use of signature whistles during separations and reunions by wild bottlenose dolphin mothers and infants. *Behav. Ecol. Sociobiol.* 33: 393–402.

Smolker, R. A., Richards, A. F., Connor, R. C., Mann, J., and Berggren, P. 1997. Sponge-carrying by Indian Ocean bottlenose dolphins: Possible tool-use by a delphinid. *Ethology* 103: 454–65.

Smolker, R. A., and Pepper, J. W. 1999. Whistle convergence among allied male bottlenose dolphins (Delphinidae, Tursiops sp.). *Ethology* 105.

Smultea, M. A. 1994. Segregation by humpback whale *(Megaptera novaeangliae)* cows with a calf in coastal habitat near the island of Hawaii. *Can. J. Zool.* 72:805–11.

Smuts, B. B. 1985. *Sex and friendship in baboons*. Chigaco: Aldine.

Smuts, B. B., and Smuts, R. W. 1993. Male aggression and sexual coercion of females in nonhuman primates and other mammals: Evidence and theoretical implications. *Adv. Stud. Behav.* 22:1–63.

Smuts, B. B., Cheney, D. L., Seyfarth, R. L., Wrangham, R. W., and Struhsaker, T. T. 1987. *Primate societies*. Chicago: University of Chicago Press.

Smythe, N. 1970. On the existence of "pursuit invitation" signals in mammals. *Am. Nat.* 104:491–94.

Snowdon, C. T. 1996. Infant care in cooperatively breeding species. In *Parental care: Evolution, mechanisms and adaptive significance,* ed. J. S. Rosenblatt and C. T. Snowdon. San Diego: Academic Press.

Sokoloff, L. 1977. Circulation and energy metabolism of the brain. In *Basic neurochemistry,* ed. G. J. Siegel. Boston: Little Brown.

Soulé, M. E. 1986. Conservation biology and the "real world." In *Conservation biology: The science of scarcity and diversity,* ed. M. E. Soulé. Sunderland, MA: Sinauer Associates.

Southwell, T. 1898. The migration of the right whale *(Balaena mysticetus). Nat. Sci. (Lond.)* 12:397–414.

Sowerby, J. 1804. Note on *Physeter bidens. Trans. Linn. Soc.* 7: 310.

Spiesberger, J. L., and Fristrup, K. M. 1990. Passive localization of calling animals and sensing their acoustic environment using acoustic tomography. *Am. Nat.* 135:107–53.

Spong, P., Bradford, J., and White, D. 1970. Field studies on the behaviour of the killer whale *(Orcinus orca).* Proceedings of the Seventh Annual Conference on Biological Sonar and Diving Mammals, 169–74.

Stacey, P. J., and Baird, R. W. 1997. Birth of a *resident* killer whale off Victoria, British Columbia, Canada. *Mar. Mamm. Sci.* 13:504–8.

Stafford, K. M., Fox, C. G., and Clark, D. S. 1998. Longrange acoustic detection and localization of blue whale calls in the northeast Pacific Ocean. *J. Acoust. Soc. Am.* 104:3616–25.

Stammbach, E. 1987. Desert, forest, and montane baboons: Multilevel societies. In *Primate societies,* ed. B. B. Smuts, D. L. Cheney, R. M. Seyfarth, R. W. Wrangham, and T. T. Struhsaker. Chicago: University of Chicago Press.

Stander, P. E. 1992. Cooperative hunting in lions: The role of the individual. *Behav. Ecol. Sociobiol.* 29:445–54.

Stearns, S. C. 1992. *The evolution of life histories.* Oxford: Oxford University Press.

Stearns, S. C., and Koella, J. 1986. The evolution of phenotypic plasticity in life-history traits: Predictions for norms of reaction for age- and size-at-maturity. *Evolution* 40:893–913.

Stejneger, L. 1883. Contributions to the history of the Commander Islands. No. 1. Notes on the natural history, including descriptions of new cetaceans. *Proc. US Natl. Mus.* 6:58–89.

Stenuit, R. 1968. *The dolphin, cousin to man.* New York: Bantam Books.

Stevens, T. A., Duffield, D. A., Asper, E. D., Hewlett, K. G., Bolz, A., Gage, L. J., and Bossart, G. D. 1989. Preliminary findings of restriction fragment differences in mitochondrial DNA among killer whales *(Orcinus orca). Can. J. Zool.* 67: 2592–95.

Stewart-Oaten, A. 1995. Rules and judgements in statistics: Three examples. *Ecology* 76:2001–9.

Stirling, I. 1973. Vocalization in the ringed seal *(Phoca hispida). J. Fish. Res. Bd. Canada* 30:1592–94.

———. 1975. Factors affecting the evolution of social behaviour in the Pinnipedia. *Rapp. P.-v. Reun. Cons. Int. Explor. Mer.* 169:205–12.

Stockley, P., and Purvis, A. 1993. Sperm competition in mammals: A comparative study of male roles and relative investment in sperm production. *Funct. Ecol.* 7:560–70.

Stoddard, P. K., and Beecher, M. D. 1983. Parental recognition of offspring in the cliff swallow. *Auk* 100:795–99.

Stommel, H. 1963. Varieties of oceanographic experience. *Science* 139:572–76.

Stone, G. S., Katona, S. K., and Tucker, E. B. 1987. History, migration and present status of humpback whales, *Megaptera novaeangliae,* at Bermuda. *Biol. Cons.* 42:133–45.

Stone, G. S., Florez-Gonzalez, L., and Katona, S. 1990. Whale migration record. *Nature* 346:705.

Stone, G., Goodyear, J., Hutt, A., and Yoshinaga, A. 1994. A new non-invasive tagging method for studying wild dolphins. *Mar. Tech. Soc. J.* 28:11–16.

Strager, H. 1995. Pod specific call repertoires and compound calls of killer whales, *Orcinus orca* Linnaeus, 1758, in the waters of northern Norway. *Can. J. Zool.* 73:1037–47.

Straley, J. M. 1990. Fall and winter occurrence of humpback whales *(Megaptera novaeangliae)* in southeastern Alaska. In *Individual recognition of cetaceans: Use of photo-identification and other techniques to estimate population parameters,* ed. P. S. Hammond, S. A. Mizroch, and G. P. Donovan, 319–23. Reports of the International Whaling Commission, special issue 12. Cambridge: International Whaling Commission.

Strassman, J. E. 1991. Costs and benefits of colony aggregation in the social wasp, *Polistes annularis. Behav. Ecol.* 2:204–9.

Stresemann, E. 1975. *Ornithology: From Aristotle to the present.* Cambridge, MA: Harvard University Press.

Strier, K. B. 1990. New World primates, new frontiers: Insights from the woolly spider monkey, or muriqui *(Brachyteles arachnoides). Int. J. Primatol.* 11:7–19.

———. 1994. Brotherhoods among atelins: Kinship, affiliation, and competition. *Behaviour* 130:151–67.

———. 1997. Behavioral ecology and conservation biology of primates and other animals. *Adv. Stud. Behav.* 26:101–58.

Stroud, C. 1996. The ethics and politics of whaling. In *The conservation of whales and dolphins: Science and practice,* ed. M. P. Simmonds and J. Hutchinson. Chichester: Wiley.

Struhsaker, T. T., and Leland, L. 1985. Infanticide in a patrilineal society of red colobus monkeys. *Z. Tierpsychol.* 69:89–132.

Sullivan, B. K. 1989. Desert environments and the structure of anuran mating systems. *J. Arid Environ.* 17:175–83.

Sumich, J. L., and Harvey, J. T. 1986. Juvenile mortality in gray whales *(Eschrichtius robustus). J. Mammal.* 67:179–82.

Sussman, R. W. 1979. Preface. In *Primate ecology: Problem-oriented field studies,* ed. R. W. Sussman. New York: John Wiley & Sons.

Swartz, S. 1986. Gray whale migratory, social and breeding behavior. In *Behaviour of whales in relation to management: Incorporating the proceedings of a workshop of the same name held in Seattle, Washington, 19–23 April 1982,* ed. G. P. Donovan, 207–29. Reports of the International Whaling Commission, special issue 8. Cambridge: International Whaling Commission.

Swartz, S. L., and Jones, M. L. 1983. Gray whale *(Eschrichtius robustus)* calf production and mortality in the winter range. *Rep. Int. Whal. Commn.* 34:503–7.

Sweeney, J. C. 1977. Difficult births and neonatal health problems in small cetaceans. In *Breeding dolphins: Present status, suggestions for the future,* ed. S. H. Ridgeway and K. Benirschke. U.S. Marine Mammal Commission Report no. MMC-76/07, Washington, DC.

Swingle, W. M., Barco, S. G., and Pitchford, T. D. 1993. Appearance of juvenile humpback whales feeding in the nearshore waters of Virginia. *Mar. Mamm. Sci.* 9:309–15.

Symington, M. M. 1990. Fission-fusion social organization in *Ateles* and *Pan*. *Int. J. Primatol.* 11:47–61.

Symmes, D., and Biben, M. 1988. Conversational vocal exchanges in squirrel monkeys. In *Primate vocal communication*, ed. D. Todt, P. Goedeking, and D. Symmes. Berlin: Springer-Verlag.

Symons, H. W., and Weston, R. D. 1958. Studies on the humpback whale *(Megaptera nodosa)* in the Bellinghausen Sea. *Norsk Hvalfangst-Tidende* 47:53–81.

Taber, S., and Thomas, P. 1982. Calf development and mother-calf spatial relationships in southern right whales. *Anim. Behav.* 30:1072–83.

Tanabe, S., and Tatsukawa, R. 1992. Chemical modernization and vulnerability of cetaceans: Increasing toxic threat of organochlorine contaminants. In *Persistent pollutants in marine ecosystems*, ed. C. H. Walker and D. R. Livingstone. Oxford: Pergamon Press.

Tanabe, S., Iwata, H., and Tatsukawa, R. 1994. Global contamination by persistent organochlorines and their toxicological impact on marine mammals. *Sci. Total Environ.* 154:163–77.

Tanaka, S. 1987. Satellite radio tracking of bottlenose dolphins *Tursiops truncatus*. *Nippon Suisan Gakkaishi* 53:1327–38.

Tarpy, C. 1979. Killer whale attack. *Nat. Geogr.* 155:542–45.

Tautz, D. 1990. Genomic fingerprinting goes simple. *Bioessays* 12:44–46.

Tavolga, M. C. 1966. Behavior of the bottlenose dolphin *(Tursiops truncatus)*: Social interactions in a captive colony. In *Whales, dolphins, and porpoises*, ed. K. S. Norris. Berkeley: University of California Press.

Tavolga, M. C., and Essapian, F. S. 1957. The behaviour of the bottle-nosed dolphin *(Tursiops truncatus)*: Mating, pregnancy, parturition and mother-infant behaviour. *Zoologica* 42(1):11–31.

Tavolga, W. N. 1964. *Marine bioacoustics*. Vol. 1. Oxford: Pergamon Press.

———. 1969. *Principles of animal behavior*. New York: Harper and Row.

———. 1983. Theoretical principles for the study of communication in cetaceans. *Mammalia* 47(1):3–26.

Tayler, C. K., and Saayman, G. S. 1972. The social organization and behavior of dolphins *(Tursiops truncatus)* and baboons *(Papio ursinus)*: Some comparisons and assessments. *Ann. Cape Prov. Mus. Nat. Hist.* 9:11–49.

Taylor, B. L., and Gerrodette, T. 1993. The uses of statistical power in conservation biology: The vaquita and the northern spotted owl. *Conserv. Biol.* 7:489–500.

Terborgh, J. 1983. *Five New World primates*. Princeton, NJ: Princeton University Press.

Tershy, B. R. 1992. Body size, diet, habitat use, and social behavior of *Balaenoptera* whales in the Gulf of California. *J. Mammal.* 73:477–86.

Thomas, J. A., Zinnel, K. C., and Ferm, L. M. 1983. Analysis of Weddell seal *(Leptonychotes weddelli)* vocalizations using underwater playbacks. *Can. J. Zool.* 61:1448–56.

Thomas, J. A., Fisher, S. R., Ferm, L. M., and Holt, R. S. 1986. Acoustic detection of cetaceans using a towed array of hydrophones. In *Behaviour of whales in relation to management: Incorporating the proceedings of a workshop of the same name held in Seattle, Washington, 19–23 April 1982*, ed. G. P. Donovan, 139–48. Reports of the International Whaling Commission, special issue 8. Cambridge: International Whaling Commission.

Thomas, P. O. 1986. Methodology for behavioural studies of cetaceans: Right whale mother-infant behaviour. In *Behaviour of whales in relation to management: Incorporating the proceedings of a workshop of the same name held in Seattle, Washington, 19–23 April 1982*, ed. G. P. Donovan, 113–19. Reports of the International Whaling Commission, special issue 8. Cambridge: International Whaling Commission.

Thomas, P. O., and Taber, S. M. 1984. Mother-infant interaction and behavioral development in southern right whales, *Eubalaena australis*. *Behaviour* 88:42–60.

Thompson, J. N. 1982. *Interaction and coevolution*. New York: John Wiley & Sons.

Thompson, P., and Mayer, S. 1996. Defining future research needs for cetacean conservation. In *The conservation of whales and dolphins: Science and practice*, ed. M. P. Simmonds and J. Hutchinson. Chichester: Wiley.

Thompson, R. K. R., and Herman, L. M. 1975. Underwater frequency discrimination in the bottlenose dolphin (1–140 kHz) and the human (1–8 kHz). *J. Acoust. Soc. Am.* 57:943–48.

Thompson, S. D. 1993. Zoo research and conservation: Beyond sperm and eggs toward the science of animal management. *Zoo Biol.* 12:155–59.

Thompson, T. J., Winn, H. E., and Perkins, P. J. 1979. Mysticete sounds. In *Behavior of marine mammals: Current perspectives in research*, vol. 3: *Cetaceans*, ed. H. E. Winn and B. L. Olla. New York: Plenum.

Thornhill, R., and Alcock, J. 1983. *The evolution of insect mating systems*. Cambridge, MA: Harvard University Press.

Thurman, G. D., and Williams, M. C. 1986. Neonatal mortality in two Indian Ocean bottlenose dolphins bred in captivity. *Aquat. Mamm.* 123:83–86.

Tikel, D., Blair, D., and Harsh, H. D. 1996. Marine mammal faeces as a source of DNA. *Mol. Ecol.* 5:456–57.

Tillman, M. F., and Breiwick, J. M. 1983. Estimates of abundance for the western North Pacific sperm whale based upon historical whaling records. In *Historical whaling records: Including the proceedings of the International Workshop on Historical Whaling Records, Sharon, MA, Sept. 12–16, 1977*, ed. M. F. Tillman and G. P. Donovan, 257–69. Reports of the International Whaling Commission, special issue 5. Cambridge: International Whaling Commission.

Tinbergen, N. 1958. *Curious naturalists*. London: Country Life.

Todd, S., Stevick, P., Lien, J., Marques, F., and Ketten, D. 1996. Behavioural effects to underwater explosions in humpback whales *(Megaptera novaeangliae)*. *Can. J. Zool.* 74:1661–72.

Tolley, K. A., Read, A. J., Wells, R. S., Urian, K. W., Scott, M. D., Irvine, A. B., and Hohn, A. A. 1995. Sexual dimorphism in wild bottlenose dolphins *(Tursiops truncatus)* from Sarasota, Florida. *J. Mammal.* 76(4):1190–98.

Tomasello, M. 1994. The question of chimpanzee culture. In *Chimpanzee cultures*, ed. R. W. Wrangham, W. C. McGrew, F. B. M. de Waal, and P. G. Heltne. Cambridge, MA: Harvard University Press.

Tormosov, D. D., Mikhaliev, Y. A., Best, P. B., Zemsky, V. A., Sekiguchi, K., and Brownell, R. L. Jr. 1998. Soviet catches of southern right whales, *Eubalaena glacialis,* 1951–1971: Biological data and conservation implications. *Biol. Conserv.* 86: 185–97.

Tovee, M. J., and Cohen-Tovee, E. M. 1993. The neural substrate of face-processing models: A review. *Cog. Neuropsychol.* 10:505–28.

Townsend, C. H. 1914. The porpoise in captivity. *Zoologica* 1: 289–99.

———. 1935. The distribution of certain whales as shown by the logbook records of American whaleships. *Zoologica* 19:1–50.

Traill, T. S. 1809. Description of a new species of whales, *Delphinus melas. Nicholson's J. Nat. Phil. Chem. Arts* 22:81–83.

Trainer, J. M., and McDonald, D. B. 1993. Vocal repertoire of the long-tailed manakin and its relation to male-male cooperation. *Condor* 95:769–81.

Triesman, M. 1975. Predation and the evolution of gregariousness. I. Models of concealment and evasion. *Anim. Behav.* 23: 779–800.

Trillmich, F. 1981. Mutual mother-pup recognition in Galápagos fur seals and sea lions: Cues used and functional significance. *Behaviour* 78:21–42.

Trivers, R. L. 1971. The evolution of reciprocal altruism. *Q. Rev. Biol.* 46:35–57.

———. 1972. Parental investment and sexual selection. In *Sexual selection and the descent of man, 1871–1971,* ed. B. Campbell. Chicago: Aldine.

———. 1974. Parent-offspring conflict. *Am. Zool.* 14:249–64.

———. 1985. *Social evolution.* Menlo Park, CA: Benjamin/Cummings.

Trivers, R. L., and Willard, D. E. 1973. Natural selection of parental ability to vary sex ratio of offspring. *Science* 179:90–92.

True, F. W. 1885. Contributions to the history of the Commander Islands. No. 5. Description of a new species of *Mesoplodon, M. stejnegeri,* obtained by Dr. Leonard Stejneger, in Bering Island. *Proc. US Natl. Mus.* 7:584–85.

———. 1890. Observations on the life history of the bottlenose porpoise. *Proc. US Natl. Mus.* 13(812):197–203.

———. 1903. On some photographs of living finback whales from Newfoundland. *Smithsonian Misc. Coll.* 45:91–94.

———. 1913. Diagnosis of a new beaked whale of the genus Mesoplodon from the coast of North Carolina. *Smithsonian Misc. Coll.* 60(25):1–2.

Turner, G. F., and Pitcher, T. J. 1986. Attack abatement: A model for group protection by combined avoidance and dilution. *Am. Nat.* 128:228–40.

Tutin, G. E. G. 1979. Mating patterns and reproductive strategies in a community of wild chimpanzees *(Pan troglodytes scweinfurthii). Behav. Ecol. Sociobiol.* 6:29–38.

Twiss, J. R., Jr., and Reeves, R. R. (eds.). 1999. *Conservation and management of marine mammals.* Washington, D.C.: Smithsonian Institution Press.

Tyack, P. L. 1981. Interactions between singing Hawaiian humpback whales and conspecifics nearby. *Behav. Ecol. Sociobiol.* 8:105–16.

———. 1982. Humpback whales respond to sounds of their neighbors. Ph.D. dissertation, Rockefeller University, New York.

———. 1983. Differential responses of humpback whales, *Megaptera novaeangliae,* to playback of song or social sounds. *Behav. Ecol. Sociobiol.* 13:49–55.

———. 1986a. Population biology, social behavior, and communication in whales and dolphins. *Trends Ecol. Evol.* 1:144–50.

———. 1986b. Whistle repertoires of two bottlenosed dolphins, *Tursiops truncatus:* Mimicry of signature whistles? *Behav. Ecol. Sociobiol.* 18:251–57.

———. 1989. Let's have less public relations and more ecology. *Oceanus* 32:103–8.

———. 1991a. If you need me, whistle. *Nat. Hist.* Aug., 60–61.

———. 1991b. Use of a telemetry device to identify which dolphin produces a sound. In *Dolphin societies: Discoveries and puzzles,* ed. K. Pryor and K. S. Norris. Berkeley: University of California Press.

———. 1993. Why ethology is necessary for the comparative study of language and communication. In *Language and communication: Comparative perspectives,* ed. H. L. Roitblat, L. M. Herman, and P. Nachtigall. Hillsdale, NJ: Lawrence Erlbaum Associates.

———. 1997. Studying how cetaceans use sound to explore their environment. In *Perspectives in ethology,* vol. 12, 251–97. New York: Plenum Press.

———. 1998. Acoustic communication under the sea. In *Animal acoustic communication: Recent technical advances,* ed. S. L. Hopp, M. J. Owren, and C. S. Evans, 163–220. Heidelberg: Springer-Verlag.

Tyack, P. L., and Sayigh, L. S. 1997. Vocal learning in cetaceans. In *Social influences on vocal development,* ed. C. T. Snowdon and M. Hausberger, 208–33. Cambridge: Cambridge University Press.

Tyack, P. L., and Whitehead, H. 1983. Male competition in large groups of wintering humpback whales. *Behaviour* 83: 132–54.

Tynan, C. T., and DeMaster, D. P. 1997. Observations and predictions of arctic climate change: Potential effects on marine mammals. *Arctic* 50:308–22.

Tyler, S. 1979. Time-sampling: A matter of convention. *Anim. Behav.* 27:801–10.

Uetz, G. W., and Hieber, C. S. 1994. Group size and predation risk in colonial web-building spiders: Analysis of attack abatement mechanisms. *Behav. Ecol.* 5:326–33.

Underwood, R. 1981. Companion preference in an eland herd. *Afr. J. Ecol.* 19:341–54.

Urban, J., and Aguayo, A. 1987. Spatial and seasonal distribution of the humpback whale, *Megaptera novaeangliae,* in the Mexican Pacific. *Mar. Mamm. Sci.* 3:333–44.

Urian, K. W., Duffield D. A., Read A. J., Wells R. S., and Shell, D. D. 1996. Seasonality of reproduction in bottlenose dolphins, *Tursiops truncatus. J. Mammal.* 77:394–403.

Urick, R. J. 1983. *Principles of underwater sound.* New York: McGraw-Hill.

———. 1986. *Ambient noise in the sea.* Los Altos, CA: Peninsula Publishing.

Valsecchi, E. 1997. Genetic studies of the humpback whale using microsatellites. Ph.D. dissertation, Cambridge University, Cambridge, England.

Vandenbergh, J. G., and Coppola, D. M. 1986. The physiology and ecology of puberty modulation by primer pheromones. *Adv. Stud. Behav.* 16:71–107.

Vedder, J. M. 1996. Levels of organochlorine contaminants in milk relative to health of bottlenose dolphins *(Tursiops truncatus)* from Sarasota, Florida. M.Sc. thesis, University of California, Santa Cruz.

Vehrencamp, S. 1983. Optimal degree of skew in cooperative societies. *Am. Zool.* 23:327–35.

Viale, D., Verneau, N., and Tison, Y. 1992. Stomach obstruction in a sperm whale beached on the Lavezzi islands: Macropollution in the Mediterranean. *J. Rech. Oceanogr.* 16:100–102.

Vidal, O., Barlow, J., Hurtado, L. A., Torre, J., Cendón, P., and Ojeda, Z. 1997. Distribution and abundance of the Amazon river dolphin *(Inia geoffrensis)* and the tucuxi *(Sotalia fluviatilis)* in the upper Amazon river. *Mar. Mamm. Sci.* 13:427–45.

Visser, I. N. 1998. Prolific body scars and collapsing dorsal fins on killer whales *(Orcinus orca)* in New Zealand waters. *Aquat. Mamm.* 24:71–81.

de Waal, F. B. M. 1987. Dynamics of social relationships. In *Primate societies,* ed. B. B. Smuts, D. Cheney, R. Seyfarth, R. W. Wrangham, and T. Struhsaker. Chicago: University of Chicago Press.

———. 1993. Reconciliation among primates: A review of empirical evidence and theoretical issues. In *Primate social conflict,* ed. W. A. Mason and S. P. Mendoza. Albany: SUNY Press.

———. 1996. *Good natured: The origins of right and wrong in humans and other animals.* Cambridge, MA.: Harvard University Press.

de Waal, F. B. M., and Luttrell, L. 1986. The similarity principle underlying social bonding among female rhesus monkeys. *Folia Primatol.* 46:215–34.

Wada, S. 1989. Latitudinal segregation of the Okhotsk Sea-West Pacific stock of minke whales. *Rep. Int. Whal. Commn.* 39:229–33.

Wada, S., Kobayashi, T., and Numachi, K. 1991. Genetic variability and differentiation of mitochondrial DNA in minke whales. In *Genetic ecology of whales and dolphins: Incorporating the proceedings of the workshop on the genetic analysis of cetacean populations,* ed. A. R. Hoelzel, 203–15. Reports of the International Whaling Commission, special issue 13. Cambridge: International Whaling Commission.

Wade, P. R. 1995. Revised estimates of incidental kill of dolphins (Delphinidae) by the purse-seine tuna fishery in the eastern tropical Pacific, 1959–1972. *Fish. Bull.* 93:345–54.

van Waerebeek, K. 1992. Population identity and general biology of the dusky dolphin *Lagenorhynchus obscurus* (Gray 1828) in the southeastern Pacific. Ph.D. thesis, University of Amsterdam.

———. 1994. A note on the status of the dusky dolphins *(Lagenorhynchus obscurus)* off Peru. In *Gillnets and cetaceans: Incorporating the proceedings of the symposium and workshop on the mortality of cetaceans in passive fishing nets and traps,* ed. W. F. Perrin, G. Donovan, and J. Barlow, 525–27. Reports of the International Whaling Commission, special issue 15. Cambridge: International Whaling Commission.

van Waerebeek, K., and Read, A. J. 1994. Reproduction of dusky dolphins, *Lagenorhynchus obscurus,* from coastal Peru. *J. Mammal.* 75:1054–62.

van Waerebeek, K., and Reyes, J. C. 1994. Post-ban small cetacean takes off Peru: A review. In *Gillnets and cetaceans: Incorporating the proceedings of the symposium and workshop on the mortality of cetaceans in passive fishing nets and traps,* ed. W. F. Perrin, G. Donovan, and J. Barlow, 503–19. Reports of the International Whaling Commission, special issue 15. Cambridge: International Whaling Commission.

van Waerebeek, K., Reyes, J. C., Read, A. J., and McKinnon, J. S. 1990. Preliminary observations of bottlenose dolphins from the Pacific coast of South America. In *The bottlenose dolphin,* ed. S. Leatherwood and R. R. Reeves. San Diego: Academic Press.

Walker, J. L., and Macko, S. A. 1999. Dietary studies of marine mammals using stable carbon and nitrogen isotopic ratios of teeth. *Mar. Mamm. Sci.* 15:314–34.

Walker, J. L., Potter, C. W., and Macko, S. A. 1999. The diets of modern and historic bottlenose dolphin populations reflected through stable isotopes. *Mar. Mamm. Sci.* 15:335–50.

Walker, L. A., Cornell, L., Dahl, K. D., Czekala, N. M., Dargen, C. M., Joseph, B., Hsueh, A. J. W., and Lasley, B. L. 1988. Urinary concentrations of ovarian steroid hormone metabolites and bioactive follicle-stimulating hormone in killer whales *(Orcinus orca)* during ovarian cycles and pregnancy. *Biol. Reprod.* 39:1013–20.

Walker, W. A. 1981. Geographical variation in morphology and biology of bottlenose dolphins *(Tursiops)* in the eastern North Pacific. National Marine Fisheries Service, Southwest Fisheries Center Administrative Report no. LJ-81-03c.

Wallace, R. L. 1994. The Marine Mammal Commission compendium of selected treaties, international agreements, and other relevant documents on marine resources, wildlife and the environment. Washington, DC: U.S. Marine Mammal Commission.

———. 1997. The Marine Mammal Commission compendium of selected treaties, international agreements, and other relevant documents on marine resources, wildlife and the environment. First update. Bethesda, MD: U.S. Marine Mammal Commission.

Wallis, G. P. 1999. Do animal mitochondrial genomes recombine? *Trends Ecol. Evol.* 14:209–10.

Walsh, M. T., Beusse, D., Bossart, G. D., Young, W. G., Odell, D. K., and Patton, G. W. 1988. Ray encounters as a mortality factor in Atlantic bottlenose dolphins *(Tursiops truncatus).* *Mar. Mamm. Sci.* 4:154–62.

Walters, E. L., Baird, R. W., and Guenther, T. J. 1992. New killer whale "pod" discovered near Victoria. *Victoria Nat.* 49(3):7–8.

Wang, D., Würsig, B., and Evans, W. 1995. Comparisons of whistles among seven odontocete species. In *Sensory systems of aquatic mammals,* ed. R. A. Kastelein, J. A. Thomas, and P. E. Nachtigall. Woerden, Netherlands: De Spil.

Wang, K. R., Payne, P. M., and Thayer, V. G. (compilers). 1994. Coastal Stock(s) of Atlantic Bottlenose Dolphin: Status Review and Management. Proceedings and Recommendations from a Workshop held in Beaufort, NC, 13–14 September 1993. NOAA Technical Memo NMFS-OPR-4.

Waples, D. M. 1995. Activity budgets of free-ranging bottlenose dolphins *(Tursiops truncatus)* in Sarasota Bay, Florida. M.Sc. thesis, University of California, Santa Cruz.

Ward, P., and Zahavi, A. 1973. The importance of certain assemblages of birds as information centres for food-finding. *Ibis* 115:517–34.

Washburn, S. L. 1951. The new physical anthropology. *Trans. NY Acad. Sci.,* ser. 2, 13:298–304.

Washburn, S. L., and DeVore, I. 1961. The social life of baboons. *Sci. Am.* 204(6):62–71.

Waters, S., and Whitehead, H. 1990. Aerial behaviour in sperm whales, *Physeter macrocephalus. Can. J. Zool.* 68:2076–82.

Watkins, W. A. 1967. The harmonic interval: Fact or artifact in spectral analysis of pulse trains. In *Marine bioacoustics,* vol. 2, ed. W. N. Tavolga. Oxford: Pergamon.

———. 1980. Acoustics and the behavior of sperm whales. In *Animal sonar systems,* ed. R.-G. Busnel and J. F. Fish. New York: Plenum Press.

———. 1981. Activities and underwater sounds of fin whales. *Sci. Rep. Whales Res. Inst.* 33:83–117.

Watkins, W. A., and Schevill, W. E. 1968. Underwater playback of their own sounds to *Leptonychotes* (Weddell seals). *J. Mammal.* 49:287–96.

———. 1972. Sound source location by arrival-times on a nonrigid three-dimensional hydrophone array. *Deep-Sea Res.* 19:691–706.

———. 1975. Sperm whales *(Physeter catodon)* react to pingers. *Deep-Sea Res.* 22:123–29.

———. 1977a. The development and testing of a radio whale tag. Technical report. Woods Hole, MA: Woods Hole Oceanographic Institution.

———. 1977b. Spatial distribution of *Physeter catodon* (sperm whales) underwater. *Deep-Sea Res.* 24:693–99.

———. 1977c. Sperm whale codas. *J. Acoust. Soc. Am.* 62:1486–90.

———. 1979. Aerial observation of feeding behavior in four baleen whales: *Eubalaena glacialis, Balaenoptera borealis, Megaptera novaeangliae,* and *Balaenoptera physalus. J. Mammal.* 60:155–63.

Watkins, W. A., and Tyack, P. T. 1991. Reaction of sperm whales *(Physeter catodon)* to tagging with implanted sonar transponder and radio tags. *Mar. Mamm. Sci.* 7:409–13.

Watkins, W. A., Schevill, W. E., and Best, P. B. 1977. Underwater sounds of *Cephalorhynchus heavisidii* (Mammalia: Cetacea). *J. Mammal.* 58:316–20.

Watkins, W. A., Moore, K. E., Wartzok, D., and Johnson, J. H. 1981. Radio tracking of finback *(Balaenoptera physalus)* and humpback *(Megaptera novaeangliae)* whales in Prince William Sound, Alaska. *Deep-Sea Res.* 28:577–88.

Watkins, W. A., Moore, K. E., and Tyack, P. 1985. Sperm whale acoustic behaviors in the southeast Caribbean. *Cetology* 49:1–15.

Watkins, W. A., Tyack, P., Moore, K. E., and Bird, J. E. 1987. The 20-Hz signals of finback *(Balaenoptera physalus). J. Acoust. Soc. Am.* 82:1901–12.

Watkins, W. A., Clark, C. W., and Dahlheim, M. E. 1988. The sounds of sperm whale calves. In *Animal sonar: Processes and performance,* ed. P. E. Nachtigall and P. W. B. Moore. New York: Plenum.

Watkins, W. A., Daher, M. A., Fristrup, K. M., Howald, T. J., and di Sciara, G. N. 1993. Sperm whales tagged with transponders and tracked underwater by sonar. *Mar. Mamm. Sci.* 9:55–67.

Watkins, W. A., Sigurjónsson, J., Wartzok, D., Maiefski, R., Howey, P. W., and Daher, M. A. 1996. Fin whale tracked by satellite off Iceland. *Mar. Mamm. Sci.* 12:564–69.

Watson, L. 1981. *Sea guide to whales of the world.* Scarborough, Ontario: Nelson.

Watts, D. P. 1990. Mountain gorilla life histories, reproductive competition, and sociosexual behavior and some implications for captive husbandry. *Zoo Biol.* 9(3):185–200.

———. 1998. Coalitionary mate guarding by male chimpanzees at Ngogo, Kibale National Park, Uganda. *Behav. Ecol. Sociobiol.* 44:43.

Wcislo, W. T. 1984. Gregarious nesting of a digger wasp as a selfish herd response to a parasitic fly (Hymenoptera: Sphecidae; Diptera: Sacrophagidae). *Behav. Ecol. Sociobiol.* 15:157–60.

Webster, M. S. 1992. Sexual dimorphism, mating system and body size in New World blackbirds (Icterinae). *Evolution* 46:1621–41.

Weilgart, L. S. 1990. Vocalizations of the sperm whale *(Physeter macrocephalus)* off the Galápagos Islands as related to behavioral and circumstantial variables. Ph.D. dissertation, Dalhousie University, Halifax, Nova Scotia.

Weilgart, L. S., and Whitehead, H. 1986. Observations of a sperm whale *(Physeter catodon)* birth. *J. Mammal.* 67:399–401.

———. 1988. Distinctive vocalizations from mature male sperm whales *(Physeter macrocephalus). Can. J. Zool.* 66:1931–37.

———. 1993. Coda vocalizations in sperm whales *(Physeter macrocephalus)* off the Galápagos Islands. *Can. J. Zool.* 71:744–52.

———. 1997. Group-specific dialects and geographical variation in coda repertoire in South Pacific sperm whales. *Behav. Ecol. Sociobiol.* 40:277–85.

Weilgart, L. S., Whitehead, H., and Payne, K. 1996. A colossal convergence. *Am. Sci.* 84:278–87.

Weinrich, M. T. 1991. Stable social associations among humpback whales *(Megaptera novaeangliae)* in the southern Gulf of Maine. *Can. J. Zool.* 69:3012–19.

———. 1995. Humpback whale competitive groups observed on a high-latitude feeding ground. *Mar. Mamm. Sci.* 11:251–54.

Weinrich, M. T., and Kuhlberg, A. E. 1991. Short-term association patterns of humpback whales *(Megaptera novaeangliae)* groups on their southern Gulf of Maine feeding grounds. *Can. J. Zool.* 69:3005–11.

Weinrich, M. T., Lambertsen, R. H., Baker, C. S., Schilling, M. R., and Belt, C. R. 1991. Behavioural responses of humpback whales *(Megaptera novaeangliae)* in the Southern Gulf of Maine to biopsy sampling. In *Genetic ecology of whales and dolphins: Incorporating the proceedings of the workshop on the genetic analysis of cetacean populations,* ed. A. R. Hoelzel, 91–97. Reports of the International Whaling Commission, special issue 13. Cambridge: International Whaling Commission.

Weinrich, M. T., Schilling, M. R., and Belt, C. R. 1992. Evidence for acquisition of a novel feeding behaviour: Lobtailing feeding in humpback whales, *Megaptera novaeangliae. Anim. Behav.* 44:1059–72.

Weller, D. W., Würsig, B., Whitehead, H., Norris, J. C., Lynn, S. K., Davis, R. W., Clauss, N., and Brown, P. 1996. Observations of an interaction between sperm whales and short-finned pilot whales in the Gulf of Mexico. *Mar. Mamm. Sci.* 12:588–94.

Wells, R. S. 1984. Reproductive behavior and hormonal correlates in Hawaiian spinner dolphins, *Stenella longirostris*. In *Reproduction in whales, dolphins and porpoises: Proceedings of the conference, Cetacean Reproduction, Estimating Parameters for Stock Assessment and Management, La Jolla, CA, 28 Nov.– 7 Dec. 1981,* ed. W. F. Perrin, R. L. Brownell Jr., and D. P. DeMaster, 465–72. Reports of the International Whaling Commission, special issue 6. Cambridge: International Whaling Commission.

———. 1986. Population structure of bottlenose dolphins: Behavioral studies of bottlenose dolphins along the central west coast of Florida. Contract Report to National Marine Fisheries Service, Southeast Fisheries Center. Contr. no. 45-WCNF-5-00366. 70 pp.

———. 1991a. Bringing up baby. *Nat. Hist.,* Aug., 56–62.

———. 1991b. Reproductive success and survivorship of free-ranging bottlenose dolphins relative to group size and stability. Abstract, Ninth Biennial Conference on the Biology of Marine Mammals, Chicago.

———. 1991c. The role of long-term study in understanding the social structure of a bottlenose dolphin community. In *Dolphin societies: Discoveries and puzzles,* ed. K. Pryor and K. S. Norris. Berkeley: University of California Press.

———. 1993a. The marine mammals of Sarasota Bay. Chapter 9 in *Sarasota Bay: 1992 Framework for Action.* Sarasota Bay National Estuary Program, 1550 Ken Thompson Parkway, Sarasota, FL 34236.

———. 1993b. Parental investment patterns of wild bottlenose dolphins. In *Proceedings of the Eighteenth International Marine Animal Trainers Association Conference,* ed. N. F. Hecker. Chicago.

———. 1993c. Why all the blubbering? *Bison* 7:11–17.

Wells, R. S., and Scott, M. D. 1990. Estimating bottlenose dolphin population parameters from individual identification and capture-release techniques. In *Individual recognition of cetaceans: Use of photo-identification and other techniques to estimate population parameters,* ed. P. S. Hammond, S. A. Mizroch, and G. P. Donovan, 407–16. Reports of the International Whaling Commission, special issue 12. Cambridge: International Whaling Commission.

———. 1997. Seasonal incidence of boat strikes on bottlenose dolphins near Sarasota, Florida. *Mar. Mamm. Sci.* 13:475–80.

———. 1999. Bottlenose dolphin—*Tursiops truncatus* (Montagu, 1821). In *Handbook of marine mammals,* vol. 6, *The second book of dolphins and porpoises,* ed. S. H. Ridgway and R. Harrison. San Diego: Academic Press.

Wells, R. S., Irvine, A. B., and Scott, M. D. 1980. The social ecology of inshore odontocetes. In *Cetacean behavior: Mechanisms and functions,* ed. L. M. Herman. New York: John Wiley & Sons.

Wells, R. S., Scott, M. D., and Irvine, A. B. 1987. The social structure of free-ranging bottlenose dolphins. In *Current mammalogy,* vol. 1., ed. H. Genoways. New York: Plenum Press.

Wells, R. S., Hansen, L. J., Baldridge, A., Dohl, T. P., Kelly, D. L., and Defran, R. H. 1990. Northward extension of the range of bottlenose dolphins along the California coast. In *The bottlenose dolphin,* ed. S. Leatherwood and R. R. Reeves. San Diego: Academic Press.

Wells, R. S., Rhinehart, H. L., Sweeney, J., Townsend, F., Casper, D., and Hansen, L. 1995a. Assessment of the health of bottlenose dolphin populations in Sarasota Bay, FL, and Matagorda Bay, TX. Abstract, Eleventh Biennial Conference on the Biology of Marine Mammals, Orlando, FL.

Wells, R. S., Urian, K. W., Read, A. J., Bassos, M. K., Carr, W. J., and Scott, M. D. 1995b. Low-level monitoring of bottlenose dolphins, *Tursiops truncatus,* in Tampa Bay, Florida: 1988–1993. Final contract report to the National Marine Fisheries Service, Southeast Fisheries Center, Miami, FL. Contract nos. 50-WCNF-7-06083 and 50-WCNF-3-06098.

Wells, R. S., Bassos, M. K., Urian, K. W., Carr, W. J., and Scott, M. D. 1996. Low-level monitoring of bottlenose dolphins, *Tursiops truncatus,* in Charlotte Harbor, Florida: 1990–1994. Final contract report to the National Marine Fisheries Service, Southeast Fisheries Center, Miami, FL. Contract no. 50-WCNF-0-06023.

Wemmer, C., and Thompson, S. 1995. A short history of scientific research in zoological gardens. In *The ark evolving: Zoos and aquariums in transition,* ed. C. M. Wemmer. Front Royal, VA: Smithsonian Institution, Conservation and Research Center.

Werth, A. J. 1989. Suction feeding in odontocetes: Water flow and head shape. *Am. Zool.* 29:92

———. 1991. Suction feeding in pilot whales: Kinematic and anatomical evidence. *Am. Zool.* 31:17A.

West-Eberhard, M. J. 1969. The social biology of polistine wasps. *Misc. Publ. Mus. Zool. Univ. Mich.* 140:1–101.

———. 1975. The evolution of social behavior by kin selection. *Q. Rev. Biol.* 50:1–33.

Western, D. 1979. Size, life history and ecology in mammals. *Afr. J. Ecol.* 17:185–204.

Westgate, A. J., Read, A. J., Berggren, P., Koopman, H. N., and Gaskin, D. E. 1995. Diving behaviour of harbour porpoises, *Phocoena phocoena. Can. J. Fish Aquat. Sci.* 52:1064–73.

White, D., Cameron, N., Spong, P., and Bradford, J. 1971. Visual acuity of the killer whale *(Orcinus orca). Exp. Neurol.* 32: 230–36.

White, T. J., Arnheim, N., and Erlich, H. A. 1989. The polymerase chain reaction. *Trends Genet.* 5:185–89.

Whitehead, H. P. 1981. The behaviour and ecology of the humpback whale in the Northwest Atlantic. Ph.D. dissertation, Cambridge University, Cambridge, England.

———. 1983. Structure and stability of humpback whale groups off Newfoundland. *Can. J. Zool.* 61:1391–97.

———. 1985a. Humpback whale breaching. *Invest. Cetacea* 17: 117–55.

———. 1985b. Humpback whale songs from the North Indian Ocean. *Invest. Cetacea* 17:157–62.

———. 1985c. Why whales leap. *Sci. Am.* 242:84–93.

———. 1987a. Social organization of sperm whales off the Galàpagos: Implications for management and conservation. *Rep. Int. Whal. Commn.* 37:195–99.

———. 1987b. Updated status of the humpback whale, *Megaptera novaeangliae*, in Canada. *Can. Field Nat.* 101:284–94.

———. 1989a. Formations of foraging sperm whales, *Physeter macrocephalus*, off the Galápagos Islands. *Can. J. Zool.* 67(9): 2131–39.

———. 1989b. *Voyage to the whales.* Toronto: Stoddart Publishing Company, Ltd.

———. 1990a. Assessing sperm whale populations using natural markings: Recent progress. In *Individual recognition of cetaceans: Use of photo-identification and other techniques to estimate population parameters,* ed. P. S. Hammond, S. A. Mizroch, and G. P. Donovan, 377–82. Reports of the International Whaling Commission, special issue 12. Cambridge: International Whaling Commission.

———. 1990b. Computer assisted individual identification of sperm whale flukes. In *Individual recognition of cetaceans: Use of photo-identification and other techniques to estimate population parameters,* ed. P. S. Hammond, S. A. Mizroch, and G. P. Donovan, 71–77. Reports of the International Whaling Commission, special issue 12. Cambridge: International Whaling Commission.

———. 1990c. Rules for roving males. *J. Theor. Biol.* 145:355–68.

———. 1993. The behaviour of mature male sperm whales on the Galápagos breeding grounds. *Can. J. Zool.* 71:689–99.

———. 1994. Delayed competitive breeding in roving males. *J. Theor. Biol.* 166:127–33.

———. 1995a. Investigating structure and temporal scale in social organizations using identified individuals. *Behav. Ecol.* 6: 199–208.

———. 1995b. Status of Pacific sperm whale stocks before modern whaling. *Rep. Int. Whal. Commn.* 45:407–12.

———. 1996a. Babysitting, dive synchrony, and indications of alloparental care in sperm whales. *Behav. Ecol. Sociobiol.* 38: 237–44.

———. 1996b. Variation in the feeding success of sperm whales: Temporal scale, spatial scale and relationship to migrations. *J. Anim. Ecol.* 65:429–38.

———. 1997. Analyzing animal social structure. *Anim. Behav.* 53:1053–67.

———. 1998. Cultural selection and genetic diversity in matrilineal whales. *Science* 282:1708–11.

———. 1999. Testing association patterns of social animals. *Anim. Behav.* 57:F26–29.

Whitehead, H., and Arnbom, T. 1987. Social organization of sperm whales off the Galápagos Islands, February-April 1985. *Can. J. Zool.* 65(4):913–19.

Whitehead, H., and Carlson, C. 1988. Social behaviour of feeding finback whales off Newfoundland: Comparisons with the sympatric humpback whale. *Can J. Zool.* 66:217–21.

Whitehead, H., and Carscadden, J. E. 1985. Predicting inshore whale abundance—whales and capelin off the Newfoundland coast. *Can. J. Fish. Aquat. Sci.* 42:976–81.

Whitehead, H., and Dufault, S. 1999. Techniques for analyzing vertebrate social structure using identified individuals: Review and recommendations. *Adv. Stud. Behav.* 28:33–74.

Whitehead, H. P., and Glass, C. 1985. Orcas (killer whales) attack humpback whales. *J. Mammal.* 66:183–85.

Whitehead, H., and Gordon, J. 1986. Methods of obtaining data for assessing and modelling sperm whale populations which do not depend on catches. In *Behaviour of whales in relation to management: Incorporating the proceedings of a workshop of the same name held in Seattle, Washington, 19–23 April 1982,* ed. G. P. Donovan, 149–65. Reports of the International Whaling Commission, special issue 8. Cambridge: International Whaling Commission.

Whitehead, H., and Kahn, B. 1992. Temporal and geographical variation in the social structure of female sperm whales. *Can. J. Zool.* 70:2145–49.

Whitehead, H. P., and Moore, M. J. 1982. Distribution and movements of West Indian humpback whales in winter. *Can. J. Zool.* 60:2203–11.

Whitehead, H., and Payne, R. 1981. New techniques for assessing population of right whales without killing them. In *Mammals in the seas,* ed. J. G. Clark. FAO Fisheries Series no. 5, vol. 3. Rome: Food and Agriculture Organization of the United Nations.

Whitehead, H., and Waters, S. 1990. Social organisation and population structure of sperm whales off the Galápagos Islands, Ecuador (1985 and 1987). In *Individual recognition of cetaceans: Use of photo-identification and other techniques to estimate population parameters,* ed. P. S. Hammond, S. A. Mizroch, and G. P. Donovan, 249–57. Reports of the International Whaling Commission, special issue 12. Cambridge: International Whaling Commission.

Whitehead, H., and Weilgart, L. 1990. Click rates from sperm whales. *J. Acoust. Soc. Am.* 87:1798–1806.

———. 1991. Patterns of visually observable behavior and vocalizations in groups of female sperm whales. *Behaviour* 118: 275–96.

Whitehead, H., Gordon, J., Mathews, E. A., and Richard, K. R. 1990. Obtaining skin samples from living sperm whales. *Mar. Mamm. Sci.* 6:316–26.

Whitehead, H., Waters, S., and Lyrholm, T. 1991. Social organization of female sperm whales and their offspring: Constant companions and casual acquaintances. *Behav. Ecol. Sociobiol.* 29:385–89.

Whitehead, H., Brennan, S., and Grover, D. 1992. Distribution and behaviour of male sperm whales on the Scotian Shelf, Canada. *Can. J. Zool.* 70:912–18.

Whitehead, H., Christal, J., and Dufault, S. 1997a. Past and distant whaling and the rapid decline of sperm whales off the Galápagos Islands. *Conserv. Biol.* 11:1387–96.

Whitehead, H., Faucher, A., Gowans, S., and McCarrey, S. 1997b. Status of the northern bottlenose whale, *Hyperoodon ampullatus*, in the Gully, Nova Scotia. *Can. Field Nat.* 111:287–92.

Whitehead, H., Gowans, S., Faucher, A., and McCarrey S. 1997c. Population of northern bottlenose whales in the Gully, Nova Scotia. *Mar. Mamm. Sci.* 13:173–85.

Whitehead, H., Dillon, M., Dufault, S., Weilgart, L., and Wright, J. 1998. Non-geographically based population structure of South Pacific sperm whales: Dialects, fluke-markings and genetics. *J. Anim. Ecol.* 67:253–62.

Wiebe, P. 1995. Developing a high-frequency system to remotely "see" plankton distributions. *Oceanus* 38:14–17.

Wiens, J. A. 1982. Song pattern variation in the sage sparrow *(Amphispiza belli):* Dialects or epiphenomenon? *Auk* 99:208–29.

Wiley, D. N., and Clapham, P. J. 1993. Does maternal condition affect the sex ratio of offspring in humpback whales? *Anim. Behav.* 46:321–24.

Wiley, R. H. 1978. Lek mating system of the sage grouse. *Sci. Am.* 238:114–25.

Williams, G. C. 1957. Pleiotropy, natural selection, and the evolution of senescence. *Evolution.* 11:398–411.

———. 1964. Measurement of consociation among fishes and comments on the evolution of schooling. *Publ. Mus. Mich. State Univ. Biol. Ser.* 2:349–84.

———. 1966. *Adaptation and natural selection.* Princeton, NJ: Princeton University Press.

Williams, T. M. 1999. The evolution of cost efficient swimming in marine mammals: Limits to energetic optimization. *Phil. Trans. R. Soc. Lond.* 354:193–201.

Williams, T. M., Friedl, W. A., Fong, M. L., Yamada, R. M., Dedivy, P., and Haun, J. E. 1992. Travel at low energetic cost by swimming and wave-riding bottlenose dolphins. *Nature* 355:821–23.

Williams, T. M., Friedl, W. A., and Haun, J. E. 1993. The physiology of bottlenose dolphins *(Tursiops truncatus):* Heart rate, metabolic rate and plasma lactate concentration during exercise. *J. Exp. Biol.* 179:31–46.

Williamson, K. 1945. The economic and ethnological importance of the Caaing Whale, *Globicephala melaena* Traill, in the Faeroe Islands. *North Western Nat.* 20:118–36.

Willson, M. F., and Pianka, E. F. 1963. Sexual selection, sex ratio and mating system. *Am. Nat.* 97:405–7.

Wilson, B. 1995. The ecology of bottlenose dolphins in the Moray Firth, Scotland: A population at the northern extreme of the species' range. Ph.D. thesis, University of Aberdeen, Scotland.

Wilson, B., Thompson, P., and Hammond, P. 1992. The ecology of bottle-nosed dolphins, *Tursiops truncatus,* in the Moray Firth. In *European research on cetaceans: Proceedings of the sixth annual conference of the European Cetacean Society,* ed. P. G. H. Evans. Cambridge: European Cetacean Society.

Wilson, B., Thompson, P., and Hammond, P. 1993. An examination of the social structure of a resident group of bottle-nosed dolphins *(Tursiops truncatus)* in the Moray Firth, N. E. Scotland. In *European research on cetaceans: Proceedings of the seventh annual conference of the European Cetacean Society,* ed. P. G. H. Evans. Cambridge: European Cetacean Society.

Wilson, B., Ross, H. M., Thompson, P. M., and Reid, R. J. 1995. Violent dolphin-porpoise interactions—fun, fear, food, or freaks? Abstract, Eleventh Biennial Conference on the Biology of Marine Mammals, Orlando, FL.

Wilson, B., Thompson, P. M., and Hammond, P. S. 1997. Habitat use by bottlenose dolphins: Seasonal distribution and stratified movement patterns in the Moray Firth, Scotland. *J. Appl. Ecol.* 34:1365–74.

Wilson, E. O. 1975. *Sociobiology: The new synthesis.* Cambridge, MA: Harvard University Press.

———. 1984. *Biophilia.* Cambridge, MA: Harvard University Press.

Wilson, F. R. 1998. *The Hand.* New York: Pantheon.

Wilson, R. P., Ryan, P. G., James, A., and Wilson, M. P. T. 1987. Conspicuous coloration may enhance prey capture in some piscivores. *Anim. Behav.* 35:1558–60.

Winn, H. E., and Olla, B. L. 1979. Introduction. In *Behavior of marine mammals: Current perspectives in research,* vol. 3, *Cetaceans.* New York: Plenum Press.

Winn, H. E., and Winn, L. K. 1978. The song of the humpback whale, *Megaptera novaeangliae,* in the West Indies. *Mar. Biol.* 47:97–114.

Winn, H. E., Edel, R. K., and Taruski, A. G. 1975. Population estimate of the humpback whale *(Megaptera novaeangliae)* in the West Indies by visual and acoustic techniques. *J. Fish. Res. Bd. Can.* 32:499–506.

Winn, H. E., Goodyear, J. D., Kenney, R. D., and Petricig, R. O. 1995. Dive patterns of tagged right whales in the Great South Channel. *Cont. Shelf Res.* 4/5:593–611.

Wisenden, B. D., and Keenleyside, M. H. A. 1992. Intraspecific brood adoption in convict cichlids: A mutual benefit. *Behav. Ecol. Sociobiol.* 31:263–69.

Withrow, D., and Angliss, R. 1992. Length frequency of bowhead whales from spring aerial photogrammetric surveys in 1985, 1986, 1989 and 1990. *Rep. Int. Whal. Commn.* 42:463–67.

Wolff, J. O. 1992. Parents suppress reproduction and stimulate dispersal in opposite-sex juvenile white-footed mice. *Nature* 359(6394):409–10.

Wong, Z., Wilson, V., Jeffreys, A. J., and Thein, S. L. 1986. Cloning a selected fragment from a human DNA "fingerprint": Isolation of an extremely polymorphic minisatellite. *Nucleic Acids Res.* 14:4605–16.

Wong, Z., Wilson, V., Patel, I., Povey, S., and Jeffreys, A. J. 1987. Characterization of a panel of highly variable minisatellites cloned from human DNA. *Ann. Hum. Genet.* 51:269–88.

Wood, F. G. 1953. Underwater sound production and concurrent behavior of captive porpoises, *Tursiops truncatus* and *Stenella plagiodon. Bull. Mar. Sci. Gulf Caribbean* 3:120–33.

———. 1973. *Marine mammals and man: The Navy's porpoises and sea lions.* New York: Robert B. Luce.

———. 1977. Birth of porpoises at Marineland, Florida, 1939 to 1969, and comments on problems involved in captive breeding of small Cetacea. In *Breeding dolphins: Present status, suggestions for the future,* ed. S. H. Ridgway and K. Benirschke. U.S. Marine Mammal Commission Report no. MMC-76/07, Washington, DC.

———. 1986. Social behavior and foraging strategies of dolphins. In *Dolphin cognition and behavior: A comparative approach,* ed. R. J. Schusterman, J. A. Thomas, and F. G. Wood. Hillsdale, NJ: Lawrence Erlbaum Associates.

Wood, F. G. Jr., Caldwell, D. K., and Caldwell, M. C. 1970. Behavioral interactions between porpoises and sharks. In *Investigations on Cetacea,* vol. 2. Berne: Brain Anatomy Institute.

Woodroffe, R., and Vincent, A. 1994. Mother's little helpers: Patterns of male care in mammals. *Trends Ecol. Evol.* 9:294–97.

Woodward, B. B. (ed.) 1903. *Catalogue of the books, manuscripts, maps, and drawings in the British Museum (Natural History).*

Vol. 2, E–K, 501–1038. London: British Museum (Natural History).

Wooster, H., Garvin, P. L., Callimahos, L. D., Lilly, J. C., Davis, W. O., and Heyden, F. J. 1966. Communication with extraterrestrial intelligence. *IEEE Spectrum* 3(3):153–63.

Worthy, G. A. J., and Edwards, E. F. 1990. Morphometric and biochemical factors affecting heat loss in a small temperate cetacean *(Phocoena phocoena)* and a small tropical cetacean *(Stenella attenuata)*. *Physiol. Zool.* 63:432–42.

Wrangham, R. W. 1979. On the evolution of ape social systems. *Soc. Sci. Info.* 18:335–68.

———. 1980a. An ecological model of female-bonded primate groups. *Behaviour* 75:262–92.

———. 1980b. Female choice of least costly males: A possible factor in the evolution of leks. *Z. Tierpsychol.* 54:357–67.

———. 1982. Mutualism, kinship, and social evolution. In *Current problems in sociobiology,* ed. Kings College Sociobiology Group. Cambridge: Cambridge University Press.

———. 1986. Ecology and social relationships in two species of chimpanzee. In *Ecological aspects of social evolution,* ed. D. I. Rubenstein and R. W. Wrangham. Princeton, NJ: Princeton University Press.

———. 1987. Evolution of social structure. In *Primate societies,* ed. B. B. Smuts, D. Cheney, R. Seyfarth, R. W. Wrangham, and T. Struhsaker. Chicago: University of Chicago Press.

———. 1993. The evolution of sexuality in chimpanzees and bonobos. *Hum. Nat.* 4(1):47–79.

Wrangham, R. W. 1999. Why are male chimpanzees more gregarious than mothers? A scramble competition hypoethesis. In *Male Primates,* ed. P. Kappeler. Cambridge: Cambridge University Press.

Wrangham, R. W., and Peterson, D. 1996. *Demonic males: Apes and the origins of human violence.* Boston: Houghton Mifflin.

Wrangham, R. W., and Rubenstein, D. I. 1986. Social evolution in birds and mammals. In *Ecological aspects of social evolution,* ed. D. I. Rubenstein and R. W. Wrangham. Princeton, NJ: Princeton University Press.

Wrangham, R. W., and Smuts, B. B. 1980. Sex differences in the behavioural ecology of chimpanzees in Gombe National Park, Tanzania. *J. Reprod. Fert. Suppl.* 28:13–31.

Wrangham, R. W., Clark, A. P., and Isabirye-Basuta, G. 1992. Female social relationships and social organization of Kibale Forest chimpanzees. In *Topics in primatology,* vol. 1., ed. T. Nishida, W. C. McGrew, P. Marler, M. Pickford, and F. B. M. de Waal. Tokyo: University of Tokyo Press.

Wrangham, R. W., Gittleman, J. L., and Chapman, C. A. 1993. Constraints on group size in primates and carnivores: Population density and day-range as assays of exploitation competition. *Behav. Ecol. Sociobiol.* 32:199–209.

Wrangham, R. W., McGrew, W. C., de Waal, F. B. M., and Heltne, P. G. 1994. *Chimpanzee cultures.* Cambridge, MA: Harvard University Press.

Wray, P., and Martin, K. R. 1983. Historical whaling records from the Western Indian Ocean. In *Historical whaling records: Including the proceedings of the International Workshop on Historical Whaling Records, Sharon, MA, Sept. 12–16, 1977,* ed. M. F. Tillman and G. P. Donovan, 213–41. Reports of the International Whaling Commission, special issue 5. Cambridge: International Whaling Commission.

Würsig, B. 1978. Occurrence and group organization of Atlantic bottlenose porpoises *(Tursiops truncatus)* in an Argentine bay. *Biol. Bull.* 154:348–59.

———. 1982. Radio tracking dusky porpoises in the South Atlantic. In *Mammals in the seas.* FAO Fisheries Series no. 5, vol. 4. Rome: Food and Agriculture Organization of the United Nations.

———. 1986. Delphinid foraging strategies. In *Dolphin cognition and behavior: A comparative approach,* ed. R. J. Schusterman, J. A. Thomas, and F. G. Wood. Hillsdale, NJ: Lawrence Erlbaum Associates.

Würsig, B., and Clark, C. 1993. Behavior. In *The bowhead whale,* ed. J. J. Burns, J. J. Montague, and C. J. Cowles, 157–99. Lawrence, KS: Society for Marine Mammalogy.

Würsig, B., and Jefferson, T. A. 1990. Methods of photo-identification for small cetaceans. In *Individual recognition of cetaceans: Use of photo-identification and other techniques to estimate population parameters,* ed. P. S. Hammond, S. A. Mizroch, and G. P. Donovan, 43–52. Reports of the International Whaling Commission, special issue 12. Cambridge: International Whaling Commission.

Würsig, B., and Würsig, M. 1977. The photographic determination of group size, composition and stability of coastal porpoises *(Tursiops truncatus)*. *Science* 198:755–56.

———. 1979. Behavior and ecology of the bottlenose dolphin, *Tursiops truncatus,* in the South Atlantic. *Fish. Bull.* 77:399–412.

———. 1980. Behavior and ecology of the dusky dolphin, *Lagenorhynchus obscurus,* in the South Atlantic. *Fish. Bull.* 77:871–90.

Würsig, B., Dorsey, E. M., Fraker, M. A., Payne, R. S., Richardson, W. J., and Wells, R. S. 1984. Behavior of bowhead whales, *Balaena mysticetus,* summering in the Beaufort Sea: Surfacing, respiration, and dive characteristics. *Can. J. Zool.* 62:1910–21.

Würsig, B., Dorsey, E. M., Fraker, M. A., Payne, R. S., and Richardson, W. J. 1985. Behavior of Bowhead whales, *Balaena mysticetus,* summering in Beaufort Sea: A description. *Fish. Bull.* 83:357–77.

Würsig, B., Dorsey, E. M., Richardson, W. J., and Wells, R. S. 1989. Feeding, aerial and play behaviour of the bowhead whale, *Balaena mysticetus,* summering in the Beaufort Sea. *Aquat. Mammal.* 15:27–37.

Würsig, B., Kieckhefer, T. R., and Jefferson, T. A. 1990. Visual displays for communication in cetaceans. In *Sensory abilities of cetaceans,* ed. J. Thomas and R. Kastelein. New York: Plenum.

Würsig, B., Cipriano, F., and Würsig, M. 1991. Dolphin movement patterns: Information from radio and theodolite tracking studies. In *Dolphin societies: Discoveries and puzzles,* ed. K. Pryor and K. S. Norris. Berkeley: University of California Press.

Würsig, B., Guerrero, J., and Silber, G. K. 1993. Social and sexual behavior of bowhead whales in fall in the Western Arctic: A re-examination of seasonal trends. *Mar. Mamm. Sci.* 9:103–10.

Würsig, B., Wells, R. S., and Norris, K. S. 1994. Food and feeding. In *The Hawaiian spinner dolphin,* ed. K. S. Norris, B. Würsig, R. S. Wells, and M. Würsig. Berkeley: University of California Press.

Wynne-Edwards, V. C. 1962. *Animal dispersion in relation to social behaviour.* Edinburgh, Scotland: Oliver and Boyd.

Yablokov, A. V. 1963. Types of colour of the Cetacea. *Bull. Moscow Soc. Nat. Biol.* 68:27–41. (Fisheries Research Board of Canada Translation Series 1239.)

———. 1972. Footnote. In *The sperm whale,* ed. A. A. Berzin. Jerusalem: Israel Program for Scientific Translations.

———. 1994. Validity of whaling data. *Nature* 367:108.

Yablokov, A. V., Zemsky, V. A., Mikhalev, Y. A., Toromsov, V. V., and Berzin, A. A. 1998. Data on Soviet whaling in the Antarctic in 1947–1972 (population aspects). *Russ. J. Ecol.* 29:38–42.

Yoshioka, M., Mohri, E., Tobayama, T., Aida, K., and Hanyu, I. 1986. Annual changes in serum reproductive hormone levels in the captive female bottle-nosed dolphins. *Bull. Jap. Soc. Sci. Fish.* 52:1939–46.

Young, S., Watt, P. J., Grover, J. P., and Thomas, D. 1994. The unselfish swarm? *J. Anim. Ecol.* 63:611–18.

Zemsky, V. A., Berzin, A. A., Mikhaliev, Y. A., and Tormosov, D. D. 1995. Soviet Antarctic pelagic whaling after WWII: Review of actual catch data. *Rep. Int. Whal. Commn.* 46:131–45.

Zenkovich, B. A. 1962. Sea mammals as observed by the round-the-world expedition of the Academy of Sciences of the USSR in 1957/58. *Norsk Hvalfangst-Tidende* 5:198–210.

Zoloth, S. R., Petersen, M. R., Beecher, M. D., Green, S., Marler, P., Moody, D. B., and Stebbins, W. 1979. Species-specific perceptual processing of vocal sounds by monkeys. *Science* 204:870–72.

Zouros, E., Freeman, K. R., Ball, A. O., and Pogson, G. H. 1992. Direct evidence for extensive paternal mitochondrial DNA inheritance in the marine mussel *Mytilus. Nature* 359:412–14.

Zuk, M., Johnson, K., Thornhill, R., and Ligon, J. D. 1990. Mechanisms of female choice in red jungle fowl. *Evolution* 44:477–85.

CITATION INDEX

Dill, L. M., 57, 62, 127, 129, 131, 132, 134, 135, 136, 139, 140, 141, 142, 143, 146, 148, 150, 151, 204, 211, 212, 216, 267
Dillon, M. C., 161, 164, 168
di Natale, A., 166, 172
di Sciara, G. N., 172
Dixon, A. F., 255
Dizon, A. E., 81, 82, 87
Dohl, T. P., 35, 102, 148, 204, 208, 210, 211, 275, 338
Doidge, D. W., 249
Dolphin, W. F., 76, 179, 180
Donoghue, M., 331
Donovan, G. P., 46, 47, 327, 331
d'Orbigny, A. C. V. D., 345
Dorsey, E. M., 35, 44, 56, 181, 190, 252, 256, 277
dos Santos, M. E., 94, 96, 104
Douglas-Hamilton, I., 20, 34, 44
Douglas-Hamilton, O., 20, 34, 44
Dover, G. A., 81, 82, 83, 134, 267
Dreher, J. J., 208, 300
Drinnan, R. L., 237
Dudgeon, D. E., 78
Dudok van Heel, W. H., 337
Dudzinski, K. M., 86
Dufault, S., 40, 67, 73, 81, 85, 86, 87, 154, 159, 161, 165, 171
Duffield, D. A., 47, 81, 82, 83, 85, 93, 95, 104, 110, 114, 120, 125, 144, 145, 263
Duffus, D. A., 152
Dunbar, M. J., 247, 252
Dunbar, R. I. M., 59, 200, 231, 255, 337
Duncan, P., 209
Dunn, D. G., 107, 119, 256
Dunning, D. J., 283
Duvernoy, G. L., 344
D'Vincent, C. G., 179, 188, 212, 218
Dziedzic, A., 24, 295

Earle, M., 311, 316
Eastcott, A., 104, 240
Eaton, R. L., 305
Eberhard, W. G., 241, 255
Economos, A. C., 245
Edds, P. L., 283
Edwards, E. F., 235
Eggert, L. S., 82

Ehrenberg, C. G., 347
Ehrlich, P. R., 330
Einhorn, R. N., 28
Eisenberg, J. F., 274
Ellifrit, D., 129, 133
Ellis, D. D., 279
Ellis, R., 16
Ellison, W. T., 78, 200, 283, 305
Emlen, J. T., 294, 295
Emlen, S. T., 192, 203, 247
Enquist, M., 252
Epple, G., 221
Essapian, F. S., 23, 24, 97, 104, 221, 239
Evans, P. G. H., 171, 222, 223, 225
Evans, W. E., 32, 37, 208, 214, 249, 280, 300, 305
Everett, W., 314
Eyre, E. J., 188

Fagen, R., 148
Fairbairn, D. J., 248
Fairbanks, L. A., 118, 221, 234
Falls, J. B., 37
Fanshawe, J. H., 208
Faucher, A., 259
Feinholz, D. M., 103
Félix, F., 94, 172
Felleman, F. L., 130, 137, 138, 140, 210
Ferrari, M. J., 47, 178, 181, 183, 185, 193, 196
Fertl, D., 94, 96, 211, 313
Fichtelius, K., 29
Findlay, K. P., 178
Fine, M. L., 291
Finley, K. J., 178, 318
Fish, J. F., 76, 284
Fisher, H. D., 77, 127, 133
Fisher, R. A., 193, 235
Fitch, W. T., 291
FitzGibbon, C. D., 208
Fletcher, S., 74, 75, 78, 80, 282, 334
Flórez-González, L., 138, 178, 180
Flower, W. H., 344
Ford, J. K. B., 35, 38, 55, 77, 127, 129, 130, 133, 134, 135, 136, 137, 138, 140, 144, 145, 146, 148, 150, 243, 268, 284, 296, 297, 298
Forney, K. A., 326
Forster, J. R., 344
Foster, M. S., 119
Foster, T. C., 204, 207, 209

Foster, W. A., 205, 206
Fowler, C. W., 231
Fragaszy, D. M., 63
Frank, L. G., 10
Frank, S. A., 203
Frankel, A. S., 76, 77, 78, 87, 188, 262, 285, 286
Frantzis, A., 318
Fraser, F. C., 14, 92, 346
Frazer, J. F. D., 173, 191
Fredericson, E., 57
Friedl, W. A., 275
Fristrup, K. M., 79, 139, 162, 283
Fruth, B., 268
Fujino, K., 194
Fuller, K., 332
Furuichi, T., 121, 268

Gabriele, C. M., 185, 189
Gadgil, M., 18
Gagneux, P., 265
Gaillard, J. M., 236
Gaimard, J. P., 346
Galdikas, B. M. F., 242
Galef, B. G. Jr., 29
Gales, N. J., 230
Gambell, R., 15, 46, 230, 234
Gamboa, G. J., 258
Gannon, D. P., 229
Gans, C., 1
Gardiner, J., 255
Gardner, M. B., 275
Gaskin, D. E., 29, 31, 37, 43, 162, 209, 214, 217, 221, 338, 339
Geist, V., 249, 310
George, J. C., 200, 233, 283, 305
Geraci, J. R., 178, 181, 196
Gerrodette, T., 324, 326, 329
Gerson, H. B., 247
Gervais, F. L. P., 344, 345, 347
Ghandour, G., 58, 59
Giard, J., 74
Gibbons, J., 311
Gifford, T., 24, 29, 31, 107, 243, 276, 277, 289, 334
Gihr, M., 102
Gill, P. C., 131, 178, 196
Gill, T., 346
Ginsberg, J. R., 55, 69, 112
Giraldeau, L.-A., 150
Gish, S. L., 302
Gittleman, J. L., 192, 234
Glass, B., 329
Glass, C., 180, 209
Glockner, D. A., 37, 176, 178, 185, 187, 285

Glockner-Ferrari, D. A., 47, 178, 181, 183, 185, 193, 196
Godin, J. G. J., 206, 208
Goldberg, T. L., 266
Goldsmith, M. L., 215
Goley, P. D., 136, 218
Gomendio, M., 242
Gompper, M. E., 200
Gonzalez, F. T., 52
Goodall, J., 2, 34, 35, 44, 46, 119, 121, 233, 265, 266
Goodall, R. N. P., 253, 321
Goodman, D., 47
Goodnight, K. F., 84
Goodson, A. D., 280
Goodyear, J. D., 75, 76
Goold, J. C., 77, 80, 159, 162, 167, 168, 282, 291, 305
Gordon, D. M., 19
Gordon, J. C. D., 35, 39, 44, 69, 77, 78, 80, 156, 157, 159, 160, 162, 165, 166, 167, 168, 172, 223, 237, 282, 291, 295, 305, 324
Gosling, L. M., 178
Gottman, J. M., 54
Gould, J. L., 9
Gould, S. J., 22, 219
Gowans, S., 93, 217, 239, 251, 259
Gray, D., 251
Gray, J. A. B., 207
Gray, J. E., 14, 343, 345, 346, 347
Greene, C. H., 335
Greenwood, P. J., 149, 182, 216
Groombridge, B., 319, 321
Gross, C. G., 300
Gross, M. R., 123, 262
Grzimek, B., 245
Guinee, L. N., 188, 292
Guinet, C., 127, 129, 131, 132, 136, 138, 139, 140, 141, 142, 144, 146, 147, 148, 243
Gulland, J. A., 160
Gulya, A. J., 278
Gunter, G., 97, 208
Gunther, E. R., 17
Gurevich, V. S., 239
Gyllensten, U., 83, 159, 161, 164, 168

Haase, B., 172
Haenel, N. J., 130, 144, 146, 239
Hafner, K., 272
Hain, J. H. W., 72, 86, 179, 210, 211, 212

Kummer, H., 30, 56, 337
Kuronuma, Y., 322
Kurt, F., 30
Kuznetsov, V. B., 275

Lacépède, B. G. E., 343, 345, 346
Lacerda, M., 94, 96
Lack, D., 19
Lagardere, J. P., 318
Lahille, F., 348
Lahvis, G. P., 95, 316
Laland, K. N., 243
Landeau, L., 206
Lansman, R. A., 83
Larsen, A. H., 177
Lavigne, D. M., 308
Law, R. J., 172, 317
Lawrence, B., 26, 27
Lawrence, M. J., 315
Laws, R. M., 20, 163, 169, 175, 191
Leaper, R., 77, 78, 156
Leatherwood, S., 29, 37, 92, 97, 98, 101, 130, 205, 211, 214, 239, 249, 277, 311, 313, 315, 316, 319, 321, 322, 323, 331
Le Boeuf, B. J., 10, 19, 29, 35, 39, 74, 193, 221, 248
LeDuc, R. G., 52, 85, 92, 93, 216, 260, 346, 348
Lee, P. C., 97, 169
Lehner, P. N., 60
Leighton, D. R., 233
Leimar, O., 236, 252
Leland, L., 119
Lescrauwaet, A.-C., 311
Lesson, R. P., 343, 347
Leuthold, W., 189
Lewontin, R. C., 219
Liberg, O., 170
Lien, J., 38, 181, 213, 327
Lilljeborg, W., 343
Lilly, J. C., 24, 25, 27, 28, 29, 30, 272, 295, 300
Lima, S. L., 207, 216
Lindhard, M., 156
Linegaugh, R. M., 26
Linnaeus, C., 343, 344, 345, 346, 348
Lissaman, P. B. S., 200
Ljungblad, D. K., 214, 287
Lockyer, C. H., 21, 46, 73, 76, 91, 92, 152, 162, 175, 180, 183, 194, 221, 228, 229, 230, 235, 248, 249
Loesche, P., 294, 295
Long, D. J., 204, 205
Longman, H. A., 344

Lopez, D., 127, 129, 130, 132, 135, 136, 142, 147, 267
Lopez, J. C., 127, 129, 130, 132, 135, 136, 142, 147, 267
Lorenz, K. Z., 189
Lott, D. F., 192
Lowry, L. F., 205
Luttrell, L., 122
Lynch, M., 83
Lynn, S. K., 102
Lyon, B. E., 294
Lyon, M., 272
Lyrholm, T., 83, 129, 130, 159, 161, 164, 168

MacGarvin, M., 318, 319
Mackintosh, N. A., 20, 22, 178, 221
MacLeod, C. D., 85, 251
Madsen, C. J., 277, 286
Major, P. F., 98, 207, 211
Makela, R., 20
Malcolm, J. R., 219, 239
Mallone, J. S., 49
Malme, C. I., 284
Mangano, A., 166
Mangel, M., 329, 330
Mann, D. A., 283, 284
Mann, J., 2, 21, 41, 42, 43, 47, 48, 49, 54, 55, 57, 58, 60, 62, 63, 64, 66, 81, 95, 97, 98, 99, 100, 101, 104, 105, 107, 108, 110, 114, 115, 117, 118, 120, 121, 122, 124, 125, 152, 163, 205, 208, 218, 221, 225, 228, 229, 230, 231, 233, 235, 236, 237, 238, 239, 240, 241, 242, 243, 283, 284, 294, 295, 296, 322, 336, 337
Manning, A., 9
Mansfield, A. W., 247
Manson, J., 46
Marino, L., 227, 336
Marler, P., 193
Marsh, H., 20, 21, 47, 102, 215, 221, 227, 230, 231, 232, 233, 235, 236, 248, 249, 260, 295, 337
Marshall, G. J., 74, 75
Martin, A. R., 161, 220, 226, 227, 230, 231, 337
Martin, K. R., 14
Martin, P., 43, 57, 58, 59
Martineau, D., 316
Mate, B. R., 37, 75, 76, 102, 156, 172, 209, 213, 214, 334

Matkin, C. O., 130, 133, 134, 138, 145, 152
Matsui, S., 236
Matthews, L. H., 174, 179, 183
Mattila, D. K., 55, 104, 178, 184, 185, 187, 189, 192, 193
Mayer, P. J, 232
Mayer, S., 326, 329, 330
Maynard Smith, J., 19, 202, 235, 236, 270, 289
Mayo, C. A., 38, 47, 174, 177, 178, 180, 181, 182, 186, 189, 192
Mayr, E., 136, 174
McBain, J., 250
McBride, A. F., 22, 23, 24, 97, 99, 100, 101, 105, 107, 204, 240, 295
McCann, C., 251
McCowan, B., 301, 302, 303
McDonald, D. B., 119, 203
McDonald, M. A., 79, 283, 304, 334
McFarland, N. W., 348
McGeer, P. L., 129
McGregor, P. K., 270, 273, 303
McHugh, J. L., 15, 17, 327
McIntyre, J., 27
McKaye, K. R., 206
McSweeney, D. J., 69, 73, 178, 187
McVay, S., 15, 19, 32, 33, 187, 274, 285, 326, 327
Mead, J. G., 85, 95, 97, 99, 103, 173, 209, 242, 249, 311, 337, 344, 348
Medrano, L., 188, 189
Medvin, M. B., 295
Medwin, H., 335
Mellinger, D. K., 79
Melville, H., 155
Mercer, M. C., 226
Merle, R., 29
Meyen, F. J. F., 347
Michaud, R., 74
Mikhalev, Y. A., 128, 131, 143, 194
Milinkovitch, M. C., 85, 154, 220
Milinski, M., 206
Millar, J. S., 169
Miller, A. M., 25, 27, 28
Miller, D. E., 294, 295
Miller, G. Sr., 345
Miller, G. W., 47, 225
Miller, K. W., 144, 145
Miller, P. J., 78, 148, 299

Miller, R. C., 206
Mitani, J. C., 122, 248, 334
Mitchell, C. L., 202
Mitchell, E. D., 14, 18, 92, 131, 152, 160, 194, 277, 310, 311
Miyashita, T., 161
Mizroch, S. A., 40, 73, 231
Mizue, K., 296
Mobley, J. R., 37, 185, 188, 189, 192, 257, 263, 276
Møhl, B., 280, 283, 289
Møller, A., 242
Møller, A. P., 241, 294
Montagu, G., 347
Monzon, F., 252, 254
Moore, J. C., 93, 344
Moore, K. E., 167, 297, 298, 304
Moore, M. J., 176, 178, 221, 257
Moore, P. W. B., 281, 306
Mooring, M. S., 200, 209
Moors, T. M., 94, 114, 119
Morgan, M. J., 206
Morgane, P. J., 275
Mori, K., 176
Morozov, D. A., 97, 212
Morris, D., 30
Morris, R. J., 73
Morton, A. B., 129, 130, 136
Morton, E. S., 270
Moscrop, A., 172
Moss, C. J., 10, 34, 35, 169, 193
Mountford, M. D., 224
Muir, D. C. G., 317
Müller, O. F., 343
Mullins, J., 156, 162
Mullis, K., 83
Mulvaney, K., 311
Murchison, A. E., 31, 281
Murphy, E. C., 294
Myrberg, A. A. Jr., 271, 283, 290
Myrick, A. C. Jr., 46, 224

Naess, A., 330
Naguib, M., 270
Napier, J. R., 233
Napier, P. H., 233
National Marine Fisheries Service, 40, 309, 315, 323
National Research Council, 325
Nelson, D. A., 289
Nemoto, T., 21, 179
Nerini, M., 210
Nestler, J. M., 283

Turner, G. F., 205
Tutin, G. E. G., 265
Tyack, P. L., 5, 9, 20, 22, 28,
 29, 31, 37, 44, 45, 49, 56,
 57, 63, 66, 67, 69, 74, 75,
 76, 78, 80, 93, 104, 117,
 125, 126, 148, 167, 176,
 185, 187, 188, 189, 190,
 192, 200, 204, 239, 242,
 255, 262, 263, 271, 273,
 276, 277, 285, 286, 287,
 288, 295, 299, 300, 301,
 302, 303, 304, 306, 307,
 308, 324, 329, 331, 336,
 337
Tyler, S., 61
Tynan, C. T., 318

Uetz, G. W., 207
Ugarte, F., 130, 132, 140, 141,
 142, 211
Underhill, L. G., 72
Underwood, R., 69
Urban, J., 176
Urian, K. W., 97
Urick, R. J., 78, 279, 280, 317

Valsecchi, E., 190
van Bénéden, P. J., 346
Vandenbergh, J. G., 243
Van der Ree, A. F., 31
van Hooff, J. A. R. A. M.,
 121, 259
Vania, J. S., 76, 284
van Lennep, E. W., 181
van Schaik, C. P., 46, 121,
 200, 218, 241, 248, 252,
 256, 259
van Utretcht, W. L., 181
van Vuren, D., 127, 135, 216
van Waerebeek, K., 93, 99,
 249, 253, 311B
Vedder, J. M., 95
Vehrencamp, S., 203
Vehrencamp, S. L., 120, 247,
 270, 288, 304
Venus, S. C., 178, 185
Viale, D., 172, 316
Vidal, O., 315
Vigne, N., 209
Vincent, A., 239
Visser, I. N., 142
Vladimirov, V. L., 128, 152
Vongraven, D., 129, 136, 146,
 210, 260

von Haast, J., 345
Vrijenhoek, R. C. 1987, 83

de Waal, F. B. M., 46, 58,
 107, 122, 203
Wada, S., 83, 186
Wade, P. R., 313
Walker, W. A., 93, 249
Wallace, R. L., 325
Wallis, G. P., 83
Walsh, M. T., 93
Walters, E. L., 137, 138
Wang, D., 296, 337
Wang, K. R., 125
Waples, D. M., 94
Ward, P., 210
Washburn, S. L., 10, 20
Waters, S., 14, 18, 57, 62,
 166, 168
Watkins, W. A., 33, 36, 37,
 66, 75, 76, 77, 78, 154,
 156, 159, 162, 166, 167,
 171, 172, 178, 179, 241,
 282, 283, 287, 288, 292,
 295, 297, 298, 299, 304,
 305, 317, 334, 337
Watson, L., 222
Watts, D. P., 265, 266
Wayne R. K., 82
Wcislo, W. T., 200
Webb, D., 66, 79, 184, 241,
 282, 283, 304
Webb, N. M., 54
Webster, M. S., 248
Weilgart, L. S., 14, 18, 41, 46,
 48, 53, 55, 56, 60, 66, 76,
 77, 101, 154, 159, 162, 164,
 165, 166, 167, 168, 169,
 173, 191, 199, 201, 209,
 210, 216, 217, 237, 239,
 241, 242, 243, 254, 255,
 258, 259, 274, 282, 291,
 295, 296, 298, 299, 304,
 305, 310, 323, 326, 335,
 336
Weiner, N., 270
Weinrich, M. T., 45, 55, 56,
 82, 174, 176, 180, 181, 184,
 185, 189, 190, 192, 194
Weller, D. W., 94, 96, 123,
 161, 165, 209
Wells, R. S., 23, 29, 31, 35,
 36, 37, 41, 42, 44, 47, 69,
 81, 82, 85, 91, 92, 93, 94,

95, 96, 97, 98, 99, 102,
 103, 104, 108, 110, 111,
 113, 114, 115, 118, 120,
 121, 123, 124, 125, 126,
 173, 221, 225, 239, 263,
 265, 266, 276, 295, 300
Wemmer, C., 30
Werth, A. J., 209
West-Eberhard, M. J., 202
Western, D., 194
Westgate, A. J., 75, 214
Weston, R. D., 175
Wheeler, J. F. G., 221
White, D., 129
White, R., 314
White, T. J., 83
Whitehead, H. P., 3, 14,
 18, 21, 22, 33, 34, 35,
 36, 39, 40, 41, 43, 45,
 46, 47, 48, 49, 53, 55,
 56, 57, 59, 60, 62, 66,
 67, 68, 69, 73, 76, 77,
 78, 81, 82, 85, 86, 87,
 93, 101, 104, 118, 135,
 155, 156, 159, 160, 161,
 162, 163, 164, 165, 166,
 167, 168, 169, 170, 171,
 172, 173, 174, 175, 176,
 178, 179, 180, 181, 184,
 185, 187, 189, 190, 191,
 193, 194, 196, 199, 200,
 201, 206, 209, 210, 214,
 216, 217, 218, 221, 224,
 225, 233, 237, 239, 241,
 242, 243, 244, 254, 255,
 257, 258, 259, 262, 273,
 274, 276, 277, 282, 285,
 288, 291, 294, 295, 296,
 298, 299, 304, 305, 310,
 316, 322, 323, 326, 332,
 333, 335, 336, 337
Whiten, A., 46, 169
Wiebe, P., 335
Wiens, J. A., 307
Wiersma, H., 280
Wiley, D. N., 45, 183, 184,
 236
Willard, D. E., 47, 183, 236
Williams, G. C., 19, 41, 207,
 232
Williams, M. C., 240
Williams, T. M., 213
Williamson, K., 14
Willson, M. F., 236

Wilson, B., 91, 94, 96, 99,
 102, 107, 112, 122, 124,
 250
Wilson, E. O., 9, 18, 19, 28,
 29, 35, 331
Wilson, F. R., 336
Wilson, R. P., 142
Winn, H. E., 17, 36, 76, 176,
 185, 187, 188
Winn, L. K., 185, 187, 188
Wisenden, B. D., 206
Withrow, D., 224
Wolf, M., 216
Wolff, J. O., 243
Wolman, A. A., 30, 195, 226,
 229, 249, 255
Wong, Z., 83
Wood, F. G. Jr., 22, 23, 26,
 27, 28, 29, 31, 44, 99, 101,
 204, 205, 208
Woodroffe, R., 239
Wooster, H., 10, 28
Worthy, G. A. J., 235
Wrangham, R. W., 41, 46,
 104, 119, 120, 121, 122,
 125, 200, 202, 203, 208,
 209, 213, 240, 247, 259,
 265, 266, 267, 269
Wray, P., 14
Würsig, B., 19, 29, 34, 35, 36,
 37, 39, 41, 44, 47, 52, 70,
 73, 91, 93, 94, 96, 101,
 108, 121, 201, 206, 210,
 211, 212, 213, 214, 257,
 277, 287
Würsig, M., 35, 36, 37, 44, 91,
 93, 94, 108, 211, 213, 214
Wynne-Edwards, V. C., 18, 19

Yablokov, A. V., 11, 182, 195,
 277, 323
York, A. E., 231
Yorke, M., 208
Yoshioka, M., 97
Young, E., 315
Young, S., 206
Young, T. P., 55, 69, 112

Zahavi, A., 210
Zemsky, V. A., 182
Zenkovich, B. A., 15
Zoloth, S. R., 79
Zouros, E., 83
Zuk, M., 310

SUBJECT INDEX

Bhulan (Ganges river dolphin, Indus river dolphin, *Platanista gangetica, Platanista minor*), 315, 345
Bicolor damselfish (*Pomacentrus partitus*), 290–91
Biodiversity, preservation of, 331–32
Biopsy sampling, 82, 159. *See also* Genetic analysis
Birth mass, 222–23, 227, 245
Births. *See also* Life history parameters; Reproduction; Seasonal
  captive, 23
  humpback whale, 181–82
  killer whale, 143–44
  singleton, 235
  size at, 224, 235
  sperm whale, 163
Biting, 289. *See also* Aggressive behavior
Black porpoise. *See* Vaquita
Black right whale. *See* Northern right whale
Blainsville's beaked whale. *See* Hubbs' beaked whale
Blasts. *See* Acousitc Tomography of Ocean Climate (ATOC) project; Explosions
Blind river dolphin. *See* Ganges river dolphin
Blowfish, *Torquigener* sp., 229
Blowholes, 154, 275
Blubber, 153, 228–29, 235. *See also* Energetics
Blue-green light, 276
Blue whale (*Balaenoptera musculus*)
  communication, 79, 255, 274, 282–83, 305
  distribution, 343
  endangered status, 319
  female mate choice, 255
  fetal growth, 226
  identification, 51–52
  penis length, 255
  predation on, 180, 204
  pregnancy, 229
  testes size, 85, 194, 254
  whaling, 15, 17–18, 21, 309–10
Blue whale units, 17–18
Boats. *See* Vessels
Body size. *See also* Life history parameters
  alliance formation, 258–259

at birth, 235
bottlenose dolphin, 92, 97, 217, 220
and communication, 291
energetics and, 213, 248
and gestation, 227
gray whale, 217
humpback whale, 174
and identification, 49
intraspecific variation in, 220
killer whale, 144
and lactation, 228
and life history parameters, 227–28, 244
and longevity, 233
and predation pressure, 217
at sexual maturity, 220–21
sexual size dimorphism, 248
sperm whale, 217
and testis size, 252
as a weapon, 249–50
Body-to-body contact, 104, 106, 166
Bonobos (*Pan paniscus*), 121, 265, 267–69
Bornean white dolphin. *See* Humpback dolphin
Boto (*Inia geoffrensis*), 226, 256, 345
Bottlenose dolphin (*Tursiops sp.*), 2–3, 91–126, 259
  acoustic fovea, 278
  affiliation, 104–6, 115
  aggressive behavior, 24, 102, 105, 107, 115, 249–50, 289
  alliances, 3, 5, 41, 111–13, 123, 203, 258–59, 268
  allomaternal behavior, 115, 239
  allopatric populations, 264
  body size, 92, 95, 97, 217, 220
  bubble streams, 277, 303
  calf foraging, 115, 228–30
  calving intervals, 46, 95, 97, 124. *See also* Interbirth intervals
  in captivity, 22–24, 27–29
  capture quotas, 40
  communication, 104–8, 211. *See also* Signature whistles
  conservation issues, 125–26
  consortships, 119, 255–56
  convergence with chimpanzees, 121, 265–66
  day range, 102, 124, 214
  diet, 97–99

dispersal, 103, 264
displays, 105, 107–8, 115, 118–19, 289
distribution, 91, 347
dominance, 105, 107, 148, 204, 243, 257
echolocation, 26, 98, 280–81
ecology, 97–104, 120–25
estrogen levels, 119
female mate choice , 119, 122, 255–56
female-female relationships, 110, 121, 123
field sites, 93–95
fisheries, 14, 46, 124, 316
fission-fusion society, 41, 91–126
focal animal studies, 47–48, 58, 117
foraging, 91, 97–99, 110, 114, 117, 122, 210–13, 228–30
genital inspections, 105
gestation, 95
group foraging, 211
groups, 95–96, 102–4, 117, 122
herding, 48, 113, 118–20, 123, 125, 242, 256
home range, 102–3, 122
identification, 35–36, 44, 47, 50–52, 71, 93–94
incidental catch, 313
infanticide, 119
infant position, 115–16, 238
inshore and offshore forms, 92–93
interbirth intervals, 46, 95, 97, 124
interspecific associations, 101–2
kidnapping, 115
lactation, 227, 237–38
life history, 95–97
locomotion, 213
low frequency calls, 211. *See also* Acoustic communication
male-female relationships, 113–15, 125
male-male relationships, 105, 111–13, 123–25, 148, 257. *See also* Alliances
mate guarding by, 288
mating strategies, 104–5, 118–20, 123–25, 255–56, 263

matrilineal bands, 41
migration, 102–3
morphology, 92, 217, 220
mortality, 97, 118, 221
mother-calf relationships, 115–19, 294–95
mounting, 104–5
nursing, 95, 97, 237
philopatry, 103, 216
predation, 99–101, 122, 204
reproduction, 95–97, 118, 120–21. *See also* Life history parameters; Mating strategies
sex ratio, 124
sexual dimorphism, 95
signature whistles, 29, 37, 41, 80, 117–18, 300–2
social behavior and communication, 104–8, 118
social relationships, 91–126
socio-sexual behavior, 104–5
sponging. *See* Tool use
super alliances, 111–12, 123–24
synchrony, 108, 115
taxonomy , 92–93, Appendix 2
testis size, 97, 254–55
tool use, 91, 98, 110, 243
ventral speckling, 92
vision, 276
vocal learning in, 5, 117–18, 305. *See also* Signature whistles; Vocalizations
vocalization, 105, 113, 211
weaning, 115, 144, 230
whaleman reports, 14
Bottlenose whale, northern (*Hyperoodon ampullatus*), 217, 227, 259, 344
Bottlenose whale, southern (*Hyperoodon planifrons*), 344
Bottlenose whale (*Hyperoodon* spp.), 231, 239, 337
Bottlenosed dolphin. *See* Bottlenose dolphin
Bouto. *See* Boto
Bowhead whale (*Balaena mysticetus*)
  aboriginal hunts, 311
  characteristics, 52
  communication, 287
  conservation, 39
  demographics, 47
  detection of ice obstacles, 200, 283

Gray whale *(continued)*
  playback experiments, 284
  predation on, 180, 218
  prey, 194
  sperm competition in, 194, 242
  testis size, 254, 255
  whaling, 11, 14
Gray's beaked whale *(Mesoplodon grayi)*, 345
Great white sharks *(Carcharodon carcharias)*, 101, 139
Greater horseshoe bat *(Rhinolophus ferrumequinum)*, 278, 305
Greenland right whale. *See* Bowhead whale
Greeting ceremonies, 134
Grind, 14
Ground squirrels, 19
Group selection models, 18–19
Groups
  behaviors, 185, 199–200
  bottlenose dolphin, 95–96, 102–4, 117, 122, 211
  on calving grounds, 256
  competitive, 185. *See* Humpback whale
  costs and benefits of, 140, 191–92, 199–202, 204, 207, 213–17
  definitions of, 55–56, 67, 95–96, 132, 164, 184, 220
  echolocation, 148
  of females, 122, 222–23, 242–43
  humpback whale, 184–86, 199–200
  hunting, 140
  identity from vocalizations, 77
  killer whale, 132, 267
  living in, 2, 199–218
  migration in, 199
  recognition of, 200–1
  selection on, 18–19, 150
  signals for maintaining cohesion, 296–300
  sperm whale, 164, 201, 210
  stability of, 36, 66
Growth
  bottlenose dolphin, 95, 97
  elephant, 163
  first year females, 222–23
  humpback whale, 181–82
  killer whale, 145–46
  neonatal rates, 236
  sperm whale, 163

Growth layer groups (GLG), 175. *See also* Age
Guarding, 257–60, 288. *See also* Herding
Gulf porpoise. *See* Vaquita
Gulf Stream beaked whale. *See* Gervais' beaked whale

Haast's beaked whale. *See* Gray's beaked whale
Habitats
  bottlenose dolphin, 102
  definitions of, 220
  degradation of, 40, 315–16, 325–36
  for female cetaceans, 222–23
  group protection from risks, 200
  of humpback whale, 178
  killer whale, 136–38
  and life history parameters, 244
  need for further study, 338
  seasonal range shifts, 103
  sperm whale, 160–61
Habituation, 44, 52–55
Hammerhead shark *(Sphyrna zygaena)*, 101
Harassment, 200, 226, 267
Harbor porpoise *(Phocoena phocoena)*
  acoustic fovea, 278
  allomaternal behavior, 239
  attacks by bottlenose dolphins, 102
  captive settings, 22
  distribution, 348
  echolocation, 280–81, 304
  female sexual maturity, 221
  ranging, 213–14
Harbor seals *(Phoca vitulina)*, 132, 140, 305
Harpoon guns, 15–16
Harris's hawks *(Parabuteo unicinctus)*, 212
Harvesting, and quotas, 17–18, 20, 40, 327. *See also* Fisheries; Whaling industry
Hastate dolphin. *See* Heaviside's dolphin
Head catalogs of right whale, 34
Head jerks, 105, 289. *See also* Displays
Hearing, 277–78, 280
Heart rates, 74
Heaviside's dolphin *(Cephalorhynchus heavisidii)*, 346

Heavy metals, 316. *See also* Pollution
Hebb, D.O., 23
Hector's beaked whale *(Mesoplodon layardii)*, 345
Hector's dolphin *(Cephalorhynchus hectori)*, 52, 148, 304, 331, 346
Helminthes, 161
Helping behavior, 24. *See also* Altruism; Cooperation
Herding. *See also* Aggressive behavior; Alliances
  bottlenose dolphin, 48, 113, 118–20, 123, 125, 212, 242, 256
  and female mate choice, 119, 242
  killer whale, 211
  and mate coercion, 256
  as mating strategy, 119–20
  as play, 118
  of prey, 211
Herring *(Alosa aestivalis)*, detection of echolocation signals, 283
Herring *(Clupea harengus)*, 130, 178
  coordinated foraging of by killer whale, 140–42, 179
Hierarchies, 170, 212. *See also* Dominance; Rank
Hinde, Robert, 66
Hinde index, 56–57
Hirundo rustica (barn swallows), 294
History, of studying cetacean societies, 2, 9–44
Hitting, 249, 289. *See also* Aggressive behavior
Home ranges, 102–3, 178
Hormones, 31, 144
Hose's dolphin. *See* Fraser's dolphin
Hotspot model, 257
Hourglass dolphin *(Lagenorhynchus cruciger)*, 5, 320–21, 346
Hubbs' beaked whale *(Mesoplodon densirostris)*, 344
Human activities. *See* Anthropogenic impacts
Human-cetacean interactions, 11, 98, 330
Humpback dolphin *(Sousa chinensis)*
  distribution, 347
  effect of incidental catch, 313

in Hong Kong harbor, 316
identification, 35, 52
live capture, 313
ranging, 213–14
Humpback whale *(Megaptera novaeangliae)*, 2, 52, 173–96
  aggressive behavior, 185, 245, 259
  annual world catch, 20
  associations, 55
  breeding grounds, 56, 194, 277
  bubble net feeding, 179–80
  calves, 15, 17, 176, 178, 183, 186, 226, 237, 239, 295–96,
  communication, 55, 66, 187–88, 274, 277–78, 285–87, 292
  comparisons with other Mysticetes, 192–195
  competitive groups, 185, 189–90, 193, 200
  competition for females, 254, 259
  conservation issues, 195–96
  copulation, 4, 190
  demographics, 47
  displays, 187, 192, 257
  distribution, 176–77, 344
  ecology, 176–81
  estrous cycles, 174, 193
  exploitation of, 309
  female–female relationships, 188–89
  female sexual maturity of, 221
  field sites, 34, 175–76
  fluking behaviors, 73
  foraging, 179–80, 184, 210
  gestation, 174, 181
  groups, 185–86, 199–200
  growth curve, 181–82
  habitat use and home range, 178
  identification of, 34–35, 49–50, 71–72, 175
  juvenile development, 186
  leks, 257
  life history parameters, 181–184
  male-female relationships, 190
  male-male relationships, 189–90
  mate choice, 242, 255, 291–92

Mesopelagic squid, 161
Mesoplodon australis. See Gray's beaked whale
Mesoplodon bidens (Sowerby's beaked whale), 344
Mesoplodon bowdoini (Andrews' beaked whale), 344
Mesoplodon dalei. See Sowerby's beaked whale
Mesoplodon densirostris (Hubbs' beaked whale), 251, 344
Mesoplodon europaeus (Gervais' beaked whale), 344
Mesoplodon floweri. See Strap-toothed whale
Mesoplodon gervaisi. See Gervais' beaked whale
Mesoplodon ginkgodens (Ginko-toothed beaked whale), 344
Mesoplodon grayi (Gray's beaked whale), 345
Mesoplodon guntheri. See Strap-toothed whale
Mesoplodon haasti. See Gray's beaked whale
Mesoplodon hotaula. See Ginko-toothed beaked whale
Mesoplodon knoxi. See Hector's beaked whale
Mesoplodon layardii. See Hector's beaked whale
Mesoplodon layardii. See Strap-toothed whale
Mesoplodon layardii (Hector's beaked whale), 345
Mesoplodon layardii (strap-toothed whale), 337, 345
Mesoplodon longirostris. See Strap-toothed whale
Mesoplodon micropterus. See Sowerby's beaked whale
Mesoplodon mirus (True's beaked whale), 345
Mesoplodon pacificus. See Indo-Pacific beaked whale
Mesoplodon peruvianus (pygmy beaked whale), 345
Mesoplodon sechellensis. See Hubbs' beaked whale
Mesoplodon sowerbyensis. See Sowerby's beaked whale
Mesoplodon sowerbyi. See Sowerby's beaked whale
Mesoplodon stejnegeri (Stejneger's beaked whale), 345
Mesoplodon thomsoni. See Strap-toothed whale

Mesoplodon traversii. See Strap-toothed whale
Methods, 45–87. *See also* Sampling methods
  for behavioral studies, 2–3, 45–64, 77, 86, 333–34
  choice of, 330
  cross-sectional studies, 46–47
  determination of duration of lactation, 227
  genetic, 49, 82–84, 86
  of interpreting communications, 273
  longitudinal studies, 46–47
  photo-identification, 34, 36, 40, 44, 69, 72–73, 93–94, 129–30, 175
MicroPascals (μPa), 279
Micropterus salmoides (large-mouth bass), 206
Migrations, 4
  baleen whale, 165–86, 176–77, 190–91, 216–17, 226
  bottlenose dolphin, 102–103
  cost-benefit analysis, 191
  gray whale, 206, 226
  groups, 199
  humpback whale, 66, 176–177, 185–186, 190–191
  maintenance of mother-offspring contact, 295
  and risk of predation, 217–18
  sperm whale, 161–62
Milk composition, 228, 237, 316. *See also* Lactation
Mimicry, 27, 274
The Mind of the Dolphin: A Nonhuman Intelligence, 28
Minke whale *(Balaenoptera acutorostrata)*
  body size, 217
  distribution, 343
  hunting of, 311, 322
  identification, 35, 52
  interbirth intervals, 231
  predation on, 180
  testis size, 255
  whaling, 15, 322
Minnows *(Phoxinus phoxinus),* 210
Mirounga leonina (elephant seals), 130
Mitochondrial diversity, 310
Mitochondrial genome, 83
Mobbing, of predators, 208
Moby Dick, 14, 155

Modelling, 40, 43
Mondon monoceros. See Narwhal
Monodontidae, taxonomy, 345
Monodontids, 209
Monogamy, 21–22, 256
Mortality
  anthropogenic impacts, 181
  bottlenose dolphin, 97, 110, 118, 221
  female cetaceans, 233–234
  of first-born offspring, 221
  humpback whale, 174, 180–181
  killer whale, 133, 144, 145, 234
  long-term studies, 47
  neonatal, 183, 221
  short-finned pilot whale, 234
  sperm whale, 163
Mother-calf relationships. *See also* Lactation; Maternal care; Nursing behaviors; Parental investment; Weaning
  bottlenose dolphin, 115–18
  communication, 293–96
  humpback whale, 186, 192, 237, 239, 295–96
  killer whale, 146
  patterns of association, 151
  recognition systems, 115–118
  separations, 237–239
  short-finned pilot whale, 295
  southern right whale, 239
  sperm whale, 164–65
Mother-son relationships, 267, 268
Mottled grampus. *See* Risso's dolphin
Mounting behaviors, 56, 104–5
Multilocus microsatellite analysis, 81
Multilocus minisatelitte fingerprinting, 83
Museau du singe, 291
Musth, 169–70
Mutualism 112, 199, 201–2
Mutualistic groups, 168
Mysticetes. *See also* Balaenidae; Balaenopteridae; Specific species
  comparisons with humpback whale, 192–95
  comparisons with odontocetes, 20, 227–31
  fasting during lactation, 237
  fat content of milk, 228

female life history parameters, 222–23
lack of social groups in, 217–18
length and life histories, 244
neonatal growth rates, 236
postpartum estrus, 230
predation on, 180
study of reproductive cycles, 20
taxonomy, 343
Myths, 29

Naked mole-rat, 10
Narwhal *(Mondon monoceros)*
  aboriginal hunts, 311, 312
  distribution, 345
  effect of selective hunting, 310
  effect of vessel noises, 317–18
  male-male competition, 247, 251–52
  tusks as displays, 251–52
Natal philopatry, 103, 215–16, 260. *See also* Philopatry
National Marine Fisheries Service, 329
Naturalists, whalers, 11–15
Negaprion brevirostris (lemon shark), 100
Nematodes, parasitism of killer whale, 142
Neobalaenidae, taxonomy, 343
Neocyamus physeteris, 20
Neonatal mortality, 183, 221
Neophocaena asiaeorientalis. See Finless porpoise
Neophocaena melas. See Finless porpoise
Neophocaena phocaenoides (finless porpoise), 313, 316, 347
Neophocaena sunameri. See Finless porpoise
Nepotism, 112, 202, 267. *See also* Kinship
Nerve endings, dermal, 275
Neural net processing, 79
New Management Procedure, 327
New York Aquarium, 22
Night, data collection, 131
Noise pollution, 5, 172, 317–18
Non-interference mutualism (NIM), 202
Non-mutualistic groups, 168

Remoras, 209
Reproduction. *See also* Life history parameters
  bottlenose dolphin, 95–97, 118, 120–21
  data from surveys, 47
  determination of success, 80
  effect of pollutants on, 316
  estimation of rates, 39, 224
  female sexual maturity, 220–21
  female strategies, 41, 153, 219–46
  humpback whale, 174, 182–83
  killer whale, 174, 182–83
  male strategies, 247–69
  maturity criteria, 157
  monogamy, 21–22, 256
  seasonality of, 97, 221, 225–26
  senescence, 46, 146, 231–33
  sex ratios, 183–84
  shifts in female foraging during, 122
  sperm whale, 157
  by subordinate individuals, 203
  suppression by dominant females, 243
Research, history of studying cetacean societies, 9–44
Respiration. *See* Breathing
Resting, 55, 122
Retinas, 276
Rhesus monkeys *(Macaca mulatta)*, 203
Rhinolophus ferrumequinum (greater horseshoe bat), 278
Right whale *(Balaena* spp.). *See also* Southern or Northern right whale
  characteristics, 52
  congregation on breeding grounds, 194
  conservation efforts, 38, 40
  copepods as prey, 194
  dependence of calves on sheltered water, 176
  identification, 34
  mating strategies, 20–21, 257
  mating system, 259
  penis length, 4, 20, 255
  sperm competition in, 194, 242
  studies on demographics, 47
  testis size, 4, 20, 85, 254

timing of sexual activity, 194
  whaling, 11, 13
Risso's dolphin *(Grampus griseus)*, 52, 251, 346
River dolphin, feeding methods, 209
Rivers, degradation of, 315
RMP, design of, 327
"Rooster struts," 108. *See also* Displays
Rorquals. *See* Baleen whale; Humpback whale; Mysticetes
Rough–toothed dolphin *(Steno bredanensis)*, 26–27, 314, 347
Roving, by odontocetes, 257–260
Rubbing behaviors 104, 106, 115, 275

Saber-toothed beaked whale. *See* Stejneger's beaked whale
Saddleback dolphin. *See* Common dolphin
Sagmatius sp. *See* Lagenorhynchus
Salmon *(Oncorhynchus* spp.), 131, 138
Sampling methods, 45–87, 333–34
  Ad libitum, 42, 63
  behavioral states versus events, 54–55, 60
  biases in choice, 58–59
  continuous, 59–60
  decisions, 41, 54–54, 57
  focal sampling, 3, 41–42, 44, 47–49, 56, 58, 62–63, 66, 117, 129, 159–60
  follow protocols and choice of, 42, 44, 47–48, 57–58, 60, 157–59
  incidental, 62
  inter–observer reliability, 54
  length of sample, 42, 58
  point sampling, 60–61
  predominant activity sampling (PAS), 61
  problems with sampling cetaceans, 64
  punch sampling, 82
  quantitative systematic observations, 58–64
  rare events, 63–64
  scan sampling, 61
  sequence, 58, 61–62
  types of, 59–63

Sanctuaries, 331
Sandbar shark *(Carcharhinus milberti)*, 100
Sand lance *(Ammodytes* spp.), 178
Sarasota. 94–95. *See also* Bottlenose dolphin field sites
Sarawak dolphin. *See* Fraser's dolphin
Satellite telemetry, 37, 75, 213–14
The Saturday Evening Post, 29
Scammon, Charles Melville, 11–13
Scamperdown whale. *See* Gray's beaked whale
Scan sampling, 61
Scarring, 85, 251
Scientific Committee, of the International Whaling Commission (IWC), 329–330
Scouting, 208, 210
Sea lions *(Otaria byronia)*, 130, 141, 206
Seasonal
  births, 181, 222–23
  breeding, 46, 169, 174, 181, 194, 221, 225–26, 256
  grouping, 216–217
  killer whale vocalizations, 132
  migrations, 190–91
  ranging, 103, 122
  vocalizations, 132
Secondary sexual characteristics, 276
Sei whale *(Balaenoptera borealis)*, 15, 41, 52, 194, 326, 343
Selection pressures, 191–92
Selfish herd hypothesis, 20, 207
Selous, 32
Semisubmersible Seasick Machine, 38
Senescence, reproductive, 231
Sensory hairs, 275
Sensory modalities, 274–283
Sentinel behaviors, 207–208, 210
Separations, mothers and calves, 117–18, 181, 237–29
Sequence sampling, 58, 61–62
Sewage, 316
Sex differences
  biased investment in pilot whales, 47
  bottlenose dolphin, 93–94
  cyamid infestation patterns, 20

humpback whale, 176
  hunting tactics, 141
  investment based on, 235–36
  locational dispersal in belugas, 81
  methods for determination, 49, 83
  and quotas, 18
  ranging patterns, 103
  sperm whale group, 157
Sex ratios
  bias in pilot whales, 47
  bottlenose dolphin, 124
  of cetaceans, 235–36
  humpback whale, 183–84
  sperm whale, 163
Sexual behaviors, social function of, 275
Sexual dimorphism. *See also* Sexual size dimorphism
  appendage sizes, 146
  bottlenose dolphin, 95
  elephant, 169
  humpback whale, 174
  and hunting ranges, 141
  and male–male competition, 248–52
  observational methods, 49
  sperm whale, 154, 168–69
Sexual maturity
  bottlenose dolphin, 95
  estimates of ages, 46
  female cetaceans, 144, 220–21, 222–23
  humpback whale, 174, 181–82
  killer whale, 145–46
  in males, 255
  sperm whale, 163
Sexual selection, 41, 247–48, 284–85. *See also* Mate choice
Sexual size dimorphism (SSD)
  allometric analysis of, 249
  allometry in cetaceans, 250
  and coercion of females, 264
  definition, 248
  and testis mass, 253, 254
Shark Bay. 95, 109
Sharks
  bull shark, 99, 204
  "cookie cutter," 209
  dusky shark, 99
  great white shark, 101, 139
  hammerhead shark, 101
  lemon shark, 100
  nurse shark, 100

and common ancestry, 133
detection of, 79–80
effect of boats on, 52
female mate choice and, 255
first structural descriptions, 27
foraging, 299, 284
harbor porpoise, 280–81, 304
humpback whale, 4, 5, 55, 187–88, 274, 277–78, 285–87, 291–92, 305
identifying of individuals using, 80
information from, 77
killer whale, 148–49, 284
low-frequency, 37, 105, 211, 274, 282–283, 288, 304–5
and mate choice, 242
and recognition systems, 292–304
seasonal comparisons, 132
separated mothers and calves, 117
sequences, 77
and social relationships, 56–57
tag recording, 74
Vulva, nerve endings near, 275

Washburn, Sherwood, 2
Washington, killer whale, 129–30
Water fleas (Bosmina longispina), 207
Wavelengths, definition, 279
Weaning
   ages of, 46

baleen whale, 227
bottlenose dolphin, 115, 144, 230
cetacean strategies for, 227–28
conflict, 230
and dependency, 95
humpback whale, 174, 181
killer whale, 144–45
sperm whale, 46
Weapons, cetacean morphology, 249–52
Whale lice (Cyamus spp.), 181
Whalemen, 11–15, 165
Whale watching, 38,152, 318, 332
Whaling industry, 10. See also International Whaling Commission
blue whale, 15, 17–18, 21, 309, 310
bowhead whale, 310
data from, 14, 20–22, 46–47
determination of reproductive history, 225
in developing countries, 311
effects of exploitation, 173, 195–96, 309–13, 331
finback whale, 20, 309
gray whale, 11, 14
humpback whale, 20, 175, 195–96, 309
and the International Whaling Commission (IWC), 17
killer whale, 152
lingering effects, 171–72

management emphasis, 18–19
minke whale, 15, 322
modern, 15–20
moratorium, 327
regulation of, 18. See also International Whaling Commission
relationship with research, 20, 32
right whale, 11, 13
selection by, 310
sperm whale, 11, 155, 170
sustainable use, 308
worldwide moratorium, 40
Whistles. See also Acoustic communication; Signature whistles
context specificity, 28
learning, 117–18
during mother-calf separations, 58, 117
and mother-offspring contact, 295
spectrograms of, 301
White-beaked dolphin (Lagenorhynchus albirostris), 52, 346
White-bellied dolphin. See Vaquita
Whitebelly dolphin. See Common dolphin
Whitefin dolphin. See Baiji
Whitefin porpoise. See Dall's porpoise
Whitefronted dolphin. See Hector's dolphin
Whiteheaded grampus. See Risso's dolphin

Whitesided porpoise. See Dall's porpoise
White whale. See Beluga whale
Wood, Forrest G., 2, 22, 27, 31
World Wildlife Fund–U.S., 332
Wynne-Edwards, V.C., 18–19

Xiphius gladius (swordfish), 172

Yangtze River dolphin. See Baiji
Years of the North Atlantic Humpback (YONAH), 175
Yellowtail (Seriola lalandei), 207
Yerkes Laboratories of Primate Biology, 23

Zebra, 34
Zebra finch, 44
ZFY, 81, 83
Ziphiidae, taxonomy, 344–345
Ziphioidea, taxonomy, 344
Ziphius australis. See Cuvier's beaked whale
Ziphius capensis. See Cuvier's beaked whale
Ziphius cavirostris (Cuvier's beaked whale), 318, 345
Ziphius chathamensis. See Cuvier's beaked whale
Ziphius indicus. See Cuvier's beaked whale
Zoos, transformation of research, 30